Bacterial Pathogenesis

A subject collection from *Cold Spring Harbor Perspectives in Medicine*

OTHER SUBJECT COLLECTIONS FROM *COLD SPRING HARBOR PERSPECTIVES IN MEDICINE*

Transplantation
Cystic Fibrosis: A Trilogy of Biochemistry, Physiology, and Therapy
Hemoglobin and Its Diseases
Addiction
Parkinson's Disease
Type 1 Diabetes
Angiogenesis: Biology and Pathology
HIV: From Biology to Prevention and Treatment
The Biology of Alzheimer Disease

SUBJECT COLLECTIONS FROM *COLD SPRING HARBOR PERSPECTIVES IN BIOLOGY*

Mitochondria
Signaling by Receptor Tyrosine Kinases
DNA Repair, Mutagenesis, and Other Responses to DNA Damage
Cell Survival and Cell Death
Immune Tolerance
DNA Replication
The Endoplasmic Reticulum
Wnt Signaling
Protein Synthesis and Translational Control
The Synapse
Extracellular Matrix Biology
Protein Homeostasis
Calcium Signaling
The Golgi
Germ Cells
The Mammary Gland as an Experimental Model
The Biology of Lipids: Trafficking, Regulation, and Function
Auxin Signaling: From Synthesis to Systems Biology
The Nucleus
Neuronal Guidance: The Biology of Brain Wiring
Cell Biology of Bacteria
Cell–Cell Junctions
Generation and Interpretation of Morphogen Gradients
Immunoreceptor Signaling
NF-κB: A Network Hub Controlling Immunity, Inflammation, and Cancer
Symmetry Breaking in Biology
The Origins of Life
The p53 Family

Bacterial Pathogenesis

A subject collection from *Cold Spring Harbor Perspectives in Medicine*

EDITED BY

Pascale Cossart
Pasteur Institute

Stanley Maloy
San Diego State University

COLD SPRING HARBOR LABORATORY PRESS
Cold Spring Harbor, New York • www.cshlpress.org

Bacterial Pathogenesis

A Subject Collection from *Cold Spring Harbor Perspectives in Medicine*
Articles online at www.perspectivesinmedicine.org

Printed in the United States of America

Executive Editor	Richard Sever
Managing Editor	Maria Smit
Project Manager	Barbara Acosta
Permissions Administrator	Carol Brown
Production Editor	Diane Schubach
Production Manager/Cover Designer	Denise Weiss
Publisher	John Inglis

Front cover artwork: Image of HeLa cells infected by *Listeria monocytogenes*. Some bacteria form actin comet tails and move in the cytoplasm of the cells. Bacteria are stained with *Listeria*-specific antibody (in red); cellular actin is stained with DAPI (in blue). Image courtesy of Edith Gouin and Pascale Cossart, Pasteur Institute, Paris.

Library of Congress Cataloging-in-Publication Data

Bacterial pathogenesis : a subject collection from Cold Spring Harbor perspectives in medicine / edited by Pascale Cossart, Pasteur Institute and Stanley Maloy, San Diego State University.
pages cm
Includes bibliographical references and index.
ISBN 978-1-936113-36-1 (hardcover : alk. paper)
1. Pathogenic bacteria. 2. Bacterial diseases--Chemotherapy--Evaluation 3. Drug resistance in microorganisms. I. Cossart, Pascale, editor of compilation. II. Maloy, Stanley R., editor of compilation.

QR175.B3365 2013
616.9'041--dc23

2013001356

10 9 8 7 6 5 4 3 2 1

All World Wide Web addresses are accurate to the best of our knowledge at the time of printing.

For a complete catalog of all Cold Spring Harbor Laboratory Press publications, visit our website at www.cshlpress.org.

Contents

PART 2: GENERAL CONCEPTS AND MECHANISMS

PART 3: THERAPEUTICS

Preface

SINCE THE EARLY 1980s, the study of bacterial pathogens and bacterial pathogenesis has received a lot of attention, and progress has been spectacular, as illustrated in this volume. The combination of various approaches, including classical genetics, molecular biology, and cell biology, over the last 20 years has led to the emergence of cellular microbiology. Genomics pushed the field even further, providing a broader view of the multitude of strategies bacteria use during a productive infection. Transcriptomics subsequently invaded the microbial world and highlighted the numerous unsuspected regulatory mechanisms that allow bacteria to optimally adapt to their hosts during infection. Finally, transgenesis and sophisticated ex vivo and in vivo models have helped to place the numerous studies of cultured cells into an in vivo context. Hence we have the necessary tools to go one step further and understand infection in the context of all the other microbes, including commensals, symbionts, and other pathogens, that are present during an infection. This opens exciting new opportunities for discovery. This book thus sets the stage for the next generation of bacterial pathogenesists!

We did not intend to cover the whole field of bacterial pathogenesis. We believe it is important to build on what we know about some well-studied pathogens, so the book begins with chapters dedicated to several important pathogens and their key features, such as the intracellular life cycles of *Chlamydia* and *Francisella*, cellular invasion by *Listeria* and *Shigella*, the pathogenesis of meningococci, and the persistence of *Helicobacter pylori* and *Salmonella*. The second part of the book transitions to chapters describing either well-established concepts or new mechanisms emerging from the study of various pathogens, including RNA-mediated regulation of virulence, the role of quorum sensing in virulence, and epigenetics and bacterial infections. Finally, such a book could not exist without chapters dedicated to vaccinology and new therapeutic approaches. In this respect, the chapter on probiotics is particularly timely.

We thank the many colleagues who agreed to write for us. We are convinced that gathering in one volume a series of chapters written by experts in their respective fields will be of great help for teaching and preparation of reviews, grant proposals, and reports. This type of book also provides an important forum for authors to express their vision, and, in this respect, we believe that this volume is a gold mine. The field is moving fast, yet there is so much we need to learn before we really understand in detail how the microbiota influence infection and how each bacterium in a population plays a role in infection. We hope that this book will be read for many years by researchers in various fields.

We thank Barbara Acosta and Richard Sever for continued support and patience during what at some stages seemed an endless project.

PASCALE COSSART
STANLEY MALOY

Bartonella and *Brucella*—Weapons and Strategies for Stealth Attack

Houchaima Ben-Tekaya[1], Jean-Pierre Gorvel[2,3,4], and Christoph Dehio[1]

[1]Focal Area Infection Biology, Biozentrum, University of Basel, 4052 Basel, Switzerland

[2]Faculté de Sciences de Luminy, Centre d'Immunologie de Marseille-Luminy, Aix Marseille Université, UM 2 Marseille Cedex, France

[3]INSERM, U 1104 Marseille, France

[4]CNRS, UMR 7280 Marseille, France

Correspondence: christoph.dehio@unibas.ch

Bartonella spp. and *Brucella* spp. are closely related α-proteobacterial pathogens that by distinct stealth-attack strategies cause chronic infections in mammals including humans. Human infections manifest by a broad spectrum of clinical symptoms, ranging from mild to fatal disease. Both pathogens establish intracellular replication niches and subvert diverse pathways of the host's immune system. Several virulence factors allow them to adhere to, invade, proliferate, and persist within various host-cell types. In particular, type IV secretion systems (T4SS) represent essential virulence factors that transfer effector proteins tailored to recruit host components and modulate cellular processes to the benefit of the bacterial intruders. This article puts the remarkable features of these two pathogens into perspective, highlighting the mechanisms they use to hijack signaling and trafficking pathways of the host as the basis for their stealthy infection strategies.

Bartonella spp. and *Brucella* spp. are Gram-negative facultative intracellular bacteria causing bartonellosis and brucellosis, respectively. These infections are typically chronic, ranging from subclinical courses to progressively debilitating and possibly life-threatening diseases. With a few exceptions in the genus *Bartonella* where the host reservoir is human, *Bartonella* spp. and *Brucella* spp. are zoonotic pathogens transmitted from mammals to humans. Wild animals are an important reservoir for the multiple species of both pathogenic genera. However, cats represent the major source for human infection by *Bartonella* (Breitschwerdt and Kordick 2000), whereas human infections by *Brucella* typically originate from livestock (Pappas et al. 2006). Therefore, individuals handling animals or infected material have a higher risk of getting infected either through direct contact or by aerosols. For *Brucella*, persons consuming infested animal products, often milk and its derivatives, are also at risk of getting infected (Pappas et al. 2006).

Via remarkably analogous mechanisms (Batut et al. 2004), *Bartonella* and *Brucella* invade various cell types using (1) cell-surface molecules that mediate host–pathogen interactions, such as adhesion proteins and receptors from

Cite this article as *Cold Spring Harb Perspect Med* doi: 10.1101/cshperspect.a010231

the bacterial and host side, respectively; and (2) type IV secretion systems (T4SS) that translocate bacterial proteins termed effectors (Backert and Meyer 2006; Llosa et al. 2009). Importantly, effectors subvert intracellular trafficking and both innate and acquired immune responses (Roop et al. 2009; Harms and Dehio 2012; von Bargen et al. 2012), thus maneuvering host processes in favor of the pathogen.

BARTONELLA AND *BRUCELLA*: CLOSELY RELATED PATHOGENS WITH DISTINCT INFECTION SEQUELS

Bartonella spp. and *Brucella* spp. evolved from a common ancestor and belong to the Rhizobiales order of the α-proteobacteria (Gillespie et al. 2011), which includes microbes that coevolved with animal or plant hosts either as pathogens or as symbionts (Fig. 1). The *Bartonella* genus is composed of 24 species sharing a core genome and subdivided based on phylogenomic analysis in four phylogenetic lineages, lineage 1 being the most ancestral lineage with *Bartonella bacilliformis* as sole species (Saenz et al. 2007; Engel and Dehio 2009; Engel et al. 2011). The *Brucella* genus includes 10 species defined according to their preferential host (Moreno et al. 2002) and divided, based on phylogenomic analysis, in four clades: *Brucella abortus–Brucella melitensis*, *Brucella suis–Brucella canis*, *Brucella ovis*, and *Brucella ceti* (Wattam et al. 2009), with *Brucella ovis* being a potential common ancestor (Foster et al. 2009).

Each species of the genera *Bartonella* or *Brucella* infects one or few closely related mammals as reservoir host(s) wherein they persist and elicit diverse infection outcomes (Fig. 1). *B. bacilliformis* and *Bartonella quintana* are an exception in having humans as the reservoir host (Maguina et al. 2009). In their mammalian reservoir hosts, *Bartonella* spp. typically cause an intraerythrocytic bacteremia with an often asymptomatic course (Chomel et al. 2009), whereas *Brucella* provoke animal brucellosis, a chronic infection characterized by abortion probably due to the infection of trophoblasts, infertility, or birth of weak infected offspring (Hignett et al. 1966; Samartino and Enright 1993). Animal brucellosis is a significant economic burden in the endemic regions that span from central Asia to South America through the Middle East and the Mediterranean region (Pappas et al. 2006; Roop et al. 2009). Transmission of *Bartonella* is mediated by blood-sucking (hematophagous) arthropod vectors (Billeter et al. 2008; Tsai et al. 2011), whereas for *Brucella*, transmission is direct between mammals via contact with bacteria-rich material (Moreno and Moriyon 2006).

The epidemiology of *Bartonella* and *Brucella* infections as well as the interaction of these bacteria with and persistence within specific host cells suggest some host specificity that determines the selective incidence in certain mammals (Dehio and Sander 1999; Breitschwerdt and Kordick 2000; Tsolis 2002). This host specificity is governed by the intraerythrocytic persistence and the prevalence and ecology of the hematophagous vectors for *Bartonella* (Harms and Dehio 2012) and possibly by effectors translocated via the T4SS and distinctive gene sets localized to islands acquired by phage-mediated integration for *Brucella* (Paulsen et al. 2002; Andersson and Kempf 2004; de Jong and Tsolis 2012). When incidentally transmitted to humans, *Bartonella* and *Brucella* cause chronic infections that are frequently misdiagnosed, misreported, and therefore underestimated (Schwartzman 1996; Velho et al. 2003; Pappas et al. 2006).

Human bartonellosis ranges from the benign self-limiting manifestation of cat scratch disease (CSD) caused by infection with *Bartonella henselae* and possibly *Bartonella clarridgeiae* (Kordick et al. 1997), to a devastating hemolytic anemia caused by *B. bacilliformis*. *B. henselae* spreads in the feline reservoir by cat fleas (*Ctenocephalides felis*) (Chomel et al. 1996) and is transmitted to humans probably via inoculation of contaminated flea feces during a cat scratch or bite (Chomel et al. 2009). Symptoms of CSD are mainly lymphadenopathy (lymph node swelling) and occasionally fatigue and fever (Lamps and Scott 2004). *B. quintana* causes trench fever, a disease with persistent bacteremia that spreads under poor hygiene conditions by body lice (*Pediculus humanus corporis*) and oc-

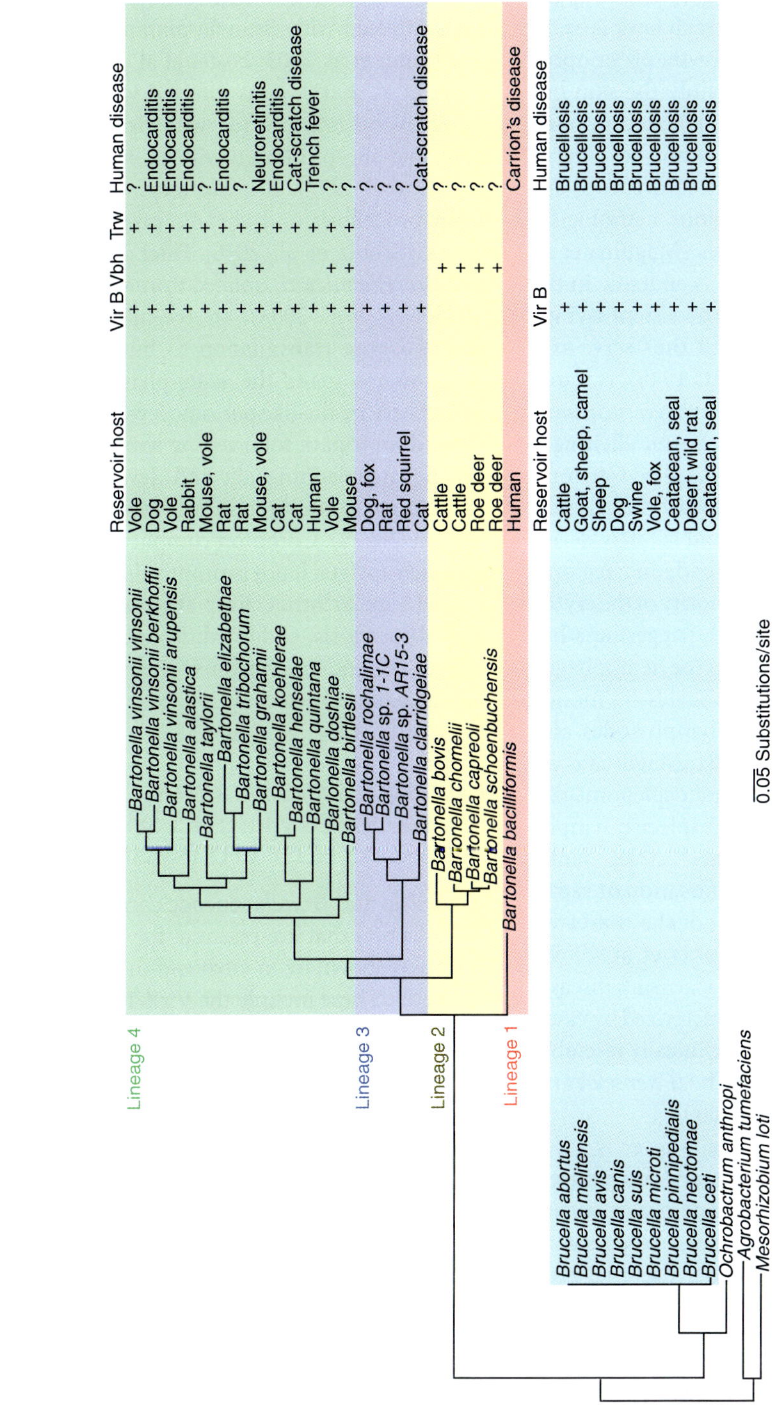

Figure 1. Phylogeny, T4SS, and epidemiology of *Bartonella* and *Brucella*. At least one type IV secretion system (T4SS; VirB, Vbh, and Trw) is present in all *Bartonella* species except for *Bartonella bacilliformis*. A related VirB T4SS is also present in *Brucella*.

casionally by cat fleas (Rolain et al. 2003). The symptoms are a 5-d cyclic fever, bone pain, and headache (Byam and Lloyd 1920; Ohl and Spach 2000). A sequel of CSD and trench fever is bacillary angiomatosis: an outgrowth of vasoproliferative tumors affecting mainly the skin but also the liver (bacillary peliosis), spleen, bone marrow, eyes, or other organs and leading to extraerythrocytic bacteremia, neuroretinitis, infective endocarditis, and various neurological symptoms such as encephalitis (Maguina et al. 2009). *B. bacilliformis*, which is endemic in the Andes, where the natives show persistent asymptomatic bacteremia and might thus serve as a reservoir for infection (Ricketts 1948), is transmitted by the sand fly (*Lutzomyia verrucarum*), and is the sole *Bartonella* pathogen eliciting a biphasic life-threatening illness called Carrion's disease. During the acute phase of this disease, which is termed Oroya fever and is rarely developed by the natives of the endemic regions, *B. bacilliformis* can infect a majority of the erythrocytes (Maguina et al. 2001), triggering a hemolytic anemia attributable to the hemophagocytosis of the infected erythrocytes by mononuclear phagocytes in spleen, lymph nodes, and liver (Walker and Winkler 1981; Maguina et al. 2009). This results in fever, hepatosplenomegaly (enlargement of the liver and spleen), lymphadenopathy, anemia (Fisman 2000; Janka 2007), and ultimately fatal suffocation or septicemia killing more than 80% of the untreated invalids (Ricketts 1948; Maguina et al. 2009). The second, chronic phase of Carrion's disease, called verruga peruana, is characterized by vasoproliferative skin lesions that clinically resemble bacillary angiomatosis caused by *B. henselae* and *B. quintana* (Maguina et al. 2009).

Human brucellosis or Malta fever is a neglected, debilitating, relapsing, and often travel-related febrile disease. It is endemic in several regions where animal brucellosis is prevalent and is estimated to be the major bacterial zoonosis worldwide. Malta fever is demanding to treat with antibiotics, and there are no vaccines for humans that prevent infection (Memish and Balkhy 2004; Young 2005; Pappas et al. 2006; Ariza et al. 2007; Pappas 2010). It is transmitted by contact with infected animals, most often goats, sheep, or camels infected with *B. melitensis*, cattle infected with *B. abortus*, or swine infected with *B. suis*. In rare cases, marine mammals infected with *Brucella pinnipedialis* or *B. ceti* (Sohn et al. 2003; Foster et al. 2007; Whatmore et al. 2008), dogs infected with *B. canis*, desert wood rats infected with *Brucella neotomae*, and sheep infected with *B. ovis* are carriers (Corbel 1997). *Brucella inopinata*, isolated from brucellosis and chronic pneumonia patients (Scholz et al. 2010; Tiller et al. 2010), and *Brucella microti*, isolated from vole and foxes (Scholz et al. 2008), are recently uncovered strains whose transmission to humans is unclear. Symptoms of the acute phase of human brucellosis are flu-like periodic fevers with muscle and joint pain followed by lymphadenopathy, hepatosplenomegaly, and development of granulomas at multiple sites of the body (Franco et al. 2007a). Common complications, some of which are fatal if not promptly diagnosed and treated, are arthritis, liver abscess formation, neurobrucellosis, endocarditis, and even abortion in the endemic regions (Khan et al. 2001; Young 2005; Franco et al. 2007b).

BARTONELLA AND *BRUCELLA*: HOMOLOGOUS VIRULENCE SYSTEMS FOR SNEAKY HOST-CELL INVASION

Bartonella and *Brucella* encode conserved virulence factors that are essential for host-cell invasion as shown by in vitro and in vivo studies (Table 1). These include the VirB T4SS and adhesion molecules.

Machineries Indispensable for Host Interactions

VirB Type IV Secretion Systems: Master Virulence Factors

VirB T4SS are central virulence factors dictating the success of infection. The VirB T4SS of *Bartonella* species (lineages 3 and 4) and *Brucella* species are related but distinct in their subunit composition: VirB2–VirB11 and VirB1–VirB12, respectively. They bridge the bacterial inner membrane, periplasm, and outer mem-

Table 1. Virulence factors of *Bartonella* and *Brucella*

	Bartonella	*Brucella*
T4SSs		
VirB	Confined to lineages 3 and 4: VirB2–VirB11 and coupling protein VirD4 Essential for infection	VirB1–VirB12 Essential for infection
Vbh	Confined to lineage 2 Putative role in infection	—
Trw	Confined to lineage 4 Mediates erythrocyte invasion Essential for persistence	—
VirB T4SS effectors		
	Beps: inhibit apoptosis; promote angiogenesis; promote invasome formation in *B. henselae*	Btp1: modulates microtubules' dynamics and TLR4-mediated signaling CstA: interacts with Sec24A PrpA: elicits T-cell-independent IL-10 secretion RicA: interacts with Rab2-GDP BPE005, BPE123, BPE275, BPE043, BvfA, VceA, VceC, BPE043, BPE152: unclear function
Regulatory systems		
Two-component regulatory systems (TC systems)	BatR/BatS Regulate essential gene transcription (i.e., *virB*)	BvrR/BvrS Regulate essential gene transcription (i.e., *virB*)
Quorum-sensing systems	—	VjbR/BlxR Regulate numerous genes (i.e., *vceA*, *vceC*)
Adhesion molecules		
Flagella	Confined to lineages 1–3 Mediate motility, erythrocytes' adhesion and invasion Escape recognition by TLR5	Act as T3SS? Essential for virulence
Trimeric autotransporters	BadA, Vomps Essential for infection	BtaE Essential for infection
Hemin-binding proteins	Hbp Essential for growth	OMP25/31
LPS		
Lipid A	Pentaacylated with long fatty acid side chain Inhibits TLR4 proinflammatory response and promotes interaction with DCs	Heptaacylated with long fatty acid side chain Inhibits TLR4 proinflammatory response
Core-polysaccharide	Uncharacterized function(s)	Inhibits complement activation and DC's maturation
O-polysaccharide	Uncharacterized function(s)	Inhibits T-cell activation

brane and extend into extracellular pili. This multiprotein complex functions as a translocation pore through which effector proteins are transferred into host cells, possibly by a piston mechanism (O'Callaghan et al. 1999; Sieira et al. 2000; Cascales and Christie 2003; Hwang and Gelvin 2004; Christie et al. 2005; Schroder and Dehio 2005; Zygmunt et al. 2006; Saenz et al. 2007; Wallden et al. 2010; Engel et al. 2011; de Jong and Tsolis 2012). As a canonical T4SS, the

Bartonella VirB system comprises an effector-recognition component: the coupling protein VirD4 (Schroder and Dehio 2005), which has no homolog in *Brucella*, and which instead expresses the unique VirB12 (O'Callaghan et al. 1999). The genes of the VirB T4SSs are encoded as operons (Berger and Christie 1994), which either locate together with effector genes (*Bartonella* lineage 4) or not (*Bartonella* lineage 3 and *Brucella*) (Schulein et al. 2005; Engel et al. 2011; de Jong and Tsolis 2012).

Because biosynthesis of the multiprotein VirB T4SS and its translocated effectors is costly for the bacterial cells, expression is tightly regulated on the transcriptional level. For instance, in *B. henselae*, the *virB* operon and the genes encoding the coupling protein and the translocated effectors are maximally induced at physiological pH (Quebatte et al. 2010), whereas the *virB* operon expression in virulent *Brucella* strains is maximally induced in acidic milieus reminiscent of the lumen of *Brucella*-containing vacuoles (BCVs) (Boschiroli et al. 2002; Rouot et al. 2003).

The well-studied family of bacterial two-component regulatory systems (TC) is composed of pairs of a histidine kinase and an interacting response regulator. The surface-exposed histidine kinase becomes autophosphorylated upon sensing of an extracellular stimulus. The phosphate group is subsequently transferred to the response regulator, which, in turn, triggers transcriptional changes (Casino et al. 2010). The VirB T4SS of *Bartonella* and *Brucella* are transcriptionally controlled by homologous TCs, that is, BatR/BatS in *Bartonella* is most active at pH7.4 (Quebatte et al. 2010), and BvrR/BvrS in *Brucella* is most active at low pH (Martinez-Nunez et al. 2010; Viadas et al. 2010; Lacerda et al. 2013).

In *Brucella*, the loci encoding the VirB T4SS and other virulence factors like flagella and outer membrane proteins are controlled by additional regulatory systems such as the quorum-sensing LuxR-type regulators VjbR and BlxR, the starvation-sensing stringent response regulator Rsh (RelA/SpoT homolog), and small regulatory RNAs (sRNA) (Roop et al. 2009; von Bargen et al. 2012).

The VirB-translocated *Bartonella* effector proteins (Bep; named BepA–BepG in *B. henselae*) display a composite domain architecture characterized by an amino-terminal effector domain and a carboxy-terminal secretion signal (Schmid et al. 2004; Dehio 2005; Schulein et al. 2005; Engel et al. 2011). The carboxy-terminal secretion signal is bipartite and composed of a Bep intracellular delivery domain (BID) of approximately 140 amino acids and a short, nonconserved, positively charged tail sequence (Schulein et al. 2005). Besides mediating T4SS-dependent translocation into host cells, some BID domains can also directly modulate host-cellular pathways. For instance, the BID domain of BepA promotes apoptosis inhibition and angiogenesis (Schmid et al. 2006; Scheidegger et al. 2009). The amino-terminal effector domain is best exemplified by the FIC (filamentation induced by cyclic AMP) domain, which is present in some effectors of lineage 3 and most effectors of lineage 4 (Engel et al. 2011). This domain mediates AMPylation, the covalent transfer on an adenyl monophosphate (AMP) moiety onto the hydroxyl side chains of host target proteins like the small GTPases of the Rho family (Palanivelu et al. 2011; Engel et al. 2012). Effectors lacking an FIC domain show duplicated BID domains and/or amino-terminal tyrosine motifs that get phosphorylated inside host cells by Src family kinase(s) (Schulein et al. 2005; Selbach et al. 2009).

The few described putative *Brucella* effectors have unclear functions and only one common feature: a nonconserved carboxy-terminal positively charged amino acid sequence considered essential for VirB T4SS-dependent translocation (von Bargen et al. 2012). Examples are (1) VceA and VceC, which belong to the VjbR regulon, consequently having their expression coregulated with the VirB T4SS (de Jong et al. 2008); (2) BPE123, whose translocation additionally depends on a Sec-secretion signal at the amino terminus; and (3) BPE005, BPE275, and BPE043 (Table 1) (Marchesini et al. 2011). Other putative effectors with indefinite positively charged amino acid carboxy-terminal sequences are (1) RicA, which is the sole effector having an assigned host partner, the guanosine diphosphate (GDP)-bound form of the small GTPase Rab2 (de Barsy et al. 2011); (2) Btp1 (TcpB in *B. melitensis*), which regulates innate immune responses via its Toll/interleukin-1 receptor (TIR) domain (Cirl et al. 2008; Salcedo et al. 2008) and possibly modulates microtubule dynamics (Radhakrishnan et al. 2011); (3) PrpA, a proline-racemase family member, which regulates immune responses (Spera et al. 2006); (4) CstA (conserved Sec24A-targeted protein A), which controls *Brucella* intracellular trafficking (de Barsy et al. 2012); and (5) BvfA, which is required for *Brucella* intracellular replication (Lavigne et al. 2005).

Instead of the VirB T4SS, some *Bartonella* species encode a VirB homologous (Vbh) T4SS (Fig. 1) (Dehio 2008). Vbh was likely the first T4SS acquired by the *Bartonella* lineage and is present in all species of lineage 2. It is believed to have been functionally replaced within lineages 3 and 4 by the subsequently acquired VirB T4SS (some lineage 4 species still encode nonfunctional remnants of Vbh) (Fig. 1). Vbh components display protein sequence identity of 40%–80% with their VirB homologs, suggesting that the two systems have only recently diverged from a common ancestor (Saenz et al. 2007). Further to the Vbh system, the *vbh* locus encodes a putative effector protein with the canonical FIC–BID architecture of the VirB-translocated Beps, advocating a potential role in virulence (Engel et al. 2011).

Adhesion Molecules: Key for Host–Pathogen Interaction

Adhesion molecules are often multifunctional membrane-exposed proteins and complexes necessary for *Bartonella* attachment to host cells (O'Rourke et al. 2011). Homologs of some of these adhesins are also present in *Brucella*, although their function is less scrutinized (von Bargen et al. 2012). They interact with host extracellular matrix (ECM) components (e.g., fibronectin, collagen) and receptors (e.g., integrins for *Bartonella* and sialic acid–bound receptors for *Brucella*) (Fig. 2). Such molecules include the Trw T4SS (for *Bartonella*), flagella, autotransporter adhesins, the ATP-binding cassette (ABC) systems, the hemin-binding proteins, and further outer membrane proteins (OMP) such as in *Bartonella* OMP89 and OMP43, a homolog of *Brucella* OMP2b (Dabo et al. 2006).

Trw T4SS. Trw is a T4SS that functionally diversified to an adhesion complex crucial for persistence of all *Bartonella* lineage 4 species (Schroder and Dehio 2005). The *trw* locus comprises an operon encoding a typical T4SS but as a particularity displays multiple tandem-repeated gene copies encoding variant forms of the extracellular pilus component. The *Bartonella* Trw T4SS is closely related to the canonical Trw conjugation system of the conjugative plasmid R388 and was thus probably gained by horizontal gene transfer (Seubert et al. 2003). The Trw T4SS mediates erythrocyte adhesion of lineage 4 species in a host-specific manner. The subsequent invasion process requires the invasion-associated locus, containing IalA and IalB, and/or the IalAB-independent erythrocyte invasion locus, Iiv (Seubert et al. 2003; Dehio 2004; Schroder and Dehio 2005; Saenz et al. 2007; Vayssier-Taussat et al. 2010; Deng et al. 2012).

Flagella. Flagella are bacterial nanomachines that mediate motility. The filament of flagella that is composed of a single protein, flagellin, is assembled by the basal body by a pathway homologous to the type 3 secretion system (T3SS) (Hueck 1998; Kirov 2003). In motile *Bartonella* species (lineages 1–3, except *Bartonella bovis*), flagella are considered to mediate erythrocyte internalization as a mechanical force or as adhesion molecules, as shown exemplarily for *B. bacilliformis* (Walker and Winkler 1981; Mernaugh and Ihler 1992). *Bartonella* species of lineage 4 that are nonmotile owing to the loss of flagella do mediate erythrocyte interaction via the Trw T4SS, which suggests that Trw has functionally replaced flagella for establishing interaction with erythrocytes (Dehio 2008). *Brucella* are nonmotile bacteria, yet flagella genes are essential for their virulence (Lestrate et al. 2003). Debatably, the basal body may function as a T3SS rather than establishing a motile nanomachine (Fretin et al. 2005).

Autotransporter Adhesins. This family mediates adhesion to nucleated cells and interac-

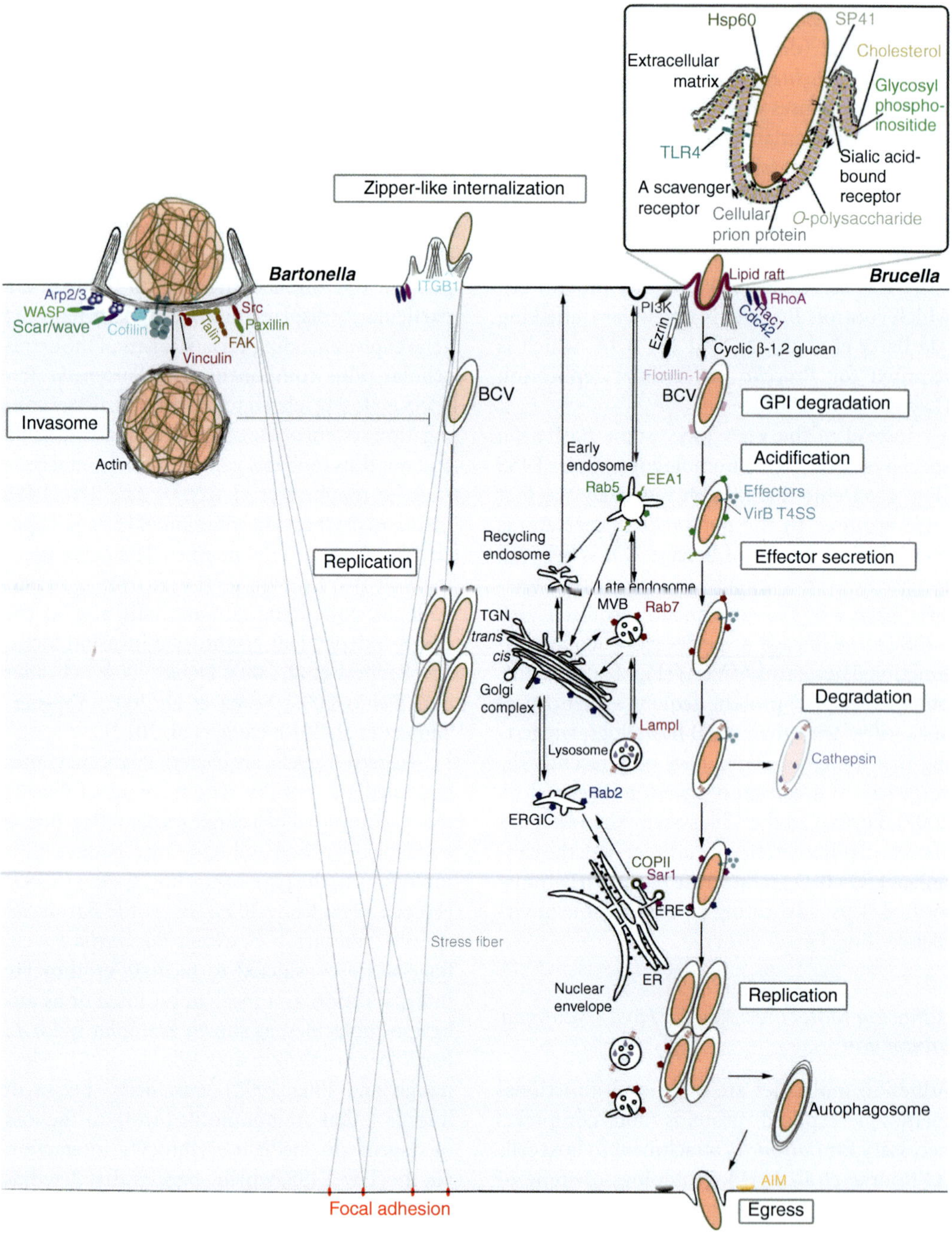

Figure 2. Intracellular life of *Bartonella* and *Brucella*. *Bartonella* and *Brucella* invade cells in vacuolar structures (BCV) or, in the case of *B. henselae*, as an aggregate in invasomes. Once internalized, these structures interact with specific host components and compartments subverting their function. AIM, Autophagy initiation markers; COPII, coatomer protein II; ER, endoplasmic reticulum; ERGIC, ER–Golgi intermediate compartment; ERES, ER exit sites; MVB, multivesicular bodies; T4SS, type IV secretion system.

tions with ECM, belongs to the type 5 secretion system (T5SS), and comprises two-partner systems (TPSs), classical monomeric autotransporter adhesins (CAAs), and trimeric autotransporter adhesins (TAAs). TPS filamentous haemagglutinins in *Bartonella* are putative adhesins encoded by multiple genes, indicating functional redundancy (O'Rourke et al. 2011). CAAs and TAAs are characterized by an amino-terminal signal peptide, a passenger domain, and a carboxy-terminal translocation domain (van Ulsen 2011). CAAs important for infection are in *Brucella* are OmaA (outer membrane autotransporter A) (Bandara et al. 2005) and BmaC (Posadas et al. 2012), and in *Bartonella* are Arp (acidic repeat protein) (Litwin et al. 2007) and Cfa (cAMP-like factor autotransporter) (Litwin and Johnson 2005), two members of the Iba (inducible *Bartonella* autotransporter) group whose expression is induced during infection. TAAs are essential for virulence and have a head–stalk–anchor structure forming modular length–adaptable trimerized cell surface-exposed filaments (Linke et al. 2006). Examples are BtaE in *Brucella* (Ruiz-Ranwez et al. 2013) and in *Bartonella*, *B. henselae* adhesin A (BadA) and the homologous *B. quintana* variably expressed outer membrane proteins (Vomps) (Zhang et al. 2004; Muller et al. 2011). *Bartonella* TAAs additionally induce angiogenesis and serum resistance (O'Rourke et al. 2011). Interestingly, BadA expression on the bacterial surface masks the activity of the VirB T4SS (Lu et al. 2013), but these two major virulence factors are differentially regulated on the transcriptional level via the TC system BatS/BatR, indicating that the two act at different stages of the infection process.

ATP-Binding Cassette Transporter Systems. ABC transporters are a family of transport systems that hydrolyze adenosine triphosphate (ATP) (Davidson and Chen 2004; Davidson et al. 2008). Examples are the heme HutABC/HmuV (Parrow et al. 2009) and iron FatBCD and Sit/YfeABCd (Battisti et al. 2006) uptake systems as well as the *ugp*-encoded glycerol 3-phosphate importers (Saenz et al. 2007) in *Bartonella*. Such an importer is in *Brucella* surface protein 41 (SP41), which probably interacts with sialylated host receptors (Castaneda-Roldan et al. 2006).

Hemin-Binding Proteins. Hemin-binding proteins (Hbps in *Bartonella* and OMP25/31 in *Brucella*) constitute a family of porin-like outer membrane proteins that bind hemin (Battisti et al. 2006; Dabo et al. 2006; Delpino et al. 2006; Caro-Hernandez et al. 2007). Hemin is essential for *Bartonella* growth (Sander et al. 2000), rendering these bacteria the most hemin-dependent microbes.

Strategies for Host Invasion

Lessons Learned from In Vitro Studies

Our understating of cellular invasion processes during *Bartonella* and *Brucella* infections is greatly based on in vitro infection models. These comprise primary cells and immortalized cell lines: professional phagocytes (e.g., macrophages, monocytes, dendritic cells), nonprofessional phagocytic cells (e.g., epithelial cells, fibroblasts, endothelial cells), erythrocytes for *Bartonella*, and trophoblasts for *Brucella* (Anderson et al. 1986; Benson et al. 1986; Kordick and Breitschwerdt 1995; Roop et al. 2009; Harms and Dehio 2012; von Bargen et al. 2012; Salcedo et al. 2013). Entry into cells leads to the establishment of BCVs (*Bartonella*/*Brucella*-containing vacuoles), structures formed when *Bartonella* and *Brucella* are taken up as single bacteria or as small bacterial clumps (Moreno and Gorvel 2004; Eicher and Dehio 2012). Some *Bartonella* species additionally enter host cells in aggregates of hundreds of bacteria, leading to the establishment of an invasion structure known as the invasome (Fig. 2) (Dehio et al. 1997).

Uptake into BCVs or Invasomes: Alternative Routes of Bacterial Entry Exemplified by B. henselae

At early stages of *B. henselae* interaction with host cells (also at subsequent stages in the case of VirB-T4SS deficiency), individual bacteria trigger a zipper-like entry mechanism involving Rho, Rac, Cdc42, tyrosine phosphorylation, and/or α5β1-integrin (Fig. 2) (Hill et al. 1992; Brouqui and Raoult 1996; Williams-Bouyer and

Hill 1999; Truttmann et al. 2011a). Once internalized, BCVs induce focal adhesion-anchored stress fibers and possibly replicate in the juxtanuclear region, where they presumably interact with the Golgi apparatus (Kempf et al. 2000; Verma et al. 2000, 2001; Verma and Ihler 2002; Kyme et al. 2005).

Once the VirB T4SS-translocated effectors reach a critical intracellular concentration, the formation of BCVs by individual bacterial uptake is blocked. Bacteria then accumulate on the cell surface, which is followed by bacterial aggregation by host-cell-driven surface transport, and subsequent internalization into a structure termed the invasome. This unique bacterial uptake structure is reminiscent of bacterial aggregates in bacillary angiomatosis lesions (Dehio et al. 1997; Dehio 2003). Invasome formation is triggered by VirB-translocated effectors via redundant pathways, which depend on either BepG or the combination of BepC and BepF. In each case, this results in the formation of a compact F-actin ring composed of condensed stress fibers, and membrane protrusions on the host-cell surface (Fig. 2) (Rhomberg et al. 2009; Truttmann et al. 2011b). This process is dependent on integrin-β1 (ITGB1) binding and activation via Talin-1 inside-out signaling and on integrin-β1-mediated outside-in signaling by focal adhesion kinase (FAK), Src, vinculin, and paxilin (Truttmann et al. 2011b). Invasome rearrangement of the F-actin cytoskeleton involves Arp2/3-dependent cortical F-actin polymerization by Rac1/Scar1/WAVE, Cdc42/WASP pathways, and additionally by cofilin-1 in the case of BepC and BepF (Rhomberg et al. 2009; Truttmann et al. 2011b,c).

Uptake of Brucella in BCVs

Contrary to *Bartonella*, the VirB T4SS of *Brucella* is not expressed during bacterial uptake and thus does not contribute to this process. Entry is mediated by surface-exposed bacterial and host membrane complexes at cholesterol- and glycosylphosphoinositide (GPI)-rich host-cell membrane domains (lipid rafts) (Pizarro-Cerda et al. 1998; Watarai et al. 2002, 2003; Kim et al. 2004; Nakato et al. 2012) in a phosphoinositide (PI)-3 kinase, actin, ezrin, Rho, Rac, Cdc42, and/or Toll-like receptor (TLR)4-dependent manner (Fig. 2) (Guzman-Verri et al. 2001; Chaves-Olarte et al. 2002; Watanabe et al. 2009). Once inside cells, BCVs' lipid raft organization is disrupted by the bacterial membrane periplasm-enriched *Brucella* virulence factor β-1,2 cyclic glucan, which promotes acidification and development of autophagosome properties (Pizarro-Cerda et al. 1998; Ugalde 1999; Arenas et al. 2000; Arellano-Reynoso et al. 2005; Starr et al. 2008). Acidification requires the recruitment of early endosome (Rab5, EEA1), late endosome (Rab7), and lysosome (Lamp1) markers, and is necessary for the activation of virulence factors such as *virB* genes (Porte et al. 1999; Boschiroli et al. 2002) essential for interactions with early secretory pathway components (Sar1/Sec24A/COPII, Rab2, and its partners) and endoplasmic reticulum-derived ribosome-rich structures (Comerci et al. 2001; Celli et al. 2003, 2005; Fugier et al. 2009) where *Brucella* replicate and persist before egressing in an autophagy initiation molecules and PI3 kinase-dependent manner (Roop et al. 2009; Starr et al. 2012).

Uptake in Erythrocytes: A Route to Persistence

A hallmark of *Bartonella* infection is invasion of erythrocytes using the Trw T4SS or flagella as adhesion factors (Harms and Dehio 2012). These processes depend on host-cell-surface glycosyl moieties, actin, spectrin, band 3 protein, glycophorin A, and/or glycophorin B (Walker and Winkler 1981; Iwaki-Egawa and Ihler 1997; Buckles and McGinnis Hill 2000; Chomel et al. 2009; Vayssier-Taussat et al. 2010; Deng et al. 2012). A potential amphiphilic small molecule, deformin, deforms the erythrocytes' membrane initiating "forced endocytosis" of *Bartonella* in vacuoles, where they shortly replicate but persist (Benson et al. 1986; Iwaki-Egawa and Ihler 1997; Derrick and Ihler 2001).

Lessons Learned from In Vivo Studies

Bartonella and *Brucella* infection spreads from the site of inoculation to distinct immune-privileged persistence niches that are essential for transmission to other hosts. During dissemination to the persistence niches, these pathogens may colonize other host niches. The identities of the primary niches remain elusive, although dendritic cells, and in the case of *Bartonella* also endothelial cells, have been proposed to represent such primary niches (Harms and Dehio 2012; von Bargen et al. 2012). The persistence niches are erythrocytes for *Bartonella* (Schulein and Dehio 2002) and the reticuloendothelial system for *Brucella* (Ficht 2003). Dissecting the sequence of events from the primary to the persistence niches is fundamental to understand the progression from an acute infection to a chronic disease. For this, clinical data and mostly laboratory in vivo models, in which different infection stages can be monitored, are essential. A widely used *Brucella* in vivo model is mouse, which largely recapitulates natural animal brucellosis (Silva et al. 2011). Several *Bartonella* species also have a small animal model reflecting infection either (1) in the reservoir host (Fig. 1), for example, rat for *Bartonella tribocorum* (Schulein and Dehio 2002) and mouse for *Bartonella birtlesii* and *Bartonella grahamii* (Koesling et al. 2001; Marignac et al. 2010); or (2) in a nonreservoir host as an incidental infection, for example, the lymphadenopathy mouse model for *B. henselae* (Kunz et al. 2008), mirroring human infection by zoonotic species (in this case, CSD). In vivo models also proved to be a significant tool in understanding immune responses during the course of infection.

BARTONELLA AND *BRUCELLA*: ANALOGOUS VIRULENCE FACTORS FOR DIVERSE APPROACHES TO MODULATE IMMUNE RESPONSES

Bartonella and *Brucella* possess conserved virulence factors essential for chronic infection. These include LPS and for some species flagella, which promote survival for longer periods in macrophages, interfere with dendritic cell (DC) maturation, and modulate immunity (Kohler et al. 2003; Roop et al. 2004; Dehio et al. 2012).

LPS Modifications: Proficiency to Escape Immune Responses

Lipopolysaccharides (LPSs) are conserved sheaths of Gram-negative bacteria composed of an endotoxin component lipid moiety (lipid A) that provides membrane anchoring, a core, and an extracellular polysaccharide chain (*O*-polysaccharide). They are pathogen-associated molecular pattern (PAMP), eliciting an immune response. However, *Bartonella* and *Brucella* LPSs have unique structural modifications that fail to display endotoxin activity (Table 1) (Moreno et al. 1981; Rasool et al. 1992; Matera et al. 2008). Similar chemical modifications aiming at detoxifying LPSs and thus dampening the immune response are used by other Gram-negative bacteria such as *Helicobacter pylori* (Miller et al. 2005).

Bartonella and *Brucella* lipid A structure is characterized by a short glycosyl backbone (diaminoglucose), a pentaacylation for *Bartonella* and a heptaacylation for *Brucella*, as well as a very long fatty acid side chain (Iriarte et al. 2004; Zahringer et al. 2004; Lapaque et al. 2006; Matera et al. 2008). This lipid A is poorly recognized by TLR4, thereby impeding a broad inflammatory response (Miller et al. 2005). *B. quintana* LPS is further a potent antagonist of TLR4 signaling (Popa et al. 2007).

Depending on the presence or absence of the *O*-polysaccharide, *Brucella* strains are, respectively, smooth human-infective or rough (Haag et al. 2010). *Brucella O*-polysaccharide inhibits T-cell activation, probably via sequestration of MHCII in large macrodomains in infected macrophages (Moreno et al. 1981; Forestier et al. 1999, 2000; Barquero-Calvo et al. 2007).

Brucella core polysaccharide inhibits DC maturation, TLR4–MD2 recognition, and complement activation, thus complement-mediated bacterial lysis (Conde-Alvarez et al. 2012). DC maturation is also inhibited by the *Brucella* T4SS effector Btp1, which, owing to its TIR domain, mimics host TLR signaling, specifically dampening TLR2 and TLR4 signaling and subsequent NF-κB activation. The Btp1 TIR domain mediates interaction with the death domain of Myd88 (myeloid differentiation factor

88) and/or TIRAP/MAL (TIR-associated protein), leading to TIRAP/MAL ubiquitination, then degradation (Cirl et al. 2008; Salcedo et al. 2008; Radhakrishnan et al. 2009; Sengupta et al. 2010; Chaudhary et al. 2012). Conversely, the glucose polymer β1,2-cyclic glucan activates DC in a TLR4 Myd88-dependent manner, thus counteracting Btp1-mediated inhibition of DC maturation (Martirosyan et al. 2012). Whether *Bartonella* has a similar DC maturation fine-tuning as *Brucella* is unclear, but *Bartonella* is hitherto known to stimulate DC maturation in a TLR2-dependent manner (Vermi et al. 2006; Pulliainen and Dehio 2012).

Flagellin Mutations: Adeptness to Escape Immune Responses

The conserved amino-terminal D1 domain of flagellin that mediates flagella assembly is also detected by TLR5, thereby triggering proinflammatory responses (Ramos et al. 2004). However, the flagellated *Bartonella* species and also a few other pathogens like *H. pylori* fail to elicit such immune responses thanks to amino acid changes in the D1 domain that dampen recognition by TLR5 and subsequent NF-κB activation. These flagellin mutations can interfere with flagella function and thus are always associated with compensatory mutations that preserve motility (Andersen-Nissen et al. 2005).

Additional Strategies to Subvert Immune Responses

Besides manipulating DC maturation and consequently T-cell-mediated immune responses, *Bartonella* and *Brucella* can also control T-cell-independent immune responses. For instance, the *Brucella* effector PrpA elicits T-cell-independent B-lymphocyte polyclonal activation and anti-inflammatory IL-10 cytokine secretion (Spera et al. 2006), which is also induced by an unknown *Bartonella* component (Matera et al. 2003, 2008). It is thus likely that both *Bartonella* and *Brucella* induce a transient and partial immune suppression to establish chronic infection (Couper et al. 2008; Harms and Dehio 2012).

Immune suppression is apparently also required for vasoproliferative tumor-like lesion formation during bacillary angiomatosis, a clinical manifestation caused by *B. henselae* or *B. quintana* infection in immunocompromised patients (Maguina et al. 2009). These lesions are characterized by small vessels lined with endothelial cells and are an amalgam of bacteria, polymorphonuclear neutrophils, macrophages, and endothelial cells, the latter two being effective producers of angiogenic factors (Cotell and Noskin 1994; Kostianovsky and Greco 1994; Manders 1996; Qian and Pollard 2010). In these lesions, *Bartonella* induces (1) autocrine angiogenesis in an angiopoetin-2, IL-8, ICAM-1, and/or Bcl-2-related protein A1-dependent manner (Cerimele et al. 2003; McCord et al. 2006); and/or (2) a VirB T4SS-dependent or -independent paracrine angiogenic loop, thus stimulating the secretion of the potent angiogenic stimulus, vascular endothelial growth factor (VEGF) (Kempf et al. 2001; Resto-Ruiz et al. 2002; Schulte et al. 2006). The VirB T4SS-dependent angiogenesis in *B. henselae* is the net effect of pro- and antiangiogenic activities of BepA/BepD and BepG, respectively (Scheidegger et al. 2009). In a Gαs-dependent manner, BepA binds adenylyl cyclase, consequently elevating cAMP levels and probably inhibiting NF-κB-dependent caspase activation and DNA fragmentation, two hallmarks of apoptosis (Kirby and Nekorchuk 2002; Schmid et al. 2006; Pulliainen et al. 2012). Similar to *Bartonella*, apoptosis suppression is a characteristic feature of *Brucella* infection. It is T4SS- and LPS-dependent and the result of (1) transcriptional regulation of pro- and antiapoptotic genes such as bcl-2 family A1 and caspase-3, and (2) inhibition of mitochondrial cytochrome *c* release and reactive oxygen species, thus dampening caspase activation (Gross et al. 2000; Eskra et al. 2003; Tolomeo et al. 2003; He et al. 2006).

CONCLUDING REMARKS

Pathogens of the closely related α-proteobacterial genera *Bartonella* and *Brucella* encode, on the one hand, homologous virulence devices and factors (i.e., VirB T4SS, LPS) and, on the other hand, specific effectors and strategies to invade, proliferate, and persist into host cells and organisms and thus engender distinctive chronic, occasionally fatal infections. Further investigation of virulence factors, deciphering their functions, and elucidating the host interaction partners and mechanisms involved in vitro and in vivo as well as in the global context of infection will be crucial to unveil the causes and processes of the persisting zoonosis. Such a goal requires concerted approaches that integrate basic sciences, veterinary sciences, and clinical medicine. This endeavor would improve diagnostics and therapeutics in animals and humans, allow the generation of animal and human vaccines, and eventually minimize outbreaks and eradicate the disease in the sporadic and endemic regions, respectively.

ACKNOWLEDGMENTS

We thank Dr. Philipp Engel for providing the phylogenetic tree of Figure 1. This study is supported by grant 31003A-132979 from the Swiss National Sciences Foundation and grant 51RT-0-126008 from SystemX.ch (to C.D.) and by The Fondation Recherche Médicale, FRM (to J.-P.G.).

REFERENCES

Andersen-Nissen E, Smith KD, Strobe KL, Barrett SL, Cookson BT, Logan SM, Aderem A. 2005. Evasion of Toll-like receptor 5 by flagellated bacteria. *Proc Natl Acad Sci* **102:** 9247–9252.

Anderson TD, Meador VP, Cheville NF. 1986. Pathogenesis of placentitis in the goat inoculated with *Brucella abortus*. I. Gross and histologic lesions. *Vet Pathol* **23:** 219–226.

Andersson SG, Kempf VA. 2004. Host cell modulation by human, animal and plant pathogens. *Int J Med Microbiol* **293:** 463–470.

Arellano-Reynoso B, Lapaque N, Salcedo S, Briones G, Ciocchini AE, Ugalde R, Moreno E, Moriyon I, Gorvel JP. 2005. Cyclic β-1,2-glucan is a *Brucella* virulence factor required for intracellular survival. *Nat Immunol* **6:** 618–625.

Arenas GN, Staskevich AS, Aballay A, Mayorga LS. 2000. Intracellular trafficking of *Brucella abortus* in J774 macrophages. *Infect Immun* **68:** 4255–4263.

Ariza J, Bosilkovski M, Cascio A, Colmenero JD, Corbel MJ, Falagas ME, Memish ZA, Roushan MR, Rubinstein E, Sipsas NV, et al. 2007. Perspectives for the treatment of brucellosis in the 21st century: The Ioannina recommendations. *PLoS Med* **4:** e317.

Backert S, Meyer TF. 2006. Type IV secretion systems and their effectors in bacterial pathogenesis. *Curr Opin Microbiol* **9:** 207–217.

Bandara AB, Sriranganathan N, Schurig GG, Boyle SM. 2005. Putative outer membrane autotransporter protein influences survival of *Brucella suis* in BALB/c mice. *Vet Microbiol* **109:** 95–104.

Barquero-Calvo E, Chaves-Olarte E, Weiss DS, Guzman-Verri C, Chacon-Diaz C, Rucavado A, Moriyon I, Moreno E. 2007. *Brucella abortus* uses a stealthy strategy to avoid activation of the innate immune system during the onset of infection. *PLoS ONE* **2:** e631.

Battisti JM, Sappington KN, Smitherman LS, Parrow NL, Minnick MF. 2006. Environmental signals generate a differential and coordinated expression of the heme receptor gene family of *Bartonella quintana*. *Infect Immun* **74:** 3251–3261.

Batut J, Andersson SG, O'Callaghan D. 2004. The evolution of chronic infection strategies in the α-proteobacteria. *Nat Rev Microbiol* **2:** 933–945.

Benson LA, Kar S, McLaughlin G, Ihler GM. 1986. Entry of *Bartonella bacilliformis* into erythrocytes. *Infect Immun* **54:** 347–353.

Berger BR, Christie PJ. 1994. Genetic complementation analysis of the *Agrobacterium tumefaciens* virB operon: *virB2* through *virB11* are essential virulence genes. *J Bacteriol* **176:** 3646–3660.

Billeter SA, Levy MG, Chomel BB, Breitschwerdt EB. 2008. Vector transmission of *Bartonella* species with emphasis on the potential for tick transmission. *Med Vet Entomol* **22:** 1–15.

Boschiroli ML, Ouahrani-Bettache S, Foulongne V, Michaux-Charachon S, Bourg G, Allardet-Servent A, Cazevieille C, Liautard JP, Ramuz M, O'Callaghan D. 2002. The *Brucella suis virB* operon is induced intracellularly in macrophages. *Proc Natl Acad Sci* **99:** 1544–1549.

Breitschwerdt EB, Kordick DL. 2000. *Bartonella* infection in animals: Carriership, reservoir potential, pathogenicity, and zoonotic potential for human infection. *Clin Microbiol Rev* **13:** 428–438.

Brouqui P, Raoult D. 1996. *Bartonella quintana* invades and multiplies within endothelial cells in vitro and in vivo and forms intracellular blebs. *Res Microbiol* **147:** 719–731.

Buckles EL, McGinnis Hill E. 2000. Interaction of *Bartonella bacilliformis* with human erythrocyte membrane proteins. *Microb Pathog* **29:** 165–174.

Byam W, Lloyd L. 1920. Trench fever: Its epidemiology and endemiology. *Proc R Soc Med* **13:** 1–27.

Caro-Hernandez P, Fernandez-Lago L, de Miguel MJ, Martin-Martin AI, Cloeckaert A, Grillo MJ, Vizcaino N. 2007. Role of the Omp25/Omp31 family in outer membrane properties and virulence of *Brucella ovis*. *Infect Immun* **75:** 4050–4061.

Cascales E, Christie PJ. 2003. The versatile bacterial type IV secretion systems. *Nat Rev Microbiol* **1:** 137–149.

Casino P, Rubio V, Marina A. 2010. The mechanism of signal transduction by two-component systems. *Curr Opin Struct Biol* **20:** 763–771.

Castaneda-Roldan EI, Ouahrani-Bettache S, Saldana Z, Avelino F, Rendon MA, Dornand J, Giron JA. 2006. Characterization of SP41, a surface protein of *Brucella* associated with adherence and invasion of host epithelial cells. *Cell Microbiol* **8:** 1877–1887.

Celli J, de Chastellier C, Franchini DM, Pizarro-Cerda J, Moreno E, Gorvel JP. 2003. *Brucella* evades macrophage killing via VirB-dependent sustained interactions with the endoplasmic reticulum. *J Exp Med* **198:** 545–556.

Celli J, Salcedo SP, Gorvel JP. 2005. *Brucella* coopts the small GTPase Sar1 for intracellular replication. *Proc Natl Acad Sci* **102:** 1673–1678.

Cerimele F, Brown LF, Bravo F, Ihler GM, Kouadio P, Arbiser JL. 2003. Infectious angiogenesis: *Bartonella bacilliformis* infection results in endothelial production of angiopoetin-2 and epidermal production of vascular endothelial growth factor. *Am J Pathol* **163:** 1321–1327.

Chaudhary A, Ganguly K, Cabantous S, Waldo GS, Micheva-Viteva SN, Nag K, Hlavacek WS, Tung CS. 2012. The *Brucella* TIR-like protein TcpB interacts with the death domain of MyD88. *Biochem Biophys Res Com* **417:** 299–304.

Chaves-Olarte E, Guzman-Verri C, Meresse S, Desjardins M, Pizarro-Cerda J, Badilla J, Gorvel JP, Moreno E. 2002. Activation of Rho and Rab GTPases dissociates *Brucella abortus* internalization from intracellular trafficking. *Cell Microbiol* **4:** 663–676.

Chomel BB, Kasten RW, Floyd-Hawkins K, Chi B, Yamamoto K, Roberts-Wilson J, Gurfield AN, Abbott RC, Pedersen NC, Koehler JE. 1996. Experimental transmission of *Bartonella henselae* by the cat flea. *J Clin Microbiol* **34:** 1952–1956.

Chomel BB, Boulouis HJ, Breitschwerdt EB, Kasten RW, Vayssier-Taussat M, Birtles RJ, Koehler JE, Dehio C. 2009. Ecological fitness and strategies of adaptation of *Bartonella* species to their hosts and vectors. *Vet Res* **40:** 29.

Christie PJ, Atmakuri K, Krishnamoorthy V, Jakubowski S, Cascales E. 2005. Biogenesis, architecture, and function of bacterial type IV secretion systems. *Annu Rev Microbiol* **59:** 451–485.

Cirl C, Wieser A, Yadav M, Duerr S, Schubert S, Fischer H, Stappert D, Wantia N, Rodriguez N, Wagner H, et al. 2008. Subversion of Toll-like receptor signaling by a unique family of bacterial Toll/interleukin-1 receptor domain-containing proteins. *Nat Med* **14:** 399–406.

Comerci DJ, Martinez-Lorenzo MJ, Sieira R, Gorvel JP, Ugalde RA. 2001. Essential role of the VirB machinery in the maturation of the *Brucella abortus*–containing vacuole. *Cell Microbiol* **3:** 159–168.

Conde-Alvarez R, Arce-Gorvel V, Iriarte M, Mancek-Keber M, Barquero-Calvo E, Palacios-Chaves L, Chacon-Diaz C, Chaves-Olarte E, Martirosyan A, von Bargen K, et al. 2012. The lipopolysaccharide core of *Brucella abortus* acts as a shield against innate immunity recognition. *PLoS Pathog* **8:** e1002675.

Corbel MJ. 1997. Brucellosis: An overview. *Emerg Infect Dis* **3:** 213–221.

Cotell SL, Noskin GA. 1994. Bacillary angiomatosis. Clinical and histologic features, diagnosis, and treatment. *Arch Intern Med* **154:** 524–528.

Couper KN, Blount DG, Riley EM. 2008. IL-10: The master regulator of immunity to infection. *J Immunol* **180:** 5771–5777.

Dabo SM, Confer AW, Saliki JT, Anderson BE. 2006. Binding of *Bartonella henselae* to extracellular molecules: Identification of potential adhesins. *Microb Pathog* **41:** 10–20.

Davidson AL, Chen J. 2004. ATP-binding cassette transporters in bacteria. *Annu Rev Biochem* **73:** 241–268.

Davidson AL, Dassa E, Orelle C, Chen J. 2008. Structure, function, and evolution of bacterial ATP-binding cassette systems. *Microbiol Mol Biol Rev* **72:** 317–364.

de Barsy M, Jamet A, Filopon D, Nicolas C, Laloux G, Rual JF, Muller A, Twizere JC, Nkengfac B, Vandenhaute J, et al. 2011. Identification of a *Brucella* spp. secreted effector specifically interacting with human small GTPase Rab2. *Cell Microbiol* **13:** 1044–1058.

de Barsy M, Mirabella A, Letesson JJ, De Bolle X. 2012. A *Brucella abortus cstA* mutant is defective for association with endoplasmic reticulum exit sites and displays altered trafficking in HeLa cells. *Microbiol* **158:** 2610–2618.

Dehio C. 2003. Recent progress in understanding *Bartonella*-induced vascular proliferation. *Curr Opin Microbiol* **6:** 61–65.

Dehio C. 2004. Molecular and cellular basis of *Bartonella* pathogenesis. *Annu Rev Microbiol* **58:** 365–390.

Dehio C. 2005. *Bartonella*–host-cell interactions and vascular tumour formation. *Nat Rev Microbiol* **3:** 621–631.

Dehio C. 2008. Infection-associated type IV secretion systems of *Bartonella* and their diverse roles in host cell interaction. *Cell Microbiol* **10:** 1591–1598.

Dehio C, Sander A. 1999. *Bartonella* as emerging pathogens. *Trends Microbiol* **7:** 226–228.

Dehio C, Meyer M, Berger J, Schwarz H, Lanz C. 1997. Interaction of *Bartonella henselae* with endothelial cells results in bacterial aggregation on the cell surface and the subsequent engulfment and internalisation of the bacterial aggregate by a unique structure, the invasome. *J Cell Sci* **110:** 2141–2154.

Dehio C, Berry C, Bartenschlager R. 2012. Persistent intracellular pathogens. *FEMS Microbiol Rev* **36:** 513.

de Jong MF, Tsolis RM. 2012. Brucellosis and type IV secretion. *Future Microbiol* **7:** 47–58.

de Jong MF, Sun YH, den Hartigh AB, van Dijl JM, Tsolis RM. 2008. Identification of VceA and VceC, two members of the VjbR regulon that are translocated into macrophages by the *Brucella* type IV secretion system. *Mol Microbiol* **70:** 1378–1396.

Delpino MV, Cassataro J, Fossati CA, Goldbaum FA, Baldi PC. 2006. *Brucella* outer membrane protein Omp31 is a haemin-binding protein. *Microbes Infect* **8:** 1203–1208.

Deng HK, Le Rhun D, Le Naour E, Bonnet S, Vayssier-Taussat M. 2012. Identification of *Bartonella* Trw host-specific receptor on erythrocytes. *PLoS ONE* **7:** e41447.

Derrick SC, Ihler GM. 2001. Deformin, a substance found in *Bartonella bacilliformis* culture supernatants, is a small, hydrophobic molecule with an affinity for albumin. *Blood Cells Mol Dis* **27:** 1013–1019.

Eicher SC, Dehio C. 2012. *Bartonella* entry mechanisms into mammalian host cells. *Cell Microbiol* **14:** 1166–1173.

Engel P, Dehio C. 2009. Genomics of host-restricted pathogens of the genus *Bartonella*. *Genome Dyn* **6:** 158–169.

Engel P, Salzburger W, Liesch M, Chang CC, Maruyama S, Lanz C, Calteau A, Lajus A, Medigue C, Schuster SC, et al.

2011. Parallel evolution of a type IV secretion system in radiating lineages of the host-restricted bacterial pathogen *Bartonella*. *PLoS Genet* **7:** e1001296.

Engel P, Goepfert A, Stanger FV, Harms A, Schmidt A, Schirmer T, Dehio C. 2012. Adenylylation control by intra- or intermolecular active-site obstruction in Fic proteins. *Nature* **482:** 107–110.

Eskra L, Mathison A, Splitter G. 2003. Microarray analysis of mRNA levels from RAW264.7 macrophages infected with *Brucella abortus*. *Infect Immun* **71:** 1125–1133.

Ficht TA. 2003. Intracellular survival of *Brucella*: Defining the link with persistence. *Vet Microbiol* **92:** 213–223.

Fisman DN. 2000. Hemophagocytic syndromes and infection. *Emerg Infect Dis* **6:** 601–608.

Forestier C, Moreno E, Meresse S, Phalipon A, Olive D, Sansonetti P, Gorvel JP. 1999. Interaction of *Brucella abortus* lipopolysaccharide with major histocompatibility complex class II molecules in B lymphocytes. *Infect Immun* **67:** 4048–4054.

Forestier C, Deleuil F, Lapaque N, Moreno E, Gorvel JP. 2000. *Brucella abortus* lipopolysaccharide in murine peritoneal macrophages acts as a down-regulator of T cell activation. *J Immunol* **165:** 5202–5210.

Foster G, Osterman BS, Godfroid J, Jacques I, Cloeckaert A. 2007. *Brucella ceti* sp. nov. and *Brucella pinnipedialis* sp. nov. for *Brucella* strains with cetaceans and seals as their preferred hosts. *Int J Syst Evol Microbiol* **57:** 2688–2693.

Foster JT, Beckstrom-Sternberg SM, Pearson T, Beckstrom-Sternberg JS, Chain PS, Roberto FF, Hnath J, Brettin T, Keim P. 2009. Whole-genome-based phylogeny and divergence of the genus *Brucella*. *J Bacteriol* **191:** 2864–2870.

Franco MP, Mulder M, Gilman RH, Smits HL. 2007a. Human brucellosis. *Lancet Infect Dis* **7:** 775–786.

Franco MP, Mulder M, Smits HL. 2007b. Persistence and relapse in brucellosis and need for improved treatment. *Trans R Soc Trop Med Hyg* **101:** 854–855.

Fretin D, Fauconnier A, Kohler S, Halling S, Leonard S, Nijskens C, Ferooz J, Lestrate P, Delrue RM, Danese I, et al. 2005. The sheathed flagellum of *Brucella melitensis* is involved in persistence in a murine model of infection. *Cell Microbiol* **7:** 687–698.

Fugier E, Salcedo SP, de Chastellier C, Pophillat M, Muller A, Arce-Gorvel V, Fourquet P, Gorvel JP. 2009. The glyceraldehyde-3-phosphate dehydrogenase and the small GTPase Rab 2 are crucial for *Brucella* replication. *PLoS Pathog* **5:** e1000487.

Gillespie JJ, Wattam AR, Cammer SA, Gabbard JL, Shukla MP, Dalay O, Driscoll T, Hix D, Mane SP, Mao C, et al. 2011. PATRIC: The comprehensive bacterial bioinformatics resource with a focus on human pathogenic species. *Infect Immun* **79:** 4286–4298.

Gross A, Terraza A, Ouahrani-Bettache S, Liautard JP, Dornand J. 2000. In vitro *Brucella suis* infection prevents the programmed cell death of human monocytic cells. *Infect Immun* **68:** 342–351.

Guzman-Verri C, Chaves-Olarte E, von Eichel-Streiber C, Lopez-Goni I, Thelestam M, Arvidson S, Gorvel JP, Moreno E. 2001. GTPases of the Rho subfamily are required for *Brucella abortus* internalization in nonprofessional phagocytes: Direct activation of Cdc42. *J Biol Chem* **276:** 44435–44443.

Haag AF, Myka KK, Arnold MF, Caro-Hernandez P, Ferguson GP. 2010. Importance of lipopolysaccharide and cyclic β-1,2-glucans in *Brucella*-mammalian infections. *Int J Microbiol* **2010:** 124509.

Harms A, Dehio C. 2012. Intruders below the radar: Molecular pathogenesis of *Bartonella* spp. *Clin Microbiol Rev* **25:** 42–78.

He Y, Reichow S, Ramamoorthy S, Ding X, Lathigra R, Craig JC, Sobral BW, Schurig GG, Sriranganathan N, Boyle SM. 2006. *Brucella melitensis* triggers time-dependent modulation of apoptosis and down-regulation of mitochondrion-associated gene expression in mouse macrophages. *Infect Immun* **74:** 5035–5046.

Hignett PG, Nagy LK, Ironside CJ. 1966. Bovine brucellosis: A study of an adult-vaccinated, *Brucella*-infected herd. I. The effect of *Brucella abortus* infection on fertility. *Vet Rec* **79:** 886–887.

Hill EM, Raji A, Valenzuela MS, Garcia F, Hoover R. 1992. Adhesion to and invasion of cultured human cells by *Bartonella bacilliformis*. *Infect Immun* **60:** 4051–4058.

Hueck CJ. 1998. Type III protein secretion systems in bacterial pathogens of animals and plants. *Microbiol Mol Biol Rev* **62:** 379–433.

Hwang HH, Gelvin SB. 2004. Plant proteins that interact with VirB2, the *Agrobacterium tumefaciens* pilin protein, mediate plant transformation. *Plant Cell* **16:** 3148–3167.

Iriarte M, Gonzalez D, Delrue RM, Monreal D, Conde R, Lopez-Goni I, Letesson JJ, Moriyon I. 2004. *Brucella* lipopolysaccharide: Structure, biosynthesis and genetics. In *Brucella: Molecular and cellular biology* (ed. Lopez-Goni I, Moriyon I), pp. 152–183. Horizon Bioscience, Norwich, UK.

Iwaki-Egawa S, Ihler GM. 1997. Comparison of the abilities of proteins from *Bartonella bacilliformis* and *Bartonella henselae* to deform red cell membranes and to bind to red cell ghost proteins. *FEMS Microbiol Lett* **157:** 207–217.

Janka GE. 2007. Hemophagocytic syndromes. *Blood Rev* **21:** 245–253.

Kempf VA, Schaller M, Behrendt S, Volkmann B, Aepfelbacher M, Cakman I, Autenrieth IB. 2000. Interaction of *Bartonella henselae* with endothelial cells results in rapid bacterial rRNA synthesis and replication. *Cell Microbiol* **2:** 431–441.

Kempf VA, Volkmann B, Schaller M, Sander CA, Alitalo K, Riess T, Autenrieth IB. 2001. Evidence of a leading role for VEGF in *Bartonella henselae*–induced endothelial cell proliferations. *Cell Microbiol* **3:** 623–632.

Khan MY, Mah MW, Memish ZA. 2001. Brucellosis in pregnant women. *Clin Infect Dis* **32:** 1172–1177.

Kim S, Watarai M, Suzuki H, Makino S, Kodama T, Shirahata T. 2004. Lipid raft microdomains mediate class A scavenger receptor-dependent infection of *Brucella abortus*. *Microb Pathog* **37:** 11–19.

Kirby JE, Nekorchuk DM. 2002. *Bartonella*-associated endothelial proliferation depends on inhibition of apoptosis. *Proc Natl Acad Sci* **99:** 4656–4661.

Kirov SM. 2003. Bacteria that express lateral flagella enable dissection of the multifunctional roles of flagella in pathogenesis. *FEMS Microbiol Lett* **224:** 151–159.

Koesling J, Aebischer T, Falch C, Schulein R, Dehio C. 2001. Cutting edge: Antibody-mediated cessation of hemotropic infection by the intraerythrocytic mouse pathogen *Bartonella grahamii*. *J Immunol* **167:** 11–14.

Kohler S, Michaux-Charachon S, Porte F, Ramuz M, Liautard JP. 2003. What is the nature of the replicative niche of a stealthy bug named *Brucella*? *Trends Microbiol* **11:** 215–219.

Kordick DL, Breitschwerdt EB. 1995. Intraerythrocytic presence of *Bartonella henselae*. *J Clin Microbiol* **33:** 1655–1656.

Kordick DL, Hilyard EJ, Hadfield TL, Wilson KH, Steigerwalt AG, Brenner DJ, Breitschwerdt EB. 1997. *Bartonella clarridgeiae*, a newly recognized zoonotic pathogen causing inoculation papules, fever, and lymphadenopathy (cat scratch disease). *J Clin Microbiol* **35:** 1813–1818.

Kostianovsky M, Greco MA. 1994. Angiogenic process in bacillary angiomatosis. *Ultrastruct Pathol* **18:** 349–355.

Kunz S, Oberle K, Sander A, Bogdan C, Schleicher U. 2008. Lymphadenopathy in a novel mouse model of *Bartonella*-induced cat scratch disease results from lymphocyte immigration and proliferation and is regulated by interferon-α/β. *Am J Pathol* **172:** 1005–1018.

Kyme PA, Haas A, Schaller M, Peschel A, Iredell J, Kempf VA. 2005. Unusual trafficking pattern of *Bartonella henselae*-containing vacuoles in macrophages and endothelial cells. *Cell Microbiol* **7:** 1019–1034.

Lacerda TL, Salcedo SP, Gorvel JP. 2013. *Brucella* T4SS: The VIP pass inside host cells. *Curr Opin Microbiol* **16:** 45–51.

Lamps LW, Scott MA. 2004. Cat-scratch disease: Historic, clinical, and pathologic perspectives. *Am J Clin Pathol* **121:** S71–S80.

Lapaque N, Forquet F, de Chastellier C, Mishal Z, Jolly G, Moreno E, Moriyon I, Heuser JE, He HT, Gorvel JP. 2006. Characterization of *Brucella abortus* lipopolysaccharide macrodomains as mega rafts. *Cell Microbiol* **8:** 197–206.

Lavigne JP, Patey G, Sangari FJ, Bourg G, Ramuz M, O'Callaghan D, Michaux-Charachon S. 2005. Identification of a new virulence factor, BvfA, in *Brucella suis*. *Infect Immun* **73:** 5524–5529.

Lestrate P, Dricot A, Delrue RM, Lambert C, Martinelli V, De Bolle X, Letesson JJ, Tibor A. 2003. Attenuated signature-tagged mutagenesis mutants of *Brucella melitensis* identified during the acute phase of infection in mice. *Infect Immun* **71:** 7053–7060.

Linke D, Riess T, Autenrieth IB, Lupas A, Kempf VA. 2006. Trimeric autotransporter adhesins: Variable structure, common function. *Trends Microbiol* **14:** 264–270.

Litwin CM, Johnson JM. 2005. Identification, cloning, and expression of the CAMP-like factor autotransporter gene (Cfa) of *Bartonella henselae*. *Infect Immun* **73:** 4205–4213.

Litwin CM, Rawlins ML, Swenson EM. 2007. Characterization of an immunogenic outer membrane autotransporter protein, Arp, of *Bartonella henselae*. *Infect Immun* **75:** 5255–5263.

Llosa M, Roy C, Dehio C. 2009. Bacterial type IV secretion systems in human disease. *Mol Microbiol* **73:** 141–151.

Lu YY, Franz B, Truttmann MC, Riess T, Gay-Fraret J, Faustmann M, Kempf VA, Dehio C. 2013. *Bartonella henselae* trimeric autotransporter adhesin BadA expression interferes with effector translocation by the VirB/D4 type IV secretion system. *Cell Microbiol* **15:** 759–778.

Maguina C, Garcia PJ, Gotuzzo E, Cordero L, Spach DH. 2001. Bartonellosis (Carrion's disease) in the modern era. *Clin Infect Dis* **33:** 772–779.

Maguina C, Guerra H, Ventosilla P. 2009. Bartonellosis. *Clin Dermatol* **27:** 271–280.

Manders SM. 1996. Bacillary angiomatosis. *Clin Dermatol* **14:** 295–299.

Marchesini MI, Herrmann CK, Salcedo SP, Gorvel JP, Comerci DJ. 2011. In search of *Brucella abortus* type IV secretion substrates: Screening and identification of four proteins translocated into host cells through VirB system. *Cell Microbiol* **13:** 1261–1274.

Marignac G, Barrat F, Chomel B, Vayssier-Taussat M, Gandoin C, Bouillin C, Boulouis HJ. 2010. Murine model for *Bartonella birtlesii* infection: New aspects. *Comp Immunol Microbiol Infect Dis* **33:** 95–107.

Martinez-Nunez C, Altamirano-Silva P, Alvarado-Guillen F, Moreno E, Guzman-Verri C, Chaves-Olarte E. 2010. The two-component system BvrR/BvrS regulates the expression of the type IV secretion system VirB in *Brucella abortus*. *J Bacteriol* **192:** 5603–5608.

Martirosyan A, Perez-Gutierrez C, Banchereau R, Dutartre H, Lecine P, Dullaers M, Mello M, Pinto Salcedo S, Muller A, Leserman L, et al. 2012. *Brucella* β 1,2 cyclic glucan is an activator of human and mouse dendritic cells. *PLoS Pathog* **8:** e1002983.

Matera G, Liberto MC, Quirino A, Barreca GS, Lamberti AG, Iannone M, Mancuso E, Palma E, Cufari FA, Rotiroti D, et al. 2003. *Bartonella quintana* lipopolysaccharide effects on leukocytes, CXC chemokines and apoptosis: A study on the human whole blood and a rat model. *Int Immunopharmacol* **3:** 853–864.

Matera G, Liberto MC, Joosten LA, Vinci M, Quirino A, Pulicari MC, Kullberg BJ, Van der Meer JW, Netea MG, Foca A. 2008. The Janus face of *Bartonella quintana* recognition by Toll-like receptors (TLRs): A review. *Eur Cytokine Netw* **19:** 113–118.

McCord AM, Resto-Ruiz SI, Anderson BE. 2006. Autocrine role for interleukin-8 in *Bartonella henselae*-induced angiogenesis. *Infect Immun* **74:** 5185–5190.

Memish ZA, Balkhy HH. 2004. Brucellosis and international travel. *J Travel Med* **11:** 49–55.

Mernaugh G, Ihler GM. 1992. Deformation factor: An extracellular protein synthesized by *Bartonella bacilliformis* that deforms erythrocyte membranes. *Infect Immun* **60:** 937–943.

Miller SI, Ernst RK, Bader MW. 2005. LPS, TLR4 and infectious disease diversity. *Nat Rev Microbiol* **3:** 36–46.

Moreno E, Gorvel JP. 2004. Invasion, intracellular trafficking and replication of *Brucella* organisms in professional and nonprofessional phagocytes. In *Brucella: Molecular and cellular biology* (ed. Moriyon I, Lopez-Goni I), pp. 287–312. Horizon Bioscience, Norwich, UK.

Moreno E, Moriyon I. 2006. The genus *Brucella*. In *Prokaryotes: A handbook on the biology of bacteria*, 3rd ed. (ed. Dworkin M, et al.), Vol. 5, pp. 315–456. Springer, New York.

Moreno E, Speth SL, Jones LM, Berman DT. 1981. Immunochemical characterization of *Brucella* lipopolysaccharides and polysaccharides. *Infect Immun* **31:** 214–222.

Moreno E, Cloeckaert A, Moriyon I. 2002. *Brucella* evolution and taxonomy. *Vet Microbiol* **90:** 209–227.

Muller NF, Kaiser PO, Linke D, Schwarz H, Riess T, Schafer A, Eble JA, Kempf VA. 2011. Trimeric autotransporter adhesin-dependent adherence of *Bartonella henselae*, *Bartonella quintana*, and *Yersinia enterocolitica* to matrix components and endothelial cells under static and dynamic flow conditions. *Infect Immun* **79:** 2544–2553.

Nakato G, Hase K, Suzuki M, Kimura M, Ato M, Hanazato M, Tobiume M, Horiuchi M, Atarashi R, Nishida N, et al. 2012. Cutting edge: *Brucella abortus* exploits a cellular prion protein on intestinal M cells as an invasive receptor. *J Immunol* **189:** 1540–1544.

O'Callaghan D, Cazevieille C, Allardet-Servent A, Boschiroli ML, Bourg G, Foulongne V, Frutos P, Kulakov Y, Ramuz M. 1999. A homologue of the *Agrobacterium tumefaciens* VirB and *Bordetella pertussis* Ptl type IV secretion systems is essential for intracellular survival of *Brucella suis*. *Mol Microbiol* **33:** 1210–1220.

Ohl ME, Spach DH. 2000. *Bartonella quintana* and urban trench fever. *Clin Infect Dis* **31:** 131–135.

O'Rourke F, Schmidgen T, Kaiser PO, Linke D, Kempf VA. 2011. Adhesins of *Bartonella* spp. *Adv Exp Med Biol* **715:** 51–70.

Palanivelu DV, Goepfert A, Meury M, Guye P, Dehio C, Schirmer T. 2011. Fic domain-catalyzed adenylylation: Insight provided by the structural analysis of the type IV secretion system effector BepA. *Prot Sci* **20:** 492–499.

Pappas G. 2010. The changing *Brucella* ecology: Novel reservoirs, new threats. *Int J Antimicrob Agents* **36:** S8–S11.

Pappas G, Papadimitriou P, Akritidis N, Christou L, Tsianos EV. 2006. The new global map of human brucellosis. *Lancet Infect Dis* **6:** 91–99.

Parrow NL, Abbott J, Lockwood AR, Battisti JM, Minnick MF. 2009. Function, regulation, and transcriptional organization of the hemin utilization locus of *Bartonella quintana*. *Infect Immun* **77:** 307–316.

Paulsen IT, Seshadri R, Nelson KE, Eisen JA, Heidelberg JF, Read TD, Dodson RJ, Umayam L, Brinkac LM, Beanan MJ, et al. 2002. The *Brucella suis* genome reveals fundamental similarities between animal and plant pathogens and symbionts. *Proc Natl Acad Sci* **99:** 13148–13153.

Pizarro-Cerda J, Meresse S, Parton RG, van der Goot G, Sola-Landa A, Lopez-Goni I, Moreno E, Gorvel JP. 1998. *Brucella abortus* transits through the autophagic pathway and replicates in the endoplasmic reticulum of nonprofessional phagocytes. *Infect Immun* **66:** 5711–5724.

Popa C, Abdollahi-Roodsaz S, Joosten LA, Takahashi N, Sprong T, Matera G, Liberto MC, Foca A, van Deuren M, Kullberg BJ, et al. 2007. *Bartonella quintana* lipopolysaccharide is a natural antagonist of Toll-like receptor 4. *Infect Immun* **75:** 4831–4837.

Porte F, Liautard JP, Kohler S. 1999. Early acidification of phagosomes containing *Brucella suis* is essential for intracellular survival in murine macrophages. *Infect Immun* **67:** 4041–4047.

Posadas DM, Ruiz-Ranwez V, Bonomi HR, Martin FA, Zorreguieta A. 2012. BmaC, a novel autotransporter of *Brucella suis*, is involved in bacterial adhesion to host cells. *Cell Microbiol* **14:** 965–982.

Pulliainen AT, Dehio C. 2012. Persistence of *Bartonella* spp. stealth pathogens: From subclinical infections to vasoproliferative tumor formation. *FEMS Microbiol Rev* **36:** 563–599.

Pulliainen AT, Pieles K, Brand CS, Hauert B, Bohm A, Quebatte M, Wepf A, Gstaiger M, Aebersold R, Dessauer CW, et al. 2012. Bacterial effector binds host cell adenylyl cyclase to potentiate Gαs-dependent cAMP production. *Proc Natl Acad Sci* **109:** 9581–9586.

Qian BZ, Pollard JW. 2010. Macrophage diversity enhances tumor progression and metastasis. *Cell* **141:** 39–51.

Quebatte M, Dehio M, Tropel D, Basler A, Toller I, Raddatz G, Engel P, Huser S, Schein H, Lindroos HL, et al. 2010. The BatR/BatS two-component regulatory system controls the adaptive response of *Bartonella henselae* during human endothelial cell infection. *J Bacteriol* **192:** 3352–3367.

Radhakrishnan GK, Yu Q, Harms JS, Splitter GA. 2009. *Brucella* TIR domain-containing protein mimics properties of the Toll-like receptor adaptor protein TIRAP. *J Biol Chem* **284:** 9892–9898.

Radhakrishnan GK, Harms JS, Splitter GA. 2011. Modulation of microtubule dynamics by a TIR domain protein from the intracellular pathogen *Brucella melitensis*. *Biochem J* **439:** 79–83.

Ramos HC, Rumbo M, Sirard JC. 2004. Bacterial flagellins: Mediators of pathogenicity and host immune responses in mucosa. *Trends Microbiol* **12:** 509–517.

Rasool O, Freer E, Moreno E, Jarstrand C. 1992. Effect of *Brucella abortus* lipopolysaccharide on oxidative metabolism and lysozyme release by human neutrophils. *Infect Immun* **60:** 1699–1702.

Resto-Ruiz SI, Schmiederer M, Sweger D, Newton C, Klein TW, Friedman H, Anderson BE. 2002. Induction of a potential paracrine angiogenic loop between human THP-1 macrophages and human microvascular endothelial cells during *Bartonella henselae* infection. *Infect Immun* **70:** 4564–4570.

Rhomberg TA, Truttmann MC, Guye P, Ellner Y, Dehio C. 2009. A translocated protein of *Bartonella henselae* interferes with endocytic uptake of individual bacteria and triggers uptake of large bacterial aggregates via the invasome. *Cell Microbiol* **11:** 927–945.

Ricketts WE. 1948. *Bartonella bacilliformis* anemia (Oroya fever): A study of 30 cases. *Blood* **3:** 1025–1049.

Rolain JM, Franc M, Davoust B, Raoult D. 2003. Molecular detection of *Bartonella quintana*, *B. koehlerae*, *B. henselae*, *B. clarridgeiae*, *Rickettsia felis*, and *Wolbachia pipientis* in cat fleas, France. *Emerg Infect Dis* **9:** 338–342.

Roop RM 2nd, Bellaire BH, Valderas MW, Cardelli JA. 2004. Adaptation of the *Brucellae* to their intracellular niche. *Mol Microbiol* **52:** 621–630.

Roop RM 2nd, Gaines JM, Anderson ES, Caswell CC, Martin DW. 2009. Survival of the fittest: How *Brucella* strains adapt to their intracellular niche in the host. *Med Microbiol Immunol* **198:** 221–238.

Rouot B, Alvarez-Martinez MT, Marius C, Menanteau P, Guilloteau L, Boigegrain RA, Zumbihl R, O'Callaghan D, Domke N, Baron C. 2003. Production of the type IV secretion system differs among *Brucella* species as revealed with VirB5- and VirB8-specific antisera. *Infect Immun* **71:** 1075–1082.

Ruiz-Ranwez V, Posadas DM, Van der Henst C, Estein SM, Arocena GM, Abdian PL, Martin FA, Sieira R, De Bolle X, Zorreguieta A. 2013. BtaE, an adhesin that belongs to the trimeric autotransporter family, is required for full virulence and defines a specific adhesive pole of *Brucella. Infect Immun* **81:** 996–1007.

Saenz HL, Engel P, Stoeckli MC, Lanz C, Raddatz G, Vayssier-Taussat M, Birtles R, Schuster SC, Dehio C. 2007. Genomic analysis of *Bartonella* identifies type IV secretion systems as host adaptability factors. *Nat Genet* **39:** 1469–1476.

Salcedo SP, Marchesini MI, Lelouard H, Fugier E, Jolly G, Balor S, Muller A, Lapaque N, Demaria O, Alexopoulou L, et al. 2008. *Brucella* control of dendritic cell maturation is dependent on the TIR-containing protein Btp1. *PLoS Pathog* **4:** e21.

Salcedo SP, Chevrier N, Santos Lacerda TL, Ben Amara A, Gerart S, Gorvel VA, de Chastellier C, Blasco JM, Mege JL, Gorvel JP. 2013. Pathogenic *Brucellae* replicate in human trophoblasts. *J Infect Dis* **207:** 1075–1083.

Samartino LE, Enright FM. 1993. Pathogenesis of abortion of bovine brucellosis. *Comp Immunol Microbiol Infect Dis* **16:** 95–101.

Sander A, Kretzer S, Bredt W, Oberle K, Bereswill S. 2000. Hemin-dependent growth and hemin binding of *Bartonella henselae. FEMS Microbiol Lett* **189:** 55–59.

Scheidegger F, Ellner Y, Guye P, Rhomberg TA, Weber H, Augustin HG, Dehio C. 2009. Distinct activities of *Bartonella henselae* type IV secretion effector proteins modulate capillary-like sprout formation. *Cell Microbiol* **11:** 1088–1101.

Schmid MC, Schulein R, Dehio M, Denecker G, Carena I, Dehio C. 2004. The VirB type IV secretion system of *Bartonella henselae* mediates invasion, proinflammatory activation and antiapoptotic protection of endothelial cells. *Mol Microbiol* **52:** 81–92.

Schmid MC, Scheidegger F, Dehio M, Balmelle-Devaux N, Schulein R, Guye P, Chennakesava CS, Biedermann B, Dehio C. 2006. A translocated bacterial protein protects vascular endothelial cells from apoptosis. *PLoS Pathog* **2:** e115.

Scholz HC, Hubalek Z, Sedlacek I, Vergnaud G, Tomaso H, Al Dahouk S, Melzer F, Kampfer P, Neubauer H, Cloeckaert A, et al. 2008. *Brucella microti* sp nov, isolated from the common vole *Microtus arvalis. Int J Syst Evol Microbiol* **58:** 375–382.

Scholz HC, Nockler K, Gollner C, Bahn P, Vergnaud G, Tomaso H, Al Dahouk S, Kampfer P, Cloeckaert A, Maquart M, et al. 2010. *Brucella inopinata* sp nov, isolated from a breast implant infection. *Int J Syst Evol Microbiol* **60:** 801–808.

Schroder G, Dehio C. 2005. Virulence-associated type IV secretion systems of *Bartonella. Trends Microbiol* **13:** 336–342.

Schulein R, Dehio C. 2002. The VirB/VirD4 type IV secretion system of *Bartonella* is essential for establishing intraerythrocytic infection. *Mol Microbiol* **46:** 1053–1067.

Schulein R, Guye P, Rhomberg TA, Schmid MC, Schroder G, Vergunst AC, Carena I, Dehio C. 2005. A bipartite signal mediates the transfer of type IV secretion substrates of *Bartonella henselae* into human cells. *Proc Natl Acad Sci* **102:** 856–861.

Schulte B, Linke D, Klumpp S, Schaller M, Riess T, Autenrieth IB, Kempf VA. 2006. *Bartonella quintana* variably expressed outer membrane proteins mediate vascular endothelial growth factor secretion but not host cell adherence. *Infect Immun* **74:** 5003–5013.

Schwartzman W. 1996. *Bartonella* (*Rochalimaea*) infections: Beyond cat scratch. *Annu Rev Med* **47:** 355–364.

Selbach M, Paul FE, Brandt S, Guye P, Daumke O, Backert S, Dehio C, Mann M. 2009. Host cell interactome of tyrosine-phosphorylated bacterial proteins. *Cell Host Microbe* **5:** 397–403.

Sengupta D, Koblansky A, Gaines J, Brown T, West AP, Zhang D, Nishikawa T, Park SG, Roop RM 2nd, Ghosh S. 2010. Subversion of innate immune responses by *Brucella* through the targeted degradation of the TLR signaling adapter, MAL. *J Immunol* **184:** 956–964.

Seubert A, Hiestand R, de la Cruz F, Dehio C. 2003. A bacterial conjugation machinery recruited for pathogenesis. *Mol Microbiol* **49:** 1253–1266.

Sieira R, Comerci DJ, Sanchez DO, Ugalde RA. 2000. A homologue of an operon required for DNA transfer in *Agrobacterium* is required in *Brucella abortus* for virulence and intracellular multiplication. *J Bacteriol* **182:** 4849–4855.

Silva TM, Costa EA, Paixao TA, Tsolis RM, Santos RL. 2011. Laboratory animal models for brucellosis research. *J Biomed Biotechnol* **2011:** 518323.

Sohn AH, Probert WS, Glaser CA, Gupta N, Bollen AW, Wong JD, Grace EM, McDonald WC. 2003. Human neurobrucellosis with intracerebral granuloma caused by a marine mammal *Brucella* spp. *Emerg Infect Dis* **9:** 485–488.

Spera JM, Ugalde JE, Mucci J, Comerci DJ, Ugalde RA. 2006. A B lymphocyte mitogen is a *Brucella abortus* virulence factor required for persistent infection. *Proc Natl Acad Sci* **103:** 16514–16519.

Starr T, Ng TW, Wehrly TD, Knodler LA, Celli J. 2008. *Brucella* intracellular replication requires trafficking through the late endosomal/lysosomal compartment. *Traffic* **9:** 678–694.

Starr T, Child R, Wehrly TD, Hansen B, Hwang S, Lopez-Otin C, Virgin HW, Celli J. 2012. Selective subversion of autophagy complexes facilitates completion of the *Brucella* intracellular cycle. *Cell Host Microbe* **11:** 33–45.

Tiller RV, Gee JE, Lonsway DR, Gribble S, Bell SC, Jennison AV, Bates J, Coulter C, Hoffmaster AR, De BK. 2010. Identification of an unusual *Brucella* strain (BO2) from a lung biopsy in a 52-year-old patient with chronic destructive pneumonia. *BMC Microbiol* **10:** 23.

Tolomeo M, Di Carlo P, Abbadessa V, Titone L, Miceli S, Barbusca E, Cannizzo G, Mancuso S, Arista S, Scarlata F. 2003. Monocyte and lymphocyte apoptosis resistance in acute and chronic brucellosis and its possible implications in clinical management. *Clin Infect Dis* **36:** 1533–1538.

Truttmann MC, Misselwitz B, Huser S, Hardt WD, Critchley DR, Dehio C. 2011a. *Bartonella henselae* engages inside-out and outside-in signaling by integrin β1 and talin1 during invasome-mediated bacterial uptake. *J Cell Sci* **124:** 3591–3602.

Truttmann MC, Rhomberg TA, Dehio C. 2011b. Combined action of the type IV secretion effector proteins BepC and BepF promotes invasome formation of *Bartonella henselae* on endothelial and epithelial cells. *Cell Microbiol* **13:** 284–299.

Truttmann MC, Rhomberg TA, Dehio C. 2011c. Combined action of the type IV secretion effector proteins BepC and BepF promotes invasome formation of *Bartonella henselae* on endothelial and epithelial cells. *Cell Microbiol* **13:** 284–299.

Tsai YL, Chang CC, Chuang ST, Chomel BB. 2011. *Bartonella* species and their ectoparasites: Selective host adaptation or strain selection between the vector and the mammalian host? *Comp Immunol Microbiol Infect Dis* **34:** 299–314.

Tsolis RM. 2002. Comparative genome analysis of the α-proteobacteria: Relationships between plant and animal pathogens and host specificity. *Proc Natl Acad Sci* **99:** 12503–12505.

Ugalde RA. 1999. Intracellular lifestyle of *Brucella* spp. Common genes with other animal pathogens, plant pathogens, and endosymbionts. *Microbes Infect* **1:** 1211–1219.

van Ulsen P. 2011. Protein folding in bacterial adhesion: Secretion and folding of classical monomeric autotransporters. *Adv Exp Med Biol* **715:** 125–142.

Vayssier-Taussat M, Le Rhun D, Deng HK, Biville F, Cescau S, Danchin A, Marignac G, Lenaour E, Boulouis HJ, Mavris M, et al. 2010. The Trw type IV secretion system of *Bartonella* mediates host-specific adhesion to erythrocytes. *PLoS Pathog* **6:** e1000946.

Velho PE, Cintra ML, Uthida-Tanaka AM, de Moraes AM, Mariotto A. 2003. What do we (not) know about the human bartonelloses? *Braz J Infect Dis* **7:** 1–6.

Verma A, Ihler GM. 2002. Activation of Rac, Cdc42 and other downstream signalling molecules by *Bartonella bacilliformis* during entry into human endothelial cells. *Cell Microbiol* **4:** 557–569.

Verma A, Davis GE, Ihler GM. 2000. Infection of human endothelial cells with *Bartonella bacilliformis* is dependent on Rho and results in activation of Rho. *Infect Immun* **68:** 5960–5969.

Verma A, Davis GE, Ihler GM. 2001. Formation of stress fibres in human endothelial cells infected with *Bartonella bacilliformis* is associated with altered morphology, impaired migration and defects in cell morphogenesis. *Cell Microbiol* **3:** 169–180.

Vermi W, Facchetti F, Riboldi E, Heine H, Scutera S, Stornello S, Ravarino D, Cappello P, Giovarelli M, Badolato R, et al. 2006. Role of dendritic cell–derived CXCL13 in the pathogenesis of *Bartonella henselae* B-rich granuloma. *Blood* **107:** 454–462.

Viadas C, Rodriguez MC, Sangari FJ, Gorvel JP, Garcia-Lobo JM, Lopez-Goni I. 2010. Transcriptome analysis of the *Brucella abortus* BvrR/BvrS two-component regulatory system. *PLoS ONE* **5:** e10216.

von Bargen K, Gorvel JP, Salcedo SP. 2012. Internal affairs: Investigating the *Brucella* intracellular lifestyle. *FEMS Microbiol Rev* **36:** 533–562.

Walker TS, Winkler HH. 1981. *Bartonella bacilliformis:* Colonial types and erythrocyte adherence. *Infect Immun* **31:** 480–486.

Wallden K, Rivera-Calzada A, Waksman G. 2010. Type IV secretion systems: Versatility and diversity in function. *Cell Microbiol* **12:** 1203–1212.

Watanabe K, Tachibana M, Kim S, Watarai M. 2009. Participation of ezrin in bacterial uptake by trophoblast giant cells. *Reprod Biol Endocrinol* **7:** 95.

Watarai M, Makino S, Fujii Y, Okamoto K, Shirahata T. 2002. Modulation of *Brucella*-induced macropinocytosis by lipid rafts mediates intracellular replication. *Cell Microbiol* **4:** 341–355.

Watarai M, Kim S, Erdenebaatar J, Makino S, Horiuchi M, Shirahata T, Sakaguchi S, Katamine S. 2003. Cellular prion protein promotes *Brucella* infection into macrophages. *J Exp Med* **198:** 5–17.

Wattam AR, Williams KP, Snyder EE, Almeida NF Jr, Shukla M, Dickerman AW, Crasta OR, Kenyon R, Lu J, Shallom JM, et al. 2009. Analysis of ten *Brucella* genomes reveals evidence for horizontal gene transfer despite a preferred intracellular lifestyle. *J Bacteriol* **191:** 3569–3579.

Whatmore AM, Dawson CE, Groussaud P, Koylass MS, King AC, Shankster SJ, Sohn AH, Probert WS, McDonald WL. 2008. Marine mammal *Brucella* genotype associated with zoonotic infection. *Emerg Infect Dis* **14:** 517–518.

Williams-Bouyer NM, Hill EM. 1999. Involvement of host cell tyrosine phosphorylation in the invasion of HEp 2 cells by *Bartonella bacilliformis*. *FEMS Microbiol Lett* **171:** 191–201.

Young E. 2005. *Brucella* species. In *Mandell, Douglas and Bennett's principles and practice of infectious diseases* (ed. Mandell BJGL, Dolin R), pp. 2669–2685. Elsevier Churchill Livingstone, Philadelphia.

Zahringer U, Lindner B, Knirel YA, van den Akker WM, Hiestand R, Heine H, Dehio C. 2004. Structure and biological activity of the short-chain lipopolysaccharide from *Bartonella henselae* ATCC 49882T. *J Biol Chem* **279:** 21046–21054.

Zhang P, Chomel BB, Schau MK, Goo JS, Droz S, Kelminson KL, George SS, Lerche NW, Koehler JE. 2004. A family of variably expressed outer-membrane proteins (Vomp) mediates adhesion and autoaggregation in *Bartonella quintana*. *Proc Natl Acad Sci* **101:** 13630–13635.

Zygmunt MS, Hagius SD, Walker JV, Elzer PH. 2006. Identification of *Brucella melitensis* 16M genes required for bacterial survival in the caprine host. *Microbes Infect* **8:** 2849–2854.

Chlamydial Intracellular Survival Strategies

Robert J. Bastidas[1], Cherilyn A. Elwell[2], Joanne N. Engel[2,3], and Raphael H. Valdivia[1]

[1]Department of Molecular Genetics and Microbiology, Center for Microbial Pathogenesis, Duke University Medical Center, Durham, North Carolina 27710

[2]Department of Medicine, University of California, San Francisco, San Francisco, California 94143

[3]Department of Microbiology and Immunology, University of California, San Francisco, San Francisco, California 94143

Correspondence: valdi001@mc.duke.edu

Chlamydia trachomatis is the most common sexually transmitted bacterial pathogen and the causative agent of blinding trachoma. Although *Chlamydia* is protected from humoral immune responses by residing within remodeled intracellular vacuoles, it still must contend with multilayered intracellular innate immune defenses deployed by its host while scavenging for nutrients. Here we provide an overview of *Chlamydia* biology and highlight recent findings detailing how this vacuole-bound pathogen manipulates host–cellular functions to invade host cells and maintain a replicative niche.

Chlamydiae are Gram-negative obligate intracellular bacterial pathogens responsible for a range of diseases of clinical and public health importance. For instance, distinct serovariants of *Chlamydia trachomatis* cause a gamut of diseases, including blinding trachoma (serovars A–C), urogenital tract infections leading to urethritis, cervicitis, and proctitis (serovars D–K), and systemic lymphogranuloma venereum (LGV) disease (serovars L1–L3) (Schachter 1999). In genital tract infections, *C. trachomatis* primarily replicates within the epithelium of the urethra of men and endocervix of women causing inflammation, edema, and mucosal discharge. Ascending uterine infections can lead to pelvic inflammatory disease (PID), tubal scarring, ectopic pregnancies, and infertility (reviewed in Bebear and de Barbeyrac 2009; Haggerty et al. 2010).

Chlamydia displays a biphasic developmental cycle (Fig. 1). The elementary body (EB) form of the bacteria attaches and invades host-epithelial cells. The nascent phagocytic compartment is rapidly modified by *Chlamydia*-derived proteins to generate a parasitophorous vacuole termed an inclusion. Within the inclusion, the EB differentiates into a reticulate body (RB), the metabolically active and replicative form of the pathogen. RBs divide by binary fission as the inclusion expands and midway through the infectious cycle begin to asynchronously differentiate back into the EB form. Newly formed EBs are eventually released by cell lysis and/or extrusion to initiate new rounds

Cite this article as *Cold Spring Harb Perspect Med* doi: 10.1101/cshperspect.a010256

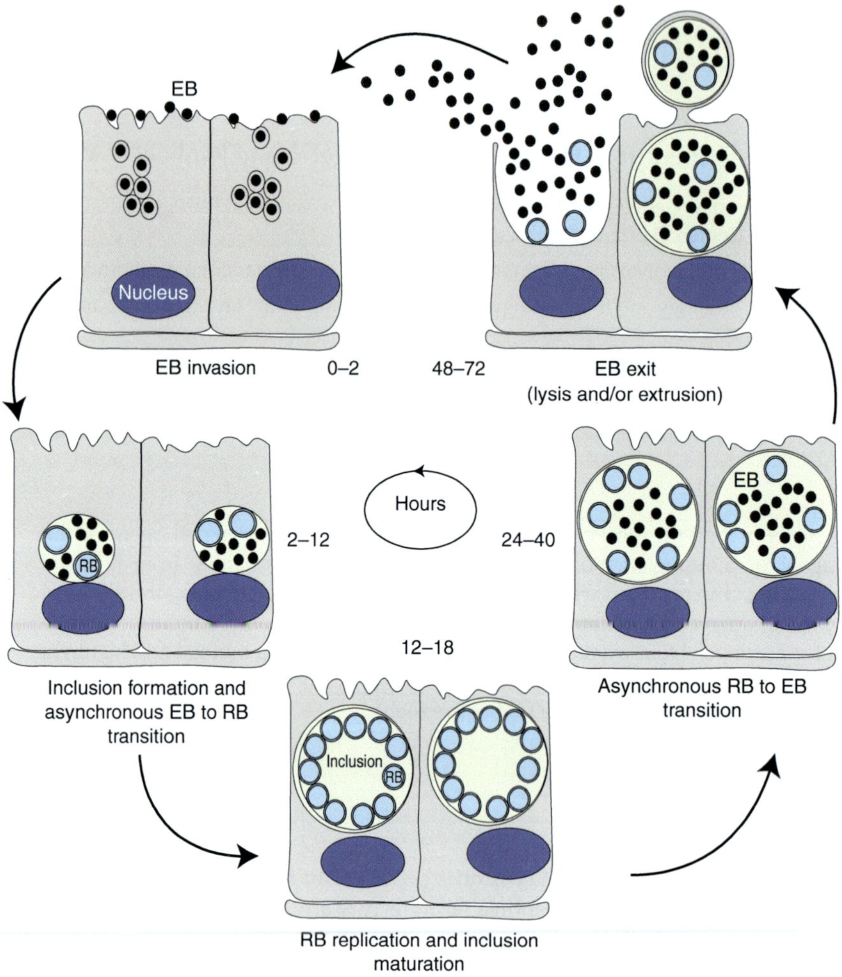

Figure 1. Life cycle of *Chlamydia.* Within the first 2 h following internalization into cells, elementary bodies (EBs) fuse to form a nascent inclusion. Between 2 and 6 h postinternalization, EBs begin to differentiate into reticulate bodies (RBs). By 12 h postinfection (hpi) RBs can be observed dividing by binary fission and by 18–24 hpi they peak in numbers. Increasing numbers of RBs differentiate back to EBs around 24 hpi and continue differentiating until lysis or release occurs ~48–72 hpi depending on the chlamydial species (Hatch 1999).

of infection (reviewed in Dautry-Varsat et al. 2005; Hybiske and Stephens 2007b).

At all stages of infection, *Chlamydia* translocates "effector" proteins directly into host membranes and cytoplasm for manipulation of host–cellular functions, including signal transduction pathways (Valdivia 2008). Here we discuss novel strategies used by *Chlamydia* to generate an environment conducive for intracellular replication and emerging avenues for dissecting chlamydial pathogenic strategies.

CELLULAR BASIS FOR THE INTERACTION OF *CHLAMYDIA* AND HOST CELLS

Mechanisms of *Chlamydia* Invasion of Epithelial Cells

Chlamydiae can invade most cultured cells, suggesting that the receptor(s) facilitating invasion are ubiquitous or that multiple receptors can be used. Binding is thought to be a two-step process for some species, involving an initial reversible interaction between the EB and the host cell mediated by heparan sulfate proteoglycans (HSPGs), followed by high-affinity irreversible binding to a secondary receptor (Dautry-Varsat et al. 2005). In addition to heparan sulfate (Chen and Stephens 1994, 1997; Chen et al. 1996; Wuppermann et al. 2001), the mannose receptor, the mannose 6-phosphate receptor, and the estrogen receptor have been proposed to act as host receptors for *Chlamydia* entry (reviewed in Cocchiaro and Valdivia 2009). Cell surface-exposed protein disulfide isomerase (PDI) has also been shown to play an important role in EB attachment and entry (Abromaitis and Stephens 2009).

Recent studies have revealed an unexpected role for growth factors and their receptors in *Chlamydia* binding (Elwell et al. 2008; Kim et al. 2011). A genome-wide loss-of-function RNAi screen in *Drosophila* S2 cells identified a potential role for the platelet-derived growth factor receptor (PDGFR), a transmembrane tyrosine kinase, in *Chlamydia* binding. In mammalian cells, depletion of PDGFRβ or inhibition with PDGFRβ-neutralizing antibodies reduces bacterial binding (Elwell et al. 2008). In subsequent studies, fibroblast growth factor 2 (FGF2) was shown to bind to EBs, where it appears to function as an HSPG-dependent bridging molecule that enhances interactions between EBs and the FGF receptor (FGFR) (Kim et al. 2011). Remarkably, bacterial binding stimulates production and release of FGF2 through a pathway that is independent of FGFR stimulation although dependent on bacterial protein synthesis and activation of the ERK1/2 signaling pathway (Kim et al. 2011). These events establish a positive-feedback loop that may further propagate bacterial binding and spreading to surrounding cells.

Multiple bacterial adhesins and ligands have been proposed to mediate invasion and their use may vary depending on both the host-cell type and the chlamydial species (Cocchiaro and Valdivia 2009). These include glycosaminoglycan (Menozzi et al. 2002), the major outer membrane protein (MOMP) (Su et al. 1996), OmcB (Fadel and Eley 2007), and PmpD (Mölleken et al. 2010). OmcB from *Chlamydia pneumoniae* but not from *C. trachomatis* has an identifiable heparan sulfate-binding domain that, when expressed on the surface of yeast, is necessary for OmcB binding to HSPGs (Moelleken and Hegemann 2008). In the absence of a system for targeted inactivation of *Chlamydia* genes, it has been difficult to confirm the relative contribution of these putative ligands to *Chlamydia* attachment and entry.

Chlamydiae entry into nonphagocytic cells is mediated by small GTPase-dependent reorganization of the actin cytoskeleton (reviewed in Hackstadt 2000; Dautry-Varsat et al. 2005; Cocchiaro and Valdivia 2009; Carabeo 2011; Scidmore 2011). Rac1 is required for *C. trachomatis* entry (Carabeo et al. 2004), whereas Cdc42 and Arf6 are also required for *Chlamydophila caviae* internalization (Subtil et al. 2004; Balana et al. 2005). Rac1 activation results in the recruitment of the actin regulators WAVE2, Abi-1, and Arp2/3, which are necessary for *C. trachomatis*-induced actin reorganization (Carabeo et al. 2007). Both chlamydial and host proteins may function synergistically to promote invasion. In the EB, the type III secretion system (a multiprotein needlelike delivery system) (reviewed in Beeckman and Vanrompay 2010), is poised for immediate discharge of effectors on contact with the host cell. At least one early secreted effector, TARP, contributes to bacterial internalization by its ability to directly nucleate actin polymerization through a WH2 actin-binding domain mimic (Jewett et al. 2006) and by recruiting the guanine nucleotide exchange factors (GEFs) Sos1 and Vav2, which activate Rac1 and signal to the actin machinery (Carabeo et al. 2007; Lane et al. 2008). This latter mechanism is only relevant in chlamydial species in which the polymorphic TARP gene (Clifton et al. 2005) can be targeted for tyrosine phosphorylation by Abl, Src, and Syk kinases (Elwell et al. 2008; Jewett et al. 2008; Lane et al. 2008; Mehlitz et al. 2008, 2010). Microinjection of cells with antibodies directed to the actin-binding domain of TARP before infection significantly reduces bacterial invasion, providing the most conclusive support for a direct role of TARP in mediating bacterial entry (Jewett et al. 2010). In addition, several host-tyrosine kinases, including PDGFR, FGFR, and Abl kinase are phosphorylated upon infection, recruited to the site of bacterial attachment, and may function redundantly in entry (Elwell et al. 2008; Kim et al. 2011). Therefore, the entry process is complex, requiring a concerted activation of growth factor receptors, cytoplasmic tyrosine kinases, small GTPases, and TARP to synergistically remodel the actin cytoskeleton and promote bacterial uptake. The actin rearrangements that occur during entry are transient and may be terminated by secreted chlamydial effectors such as CT166, which glucosylates Rac1 (Thalmann et al. 2010), or CT694, which interacts and colocalizes with AHNAK, an actin-binding protein (Hower et al. 2009). Additional host factors that contribute to uptake into nonphagocytic cells include clathrin (Boleti et al. 1999; Hybiske and Stephens 2007a) and cholesterol-rich microdomains (Norkin et al. 2001; Jutras et al. 2003; Stuart et al. 2003; Gabel et al. 2004).

Creating a Unique Niche

Following entry, EBs are sequestered within a membrane-bound compartment, termed the inclusion. The inclusion then quickly dissociates from the endolysosomal pathway, traffics along microtubules to a peri-Golgi location, and avoids fusion with lysosomes (reviewed in Hackstadt 2000). Early studies failed to identify host proteins on the inclusion membrane, leading to a model wherein the inclusion was viewed as an isolated compartment with little to no interactions with other host-cell organelles (reviewed in Fields and Hackstadt 2002). However, recent studies reveal a complex set of interactions between the chlamydial inclusion and host-cell trafficking pathways to facilitate acquisition of essential host-derived nutrients, such as amino acids, lipids, and iron, while limiting detection by the innate immune system. The list of host proteins that associate with the inclusion is rapidly growing; however, the mechanism by which these proteins are recruited to the inclusion has only been elucidated in a few cases.

Transport to the Peri-Golgi Region

Soon after entry, nascent inclusions are transported along microtubules to the microtubule organizing center (MTOC) in a dynein-dependent but dynactin-independent manner (Clausen et al. 1997; Grieshaber et al. 2003; Grieshaber et al. 2006), suggesting that one or more unknown bacterial effectors mimic the cargo-binding activity of dynactin and tether the inclusion to dynein and/or to centrosomes. Src family kinases are required in human-adapted *Chlamydia* strains for inclusions to migrate to the MTOC and for intracellular growth, even though they are dispensable for binding and entry (Mital and Hackstadt 2011). Recent studies indicate that the inclusion membrane is not homogenous and that microdomains consisting of a subset of inclusion proteins (Incs) (IncB, Inc101, Inc222, and Inc850), active Src family kinases, and cholesterol, associate with centrosomes and dynein (Mital et al. 2010; Mital and Hackstadt 2011). One or more of these Incs may recruit active Src family kinases to the inclusion to promote microtubule nucleation, dynein-dependent movement, and sphingomyelin acquisition (Mital et al. 2010; Mital and Hackstadt 2011). The inclusion maintains a dynein-dependent association with the centrosome throughout the cell cycle, resulting in supernumerary centrosomes, abnormal spindle poles, and segregation defects (Grieshaber et al. 2006; Johnson et al. 2009). These findings are intriguing in light of the association of *Chlamydia* and human papilloma virus-associated cervical cancer (Wallin et al. 2002).

Interactions with Regulators of Membrane-Trafficking Pathways

The inclusion interacts with a number of host-cell molecules that are involved in membrane trafficking, fusion, and organelle identity. A subset of these are Rab GTPases including Rabs 4, 11, and 14, which are associated with recycling endosomes, and Rabs 1, 6, and 10, which are associated with ER-Golgi traffic (Rzomp et al. 2006; Brumell and Scidmore 2007; Moorhead et al. 2007, 2010; Capmany and Damiani 2010). Rabs 6 and 10 are recruited to inclusions in a species-specific manner, whereas recruitment of Rab1, 4, and 11 appear to be ubiquitous to all *Chlamydiae* (Rzomp et al. 2003). The

C. trachomatis Inc protein CT229 has been identified as a Rab4-interacting protein, although it is not conserved in other strains or serovars that recruit Rab4, suggesting redundancy in Inc protein function (Rzomp et al. 2006). In the case of *C. pneumoniae*, Cpn0585 is thought to be responsible for the selective recruitment of Rab1, 10, and 11 (Cortes et al. 2007). In some instances, Rab-binding proteins are directly recruited to the inclusion membrane. Bicaudal D1 (BICD1), a Rab6-interacting protein, is recruited in a Rab6-independent manner, suggesting that an Inc protein may directly recruit BICD1 (Moorhead et al. 2007). The recruitment of Rabs and their associated proteins likely promotes selective interaction/fusion with host vesicles containing essential nutrients. Rab6, 11, and 14 play a role in the acquisition of sphingomyelin by the inclusion (Rejman Lipinski et al. 2009; Capmany and Damiani 2010), whereas Rab4 and Rab11 participate in iron acquisition by regulating a slow transferrin recycling pathway intercepted by the inclusion (Rzomp et al. 2003, 2006; Ouellette and Carabeo 2010). Rab6 and Rab11 facilitate sphingomyelin transport to the inclusion by regulating fragmentation of the Golgi into ministacks during infection (Heuer et al. 2009; Rejman Lipinski et al. 2009).

Phosphoinositides are also key determinants of membrane identity and vesicle fusion (Poccia and Larijani 2009), and several proteins associated with phosphoinositol-4-phosphate (PI4P) metabolism are recruited to the inclusion. These include oculocerebrorenal syndrome of Lowe protein 1 (OCRL1), phosphatidylinositol 4-kinase (PI4KIIα), and Arf1 (a GTPase that recruits PI4K enzymes and PI4P-binding proteins to the Golgi) (Moorhead et al. 2010; Elwell et al. 2011). Depletion of OCRL1, PI4KIIα, and Arf1 reduces inclusion formation and the production of infectious progeny, suggesting that generation of PI4P at the inclusion plays an important role in infection (Moorhead et al. 2010). Recruitment of Arf1 is intriguing, as it also regulates actin dynamics and vesicle trafficking at the Golgi (D'Souza-Schorey and Chavrier 2006), and thus may play multiple roles during infection.

In addition to recruiting Rab GTPases, *Chlamydia* may regulate fusion with host vesicles by recruiting host SNARES, which are key components of intracellular fusion machinery (Sudhof and Rothman 2009). In support of this notion, at least two Golgi-specific SNARES, including Syntaxin 6 and GS15 (Moore et al. 2011; Pokrovskaya et al. 2012), as well as vesicles containing the endocytic SNARES, Vamp3, Vamp7, and Vamp8, are recruited to the inclusion (Delevoye et al. 2008; Paumet et al. 2009). Interestingly, a subset of Incs contains motifs that display similarities to eukaryotic SNAREs, including IncA, CT813, and CT223 (Delevoye et al. 2008; Paumet et al. 2009). It is speculated that these SNARE-like domains pair with host-SNARE proteins on target vesicles to assemble a four-helix bundle that provides the energy to promote fusion of opposing membranes. Indeed, IncA, along with CT813, binds to host SNAREs, Vamp3, 7, and 8, suggesting that host SNAREs may work in concert with IncA to regulate membrane fusion (Delevoye et al. 2004, 2008; Paumet et al. 2009). However, in vitro liposome fusion assays indicate that instead of promoting fusion with endocytic compartments, IncA acts as an inhibitory SNARE to limit fusion with these compartments (Paumet et al. 2009). IncA has also been shown to participate in homotypic fusion of individual inclusions, as naturally occurring IncA-deficient variants or infected cells microinjected with anti-IncA antibodies display an aberrant multilobed inclusion structure (Hackstadt et al. 1999; Delevoye et al. 2004; Xia et al. 2005).

Interactions with Host-Cell Organelles for Acquisition of Host Nutrients

Early studies established a close association of the *Chlamydia* inclusion with the Golgi apparatus from which it intercepts sphingomyelin and cholesterol-containing exocytic vesicles by a Brefeldin A (BFA)-sensitive vesicular-trafficking pathway (Hackstadt et al. 1996; Carabeo et al. 2003). BFA inhibits activation of Arf1 (a key player in vesicle formation) by targeting its activators, the GEFs GBF1 and BIG1/2 (Claude et al. 1999; Mansour et al. 1999; Yamaji et al.

2000; Kawamoto et al. 2002), which are localized to the *cis*- and *trans*-Golgi, respectively, and have distinct functions at the Golgi (Manolea et al. 2008). Recent studies reveal that *Chlamydia* uses GBF1 (but not BIG1/2) to acquire sphingomyelin, which is necessary for inclusion membrane growth and stability but not bacterial replication (Elwell et al. 2011). Nonvesicular trafficking may also play a major role in lipid acquisition by *Chlamydia* (Derre et al. 2011; Elwell et al. 2011). Ceramide transfer protein (CERT), a cytosolic protein that transports ceramide from the ER to the *trans*-Golgi (Hanada 2010), is recruited to the inclusion membrane, possibly by interaction with the inclusion membrane protein IncD (Derre et al. 2011), and is involved in sphingomyelin acquisition (Elwell et al. 2011). Depletion of CERT significantly reduces the production of infectious progeny (Derre et al. 2011; Elwell et al. 2011). It is speculated that CERT mediates the transfer of ceramide from the ER to the inclusion, perhaps at ER-inclusion membrane contact sites, to generate specialized metabolic and/or signaling platforms (Derre et al. 2011). In addition, at least one of the two sphingomyelin synthase enzymes, which converts ceramide to sphingomyelin (SM) (Huitema et al. 2004), is recruited to the inclusion membrane (Elwell et al. 2011), raising the possibility that ceramide may also be converted to SM at the inclusion membrane, thereby creating a localized SM biosynthetic factory at the inclusion membrane. These observations shed some light on the paradox that although host-sphingomyelin biosynthesis is required for progeny production and inclusion biogenesis (van Ooij et al. 2000; Robertson et al. 2009), blocking vesicular transport with BFA has no inhibitory effect on the production of infectious progeny (Hackstadt et al. 1996). Acquisition of sphingomyelin by BFA-sensitive vesicular and nonvesicular pathways may play distinct functional roles, with BFA-sensitive vesicular transport participating in inclusion biogenesis and nonvesicular transport playing a key role in bacterial replication (Elwell et al. 2011). Host glycerophospholipids, such as phosphatidylinositol and phosphatidylcholine, are also acquired through a nonvesicular transport pathway by a process that requires activation of ERK and cytosolic phospholipase A2 (Wylie et al. 1997; Hatch and McClarty 1998; Su et al. 2004).

In addition to the Golgi, the inclusion interacts with other host-cell organelles, including (1) multivesicular bodies (MVBs) (Beatty 2006, 2008; Robertson et al. 2009), which may also serve as a source of sphingolipids and cholesterol, (2) lipid droplets, which are translocated into the inclusion lumen following their capture by the chlamydial proteins Lda1 and/or Lda3 and which may serve as a source of neutral lipids (Kumar et al. 2006; Cocchiaro et al. 2008), (3) mitochondria (Matsumoto et al. 1991; Derre et al. 2007), and (4) lysosomes, which may be a source of essential amino acids derived from host-protein degradation (Ouellette et al. 2011). Together, these interactions (summarized in Fig. 2) likely facilitate acquisition of nutrients for bacterial replication as well as for inclusion membrane stability and expansion.

Exiting the Host Cell

At the end of the developmental cycle, EBs are released by two mutually exclusive mechanisms: cell lysis and extrusion (Hybiske and Stephens 2007b). Cell lysis involves the sequential disruption of the inclusion and cellular membranes by cysteine proteases (Hybiske and Stephens 2007b). In contrast, the extrusion of inclusions requires actin polymerization, myosin, the Rho family GTPase RhoA, and leaves the host cell intact (Hybiske and Stephens 2007b). EB exit from *C. trachomatis* serovar E-infected cells is accompanied by lysosome-mediated repair of the plasma membrane (Beatty 2007).

MODIFYING THE HOST RESPONSE

Detection of *Chlamydia* by the Host

As with most bacteria, *Chlamydia* infections are detected by host pattern recognition receptors (PRRs) that recognize chlamydial LPS via TLR4 (Ingalls et al. 1995; Prebeck et al. 2001, 2003; Heine et al. 2003) and heat shock protein Hsp60 through TLR2 and TLR4 (Kol et al. 1999,

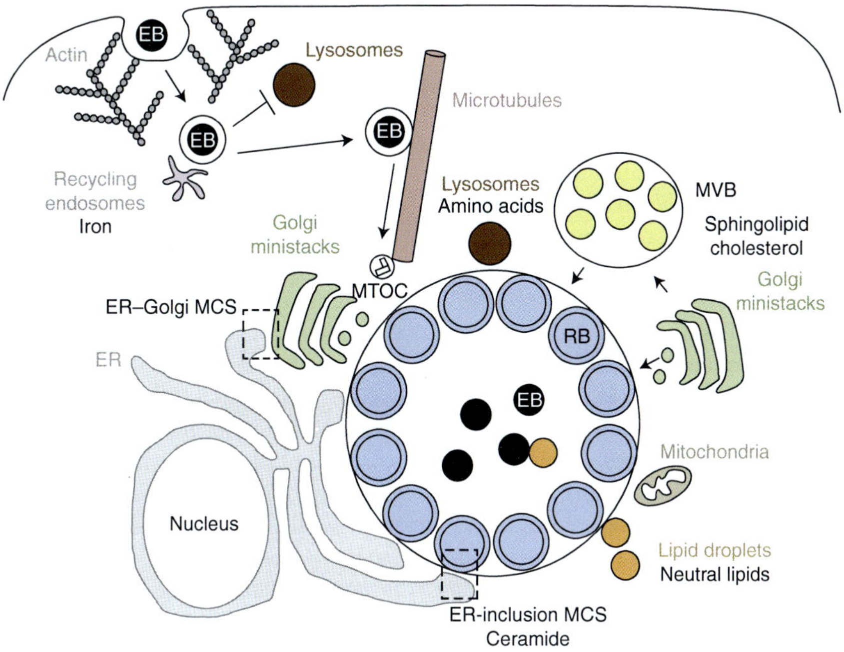

Figure 2. *Chlamydia* and host–cellular interactions. *Chlamydia* enters host cells by an actin-dependent mechanism that involves both host and bacterial factors. The EBs are enclosed within a membrane-bound inclusion, which becomes modified by insertion of bacterial inclusion proteins (Incs) that prevent fusion with lysosomes and promote interactions with recycling endosomes for iron acquisition. The nascent inclusion is transported along microtubules to the MTOC and EBs transition into RBs. The inclusion expands to accommodate the replicating RBs. During this time, the inclusion interacts, likely via bacterial effectors, with multiple host-cell organelles, including fragmented Golgi ministacks, the ER, lipid droplets, MVBs, and the mitochondria. Delivery of essential host lipids to the inclusion involves vesicular trafficking from Golgi ministacks and MVBs as well as nonvesicular trafficking from lipid droplets and membrane contact sites (MCS) formed between the ER and the inclusion. Although the inclusion does not fuse with lysosomes, *Chlamydia* may import essential amino acids derived from host-protein degradation within lysosomes.

2000; Vabulas et al. 2001; Bulut et al. 2002; Costa et al. 2002; Bulut et al. 2009). TLR2 appears to be the predominant receptor required for an inflammatory response to infection (Prebeck et al. 2001; Darville et al. 2003; O'Connell et al. 2006). Interestingly, TLR2 and its adaptor MyD88 localize to the periphery of the chlamydial inclusion during active infection, suggesting that TLR2 may signal intracellularly during infection (Mackern-Oberti et al. 2006; O'Connell et al. 2006).

In addition to TLRs, cytosolic PRRs recognize chlamydial PAMPS. RNAi-mediated knockdown of NOD1, or its downstream adaptor RIP2, in infected epithelial and endothelial cells ablates expression of inflammatory cytokines (Opitz et al. 2005; Welter-Stahl et al. 2006; Buchholz and Stephens 2008). Similarly, $Nod1^{-/-}$ and $Rip2^{-/-}$ mice show significantly delayed clearance of *C. pneumoniae* from their lungs (Shimada et al. 2009). Presumably, Nod1 is activated by trace amounts of peptidoglycans (PGNs), and even though PGN components such as *N*-acetylmuramic acid have not been detected in *Chlamydiae* above trace levels, the full complement of genes required for PGN biosynthesis and assembly are present in *Chlamydiae* genomes (Stephens et al. 1998; Kalman et al. 1999; Read et al. 2000). Interestingly, in place of an abundant layer of PGNs, the cell envelope of EBs is comprised of a highly cross-linked latticework of outer membrane proteins

(MOMP, OmcA, OmcB, and others) (reviewed in Hatch 1996), many of which are highly immunogenic (Wang et al. 2010) and could possibly be recognized as PAMPs by cytosolic receptors. In addition, infected cells induce type I IFNs, which can activate cell-autonomous resistance mechanisms to *Chlamydia* (Bernstein-Hanley et al. 2006). The ligands responsible for inducing the expression of type I IFNs are unknown, but the recently reported role for STING in IFN expression (Prantner et al. 2010) during *Chlamydia* infection point to a role for bacterial nucleic acids as potential ligands.

Antimicrobial molecules such as reactive oxygen species are also synthesized in response to infections (reviewed in D'Autreaux and Toledano 2007), and K^+ efflux triggered by *Chlamydia* infection activates reactive oxygen species (ROS) production (Abdul-Sater et al. 2009; Boncompain et al. 2010; He et al. 2010). ROS production also leads to activation of caspase 1 through the NLRP3-ASC inflammasome and processing of the proinflammatory cytokines IL-1β and IL-18 in *Chlamydia*-infected cells (Rothermel et al. 1989; Entrican et al. 1999; Lu et al. 2000; Gervassi et al. 2004; Abdul-Sater et al. 2009; He et al. 2010).

Initiating Innate Immune Responses

Early during *Chlamydia* infection there is an acute localized inflammatory response largely mediated by polymorphonuclear leukocytes (PMNs) and mononuclear leukocytes, recruited by cytokines and chemokines released by infected epithelial cells (Kelly and Rank 1997; Rank et al. 2000, 2008). Proinflammatory cytokines, chemokines, and interferons such as IL-1α, IL-1β, GM-CSF, IL-8, and IFN-α and γ are secreted by *Chlamydia*-infected cells in in vitro culture systems (Table 1). Activation of MEK-ERK (independent from the p38 and Jun amino-terminal MAPK pathways) and Nod1 signaling pathways, as well as nuclear IL-1α, contribute to cytokine production by infected epithelial cells (Buchholz and Stephens 2007, 2008; Cheng et al. 2008).

Cytokine/chemokine profiles have also been measured in vivo using a mouse model of genital tract infections with *Chlamydia muridarum* (Table 1). As early as 3 h postinfection, genes encoding chemokines (CCL20, CCL3, and CCL24) that recruit immature dendritic cells (DCs) are up-regulated, as well as those encoding for TNF-α and C3 (alternative complement). Therefore, early responses prime the site of infection for immature DC infiltration, which is an important step of the adaptive response (Rank et al. 2010). By 12 h postinfection, expression of chemokines that are chemotactic for natural killer (NK) cells (CXCL9-11, CCL2-4, CCL7, CCL8, and CCL12) is significantly induced, which is in agreement with the observed recruitment of NK cells to the site of infection (Tseng and Rank 1998; Rank et al. 2010). Not surprisingly, the proinflammatory cytokines IL-1α, IL-1β, IL-1F6, and IL-1F8 are also expressed as are PMN-recruiting chemokines (CXCL1, CXCL5, and CXCL15) (Rank et al. 2010).

The recruitment of immune cells that mediate innate and adaptive immune responses to *Chlamydia* infections are, under most circumstances, sufficient for bacterial clearance during primary infections, which may explain why most infections are asymptomatic. However, inflammatory responses resulting from recurring infections or when bacteria access sterile sites (i.e., the upper genital tract) contribute to the detrimental scarring and pathology observed in some infected individuals (Stephens 2003; Darville and Hiltke 2010).

Subversion of Host-Innate Immune Response

Despite the inflammatory response generated by the host during primary *Chlamydia* infection, in some individuals *C. trachomatis* infections can persist asymptomatically for months before being spontaneously cleared (Parks et al. 1997; Golden et al. 2000). How infection can persist for so long is not well understood. *Chlamydia* uses a wide range of strategies that may potentially subvert the host-immune responses, including interfering with the function of the transcription factor NF-κB (p65/RelA), a central regulator of immune responses. NF-κB functions downstream from PPRs and proinflammatory cytokine receptors such as the

Table 1. Chemokines and cytokines produced during chlamydial infection

	Mouse genital tract *Chlamydia muridarum*			Cell types			
				C.trachomatis		*C.pneumoniae*	
	3 hpi	12 hpi	24 hpi	HeLa	PBMC	PBMC	BMM
Chemokine receptors							
CCR1	■	■					
CCR2	■	■					
CCR6	■	■					
Chemokines							
CCL2 (MCP-1)		■	■		■		
CCL3 (MIP-1α)		■	■				
CCL4 (MIP-1β)		■	■				
CCL5 (RANTES)		■	■				
CCL20 (MIP-3α)	■		■				
CCL24	■	■					
CXCL1 (GRO-α)		■	■	■			
CXCL4 and 5		■	■				
CXCL8 (IL8 [h], MIP-2 [m])				■			
CXCL9 (MIG)		■	■				
CXCL10 (IP-10)		■	■				
CXCL11 (I-TAC)		■	■				
Cytokines							
IFN-α						■	■
IFN-γ			■		■		■
IL-1α			■	■			■
IL-1β			■		■	■	
IL-6				■	■	■	■
IL-10	NE	NE	NE		■	■	
IL-11				■	■		
IL-12						■	■
IL-13	■	■					
IL-18	NE	NE	NE		■	■	
TGF-β1		■		■			
TNF-α	■	■	■		■	■	■
GM-CSF				■			

Induced chemokines and cytokines are represented as gray boxes.

Data adapted from Kaukoranta-Tolvanen et al. 1996; Rasmussen et al. 1997; Dessus-Babus et al. 2000; Geng et al. 2000; Netea et al. 2004; Rothfuchs et al. 2004; Rank et al. 2010.

PBMC, peripheral blood mononuclear cell; BMM, murine bone marrow-derived macrophage; IL-8 (h), human CXCL8 homolog; MIP-2 (m), murine CXCL8 homolog; NE, not expressed.

TLR/IL-1R and tumor necrosis factor (TNF) family of receptors. Signaling is initiated by adaptor proteins belonging to a family of TNF receptor-associated factors (TRAF) such as TRAF2 and TRAF6. In the canonical pathway, a family of IκB proteins maintains NF-κB homo- or heterodimers sequestered in the cytoplasm. Activation of the pathway triggers phosphorylation of IκB and subsequent ubiquitination and proteasome-dependent degradation of IκB proteins culminating in NF-κB release, nuclear translocation, and expression of hundreds of genes, including innate immune response factors (reviewed in Chen 2005).

Chlamydia blocks degradation of the NF-κB retention factor IκBα and nuclear translocation of NF-κB (Lad et al. 2007a). This activity has been attributed to *Chlamydia* proteins possessing deubiquitinating (DUB) and deneddylating activity, ChlaDub1 and ChlaDub2, which may potentially interfere with host-ubiquitination pathways (Misaghi et al. 2006). Ectopic expression of ChlaDub1 in HeLa cells impaired ubiquitination and degradation of IκBα on stimulation with a variety of NF-κB agonists including TNF-α and IL-1β (Le Negrate et al. 2008). ChlaDub1 did not affect ubiquitination of TRAF2 or TRAF6, or IκBα phosphorylation, indicating that ChlaDub1 activity does not interfere with upstream components of the pathway. However, ChlaDub1 coprecipitates with IκBα suggesting that ChlaDub1 activity may very well be directed at IκBα (Le Negrate et al. 2008).

Chlamydia can also block NF-κB activation directly through the proteolysis of the p65/RelA subunit of NF-κB (Lad et al. 2007a). Initially this activity was attributed to the chlamydial tail-specific (TSP) endoprotease CT441 (Lad et al. 2007a,b). Although CT441 shows robust human p65 proteolytic activity, it is unclear whether CT441 can access the cytoplasm of the host cell where it could engage its target. Instead, a chlamydial protease named CPAF (chlamydial protease-like activity factor) has been recently implicated in the degradation of p65 during infection (Christian et al. 2010). Although CPAF is an extensively characterized protease with numerous potential substrates of importance to innate immunity (reviewed in Zhong 2011), a recent report suggests that several proteins targeted by CPAF, including the p65/RelA subunit of NF-κB, RFX5, Bim, and Puma (discussed below), may not be bona fide CPAF substrates in in vivo settings (Chen et al. 2012).

Chlamydia has also evolved a mechanism to limit the recognition of PAMPs by ensuring the stability of the inclusion during the intracellular stage of infections. This is achieved by reorganizing actin and intermediate filaments at the periphery of the inclusion (Kumar and Valdivia 2008). Disruption of these cytoskeletal elements results in inclusion membrane instability and leakage of inclusion luminal contents into the host-cell cytoplasm. As a result, there is an increase in IL-8 gene transcription (Kumar and Valdivia 2008).

CPAF may also play a role in limiting the presentation to immune cells of chlamydial antigens by major histocompatibility complexes (MHCs). MHC class I and II complexes present bacterial antigens to $CD8^+$ and $CD4^+$ T lymphocytes, respectively (reviewed in Germain 1994). To limit detection by $CD8^+$ and $CD4^+$ T cells, *Chlamydia* promotes CPAF-dependent degradation of the transcription factors RFX5 and USF-1, two upstream transcriptional regulators of IFN-γ-induced MHC I and II genes (Zhong et al. 1999, 2000). CPAF also stimulates degradation of CD1d, an MHC-like protein complex that stimulates NK and NKT immune cells (Kawana et al. 2007, 2008). In parallel, CPAF also cleaves a subset of chlamydial secreted effectors and inclusion membrane proteins, potentially curtailing their availability as substrates for antigen presentation (Jorgensen et al. 2011). These multiple potential mechanisms used by *Chlamydia* to dampen immune responses are summarized in Figure 3.

Maintaining the Host Cell Alive

Cell death on stimulation by death-inducing ligands secreted by immune cells or in response to cellular cytopathic signals is a common cell-autonomous response to infection (Ying et al. 2006). *Chlamydia* interferes with multiple proapoptotic pathways as well as potential necrotic cell death to guarantee survival within host cells (Fan et al. 1998; Fischer et al. 2001; Rajalingam et al. 2001; Yu et al. 2010). A prominent mechanism for preventing cell death includes CPAF-mediated degradation of the proapoptotic BH3-only proteins Bad, Bim, and Puma (Fischer et al. 2004; Dong et al. 2005; Ying et al. 2005; Pirbhai et al. 2006). This activity is proposed to be responsible for the reduced activation of Bax and Bak and subsequent block in cytochrome *c* release observed during infection (Fan et al. 1998; Fischer et al. 2004; Xiao et al. 2004). However, new observations indicate that Bim and Puma may not be targeted for degradation during infection (Chen et al. 2012). These differences will

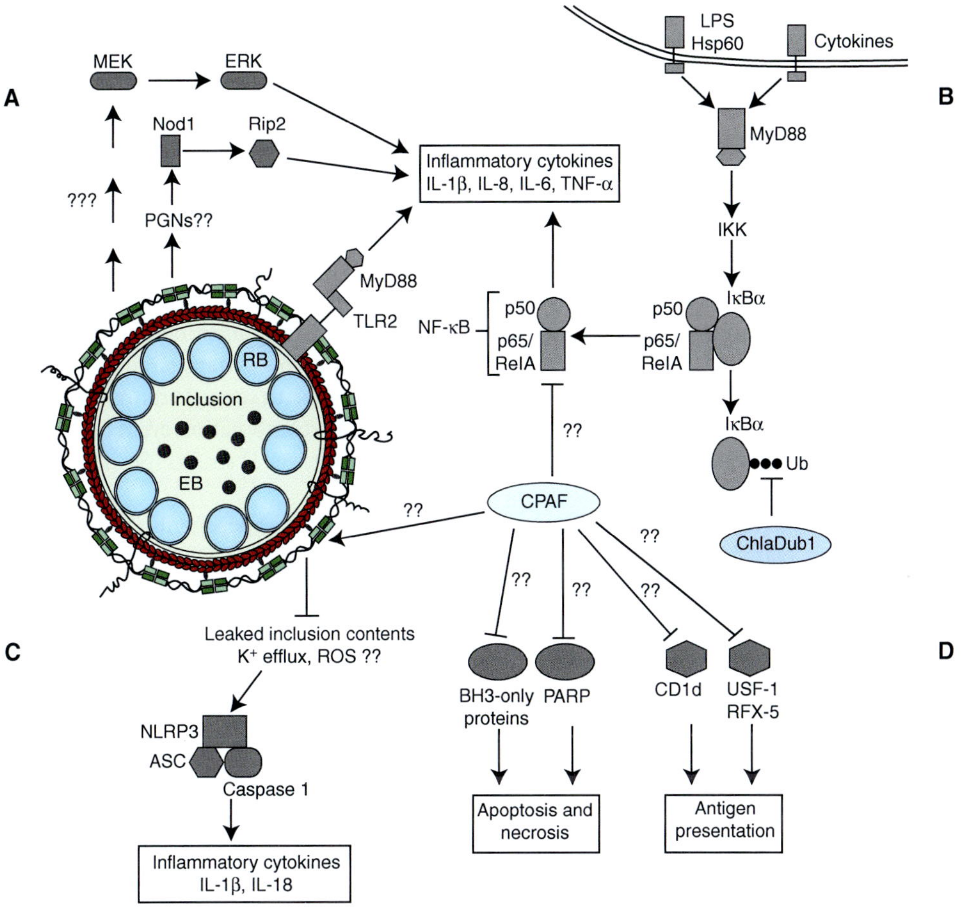

Figure 3. Current model of chlamydial effector interactions with host-innate immune signaling pathways. (*A*) TLR2 and MyD88 are trafficked to the inclusion by unknown mechanisms and promote expression of inflammatory cytokines. Nod1 and MEK-ERK MAPK signaling pathways also stimulate inflammatory cytokine gene expression. (*B*) Chlamydial microbial products and cytokines secreted by recruited immune cells and neighboring infected cells activate Toll-like and cytokine receptors. Receptor signaling activates NF-κB signaling and expression of proinflammatory cytokines/chemokines. Established inclusions impede NF-κB signaling by reversing IκBα ubiquitination through the activity of secreted ChlaDub1 and possibly by CPAF-mediated degradation of p65/RelA. (*C*) A cytoskeleton cage comprised of actin (red) and intermediate filaments (green) envelops the established inclusion. Remodeling of the cage curtails inclusion contents from activating cytosolic PRRs, possibly the NLRP3-ASC-caspase 1 inflammasome. However, unknown chlamydial agonists can activate the NLRP3-ASC inflammasome and caspase 1-dependent processing of inflammatory cytokines. (*D*) Additional mechanisms facilitating host survival may include the degradation of BH3-only proteins and PARP (Yu et al. 2010) and blocking death-inducing signaling pathways presumably originating from the TNF family of receptors and cytokine receptors. CPAF activity may also be deployed by *Chlamydia* to subvert antigen presentation.

need to be reconciled in the future. In parallel, infection also leads to stabilization of IAP2 (inhibitor of apoptosis protein 2) and up-regulation of the prosurvival factor Mcl-1 (myeoloid cell leukemia) (Rajalingam et al. 2006, 2008; Sharma et al. 2011). Other mechanisms potentially include sequestration of Bad at the inclusion via 14-3-3β, and of protein kinase C-δ (PKCδ) through binding to diacylglycerol-enriched membranes in the vicinity of the inclusion (Scidmore and Hackstadt 2001; Tse et al. 2005; Verbeke et al. 2006).

In addition, *Chlamydia* contends with the antimicrobial function of IFN-γ, which is secreted by infiltrating immune cells. IFN-γ induces tryptophan depletion by increasing host-indoleamine 2,3-dioxygenase (IDO) expression (reviewed in Taylor and Feng 1991). Tryptophan depletion is detrimental to *Chlamydia* because most strains are tryptophan auxotrophs. In response to this stress, *Chlamydia* transitions into quiescent noninfectious aberrant structures termed the "persistent" form (Beatty et al. 1994; Schoborg 2011). Interestingly, cells infected with these chlamydial forms are still resistant to various forms of apoptosis, using unknown mechanisms that are largely independent of CPAF (Dean and Powers 2001; Rodel et al. 2011).

Infection activates the MAPK and PI3K prosurvival signaling pathways. The ERK MAPK signaling pathway appears to be a central regulator for bacterial nutrient acquisition, expression of antiapoptotic factors, and synthesis of proinflammatory cytokines (Su et al. 2004; Buchholz and Stephens 2007; Rajalingam et al. 2008). Epithelial cells display sustained activation of upstream components of the ERK signaling module including RAS and RAF (Su et al. 2004; Gurumurthy et al. 2010; Vignola et al. 2010). However, this activation of ERK is uncoupled from activated RAS and RAF during infection (Gurumurthy et al. 2010; Vignola et al. 2010), suggesting multiple signaling pathways are engaged to further *Chlamydia* replication and intracellular survival. Avoidance of cell death and promoting self-survival are complex and very likely temporally regulated strategies deployed during infection and reinfection.

GENETICS AND GENOMICS OF *Chlamydia*

The genomes of a wide range of *Chlamydiae* serovars and species have been sequenced. Chlamydial genomes are small (<2 Mb), with few repetitive elements and pseudogenes, reflecting their long-term coadaptation to life within eukaryotic hosts (Collingro et al. 2011). Indeed, many of the genes required for the biosynthesis of key metabolites have been lost, presumably as these are now directly acquired from the host. The dearth of genes encoding DNA competence, conjugation, phages, or DNA modification systems in pathogenic *Chlamydiae* is consistent with that of an organism undergoing very limited genetic exchange (Stephens et al. 1998). Surprisingly, recent observations indicate that *Chlamydia* can exchange DNA efficiently with other strains. A genomic analysis of clinical *Chlamydia* isolates provided the first evidence of chimeric strains (Gomes et al. 2007). Such in vivo recombinant strains can have significant clinical implications. For instance, a hypervirulent *C. trachomatis* strain (L_2c) isolated from a patient with severe hemorrhagic proctitis represents a recombination event between serovar L2 and serovar D strains (Somboonna et al. 2011).

In landmark experiments, Robert DeMars provided evidence of DNA exchange among *Chlamydiae* in the laboratory (Demars et al. 2007; DeMars and Weinfurter 2008). In cells coinfected with *C. trachomatis* serovars that were naturally resistant to ofloxacin or to rifampin, doubly drug-resistant recombinant *Chlamydia* arose at $\sim 10^4$ times higher frequency than that of spontaneous mutations, and the genomes of these recombinants were chimeric recombinants of the parental strains. Whole genome sequencing of recombinants from additional crosses provided evidence that recombination events ranged from 40 kb to 790 kb and that the sizes of recombination sites range from a short 11-bp sequence to a 366-bp stretch (Suchland et al. 2009). The mechanisms underlying the release of DNA, uptake, and recombination are largely unknown.

Introduction of recombinant DNA into *Chlamydia* has been of limited success until very recently. Early efforts achieved transient transformation of chimeric *C. trachomatis* (pCT) plasmids (Tam et al. 1994). In later experiments, *Chlamydia psittaci* was transformed with allelic exchange vectors that targeted the 16S rRNA gene, and the investigators were able to show successful replacement of 16S rRNA loci with the introduced DNA (Binet and Maurelli 2009). Very recently, Clarke and colleagues reported successful stable transformation of *C. trachomatis* with a pCT shuttle plasmid by calcium chloride-mediated transformation of EBs and selection for resistance to β lactams (Wang et al. 2011). These stable transformants can express a variety of recombinant genes including the green fluorescent protein. Although in its present form the transformation and selection procedures are not very efficient, these experiments represent a significant breakthrough in our ability to experimentally manipulate these organisms.

Despite the lack of molecular genetic tools to manipulate *Chlamydiae*, significant advances have been made to experimentally address the role of individual genes in pathogenesis. This includes gain-of-function approaches, such as expressing *Chlamydia* genes in heterologous systems to identify proteins that modulate host functions (Sisko et al. 2006) or to identify proteins that can be secreted by surrogate type III secretion systems (Subtil et al. 2001). Loss-of-function studies have also recently become possible with the implementation of targeting-induced local lesions in genomes (TILLING), an approach used extensively in plant and zebrafish to screen for mutants in genes of interest (reviewed in Stemple 2004). In this method, PCR fragments that span the gene in which one wishes to identify mutations are generated from wild-type and mutagenized strains. DNA heteroduplexes are then generated by annealing these fragments, and if the gene amplified from the mutants strains harbors a nucleotide variant in the target gene, the annealed PCR products will generate a mismatch that can be recognized and cleaved by the endonuclease CEL I (reviewed in Stemple 2004). Kari and colleagues applied TILLING in EMS-mutagenized *Chlamydia* to identify mutations, including null alleles, in the tryptophan synthase operon *trpBA* (Kari et al. 2011). Such approaches enable the implementation of reverse genetic approaches by identifying strains with specific mutations. In a complementary approach, a system that uses the concept of "phenotype sequencing" (Harper et al. 2011) has been recently implemented in *Chlamydia* to perform forward genetic screens. In this approach, the genomes of EMS-derived mutants with similar phenotypes ("granular" plaque morphologies) were sequenced and a common gene (*glgB*-encoding glycogen branching enzyme) was found to be mutated in a majority of strains (Nguyen and Valdivia 2012). By taking advantage of genetic exchange among *Chlamydia* strains during coinfections, recombinant strains were generated in which a *glgB* mutation was the only gene lesion inherited from the parental mutant strain, thus unambiguously linking this mutation to a glycogen granule accumulation plaque morphotype. The ability to screen and select for mutants, mapping the underlying genetic lesions and generating congenic strains by recombination now opens up a broad range of possibilities for the genetic analysis of *Chlamydia* virulence.

CONCLUDING REMARKS

The obligate intracellular lifestyle of *Chlamydia* offers a unique opportunity to understand how the host-immune system contends with invading pathogens. *Chlamydia* species translocate a staggering number of protein effectors into the mammalian host cell to coopt cellular processes. Elucidating the function of these effectors will lead to new insights as to how pathogens subvert their hosts and on the function of the cellular processes they target. Despite the absence of classical genetic tools to study these ancient organisms, recent advances in deep sequencing technologies, the implementation of high-throughput RNAi screens, and comparative genomics approaches have broadened our understanding of the biology of these pathogens. Observations of widespread DNA exchange among *Chlamydia* organisms in experimental settings coupled with exciting new reports of *Chlamydia* plasmid transformation techniques will likely lead to major advances in our molecular understanding of chlamydial pathogenesis in the near future.

ACKNOWLEDGMENTS

We thank J.D. Dunn for comments on the manuscript. We apologize to those investigators whose work we were unable to cite because of space constraints. Work in our laboratories is supported by National Institutes of Health (NIH) R01 AI07377 (J.N.E. and C.A.E.) and

NIH grants R01AI081694 and R01AI081694-S1 (R.H.V. and R.J.B.). R.H.V. is a Burroughs Wellcome Scholar in the Pathogenesis of Infectious Diseases.

REFERENCES

Abdul-Sater AA, Koo E, Hacker G, Ojcius DM. 2009. Inflammasome-dependent caspase-1 activation in cervical epithelial cells stimulates growth of the intracellular pathogen *Chlamydia trachomatis*. *J Biol Chem* **284:** 26789–26796.

Abromaitis S, Stephens RS. 2009. Attachment and entry of *Chlamydia* have distinct requirements for host protein disulfide isomerase. *PLoS Pathog* **5:** e1000357.

Balana ME, Niedergang F, Subtil A, Alcover A, Chavrier P, Dautry-Varsat A. 2005. ARF6 GTPase controls bacterial invasion by actin remodelling. *J Cell Sci* **118:** 2201–2210.

Beatty WL. 2006. Trafficking from CD63-positive late endocytic multivesicular bodies is essential for intracellular development of *Chlamydia trachomatis*. *J Cell Sci* **119:** 350–359.

Beatty WL. 2007. Lysosome repair enables host cell survival and bacterial persistence following *Chlamydia trachomatis* infection. *Cell Microbiol* **9:** 2141–2152.

Beatty WL. 2008. Late endocytic multivesicular bodies intersect the chlamydial inclusion in the absence of CD63. *Infect Immun* **76:** 2872–2881.

Beatty WL, Belanger TA, Desai AA, Morrison RP, Byrne GI. 1994. Tryptophan depletion as a mechanism of γ interferon-mediated chlamydial persistence. *Infect Immun* **62:** 3705–3711.

Bebear C, de Barbeyrac B. 2009. Genital *Chlamydia trachomatis* infections. *Clin Microbiol Infect* **15:** 4–10.

Beeckman DS, Vanrompay DC. 2010. Bacterial secretion systems with an emphasis on the chlamydial type III secretion system. *Curr Issues Mol Biol* **12:** 17–42.

Bernstein-Hanley I, Coers J, Balsara ZR, Taylor GA, Starnbach MN, Dietrich WF. 2006. The p47 GTPases Igtp and Irgb10 map to the *Chlamydia trachomatis* susceptibility locus Ctrq-3 and mediate cellular resistance in mice. *Proc Natl Acad Sci* **103:** 14092–14097.

Binet R, Maurelli AT. 2009. Transformation and isolation of allelic exchange mutants of *Chlamydia psittaci* using recombinant DNA introduced by electroporation. *Proc Natl Acad Sci* **106:** 292–297.

Boleti H, Benmerah A, Ojcius DM, Cerf-Bensussan N, Dautry-Varsat A. 1999. *Chlamydia* infection of epithelial cells expressing dynamin and Eps15 mutants: Clathrin-independent entry into cells and dynamin-dependent productive growth. *J Cell Sci* **112** (Pt 10): 1487–1496.

Boncompain G, Schneider B, Delevoye C, Kellermann O, Dautry-Varsat A, Subtil A. 2010. Production of reactive oxygen species is turned on and rapidly shut down in epithelial cells infected with *Chlamydia trachomatis*. *Infect Immun* **78:** 80–87.

Brumell JH, Scidmore MA. 2007. Manipulation of rab GTPase function by intracellular bacterial pathogens. *Microbiol Mol Biol Rev* **71:** 636–652.

Buchholz KR, Stephens RS. 2007. The extracellular signal-regulated kinase/mitogen-activated protein kinase pathway induces the inflammatory factor interleukin-8 following *Chlamydia trachomatis* infection. *Infect Immun* **75:** 5924–5929.

Buchholz KR, Stephens RS. 2008. The cytosolic pattern recognition receptor NOD1 induces inflammatory interleukin-8 during *Chlamydia trachomatis* infection. *Infect Immun* **76:** 3150–3155.

Bulut Y, Faure E, Thomas L, Karahashi H, Michelsen KS, Equils O, Morrison SG, Morrison RP, Arditi M. 2002. Chlamydial heat shock protein 60 activates macrophages and endothelial cells through Toll-like receptor 4 and MD2 in a MyD88-dependent pathway. *J Immunol* **168:** 1435–1440.

Bulut Y, Shimada K, Wong MH, Chen S, Gray P, Alsabeh R, Doherty TM, Crother TR, Arditi M. 2009. Chlamydial heat shock protein 60 induces acute pulmonary inflammation in mice via the Toll-like receptor 4- and MyD88-dependent pathway. *Infect Immun* **77:** 2683–2690.

Capmany A, Damiani MT. 2010. *Chlamydia trachomatis* intercepts Golgi-derived sphingolipids through a Rab14-mediated transport required for bacterial development and replication. *PLoS ONE* **5:** e14084.

Carabeo R. 2011. Bacterial subversion of host actin dynamics at the plasma membrane. *Cell Microbiol* **13:** 1460–1469.

Carabeo RA, Mead DJ, Hackstadt T. 2003. Golgi-dependent transport of cholesterol to the *Chlamydia trachomatis* inclusion. *Proc Natl Acad Sci* **100:** 6771–6776.

Carabeo RA, Grieshaber SS, Hasenkrug A, Dooley C, Hackstadt T. 2004. Requirement for the Rac GTPase in *Chlamydia trachomatis* invasion of non-phagocytic cells. *Traffic* **5:** 418–425.

Carabeo RA, Dooley CA, Grieshaber SS, Hackstadt T. 2007. Rac interacts with Abi-1 and WAVE2 to promote an Arp2/3-dependent actin recruitment during chlamydial invasion. *Cell Microbiol* **9:** 2278–2288.

Chen ZJ. 2005. Ubiquitin signalling in the NF-κB pathway. *Nat Cell Biol* **7:** 758–765.

Chen JC, Stephens RS. 1994. Trachoma and LGV biovars of *Chlamydia trachomatis* share the same glycosaminoglycan-dependent mechanism for infection of eukaryotic cells. *Mol Microbiol* **11:** 501–507.

Chen JC, Stephens RS. 1997. *Chlamydia trachomatis* glycosaminoglycan-dependent and independent attachment to eukaryotic cells. *Microb Pathog* **22:** 23–30.

Chen JC, Zhang JP, Stephens RS. 1996. Structural requirements of heparin binding to *Chlamydia trachomatis*. *J Biol Chem* **271:** 11134–11140.

Chen AL, Johnson KA, Lee JK, Sütterlin C, Tan M. 2012. CPAF: A chlamydial protease in search of an authentic substrate. *PLoS Pathog* **8:** e1002842.

Cheng W, Shivshankar P, Zhong Y, Chen D, Li Z, Zhong G. 2008. Intracellular interleukin-1α mediates interleukin-8 production induced by *Chlamydia trachomatis* infection via a mechanism independent of type I interleukin-1 receptor. *Infect Immun* **76:** 942–951.

Christian J, Vier J, Paschen SA, Hacker G. 2010. Cleavage of the NF-κB family protein p65/RelA by the chlamydial protease-like activity factor (CPAF) impairs proinflammatory signaling in cells infected with *Chlamydiae*. *J Biol Chem* **285:** 41320–41327.

Claude A, Zhao BP, Kuziemsky CE, Dahan S, Berger SJ, Yan JP, Armold AD, Sullivan EM, Melancon P. 1999. GBF1: A novel Golgi-associated BFA-resistant guanine nucleotide exchange factor that displays specificity for ADP-ribosylation factor 5. *J Cell Biol* **146:** 71–84.

Clausen JD, Christiansen G, Holst HU, Birkelund S. 1997. *Chlamydia trachomatis* utilizes the host cell microtubule network during early events of infection. *Mol Microbiol* **25:** 441–449.

Clifton DR, Dooley CA, Grieshaber SS, Carabeo RA, Fields KA, Hackstadt T. 2005. Tyrosine phosphorylation of the chlamydial effector protein Tarp is species specific and not required for recruitment of actin. *Infect Immun* **73:** 3860–3868.

Cocchiaro JL, Valdivia RH. 2009. New insights into *Chlamydia* intracellular survival mechanisms. *Cell Microbiol* **11:** 1571–1578.

Cocchiaro JL, Kumar Y, Fischer ER, Hackstadt T, Valdivia RH. 2008. Cytoplasmic lipid droplets are translocated into the lumen of the *Chlamydia trachomatis* parasitophorous vacuole. *Proc Natl Acad Sci* **105:** 9379–9384.

Collingro A, Tischler P, Weinmaier T, Penz T, Heinz E, Brunham RC, Read TD, Bavoil PM, Sachse K, Kahane S, et al. 2011. Unity in variety-the pan-genome of the *Chlamydiae*. *Molec Biol Evol* **28:** 3253–3270.

Cortes C, Rzomp KA, Tvinnereim A, Scidmore MA, Wizel B. 2007. *Chlamydia pneumoniae* inclusion membrane protein Cpn0585 interacts with multiple Rab GTPases. *Infect Immun* **75:** 5586–5596.

Costa CP, Kirschning CJ, Busch D, Durr S, Jennen L, Heinzmann U, Prebeck S, Wagner H, Miethke T. 2002. Role of chlamydial heat shock protein 60 in the stimulation of innate immune cells by *Chlamydia pneumoniae*. *Eur J Immunol* **32:** 2460–2470.

Darville T, Hiltke TJ. 2010. Pathogenesis of genital tract disease due to *Chlamydia trachomatis*. *J Infect Dis* **201** (Suppl 2)**:** S114–S125.

Darville T, O'Neill JM, Andrews CW Jr, Nagarajan UM, Stahl L, Ojcius DM. 2003. Toll-like receptor-2, but not Toll-like receptor-4, is essential for development of oviduct pathology in chlamydial genital tract infection. *J Immunol* **171:** 6187–6197.

D'Autreaux B, Toledano MB. 2007. ROS as signalling molecules: Mechanisms that generate specificity in ROS homeostasis. *Nat Rev Mol Cell Biol* **8:** 813–824.

Dautry-Varsat A, Subtil A, Hackstadt T. 2005. Recent insights into the mechanisms of *Chlamydia* entry. *Cell Microbiol* **7:** 1714–1722.

Dean D, Powers VC. 2001. Persistent *Chlamydia trachomatis* infections resist apoptotic stimuli. *Infect Immun* **69:** 2442–2447.

Delevoye C, Nilges M, Dautry-Varsat A, Subtil A. 2004. Conservation of the biochemical properties of IncA from *Chlamydia trachomatis* and *Chlamydia caviae*: Oligomerization of IncA mediates interaction between facing membranes. *J Biol Chem* **279:** 46896–46906.

Delevoye C, Nilges M, Dehoux P, Paumet F, Perrinet S, Dautry-Varsat A, Subtil A. 2008. SNARE protein mimicry by an intracellular bacterium. *PLoS Pathog* **4:** e1000022.

DeMars R, Weinfurter J. 2008. Interstrain gene transfer in *Chlamydia trachomatis* in vitro: Mechanism and significance. *J Bacteriol* **190:** 1605–1614.

DeMars R, Weinfurter J, Guex E, Lin J, Potucek Y. 2007. Lateral gene transfer in vitro in the intracellular pathogen *Chlamydia trachomatis*. *J Bacteriol* **189:** 991–1003.

Derre I, Pypaert M, Dautry-Varsat A, Agaisse H. 2007. RNAi screen in *Drosophila* cells reveals the involvement of the Tom complex in *Chlamydia* infection. *PLoS Pathog* **3:** 1446–1458.

Derre I, Swiss R, Agaisse H. 2011. The lipid transfer protein CERT interacts with the *Chlamydia* inclusion protein IncD and participates to ER-*Chlamydia* inclusion membrane contact sites. *PLoS Pathog* **7:** e1002092.

Dessus-Babus S, Knight ST, Wyrick PB. 2000. Chlamydial infection of polarized HeLa cells induces PMN chemotaxis but the cytokine profile varies between disseminating and non-disseminating strains. *Cell Microbiol* **2:** 317–327.

Dong F, Pirbhai M, Xiao Y, Zhong Y, Wu Y, Zhong G. 2005. Degradation of the proapoptotic proteins Bik, Puma, and Bim with Bcl-2 domain 3 homology in *Chlamydia trachomatis*-infected cells. *Infect Immun* **73:** 1861–1864.

D'Souza-Schorey C, Chavrier P. 2006. ARF proteins: Roles in membrane traffic and beyond. *Nat Rev Mol Cell Biol* **7:** 347–358.

Elwell CA, Ceesay A, Kim JH, Kalman D, Engel JN. 2008. RNA interference screen identifies Abl kinase and PDGFR signaling in *Chlamydia trachomatis* entry. *PLoS Pathog* **4:** e1000021.

Elwell CA, Jiang S, Kim JH, Lee A, Wittmann T, Hanada K, Melancon P, Engel JN. 2011. *Chlamydia trachomatis* coopts GBF1 and CERT to acquire host sphingomyelin for distinct roles during intracellular development. *PLoS Pathog* **7:** e1002198.

Entrican G, Wilkie R, McWaters P, Scheerlinck J, Wood PR, Brown J. 1999. Cytokine release by ovine macrophages following infection with *Chlamydia psittaci*. *Clin Exp Immunol* **117:** 309–315.

Fadel S, Eley A. 2007. *Chlamydia trachomatis* OmcB protein is a surface-exposed glycosaminoglycan-dependent adhesin. *J Med Microbiol* **56:** 15–22.

Fan T, Lu H, Hu H, Shi L, McClarty GA, Nance DM, Greenberg AH, Zhong G. 1998. Inhibition of apoptosis in *Chlamydia*-infected cells: Blockade of mitochondrial cytochrome c release and caspase activation. *J Exp Med* **187:** 487–496.

Fields KA, Hackstadt T. 2002. The chlamydial inclusion: Escape from the endocytic pathway. *Annu Rev Cell Dev Biol* **18:** 221–245.

Fischer SF, Schwarz C, Vier J, Hacker G. 2001. Characterization of antiapoptotic activities of *Chlamydia pneumoniae* in human cells. *Infect Immun* **69:** 7121–7129.

Fischer SF, Harlander T, Vier J, Hacker G. 2004. Protection against CD95-induced apoptosis by chlamydial infection at a mitochondrial step. *Infect Immun* **72:** 1107–1115.

Gabel BR, Elwell C, van Ijzendoorn SC, Engel JN. 2004. Lipid raft-mediated entry is not required for *Chlamydia trachomatis* infection of cultured epithelial cells. *Infect Immun* **72:** 7367–7373.

Geng Y, Shane RB, Berencsi K, Gonczol E, Zaki MH, Margolis DJ, Trinchieri G, Rook AH. 2000. *Chlamydia pneumoniae* inhibits apoptosis in human peripheral blood mononuclear cells through induction of IL-10. *J Immunol* **164:** 5522–5529.

Germain RN. 1994. MHC-dependent antigen processing and peptide presentation: Providing ligands for T lymphocyte activation. *Cell* **76:** 287–299.

Gervassi A, Alderson MR, Suchland R, Maisonneuve JF, Grabstein KH, Probst P. 2004. Differential regulation of inflammatory cytokine secretion by human dendritic cells upon *Chlamydia trachomatis* infection. *Infect Immun* **72:** 7231–7239.

Golden MR, Schillinger JA, Markowitz L, St Louis ME. 2000. Duration of untreated genital infections with *Chlamydia trachomatis*: A review of the literature. *Sex Transm Dis* **27:** 329–337.

Gomes JP, Bruno WJ, Nunes A, Santos N, Florindo C, Borrego MJ, Dean D. 2007. Evolution of *Chlamydia trachomatis* diversity occurs by widespread interstrain recombination involving hotspots. *Genome Res* **17:** 50–60.

Grieshaber SS, Grieshaber NA, Hackstadt T. 2003. *Chlamydia trachomatis* uses host cell dynein to traffic to the microtubule-organizing center in a p50 dynamitin-independent process. *J Cell Sci* **116:** 3793–3802.

Grieshaber SS, Grieshaber NA, Miller N, Hackstadt T. 2006. *Chlamydia trachomatis* causes centrosomal defects resulting in chromosomal segregation abnormalities. *Traffic* **7:** 940–949.

Gurumurthy RK, Maurer AP, Machuy N, Hess S, Pleissner KP, Schuchhardt J, Rudel T, Meyer TF. 2010. A loss-of-function screen reveals Ras- and Raf-independent MEK-ERK signaling during *Chlamydia trachomatis* infection. *Sci Signal* **3:** ra21.

Hackstadt T. 2000. Redirection of host vesicle trafficking pathways by intracellular parasites. *Traffic* **1:** 93–99.

Hackstadt T, Rockey DD, Heinzen RA, Scidmore MA. 1996. *Chlamydia trachomatis* interrupts an exocytic pathway to acquire endogenously synthesized sphingomyelin in transit from the Golgi apparatus to the plasma membrane. *EMBO J* **15:** 964–977.

Hackstadt T, Scidmore-Carlson MA, Shaw EI, Fischer ER. 1999. The *Chlamydia trachomatis* IncA protein is required for homotypic vesicle fusion. *Cell Microbiol* **1:** 119–130.

Haggerty CL, Gottlieb SL, Taylor BD, Low N, Xu F, Ness RB. 2010. Risk of sequelae after *Chlamydia trachomatis* genital infection in women. *J Infect Dis* **201** (Suppl 2): S134–S155.

Hanada K. 2010. Intracellular trafficking of ceramide by ceramide transfer protein. *Proc Jpn Acad Ser B Phys Biol Sci* **86:** 426–437.

Harper MA, Chen Z, Toy T, Machado IM, Nelson SF, Liao JC, Lee CJ. 2011. Phenotype sequencing: Identifying the genes that cause a phenotype directly from pooled sequencing of independent mutants. *PLoS ONE* **6:** e16517.

Hatch TP. 1996. Disulfide cross-linked envelope proteins: The functional equivalent of peptidoglycan in chlamydiae? *J Bacteriol* **178:** 1–5

Hatch TP. 1999. Developmental biology. In *Chlamydia: Intracellular biology, pathogenesis, and immunity* (ed. Stephens RS), pp. 29–68. American Society for Microbiology, Washington, DC.

Hatch GM, McClarty G. 1998. Phospholipid composition of purified *Chlamydia trachomatis* mimics that of the eucaryotic host cell. *Infect Immun* **66:** 3727–3735.

He X, Mekasha S, Mavrogiorgos N, Fitzgerald KA, Lien E, Ingalls RR. 2010. Inflammation and fibrosis during *Chlamydia pneumoniae* infection is regulated by IL-1 and the NLRP3/ASC inflammasome. *J Immunol* **184:** 5743–5754.

Heine H, Muller-Loennies S, Brade L, Lindner B, Brade H. 2003. Endotoxic activity and chemical structure of lipopolysaccharides from *Chlamydia trachomatis* serotypes E and L2 and *Chlamydophila psittaci* 6BC. *Eur J Biochem* **270:** 440–450.

Heuer D, Rejman Lipinski A, Machuy N, Karlas A, Wehrens A, Siedler F, Brinkmann V, Meyer TF. 2009. *Chlamydia* causes fragmentation of the Golgi compartment to ensure reproduction. *Nature* **457:** 731–735.

Hower S, Wolf K, Fields KA. 2009. Evidence that CT694 is a novel *Chlamydia trachomatis* T3S substrate capable of functioning during invasion or early cycle development. *Mol Microbiol* **72:** 1423–1437.

Huitema K, van den Dikkenberg J, Brouwers JF, Holthuis JC. 2004. Identification of a family of animal sphingomyelin synthases. *EMBO J* **23:** 33–44.

Hybiske K, Stephens RS. 2007a. Mechanisms of *Chlamydia trachomatis* entry into nonphagocytic cells. *Infect Immun* **75:** 3925–3934.

Hybiske K, Stephens RS. 2007b. Mechanisms of host cell exit by the intracellular bacterium *Chlamydia*. *Proc Natl Acad Sci* **104:** 11430–11435.

Ingalls RR, Rice PA, Qureshi N, Takayama K, Lin JS, Golenbock DT. 1995. The inflammatory cytokine response to *Chlamydia trachomatis* infection is endotoxin mediated. *Infect Immun* **63:** 3125–3130.

Jewett TJ, Fischer ER, Mead DJ, Hackstadt T. 2006. Chlamydial TARP is a bacterial nucleator of actin. *Proc Natl Acad Sci* **103:** 15599–15604.

Jewett TJ, Dooley CA, Mead DJ, Hackstadt T. 2008. *Chlamydia trachomatis* Tarp is phosphorylated by Src family tyrosine kinases. *Biochem Biophys Res Commun* **371:** 339–344.

Jewett TJ, Miller NJ, Dooley CA, Hackstadt T. 2010. The conserved Tarp actin binding domain is important for chlamydial invasion. *PLoS Pathog* **6:** e1000997.

Johnson KA, Tan M, Sutterlin C. 2009. Centrosome abnormalities during a *Chlamydia trachomatis* infection are caused by dysregulation of the normal duplication pathway. *Cell Microbiol* **11:** 1064–1073.

Jorgensen I, Bednar MM, Amin V, Davis BK, Ting JP, McCafferty DG, Valdivia RH. 2011. The *Chlamydia* protease CPAF regulates host and bacterial proteins to maintain pathogen vacuole integrity and promote virulence. *Cell Host Microbe* **10:** 21–32.

Jutras I, Abrami L, Dautry-Varsat A. 2003. Entry of the lymphogranuloma venereum strain of *Chlamydia trachomatis* into host cells involves cholesterol-rich membrane domains. *Infect Immun* **71:** 260–266.

Kalman S, Mitchell W, Marathe R, Lammel C, Fan J, Hyman RW, Olinger L, Grimwood J, Davis RW, Stephens RS. 1999. Comparative genomes of *Chlamydia pneumoniae* and *C. trachomatis*. *Nat Genet* **21:** 385–389.

Kari L, Goheen MM, Randall LB, Taylor LD, Carlson JH, Whitmire WM, Virok D, Rajaram K, Endresz V, McClarty G, et al. 2011. Generation of targeted *Chlamydia trachomatis* null mutants. *Proc Natl Acad Sci* **108:** 7189–7193.

Kaukoranta-Tolvanen SS, Teppo AM, Laitinen K, Saikku P, Linnavuori K, Leinonen M. 1996. Growth of *Chlamydia pneumoniae* in cultured human peripheral blood mononuclear cells and induction of a cytokine response. *Microb Pathog* **21:** 215–221.

Kawamoto K, Yoshida Y, Tamaki H, Torii S, Shinotsuka C, Yamashina S, Nakayama K. 2002. GBF1, a guanine nucleotide exchange factor for ADP-ribosylation factors, is localized to the *cis*-Golgi and involved in membrane association of the COPI coat. *Traffic* **3:** 483–495.

Kawana K, Quayle AJ, Ficarra M, Ibana JA, Shen L, Kawana Y, Yang H, Marrero L, Yavagal S, Greene SJ, et al. 2007. CD1d degradation in *Chlamydia trachomatis*-infected epithelial cells is the result of both cellular and chlamydial proteasomal activity. *J Biol Chem* **282:** 7368–7375.

Kawana K, Matsumoto J, Miura S, Shen L, Kawana Y, Nagamatsu T, Yasugi T, Fujii T, Yang H, Quayle AJ, et al. 2008. Expression of CD1d and ligand-induced cytokine production are tissue specific in mucosal epithelia of the human lower reproductive tract. *Infect Immun* **76:** 3011–3018.

Kelly KA, Rank RG. 1997. Identification of homing receptors that mediate the recruitment of CD4 T cells to the genital tract following intravaginal infection with *Chlamydia trachomatis*. *Infect Immun* **65:** 5198–5208.

Kim JH, Jiang S, Elwell CA, Engel JN. 2011. *Chlamydia trachomatis* co-opts the FGF2 signaling pathway to enhance infection. *PLoS Pathog* **7:** e1002285.

Kol A, Bourcier T, Lichtman AH, Libby P. 1999. Chlamydial and human heat shock protein 60s activate human vascular endothelium, smooth muscle cells, and macrophages. *J Clin Invest* **103:** 571–577.

Kol A, Lichtman AH, Finberg RW, Libby P, Kurt-Jones EA. 2000. Cutting edge: Heat shock protein (HSP) 60 activates the innate immune response: CD14 is an essential receptor for HSP60 activation of mononuclear cells. *J Immunol* **164:** 13–17.

Kumar Y, Valdivia RH. 2008. Actin and intermediate filaments stabilize the *Chlamydia trachomatis* vacuole by forming dynamic structural scaffolds. *Cell Host Microbe* **4:** 159–169.

Kumar Y, Cocchiaro J, Valdivia RH. 2006. The obligate intracellular pathogen *Chlamydia trachomatis* targets host lipid droplets. *Curr Biol* **16:** 1646–1651.

Lad SP, Li J, da Silva Correia J, Pan Q, Gadwal S, Ulevitch RJ, Li E. 2007a. Cleavage of p65/RelA of the NF-κB pathway by *Chlamydia*. *Proc Natl Acad Sci* **104:** 2933–2938.

Lad SP, Yang G, Scott DA, Wang G, Nair P, Mathison J, Reddy VS, Li E. 2007b. Chlamydial CT441 is a PDZ domain-containing tail-specific protease that interferes with the NF-κB pathway of immune response. *J Bacteriol* **189:** 6619–6625.

Lane BJ, Mutchler C, Al Khodor S, Grieshaber SS, Carabeo RA. 2008. Chlamydial entry involves TARP binding of guanine nucleotide exchange factors. *PLoS Pathog* **4:** e1000014.

Le Negrate G, Krieg A, Faustin B, Loeffler M, Godzik A, Krajewski S, Reed JC. 2008. ChlaDub1 of *Chlamydia trachomatis* suppresses NF-κB activation and inhibits IκBα ubiquitination and degradation. *Cell Microbiol* **10:** 1879–1892.

Lu H, Shen C, Brunham RC. 2000. *Chlamydia trachomatis* infection of epithelial cells induces the activation of caspase-1 and release of mature IL-18. *J Immunol* **165:** 1463–1469.

Mackern-Oberti JP, Maccioni M, Cuffini C, Gatti G, Rivero VE. 2006. Susceptibility of prostate epithelial cells to *Chlamydia muridarum* infection and their role in innate immunity by recruitment of intracellular Toll-like receptors 4 and 2 and MyD88 to the inclusion. *Infect Immun* **74:** 6973–6981.

Manolea F, Claude A, Chun J, Rosas J, Melancon P. 2008. Distinct functions for Arf guanine nucleotide exchange factors at the Golgi complex: GBF1 and BIGs are required for assembly and maintenance of the Golgi stack and trans-Golgi network, respectively. *Mol Biol Cell* **19:** 523–535.

Mansour SJ, Skaug J, Zhao XH, Giordano J, Scherer SW, Melancon P. 1999. p200 ARF-GEP1: A Golgi-localized guanine nucleotide exchange protein whose Sec7 domain is targeted by the drug brefeldin A. *Proc Natl Acad Sci* **96:** 7968–7973.

Matsumoto A, Bessho H, Uehira K, Suda T. 1991. Morphological studies of the association of mitochondria with chlamydial inclusions and the fusion of chlamydial inclusions. *J Electron Microsc (Tokyo)* **40:** 356–363.

Mehlitz A, Banhart S, Hess S, Selbach M, Meyer TF. 2008. Complex kinase requirements for *Chlamydia trachomatis* Tarp phosphorylation. *FEMS Microbiol Lett* **289:** 233–240.

Mehlitz A, Banhart S, Maurer AP, Kaushansky A, Gordus AG, Zielecki J, Macbeath G, Meyer TF. 2010. Tarp regulates early *Chlamydia*-induced host cell survival through interactions with the human adaptor protein SHC1. *J Cell Biol* **190:** 143–157.

Menozzi FD, Pethe K, Bifani P, Soncin F, Brennan MJ, Locht C. 2002. Enhanced bacterial virulence through exploitation of host glycosaminoglycans. *Mol Microbiol* **43:** 1379–1386.

Misaghi S, Balsara ZR, Catic A, Spooner E, Ploegh HL, Starnbach MN. 2006. *Chlamydia trachomatis*-derived deubiquitinating enzymes in mammalian cells during infection. *Mol Microbiol* **61:** 142–150.

Mital J, Hackstadt T. 2011. Diverse requirements for Src-family tyrosine kinases distinguish chlamydial species. *mBio* **2:** e00031–11.

Mital J, Miller NJ, Fischer ER, Hackstadt T. 2010. Specific chlamydial inclusion membrane proteins associate with active Src family kinases in microdomains that interact with the host microtubule network. *Cell Microbiol* **12:** 1235–1249.

Moelleken K, Hegemann JH. 2008. The *Chlamydia* outer membrane protein OmcB is required for adhesion and exhibits biovar-specific differences in glycosaminoglycan binding. *Mol Microbiol* **67:** 403–419.

Mölleken K, Schmidt E, Hegemann JH. 2010. Members of the Pmp protein family of *Chlamydia pneumoniae* mediate adhesion to human cells via short repetitive peptide motifs. *Mol Microbiol* **78:** 1004–1017.

Moore ER, Mead DJ, Dooley CA, Sager J, Hackstadt T. 2011. The trans-Golgi SNARE syntaxin 6 is recruited to the chlamydial inclusion membrane. *Microbiology* **157:** 830–838.

Moorhead AR, Rzomp KA, Scidmore MA. 2007. The Rab6 effector Bicaudal D1 associates with *Chlamydia trachomatis* inclusions in a biovar-specific manner. *Infect Immun* **75:** 781–791.

Moorhead AM, Jung JY, Smirnov A, Kaufer S, Scidmore MA. 2010. Multiple host proteins that function in phosphatidylinositol-4-phosphate metabolism are recruited to the chlamydial inclusion. *Infect Immun* **78:** 1990–2007.

Netea MG, Kullberg BJ, Jacobs LE, Verver-Jansen TJ, van der Ven-Jongekrijg J, Galama JM, Stalenhoef AF, Dinarello CA, Van der Meer JW. 2004. *Chlamydia pneumoniae* stimulates IFN-γ synthesis through MyD88-dependent, TLR2- and TLR4-independent induction of IL-18 release. *J Immunol* **173:** 1477–1482.

Nguyen BD, Valdivia RH. 2012. Virulence determinants in the obligate intracellular pathogen *Chlamydia trachomatis* revealed by forward genetic approaches. *Proc Natl Acad Sci* **109:** 1263–1268.

Norkin LC, Wolfrom SA, Stuart ES. 2001. Association of caveolin with *Chlamydia trachomatis* inclusions at early and late stages of infection. *Exp Cell Res* **266:** 229–238.

O'Connell CM, Ionova IA, Quayle AJ, Visintin A, Ingalls RR. 2006. Localization of TLR2 and MyD88 to *Chlamydia trachomatis* inclusions. Evidence for signaling by intracellular TLR2 during infection with an obligate intracellular pathogen. *J Biol Chem* **281:** 1652–1659.

Opitz B, Forster S, Hocke AC, Maass M, Schmeck B, Hippenstiel S, Suttorp N, Krull M. 2005. Nod1-mediated endothelial cell activation by *Chlamydophila pneumoniae*. *Circ Res* **96:** 319–326.

Ouellette SP, Carabeo RA. 2010. A functional slow recycling pathway of transferrin is required for growth of *Chlamydia*. *Front Microbiol* **1:** 112.

Ouellette SP, Dorsey FC, Moshiach S, Cleveland JL, Carabeo RA. 2011. *Chlamydia* species-dependent differences in the growth requirement for lysosomes. *PLoS ONE* **6:** e16783.

Parks KS, Dixon PB, Richey CM, Hook EW 3rd. 1997. Spontaneous clearance of *Chlamydia trachomatis* infection in untreated patients. *Sex Transm Dis* **24:** 229–235.

Paumet F, Wesolowski J, Garcia-Diaz A, Delevoye C, Aulner N, Shuman HA, Subtil A, Rothman JE. 2009. Intracellular bacteria encode inhibitory SNARE-like proteins. *PLoS ONE* **4:** e7375.

Pirbhai M, Dong F, Zhong Y, Pan KZ, Zhong G. 2006. The secreted protease factor CPAF is responsible for degrading pro-apoptotic BH3-only proteins in *Chlamydia trachomatis*-infected cells. *J Biol Chem* **281:** 31495–31501.

Poccia D, Larijani B. 2009. Phosphatidylinositol metabolism and membrane fusion. *Biochem J* **418:** 233–246.

Pokrovskaya ID, Szwedo JW, Goodwin A, Lupashina TV, Nagarajan UM, Lupashin VV. 2012. *Chlamydia trachomatis* hijacks intra-Golgi COG complex-dependent vesicle trafficking pathway. *Cell Microbiol* **14:** 656–668.

Prantner D, Darville T, Nagarajan UM. 2010. Stimulator of IFN gene is critical for induction of IFN-β during *Chlamydia muridarum* infection. *J Immunol* **184:** 2551–2560.

Prebeck S, Kirschning C, Durr S, da Costa C, Donath B, Brand K, Redecke V, Wagner H, Miethke T. 2001. Predominant role of toll-like receptor 2 versus 4 in *Chlamydia pneumoniae*-induced activation of dendritic cells. *J Immunol* **167:** 3316–3323.

Prebeck S, Brade H, Kirschning CJ, da Costa CP, Durr S, Wagner H, Miethke T. 2003. The Gram-negative bacterium *Chlamydia trachomatis* L2 stimulates tumor necrosis factor secretion by innate immune cells independently of its endotoxin. *Microbes Infect* **5:** 463–470.

Rajalingam K, Al-Younes H, Muller A, Meyer TF, Szczepek AJ, Rudel T. 2001. Epithelial cells infected with *Chlamydophila pneumoniae* (*Chlamydia pneumoniae*) are resistant to apoptosis. *Infect Immun* **69:** 7880–7888.

Rajalingam K, Sharma M, Paland N, Hurwitz R, Thieck O, Oswald M, Machuy N, Rudel T. 2006. IAP-IAP complexes required for apoptosis resistance of *C. trachomatis*-infected cells. *PLoS Pathog* **2:** e114.

Rajalingam K, Sharma M, Lohmann C, Oswald M, Thieck O, Froelich CJ, Rudel T. 2008. Mcl-1 is a key regulator of apoptosis resistance in *Chlamydia trachomatis*-infected cells. *PLoS ONE* **3:** e3102.

Rank RG, Bowlin AK, Kelly KA. 2000. Characterization of lymphocyte response in the female genital tract during ascending Chlamydial genital infection in the guinea pig model. *Infect Immun* **68:** 5293–5298.

Rank RG, Whittimore J, Bowlin AK, Dessus-Babus S, Wyrick PB. 2008. *Chlamydiae* and polymorphonuclear leukocytes: Unlikely allies in the spread of chlamydial infection. *FEMS Immunol Med Microbiol* **54:** 104–113.

Rank RG, Lacy HM, Goodwin A, Sikes J, Whittimore J, Wyrick PB, Nagarajan UM. 2010. Host chemokine and cytokine response in the endocervix within the first developmental cycle of *Chlamydia muridarum*. *Infect Immun* **78:** 536–544.

Rasmussen SJ, Eckmann L, Quayle AJ, Shen L, Zhang YX, Anderson DJ, Fierer J, Stephens RS, Kagnoff MF. 1997. Secretion of proinflammatory cytokines by epithelial cells in response to *Chlamydia* infection suggests a central role for epithelial cells in chlamydial pathogenesis. *J Clin Invest* **99:** 77–87.

Read TD, Brunham RC, Shen C, Gill SR, Heidelberg JF, White O, Hickey EK, Peterson J, Utterback T, Berry K, et al. 2000. Genome sequences of *Chlamydia trachomatis* MoPn and *Chlamydia pneumoniae* AR39. *Nucleic Acids Res* **28:** 1397–1406.

Rejman Lipinski A, Heymann J, Meissner C, Karlas A, Brinkmann V, Meyer TF, Heuer D. 2009. Rab6 and Rab11 regulate *Chlamydia trachomatis* development and golgin-84-dependent Golgi fragmentation. *PLoS Pathog* **5:** e1000615.

Robertson DK, Gu L, Rowe RK, Beatty WL. 2009. Inclusion biogenesis and reactivation of persistent *Chlamydia trachomatis* requires host cell sphingolipid biosynthesis. *PLoS Pathog* **5:** e1000664.

Rodel J, Grosse C, Yu H, Wolf K, Otto GP, Liebler-Tenorio E, Forsbach-Birk V, Straube E. 2011. Persistent *Chlamydia trachomatis* infection of HeLa cells mediates apoptosis resistance through a CPAF-independent mechanism and induces HMGB1 release. *Infect Immun* **80:** 195–205.

Rothermel CD, Schachter J, Lavrich P, Lipsitz EC, Francus T. 1989. *Chlamydia trachomatis*-induced production of interleukin-1 by human monocytes. *Infect Immun* **57:** 2705–2711.

Rothfuchs AG, Trumstedt C, Wigzell H, Rottenberg ME. 2004. Intracellular bacterial infection-induced IFN-γ is critical but not solely dependent on Toll-like receptor 4-myeloid differentiation factor 88-IFN-αβ-STAT1 signaling. *J Immunol* **172:** 6345–6353.

Rzomp KA, Scholtes LD, Briggs BJ, Whittaker GR, Scidmore MA. 2003. Rab GTPases are recruited to chlamydial inclusions in both a species-dependent and species-independent manner. *Infect Immun* **71:** 5855–5870.

Rzomp KA, Moorhead AR, Scidmore MA. 2006. The GTPase Rab4 interacts with *Chlamydia trachomatis* inclusion membrane protein CT229. *Infect Immun* **74:** 5362–5373.

Schachter J. 1999. Infection and disease epidemiology. In *Chlamydia: Intracellular biology, pathogenesis, and immunity* (ed. Stephens RS), pp. 139–170. American Society for Microbiology, Washington, DC.

Schoborg RV. 2011. *Chlamydia* persistence—A tool to dissect *Chlamydia*-host interactions. *Microbes Infect* **13:** 649–662.

Scidmore MA. 2011. Recent advances in *Chlamydia* subversion of host cytoskeletal and membrane trafficking pathways. *Microbes Infect* **13:** 527–535.

Scidmore MA, Hackstadt T. 2001. Mammalian 14-3-3β associates with the *Chlamydia trachomatis* inclusion membrane via its interaction with IncG. *Mol Microbiol* **39:** 1638–1650.

Sharma M, Machuy N, Bohme L, Karunakaran K, Maurer AP, Meyer TF, Rudel T. 2011. HIF-1α is involved in mediating apoptosis resistance to *Chlamydia trachomatis*-infected cells. *Cell Microbiol* **13:** 1573–1585.

Shimada K, Chen S, Dempsey PW, Sorrentino R, Alsabeh R, Slepenkin AV, Peterson E, Doherty TM, Underhill D, Crother TR, et al. 2009. The NOD/RIP2 pathway is essential for host defenses against *Chlamydophila pneumoniae* lung infection. *PLoS Pathog* **5:** e1000379.

Sisko JL, Spaeth K, Kumar Y, Valdivia RH. 2006. Multifunctional analysis of *Chlamydia*-specific genes in a yeast expression system. *Mol Microbiol* **60:** 51–66.

Somboonna N, Wan R, Ojcius DM, Pettengill MA, Joseph SJ, Chang A, Hsu R, Read TD, Dean D. 2011. Hypervirulent *Chlamydia trachomatis* clinical strain is a recombinant between lymphogranuloma venereum (L_2) and D lineages. *mBio* **2:** e00045–11.

Stemple DL. 2004. TILLING—A high-throughput harvest for functional genomics. *Nat Rev Genet* **5:** 145–150.

Stephens RS. 2003. The cellular paradigm of chlamydial pathogenesis. *Trends Microbiol* **11:** 44–51.

Stephens RS, Kalman S, Lammel C, Fan J, Marathe R, Aravind L, Mitchell W, Olinger L, Tatusov RL, Zhao Q, et al. 1998. Genome sequence of an obligate intracellular pathogen of humans: *Chlamydia trachomatis*. *Science* **282:** 754–759.

Stuart ES, Webley WC, Norkin LC. 2003. Lipid rafts, caveolae, caveolin-1, and entry by *Chlamydiae* into host cells. *Exp Cell Res* **287:** 67–78.

Su H, Raymond L, Rockey DD, Fischer E, Hackstadt T, Caldwell HD. 1996. A recombinant *Chlamydia trachomatis* major outer membrane protein binds to heparan sulfate receptors on epithelial cells. *Proc Natl Acad Sci* **93:** 11143–11148.

Su H, McClarty G, Dong F, Hatch GM, Pan ZK, Zhong G. 2004. Activation of Raf/MEK/ERK/cPLA2 signaling pathway is essential for chlamydial acquisition of host glycerophospholipids. *J Biol Chem* **279:** 9409–9416.

Subtil A, Parsot C, Dautry-Varsat A. 2001. Secretion of predicted Inc proteins of *Chlamydia pneumoniae* by a heterologous type III machinery. *Mol Microbiol* **39:** 792–800.

Subtil A, Wyplosz B, Balana ME, Dautry-Varsat A. 2004. Analysis of *Chlamydia caviae* entry sites and involvement of Cdc42 and Rac activity. *J Cell Sci* **117:** 3923–3933.

Suchland RJ, Sandoz KM, Jeffrey BM, Stamm WE, Rockey DD. 2009. Horizontal transfer of tetracycline resistance among *Chlamydia* spp. in vitro. *Antimicrob Agents Chemother* **53:** 4604–4611.

Sudhof TC, Rothman JE. 2009. Membrane fusion: Grappling with SNARE and SM proteins. *Science* **323:** 474–477.

Tam JE, Davis CH, Wyrick PB. 1994. Expression of recombinant DNA introduced into *Chlamydia trachomatis* by electroporation. *Can J Microbiol* **40:** 583–591.

Taylor MW, Feng GS. 1991. Relationship between interferon-γ, indoleamine 2,3-dioxygenase, and tryptophan catabolism. *FASEB J* **5:** 2516–2522.

Thalmann J, Janik K, May M, Sommer K, Ebeling J, Hofmann F, Genth H, Klos A. 2010. Actin re-organization induced by *Chlamydia trachomatis* serovar D-evidence for a critical role of the effector protein CT166 targeting Rac. *PLoS ONE* **5:** e9887.

Tse SM, Mason D, Botelho RJ, Chiu B, Reyland M, Hanada K, Inman RD, Grinstein S. 2005. Accumulation of diacylglycerol in the *Chlamydia* inclusion vacuole: Possible role in the inhibition of host cell apoptosis. *J Biol Chem* **280:** 25210–25215.

Tseng CT, Rank RG. 1998. Role of NK cells in early host response to chlamydial genital infection. *Infect Immun* **66:** 5867–5875.

Vabulas RM, Ahmad-Nejad P, da Costa C, Miethke T, Kirschning CJ, Hacker H, Wagner H. 2001. Endocytosed HSP60s use toll-like receptor 2 (TLR2) and TLR4 to activate the toll/interleukin-1 receptor signaling pathway in innate immune cells. *J Biol Chem* **276:** 31332–31339.

Valdivia RH. 2008. *Chlamydia* effector proteins and new insights into chlamydial cellular microbiology. *Curr Opin Microbiol* **11:** 53–59.

van Ooij C, Kalman L, van I, Nishijima M, Hanada K, Mostov K, Engel JN. 2000. Host cell-derived sphingolipids are required for the intracellular growth of *Chlamydia trachomatis*. *Cell Microbiol* **2:** 627–637.

Verbeke P, Welter-Stahl L, Ying S, Hansen J, Hacker G, Darville T, Ojcius DM. 2006. Recruitment of BAD by the *Chlamydia trachomatis* vacuole correlates with host-cell survival. *PLoS Pathog* **2:** e45.

Vignola MJ, Kashatus DF, Taylor GA, Counter CM, Valdivia RH. 2010. cPLA2 regulates the expression of type I interferons and intracellular immunity to *Chlamydia trachomatis*. *J Biol Chem* **285:** 21625–21635.

Wallin KL, Wiklund F, Luostarinen T, Angstrom T, Anttila T, Bergman F, Hallmans G, Ikaheimo I, Koskela P, Lehtinen M, et al. 2002. A population-based prospective study of *Chlamydia trachomatis* infection and cervical carcinoma. *Int J Cancer J* **101:** 371–374.

Wang J, Zhang Y, Lu C, Lei L, Yu P, Zhong G. 2010. A genome wide profiling of the humoral immune response to *Chlamydia trachomatis* infection reveals vaccine candidate antigens expressed in humans. *J Immunol* **185:** 1670–1680.

Wang Y, Kahane S, Cutcliffe LT, Skilton RJ, Lambden PR, Clarke IN. 2011. Development of a transformation system for *Chlamydia trachomatis*: Restoration of glycogen biosynthesis by acquisition of a plasmid shuttle vector. *PLoS Pathog* **7:** e1002258.

Welter-Stahl L, Ojcius DM, Viala J, Girardin S, Liu W, Delarbre C, Philpott D, Kelly KA, Darville T. 2006. Stimulation of the cytosolic receptor for peptidoglycan, Nod1, by infection with *Chlamydia trachomatis* or *Chlamydia muridarum*. *Cell Microbiol* **8:** 1047–1057.

Wuppermann FN, Hegemann JH, Jantos CA. 2001. Heparan sulfate-like glycosaminoglycan is a cellular receptor for *Chlamydia pneumoniae*. *J Infect Dis* **184:** 181–187.

Wylie JL, Hatch GM, McClarty G. 1997. Host cell phospholipids are trafficked to and then modified by *Chlamydia trachomatis*. *J Bacteriol* **179:** 7233–7242.

Xia M, Suchland RJ, Bumgarner RE, Peng T, Rockey DD, Stamm WE. 2005. *Chlamydia trachomatis* variant with nonfusing inclusions: Growth dynamic and host-cell transcriptional response. *J Infect Dis* **192:** 1229–1236.

Xiao Y, Zhong Y, Greene W, Dong F, Zhong G. 2004. *Chlamydia trachomatis* infection inhibits both Bax and Bak activation induced by staurosporine. *Infect Immun* **72:** 5470–5474.

Yamaji R, Adamik R, Takeda K, Togawa A, Pacheco-Rodriguez G, Ferrans VJ, Moss J, Vaughan M. 2000. Identification and localization of two brefeldin A-inhibited guanine nucleotide-exchange proteins for ADP-ribosylation factors in a macromolecular complex. *Proc Natl Acad Sci* **97:** 2567–2572.

Ying S, Seiffert BM, Hacker G, Fischer SF. 2005. Broad degradation of proapoptotic proteins with the conserved Bcl-2 homology domain 3 during infection with *Chlamydia trachomatis*. *Infect Immun* **73:** 1399–1403.

Ying S, Fischer SF, Pettengill M, Conte D, Paschen SA, Ojcius DM, Hacker G. 2006. Characterization of host cell death induced by *Chlamydia trachomatis*. *Infect Immun* **74:** 6057–6066.

Yu H, Schwarzer K, Forster M, Kniemeyer O, Forsbach-Birk V, Straube E, Rodel J. 2010. Role of high-mobility group box 1 protein and poly(ADP-ribose) polymerase 1 degradation in *Chlamydia trachomatis*-induced cytopathicity. *Infect Immun* **78:** 3288–3297.

Zhong G. 2011. *Chlamydia trachomatis* secretion of proteases for manipulating host signaling pathways. *Front Microbiol* **2:** 14.

Zhong G, Fan T, Liu L. 1999. *Chlamydia* inhibits interferon γ-inducible major histocompatibility complex class II expression by degradation of upstream stimulatory factor 1. *J Exp Med* **189:** 1931–1938.

Zhong G, Liu L, Fan T, Fan P, Ji H. 2000. Degradation of transcription factor RFX5 during the inhibition of both constitutive and interferon γ-inducible major histocompatibility complex class I expression in *Chlamydia*-infected cells. *J Exp Med* **191:** 1525–1534.

In Vitro and In Vivo Model Systems for Studying Enteropathogenic *Escherichia coli* Infections

Robyn J. Law[1,2], Lihi Gur-Arie[3], Ilan Rosenshine[3], and B. Brett Finlay[1,2,4]

[1]Michael Smith Laboratories, University of British Columbia, Vancouver, British Columbia V6T 1Z4, Canada

[2]Department of Microbiology and Immunology, University of British Columbia, Vancouver, British Columbia V6T 1Z4, Canada

[3]Department of Microbiology and Molecular Genetics, IMRIC, The Hebrew University of Jerusalem, Faculty of Medicine, Jerusalem, Israel

[4]Department of Biochemistry and Molecular Biology, University of British Columbia, Vancouver, British Columbia V6T 1Z4, Canada

Correspondence: bfinlay@msl.ubc.ca

Enteropathogenic *Escherichia coli* (EPEC) and enterohemorrhagic *E. coli* (EHEC) belong to a group of bacteria known as attaching and effacing (A/E) pathogens that cause disease by adhering to the lumenal surfaces of their host's intestinal epithelium. EPEC and EHEC are major causes of infectious diarrhea that result in significant childhood morbidity and mortality worldwide. Recent advances in in vitro and in vivo modeling of these pathogens have contributed to our knowledge of how EPEC and EHEC attach to host cells and subvert host-cell signaling pathways to promote infection and cause disease. A more detailed understanding of how these pathogenic microbes infect their hosts and how the host responds to infection could ultimately lead to new therapeutic strategies to help control these significant enteric pathogens.

Escherichia coli is normally a harmless commensal of the human intestinal flora but some strains have acquired the ability to cause disease in humans ranging from meningitis to urinary tract infections and diarrhea. Six of the eight well-studied *E. coli* pathovars are diarrheagenic, and although each pathovar has a distinct mechanism of disease, many virulence strategies are shared between them (Croxen and Finlay 2010). Enteropathogenic *E. coli* (EPEC) is a leading cause of watery and potentially fatal diarrhea in infants that is often accompanied by vomiting, fever, and dehydration (Nataro and Kaper 1998; Kaper et al. 2004; Croxen and Finlay 2010). EPEC poses a substantial health concern mainly for children in developing countries (Kaper et al. 2004) but has also been linked to sporadic cases in daycare facilities in North America and Europe (Nataro and Kaper 1998; Goosney et al. 2000; Jenkins et al. 2003). Enterohemorrhagic *E. coli* (EHEC) causes more serious diarrhea than EPEC, provoking a variety of symptoms that range in severity from acute gastroenteritis to hemorrhagic colitis and in the most severe cases, hemolytic uremic syndrome (HUS) (Goosney et al. 2000), with progression

Cite this article as *Cold Spring Harb Perspect Med* doi: 10.1101/cshperspect.a009977

to HUS occurring in 15% of childhood cases (Tarr et al. 2005). The most prevalent EHEC serotype causing outbreaks in North America is O157:H7 with cattle serving as the primary reservoir for this highly infectious organism, thus making EHEC contamination a major issue in food and water safety (Croxen and Finlay 2010). The annual hospitalization rate for laboratory-confirmed EHEC O157 infections is 46.2% compared with that of non-O157 Shiga toxin-producing *E. coli* infections with a rate of only 12.8% (Scallan et al. 2011).

EPEC and EHEC belong to a group of extracellular pathogens characterized by their ability to form distinctive pedestal-like structures on the surfaces of intestinal epithelial cells (IEC) (Goosney et al. 2000). This characteristic lesion, known as the attaching and effacing (A/E) lesion, is produced through subversion of host-cell actin dynamics mediated by the bacterial translocated intimin receptor, Tir (Kalman et al. 1999; Gruenheid et al. 2001). EPEC and EHEC are normally noninvasive enteric pathogens, and their intimate attachment to intestinal surfaces is a hallmark of A/E pathogen infection requiring the bacterial outer membrane protein, intimin, and the secreted effector, Tir (Kenny et al. 1997). EPEC and EHEC deliver effector proteins into host cells via a type III secretion system (T3SS) encoded on a conserved pathogenicity island known as the locus of enterocyte effacement (LEE) (McDaniel et al. 1995). The LEE is found in all A/E pathogens and is absolutely required for their pathogenesis (Kenny 2002). Encoded within the LEE are regulatory elements, seven secreted effectors and their related chaperones, as well as the T3SS apparatus itself (Frankel et al. 1998b; Deng et al. 2004). In addition to Tir, the T3SS translocates an assortment of effector proteins into host cells, which subvert cellular processes to promote A/E pathogen infection (Kenny 2002; Dean et al. 2005; Wong et al. 2011). To date, several LEE-encoded effectors, as well as a number of non-LEE (Nle)-encoded effectors (Deng et al. 2004), have been identified in the genomes of EPEC and EHEC (Tobe et al. 2006; Coburn et al. 2007; Iguchi et al. 2009).

Effector proteins perform pivotal roles during T3SS-mediated disease by subverting host defenses and cellular trafficking and by manipulating the host cytoskeleton (Fig. 1) (Coburn et al. 2007). The seven LEE-encoded type III secreted (T3S) effectors as well as some of the Nle effectors are well conserved across A/E pathogen species, whereas the number and type of other Nle effectors varies significantly (Dean and Kenny 2009). The prototypic EPEC strain E2348/69 encodes 23 effectors (Iguchi et al. 2009; Deng et al. 2012), whereas certain serotypes of EHEC encode as many as 41 (Tobe et al. 2006; Ogura et al. 2009). Likewise, there is considerable variation in the number of effectors found among A/E animal pathogens such as *Citrobacter rodentium* (Deng et al. 2010), which colonizes mice (Luperchio and Schauer 2001). The variation in effector repertoire among A/E pathogens likely reflects differences in virulence phenotypes and adaptation to diverse host environments (Tobe et al. 2006; Deng et al. 2010).

EPEC and EHEC are strictly human enteric pathogens, but the availability of surrogate animal models, in addition to the advent of tissue culture, has made it possible to study the complex interactions between these pathogens and their hosts (Vallance et al. 2004). In vitro models such as IEC in tissue culture and in vitro organ culture (IVOC) have been used extensively to define, at the molecular level, T3SS-related proteins necessary for effector secretion and A/E lesion formation, as well as effector localization, function, and protein interactions within host cells. Alternatively, in vivo models including nematodes, mice, rabbits, as well as studies in human volunteers, provide essential information with respect to virulence, host immunity and how the inflammatory response contributes to the pathology of A/E pathogen-mediated disease, and the interplay with the microbiota. Current research into the mechanisms of EPEC/EHEC-induced diarrheal disease and the inflammatory response of the host has extended our knowledge of how A/E pathogens infect their hosts and how the immune response they elicit contributes to disease pathogenesis. Further elucidation of the molecular and innate immune mechanisms behind EPEC/EHEC pathogenesis through in vitro and in vivo modeling is crucial to better understand the viru-

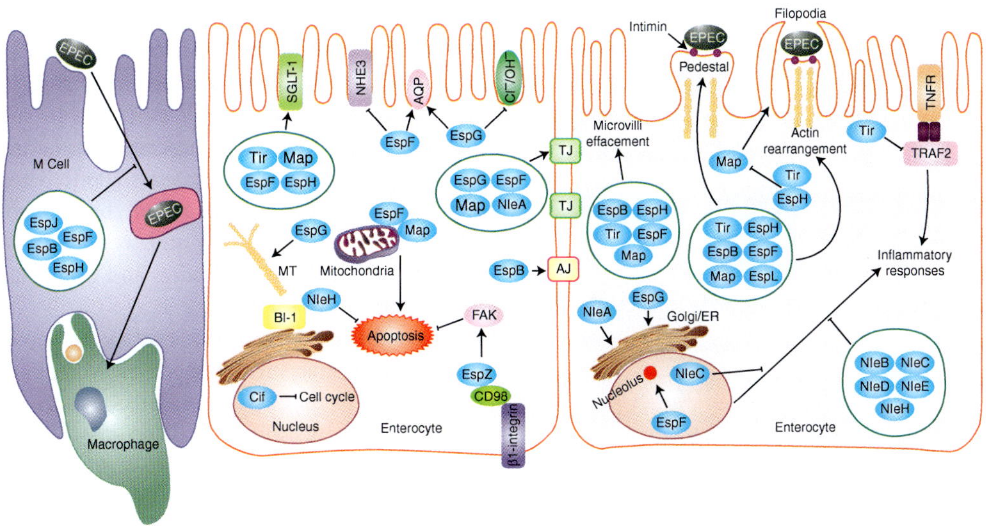

Figure 1. Major T3SS effectors of EPEC/EHEC and their targets within host cells. Adapted and updated with additional information from Dean and Kenny (2009) to show the multifunctional nature of effectors common to EPEC and EHEC and their overlapping functions. TJ, tight junction; AJ, adherens junction; ER, endoplasmic reticulum; MT, microtubules; SGLT-1, sodium glucose cotransporter; NHE3, sodium hydrogen exchanger; AQP, aquaporin; Cl^-/OH^-, Cl^-/OH^- transporter; FAK, focal adhesion kinase; BI-1, BAX inhibitor-1; TNFR, tumor necrosis factor receptor; TRAF2, TNF-associated factor 2.

lence strategies used by these organisms in causing disease.

IN VITRO INFECTION MODELS

Interaction of EPEC with tissue culture cells represents a simple, yet extremely powerful model system for studying A/E pathogens (Fig. 2). EPEC exhibits tight specificity to IEC of the pediatric small intestine during natural infection. Intriguingly, under tissue culture conditions, this selectivity is lost and EPEC interacts with cells originating from a large variety of tissues and species. Although this precludes the use of cell lines to study species specificity and tissue tropism of EPEC, it opens up the opportunity for using specialized cell lines, including those containing specific mutations, to address the role of specific host genes and proteins in a given process. Many features of commonly used cell lines have been characterized, making them an effective tool for studying the role of a given host cell protein in response to bacterial infection.

Although nonpolarized cell lines, such as HeLa cells, are very different from polarized cells of the intestinal epithelium, they have been used extensively to study basic aspects of the EPEC–host interaction. The best evidence in support of adopting such cell lines as model systems to study EPEC is the dissection of actin pedestal biogenesis, including the elucidation of the involved bacterial and host proteins. Early studies showed the process of actin pedestal formation during infection of tissue culture cells (Knutton et al. 1989; Rosenshine et al. 1996), which provided a solid foundation for several seminal discoveries based on the molecular mechanisms of pedestal formation during infection with A/E pathogens (reviewed in Frankel and Phillips 2008 and Campellone 2010). Additional examples of the usefulness of nonpolarized cells as models for studying A/E pathogen infection are their use in studies aimed at elucidating the process of T3SS-dependent protein translocation into host cells as well as functional analyses of the injected effectors. These include genetic analysis defining components that are required

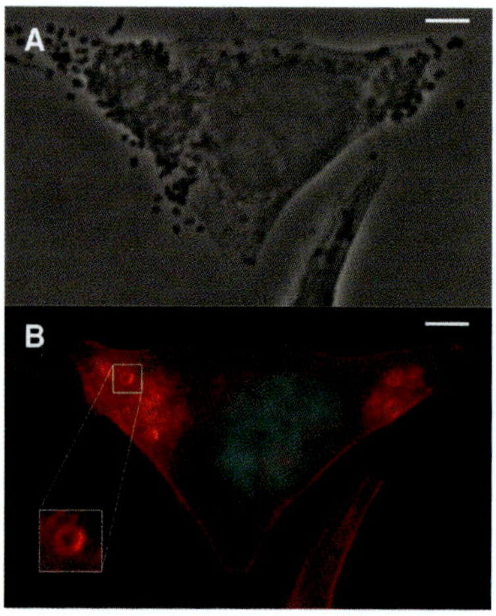

Figure 2. The use of cell culture models to study EPEC pathogenicity. HeLa cells were infected for 3 h with EPEC and visualized using phase contrast (*A*), or phalloidin rhodamine (red) and DAPI (cyan) staining (*B*). The dense F-actin staining (red) is representative of actin pedestals formed beneath the infecting bacteria. Magnification, 100×; scale bar, 5 μm.

for T3SS assembly and/or activity (reviewed in Cornelis 2006), and analysis of the injection process (Mills et al. 2008). Studies of the biochemical activity of injected effectors also benefit from using tissue culture cells as they are easy to grow and manipulate. This is nicely exemplified by analysis of the function of Tir, as well as most other studied effectors (Nieto-Pelegrin and Martinez-Quiles 2009; Nadler et al. 2010; Baruch et al. 2011). Indeed, simple nonpolarized cell lines provide an effective model system to study basic aspects of T3SS function and the immediate interface between bacterial and host proteins.

Nonpolarized cell lines are not as effective for studying the influence of EPEC/EHEC on the physiological properties of IEC, such as barrier function or maintenance of the brush border. This is particularly true in cases in which the effect on barrier function is secondary. For instance, the effector Map appears to function as a guanine nucleotide exchange factor of Cdc42 GTPase, an activity that results in barrier function disruption (Dean and Kenny 2004). Cell lines that mimic polarization by formation of tight junctions and brush border, including MDCK, Caco-2, T84, and HT29, provide instrumental models for describing how EPEC disrupts epithelial barrier function during infection and for identifying effectors that are specifically involved in this process (McNamara et al. 2001, and reviewed in Sears 2000). These cell lines offer the advantages of convenience, rapid growth, uniformity, and availability of genetic tools such as antibodies and genetically manipulated cell lines.

In addition to interactions with enterocytes, EPEC also interacts with phagocytic M cells and dendritic cells that extend protrusions into the gut lumen where EPEC may encounter infiltrating innate immune cells (Inman and Cantey 1983; Martinez-Argudo et al. 2007; Vossenkämper et al. 2010). These particular immune cell types are modeled by macrophage-like cell lines such as J774 and U937, and using these models, it was found that EPEC resists uptake by professional phagocytes (Goosney et al. 1999). Macrophage-like cell lines also enabled the identification of EspF, EspJ, and EspH effector functions, in targeting independent aspects of the phagocytic function of mammalian macrophages (Quitard et al. 2006; Marchès et al. 2008; Dong et al. 2010). Macrophage-like cells were also used to show that EspT plays a role in triggering the production of central immune mediators (Raymond et al. 2011).

The physiological relevance of the findings discovered in cultured cell lines needs confirmation in more biologically relevant models. For instance, the use of IVOC has shown that pathways of A/E lesion formation based on cultured cell lines do not necessarily apply in organ culture (Schüller et al. 2007; Bai et al. 2008). The IVOC system is the most complex in vitro model to study EPEC infection (Knutton et al. 1987). This model includes infection of freshly obtained human intestinal biopsies kept in tissue culture media under high oxygen pressure to delay ischemia and cell death. The advantage of IVOC is that the infected tissue is as close as one can get in vitro to native live tissue. In a recent study, an improved polarized version of this model was presented in which the apical surface of the tissue biopsy was affixed to an acrylic glass disk to seal the mucosal side of the

 Cite this article as *Cold Spring Harb Perspect Med* doi: 10.1101/cshperspect.a009977

sample and decrease bacterial leakage to the basolateral surface (Schüller et al. 2009). Thus, the polarized IVOC model allows specific apical EPEC infection that better mimics the in vivo situation. Like other in vitro systems, experimentation time is limited to several hours until the tissue dies and, compared with other in vitro models, the use of IVOC for experimental infection is technically challenging and requires coordination with a clinic to obtain fresh tissue. In addition, the variability in sample properties between donors and results is far greater than in cell line models. The IVOC system has been used to study EPEC-host specificity and tissue tropism (Girard et al. 2005; Mundy et al. 2007), Tir-intimin-dependent colonization and A/E lesion formation (Frankel et al. 1998a; Schüller et al. 2007; Frankel and Phillips 2008). The polarized IVOC method has shown that apical EPEC infection of duodenal mucosa results in a flagellin-dependent increase in IL-8 levels (Schüller et al. 2009).

The use of cell lines for pathogenic *E. coli* research is very common and extensive. Cell lines simplify very complex in vivo processes, and their use may significantly shorten the duration of research compared with in vivo models. EPEC is a human-specific pathogen and the use of human cell lines enables the study of direct interactions between the bacterium and its natural host cells, making IEC lines an excellent tool to study A/E pathogen–host-cell dynamics. Yet, one must consider that the biological relevance of results obtained via the use of cultured cell lines requires further validation in a relevant in vivo model. In vitro systems lack additional host cell types, immune interactions, natural microflora, and the specific physiological conditions that influence host–microbe interactions in the small intestine.

IN VIVO INFECTION MODELS

Given the narrow host range of human EPEC, studies exploring the role of genes involved in A/E pathogen virulence and lesion formation have been limited by a lack of in vivo models reflecting the natural disease process in humans. The primary site of infection by EPEC and EHEC are epithelial cells of the small intestine and colon, respectively. Although much information has been gathered through in vitro analyses with respect to localization and function of T3S effector proteins, in vivo systems are the only way of assessing A/E pathogen virulence within the context of the complete intestinal environment. A/E bacteria are able to infect diverse animal species (Wales et al. 2005) from the invertebrate *Caenorhabditis elegans* to mice, rabbits, pigs, and cattle. As such, the diversity of available animal models has provided the research community with a wealth of potential hosts with which to simulate the various aspects of human infection.

C. elegans is a small free-living nematode that has traditionally been used for the study of development, neurobiology, and aging (Brenner 1974; Larsen et al. 1995), but has garnered interest as a model to study microbial virulence strategies and conserved mechanisms of innate immunity (Kurz and Ewbank 2000; Schulenburg et al. 2004; Hilbi et al. 2007). The simplicity of the nematode provides an inexpensive and uncomplicated model in which to follow the infection process. In 1999, Tan et al. first reported the utility of *C. elegans* as a model for feeding-based pathogenicity studies using the opportunistic pathogen *Pseudomonas aeruginosa* (Tan et al. 1999). Like *P. aeruginosa*, EPEC has been shown to both infect and kill *C. elegans* via a "slow killing" mechanism resulting from bacterial accumulation in the nematode intestine, as well as to form microcolonies similar to those formed during EPEC infection of cultured epithelial cells (Mellies et al. 2006). On the other hand, EPEC has also been reported to cause paralysis and death of *C. elegans* via a "fast killing" mechanism involving secretion of diffusible exotoxins (Anyanful et al. 2005; Bhatt et al. 2011). The usefulness of the nematode as a simple animal model for the study of EPEC pathogenesis is validated by the fact that mutation-based analysis of specific genes required for full virulence of EPEC in mammalian systems are similarly required for maximum pathogenicity in the nematode. For example, the global virulence regulator Ler is essential for virulence of rabbit enteropathogenic *E. coli* (REPEC) and

C. rodentium in the weaned rabbit and mouse models of infection, respectively (Deng et al. 2004; Zhu et al. 2006), and is similarly required for colonization of *C. elegans* by EPEC (Mellies et al. 2006). In addition, EPEC is shown to elicit a protective response in *C. elegans* through induction of genes regulating nematode innate immunity in response to pathogen exposure (Anyanful et al. 2009). *C. elegans* provides a straightforward in vivo system in which to study EPEC attachment and colonization factors (Mellies et al. 2006), however, practical limitations to this model do exist as *C. elegans* cannot survive at the optimal temperature required for expression of the virulence machinery of A/E pathogens (Umanski et al. 2002). Furthermore, administration of precise bacterial inocula can be technically difficult (Mylonakis and Aballay 2005). Additionally, although some degree of conservation exists between innate immunity pathways of the nematode and those of higher organisms (2003), and nonmammalian vertebrate models like the zebrafish have been used to study neutrophil modulation during EHEC infection (Szabady et al. 2009), mammalian models such as mice and rabbits are more representative of the inflammatory response seen during human EPEC/EHEC infection.

Although EPEC and EHEC display narrow host specificity and are primarily human pathogens (Goosney et al. 2000), mouse models of infection have also been reported. Savkovic et al. found that infection of C57BL/6 mice with EPEC resulted in colonization and adherence to the intestinal epithelium as well as distortion and loss of intestinal microvilli concomitant with actin accumulation at the site of EPEC infection (Savkovic et al. 2005), all of which is consistent with A/E lesion formation. Recent studies using this model to investigate changes in tight junction morphology and epithelial barrier function in mice as a result of EPEC infection found that infections were accompanied by an inflammatory response in the intestines of infected mice (Zhang et al. 2010a,b). In both cases, wild-type EPEC induced inflammatory cell infiltration and tissue damage similar to what has been previously reported (Savkovic et al. 2005). Conversely, several EPEC derivatives were unable to colonize Swiss NIH mice (Frankel et al. 1996) and Klapproth et al. found that EPEC was rapidly cleared from the intestines of infected C57BL/6 mice with no evidence of A/E lesion formation (Klapproth et al. 2005). Similarly, a study comparing the colonization dynamics of C3H/HeJ and C57BL/6 mice infected with EPEC, EHEC, or *C. rodentium* found very low levels of EPEC/EHEC colonization with no apparent signs of pathology (Mundy et al. 2006). This was in contrast to mice infected with *C. rodentium* that showed high levels of colonization accompanied by increased intestinal weights and visible thickening of the colon (Mundy et al. 2006). These data suggest a limited application for the model due to the inconsistency in EPEC colonization levels and manifestation of A/E pathogen-associated hallmarks of disease among several laboratory mouse strains.

Streptomycin-treated mouse models have also been used to study the colonization dynamics of EPEC and EHEC by eliminating or reducing normal gut microbial communities to avoid competitive exclusion and facilitate colonization (Royan et al. 2010; Mohawk and O'Brien 2011). Bioluminescent strains of EPEC and EHEC show tight association of the bacteria with cecal and colonic tissues of streptomycin-treated mice accompanied by mild inflammation (Roxas et al. 2010; Rhee et al. 2011) and, in the case of EHEC, microvilli effacement (Roxas et al. 2010). Germ-free mice have been used to study toxin suppression and production of other virulence factors by EHEC as well as the use of probiotics for protection against colonization and EHEC-mediated disease (Takahashi et al. 2004; Eaton et al. 2008, 2011). Alternatively, human xenografts of the small and large intestine in severe combined immunodeficiency mice have recently been proposed as a model to evaluate T3S effector function in native tissue and to investigate tissue tropism (Golan et al. 2011). EHEC is able to specifically interact with human colonic xenografts, inducing A/E lesions and tissue damage dependent on a functional T3SS (Golan et al. 2011). This type of model has also been used to show that H7 flagellin is a major factor involved in increases of proinflammatory chemokines in response to EHEC infection (Miyamoto et al. 2006). An advantage of this model is the ability to study the immunogenic reaction of A/E pathogens in the absence of other microbial products. However, both the xenograft and streptomycin pretreatment models are limited in their capacity to assess A/E pathogen infection under physiologically relevant conditions that take into account the complete intestinal mucosa and interactions with the resident microflora.

A great deal of in vivo research on A/E pathogens has been performed using natural animal pathogens related to EPEC and EHEC. Infection of rabbits and mice with REPEC and *C. rodentium*, respectively, has been used to study the contributions of many translocated effectors and T3SS-related genes to EPEC virulence and disease. REPEC encompasses a number of serotypes of A/E lesion-forming, diarrheagenic *E. coli* that share many virulence characteristics with human EPEC and are capable of causing diarrheal disease in weaned rabbits (Goosney et al. 2000). Similar to human-specific EPEC strains, REPEC induces A/E lesions and intestinal pathology without invasion of the epithelia or production of enterotoxins (Robins-Browne et al. 1994b). The REPEC LEE has been characterized and shares a high degree of homology with the LEE region of EPEC (Tauschek et al. 2002), thereby facilitating the discovery of LEE genes required for A/E lesion formation and disease (Abe et al. 1998; Marchès et al. 2000).

The LEE of *C. rodentium* is also similar to those of EPEC and EHEC (Deng et al. 2001) and is required to produce A/E lesions in mouse intestinal cells (Schauer and Falkow 1993). Functional conservation of Tir and intimin from EPEC and *C. rodentium* has been shown and confirms the usefulness of the *C. rodentium*-mouse model to study human infections with EPEC and EHEC (Fig. 3) (Frankel et al. 1996; Deng et al. 2003). However, instead of causing obvious diarrhea as seen with EPEC, EHEC, and REPEC, *C. rodentium* infection results in colonic inflammation and hyperplasia (Barthold et al. 1978). Despite this difference, mouse infections with *C. rodentium* have contributed significantly to our understanding of A/E pathogen virulence and mechanisms of disease in vivo. Using the mouse model 33 virulence factors were identified during a screen of all 41 genes in the *C. rodentium* LEE (Deng et al. 2004). A comprehensive analysis of the *C. rodentium* secretome has since been performed, effectively confirming secretion of many previously identified effectors while expanding the known repertoire of T3S proteins among A/E pathogens (Deng et al. 2010). Although much of the in vivo work on A/E pathogens has focused on the virulence roles of secreted proteins, others have attempted to determine tissue localization of effectors such as Tir (Deng et al. 2003), in an effort to uncover function. Infection of mice with *C. rodentium* mutants confirms a critical role in colonization and virulence of the LEE-encoded effectors Tir, EspB, and EspZ, as well as the non-LEE-encoded effectors NleA and NleB (Newman et al. 1999; Deng et al. 2003, 2004; Gruenheid et al. 2004; Mundy et al. 2004; Kelly et al. 2006), whereas others are thought to contribute collectively (Dean and Kenny 2009).

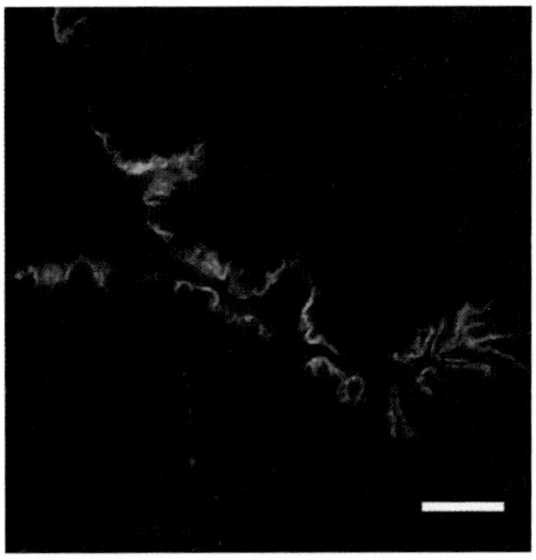

Figure 3. *C. rodentium* colonizes the apical surfaces of mouse intestinal epithelial cells (IEC) in vivo. Mice were orally infected with *C. rodentium* and colonic tissues were harvested and subjected to immunofluorescent staining at 6 days postinfection. Immunostaining was performed using a *C. rodentium*-specific anti-Tir antibody (green) with DAPI (blue) as a counterstain for host-cell nuclei. Magnification, 40×; scale bar, 50 μm. For further details, see Wlodarska et al. (2011).

A/E pathogens encode a series of secreted proteins believed to interfere with various host inflammatory signaling pathways (Dean and Kenny 2009; Wong et al. 2011). Thus, another significant advantage to the use of in vivo models is the ability to assess host-generated immune responses to bacterial infections. The *C. rodentium* mouse model has been widely used to explore the host immune response to pathogenic *E. coli* infections (Luperchio and Schauer 2001; Borenshtein et al. 2008; Kum et al. 2010). In susceptible mice, *C. rodentium* induces colonic hyperplasia and intestinal inflammation and mice often die as a result of infection, whereas less susceptible mouse strains are generally able to clear these infections (Luperchio and Schauer 2001) and develop a protective response against subsequent infection (Ghaem-Maghami et al. 2001). The role of IL-22 in regulating early control of host defense in response to *C. rodentium* infection has been shown (Zheng et al. 2008; Ota et al. 2011). Direct induction of Reg antimicrobial proteins requires IL-22 in the colonic epithelium, and survival of IL-22 knockout mice infected with *C. rodentium* is greatly improved in the presence of exogenous mouse or human RegIIIγ (Zheng et al. 2008). Mice that lack adaptive immunity of B and T lymphocytes are unable to efficiently clear *C. rodentium* infection (Vallance et al. 2002; MacDonald et al. 2003), as are mice lacking the p50 subunit of nuclear factor κB (NF-κB) (Dennis et al. 2008) and the Toll-like receptor adapter protein MyD88 (Lebeis et al. 2007; Gibson et al. 2008). Furthermore, sensitivity to *C. rodentium* is strongly influenced by the presence of segmented filamentous bacteria of the intestinal microbiota (Ivanov et al. 2009). Taken together, these studies provide clear evidence that both innate and adaptive immune pathways, as well as components of the microbiota, have a protective role against A/E pathogen infection and future in vivo work should help to shed more light on how different branches of the immune system respond to these infections.

Recent progress in the identification of T3SS structural proteins and secreted effectors has advanced our knowledge of A/E pathogenesis substantially at the molecular level, but the relevance of these proteins in EPEC and EHEC and their functions during human infection can only be validated through the use of human volunteers. Adult volunteers have been used to verify the roles of intimin, EspB, and bundle-forming pili (BFP) in EPEC pathogenesis (Donnenberg et al. 1993; Bieber et al. 1998; Tacket et al. 2000). BFP is an EPEC-specific virulence factor but intimin and EspB are found in the genomes of diverse A/E pathogen species (Tauschek et al. 2002; Iguchi et al. 2009; Ogura et al. 2009; Petty et al. 2010). Intimin and EspB are essential for A/E lesion formation in tissue culture (Foubister et al. 1994; Abe et al. 1997) and intestinal damage in rabbits and mice infected with REPEC (Abe et al. 1998; Marchès et al. 2000) or *C. rodentium*, respectively (Schauer and Falkow 1993; Newman et al. 1999; Deng et al. 2004). Intestinal biopsies taken from human volunteers infected with wild-type EPEC show destruction of the epithelial brush border (Tacket et al. 2000), a phenotype that is reproducible in tissue culture (Knutton et al. 1987). All but one individual developed diarrhea after consumption of the wild-type strain (Tacket et al. 2000), whereas only one instance of diarrhea occurred in 10 individuals who consumed the *espB* mutant (Tacket et al. 2000) and four out of 11 who consumed the intimin mutant (Donnenberg et al. 1993). Due to the relative inability of most EPEC strains to colonize nonhuman hosts and cause diarrhea, clinical trials using adult volunteers are important to further our knowledge of EPEC pathogenesis in humans and to determine the role of specific virulence factors in EPEC-mediated disease (Tacket et al. 2000). However, the use of human volunteer studies is not without its drawbacks as the diarrhea that develops is often not reproducible (Donnenberg et al. 1993) and susceptibility among individual hosts is likely to vary based on differences in immune response and prior exposure to the pathogen. Additionally, the tendency for EPEC and EHEC to infect and cause disease in young children (Kaper et al. 2004) and the varying degree of disease severity associated with EHEC infections (Nataro and Kaper 1998) precludes the use of appropriate volunteer populations, thus making the identification of host resistance genes difficult. As a result, surrogate models using related A/E pathogens and their natural animal hosts have given way to important advances in the pathogenesis of EPEC and EHEC (Vallance et al. 2004).

COMPARISON OF MODELS

No model system currently in use for the study of EPEC/EHEC pathogenesis is able to completely replicate the natural infection process in humans. However, each model does have its particular advantages and it is the combination of all these models that has advanced our knowledge on A/E pathogens tremendously over the last 20 years (Fig. 4). An area of increasing interest in A/E pathogen research involves identifying cellular receptors for T3S effectors once inside host cells. Tissue culture methods have been used successfully to identify a few host targets and functions for a number of secreted effectors. Infection of immortalized epithelial cells provides a simple and convenient means to assess the direct relationship between the infecting organism and the intestinal epithelia in the absence of additional host and microbial factors that would normally be present within the lumen. In comparison to nonpolarized epithelial cells, Caco-2 and T84 cells are capable of mimicking polarized enterocytes of the small and large intestine, respectively, and have additionally been used to study the effects of EPEC infection on tight junction integrity and overall intestinal barrier function (McNamara et al. 2001; Dean and Kenny 2004; Thanabalasuriar et al. 2010). Alternatively, polarized intestinal IVOC systems also provide an excellent in vitro model for studying EPEC infection of the intestinal mucosa because A/E pathogen exposure can be restricted to mucosal tissue surfaces, thereby simulating natural infection (Schüller et al. 2009). Because polarized IVOC maintains tissue structure and membrane integrity similar to in vivo systems, this method is far more conducive to determining the significance of effector protein distribution during infection and modeling in vivo colonization dynamics than infection of cell lines. However, human intestinal specimens are not as readily available to most laboratories as standard tissue culture cell lines (Vallance et al. 2004). Additionally, the short life span of cultured organ sections and the inability of in vitro systems to metabolize

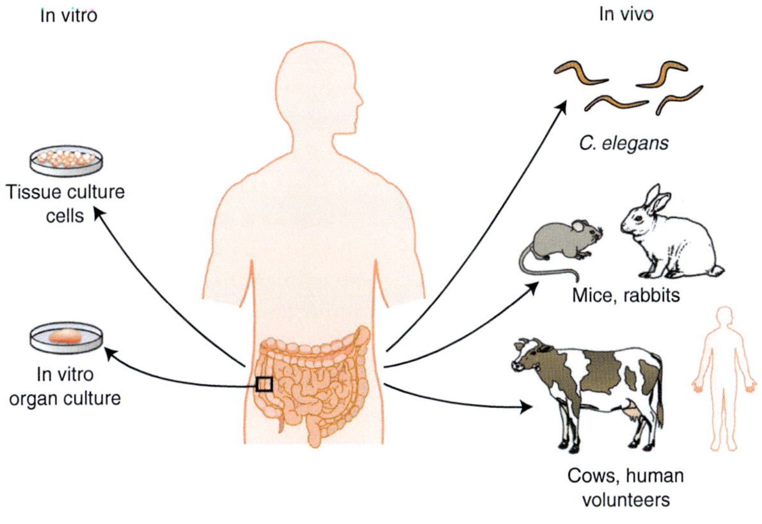

Figure 4. The various model systems currently available to study EPEC/EHEC infection biology. In vitro models such as tissue culture cells and in vitro organ culture (IVOC) are best suited to assess bacteria–host-cell interactions such as pedestal formation, signal transduction, effector translocation, and effector function. Alternatively, in vivo models are more appropriate for studies on bacterial virulence and host immune response.

drug candidates render these models ineffective for drug action studies in comparison to whole animal models.

Cell culture methods provide important information with respect to the cellular biology of T3S effector function, but are of limited use in studying the interaction of A/E pathogens with the host immune system and microbiota. Related A/E animal pathogens, infecting their natural hosts, have come to serve as effective models for EPEC/EHEC pathogenesis although not all are capable of causing equivalent intestinal disease in their animal hosts. For example, mice infected with *C. rodentium* tend to develop colonic hyperplasia accompanied by soft stool and rectal prolapse (Luperchio and Schauer 2001), whereas infection of children by EPEC can result in copious and often fatal watery diarrhea (Nataro and Kaper 1998). As natural pathogens of rabbits and mice, REPEC and *C. rodentium* provide highly relevant surrogate models to study EPEC infection in humans. Rabbits naturally develop severe and often lethal diarrhea in response to REPEC infection (Robins-Browne et al. 1994a) and with the exception of initial host-cell attachment, REPEC displays the same pathological characteristics, host age, and tissue specificity as human EPEC (Robins-Browne et al. 1994b), making this one of the most applicable models for studying EPEC infection. Limitations to this model do exist, however, in that few genetic tools are available to evaluate the host response to REPEC infection, and due to the significant mortality resulting from this infection, modest differences in virulence are difficult to detect (Vallance et al. 2004). *C. rodentium*, on the other hand, naturally infects mice of all ages and offers the convenience of several genetic and immunological tools to study infection and host response, including a variety of genetically manipulated mouse strains (Vallance et al. 2004). In addition, increasing interest on the essential role of the intestinal microbiota in maintaining overall gut homeostasis as well as protection against enteric pathogens offers another level of complexity to A/E pathogen infection that can only be assessed in vivo. Disruption or elimination of normal gut microbial communities could lead to altered epithelial cell responses and host susceptibility to infection (Sekirov and Finlay 2009; Wlodarska et al. 2011). Thus, a more detailed understanding of the mechanisms behind the host–pathogen–microbiota interaction should lead to an improved therapeutic approach in managing A/E pathogen infection. Animal models provide a convenient means to study A/E pathogens, yet the symptoms and severity of diarrheal disease can vary significantly between animal hosts. Although the clinical outcome of EPEC or EHEC infection in humans differs from that of *C. rodentium* in mice, this model remains the best for studies on host immune response to A/E pathogen infection. However, heavier costs associated with housing and handling of animals, as well as increased complexity of the gastrointestinal environment in comparison to in vitro infection models, are major limitations of in vivo systems (Table 1).

Currently, there is no vaccine available to humans to prevent infection by EPEC or EHEC, and the use of antibiotics for treatment of EHEC infections is controversial as certain antibiotics may exacerbate disease symptoms (Panos et al. 2006), as well as induce Shiga toxin production. Numerous EHEC outbreaks have been traced back to ingestion of contaminated beef and vegetable products (Nataro and Kaper 1998; Berger et al. 2010). Although a major environmental reservoir for EPEC has not yet been determined, it is generally accepted that transmission to humans occurs through contaminated food and water sources (Nataro and Kaper 1998). As a result, reducing carriage and transmission of EHEC in cattle as well as bacterial attachment to and colonization of produce are important areas of A/E pathogen research. A logical approach to preventing infections in humans has therefore been to immunize cattle to reduce colonization and fecal shedding to the environment. In vivo models such as mice, pigs, and cattle have all been used to study essential virulence factors conserved between EPEC and EHEC as potential vaccine candidates (Dean-Nystrom et al. 2002; Potter et al. 2004; McNeilly et al. 2010; Misyurina et al. 2010; Mohawk and O'Brien 2011). Thus, the EPEC/EHEC T3SS

Table 1. Summary of in vitro and in vivo infection models

	Advantages	Disadvantages	Applications
In vitro models			
Cell lines	Simplicity, uniformity, rapid growth, availability of human cells, and knockout cell lines	Infection times limited by bacterial overgrowth, limited ability to assess host response, do not necessarily represent physiology in vivo	Effector function and translocation, pedestal formation, bacteria–host-cell interactions
IVOC	Enables research in human tissues, maintains tissue structure/membrane integrity, similar to in vivo system	Low availability of human tissue biopsies, short life span of tissue limits experimentation, variation between samples	Host specificity, tissue tropism, colonization dynamics
In vivo models			
Invertebrate	Simple, inexpensive, rapid generation time, availability of knockout mutants	Cannot survive at mammalian body temperature, no adaptive immune system, difficult to deliver precise inocula	Colonization and attachment factors, microbial virulence
Mammalian	Availability of knockout mice, can assess innate and adaptive immune responses, can assess pathogen virulence	Increased cost and complexity, subject variability, no model replicates all aspects of human infection	Microbial virulence, host immune response, effects of host microbiota, vaccine trials

See text for references. IVOC, in vitro organ culture.

and secreted proteins offer attractive targets for development of novel vaccines and therapeutics whose effectiveness can only be evaluated through in vivo experimentation.

CONCLUDING REMARKS

Diarrheagenic *E. coli* present a significant risk to human health worldwide, and due to the severity of symptoms associated with infections caused by EPEC and EHEC, few opportunities have presented themselves to study host susceptibility to these microbes and disease pathogenesis in the human host. For this reason, a variety of infection models have been developed to provide insight into our current understanding of the virulence mechanisms employed by A/E bacterial pathogens and to explore the molecular mechanisms that contribute to virulence of pathogenic *E. coli*. In vitro systems are ideal for assessing the cellular biology of T3S effector function and the bacterial–host-cell interactions that contribute to initial attachment and colonization. In contrast, in vivo infection models are crucial to identify and differentiate between virulence factors that may be common to all A/E pathogens and those that are specific to EPEC and EHEC, as well as to determine the basis for host resistance or susceptibility to these pathogens to properly control the public health burdens that they impose (Vallance et al. 2004).

New and rapidly evolving genomic tools are emerging, allowing for novel revelations of how differences in gut microbial composition and variability of the human metagenome among individuals correlate with human health and disease states (Arumugam et al. 2011; Pflughoeft and Versalovic 2011). In the future, global surveys of the human metagenome should address how the composition of the microflora and the genome of the individual contribute to host susceptibility to A/E pathogens, whether in a carrier state or during acute infection. Studies of this nature should provide information that will lead to a better understanding of the infection process as it relates to host–

pathogen–microbiota interplay. Currently, no in silico models exist to directly simulate EPEC or EHEC infection (Vallance et al. 2004), however, publication of the EPEC E2348/69 (Iguchi et al. 2009), EHEC O157:H7 (Hayashi et al. 2001; Perna et al. 2001), and *C. rodentium* (Petty et al. 2010) genome sequences has facilitated identification of conserved and unique T3SS-related proteins and effectors and should contribute to the development of microarray databases and other computer-based technologies to identify and study A/E bacterial pathogens.

Major advances in A/E pathogen research have been made over the years through the use of different EPEC/EHEC infection models. Novel adherence factors and secreted effectors have been identified and for many of these proteins, their mechanism of action and host receptors have been determined. However, to fully understand how these molecular interactions contribute to disease, in vivo studies are required to show a direct impact by A/E pathogen virulence. A more detailed understanding of how EPEC and EHEC exploit cellular functions as part of the disease process could ultimately lead to potential targets for new therapeutics to help control these significant enteric pathogens.

ACKNOWLEDGMENTS

We thank Y. Litvak from the Hebrew University of Jerusalem and M. Wlodarska from the University of British Columbia for providing fluorescent images. We also thank members of the Finlay and Rosenshine laboratories for constructive comments and critical reading of the manuscript. Work in B.B.F.'s laboratory is supported by an operating grant from the Canadian Institute of Health Research (CIHR). Work in I.R.'s laboratory is supported by grants from the Israel Science Academy of Science and Humanities, the Israeli Ministry of Health, and the German Israeli Science Foundation (GIF). R.J.L. is supported by a Natural Sciences and Engineering Research Council of Canada (NSERC) Postgraduate scholarship. I.R. is an Etta Rosensohn Professor of Bacteriology. B.B.F. is the UBC Peter Wall Distinguished Professor.

REFERENCES

Abe A, Kenny B, Stein M, Finlay BB. 1997. Characterization of two virulence proteins secreted by rabbit enteropathogenic *Escherichia coli*, EspA and EspB, whose maximal expression is sensitive to host body temperature. *Infect Immun* **65:** 3547–3555.

Abe A, Heczko U, Hegele RG, Finlay BB. 1998. Two enteropathogenic *Escherichia coli* type III secreted proteins, EspA and EspB, are virulence factors. *J Exp Med* **188:** 1907–1916.

Anyanful A, Dolan-Livengood JM, Lewis T, Sheth S, Dezalia MN, Sherman MA, Kalman LV, Benian GM, Kalman D. 2005. Paralysis and killing of *Caenorhabditis elegans* by enteropathogenic *Escherichia coli* requires the bacterial tryptophanase gene. *Mol Microbiol* **57:** 988–1007.

Anyanful A, Easley KA, Benian GM, Kalman D. 2009. Conditioning protects *C. elegans* from lethal effects of enteropathogenic *E. coli* by activating genes that regulate lifespan and innate immunity. *Cell Host Microbe* **5:** 450–462.

Arumugam M, Raes J, Pelletier E, Le Paslier D, Yamada T, Mende DR, Fernandes GR, Tap J, Bruls T, Batto JM, et al. 2011. Enterotypes of the human gut microbiome. *Nature* **473:** 174–180.

Bai L, Schüller S, Whale A, Mousnier A, Marches O, Wang L, Ooka T, Heuschkel R, Torrente F, Kaper JB, et al. 2008. Enteropathogenic *Escherichia coli* O125:H6 triggers attaching and effacing lesions on human intestinal biopsy specimens independently of Nck and TccP/TccP2. *Infect Immun* **76:** 361–368.

Barthold SW, Coleman GL, Jacoby RO, Livestone EM, Jonas AM. 1978. Transmissible murine colonic hyperplasia. *Vet Pathol* **15:** 223–236.

Baruch K, Gur-Arie L, Nadler C, Koby S, Yerushalmi G, Ben-Neriah Y, Yogev O, Shaulian E, Guttman C, Zarivach R, et al. 2011. Metalloprotease type III effectors that specifically cleave JNK and NF-κB. *EMBO J* **30:** 221–231.

Berger CN, Sodha SV, Shaw RK, Griffin PM, Pink D, Hand P, Frankel G. 2010. Fresh fruit and vegetables as vehicles for the transmission of human pathogens. *Environ Microbiol* **12:** 2385–2397.

Bhatt S, Anyanful A, Kalman D. 2011. CsrA and TnaB coregulate tryptophanase activity to promote exotoxin-induced killing of *Caenorhabditis elegans* by enteropathogenic *Escherichia coli*. *J Bacteriol* **193:** 4516–4522.

Bieber D, Ramer SW, Wu CY, Murray WJ, Tobe T, Fernandez R, Schoolnik GK. 1998. Type IV pili, transient bacterial aggregates, and virulence of enteropathogenic *Escherichia coli*. *Science (New York, NY)* **280:** 2114–2118.

Borenshtein D, McBee ME, Schauer DB. 2008. Utility of the *Citrobacter rodentium* infection model in laboratory mice. *Curr Opin Gastroenterol* **24:** 32–37.

Brenner S. 1974. The genetics of *Caenorhabditis elegans*. *Genetics* **77:** 71–94.

Campellone KG. 2010. Cytoskeleton-modulating effectors of enteropathogenic and enterohaemorrhagic *Escherichia coli*: Tir, EspFU and actin pedestal assembly. *FEBS J* **277:** 2390–2402.

Coburn B, Sekirov I, Finlay BB. 2007. Type III secretion systems and disease. *Clin Microbiol Rev* **20:** 535–549.

Cornelis GR. 2006. The type III secretion injectisome. *Nat Rev Microbiol* **4:** 811–825.

Croxen MA, Finlay BB. 2010. Molecular mechanisms of *Escherichia coli* pathogenicity. *Nat Rev Microbiol* **8:** 26–38.

Dean P, Kenny B. 2004. Intestinal barrier dysfunction by enteropathogenic *Escherichia coli* is mediated by two effector molecules and a bacterial surface protein. *Mol Microbiol* **54:** 665–675.

Dean P, Kenny B. 2009. The effector repertoire of enteropathogenic *E. coli*: Ganging up on the host cell. *Curr Opin Microbiol* **12:** 101–109.

Dean P, Maresca M, Kenny B. 2005. EPEC's weapons of mass subversion. *Curr Opin Microbiol* **8:** 28–34.

Dean-Nystrom EA, Gansheroff LJ, Mills M, Moon HW, O'Brien AD. 2002. Vaccination of pregnant dams with intimin(O157) protects suckling piglets from *Escherichia coli* O157:H7 infection. *Infect Immun* **70:** 2414–2418.

Deng W, Li Y, Vallance BA, Finlay BB. 2001. Locus of enterocyte effacement from *Citrobacter rodentium*: Sequence analysis and evidence for horizontal transfer among attaching and effacing pathogens. *Infect Immun* **69:** 6323–6335.

Deng W, Vallance BA, Li Y, Puente JL, Finlay BB. 2003. *Citrobacter rodentium* translocated intimin receptor (Tir) is an essential virulence factor needed for actin condensation, intestinal colonization and colonic hyperplasia in mice. *Mol Microbiol* **48:** 95–115.

Deng W, Puente JL, Gruenheid S, Li Y, Vallance BA, Vázquez A, Barba J, Ibarra JA, O'Donnell P, Metalnikov P, et al. 2004. Dissecting virulence: Systematic and functional analyses of a pathogenicity island. *Proc Natl Acad Sci* **101:** 3597–3602.

Deng W, de Hoog CL, Yu HB, Li Y, Croxen MA, Thomas NA, Puente JL, Foster LJ, Finlay BB. 2010. A comprehensive proteomic analysis of the type III secretome of *Citrobacter rodentium*. *J Biol Chem* **285:** 6790–6800.

Deng W, Yu HB, de Hoog CL, Stoynov N, Li Y, Foster LJ, Finlay BB. 2012. Quantitative proteomic analysis of type III secretome of Enteropathogenic *E. coli* reveals an expanded effector repertoire for attaching/effacing bacterial pathogens. *Mol Cell Proteomics* **11:** 692–709.

Dennis A, Kudo T, Kruidenier L, Girard F, Crepin VF, MacDonald TT, Frankel G, Wiles S. 2008. The p50 subunit of NF-κB is critical for *in vivo* clearance of the noninvasive enteric pathogen *Citrobacter rodentium*. *Infect Immun* **76:** 4978–4988.

Dong N, Liu L, Shao F. 2010. A bacterial effector targets host DH-PH domain RhoGEFs and antagonizes macrophage phagocytosis. *EMBO J* **29:** 1363–1376.

Donnenberg MS, Tacket CO, James SP, Losonsky G, Nataro JP, Wasserman SS, Kaper JB, Levine MM. 1993. Role of the *eaeA* gene in experimental enteropathogenic *Escherichia coli* infection. *J Clin Invest* **92:** 1412–1417.

Eaton KA, Friedman DI, Francis GJ, Tyler JS, Young VB, Haeger J, Abu-Ali G, Whittam TS. 2008. Pathogenesis of renal disease due to enterohemorrhagic *Escherichia coli* in germ-free mice. *Infect Immun* **76:** 3054–3063.

Eaton KA, Honkala A, Auchtung TA, Britton RA. 2011. Probiotic *Lactobacillus reuteri* ameliorates disease due to enterohemorrhagic *Escherichia coli* in germfree mice. *Infect Immun* **79:** 185–191.

Foubister V, Rosenshine I, Donnenberg MS, Finlay BB. 1994. The *eaeB* gene of enteropathogenic *Escherichia coli* is necessary for signal transduction in epithelial cells. *Infect Immun* **62:** 3038–3040.

Frankel G, Phillips AD. 2008. Attaching effacing *Escherichia coli* and paradigms of Tir-triggered actin polymerization: Getting off the pedestal. *Cell Microbiol* **10:** 549–556.

Frankel G, Phillips AD, Novakova M, Field H, Candy DC, Schauer DB, Douce G, Dougan G. 1996. Intimin from enteropathogenic *Escherichia coli* restores murine virulence to a *Citrobacter rodentium eaeA* mutant: Induction of an immunoglobulin A response to intimin and EspB. *Infect Immun* **64:** 5315–5325.

Frankel G, Philips AD, Novakova M, Batchelor M, Hicks S, Dougan G. 1998a. Generation of *Escherichia coli* intimin derivatives with differing biological activities using site-directed mutagenesis of the intimin C-terminus domain. *Mol Microbiol* **29:** 559–570.

Frankel G, Phillips AD, Rosenshine I, Dougan G, Kaper JB, Knutton S. 1998b. Enteropathogenic and enterohaemorrhagic *Escherichia coli*: More subversive elements. *Mol Microbiol* **30:** 911–921.

Ghaem-Maghami M, Simmons CP, Daniell S, Pizza M, Lewis D, Frankel G, Dougan G. 2001. Intimin-specific immune responses prevent bacterial colonization by the attaching-effacing pathogen *Citrobacter rodentium*. *Infect Immun* **69:** 5597–5605.

Gibson DL, Ma C, Bergstrom KSB, Huang JT, Man C, Vallance BA. 2008. MyD88 signalling plays a critical role in host defence by controlling pathogen burden and promoting epithelial cell homeostasis during *Citrobacter rodentium*-induced colitis. *Cell Microbiol* **10:** 618–631.

Girard F, Batisson I, Frankel GM, Harel J, Fairbrother JM. 2005. Interaction of enteropathogenic and Shiga toxin-producing *Escherichia coli* and porcine intestinal mucosa: Role of intimin and Tir in adherence. *Infect Immun* **73:** 6005–6016.

Golan L, Gonen E, Yagel S, Rosenshine I, Shpigel NY. 2011. Enterohemorrhagic *Escherichia coli* induce attaching and effacing lesions and hemorrhagic colitis in human and bovine intestinal xenograft models. *Dis Model Mech* **4:** 86–94.

Goosney DL, Celli J, Kenny B, Finlay BB. 1999. Enteropathogenic *Escherichia coli* inhibits phagocytosis. *Infect Immun* **67:** 490–495.

Goosney DL, Gruenheid S, Finlay BB. 2000. Gut feelings: Enteropathogenic *E. coli* (EPEC) interactions with the host. *Ann Rev Cell Dev Biol* **16:** 173–189.

Gruenheid S, DeVinney R, Bladt F, Goosney D, Gelkop S, Gish GD, Pawson T, Finlay BB. 2001. Enteropathogenic *E. coli* Tir binds Nck to initiate actin pedestal formation in host cells. *Nat Cell Biol* **3:** 856–859.

Gruenheid S, Sekirov I, Thomas NA, Deng W, O'Donnell P, Goode D, Li Y, Frey EA, Brown NF, Metalnikov P, et al. 2004. Identification and characterization of NleA, a non-LEE-encoded type III translocated virulence factor of enterohaemorrhagic *Escherichia coli* O157:H7. *Mol Microbiol* **51:** 1233–1249.

Hayashi T, Makino K, Ohnishi M, Kurokawa K, Ishii K, Yokoyama K, Han CG, Ohtsubo E, Nakayama K,

Murata T, et al. 2001. Complete genome sequence of enterohemorrhagic *Escherichia coli* O157:H7 and genomic comparison with a laboratory strain K-12. *DNA Res* **8:** 11–22.

Hilbi H, Weber SS, Ragaz C, Nyfeler Y, Urwyler S. 2007. Environmental predators as models for bacterial pathogenesis. *Environ Microbiol* **9:** 563–575.

Iguchi A, Thomson NR, Ogura Y, Saunders D, Ooka T, Henderson IR, Harris D, Asadulghani M, Kurokawa K, Dean P, et al. 2009. Complete genome sequence and comparative genome analysis of enteropathogenic *Escherichia coli* O127:H6 strain E2348/69. *J Bacteriol* **191:** 347–354.

Inman LR, Cantey JR. 1983. Specific adherence of *Escherichia coli* (strain RDEC-1) to membranous (M) cells of the Peyer's patch in *Escherichia coli* diarrhea in the rabbit. *J Clin Invest* **71:** 1–8.

Ivanov II, Atarashi K, Manel N, Brodie EL, Shima T, Karaoz U, Wei D, Goldfarb KC, Santee CA, Lynch SV, et al. 2009. Induction of intestinal Th17 cells by segmented filamentous bacteria. *Cell* **139:** 485–498.

Jenkins C, Lawson AJ, Cheasty T, Willshaw GA, Wright P, Dougan G, Frankel G, Smith HR. 2003. Subtyping intimin genes from enteropathogenic *Escherichia coli* associated with outbreaks and sporadic cases in the United Kingdom and Eire. *Mol Cell Probe* **17:** 149–156.

Kalman D, Weiner OD, Goosney DL, Sedat JW, Finlay BB, Abo A, Bishop JM. 1999. Enteropathogenic *E. coli* acts through WASP and Arp2/3 complex to form actin pedestals. *Nat Cell Biol* **1:** 389–391.

Kaper JB, Nataro JP, Mobley HL. 2004. Pathogenic *Escherichia coli*. *Nat Rev Microbiol* **2:** 123–140.

Kelly M, Hart E, Mundy R, Marchès O, Wiles S, Badea L, Luck S, Tauschek M, Frankel G, Robins-Browne RM, et al. 2006. Essential role of the type III secretion system effector NleB in colonization of mice by *Citrobacter rodentium*. *Infect Immun* **74:** 2328–2337.

Kenny B. 2002. Mechanism of action of EPEC type III effector molecules. *Int J Med Microbiol* **291:** 469–477.

Kenny B, DeVinney R, Stein M, Reinscheid DJ, Frey EA, Finlay BB. 1997. Enteropathogenic *E. coli* (EPEC) transfers its receptor for intimate adherence into mammalian cells. *Cell* **91:** 511–520.

Klapproth JM, Sasaki M, Sherman M, Babbin B, Donnenberg MS, Fernandes PJ, Scaletsky ICA, Kalman D, Nusrat A, Williams IR. 2005. *Citrobacter rodentium lifA/efa1* is essential for colonic colonization and crypt cell hyperplasia *in vivo*. *Infect Immun* **73:** 1441–1451.

Knutton S, Lloyd DR, McNeish AS. 1987. Adhesion of enteropathogenic *Escherichia coli* to human intestinal enterocytes and cultured human intestinal mucosa. *Infect Immun* **55:** 69–77.

Knutton S, Baldwin T, Williams PH, McNeish AS. 1989. Actin accumulation at sites of bacterial adhesion to tissue culture cells: Basis of a new diagnostic test for enteropathogenic and enterohemorrhagic *Escherichia coli*. *Infect Immun* **57:** 1290–1298.

Kum WWS, Lo BC, Deng W, Ziltener HJ, Finlay BB. 2010. Impaired innate immune response and enhanced pathology during *Citrobacter rodentium* infection in mice lacking functional P-selectin. *Cell Microbiol* **12:** 1250–1271.

Kurz CL, Ewbank JJ. 2000. *Caenorhabditis elegans* for the study of host-pathogen interactions. *Trends Microbiol* **8:** 142–144.

Kurz CL, Ewbank JJ. 2003. *Caenorhabditis elegans*: An emerging genetic model for the study of innate immunity. *Nat Rev Genet* **4:** 380–390.

Larsen PL, Albert PS, Riddle DL. 1995. Genes that regulate both development and longevity in *Caenorhabditis elegans*. *Genetics* **139:** 1567–1583.

Lebeis SL, Bommarius B, Parkos CA, Sherman MA, Kalman D. 2007. TLR signaling mediated by MyD88 is required for a protective innate immune response by neutrophils to *Citrobacter rodentium*. *J Immunol* **179:** 566–577.

Luperchio SA, Schauer DB. 2001. Molecular pathogenesis of *Citrobacter rodentium* and transmissible murine colonic hyperplasia. *Microbes Infect* **3:** 333–340.

MacDonald TT, Frankel G, Dougan G, Goncalves NS, Simmons C. 2003. Host defences to *Citrobacter rodentium*. *Int J Med Microbiol* **293:** 87–93.

Marchès O, Nougayrède JP, Boullier S, Mainil J, Charlier G, Raymond I, Pohl P, Boury M, De Rycke J, Milon A, et al. 2000. Role of Tir and intimin in the virulence of rabbit enteropathogenic *Escherichia coli* serotype O103:H2. *Infect Immun* **68:** 2171–2182.

Marchès O, Covarelli V, Dahan S, Cougoule C, Bhatta P, Frankel G, Caron E. 2008. EspJ of enteropathogenic and enterohaemorrhagic *Escherichia coli* inhibits opsono-phagocytosis. *Cell Microbiol* **10:** 1104–1115.

Martinez-Argudo I, Sands C, Jepson MA. 2007. Translocation of enteropathogenic *Escherichia coli* across an *in vitro* M cell model is regulated by its type III secretion system. *Cell Microbiol* **9:** 1538–1546.

McDaniel TK, Jarvis KG, Donnenberg MS, Kaper JB. 1995. A genetic locus of enterocyte effacement conserved among diverse enterobacterial pathogens. *Proc Natl Acad Sci* **92:** 1664–1668.

McNamara BP, Koutsouris A, O'Connell CB, Nougayréde JP, Donnenberg MS, Hecht G. 2001. Translocated EspF protein from enteropathogenic *Escherichia coli* disrupts host intestinal barrier function. *J Clin Invest* **107:** 621–629.

McNeilly TN, Mitchell MC, Rosser T, McAteer S, Low JC, Smith DGE, Huntley JF, Mahajan A, Gally DL. 2010. Immunization of cattle with a combination of purified intimin-531, EspA and Tir significantly reduces shedding of *Escherichia coli* O157:H7 following oral challenge. *Vaccine* **28:** 1422–1428.

Mellies JL, Barron AMS, Haack KR, Korson AS, Oldridge DA. 2006. The global regulator Ler is necessary for enteropathogenic *Escherichia coli* colonization of *Caenorhabditis elegans*. *Infect Immun* **74:** 64–72.

Mills E, Baruch K, Charpentier X, Kobi S, Rosenshine I. 2008. Real-time analysis of effector translocation by the type III secretion system of enteropathogenic *Escherichia coli*. *Cell Host Microbe* **3:** 104–113.

Misyurina O, Asper DJ, Deng W, Finlay BB, Rogan D, Potter AA. 2010. The role of Tir, EspA, and NleB in the colonization of cattle by Shiga toxin producing *Escherichia coli* O26:H11. *Can J Microbiol* **56:** 739–747.

Miyamoto Y, Iimura M, Kaper JB, Torres AG, Kagnoff MF. 2006. Role of Shiga toxin versus H7 flagellin in enterohaemorrhagic *Escherichia coli* signalling of human colon epithelium *in vivo*. *Cell Microbiol* **8:** 869–879.

Mohawk KL, O'Brien AD. 2011. Mouse models of *Escherichia coli* O157:H7 infection and shiga toxin injection. 2011. *J Biomed Biotechnol* **2011:** 258185.

Mundy R, Petrovska L, Smollett K, Simpson N, Wilson RK, Yu J, Tu X, Rosenshine I, Clare S, Dougan G, et al. 2004. Identification of a novel *Citrobacter rodentium* type III secreted protein, EspI, and roles of this and other secreted proteins in infection. *Infect Immun* **72:** 2288–2302.

Mundy R, Girard F, FitzGerald AJ, Frankel G. 2006. Comparison of colonization dynamics and pathology of mice infected with enteropathogenic *Escherichia coli*, enterohaemorrhagic *E. coli* and *Citrobacter rodentium*. *FEMS Microbiol Lett* **265:** 126–132.

Mundy R, Schüller S, Girard F, Fairbrother JM, Phillips AD, Frankel G. 2007. Functional studies of intimin *in vivo* and *ex vivo*: Implications for host specificity and tissue tropism. *Microbiology* **153:** 959–967.

Mylonakis E, Aballay A. 2005. Worms and flies as genetically tractable animal models to study host-pathogen interactions. *Infect Immun* **73:** 3833–3841.

Nadler C, Baruch K, Kobi S, Mills E, Haviv G, Farago M, Alkalay I, Bartfeld S, Meyer TF, Ben-Neriah Y, et al. 2010. The type III secretion effector NleE inhibits NF-κB activation. *PLoS Pathog* **6:** e1000743.

Nataro JP, Kaper JB. 1998. Diarrheagenic *Escherichia coli*. *Clin Microbiol Rev* **11:** 142–201.

Newman JV, Zabel BA, Jha SS, Schauer DB. 1999. *Citrobacter rodentium espB* is necessary for signal transduction and for infection of laboratory mice. *Infect Immun* **67:** 6019–6025.

Nieto-Pelegrin E, Martinez-Quiles N. 2009. Distinct phosphorylation requirements regulate cortactin activation by TirEPEC and its binding to N-WASP. *Cell Commun Signal* **7:** 11.

Ogura Y, Ooka T, Iguchi A, Toh H, Asadulghani M, Oshima K, Kodama T, Abe H, Nakayama K, Kurokawa K, et al. 2009. Comparative genomics reveal the mechanism of the parallel evolution of O157 and non-O157 enterohemorrhagic *Escherichia coli*. *Proc Natl Acad Sci* **106:** 17939–17944.

Ota N, Wong K, Valdez P, Zheng Y, Crellin NK, Diehl L, Ouyang W. 2011. IL-22 bridges the lymphotoxin pathway with the maintenance of colonic lymphoid structures during infection with *Citrobacter rodentium*. *Nat Immun* **12:** 941–948.

Panos GZ, Betsi GI, Falagas ME. 2006. Systematic review: Are antibiotics detrimental or beneficial for the treatment of patients with *Escherichia coli* O157:H7 infection? *Aliment Pharmacol Ther* **24:** 731–742.

Perna NT, Plunkett G, Burland V, Mau B, Glasner JD, Rose DJ, Mayhew GF, Evans PS, Gregor J, Kirkpatrick HA, et al. 2001. Genome sequence of enterohaemorrhagic *Escherichia coli* O157:H7. *Nature* **409:** 529–533.

Petty NK, Bulgin R, Crepin VF, Cerdeño-Tárraga AM, Schroeder GN, Quail MA, Lennard N, Corton C, Barron A, Clark L, et al. 2010. The *Citrobacter rodentium* genome sequence reveals convergent evolution with human pathogenic *Escherichia coli*. *J Bacteriol* **192:** 525–538.

Pflughoeft KJ, Versalovic J. 2011. Human microbiome in health and disease. *Annu Rev Pathol* **7:** 99–122.

Potter AA, Klashinsky S, Li Y, Frey E, Townsend H, Rogan D, Erickson G, Hinkley S, Klopfenstein T, Moxley RA, et al. 2004. Decreased shedding of *Escherichia coli* O157:H7 by cattle following vaccination with type III secreted proteins. *Vaccine* **22:** 362–369.

Quitard S, Dean P, Maresca M, Kenny B. 2006. The enteropathogenic *Escherichia coli* EspF effector molecule inhibits PI-3 kinase-mediated uptake independently of mitochondrial targeting. *Cell Microbiol* **8:** 972–981.

Raymond B, Crepin VF, Collins JW, Frankel G. 2011. The WxxxE effector EspT triggers expression of immune mediators in an Erk/JNK and NF-κB-dependent manner. *Cell Microbiol* **13:** 1881–1893.

Rhee KJ, Cheng H, Harris A, Morin C, Kaper JB, Hecht G. 2011. Determination of spatial and temporal colonization of enteropathogenic *E. coli* and enterohemorrhagic *E. coli* in mice using bioluminescent *in vivo* imaging. *Gut Microbes* **2:** 34–41.

Robins-Browne RM, Tokhi AM, Adams LM, Bennett-Wood V. 1994a. Host specificity of enteropathogenic *Escherichia coli* from rabbits: Lack of correlation between adherence *in vitro* and pathogenicity for laboratory animals. *Infect Immun* **62:** 3329–3336.

Robins-Browne RM, Tokhi AM, Adams LM, Bennett-Wood V, Moisidis AV, Krejany EO, O'Gorman LE. 1994b. Adherence characteristics of attaching and effacing strains of *Escherichia coli* from rabbits. *Infect Immun* **62:** 1584–1592.

Rosenshine I, Ruschkowski S, Stein M, Reinscheid DJ, Mills SD, Finlay BB. 1996. A pathogenic bacterium triggers epithelial signals to form a functional bacterial receptor that mediates actin pseudopod formation. *EMBO J* **15:** 2613–2624.

Roxas JL, Koutsouris A, Bellmeyer A, Tesfay S, Royan S, Falzari K, Harris A, Cheng H, Rhee KJ, Hecht G. 2010. Enterohemorrhagic *E. coli* alters murine intestinal epithelial tight junction protein expression and barrier function in a Shiga toxin independent manner. *Lab Invest* **90:** 1152–1168.

Royan SV, Jones RM, Koutsouris A, Roxas JL, Falzari K, Weflen AW, Kim A, Bellmeyer A, Turner JR, Neish AS, et al. 2010. Enteropathogenic *E. coli* non-LEE encoded effectors NleH1 and NleH2 attenuate NF-κB activation. *Mol Microbiol* **78:** 1232–1245.

Savkovic SD, Villanueva J, Turner JR, Matkowskyj KA, Hecht G. 2005. Mouse model of enteropathogenic *Escherichia coli* infection. *Infect Immun* **73:** 1161–1170.

Scallan E, Hoekstra RM, Angulo FJ, Tauxe RV, Widdowson MA, Roy SL, Jones JL, Griffin PM. 2011. Foodborne illness acquired in the United States—Major pathogens. *Emerg Infect Dis* **17:** 7–15.

Schauer DB, Falkow S. 1993. The *eae* gene of *Citrobacter freundii* biotype 4280 is necessary for colonization in transmissible murine colonic hyperplasia. *Infect Immun* **61:** 4654–4661.

Schulenburg H, Kurz CL, Ewbank JJ. 2004. Evolution of the innate immune system: The worm perspective. *Immunol Rev* **198:** 36–58.

Schüller S, Chong Y, Lewin J, Kenny B, Frankel G, Phillips AD. 2007. Tir phosphorylation and Nck/N-WASP recruitment by enteropathogenic and enterohaemorrhagic *Escherichia coli* during *ex vivo* colonization of human intestinal mucosa is different to cell culture models. *Cell Microbiol* **9:** 1352–1364.

Schüller S, Lucas M, Kaper JB, Girón JA, Phillips AD. 2009. The *ex vivo* response of human intestinal mucosa to enteropathogenic *Escherichia coli* infection. *Cell Microbiol* **11:** 521–530.

Sears CL. 2000. Molecular physiology and pathophysiology of tight junctions V. assault of the tight junction by enteric pathogens. *Am J Physiol Gastrointest Liver Physiol* **279:** G1129–G1134.

Sekirov I, Finlay BB. 2009. The role of the intestinal microbiota in enteric infection. *J Physiol* **587:** 4159–4167.

Szabady RL, Lokuta MA, Walters KB, Huttenlocher A, Welch RA. 2009. Modulation of neutrophil function by a secreted mucinase of *Escherichia coli* O157:H7. *PLoS Pathog* **5:** e1000320.

Tacket CO, Sztein MB, Losonsky G, Abe A, Finlay BB, McNamara BP, Fantry GT, James SP, Nataro JP, Levine MM, et al. 2000. Role of EspB in experimental human enteropathogenic *Escherichia coli* infection. *Infect Immun* **68:** 3689–3695.

Takahashi M, Taguchi H, Yamaguchi H, Osaki T, Komatsu A, Kamiya S. 2004. The effect of probiotic treatment with Clostridium butyricum on enterohemorrhagic *Escherichia coli* O157:H7 infection in mice. *FEMS Immunol Med Microbiol* **41:** 219–226.

Tan MW, Mahajan-Miklos S, Ausubel FM. 1999. Killing of *Caenorhabditis elegans* by *Pseudomonas aeruginosa* used to model mammalian bacterial pathogenesis. *Proc Natl Acad Sci* **96:** 715–720.

Tarr PI, Gordon CA, Chandler WL. 2005. Shiga-toxin-producing *Escherichia coli* and haemolytic uraemic syndrome. *Lancet* **365:** 1073–1086.

Tauschek M, Strugnell RA, Robins-Browne RM. 2002. Characterization and evidence of mobilization of the LEE pathogenicity island of rabbit-specific strains of enteropathogenic *Escherichia coli*. *Mol Microbiol* **44:** 1533–1550.

Thanabalasuriar A, Koutsouris A, Weflen A, Mimee M, Hecht G, Gruenheid S. 2010. The bacterial virulence factor NleA is required for the disruption of intestinal tight junctions by enteropathogenic *Escherichia coli*. *Cell Microbiol* **12:** 31–41.

Tobe T, Beatson SA, Taniguchi H, Abe H, Bailey CM, Fivian A, Younis R, Matthews S, Marches O, Frankel G, et al. 2006. An extensive repertoire of type III secretion effectors in *Escherichia coli* O157 and the role of lambdoid phages in their dissemination. *Proc Natil Acad Sci* **103:** 14941–14946.

Umanski T, Rosenshine I, Friedberg D. 2002. Thermoregulated expression of virulence genes in enteropathogenic *Escherichia coli*. *Microbiology* **148:** 2735–2744.

Vallance BA, Deng W, Knodler LA, Finlay BB. 2002. Mice lacking T and B lymphocytes develop transient colitis and crypt hyperplasia yet suffer impaired bacterial clearance during *Citrobacter rodentium* infection. *Infect Immun* **70:** 2070–2081.

Vallance BA, Khan MA, Deng W, Gruenheid S, Finlay BB. 2004. Modeling enteropathogenic and enterohemorrhagic *E. coli* infections and disease. *Drug Discovery Today* **1:** 73–79.

Vossenkämper A, Marchès O, Fairclough PD, Warnes G, Stagg AJ, Lindsay JO, Evans PC, Luong LA, Croft NM, Naik S, et al. 2010. Inhibition of NF-κB signaling in human dendritic cells by the enteropathogenic *Escherichia coli* effector protein NleE. *J Immunol* **185:** 4118–4127.

Wales AD, Woodward MJ, Pearson GR. 2005. Attaching-effacing bacteria in animals. *J Comp Pathol* **132:** 1–26.

Wlodarska M, Willing B, Keeney KM, Menendez A, Bergstrom KS, Gill N, Russell SL, Vallance BA, Finlay BB. 2011. Antibiotic treatment alters the colonic mucus layer and predisposes the host to exacerbated *Citrobacter rodentium*-induced colitis. *Infect Immun* **79:** 1536–1545.

Wong ARC, Pearson JS, Bright MD, Munera D, Robinson KS, Lee SF, Frankel G, Hartland EL. 2011. Enteropathogenic and enterohaemorrhagic *Escherichia coli*: Even more subversive elements. *Mol Microbiol* **80:** 1420–1438.

Zhang Q, Li Q, Wang C, Li N, Li J. 2010a. Redistribution of tight junction proteins during EPEC infection *in vivo*. *Inflammation* **35:** 23–32.

Zhang Q, Li Q, Wang C, Liu X, Li N, Li J. 2010b. Enteropathogenic *Escherichia coli* changes distribution of occludin and ZO-1 in tight junction membrane microdomains *in vivo*. *Microb Pathogenesis* **48:** 28–34.

Zheng Y, Valdez PA, Danilenko DM, Hu Y, Sa SM, Gong Q, Abbas AR, Modrusan Z, Ghilardi N, de Sauvage FJ, et al. 2008. Interleukin-22 mediates early host defense against attaching and effacing bacterial pathogens. *Nat Med* **14:** 282–289.

Zhu C, Feng S, Thate TE, Kaper JB, Boedeker EC. 2006. Towards a vaccine for attaching/effacing *Escherichia coli*: A LEE encoded regulator (*ler*) mutant of rabbit enteropathogenic *Escherichia coli* is attenuated, immunogenic, and protects rabbits from lethal challenge with the wild-type virulent strain. *Vaccine* **24:** 3845–3855.

Genome Dynamics in *Legionella*: The Basis of Versatility and Adaptation to Intracellular Replication

Laura Gomez-Valero and Carmen Buchrieser

Institut Pasteur, Biologie des Bactéries Intracellulaires and CNRS UMR 3525, 75724 Paris, France

Correspondence: cbuch@pasteur.fr

Legionella pneumophila is a bacterial pathogen present in aquatic environments that can cause a severe pneumonia called Legionnaires' disease. Soon after its recognition, it was shown that *Legionella* replicates inside amoeba, suggesting that bacteria replicating in environmental protozoa are able to exploit conserved signaling pathways in human phagocytic cells. Comparative, evolutionary, and functional genomics suggests that the *Legionella*–amoeba interaction has shaped this pathogen more than previously thought. A complex evolutionary scenario involving mobile genetic elements, type IV secretion systems, and horizontal gene transfer among *Legionella*, amoeba, and other organisms seems to take place. This long-lasting coevolution led to the development of very sophisticated virulence strategies and a high level of temporal and spatial fine-tuning of bacteria host–cell interactions. We will discuss current knowledge of the evolution of virulence of *Legionella* from a genomics perspective and propose our vision of the emergence of this human pathogen from the environment.

Legionellosis or Legionnaires' disease is a severe pneumonia caused by bacteria belonging to the genus *Legionella*. The first member of this genus, *Legionella pneumophila,* was recognized as a human pathogen during an outbreak of severe pneumonia in 1976 in Philadelphia, PA (Fraser et al. 1977; McDade et al. 1977). Since then, outbreaks occur every year, in particular in the industrialized world, and legionellosis remains an up-to-date health problem. Thirty-five years after the discovery of *Legionella pneumophila*, the genus *Legionella* has grown considerably as it now contains 59 species (http://old.dsmz.de/microorganisms/bacterial_nomenclature_info.php?genus=Legionella), all of which are ubiquitous, environmental bacteria. Although 24 different species of *Legionella* have been isolated at least once from humans, *L. pneumophila* is the major cause of Legionnaires' disease worldwide as it causes more than 90% of the diagnosed cases. *Legionella longbeachae* is the second cause of legionellosis, responsible for 3.9% of the cases (Yu et al. 2002; Newton et al. 2010). Interestingly, the epidemiology is different in Australia and New Zealand, where *L. pneumophila* accounts for only 45.7% but *L. longbeachae* for 30.4% of the cases. Thus, research focuses mainly on *L. pneumophila* and *L. longbeachae*. In contrast, little or nothing is known about the other *Legionella* species with respect to

Cite this article as *Cold Spring Harb Perspect Med* doi: 10.1101/cshperspect.a009993

their environmental distribution, ecology, or the capacity to trigger human disease. All *Legionella* species, except *L. longbeachae*, which is present mainly in compost and potting mix, are found in freshwater environments (Whiley and Bentham 2011).

Legionella are intracellular bacteria whose natural hosts are aquatic protozoa, in which these bacteria replicate and are protected from harsh environmental conditions (Rowbotham 1980). *Legionella* are able to parasitize at least 20 different species of amoeba, two species of ciliated protozoa, and one slime mold (Lau and Ashbolt 2009), but are associated most frequently with amoeba belonging to the genera *Acanthamoeba, Hartmanella,* and *Naegleria* (Fields 1996). Interestingly, not all *Legionella* will grow in the same amoebal host (Rowbotham 1986) and certain amoeba seem also to be resistant to *Legionell*a infection (Atlan et al. 2012). Thus, *Legionella* are broad host-range protozoan parasites but some species, in particular *L. pneumophila*, are also pathogens of humans. The vehicle of transmission to humans is aerosols loaded with *Legionella.* Changes in the human lifestyle in the last decade led to an increased number of artificial water systems such as air conditioning units, cooling towers, showers, and fountains, where water is given off in a fine spray providing the bacterium with the possibility to reach humans as an accidental host. However, although *Legionella* is able to infect humans, human infection is a dead end for *Legionella* replication, as person-to-person transmission has never been reported. Therefore, the evolution of virulence traits in *L. pneumophila* and *L. longbeachae* has resulted from the interaction with environmental protozoa.

The capacity of *Legionella* to infect its natural hosts, amoeba or ciliated protozoa but also human macrophages, indicates that the mechanisms developed by this bacterium to infect low eukaryotes can also be used for infection of cells of high eukaryotes like humans as was proposed the first time by Rowbotham in 1980 (Rowbotham 1980). This also suggests that virulence factors of *Legionella* probably target conserved eukaryotic pathways, allowing *Legionella* to infect many phagocytic eukaryotic cells like different aquatic protozoa, ciliates, and human cells. Therefore, the identification of what are these factors, how they have been acquired, and what are their functions, is fundamental to decipher host–pathogen interactions and the virulence strategies used by intracellular pathogens.

In recent years, it has been shown that several intracellular bacterial pathogens are able to enter and replicate in environmental amoebae, similar to what is known for *Legionella* (Molmeret et al. 2005). This observation has led to the idea that free-living amoebae are the "Trojan horses" of the microbial world, as they have a role in survival, replication, and transmission of bacteria (Winiecka-Krusnell and Linder 1999). The interaction between both organisms provides bacteria with the capacity to infect human cells and to become human pathogens. Furthermore, genome analyses and comparative genomics of amoeba-associated bacteria revealed that common virulence factors are present in bacteria living with amoebae (Schmitz-Esser et al. 2010). This finding suggested that the genes coding for these virulence factors may be interchanged and that they may thus be part of mobile genetic elements. The complete set of mobile genetic elements of a genome is commonly called "mobilome" (Siefert 2009), a term we will use hereafter. Particularly in bacteria living with amoebae, this mobilome seems to be highly dynamic. Because *L. pneumophila* is probably the best-studied amoeba-associated bacterium, it has become a model to study the role of amoebal hosts in bacterial pathogenesis and the acquisition of virulence factors from this host–pathogen relationship.

In this review, we will present the state-of-the-art of the evolution of virulence in *Legionella* from a genomics perspective and how current data increased our knowledge about *Legionella* virulence evolution and host–pathogen interaction and acquisition of virulence factors in general.

THE *LEGIONELLA* GENOMES ARE GENETICALLY DIVERSE

L. pneumophila strains Philadelphia, Paris, and Lens were the first *Legionella* genomes sequenced (Cazalet et al. 2004; Chien et al. 2004). Each strain contains one circular chromosome that is characterized by an average G + C content of about 38% and by a size between 3.3 and 3.5 Mb. In each strain ~3000 protein-coding genes are predicted, representing a coding capacity of 88% (Table 1A). Comparative genomics revealed that ~300 genes (~10%) are specific for each strain. Taking into account that these strains belong to the same species and serogroup, the genomic diversity among them is relatively high. Further analysis of the gene content of 217 *L. pneumophila* strains using a *Legionella* DNA array confirmed that the gene content of the *L. pneumophila* genome is highly diverse (Cazalet et al. 2008). Both studies suggested that the main source of diversity among the *Legionella* genomes is mobile genetic elements and horizontal gene transfer among *L. pneumophila* strains, but also among strains belonging to different *Legionella* species, other bacterial species, and most surprisingly, probably also between *Legionella* and their eukaryotic hosts (Cazalet et al. 2004, 2008).

Recent years have seen five new *L. pneumophila* genomes sequenced providing the possibility to study their diversity and evolution in depth (Table 1A) (Steinert et al. 2007; D'Auria et al. 2010; Schroeder et al. 2010; Gomez-Valero et al. 2011b). Comparative analyses of the eight available genomes showed that the core genome contains 2405 orthologous groups of genes, and each genome when compared with the seven others contains between 154 and 271 strain-specific genes (Fig. 1). Furthermore, more than 1000 genes are not present in all eight genomes and thus also belong to the flexible gene content or accessory genome. In contrast, the *L. long-*

Table 1. General features of the *Legionella* genome sequences published

	(A) *L. pneumophila*							
	Paris	Lens	Philadelphia	Corby	Alcoy	130b	Lorraine	HL06041035
Chromosome size (Mb)	3.5	3.3	3.4	3.6	3.5	3.5	3.5	3.5
G + C content (%)	38	38	38	38	38	38	38	38
Number of genes	3178	3034	3083	3290	3197	3293	3170	3184
Number of protein-coding genes	3079	2921	2999	3193	3191	3288	3080	3079
Pseudogenes	45	59	32	44	?	?	37	53
tRNA	43	43	43	44	?	42	44	43
16S/23S/5S	3/3/3	3/3/3	3/3/3	3/3/3	3/3/3	3/3/3	3/3/3	3/3/3
Coding density (%)	87	87	88	87	86		87	87
Plasmids	1	1	0	0	0	0	1	0

	(B) *L. longbeachae*				
	NSW 150#	D-4968+	ATCC39462+	98072+	C-4E7+
Chromosome size (Kb)	4.0	4.0	4.1	4.0	3.9
G + C content (%)	37.1	37.0	37.0	37.0	37
Number of genes	3660	3557	–	–	–
16S/23S/5S	4/4/4	4/4/4	4/4/4	4/4/4	4/4/4
No. of contigs > 0.5–300 kb	Complete	13	64	65	63
N50 contig size[a]	Complete	–	138 kb	129 kb	134 kb
Percentage of coverage[b]	100%	96.3	96.3	93.4	93.1
Number of SNP with NSW150	–	1900	1611	16 853	16 820
Plasmids	1	1	0	1	1

[a]N50 contig size, calculated by ordering all contig sizes and adding the lengths (starting from the longest contig) until the summed length exceeds 50% of the total length of all contigs.

[b]For SNP detection: #, completed sequence; +, draft sequence, not completed; SNP, single nucleotide polymorphism.

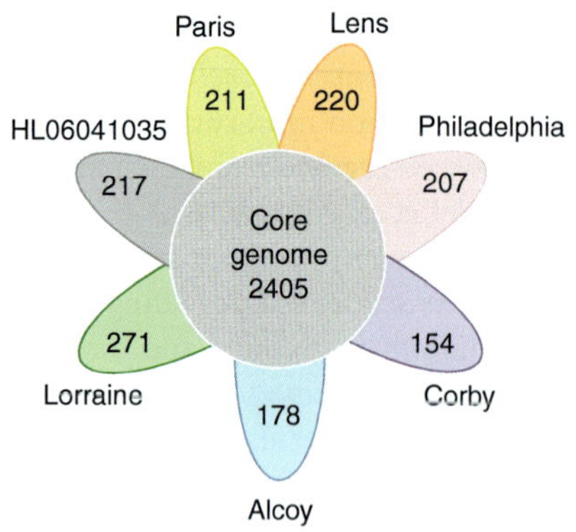

Figure 1. Shared and specific gene content of eight *L. pneumophila* genomes. Each petal represents a genome with an associated color. The number in the center of the diagram represents the orthologous groups of genes shared by all the genomes. The number inside of each individual petal corresponds to the specific genes of each genome with nonorthologous genes in any of the other genomes. The orthologous groups were defined using the program PanOCT (Fouts et al. 2012) (the parameters used were *e*-value $\leq 1 \times 10^{-5}$, percent identity ≥ 35, and length of the match ≥ 75).

beachae genomes, five of which have been sequenced to date (Cazalet et al. 2010; Kozak et al. 2010), have a slightly larger genome size of 3.9–4.1 Mb and show much lower diversity among them (Table 1B). When comparing two strains of *L. longbeachae* belonging to different serogroups (Sg), the overall DNA polymorphism is less than 0.4%, whereas this polymorphism is ~2% between *L. pneumophila* strains of the same Sg (Cazalet et al. 2010). Interestingly, despite the fact that *L. pneumophila* and *L. longbeachae* cause the same human disease, high divergence of these two species was observed. Only 2290 (65.2%) *L. longbeachae* genes are orthologous to *L. pneumophila* genes, whereas 1222 (34.8%) are *L. longbeachae* specific with respect to *L. pneumophila* Paris, Lens, Philadelphia, and Corby (Cazalet et al. 2010; Gomez-Valero et al. 2011b). The different genome analyses, when combined, revealed that the accessory genome of *Legionella*, containing approximately 3500 genes when eight *L. pneumophila* and five *L. longbeachae* genomes are analyzed, is large and, most interestingly, the main part seems to be transferable, thus constituting the mobilome of the genus.

THE MOBILOME OF *Legionella* HAS A DIVERSE ORIGIN

The *Legionella* mobilome comprises plasmids, integrative conjugative elements, insertion sequences, and genomic-island-like regions like most of the prokaryotic genomes. However, it also has a particular feature, which is a large array of so-called eukaryotic-like genes that were identified during the *L. pneumophila* genome analyses of strains Paris and Lens in 2004 (Cazalet et al. 2004). These genes code for proteins that show the strongest similarity, not to prokaryotic but to eukaryotic proteins, or encode motifs known to be implicated in protein–protein interactions, which are present only or primarily in eukaryotes. The eukaryotic-like proteins were predicted to mimic host proteins to allow intracellular replication of *Legionella* and are thus good candidates for being implicated in host–pathogen interactions (Cazalet et al. 2004). Interestingly, many of these potential virulence factors never had been identified in a prokaryotic genome before the discovery in *L. pneumophila*, suggesting that *L. pneumophila* may have developed specific mechanisms to cross talk with its eukaryotic hosts (Bruggemann et al. 2006). The presence of a high number of eukaryotic-like proteins in the genus *Legionella* was further confirmed when analyzing different *L. pneumophila* and *L. longbeachae* genomes (de Felipe et al. 2005; Cazalet et al. 2010; Kozak et al. 2010; Schroeder et al. 2010; Gomez-Valero et al. 2011b). Thus, the *Legionella* genomes reflect the interaction of this bacterium with amoeba, which is probably the driving force in the evolution of *Legionella* pathogenicity.

Although few eukaryotic-like genes were then also identified in other bacteria like *Rickettsia belli* or *Wolbachia* (Ogata et al. 2006; Walker et al. 2007), *Legionella* is the bacterium with the widest variety of eukaryotic-like proteins and eukaryotic domain-carrying proteins; a finding that still holds true today, despite the hundreds of different bacterial genome sequences that have been sequenced and analyzed since then. In 2004, more than 60 of these proteins were described in the genomes of strains Paris and Lens (Cazalet et al. 2004) and in 2005, in strain Philadelphia (de Felipe et al. 2005). Later, the analysis of the distribution of these genes among more than 200 *L. pneumophila* strains using DNA arrays showed that the conservation of certain eukaryotic-like genes within the species *L. pneumophila* was high, indicating strong environmental selection pressures for their preservation (Cazalet et al. 2008).

Recently, the sequencing and analyses of five additional *L. pneumophila* genomes revealed that over half of the eukaryotic-like proteins are present in all eight *L. pneumophila* strains (Gomez-Valero et al. 2011a,b). We thus propose that these conserved eukaryotic-like proteins have probably allowed the ancestor of *L. pneumophila* to better adapt to the intracellular environment, but now most of them are evolving as part of the core genome. Interestingly, approximately 30% of the proteins containing eukaryotic motifs are also present in the species *L. longbeachae* (Cazalet et al. 2010; Gomez-Valero et al. 2011a). These proteins might be part of the core of eukaryotic-like proteins essential for the genus *Legionella* for its intracellular replication. However, simultaneously, a flexible pool of eukaryotic-like proteins is present in the *Legionella* genomes. It represents more than half of the eukaryotic-like proteins identified to date in different *L. pneumophila* strains. The fact that even phylogenetically very closely related strains like the *L. pneumophila* strains HL06041035 and Paris present some differences in their repertoire of eukaryotic-like proteins suggests that the acquisition and loss of these genes is an ongoing process that has taken place multiple times and further underlines the high dynamics and plasticity of the *Legionella* genomes. In addition to horizontal gene transfer a process of convergent evolution might be important in the evolution of *Legionella* as, although many of the proteins containing eukaryotic motifs are not conserved between the species *L. pneumophila* and *L. longbeachae*, the same eukaryotic motifs like F-box, U-box, ankyrin, or serine thereonine protein kinase (STPK) motifs are found in both species (Cazalet et al. 2010). A similar observation has been made when the genome sequence of the amoeba symbiont "*Candidatus Amoebophilus asiaticus*" has been analyzed. Interestingly, many of the eukaryotic motifs identified in *Legionella* are also present in this bacterium (Schmitz-Esser et al. 2010). Most interestingly, an enrichment analysis comparing the fraction of all functional protein domains among 514 bacterial proteomes revealed that all genomes of bacteria replicating in amoeba were enriched in protein domains that are predominantly found in eukaryotic proteins (Schmitz-Esser et al. 2010). The most highly enriched domains were ANK repeats, LRR, SEL1 repeats, and F- and U-box domains, all of which are also present in *L. pneumophila* and *L. longbeachae.* Taken together, genome analyses of different bacteria associated with amoeba undertaken since the *L. pneumophila* genome sequence had been published in 2004, showed that bacteria that can exploit amoebae as hosts share a set of eukaryotic domain proteins that are most probably important for bacteria–host–cell interactions, despite their different lifestyles and their large phylogenetic diversity. Thus, a global, convergent adaptation mechanism to intracellular parasitism seems to exist in amoeba-associated bacteria.

EUKARYOTIC-LIKE PROTEINS OF *LEGIONELLA* ARE VIRULENCE FACTORS THAT MIMIC HOST PROTEINS

One important prediction from the genome analysis of strains Paris and Lens was that the eukaryotic-like proteins of *Legionella* are secreted virulence factors able to interfere with many different host-signaling pathways by mimicking host proteins (Cazalet et al. 2004; Bruggemann et al. 2006). Therefore, many groups started functional studies focusing on these proteins and within a short period of time the importance of the eukaryotic-like proteins for the virulence of *Legionella* was shown (de Felipe et al. 2008; Nora et al. 2009; Hubber and Roy 2010). The main secretion system of *Legionella*, essential for intracellular growth, is the type IVB Dot/Icm secretion system (Marra and Shuman 1992; Berger and Isberg 1993). Using many different methods it has been shown that at least 275 *L. pneumophila* proteins are secreted by this system (Campodonico et al. 2005; de Felipe et al. 2005; Shohdy et al. 2005; de Felipe et al. 2008; Burstein et al. 2009; Heidtman et al. 2009; Zhu et al. 2011), which represents nearly 10% of the protein-coding genes predicted in the *L. pneumophila* genomes. At least 75 of these are eukaryotic-like proteins or carry eukaryotic domains and, for certain of these, it has also been shown experimentally that they interfere with host–cell signaling pathways (Table 2).

The diverse motifs identified, like ankyrin repeats, F-box and U-box domains, STPK domains, coil-coiled domains, SET, and Sel1 or Sec-7 domains indicate also very diverse functional roles of these proteins. This fact is, for example, reflected in the very high number of different posttranslational modifications *Legionella* is able to induce in its hosts. To date, it has been shown that *L. pneumophila* proteins are able to ubiquitinate, phosphorylate, lipidate, glycosylate, AMPylate, DeAMPylate, phosphocholinate, and dephosphoryl-cholinate specific host proteins to modulate multiple host pathways to the pathogen's advantage (Rolando and Buchrieser 2012). Recently, a new posttranslational modification induced by *L. pneumophila* was reported. The *Legionella*-secreted effector, RomA, was shown to be a SET domain-containing methyltransferase that uniquely modifies host chromatin to repress gene expression and promote efficient intracellular bacterial replication (Rolando et al. 2013). Interestingly, a combination of eukaryotic domains was predicted in several of the eukaryotic-like proteins, leading to a modular protein structure. One example is AnkB/LegAU13/Ceg27/Lpg2144/Lpp2082 that contains three different eukaryotic protein domains: an F-box domain, an ankyrin domain, and a CaaX motif (Table 2). The F-box domain interacts with the SKP1 component of the SCF1 ubiquitin ligase complex and allows *Legionella* to exploit the host ubiquitination machinery, whereas the ankyrin motif, a widespread protein–protein interaction motif, probably interacts with the cellular target like Parvin B described for strain Paris (Price et al. 2009; Lomma et al. 2010). The F-box domain, as well as the two ankyrin protein–protein interaction domains, is essential for the biological function of AnkB. The CaaX motif exploits the host prenyltransferase machinery to facilitate membrane localization of *L. pneumophila* effector proteins (Ivanov et al. 2010; Price et al. 2010). Another modular protein is SdhA/Lpg0376, a Dot/Icm-secreted virulence factor that contains a GRIP and a coiled-coil domain (Table 2). SdhA is required for growth within macrophages and protection from host–cell death (Laguna et al. 2006). Furthermore, the *Legionella*-containing vacuole is actively stabilized by the SdhA protein during intracellular replication, which affords the bacterium protection from cytosolic host factors that degrade bacteria and initiate immune responses (Creasey and Isberg 2012). AnkY/LegA9/Lpg0402 is also a modular protein as it contains an STPK domain and an ankyrin motif. However, to date, its functional role and putative implication in virulence has not been shown.

A specific example for the amazing temporal and spatial fine-tuning *Legionella* has evolved in its interactions with the host cell is the Dot/Icm-secreted effector LubX/Lpg2830. LubX (for *Legionella* U-box) is a protein containing two U-box domains (U-box1 and U-box2) similar to eukaryotic E3 ubiquitin ligases that function as a ubiquitin ligase in conjunction with host UbcH5a or UbcH5c E2 enzymes and mediates polyubiquitination of cellular Clk1 (Cdc2-like kinase) (Kubori et al. 2008). U-box1 seems to be critical for ubiquitin ligation, whereas the U-box2 domain interacts with the substrate. Recently, it was shown that the secreted *L. pneumophila* effector protein SidH is a second target of LubX within host cells. LubX directly binds and polyubiquitinates SidH in vitro and mediates its proteasomal degradation in infected cells. LubX is the first example of a bacterial metaeffector that regulates spatially and temporally the expression level of another effector (Kubori et al. 2010). The identification of this mechanism points to the fact that there might be many different levels of regulation among the large number of *Legionella* effectors as well as between the effectors and their host targets.

 Cite this article as *Cold Spring Harb Perspect Med* doi: 10.1101/cshperspect.a009993

Table 2. List of selected substrates of the Dot/Icm secretion system in *L. pneumophila* and *L. longbeachae* strains

	Phila	*L. pneumophila* strains: 130b	Paris	Lens	Corby	Alcoy	Lorraine	HL06041035	*L. longbeachae* strains: NSW150	D-4968	ATCC39462	98072	C-4E7	Name	Product or associated motif	Function
Legionella core genome	*lpg0103*								*llo3312*					*vipF*	Amino-terminal acetyltransferase, GNAT	Putative modulation of host protein trafficking
	lpg0257								*llo2362*					*sdeA*	Multidrug resistance protein	Adhesion, invasion
	lpg0376								*llo0548*					*sdhA*	GRIP, coiled-coil	Putative antiapoptotic
	lpg0405								*llo2845*					-	Spectrin	Unknown
	lpg0422								*llo2801*					*legY*	Putative glucan 1,4-α-glucosidase	Unknown
	lpg0483								*llo2705*					*ankC/legA12*	Ankyrin	Unknown
	lpg0515								*llo3224*					*legD2*	Phytanoyl-CoA dioxygenase	Unknown
	lpg1426								*llo1791*					*vpdC*	Patatin	Unknown
	lpg1483								*llo1682*					*legK1*	STPK	Modulation of host immune signaling
	lpg1661								*llo1691*					-	Putative *N*-acetyl transferase	Unknown
	lpg1950								*llo1397*					*ralF*	Sec7 domain	Protein recruitment to the LCV
	lpg2222								*llo1443*					*lpnE*	Putative β-lactamase (sel1 and TTR domains)	Host cell entry
	lpg2298								*llo1707*					*ylfA/legC7*	Coiled-coil	Putative modulation of host protein trafficking
	lpg2300								*llo0584*					*ankH/legA3/ankW*	Ankyrin, NF-κB inhibitor	Intracelullar proliferation
	lpg2322								*llo0570*					*ankK/legA5*	Ankyrin	Unknown
	lpg2456								*llo0365*					*ankD/legA15*	Ankyrin	Unknown
	lpg2556								*llo2218*					*legK3*	STPK	Unknown
	lpg2832								*llo0214*					-	Hypothetical protein	Unknown
	lpg2936								*llo0081*					-	rRNA methyltransferase E	Unknown
L. pneumophila core genome	*lpg0038*													*ankQ/legA10*	Ankyrin	Unknown
	lpg0403													*ankG/ankZ/ legA7*	Ankyrin	Unknown
	lpg0436													*ankJ/legA11*	Ankyrin	Intracellular proliferation
	lpg0695													*ankN/ankX/legA8*	Ankyrin	Modulation of vesicular transport
	lpg1488													*lgt3/legc5*	Coiled-coil	Cytotoxicity
	lpg1701													*ppeA/legC3*	Coiled-coil	Putative modulation of host protein trafficking
	lpg1718													*ankI/legAS4*	Ankyrin methyltransferase	Unknown
	lpg1884													*ylfB/legC2*	Coiled-coil	Putative modulation of host protein trafficking
	lpg1962													*lirB/ppiB/rotA*	Rotamase	Unknown
	lpg1976													*pieG/legG1*	Regulator of chromosome condensation	Unknown
	lpg1978													*setA*	Putative glucosyltransferase	Putative modulation of host protein trafficking
	lpg2137													*legK2*	Ca-depPK	ER recruitment to the LCV
	lpg2144													*ankB/legAU13/ ceg27*	Ankyrin, F-box, CaaX	Recruitment of polyubiquitinated proteins to the LCV
	lpg2176													*legS2*	Sphingosine-1-phosphate lyase	Autophagy
	lpg2215													*legA2*	Ankyrin	Unknown
	lpg2410													*vpdA*	Patatin domain	Unknown
	lpg2452													*ankF/legA14/ceg31*	Ankyrin	Unknown
	lpg2793													*lepA*	Effector protein A	Release of bacteria
	lpg2999													*legP*	Astacin protease	Unknown

Continued

Table 2. *Continued*

	Phila	*L. pneumophila* strains							*L. longbeachae* strains					Name	Product or associated motif	Function
		130b	Paris	Lens	Corby	Alcoy	Lorraine	HL06041035	NSW150	D-4968	ATCC39462	98072	C-4E7			
L. longbeachae core genome									*llo0037*	■	■	■	■	-	Ankyrin	Unknown
									llo1168	■	■	■	■	-	Ankyrin	Unknown
									llo1371	■	■	■	■	-	Ankyrin, coiled-coil	Unknown
									llo2668	■	■	■	■	-	Ankyrin	Unknown
									llo3081	■	■	■	■	-	Ankyrin, patatin-like phospholipase	Unknown
									llo3093	■	■	■	■	-	Ankyrin, STPK	Unknown
									llo0114	■	■	■	■	-	LRR	Unknown
									llo1681	■	■	■	■	-	STPK	Unknown
									llo1427	■	■	■	■	-	F-Box	Unknown
									llo0448	■	■	■	■	-	U-Box	Unknown
									llo2643	■	■	■	■	-	PPR, Coiled-coil	Unknown
									llo2200	■	■	■	■	-	TTL	Unknown
									llo2327	■	■	■	■	-	SH2	Unknown
									llo2352	■	■	■	■	-	PAM2	Unknown
									llo1196	■	■	■	■	-	Snare	Unknown
									llo0793	■	■	■	■	-	Phosphatidylinositol-4-phosphate 5-kinase	Unknown
									llo2329	■	■	■	■	-	Ras-related small GTPase, Miro-like domain	Unknown
									llo2249	■	■	■	■	-	Miro-like domains	Unknown
									llo1892	■	■	■	■	-	Putative immunoglobulin I-set domain	Unknown
Legionella accessory genome									*llo0990*	■	■			-	Ankyrin	Unknown
									llo3116					-	LRR	Unknown
									llo3118					-	LRR	Unknown
	lpg0041				■	■		■						-	Putative metalloprotease	Unknown
	lpg0171	■	■	■			■	■						*legU1*	F-box motif	Interaction with host ubiquitination machinery
	lpg0402							■						*ankY/legA9*	Ankyrin, STPK	Unknown
	lpg1328		■	■	■	■	■	■						*legT*	Thaumatin domain	Unknown
	lpg1355		■					■						*sidG*	Coiled-coil	Interaction with dot/icms system—translocation
	lpg1684				■	■		■						-	Unknown	Unknown
	lpg1890	■		■	■	■	■	■						*legLC8*	LRR, coiled-coil	Unknown
	lpg1947	■	■	■			■							*lem16*	Spectrin	Unknown
	lpg1948						■	■						*legLC4*	LRR, coiled-coil	Unknown
	lpg2224							■						*ppgA*	Regulator of chromosome condensation	Unknown
	lpg2400	■		■			■							*legL7*	LRR	Unknown
	lpg2830	■	■				■	■						*lubX/legU2*	U box motif	Polyubiquitination of host proteins
	lpg2862						■							*Lgt2/legC8*	Glucosyltransferase	Cytotoxicity

Filled squares represent presence, white squares absence of a gene. ER, endoplasmatic reticulum; LCV, *Legionella*-containing vacuole; STPK, seronine threonine protein kinase; LLR, leucine rich repeat; TTL, tubulin-tyrosine ligase; SH2, Src homology 2; PAM, PCI/PINT-associated module; Rab1 GAP, GTPase-activating protein; TTR, tetratricopeptide repeats; Ca-depPK, calmodulin-dependent protein kinase.

EUKARYOTIC-LIKE PROTEINS WITNESS THE LONG-LASTING COEVOLUTION OF *Legionella* WITH EUKARYOTIC CELLS

The presence of these many eukaryotic-like proteins in the genome of amoeba-associated bacteria raises the question of the origin of this part of the genome. Are these eukaryotic-like genes laterally acquired or are they the result of adaptive changes generating motifs similar to the eukaryotic ones? The latter is known as "convergent evolution" in contrast to horizontal gene transfer (HGT). Although both can lead to similar proteins, similarity resulting from HGT is detectable at the primary sequence level, whereas convergent evolution more frequently leads to similar tridimensional structures but not necessarily to similar amino acid sequences. *Legionella* possesses all necessary factors for acquiring genes laterally as these bacteria are naturally competent and a complete recombination machinery is present, thus it is conceivable that eukaryotic-like proteins have been acquired by HGT. To analyze this question, different phylogenetic analyses of some of these eukaryotic-like proteins have been performed revealing a clustering of the *Legionella* eukaryotic-like proteins with eukaryotic sequences, further supporting the hypothesis that these were acquired by HGT from eukaryotic organisms like amoeba (de Felipe et al. 2005; Nora et al. 2009; Lurie-Weinberger et al. 2010; Gomez-Valero et al. 2011a). An example is the enzyme arylamine *N*-acetyltransferase (NAT) encoded by *lpp2516* in strain Paris. The *L. pneumophila* NAT is an atypical member of the arylamine *N*-acetyltransferase family that allows *L. pneumophila* to detoxify aromatic amine chemicals and to grow in their presence (Kubiak et al. 2012). This enzyme probably allows *L. pneumophila* that is exposed to many chemicals in its natural and man-made environments, to better adapt and survive in these environments (Kubiak et al. 2012). Phylogenetic reconstruction shown in Figure 2A indicates that horizontal acquisition of this gene from ciliates is the most probable origin of this enzyme in *Legionella* (Kubiak et al. 2012).

Another convincing example is the eukaryote-like sphingosine-1-phosphate lyase of *L. pneumophila* that is encoded by *lpp2128* in strain Paris. It has been shown that the *L. pneumophila* Spl is a Dot/Icm-secreted effector that is able to complement the sphingosine-sensitive phenotype of *Saccharomyces cervisiae* and that colocalizes with host–cell mitochondria (Degtyar et al. 2009). Figure 2B shows a phylogenetic reconstruction of Spl sequences that we have undertaken. The *L. pneumophila* Spl protein sequence clearly falls into the eukaryotic clade of SPL sequences and those of aquatic protozoa are the closest phylogenetic relatives, suggesting acquisition of this gene by HGT from an amoeba host. Most interestingly, *Legionella* also encodes proteins not present in any prokaryote sequenced to date, as amoebae are the only organisms in which we identified sequences with significant similarity. An example is Lpg1684 that codes a protein of unknown function and belongs to the eukaryotic D123 protein family. Eukaryotic-like proteins may have been acquired also from other eukaryotic organisms like viruses. For example, the eukaryotic-like protein Lpg2416 is an ankyrin-containing protein whose best homolog is present in the *Acanthamoeba polyphaga* mimivirus, a giant virus infecting *Acanthamoeba* that are also the hosts of *Legionella* (Lurie-Weinberger et al. 2010). Furthermore, recent studies have shown that these eukaryotic-like proteins can also be exchanged among different amoeba-related bacteria, adding complexity to the possible evolutionary scenarios. For example, the sequencing and analysis of the genome of *L. drancourtii* have identified numerous cases of exchange between the intracellular bacteria of the order Legionellales and Chlamydiales (Gimenez et al. 2011). However, for several eukaryotic-like proteins, a clear description of their phylogenetic history remains difficult, as the acquisition events might have been too ancient for the phylogenetic signal in the sequences to be preserved. Moreover, in some of these proteins, the eukaryotic sequence is restricted to a motif and therefore to a very short region, for which phylogenetic analyses are less evident. Furthermore, only few genome sequences of aquatic amoeba and other protozoa that are most probably the main donors are currently present in the databases. Only

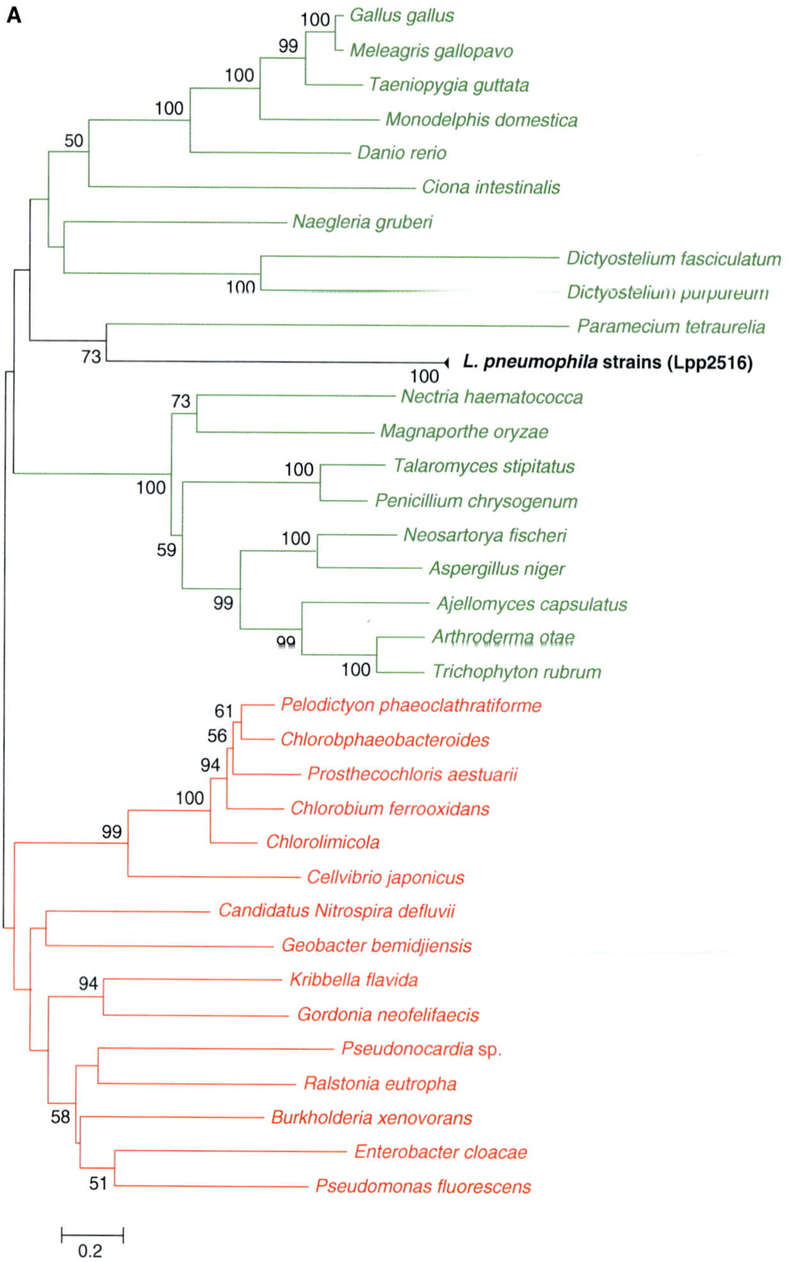

Figure 2. Phylogenetic tree of selected eukaryotic-like proteins of *Legionella* spp. (*A*) Phylogeny of the enzyme arylamine *N*-acetyltransferase (NAT) (Lpp2516) of *L. pneumophila*. (*B*) Phylogeny of the sphingosine-1 phosphate lyse Lpp2128 of *L. pneumophila*. Homolog sequences were recruited from a database containing only completed genome sequences. Selected representatives of all eukaryotic groups and one representative of each bacterial species are included in the analyses. After blastp only significant hits were recruited (*e*-value $<10 \times 10^{-4}$), and only one hit for each species was retained. The alignments were performed with T-coffee (expresso) for Lpp2616 and Muscle for Lpp2128, and followed by manual curation. Phylogenies were reconstructed using a distance method (NJ) with 1000 bootstrap replicates. The corresponding support values are shown in each node (values lower than 50 are not represented). Bars represent 20% and 10% of estimated phylogenetic divergence, respectively. (*Legend continues on following page.*)

 Cite this article as *Cold Spring Harb Perspect Med* doi: 10.1101/cshperspect.a009993

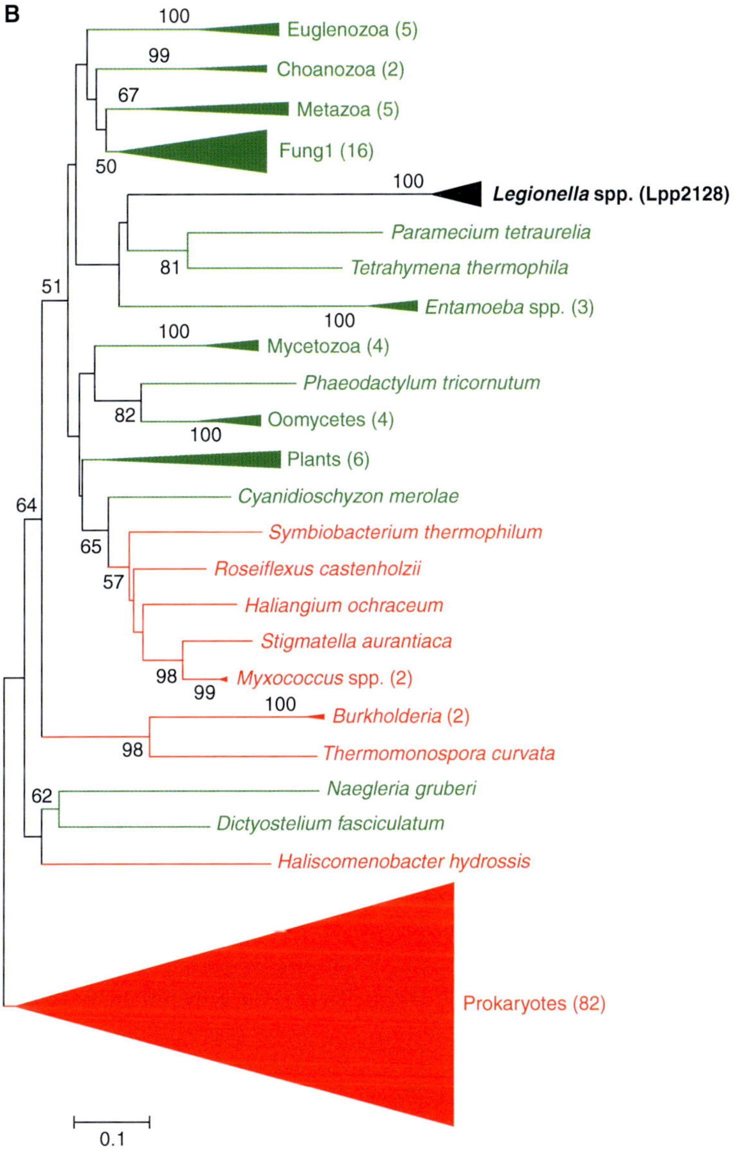

Figure 2. (*Continued*). The clustering of *Legionella* and eukaryotes was confirmed in both cases by in-parallel likelihood reconstructions.

once the gap of genome sequences of aquatic protozoa in the databases is filled will we be able to fully understand the origin and history of these eukaryotic-like proteins.

Independently of the origin of these eukaryotic-like genes, their mechanism of integration remains unknown. How are these eukaryotic genes without any similarity to prokaryotic sequences integrated in the *Legionella* genome? Many eukaryotic genes have introns, but their prokaryotic counterparts do not. How can these introns be processed or how did they "disappear?" It is tempting to propose that this happened not via HGT but via "horizontal RNA transfer." Indeed, the *L. pneumophila* genomes encode a reverse transcriptase that could be implicated in this process. However, the integration mechanism remains to be discovered.

Moreover, there is evidence for the presence of a secretion signal in the carboxy terminal of the Dot/Icm substrates (Nagai et al. 2005; Kubori et al. 2008; Huang et al. 2010); therefore, acquisition of this signal is also a challenge for these proteins to be secreted. Thus, although we start to discern how these eukaryotic proteins have appeared in the *Legionella* genomes, many questions with respect to the detailed mechanism of acquisition and integration remain unanswered.

THE *LEGIONELLA* ACCESSORY GENES ARE PART OF THE MOBILOME

The most evident component of the *Legionella* accessory genome is the plasmids. The strains Paris, Lens, and Lorraine each contain a plasmid of 132, 60, and 150 kb, respectively (Cazalet et al. 2004; Chien et al. 2004). Likewise, *L. longbeachae* strain NSW150 possesses a plasmid of 72 kb, and also strain D-4968 seems to contain a plasmid (Kozak et al. 2010). These replicons show a heterogeneous distribution among the different genomes, as the same plasmids or partially similar plasmids are present in different strains (Cazalet et al. 2008). Likewise, a comparative analysis of these plasmids revealed that the 30.4 kb Tra region present in the plasmid of *L. pneumophila* strain Paris shows much higher identity with the Tra region located on the *L. longbeachae* plasmid, than with those of the other *L. pneumophila* strains sequenced (Gomez-Valero et al. 2011b). Interestingly, regions coding proteins homologous to Tra proteins probably coding for a conjugative machinery are present in all *Legionella* plasmids. They are similar to F-type T4SS that allow the synthesis of a long and flexible pilus for conjugation in liquid and solid media (Lawley et al. 2003). However, such a system is also found in a chromosomal localization in the *L. pneumophila* strain Philadelphia and in *L. longbeachae* strain NSW150. In both strains, this region is inserted in a tRNA gene next to an integrase and it is bordered by flanking repeats. The presence of these elements suggests that these T4SSs are mobile and that their heterogeneous distribution is the result of the lateral movement of these plasmids. Furthermore, these conjugative elements can be found in different plasmids or can be completely or partially present in the chromosome, indicating that they might have the capacity to integrate and excise from the *Legionella* genomes. Indeed, this has been shown for the region carrying the Lvh genes. Lvh is a genomic-island-like region that encodes a T4ASS implicated in conjugation and in virulence related phenotypes under laboratory conditions mimicking the spread of Legionnaires' disease from environmental niches (Ridenour et al. 2003; Bandyopadhyay et al. 2007). This region can be integrated in the chromosome but can also excise in a site-specific manner to exist as a low copy number plasmid that is undoubtedly transferable among strains (Doleans-Jordheim et al. 2006). Indeed, it is present in five of the eight sequenced *L. pneumophila* strains, in *L. longbeachae* strain D-4968 (Kozak et al. 2010), and according to hybridization results also partly in *Legionella rubrilucens* (Cazalet et al. 2008). Interestingly, in the *Legionella* strains where the Lvh T4SS is not present, another Tra system, a P-type T4SS that codes for short and rigid pili that allow surface mating for conjugation, is present in the same chromosomal position (Gomez-Valero et al. 2011b). This remarkable diversity and mobility of conjugative elements probably coding T4SSs in *Legionella* and their presence in all strains, suggests a role in the bacterial adaptation to the different environments and hosts.

An intriguing feature of these probably mobile regions is their association with genes coding for homologs of a putative phage repressor protein (PrpA) and for homologs of LvrA, LvrB, and LvrC, first described for the Lvh region of *L. pneumophila*. LvrC is a homolog of CsrA, an RNA-binding protein crucial for the regulation of the switch between the replicative and transmissive phases of *L. pneumophila* (Molofsky and Swanson 2004). It is tempting to assume that the CsrA homolog encoded on each of these regions is implicated in the regulation of the mobility and/or the integration or excision of these islands. Perhaps under certain conditions it is advantageous for *Legionella* to have multiple copies of these plasmids to achieve a higher expression of these genes or for promoting high frequencies of DNA interchange to rapidly adapt to new conditions. This idea is further underlined by the observation that the presence of homologous Lvr loci is not only limited to the Lvh or Tra regions, but they are also present in other genomic locations that seem to be mobile. Thus, large parts of the accessory genome of *Legionella* are mobile, and it seems that their mobility is regulated by a particular mechanism that still remains elusive.

CONCLUSIONS: THE *Legionella* ACCESSORY GENOME IS PART OF A GLOBAL MOBILOME CONFERRING VERSATILITY

One of the best-known characteristics of obligate, intracellular bacteria is a reduction of the genome size as many genes become dispensable given the stability of the intracellular environment they inhabit. In addition, there is a minor probability of DNA exchange as they live in a closed niche (Moya et al. 2008). Extreme examples are the symbiontic bacteria *Buchnera aphidicola*, an insect endosymbiont found in the aphid *Cinaria cedri* with a genome size of 416 Kb coding solely for 357 protein-coding genes (Pérez-Brocal et al. 2006) or the obligate intracellular parasite *Rickettsia prowazekii*, the causative agent of epidemic typhus with a 1.1 Mb genome (Andersson et al. 1998). However, this does not hold true for amoeba-associated bacteria like *Legionella*, as no signs that they follow a reductive evolutionary path are present in their genomes, although *Legionella* are probably replicating exclusively intracellularly. Furthermore, *Legionella* have highly dynamic genomes and a large flexible gene pool is present in the genus comprising plasmids, mobile genetic elements, and genes probably transferred from their hosts. This large mobilome confers to these bacteria their high versatility and capacity to adapt to many different environmental conditions and

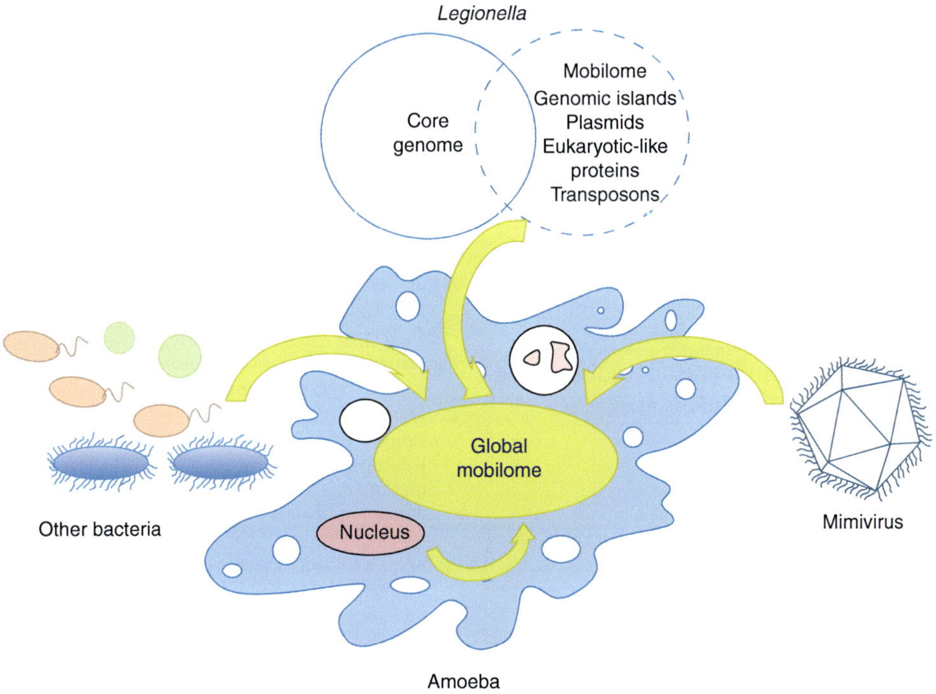

Figure 3. Model of the genetic flow inside amoeba. The accessory genome of *Legionella*, other intracellular bacteria, mimivirus, and amoeba are contributing to a "global mobilome" that is accessible to the different organisms living inside amoeba to adapt to different conditions and allows versatility in the host–pathogen interactions.

to infect diverse hosts like amoeba, protists, probably algae, and humans. Thus, the intracellular environment provided by amoeba is different from the environment in which symbionts like (e.g., *Buchneria* or *Rickettsia* live), probably because amoeba may host different bacteria and viruses at the same time, allowing exchange of genetic material among a variety of organisms. We thus propose that amoeba and aquatic protozoa form communities with a large pool of exchangeable genetic material, and have a large common accessory genome, derived from the host and from the diverse intracellular organisms that inhabit them as parasites or symbionts (Fig. 3). This big "global mobilome" allows the constant acquisition of DNA and provides a large repertoire of diverse functions for adapting to changing environmental conditions. This large mobile gene pool combined with the high genome dynamics provide the basis of versatility and adaptability in *Legionella* and allow successful intracellular parasitism in many different hosts.

ACKNOWLEDGMENTS

This work received financial support from the Institut Pasteur, the Centre National de la Recherche (CNRS), the Institut Carnot-Pasteur MI, the Fondation pour la Recherche Médicale (FRM), grant No. DEQ20120323697, ANR-10-LABX-62-IBEID, and the ANR-10-PATH-004 project "MobilGenomics," in the frame of ERA-NET PathoGenoMics. We are most grateful to Philippe Glaser for critical reading of the manuscript and for fruitful discussions.

REFERENCES

Andersson SG, Zomorodipour A, Andersson JO, Sicheritz-Pontén T, Alsmark UC, Podowski RM, Näslund AK, Eriksson AS, Winkler HH, Kurland CG. 1998. The genome sequence of *Rickettsia prowazekii* and the origin of mitochondria. *Nature* **396:** 133–140.

Atlan D, Coupat-Goutaland B, Risler A, Reyrolle M, Souchon M, Briolay J, Jarraud S, Doublet P, Pélandakis M. 2012. *Micriamoeba tesseris* nov. gen. nov. sp: A new taxon of free-living small-sized *Amoebae* non-permissive to virulent *Legionellae*. *Protist* **163:** 888–902.

Bandyopadhyay P, Liu S, Gabbai CB, Venitelli Z, Steinman HM. 2007. Environmental mimics and the Lvh type IVA secretion system contribute to virulence-related phenotypes of *Legionella pneumophila*. *Infect Immun* **75:** 723–735.

Berger KH, Isberg RR. 1993. Two distinct defects in intracellular growth complemented by a single genetic locus in *Legionella pneumophila*. *Mol Microbiol* **7:** 7–19.

Bruggemann H, Cazalet C, Buchrieser C. 2006. Adaptation of *Legionella pneumophila* to the host environment: Role of protein secretion, effectors and eukaryotic-like proteins. *Curr Opin Microbiol* **9:** 86–94.

Burstein D, Zusman T, Degtyar E, Viner R, Segal G, Pupko T. 2009. Genome-scale identification of *Legionella pneumophila* effectors using a machine learning approach. *PLoS Pathog* **5:** e1000508.

Campodonico EM, Chesnel L, Roy CR. 2005. A yeast genetic system for the identification and characterization of substrate proteins transferred into host cells by the *Legionella pneumophila* Dot/Icm system. *Mol Microbiol* **56:** 918–933.

Cazalet C, Rusniok C, Bruggemann H, Zidane N, Magnier A, Ma L, Tichit M, Jarraud S, Bouchier C, Vandenesch F, et al. 2004. Evidence in the *Legionella pneumophila* genome for exploitation of host cell functions and high genome plasticity. *Nat Genet* **36:** 1165–1173.

Cazalet C, Jarraud S, Ghavi-Helm Y, Kunst F, Glaser P, Etienne J, Buchrieser C. 2008. Multigenome analysis identifies a worldwide distributed epidemic *Legionella pneumophila* clone that emerged within a highly diverse species. *Genome Res* **18:** 431–441.

Cazalet C, Gomez-Valero L, Rusniok C, Lomma M, Dervins-Ravault D, Newton HJ, Sansom FM, Jarraud S, Zidane N, Ma L, et al. 2010. Analysis of the *Legionella longbeachae* genome and transcriptome uncovers unique strategies to cause Legionnaires' disease. *PLoS Genet* **6:** e1000851.

Chien M, Morozova I, Shi S, Sheng H, Chen J, Gomez SM, Asamani G, Hill K, Nuara J, Feder M, et al. 2004. The genomic sequence of the accidental pathogen *Legionella pneumophila*. *Science* **305:** 1966–1968.

Creasey EA, Isberg RR. 2012. The protein SdhA maintains the integrity of the *Legionella*-containing vacuole. *Proc Natl Acad Sci* **109:** 3481–3486.

D'Auria G, Jimenez-Hernandez N, Peris-Bondia F, Moya A, Latorre A. 2010. *Legionella pneumophila* pangenome reveals strain-specific virulence factors. *BMC Genomics* **11:** 181.

de Felipe KS, Pampou S, Jovanovic OS, Pericone CD, Ye SF, Kalachikov S, Shuman HA. 2005. Evidence for acquisition of *Legionella* type IV secretion substrates via interdomain horizontal gene transfer. *J Bacteriol* **187:** 7716–7726.

de Felipe KS, Glover RT, Charpentier X, Anderson OR, Reyes M, Pericone CD, Shuman HA. 2008. *Legionella* eukaryotic-like type IV substrates interfere with organelle trafficking. *PLoS Pathog* **4:** e1000117.

Degtyar E, Zusman T, Ehrlich M, Segal G. 2009. A *Legionella* effector acquired from protozoa is involved in sphingolipids metabolism and is targeted to the host cell mitochondria. *Cell Microbiol* **11:** 1219–1235.

Doleans-Jordheim A, Akermi M, Ginevra C, Cazalet C, Kay E, Schneider D, Buchrieser C, Atlan D, Vandenesch F, Etienne J, et al. 2006. Growth-phase-dependent mobility of the lvh-encoding region in *Legionella pneumophila* strain Paris. *Microbiology* **152:** 3561–3568.

Fields BS. 1996. The molecular ecology of *Legionellae*. *Trends Microbiol* **4:** 286–290.

Fouts DE, Brinkac L, Beck E, Inman J, Sutton G. 2012. PanOCT: Automated clustering of orthologs using conserved gene neighborhood for pan-genomic analysis of bacterial strains and closely related species. *Nucleic Acids Res* **40:** e172.

Fraser DW, Tsai TR, Orenstein W, Parkin WE, Beecham HJ, Sharrar RG, Harris J, Mallison GF, Martin SM, McDade JE, et al. 1977. Legionnaires' disease: Description of an epidemic of pneumonia. *N Engl J Med* **297:** 1189–1197.

Gimenez G, Bertelli C, Moliner C, Robert C, Raoult D, Fournier PE, Greub G. 2011. Insight into cross-talk between intra-amoebal pathogens. *BMC Genomics* **12:** 542.

Gomez-Valero L, Rusniok C, Cazalet C, Buchrieser C. 2011a. Comparative and functional genomics of *Legionella* identified eukaryotic like proteins as key players in host–pathogen interactions. *Front Microbiol* **2:** 208.

Gomez-Valero L, Rusniok C, Jarraud S, Vacherie B, Rouy Z, Barbe V, Medigue C, Etienne J, Buchrieser C. 2011b. Extensive recombination events and horizontal gene transfer shaped the *Legionella pneumophila* genomes. *BMC Genomics* **12:** 536.

Heidtman M, Chen EJ, Moy MY, Isberg RR. 2009. Large-scale identification of *Legionella pneumophila* Dot/Icm substrates that modulate host cell vesicle trafficking pathways. *Cell Microbiol* **11:** 230–248.

Huang L, Boyd D, Amyot WM, Hempstead AD, Luo ZQ, O'Connor TJ, Chen C, Machner M, Montminy T, Isberg RR. 2010. The E Block motif is associated with *Legionella pneumophila* translocated substrates. *Cell Microbiol* **13:** 227–245.

Hubber A, Roy CR. 2010. Modulation of host cell function by *Legionella pneumophila* type IV effectors. *Annu Rev Cell Dev Biol* **26:** 261–283.

Ivanov SS, Charron G, Hang HC, Roy CR. 2010. Lipidation by the host prenyltransferase machinery facilitates membrane localization of *Legionella pneumophila* effector proteins. *J Biol Chem* **285:** 34686–34698.

Kozak NA, Buss M, Lucas CE, Frace M, Govil D, Travis T, Olsen-Rasmussen M, Benson RF, Fields BS. 2010. Virulence factors encoded by *Legionella longbeachae* identified on the basis of the genome sequence analysis of clinical isolate D-4968. *J Bacteriol* **192:** 1030–1044.

Kubiak X, Dervins-Ravault D, Pluvinage B, Chaffotte AF, Gomez-Valero L, Dairou J, Busi F, Dupret JM, Buchrieser C, Rodrigues-Lima F. 2012. Characterization of an acetyltransferase that detoxifies aromatic chemicals in *Legionella pneumophila*. *Biochem J* **445:** 219–229.

Kubori T, Hyakutake A, Nagai H. 2008. *Legionella* translocates an E3 ubiquitin ligase that has multiple U-boxes with distinct functions. *Mol Microbiol* **67:** 1307–1319.

Kubori T, Shinzawa N, Kanuka H, Nagai H. 2010. *Legionella* metaeffector exploits host proteasome to temporally regulate cognate effector. *PLoS Pathog* **6:** e1001216.

Laguna RK, Creasey EA, Li Z, Valtz N, Isberg RR. 2006. A *Legionella pneumophila*-translocated substrate that is required for growth within macrophages and protection from host cell death. *Proc Natl Acad Sci* **103:** 18745–18750.

Lau HY, Ashbolt NJ. 2009. The role of biofilms and protozoa in *Legionella* pathogenesis: Implications for drinking water. *J Appl Microbiol* **107:** 368–378.

Lawley TD, Klimke WA, Gubbins MJ, Frost LS. 2003. F factor conjugation is a true type IV secretion system. *FEMS Microbiol Lett* **224:** 1–15.

Lomma M, Dervins-Ravault D, Rolando M, Nora T, Newton HJ, Sansom FM, Sahr T, Gomez-Valero L, Jules M, Hartland EL, et al. 2010. The *Legionella pneumophila* F-box protein Lpp2082 (AnkB) modulates ubiquitination of the host protein parvin B and promotes intracellular replication. *Cell Microbiol* **12:** 1272–1291.

Lurie-Weinberger MN, Gomez-Valero L, Merault N, Glockner G, Buchrieser C, Gophna U. 2010. The origins of eukaryotic-like proteins in *Legionella pneumophila*. *Int J Med Microbiol* **300:** 470–481.

Marra A, Shuman HA. 1992. Genetics of *Legionella pneumophila* virulence. *Annu Rev Genet* **26:** 51–69.

McDade JE, Shepard CC, Fraser DW, Tsai TR, Redus MA, Dowdle WR. 1977. Legionnaires' disease: Isolation of a bacterium and demonstration of its role in other respiratory disease. *N Engl J Med* **297:** 1197–1203.

Molmeret M, Horn M, Wagner M, Santic M, Abu Kwaik Y. 2005. Amoebae as training grounds for intracellular bacterial pathogens. *Appl Environ Microbiol* **71:** 20–28.

Molofsky AB, Swanson MS. 2004. Differentiate to thrive: Lessons from the *Legionella pneumophila* life cycle. *Mol Microbiol* **53:** 29–40.

Moya A, Pereto J, Gil R, Latorre A. 2008. Learning how to live together: Genomic insights into prokaryote-animal symbioses. *Nat Rev Genet* **9:** 218–229.

Nagai H, Cambronne ED, Kagan JC, Amor JC, Kahn RA, Roy CR. 2005. A C-terminal translocation signal required for Dot/Icm-dependent delivery of the *Legionella* RalF protein to host cells. *Proc Natl Acad Sci* **102:** 826–831.

Newton HJ, Ang DK, van Driel IR, Hartland EL. 2010. Molecular pathogenesis of infections caused by *Legionella pneumophila*. *Clin Microbiol Rev* **23:** 274–298.

Nora T, Lomma M, Gomez-Valero L, Buchrieser C. 2009. Molecular mimicry: An important virulence strategy employed by *Legionella pneumophila* to subvert host functions. *Future Microbiol* **4:** 691–701.

Ogata H, La Scola B, Audic S, Renesto P, Blanc G, Robert C, Fournier PE, Claverie JM, Raoult D. 2006. Genome sequence of *Rickettsia bellii* illuminates the role of amoebae in gene exchanges between intracellular pathogens. *PLoS Genet* **2:** e:76.

Pérez-Brocal V, Gil R, Ramos S, Lamelas A, Postigo M, Michelena JM, Silva FJ, Moya A, Latorre A. 2006. A small microbial genome: The end of a long symbiotic relationship? *Science* **314:** 312–313.

Price CT, Al-Khodor S, Al-Quadan T, Santic M, Habyarimana F, Kalia A, Kwaik YA. 2009. Molecular mimicry by an F-box effector of *Legionella pneumophila* hijacks a conserved polyubiquitination machinery within macrophages and protozoa. *PLoS Pathog* **5:** e1000704.

Price CT, Al-Quadan T, Santic M, Jones SC, Abu Kwaik Y. 2010. Exploitation of conserved eukaryotic host cell farnesylation machinery by an F-box effector of *Legionella pneumophila*. *J Exp Med* **207:** 1713–1726.

Ridenour DA, Cirillo SL, Feng S, Samrakandi MM, Cirillo JD. 2003. Identification of a gene that affects the efficiency of host cell infection by *Legionella pneumophila* in a temperature-dependent fashion. *Infect Immun* **71:** 6256–6263.

Rolando M, Buchrieser C. 2012. Post-translational modifications of host proteins by *Legionella pneumophila*: A sophisticated survival strategy. *Future Microbiol* **7:** 369–381.

Rolando M, Sanulli S, Rusniok C, Gomez-Valero L, Bertholet C, Sahr T, Margueron R, Buchrieser C. 2013. Legionella pneumophila effector RomA uniquely modifies host chromatin to repress gene expression and promote intracellular bacterial replication. *Cell Host Microbe* **13:** 395–405.

Rowbotham TJ. 1980. Preliminary report on the pathogenicity of *Legionella pneumophila* for freshwater and soil amoebae. *J Clin Pathol* **33:** 1179–1183.

Rowbotham TJ. 1986. Current views on the relationships between *amoebae, legionellae* and man. *Isr J Med Sci* **22:** 678–689.

Schmitz-Esser S, Tischler P, Arnold R, Montanaro J, Wagner M, Rattei T, Horn M. 2010. The genome of the amoeba symbiont "*Candidatus Amoebophilus asiaticus*" reveals common mechanisms for host cell interaction among amoeba-associated bacteria. *J Bacteriol* **192:** 1045–1057.

Schroeder GN, Petty NK, Mousnier A, Harding CR, Vogrin AJ, Wee B, Fry NK, Harrison TG, Newton HJ, Thomson NR, et al. 2010. *Legionella pneumophila* strain 130b possesses a unique combination of type IV secretion systems and novel Dot/Icm secretion system effector proteins. *J Bacteriol* **192:** 6001–6016.

Shohdy N, Efe JA, Emr SD, Shuman HA. 2005. Pathogen effector protein screening in yeast identifies *Legionella* factors that interfere with membrane trafficking. *Proc Natl Acad Sci* **102:** 4866–4871.

Siefert JL. 2009. Defining the mobilome. *Methods Mol Biol* **532:** 13–27.

Steinert M, Heuner K, Buchrieser C, Albert-Weissenberger C, Glockner G. 2007. *Legionella* pathogenicity: Genome structure, regulatory networks and the host cell response. *Int J Med Microbiol* **297:** 577–587.

Walker T, Klasson L, Sebaihia M, Sanders MJ, Thomson NR, Parkhill J, Sinkins SP. 2007. Ankyrin repeat domain-encoding genes in the wPip strain of *Wolbachia* from the Culex pipiens group. *BMC Biol* **5:** 39.

Whiley H, Bentham R. 2011. *Legionella longbeachae* and legionellosis. *Emerg Infect Dis* **17:** 579–583.

Winiecka-Krusnell J, Linder E. 1999. Free-living amoebae protecting *Legionella* in water: The tip of an iceberg? *Scand J Infect Dis* **31:** 383–385.

Yu VL, Plouffe JF, Pastoris MC, Stout JE, Schousboe M, Widmer A, Summersgill J, File T, Heath CM, Paterson DL, et al. 2002. Distribution of *Legionella* species and serogroups isolated by culture in patients with sporadic community-acquired legionellosis: An international collaborative survey. *J Infect Dis* **186:** 127–128.

Zhu W, Banga S, Tan Y, Zheng C, Stephenson R, Gately J, Luo ZQ. 2011. Comprehensive identification of protein substrates of the Dot/Icm type IV transporter of *Legionella pneumophila*. *PLoS ONE* **6:** e17638.

Mechanisms of *Francisella tularensis* Intracellular Pathogenesis

Jean Celli[1] and Thomas C. Zahrt[2]

[1]Laboratory of Intracellular Parasites, Rocky Mountain Laboratories, National Institute of Allergy and Infectious Diseases, National Institutes of Health, Hamilton, Montana 59840

[2]Department of Microbiology and Molecular Genetics, Center for Infectious Disease Research, Medical College of Wisconsin, Milwaukee, Wisconsin 53226

Correspondence: tzahrt@mcw.edu

Francisella tularensis is a zoonotic intracellular pathogen and the causative agent of the debilitating febrile illness tularemia. Although natural infections by *F. tularensis* are sporadic and generally localized, the low infectious dose, with the ability to be transmitted to humans via multiple routes and the potential to cause life-threatening infections, has led to concerns that this bacterium could be used as an agent of bioterror and released intentionally into the environment. Recent studies of *F. tularensis* and other closely related *Francisella* species have greatly increased our understanding of mechanisms used by this organism to infect and cause disease within the host. Here, we review the intracellular life cycle of *Francisella* and highlight key genetic determinants and/or pathways that contribute to the survival and proliferation of this bacterium within host cells.

Francisella are nonmotile, encapsulated, Gram-negative coccobacilli and are facultative intracellular pathogens of humans and many animals. The genus consists of three recognized species: *Francisella tularensis*, *Francisella novicida*, and *Francisella philomiragia*. *F. tularensis* is highly infectious and causes a potentially debilitating febrile illness known as tularemia. *F. novicida* and *F. philomiragia* are rarely pathogenic in man and usually only in individuals who are severely immunocompromised. Although *F. novicida* has a limited ability to cause disease in humans, this organism continues to serve as an important surrogate model to study aspects of *F. tularensis* pathogenesis and host response to infection owing to its reduced biosafety requirements, the conserved nature of its genome relative to pathogenic *F. tularensis* derivatives, its apparently similar intracellular life cycle, and its ability to cause a tularemia-like disease in in vivo model systems of infection.

F. tularensis is transmitted from infected animals to humans by multiple routes and can cause disease of varying severities depending on the portal of entry, infectious dose, and subspecies (biovar) of the infecting strain. Person-to-person transmission of *F. tularensis* has not yet been reported. *F. tularensis* subspecies *tularensis* is the most infectious biovar ($ID_{50} < 10$ cfu) and is responsible for most cases of tularemia in North America (Saslaw et al. 1961a). This subspecies causes the most severe disease symptoms

Cite this article as *Cold Spring Harb Perspect Med* doi: 10.1101/cshperspect.a010314

and has mortality rates approaching 60% if untreated (Saslaw et al. 1961a,b; Dienst 1963). Type A strain Schu S4 is the most commonly studied isolate from this subspecies. *F. tularensis* subspecies *holarctica* has an infectious dose $<10^3$ cfu and is the primary cause of tularemia in Europe and other regions in the Northern Hemisphere. Infections by this subspecies are generally associated with milder disease symptoms and are rarely fatal. The live vaccine strain (LVS) is an attenuated isolate derived from this subspecies and was developed in the former Soviet Union. However, it is not licensed for use in the United States. The remaining biovar, *F. tularensis* subspecies *mediastica*, only occasionally causes disease in humans.

Infection by *F. tularensis* occurs primarily after inadvertent exposure to infected wildlife species, most frequently rodents, hares, and rabbits. Transmission to humans occurs via direct contact, through arthropod or insect vectors, by ingestion of contaminated material(s), or by inhalation of aerosolized organisms. Regardless of the entry route, *F. tularensis* can disseminate from the initial infection site to the lungs where it can cause respiratory tularemia, the most severe form of the disease. The low infectious dose, with the ability to be transmitted to humans via multiple routes, and potential to cause life-threatening illness has resulted in the designation of *F. tularensis* by the United States Centers for Infectious Disease Control and Prevention as a Category A Select Agent with potential to be weaponized and/or intentionally released into the environment. These characteristics have resulted in a renewed interest in the study of *Francisella*, including characterization of the *F. tularensis* life cycle, and identification of bacterial and/or host determinants important for aspects of its pathogenesis.

OVERVIEW OF THE *Francisella* LIFE CYCLE

Although *Francisella* shows an extracellular phase during bacteriemia in mice (Forestal et al. 2007), survival and replication within host cells is thought to be a key aspect of its life cycle. This is exemplified by the ability of various strains of *F. tularensis* subsp. *tularensis* and *holarctica* and of *F. novicida* to enter, survive, and proliferate within a variety of host-cell types, including macrophages, dendritic cells, polymorphonuclear neutrophils, hepatocytes, endothelial, and type II alveolar lung epithelial cells (Oyston et al. 2004; McCaffrey and Allen 2006; Hall et al. 2007, 2008). Because intracellular proliferation is essential to *Francisella* virulence, much research has focused on understanding and characterizing specific steps in the intracellular cycle of this bacterium. It has become clear that *Francisella* survival and proliferation strategies rely on physical escape from its original phagosome and replication in the host-cell cytosol (Fig. 1), making this bacterium a typical cytosol-dwelling pathogen.

Francisella ENTRY INTO MAMMALIAN CELLS

Although entry into nonphagocytic cells remains to be further defined, phagocytosis of *Francisella* by macrophages has been extensively studied and involves the engagement of different phagocytic receptors depending on the bacterium's opsonization conditions. The mannose receptor (MR) plays a significant role in nonopsonic uptake of *F. novicida* and *F. tularensis* strains by either human monocyte-derived macrophages (MDMs), murine bone marrow-derived macrophages (BMMs), or J774A.1 macrophage-like cells (Balagopal et al. 2006; Schulert and Allen 2006; Geier and Celli 2011). Additional, yet-to-be-identified receptors are also likely engaged by nonopsonized *Francisella*. Serum opsonization, which markedly enhances *Francisella* uptake (Clemens et al. 2004, 2005; Balagopal et al. 2006; Schulert and Allen 2006; Geier and Celli 2011), mostly redirects the bacterium to the complement receptor CR3 in human and murine macrophages, human neutrophils, and dendritic cells (Balagopal et al. 2006; Ben Nasr et al. 2006; Schulert and Allen 2006; Barker et al. 2009a). The scavenger receptor A (SR-A) (Pierini 2006; Geier and Celli 2011), Fcγ receptors (Balagopal et al. 2006), nucleolin (Barel et al. 2008), and the lung surfactant protein A (SP-A) (Balagopal et al. 2006) have also been implicated to various degrees in uptake of serum-opsonized *Francisella* by murine or human macrophages.

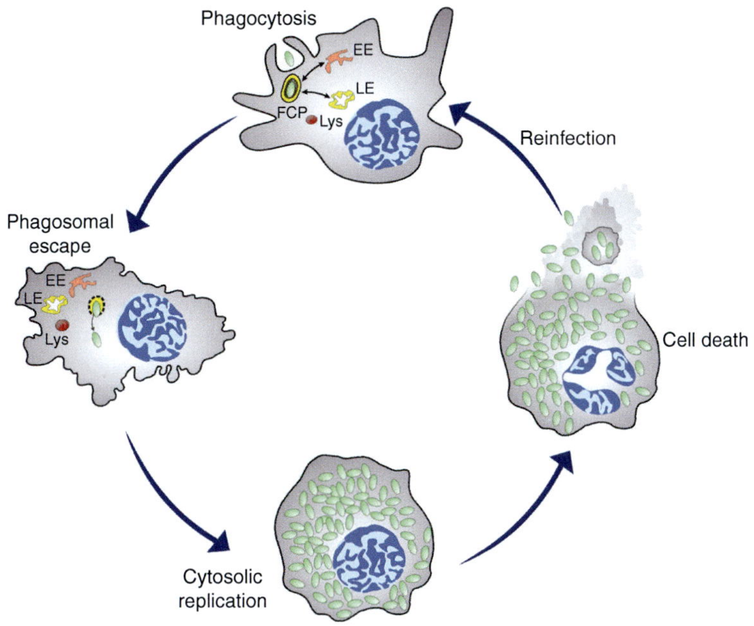

Figure 1. Model of the *Francisella* intracellular cycle depicting stages that are common to murine and human phagocytes. Upon phagocytosis, bacteria reside in an early phagosome (FCP) that interacts with early (EE) and late (LE) endocytic compartments but not lysosomes (Lys). Bacteria rapidly disrupt the FCP membrane and reach the cytosol where they undergo extensive replication, a process followed by cell death, bacterial release, and subsequent infection.

Conversely, Fcγ receptors are the main phagocytic receptors engaged during uptake of antibody-opsonized *Francisella*, which is also enhanced compared with nonopsonic conditions (Balagopal et al. 2006; Geier and Celli 2011). Opsonophagocytosis of microbes is generally associated with enhanced uptake, as is the case for *Francisella*. This suggests that this pathogen exposes a limited number of ligands to nonopsonic phagocytic receptors, possibly providing the bacterium with means to limit uptake by bactericidal phagocytes. Additionally, such bacterial ligands may engage receptors that trigger less bactericidal pathways, thereby ensuring better intracellular survival. Consistent with this hypothesis, intracellular proliferation of nonopsonic SchuS4 in murine BMMs is superior to that of opsonized bacteria (Geier and Celli 2011).

THE *Francisella*-CONTAINING PHAGOSOME

Following uptake, *Francisella* resides within a phagosome, the *Francisella*-containing phagosome (FCP), an initial vacuolar compartment along the endocytic degradative pathway that is normally subjected to progressive maturation into a bactericidal phagolysosome. Newly formed FCPs sequentially acquire markers of early endosomes and late endosomes, such as EEA-1, CD63, LAMP-1, LAMP-2, and Rab7 (Clemens et al. 2004, 2009; Santic et al. 2005a, 2008; Checroun et al. 2006; Chong et al. 2008; Wehrly et al. 2009), indicative of a normal maturation process. Yet, such interactions do not proceed toward fusion with lysosomes, as FCPs do not seem to accumulate lysosomal luminal hydrolases, such as cathepsin D, or lysosomal tracers (Anthony et al. 1991; Clemens et al. 2004; Santic et al. 2005b; Bonquist et al. 2008) and bacteria physically disrupt the phagosomal membrane and escape into the host-cell cytosol (see below). Another important feature of phagosomal maturation is the progressive acidification of the phagosomal lumen, via the recruitment of the vacuolar ATPase (v-ATPase), which is both a requirement and a consequence of phagosomal maturation (Huynh and Grinstein 2007). Whether the FCP becomes acidified or not before disruption remains contentious. Some studies have shown that FCPs containing either LVS, Schu S4, or a *F. tularensis* clinical isolate resist acidification and acquire limited amounts of v-ATPase (Clemens et al. 2004; Bonquist et al. 2008; Cremer et al. 2009), whereas others report that FCPs containing either *F. novicida* strain U112 or SchuS4 become acidified and acquire the v-ATPase proton pump before phagosomal disruption (Chong et al. 2008; Santic et al. 2008). Additional controversy resides in the fact that studies that found that FCPs become acidified show a role for acidification in optimal phagosomal escape (Chong et al. 2008; Santic et al. 2008), whereas a study in which FCPs were found to resist acidification did not observe any effect of acidification inhibitors (Clemens et al. 2009). It is noteworthy that acidified FCPs are observed using nonopsonic infection conditions, whereas nonacidified FCPs are generated using serum-opsonized *Francisella*, suggesting that the mode of uptake may affect acidification of the FCP and account for discrepancy in the literature. Although FCP intraluminal pH remains to be measured under different opsonization conditions, these studies nonetheless suggest that physicochemical cues within the FCP contribute to *Francisella* intracellular fate. This reinforces the importance of the FCP in the *Francisella* intracellular cycle, in agreement with the fact that it constitutes the site of intracellular induction of virulence genes encoded within the *Francisella* pathogenicity island (FPI), a genetic loci that has been predicted to encode a type VI secretion system (Nano et al. 2004; Chong et al. 2008; Wehrly et al. 2009).

REACTIVE OXYGEN SPECIES AND INHIBITION OF NADPH OXIDASE ACTIVATION

A major bactericidal function of phagosomes is production of reactive oxygen species (ROS) during the oxidative burst, a process mediated by recruitment and assembly of a multiprotein complex called the NADPH oxidase onto the phagosomal membrane. This enzyme mediates the conversion of molecular oxygen to superoxide (O_2^-) anions, an ROS that can be subsequently converted to other reactive species including hydrogen peroxide (H_2O_2), hypochlorous acid (bleach), and peroxynitrite ($ONOO^-$) within certain cell types or under certain conditions (Zahrt and Deretic 2002). The NADPH oxidase complex consists of two integral membrane proteins (gp91^{phox} and p22^{phox}) that together form flavocytochrome b_{558}, and various soluble components that are translocated to the phagosomal membrane on activation. Flavocytochrome b_{558} contains the redox center of NADPH oxidase and is the docking site for cytosolic subunits of the enzyme complex including p47^{phox}, p40^{phox}, and p67^{phox}. Translocation of these determinants to flavocytochrome b_{558} requires phosphorylation of p47^{phox} and p40^{phox}; activation of NADPH oxidase also requires association with a small G protein such as Rac1 or Rac2 (Babior 2004).

To ensure their survival, many bacterial pathogens have evolved strategies to resist and/or interfere with ROS generation. For *Francisella*, these survival strategies have been best elucidated in human polymorphonuclear neutrophils (PMNs) (McCaffrey and Allen 2006; Allen and McCaffrey 2007; Buchan et al. 2009; Schulert et al. 2009; McCaffrey et al. 2010; Mohapatra et al. 2010), a cell type normally encountered by the bacterium in vivo and a potent producer of NADPH oxidase-generated O_2^- (Allen and McCaffrey 2007). In these cells, *F. tularensis* and *F. novicida* have been shown to actively disrupt ROS generation by altering at least two different steps in NADPH oxidase activation: (1) by inhibiting assembly of flavocytochrome b_{558} in the phagosomal membrane (McCaffrey et al. 2010), and/or (2) by inhibiting the phosphorylation of p47^{phox} and p40^{phox} (McCaffrey et al. 2010; Mohapatra et al. 2010). Inhibition of NADPH oxidase assembly by *F. tularensis* is dependent on opsonization, and is observed with strains from both *F. tularensis* subspecies *tularensis* and *holarctica*, including the LVS (McCaffrey et al. 2010; Geier and Celli 2011). Interestingly, *F. tularensis* also possesses the unique ability to inhibit NADPH oxidase activity postcomplex assembly, an event that limits subsequent PMN activation by other stimuli and diminishes ROS production following phagocytosis of antibody-opsonized *F. tularensis* (McCaffrey et al. 2010). Regardless of mechanism(s), inhibition of NADPH oxidase activity enhances survival of both *F. tularensis* and *F. novicida* in this cell type (McCaffrey and Allen 2006; McCaffrey et al. 2010; Mohapatra et al. 2010; Geier and Celli 2011). Similarly, monocytes or MDMs that have been infected with *F. tularensis* or *F. novicida* also fail to trigger a robust and/or productive oxidative burst (Schulert et al. 2009; Mohapatra et al. 2010; Geier and Celli 2011). However, in contrast to PMNs, the inability of monocytes and MDMs to produce ROS following infection is likely owing to *Francisella* engagement of and/or internalization by receptors not coupled to NADPH oxidase activation (Schulert et al. 2009). Consistent with this observation, intracellular growth of *F. tularensis* or *F. novicida* is similar following ex vivo infection of monocytes and/or MDMs from wild-type mice compared with $p47^{phox-/-}$ mice (Lindgren et al. 2005; Mohapatra et al. 2010). Furthermore, IgG-opsonization of *F. tularensis* results in NADPH oxidase-dependent ROS generation in wild-type murine BMMs, an affect that is not observed in macrophages lacking $gp91^{phox}$ (Geier and Celli 2011).

Although the ability of *Francisella* to inhibit NADPH oxidase activity in neutrophils in vitro has been well established, the importance of neutrophils and/or NADPH oxidase activity in this and other cell types in vivo remains less clear. Antibody-mediated depletion of neutrophils from mice, or targeted recruitment of neutrophils to the lungs using intranasally delivered recombinant MIP-2, a neutrophil-chemotactic chemokine, does not affect bacterial burden or time-to-death kinetics in mice that have been infected with *F. tularensis* compared with untreated control animals (KuoLee et al. 2011). Similarly, $gp91^{phox-/-}$ mice show only slight differences in time-to-death and overall bacterial burden in target organs of infection compared with wild-type animals following respiratory infection with *F. tularensis* (KuoLee et al. 2011). Thus, although *Francisella* may possess specific strategies to inhibit, counteract, or bypass the production of ROS by neutrophils and other phagocytes, the importance of the NADPH oxidase in controlling infection by pathogenic biovars of *Francisella* appears limited in murine tularemia.

Several bacterial determinants have been implicated in the resistance of *Francisella* to ROS. For example, *Francisella* encodes a number of putative acid phosphatases, enzymes that hydrolyze monoesters at acidic pH, that may inhibit the respiratory burst. Purified AcpA, the major contributor of acid phosphatase activity in *Francisella*, is able to inhibit the oxidative burst in porcine neutrophils that have been activated with exogenous stimulants (Reilly et al. 1996). Recent evidence suggests that AcpA is secreted by *F. tularensis* and *F. novicida* strains in vitro (Dai et al. 2011), and is detectable in the cytosol of macrophages infected with *F. novicida* shortly after infection (Dai et al. 2011). These findings indicate that AcpA would be in a physical location (near the phagosomal membrane or in the cytosol) to interact with membrane-bound or soluble components of the NADPH oxidase (Dai et al. 2011). Consistent with these observations, an *F. novicida* mutant lacking AcpA is partially compromised in its ability to suppress the oxidative burst following infection of human neutrophils or MDMs (Mohapatra et al. 2010), a phenotype that becomes more pronounced as additional predicted acid phosphatases (*acpB*, *acpC*, and *hap*) are deleted from the bacterium (Mohapatra et al. 2010). Furthermore, loss of AcpA and/or other acid phosphatases reduces the ability of *F. novicida* to survive following infection of these and other cell types in vitro (Mohapatra et al. 2007, 2008, 2010), and in vivo following intraperitoneal or intranasal infection of mice (Mohapatra et al. 2007, 2008). The importance of acid phosphatases in resistance of *F. novicida* to ROS can also be seen following treatment of neutrophils and MDMs with reagents that block assembly and/or function of NADPH oxidase, or in macrophages from $p47^{phox-/-}$ mice, where the ability of *F. novicida* $\Delta acpA$ and/or $\Delta acpABCH$ mutants to survive intracellularly is improved (Mohapatra et al. 2010). Additionally, ascorbate treatment of murine or human macrophages infected with the LVS blocks acid phosphatase production and inhibits intracellular growth of the bacterium, a phenotype that is dependent on production of AcpA (McRae et al. 2009). Thus, AcpA and other predicted acid phosphatases appear to play an important role in NADPH oxidase inhibition and resistance to ROS following infection by *F. novicida*.

It is, however, important to note that AcpA and/or the other acid phosphatases appear to play little if any role in the inhibition of NADPH oxidase activity or resistance to ROS in phagocytes that have been infected with human-virulent strains of *F. tularensis*. An *acpA* mutant generated in Schu S4 does not trigger a respiratory burst and shows similar growth characteristics to the wild-type parent following infection of human PMNs, even though this bacterium lacks detectable acid phosphatase activity (McCaffrey et al. 2010). Similarly, deletion of *acpA*, *acpB*, or *acpC* alone or sequentially does not alter growth and/or trafficking characteristics of Schu S4 in vitro in murine BMMs or human MDMs, or in vivo following intranasal infection of mice (Child et al. 2010). Thus, acid phosphatases may play a different role in *F. novicida* compared with *F. tularensis*, or strains of *Francisella* that are pathogens of humans may possess/use additional determinants/strategies for ROS resistance that are not conserved or functional in *F. novicida*.

Apart from acid phosphatases, other determinants have also been implicated in *Francisella* resistance to ROS. A number of determinants involved in pyrimidine biosynthesis (*carA*, *carB*, and *pyrB*) mediate resistance to ROS in *F. tularensis* LVS, although it is unclear whether these genes act indirectly by simply enhancing the fitness of the bacterium (Schulert et al. 2009). FevR, a transcription factor that regulates expression of $\sim$100 genes including those from the FPI (Brotcke and Monack 2008; Meibom et al. 2009), is also required for the inhibition of NADPH oxidase activation and subsequent ROS production in PMNs by both *F. tularensis* LVS (Buchan et al. 2009) and *F. tularensis* subspecies *tularensis* (McCaffrey et al. 2010). *FTN1133* (*FTL0803*) encodes a hypothetical gene product that is required for resistance of *F. novicida* and *F. tularensis* LVS to organic hydroperoxides (Llewellyn et al. 2011). Mutants lacking this enzyme are attenuated for survival in macrophages in vitro and show reduced burdens in mice following intradermal infection, a phenotype that is partially or completely ablated in cells or animals lacking gp91^{phox} (Llewellyn et al. 2011). *F. tularensis* also encodes at least two superoxide dismutases that mediate the detoxification of superoxide to H_2O_2 and O_2. *sodB*, encoding an FeSOD, and *sodC*, encoding CuZn SOD, mediate resistance to paraquat, H_2O_2, and/or pyrogallol in *F. tularensis* LVS (Bakshi et al. 2006; Melillo et al. 2009), and are required for virulence of *F. tularensis* LVS in macrophages and mice (Bakshi et al. 2006). The attenuation in virulence observed with these strains in vitro and in vivo is attributable to NADPH oxidase, as addition of NADPH oxidase inhibitors or utilization of cells or mice unable to make this enzyme complex fail to control infection (Melillo et al. 2009). Finally, *F. tularensis* LVS mutants lacking KatG, a catalase, show increased sensitivity to H_2O_2 in vitro and are attenuated for virulence in mice in vivo (Lindgren et al. 2007). Interestingly, a similar phenotype is not observed in strains from subspecies *tularensis* (Lindgren et al. 2007). Thus, *Francisella* likely uses multiple strategies and a variety of determinants to mediate resistance to ROS.

Francisella PHAGOSOMAL ESCAPE

Several laboratories have described phagosomal escape by various strains of *F. tularensis* or *F. novicida* in either human or murine macrophages, and other cell types (Golovliov et al. 2003; Clemens et al. 2004; Santic et al. 2005a, 2008; Checroun et al. 2006; Chong et al. 2008), clearly indicating that this process is conserved and is a hallmark of the *Francisella* intracellular life cycle. The study of phagosomal escape kinetics have nonetheless brought controversy to the field, as early studies have reported phagosomal disruption and bacterial access to the cytosol that range anywhere from 1 h to 8 h post-infection (Golovliov et al. 2003; Clemens et al.

2004; Santic et al. 2005a; Checroun et al. 2006; McCaffrey and Allen 2006). Such large variations in kinetics are likely attributed to technical differences in the assays and criteria used to assess phagosomal disruption, and also to variations in the models studied between laboratories where different species and strains of *Francisella*, and different macrophage models were examined. In support of the latter, more recent side-by-side comparisons of phagosomal escape kinetics of *F. tularensis* and *F. novicida* strains have highlighted differences in this process (Chong et al. 2008). For example, more rapid phagosomal escape is observed under nonopsonic conditions (Golovliov et al. 2003; Lindgren et al. 2004; Checroun et al. 2006; Chong et al. 2008; Santic et al. 2008; Barker et al. 2009a; Wehrly et al. 2009; Child et al. 2010; Edwards et al. 2010), whereas slower escape kinetics occur when opsonic conditions are used (Clemens et al. 2004; McCaffrey and Allen 2006; Schulert et al. 2009). Recent systematic comparisons of phagosomal escape kinetics by nonopsonic and opsonized SchuS4 in murine macrophages have clarified this issue further by demonstrating that opsonization of *Francisella* with either complement or antibodies targets bacteria to phagocytic pathways that restrict the extent and timing of phagosomal escape (Geier and Celli 2011). Although the basis for restricted phagosomal escape on serum opsonization is unclear, that occurring on antibody opsonization and bacterial targeting to Fcγ receptors depends on the rapid and transient NADPH oxidase-dependent oxidative burst (Geier and Celli 2011) that the bacteria do not seem to be able to prevent, further confirming that conditions encountered by phagocytosed *Francisella* within the FCP can influence its intracellular fate.

Francisella phagosomal escape is requisite to intracellular proliferation, as exemplified by the inability of numerous phagosomal escape-deficient mutants to grow within macrophages (Lindgren et al. 2004; Santic et al. 2005b; Bonquist et al. 2008; Barker et al. 2009b; Wehrly et al. 2009; Broms et al. 2012). Given the important role of phagosomal escape in the *Francisella* intracellular cycle, much effort has been made to identify bacterial factors that contribute to this process. Unlike *Listeria monocytogenes* or *Shigella flexneri*, *Francisella* does not encode lipid hydrolases or phospholipases classically involved in phagosomal membrane disruption (Larsson et al. 2005). However, *F. tularensis* does appear to express a type VI-like secretion system (encoded by the FPI) (Nano et al. 2004), and numerous genes that are expressed from this genetic locus or that regulate expression from this locus are important for this phagosomal egress (reviewed in Chong et al. 2008). These include *iglC* and *iglD* (Lindgren et al. 2004; Santic et al. 2005b; Bonquist et al. 2008; Chong et al. 2008), *pdpA* (Schmerk et al. 2009), *iglI* and *iglJ* (Barker et al. 2009b; McCaffrey et al. 2010), *vgrG* (Barker et al. 2009b; Broms et al. 2012), *dotU* (Broms et al. 2012), *mglA* (Baron and Nano 1998; Santic et al. 2005b; Bonquist et al. 2008), *fevR* (Brotcke and Monack 2008; Buchan et al. 2009; Wehrly et al. 2009), and *migR* (Buchan et al. 2009). However, other determinants not linked to the FPI have also been implicated in phagosomal escape by *Francisella* including the acid phosphatases of *F. novicida* (AcpA, AcpB, AcpC, and Hap) (Mohapatra et al. 2008), pyrimidine biosynthetic genes (*carA*, *carB*, and *pyrB*) (Schulert et al. 2009), as well as several genes of unknown function including *FTT1103* (Qin and Mann 2006; Qin et al. 2009) and *FTT1676* (Wehrly et al. 2009). Thus, it remains unclear whether *Francisella* use a single mechanism for escape from the phagosome; further delineation of this process, including the genes required, remains a top priority in the field given its importance for subsequent stages of *Francisella* survival.

Francisella REPLICATION IN THE CYTOSOL

Besides phagosomal escape, the ability of *Francisella* to proliferate within the host-cell cytosol is another key aspect of its intracellular life cycle, as exemplified by the avirulence in mice of the large number of identified replication-deficient mutants (Brotcke et al. 2006; Tempel et al. 2006; Su et al. 2007; Weiss et al. 2007; Alkhuder et al. 2009; Wehrly et al. 2009). Although the cytosol may be viewed as permissive for bacterial proliferation, compared with a vacuole along the endocytic degradative pathway, only cytosol-adapted pathogens can readily proliferate within this host-cell compartment (O'Riordan and Portnoy 2002), through the expression of dedicated factors and interactions with host components that either favor or antagonize bacterial growth. Because phagosomal escape is a prerequisite to cytosolic replication, identifying bacterial factors that are specifically required for growth in the cytosol (and not for phagosomal escape) via mutagenesis and in vitro and/or in vivo screens has been difficult. In general, identification of specific *Francisella* cytosolic replication-defective mutants has required use of more specialized secondary readouts including those based on electron and/or immunofluorescence microscopy. To date, only purine biosynthetic genes (*purMCD*) (Pechous et al. 2006, 2008), a γ-glutamyl transpeptidase (*ggt*) (Alkhuder et al. 2009), and several genes of unknown function including *FTT0369c/dipA* (Wehrly et al. 2009; Chong et al. 2012), *FTT0989* (Brotcke et al. 2006), and *ripA* (Fuller et al. 2008), have been identified as being specifically required for cytosolic replication by *Francisella*. However, it is likely that there remain a large number of additional gene products that also contribute to this process (reviewed in Chong et al. 2008).

Although *Francisella* have evolved mechanisms to adapt to a cytosolic lifestyle, some host factors have been identified that either contribute or interfere with cytosolic proliferation. Recently, Akimana et al. have identified through a genome-wide RNAi screen in *Drosophila melanogaster* S2R$^+$ cells the type III PI4-kinase α subunit PI4KC and the ubiquitin-specific peptidase USP22 as required for cytosolic replication of *F. novicida* (Akimana et al. 2010). Whether the same host proteins are required for replication of virulent *F. tularensis* and what their function in bacterial replication is remains to be established. Nonetheless, these findings argue that this bacterium modulates specific host pathways associated with lipid signaling and ubiquitin systems to foster replication.

Not surprisingly, innate immune mechanisms have been shown to negatively control intracellular proliferation of *Francisella*. The cytokine interferon (IFN)-γ is essential to control primary infections (Leiby et al. 1992; Elkins et al. 1996), and has long been known to restrict intracellular growth of various *Francisella* strains in different host-cell models (Anthony et al. 1992; Fortier et al. 1992; Polsinelli et al. 1994), although the effector mechanisms are unclear. IFN-γ control of either LVS or Schu S4 intracellular growth in murine peritoneal exudate cells (PECs) seems to depend on generation of nitric oxide by the inducible nitric oxide synthase (iNOS) (Lindgren et al. 2005, 2007), but not in murine alveolar macrophages (Polsinelli et al. 1994), nor in either human blood-derived or murine bone marrow-derived primary macrophages (Edwards et al. 2010). Nonetheless, independent studies have established that cytosolic replication rather than phagosomal escape of *F. tularensis* is the target of IFN-γ, as this cytokine does not affect phagosomal escape of either LVS or SchuS4 in either J774A.1 cells and human or murine primary macrophages (Bonquist et al. 2008; Edwards et al. 2010), but restricts cytosolic proliferation (Edwards et al. 2010). These findings are, however, inconsistent with those of Santic et al. (2005a), who reported that IFN-γ activation of human blood-derived macrophages prevented phagosomal escape of *F. novicida*. Although variations in the experimental designs of these studies may account for such contrasting conclusions, and further work is required to reconcile these results and clarify the effector mechanisms induced by IFN-γ, these findings clearly illustrate that phagocytes express specific mechanisms that control cytosolic proliferation of *Francisella*. In addition to the effect of IFN-γ on *Francisella* intracellular growth, alternative activation of macrophages by mast cells via interleukin-4 (IL-4) controls intramacrophage growth of LVS (Ketavarapu et al. 2008). Although increased ATP production, prolonged FCP acidification, and up-regulation of the MR have been invoked in this control (Rodriguez et al. 2010), whether IL-4 affects *Francisella* phagosomal escape and/or cytosolic replication remains to be examined in more depth.

Francisella AND INNATE IMMUNE RECOGNITION

Francisella phagocytosis by macrophages and intracellular trafficking through vacuolar and cytosolic compartments likely subjects the bacterium to innate immune recognition by various pattern recognition receptors (PRRs) either located on the cell surface or endosomes, such as the Toll-like receptor (TLRs), or in the cytosolic compartment such as Nod-like receptors (NLRs). TLRs or NLRs typically recognize pathogen-associated molecular patterns (PAMPs) or danger-associated molecular patterns (DAMPs) to trigger proinflammatory responses, among which the activation of the inflammasome, a cytosolic molecular complex, orchestrates caspase-1 activation, processing and release of the proinflammatory cytokines IL-1β and IL-18, and cell death via pyroptosis.

Because of the unique structure of its lipid A moiety (Hajjar et al. 2006), *Francisella* LPS shows very low endotoxicity and the bacterium does not induce TLR4-mediated signaling. Instead, LVS significantly stimulates TLR2-mediated signaling resulting in proinflammatory cytokine production (Katz et al. 2006; Cole et al. 2007), although it is also capable of subsequently suppressing TLR-mediated signaling and secretion of proinflammatory cytokines induced by other agonists (Telepnev et al. 2003). This ability of *Francisella* to suppress or dampen proinflammatory responses has been extended to the virulent subspecies *tularensis* in human dendritic cells, where infection impairs their activation and function (Bosio and Dow 2005; Chase et al. 2009), further supporting the notion that *Francisella* interferes with immune detection and responses.

INFLAMMASOME ACTIVATION

Given its cytosolic location, *Francisella* has become a model pathogen to study the function of inflammasomes, which has also expanded our understanding of how this bacterium may interfere with immune recognition. Release of *F. novicida* in the macrophage cytosol is sensed in a type I IFN-dependent manner (Henry et al. 2007), and activates the absent in melanoma 2 (AIM2)-containing inflammasome (Fernandes-Alnemri et al. 2010; Rathinam et al. 2010; Jones et al. 2011), a type I IFN-inducible complex whose activation induces secretion of the proinflammatory cytokines IL1β and pro-IL18 and pyroptosis (Mariathasan et al. 2005). AIM2 inflammasome activation is potentiated by TLR2 signaling (Jones et al. 2010) and is required in vivo in mice to control bacterial burden by eliminating infected macrophages that constitute the *Francisella* proliferation niche (Mariathasan et al. 2005; Fernandes-Alnemri et al. 2010; Jones et al. 2011). Consistent with AIM2 recognizing cytosolic double-stranded DNA (Rathinam et al. 2010), *F. novicida* activation of the AIM2 inflammasome is associated with bacterial release of DNA through lysis in the cytosol (Jones et al. 2011). These studies have focused on murine systems and the nonhuman pathogenic *F. novicida* species, so an important question that remains unanswered is whether the conclusions drawn from these studies also apply to humans and virulent strains. For example, recent evidence indicates that infection of human cells with *F. novicida*, LVS, and the highly virulent *F. tularensis* Schu S4 strains activate the NLRP3 inflammasome in addition to the AIM2 inflammasome (Atianand et al. 2011), suggesting some functional differences in the innate immune complexes involved in *Francisella* recognition between mice and humans. Moreover, several studies using *F. tularensis* subspecies suggest that infections of BMMs with the LVS strain (Huang et al. 2010; Ulland et al. 2010), or of human primary macrophages with virulent Schu S4 (Lindemann et al. 2010) do induce little cytotoxicity compared with *F. novicida*, raising the question of relevance of cytotoxicity in infections with virulent *tularensis* strains.

Given the high adaptation of *Francisella* to the cytosol and the potential of proinflammatory pathways to restrict its proliferation, this bacterium has likely evolved mechanisms to counteract inflammasome activation to preserve its replication niche. In support of this concept, many laboratories have identified hypercytotoxic mutants of *F. novicida* (Brotcke et al. 2006; Hager et al. 2006; Weiss et al. 2007; Lai et al. 2010), LVS (Huang et al. 2010; Ulland et al. 2010), and Schu S4 (Lindemann et al. 2010), which typically induced increased inflammasome-dependent cell death and secretion of proinflammatory cytokines in either mouse or human macrophages. Intriguingly, the hypercytotoxic phenotype results from the interruption of genes encoding a variety of proteins, such as the oligopeptide permease OppB (Brotcke et al. 2006), the metallopeptidase PepO (Brotcke et al. 2006; Hager et al. 2006), the inner membrane protein RipA (Huang et al. 2010), the IclR family transcriptional factor FTT0784 and an unknown protein FTT0584 (Weiss et al. 2007), and several proteins involved in biosynthesis of type IV pili, the LPS and the O-antigen polysaccharidic capsule (Huang et al. 2010; Ulland et al. 2010), bringing some confusion as to whether *Francisella* specifically suppress inflammasome activation. The hypothesis of *Francisella* evasion of inflammasome activation was recently challenged by Peng et al., who showed that many previously identified hypercytotoxic mutants of *F. novicida* and LVS are deficient in membrane-associated proteins and lyse more extensively in the cytosol than wild-type bacteria, consequently triggering increased inflammasome activation, secretion of proinflammatory cytokines, and pyroptosis (Peng et al. 2011). Although this study does not completely rule out the existence of *Francisella* inflammasome suppression factors, it elegantly argues that most genes identified as potential suppressors of inflammasome activation do not act specifically and that the hypercytotoxic phenotype is in many cases an indirect consequence of bacterial cell wall fragilization and enhanced release of PAMPs. Because this study has been performed in *F. novicida* and LVS, it will also be interesting to verify that these conclusions also apply to virulent strains.

XENOPHAGIC CAPTURE

Another intracellular innate immune defense mechanism is xenophagy, a selective process of capture of intracellular microorganisms within a double-membrane-bound vacuole, the autophagosome, for delivery to and degradation into the lysosomal compartment (Levine et al. 2011). *Francisella* cytosolic location makes this pathogen an ideal substrate for selective autophagic recognition and capture, a process that shares many molecular machineries with nonselective, canonical autophagy. Yet, surprisingly, this bacterium is able to replicate within murine and human macrophages over long periods of time (>18 h) without evidence of a xenophagic response, suggesting it is either capable of avoiding recognition or inhibiting autophagic processes. A belated autophagic response has nonetheless been observed in murine BMMs that encloses a fraction of cytosolic *Francisella* into large vacuoles with autophagic features (Checroun et al. 2006), but the role of these vacuoles and their relevance is unclear, as they do not form in human macrophages (Akimana et al. 2010; Edwards et al. 2010) and may not constitute a bona fide stage of the bacterium's intracellular cycle. Current evidence that *Francisella* may interfere with the autophagy pathway stems from the down-regulation of several autophagy-related genes, such as *beclin1*, *ATG5*, *ATG12*, *ATG16L*, *ATG7*, and *ATG4a* in either Schu S4- or *F. novicida*-infected human monocytes (Butchar et al. 2008; Cremer et al. 2009), which suggests that *Francisella* may suppress an autophagic response at the gene expression level. Yet, it remains to be shown whether down-regulation of some autophagy gene during infection is sufficient to block this constitutive pathway rapidly enough to prevent bacteria that have reached the cytosol from being targeted. Interestingly, a Δ*dipA* (FTT0369c) deletion mutant of SchuS4, which is deficient in cytosolic replication, is eventually captured and cleared by autophagy following a loss of viability in the cytosol (Chong et al. 2012), suggesting that viable *Francisella* may possess mechanisms that prevent their recognition by the autophagic machinery. Additionally, whether *Francisella* infection negatively modulates nonselective autophagy also needs to be examined to further understand *Francisella* potential interference with this degradative pathway.

CONCLUDING REMARKS

The ability of *Francisella* to survive and replicate within phagocytes and other host cells following infection is a key aspect of its life cycle and represents an essential step required by this bacterium to cause disease within the host. Alterations in mechanisms that mediate this ability, including the engagement of appropriate receptors, resistance to ROS, escape from the phagosome, replication within the cytosol, and/or avoidance of innate immune recognition or other defense mechanisms, can lead to altered survival characteristics and potential attenuation of *F. tularensis* in virulence in vitro and/or in vivo. Continued study of the genetic determinants contributing to these mechanisms will be essential for the development of therapeutics or vaccines that are able to specifically inhibit or prevent disease caused by this bacterium.

ACKNOWLEDGMENTS

The authors apologize to those investigators in the field whose excellent work we were not able to reference. We are grateful to Anita Mora at the Rocky Mountain Laboratory for her assistance with graphic art. T.C.Z. acknowledges support from the National Institutes of Health, National Institute of Allergy and Infectious Diseases (NIAID) (1R21 AI097597-01A1), and the NIAID's Regional Center of Excellence for Biodefense and Emerging Infectious Diseases Research Program. Membership within and support from the Region V Great Lakes Regional Center of Excellence (National Institutes of Health Award 2-U54-AI-057153) is also acknowledged. J.C. acknowledges support from the Intramural Research Program of the National Institutes of Health, National Institute of Allergy and Infectious Diseases.

REFERENCES

Akimana C, Al-Khodor S, Abu Kwaik Y. 2010. Host factors required for modulation of phagosome biogenesis and proliferation of *Francisella tularensis* within the cytosol. *PLoS ONE* **5:** e11025.

Alkhuder K, Meibom KL, Dubail I, Dupuis M, Charbit A. 2009. Glutathione provides a source of cysteine essential for intracellular multiplication of *Francisella tularensis*. *PLoS Pathog* **5:** e1000284.

Allen LA, McCaffrey RL. 2007. To activate or not to activate: Distinct strategies used by *Helicobacter pylori* and *Francisella tularensis* to modulate the NADPH oxidase and survive in human neutrophils. *Immunol Rev* **219:** 103–117.

Anthony LD, Burke RD, Nano FE. 1991. Growth of *Francisella* spp. in rodent macrophages. *Infect Immun* **59:** 3291–3296.

Anthony LS, Morrissey PJ, Nano FE. 1992. Growth inhibition of *Francisella tularensis* live vaccine strain by IFN-γ-activated macrophages is mediated by reactive nitrogen intermediates derived from L-arginine metabolism. *J Immunol* **148:** 1829–1834.

Atianand MK, Duffy EB, Shah A, Kar S, Malik M, Harton JA. 2011. *Francisella tularensis* reveals a disparity between human and mouse NLRP3 inflammasome activation. *J Biol Chem* **286:** 39033–39042.

Babior BM. 2004. NADPH oxidase. *Curr Opin Immunol* **16:** 42–47.

Bakshi CS, Malik M, Regan K, Melendez JA, Metzger DW, Pavlov VM, Sellati TJ. 2006. Superoxide dismutase B gene (sodB)-deficient mutants of *Francisella tularensis* demonstrate hypersensitivity to oxidative stress and attenuated virulence. *J Bacteriol* **188:** 6443–6448.

Balagopal A, MacFarlane AS, Mohapatra N, Soni S, Gunn JS, Schlesinger LS. 2006. Characterization of the receptor-ligand pathways important for entry and survival of *Francisella tularensis* in human macrophages. *Infect Immun* **74:** 5114–5125.

Barel M, Hovanessian AG, Meibom K, Briand JP, Dupuis M, Charbit A. 2008. A novel receptor-ligand pathway for entry of *Francisella tularensis* in monocyte-like THP-1 cells: Interaction between surface nucleolin and bacterial elongation factor Tu. *BMC Microbiol* **8:** 145.

Barker JH, McCaffrey RL, Baman NK, Allen LA, Weiss JP, Nauseef WM. 2009a. The role of complement opsonization in interactions between *F. tularensis* subsp. novicida and human neutrophils. *Microbes Infect* **11:** 762–769.

Barker JR, Chong A, Wehrly TD, Yu JJ, Rodriguez SA, Liu J, Celli J, Arulanandam BP, Klose KE. 2009b. The *Francisella tularensis* pathogenicity island encodes a secretion system that is required for phagosome escape and virulence. *Mol Microbiol* **74:** 1459–1470.

Baron GS, Nano FE. 1998. MglA and MglB are required for the intramacrophage growth of *Francisella novicida*. *Mol Microbiol* **29:** 247–259.

Ben Nasr A, Haithcoat J, Masterson JE, Gunn JS, Eaves-Pyles T, Klimpel GR. 2006. Critical role for serum opsonins and complement receptors CR3 (CD11b/CD18) and CR4 (CD11c/CD18) in phagocytosis of *Francisella tularensis* by human dendritic cells (DC): Uptake of *Francisella* leads to activation of immature DC and intracellular survival of the bacteria. *J Leukoc Biol* **80:** 774–786.

Bonquist L, Lindgren H, Golovliov I, Guina T, Sjostedt A. 2008. MglA and Igl proteins contribute to the modulation of *Francisella tularensis* live vaccine strain-containing phagosomes in murine macrophages. *Infect Immun* **76:** 3502–3510.

Bosio CM, Dow SW. 2005. *Francisella tularensis* induces aberrant activation of pulmonary dendritic cells. *J Immunol* **175:** 6792–6801.

Broms JE, Meyer L, Lavander M, Larsson P, Sjostedt A. 2012. DotU and VgrG, core components of type VI secretion systems, are essential for *Francisella* LVS pathogenicity. *PLoS ONE* **7:** e34639.

Brotcke A, Monack DM. 2008. Identification of fevR, a novel regulator of virulence gene expression in *Francisella novicida*. *Infect Immun* **76:** 3473–3480.

Brotcke A, Weiss DS, Kim CC, Chain P, Malfatti S, Garcia E, Monack DM. 2006. Identification of MglA-regulated genes reveals novel virulence factors in *Francisella tularensis*. *Infect Immun* **74:** 6642–6655.

Buchan BW, McCaffrey RL, Lindemann SR, Allen LA, Jones BD. 2009. Identification of migR, a regulatory element of the *Francisella tularensis* live vaccine strain iglABCD virulence operon required for normal replication and trafficking in macrophages. *Infect Immun* **77:** 2517–2529.

Butchar JP, Cremer TJ, Clay CD, Gavrilin MA, Wewers MD, Marsh CB, Schlesinger LS, Tridandapani S. 2008. Microarray analysis of human monocytes infected with *Francisella tularensis* identifies new targets of host response subversion. *PLoS ONE* **3:** e2924.

Chase JC, Celli J, Bosio CM. 2009. Direct and indirect impairment of human dendritic cell function by virulent *Francisella tularensis* Schu S4. *Infect Immun* **77:** 180–195.

Checroun C, Wehrly TD, Fischer ER, Hayes SF, Celli J. 2006. Autophagy mediated reentry of *Francisella tularensis* into the endocytic compartment after cytoplasmic replication. *Proc Natl Acad Sci* **103:** 14578–14583.

Child R, Wehrly TD, Rockx-Brouwer D, Dorward DW, Celli J. 2010. Acid phosphatases do not contribute to the pathogenesis of type A *Francisella tularensis*. *Infect Immun* **78:** 59–67.

Chong A, Wehrly TD, Nair V, Fischer ER, Barker JR, Klose KE, Celli J. 2008. The early phagosomal stage of *Francisella tularensis* determines optimal phagosomal escape and *Francisella* pathogenicity island protein expression. *Infect Immun* **76:** 5488–5499.

Chong A, Wehrly TD, Child R, Hansen B, Hwang S, Virgin HW, Celli J. 2012. Cytosolic clearance of replication-deficient mutants reveals *Francisella tularensis* interactions with the autophagic pathway. *Autophagy* **8:** 1342–1356.

Clemens DL, Lee BY, Horwitz MA. 2004. Virulent and avirulent strains of *Francisella tularensis* prevent acidification and maturation of their phagosomes and escape into the cytoplasm in human macrophages. *Infect Immun* **72:** 3204–3217.

Clemens DL, Lee BY, Horwitz MA. 2005. *Francisella tularensis* enters macrophages via a novel process involving pseudopod loops. *Infect Immun* **73:** 5892–5902.

Clemens DL, Lee BY, Horwitz MA. 2009. *Francisella tularensis* phagosomal escape does not require acidification of the phagosome. *Infect Immun* **77:** 1757–1773.

Cole LE, Shirey KA, Barry E, Santiago A, Rallabhandi P, Elkins KL, Puche AC, Michalek SM, Vogel SN. 2007. Toll-like receptor 2-mediated signaling requirements for *Francisella tularensis* live vaccine strain infection of murine macrophages. *Infect Immun* **75:** 4127–4137.

Cremer TJ, Amer A, Tridandapani S, Butchar JP. 2009. *Francisella tularensis* regulates autophagy-related host cell signaling pathways. *Autophagy* **5:** 125–128.

Dai S, Mohapatra NP, Schlesinger LS, Gunn JS. 2011. The acid phosphatase AcpA is secreted in vitro and in macrophages by *Francisella* spp. *Infect Immun* **80:** 1088–1097.

Dienst Jr. FTJr. 1963. Tularemia: A perusal of three hundred thirty-nine cases. *J La State Med Soc* **115:** 114–127.

Edwards JA, Rockx-Brouwer D, Nair V, Celli J. 2010. Restricted cytosolic growth of *Francisella tularensis* subsp. tularensis by IFN-γ activation of macrophages. *Microbiology* **156:** 327–339.

Elkins KL, Rhinehart-Jones TR, Culkin SJ, Yee D, Winegar RK. 1996. Minimal requirements for murine resistance to infection with *Francisella tularensis* LVS. *Infect Immun* **64:** 3288–3293.

Fernandes-Alnemri T, Yu JW, Juliana C, Solorzano L, Kang S, Wu J, Datta P, McCormick M, Huang L, McDermott E, et al. 2010. The AIM2 inflammasome is critical for innate immunity to *Francisella tularensis*. *Nat Immunol* **11:** 385–393.

Forestal CA, Malik M, Catlett SV, Savitt AG, Benach JL, Sellati TJ, Furie MB. 2007. *Francisella tularensis* has a significant extracellular phase in infected mice. *J Infect Dis* **196:** 134–137.

Fortier AH, Polsinelli T, Green SJ, Nacy CA. 1992. Activation of macrophages for destruction of *Francisella tularensis*: Identification of cytokines, effector cells, and effector molecules. *Infect Immun* **60:** 817–825.

Fuller JR, Craven RR, Hall JD, Kijek TM, Taft-Benz S, Kawula TH. 2008. RipA, a cytoplasmic membrane protein conserved among *Francisella* species, is required for intracellular survival. *Infect Immun* **76:** 4934–4943.

Geier H, Celli J. 2011. Phagocytic receptors dictate phagosomal escape and intracellular proliferation of *Francisella tularensis*. *Infect Immun* **79:** 2204–2214.

Golovliov I, Baranov V, Krocova Z, Kovarova H, Sjostedt A. 2003. An attenuated strain of the facultative intracellular bacterium *Francisella tularensis* can escape the phagosome of monocytic cells. *Infect Immun* **71:** 5940–5950.

Hager AJ, Bolton DL, Pelletier MR, Brittnacher MJ, Gallagher LA, Kaul R, Skerrett SJ, Miller SI, Guina T. 2006. Type IV pili-mediated secretion modulates *Francisella* virulence. *Mol Microbiol* **62:** 227–237.

Hajjar AM, Harvey MD, Shaffer SA, Goodlett DR, Sjostedt A, Edebro H, Forsman M, Bystrom M, Pelletier M, Wilson CB, et al. 2006. Lack of in vitro and in vivo recognition of *Francisella tularensis* subspecies lipopolysaccharide by Toll-like receptors. *Infect Immun* **74:** 6730–6738.

Hall JD, Craven RR, Fuller JR, Pickles RJ, Kawula TH. 2007. *Francisella tularensis* replicates within alveolar type II epithelial cells in vitro and in vivo following inhalation. *Infect Immun* **75:** 1034–1039.

Hall JD, Woolard MD, Gunn BM, Craven RR, Taft-Benz S, Frelinger JA, Kawula TH. 2008. Infected-host-cell repertoire and cellular response in the lung following inhalation of *Francisella tularensis* Schu S4, LVS, or U112. *Infect Immun* **76:** 5843–5852.

Henry T, Brotcke A, Weiss DS, Thompson LJ, Monack DM. 2007. Type I interferon signaling is required for activation of the inflammasome during *Francisella* infection. *J Exp Med* **204:** 987–994.

Huang MT, Mortensen BL, Taxman DJ, Craven RR, Taft-Benz S, Kijek TM, Fuller JR, Davis BK, Allen IC, Brickey WJ, et al. 2010. Deletion of ripA alleviates suppression of the inflammasome and MAPK by *Francisella tularensis*. *J Immunol* **185:** 5476–5485.

Huynh KK, Grinstein S. 2007. Regulation of vacuolar pH and its modulation by some microbial species. *Microbiol Mol Biol Rev* **71:** 452–462.

Jones JW, Kayagaki N, Broz P, Henry T, Newton K, O'Rourke K, Chan S, Dong J, Qu Y, Roose-Girma M, et al. 2010. Absent in melanoma 2 is required for innate immune recognition of *Francisella tularensis*. *Proc Natl Acad Sci* **107:** 9771–9776.

Jones JW, Broz P, Monack DM. 2011. Innate immune recognition of *Francisella tularensis*: Activation of type-I interferons and the inflammasome. *Front Microbiol* **2:** 16.

Katz J, Zhang P, Martin M, Vogel SN, Michalek SM. 2006. Toll-like receptor 2 is required for inflammatory responses to *Francisella tularensis* LVS. *Infect Immun* **74:** 2809–2816.

Ketavarapu JM, Rodriguez AR, Yu JJ, Cong Y, Murthy AK, Forsthuber TG, Guentzel MN, Klose KE, Berton MT, Arulanandam BP. 2008. Mast cells inhibit intramacrophage *Francisella tularensis* replication via contact and secreted products including IL-4. *Proc Natl Acad Sci* **105:** 9313–9318.

KuoLee R, Harris G, Conlan JW, Chen W. 2011. Role of neutrophils and NADPH phagocyte oxidase in host defense against respiratory infection with virulent *Francisella tularensis* in mice. *Microbes Infect* **13:** 447–456.

Lai XH, Shirley RL, Crosa L, Kanistanon D, Tempel R, Ernst RK, Gallagher LA, Manoil C, Heffron F. 2010. Mutations of *Francisella novicida* that alter the mechanism of its phagocytosis by murine macrophages. *PLoS ONE* **5:** e11857.

Larsson P, Oyston PC, Chain P, Chu MC, Duffield M, Fuxelius HH, Garcia E, Halltorp G, Johansson D, Isherwood KE, et al. 2005. The complete genome sequence of *Francisella tularensis*, the causative agent of tularemia. *Nat Genet* **37:** 153–159.

Leiby DA, Fortier AH, Crawford RM, Schreiber RD, Nacy CA. 1992. In vivo modulation of the murine immune response to *Francisella tularensis* LVS by administration of anticytokine antibodies. *Infect Immun* **60:** 84–89.

Levine B, Mizushima N, Virgin HW. 2011. Autophagy in immunity and inflammation. *Nature* **469:** 323–335.

Lindemann SR, Peng K, Long ME, Hunt JR, Apicella MA, Monack DM, Allen LA, Jones BD. 2010. *Francisella tularensis* Schu S4 O-antigen and capsule biosynthesis gene mutants induce early cell death in human macrophages. *Infect Immun* **79:** 581–594.

Lindgren H, Golovliov I, Baranov V, Ernst RK, Telepnev M, Sjostedt A. 2004. Factors affecting the escape of *Francisella tularensis* from the phagolysosome. *J Med Microbiol* **53:** 953–958.

Lindgren H, Stenman L, Tarnvik A, Sjostedt A. 2005. The contribution of reactive nitrogen and oxygen species to the killing of *Francisella tularensis* LVS by murine macrophages. *Microbes Infect* **7:** 467–475.

Lindgren H, Shen H, Zingmark C, Golovliov I, Conlan W, Sjostedt A. 2007. Resistance of *Francisella tularensis* strains against reactive nitrogen and oxygen species with special reference to the role of KatG. *Infect Immun* **75:** 1303–1309.

Llewellyn AC, Jones CL, Napier BA, Bina JE, Weiss DS. 2011. Macrophage replication screen identifies a novel *Francisella* hydroperoxide resistance protein involved in virulence. *PLoS ONE* **6:** e24201.

Mariathasan S, Weiss DS, Dixit VM, Monack DM. 2005. Innate immunity against *Francisella tularensis* is dependent on the ASC/caspase-1 axis. *J Exp Med* **202:** 1043–1049.

McCaffrey RL, Allen LA. 2006. *Francisella tularensis* LVS evades killing by human neutrophils via inhibition of the respiratory burst and phagosome escape. *J Leukoc Biol* **80:** 1224–1230.

McCaffrey RL, Schwartz JT, Lindemann SR, Moreland JG, Buchan BW, Jones BD, Allen LA. 2010. Multiple mechanisms of NADPH oxidase inhibition by type A and type B *Francisella tularensis*. *J Leukoc Biol* **88:** 791–805.

McRae S, Pagliai FA, Mohapatra NP, Gener A, Mahmou AS, Gunn JS, Lorca GL, Gonzalez CF. 2009. Inhibition of AcpA phosphatase activity with ascorbate attenuates *Francisella tularensis* intramacrophage survival. *J Biol Chem* **285:** 5171–5177.

Meibom KL, Barel M, Charbit A. 2009. Loops and networks in control of *Francisella tularensis* virulence. *Future Microbiol* **4:** 713–729.

Melillo AA, Mahawar M, Sellati TJ, Malik M, Metzger DW, Melendez JA, Bakshi CS. 2009. Identification of *Francisella tularensis* live vaccine strain CuZn superoxide dismutase as critical for resistance to extracellularly generated reactive oxygen species. *J Bacteriol* **191:** 6447–6456.

Mohapatra NP, Balagopal A, Soni S, Schlesinger LS, Gunn JS. 2007. AcpA is a *Francisella* acid phosphatase that affects intramacrophage survival and virulence. *Infect Immun* **75:** 390–396.

Mohapatra NP, Soni S, Reilly TJ, Liu J, Klose KE, Gunn JS. 2008. Combined deletion of four *Francisella novicida* acid phosphatases attenuates virulence and macrophage vacuolar escape. *Infect Immun* **76:** 3690–3699.

Mohapatra NP, Soni S, Rajaram MV, Dang PM, Reilly TJ, El-Benna J, Clay CD, Schlesinger LS, Gunn JS. 2010. *Francisella* acid phosphatases inactivate the NADPH oxidase in human phagocytes. *J Immunol* **184:** 5141–5150.

Nano FE, Zhang N, Cowley SC, Klose KE, Cheung KK, Roberts MJ, Ludu JS, Letendre GW, Meierovics AI, Stephens G, et al. 2004. A *Francisella tularensis* pathogenicity island required for intramacrophage growth. *J Bacteriol* **186:** 6430–6436.

O'Riordan M, Portnoy DA. 2002. The host cytosol: Front-line or home front? *Trends Microbiol* **10:** 361–364.

Oyston PC, Sjostedt A, Titball RW. 2004. Tularaemia: Bioterrorism defence renews interest in *Francisella tularensis*. *Nat Rev Microbiol* **2:** 967–978.

Pechous R, Celli J, Penoske R, Hayes SF, Frank DW, Zahrt TC. 2006. Construction and characterization of an attenuated purine auxotroph in a *Francisella tularensis* live vaccine strain. *Infect Immun* **74:** 4452–4461.

Pechous RD, McCarthy TR, Mohapatra NP, Soni S, Penoske RM, Salzman NH, Frank DW, Gunn JS, Zahrt TC. 2008. A *Francisella tularensis* Schu S4 purine auxotroph is highly attenuated in mice but offers limited protection against homologous intranasal challenge. *PLoS ONE* **3:** e2487.

Peng K, Broz P, Jones J, Joubert LM, Monack D. 2011. Elevated AIM2-mediated pyroptosis triggered by hypercytotoxic *Francisella* mutant strains is attributed to increased intracellular bacteriolysis. *Cell Microbiol* **13:** 1586–1600.

Pierini LM. 2006. Uptake of serum-opsonized *Francisella tularensis* by macrophages can be mediated by class A scavenger receptors. *Cell Microbiol* **8:** 1361–1370.

Polsinelli T, Meltzer MS, Fortier AH. 1994. Nitric oxide-independent killing of *Francisella tularensis* by IFN-γ-stimulated murine alveolar macrophages. *J Immunol* **153:** 1238–1245.

Qin A, Mann BJ. 2006. Identification of transposon insertion mutants of *Francisella tularensis tularensis* strain Schu S4 deficient in intracellular replication in the hepatic cell line HepG2. *BMC Microbiol* **6:** 69.

Qin A, Scott DW, Thompson JA, Mann BJ. 2009. Identification of an essential *Francisella tularensis* subsp. tularensis virulence factor. *Infect Immun* **77:** 152–161.

Rathinam VA, Jiang Z, Waggoner SN, Sharma S, Cole LE, Waggoner L, Vanaja SK, Monks BG, Ganesan S, Latz E, et al. 2010. The AIM2 inflammasome is essential for host defense against cytosolic bacteria and DNA viruses. *Nat Immunol* **11:** 395–402.

Reilly TJ, Baron GS, Nano FE, Kuhlenschmidt MS. 1996. Characterization and sequencing of a respiratory burst-inhibiting acid phosphatase from *Francisella tularensis*. *J Biol Chem* **271:** 10973–10983.

Rodriguez AR, Yu JJ, Murthy AK, Guentzel MN, Klose KE, Forsthuber TG, Chambers JP, Berton MT, Arulanandam BP. 2010. Mast cell/IL-4 control of *Francisella tularensis* replication and host cell death is associated with increased ATP production and phagosomal acidification. *Mucosal Immunol* **4:** 217–226.

Santic M, Molmeret M, Abu Kwaik Y. 2005a. Modulation of biogenesis of the *Francisella tularensis* subsp. novicida-containing phagosome in quiescent human macrophages and its maturation into a phagolysosome upon activation by IFN-γ. *Cell Microbiol* **7:** 957–967.

Santic M, Molmeret M, Klose KE, Jones S, Kwaik YA. 2005b. The *Francisella tularensis* pathogenicity island protein IglC and its regulator MglA are essential for modulating phagosome biogenesis and subsequent bacterial escape into the cytoplasm. *Cell Microbiol* **7:** 969–979.

Santic M, Asare R, Skrobonja I, Jones S, Abu Kwaik Y. 2008. Acquisition of the vacuolar ATPase proton pump and phagosome acidification are essential for escape of *Francisella tularensis* into the macrophage cytosol. *Infect Immun* **76:** 2671–2677.

Saslaw S, Eigelsbach HT, Prior JA, Wilson HE, Carhart S. 1961a. Tularemia vaccine study. II. Respiratory challenge. *Arch Intern Med* **107:** 702–714.

Saslaw S, Eigelsbach HT, Wilson HE, Prior JA, Carhart S. 1961b. Tularemia vaccine study. I. Intracutaneous challenge. *Arch Intern Med* **107:** 689–701.

Schmerk CL, Duplantis BN, Howard PL, Nano FE. 2009. A *Francisella novicida* pdpA mutant exhibits limited intracellular replication and remains associated with the lysosomal marker LAMP-1. *Microbiology* **155:** 1498–1504.

Schulert GS, Allen LA. 2006. Differential infection of mononuclear phagocytes by *Francisella tularensis*: Role of the macrophage mannose receptor. *J Leukoc Biol* **80:** 563–571.

Schulert GS, McCaffrey RL, Buchan BW, Lindemann SR, Hollenback C, Jones BD, Allen LA. 2009. *Francisella tularensis* genes required for inhibition of the neutrophil respiratory burst and intramacrophage growth identified by random transposon mutagenesis of strain LVS. *Infect Immun* **77:** 1324–1336.

Su J, Yang J, Zhao D, Kawula TH, Banas JA, Zhang JR. 2007. Genome-wide identification of *Francisella tularensis* virulence determinants. *Infect Immun* **75:** 3089–3101.

Telepnev M, Golovliov I, Grundstrom T, Tarnvik A, Sjöstedt A. 2003. *Francisella tularensis* inhibits Toll-like receptor-mediated activation of intracellular signalling and secretion of TNF-α and IL-1 from murine macrophages. *Cell Microbiol* **5:** 41–51.

Tempel R, Lai XH, Crosa L, Kozlowicz B, Heffron F. 2006. Attenuated *Francisella novicida* transposon mutants protect mice against wild-type challenge. *Infect Immun* **74:** 5095–5105.

Ulland TK, Buchan BW, Ketterer MR, Fernandes-Alnemri T, Meyerholz DK, Apicella MA, Alnemri ES, Jones BD, Nauseef WM, Sutterwala FS. 2010. Cutting edge: Mutation of *Francisella tularensis* mviN leads to increased macrophage absent in melanoma 2 inflammasome activation and a loss of virulence. *J Immunol* **185:** 2670–2674.

Wehrly TD, Chong A, Virtaneva K, Sturdevant DE, Child R, Edwards JA, Brouwer D, Nair V, Fischer ER, Wicke L, et al. 2009. Intracellular biology and virulence determinants of *Francisella tularensis* revealed by transcriptional profiling inside macrophages. *Cell Microbiol* **11:** 1128–1150.

Weiss DS, Brotcke A, Henry T, Margolis JJ, Chan K, Monack DM. 2007. In vivo negative selection screen identifies genes required for *Francisella* virulence. *Proc Natl Acad Sci* **104:** 6037–6042.

Zahrt TC, Deretic V. 2002. Reactive nitrogen and oxygen intermediates and bacterial defenses: Unusual adaptations in Mycobacterium tuberculosis. *Antioxid Redox Signal* **4:** 141–159.

Echoes of a Distant Past: The *cag* Pathogenicity Island of *Helicobacter pylori*

Nicola Pacchiani[1], Stefano Censini[1], Ludovico Buti[2], and Antonello Covacci[1]

[1]Systems Biology Unit, Novartis Vaccines and Diagnostics, 53100 Siena, Italy

[2]Ludwig Institute for Cancer Research Oxford Branch, Oxford, United Kingdom

Correspondence: antonello.covacci@novartis.com

This review discusses the multiple roles of the CagA protein encoded by the *cag* pathogenicity island of *Helicobacter pylori* and highlights the CagA degradation activities on p53. By subverting the p53 tumor suppressor pathway CagA induces a strong antiapoptotic effect. *Helicobacter pylori* infection has been always associated with an increased risk of gastric cancer. The pro-oncogenic functions of CagA also target the tumor suppressor ASPP2. In the absence of tumor suppressor genes, cells survive and proliferate at times and in places where their survival and proliferation are inappropriate.

Helicobacter pylori (*H. pylori*) is a Gram-negative, helix-shaped, and microaerophilic bacterium that colonizes the mucosa of the stomach lifelong and has succeeded in living in a hostile niche where most other microorganisms fail to thrive, and in which competition is limited or absent. It is one of the most common human infections, affecting half of the world population and it is the causative agent of peptic ulcer disease (PUD), chronic gastritis, and is associated with an increased risk of gastric adenocarcinoma and MALT lymphoma. More than 80% of individuals are asymptomatic. During their lives, 10%–20% are at risk of PUD, 1%–3% will develop a gastric adenocarcinoma, and 0.1% MALT lymphoma. Infections are more frequent in the developing world than in Western Europe, North America, and Australasia. The age of acquisition is critical because it is associated with more severe outcomes. Transmission is by fecal–oral or oral–oral routes; it occurs within families in developed nations. In developing countries it is community acquired.

H. pylori was acquitted by a host jump from an unknown source in Africa, where two discrete superlineages differentiated. One of the superlineages was predominantly associated until very recently with San (hpAfrica2) and large felines (Hac), whereas the second is widespread throughout Africa (hpAfrica1, hpNEAfrica) and accompanied anatomically modern humans during their first "out of Africa" migration. Subsequent migration resulted in the Asian and Oceanic lineages hpAsia2, hpAsia, and hpSahul. After new migratory waves ancestors of the African hpNEAfrica and/or the Asian hpAsia2 populations resulted in the admixed hpEurope population, which then became the predominant population of extant *H. pylori* in Europe, the middle East, and western Asia (Moodley et al. 2012). The distribution of human languages is quite a sensitive indicator of

Cite this article as *Cold Spring Harb Perspect Med* doi: 10.1101/cshperspect.a010355

the dispersal of modern human beings around the globe. When cluster analysis was applied to a set of randomly selected gene samples, six *H. pylori* strains isolated from East Asia clustered with a strain from Peru. It turned out that the Peruvian isolate "was from an ethnic Japanese living in Peru" (Achtman et al. 1999). DNA fingerprinting analysis in Maori and Pacific Islanders by O'Toole *et al.* suggested that Polynesians are infected by a group of *H. pylori* strains that are distinct from European New Zealanders (Campbell et al. 1997). Likewise, isolates from Australian Aborigines are distinct from European origin Australians and indicated that the bacterial lineages first arrived in Australia with the earliest human migrations. New results lend support for two distinct waves of migrations into the Pacific. First are the early migrations to New Guinea and Australia accompanied by hpSahul and, second, a much later dispersal of hspMaori from Taiwan through the Pacific by the Malayo-Polynesian-speaking Lapita culture. Each sampling area yielded either hpSahul or hspMaori haplotypes, but not both (Moodley et al. 2009).

The major determinant of virulence is the *cag* pathogenicity island (PAI), a chromosomal segment of 40 kb containing 30 genes (Blum et al. 1994; Covacci et al. 1999). In comparison with partially deleted *cag* PAI strains, those with a functional intact *cag* PAI would increase the risk of gastric carcinoma 10-fold in infected subjects (Censini et al. 1996; Akopyants et al. 1998; Nguyen et al. 2008). Encoded in the *cag* PAI are the CagA antigen (Covacci et al. 1993; Tummuru et al. 1993) and the *cag* type IV secretion system (TFSS) (Covacci and Rappuoli 2000). The TFSS can be seen as a surface organelle, membrane-sheeted, forming a conduit for translocation of the substrate, the CagA protein (Rohde et al. 2003). Recent data suggest that $\alpha5\beta1$ integrin acts a receptor for CagA translocation (Kaplan-Türköz et al. 2012).

THE CagA PROTEIN

We are updating our knowledge on both the type IV secretion system of *H. pylori* and on a specific role of CagA in the p53 pathway. Together with essential notions on the microorganism we will provide an insight into CagA-host interactions. Furthermore, new data on the TFSS of a conjugative plasmid will help to explain the first phase of CagA secretion. Great progress has been made on the study of co-evolution of *H. pylori* with its human host and the use of *H. pylori* as a marker for the study of human migrations. We are approaching a phase in which genome-wide association studies will merge with data about migration of humans infected by *H. pylori*, providing a scaffold to understand language dispersal and anthropology as well as transmission of the infectious agents within families, persistent infection, and tumor progression. The last part will concentrate on the cellular mechanisms exploited by CagA in epithelial damage (read lesions) and the earliest phase of tumorigenic progression in concert with the tumor suppressor ASPP2 and p53 degradation.

Scanning electron microscopy images of human gastric mucosa revealed the presence of large amounts of *H. pylori* cells at the level of junctions (Steer 1984; Hazell et al. 1986). In addition, during coculture of bacterial and epithelial cells of animal or human origin massive elongations of the host cells were observed and named "hummingbird phenotype." CagA molecules, once translocated via the cag TFSS into the host cells, are tyrosine-phosphorylated within the repetitive sequence motif EPIYA and this activation has a central role in inducing changes in the host-cell morphology (Stein et al. 2000).

CagA tyrosine phosphorylation initiates host-cell signaling events via interaction with a SH2 or SH3 domain. This leads to the induction of a signaling cascade that mimics growth factor-like responses. In addition, inhibitors specific for the Src kinase family abolish CagA tyrosine phosphorylation in vitro and in tissue culture infection experiments, and two members of the Src kinase family, c-Src and Lyn, are the major CagA kinases (Stein et al. 2002). Src family kinases are strongly implicated in the development, growth, progression, and metastasis of a number of human cancers. They belong to the family of nonreceptor kinases and are posttranslationally modified through covalent attachment of a 14-carbon fatty acid moiety, myristate. Together with six basic amino acid residues the myristylation acts as the amino-terminal signal for membrane anchorage of the Src kinases. The main target is the plasma membrane, but perinuclear and endosomal membranes are also targeted. In the plasma membrane, Src kinases are found associated with focal adhesions, the actin cytoskeleton and adherens junctions. Src responses include mitogenic signaling, through association with growth factor receptors, cell adhesion and spreading, cell migration, and regulation of cell–cell contact formation (Segal et al. 1999; Asahi et al. 2000; Backert et al. 2000; Odenbreit et al. 2000).

CagA and Host–Cell Interactions

H. pylori adheres to cells in the proximity of the apical–junctional complex. The epithelial apical–junctional complex forms a network of transmembrane, scaffolding, and signaling proteins, and serves as a barrier, adhesion site, and pathways network to control cell polarity, proliferation, and differentiation processes. Simultaneous visualization of CagA, ZO-1, and *H. pylori* showed that injected CagA colocalized with ZO-1 at sites of bacterial attachment, as well as at cell–cell contacts. In Madin-Darby canine kidney (MDCK) monolayers infected with wild-type *H. pylori,* individual junctions showed barrier dysfunction with leakage of ruthenium red into the basolateral space. MDCK cells elongated and extended processes between adjacent cells, resulting in dysplastic cell shapes. Intraepithelial dysplasia depended on the injection of CagA and was reversible if monolayers were treated with antibiotics. CagA interferes with the assembly of functional tight junctions and alters the composition of junctional protein complexes (Amieva et al. 2003). The number of copies of the repeated region is strain specific; size variability of CagA ranges from 128 kDa (absence of 102 bp repeat) to 144 kDa (four 102 bp repeats in tandem) (Fig. 1) (Xiang et al. 1995). CagA molecules lacking the EPIYA motifs are translocated, but not phosphorylated. Unlike the parental strain the mutants failed to induce signaling events that govern cellular elongation and suggest that tyrosine phosphorylation is critical for CagA activation (Hayashi et al. 2012).

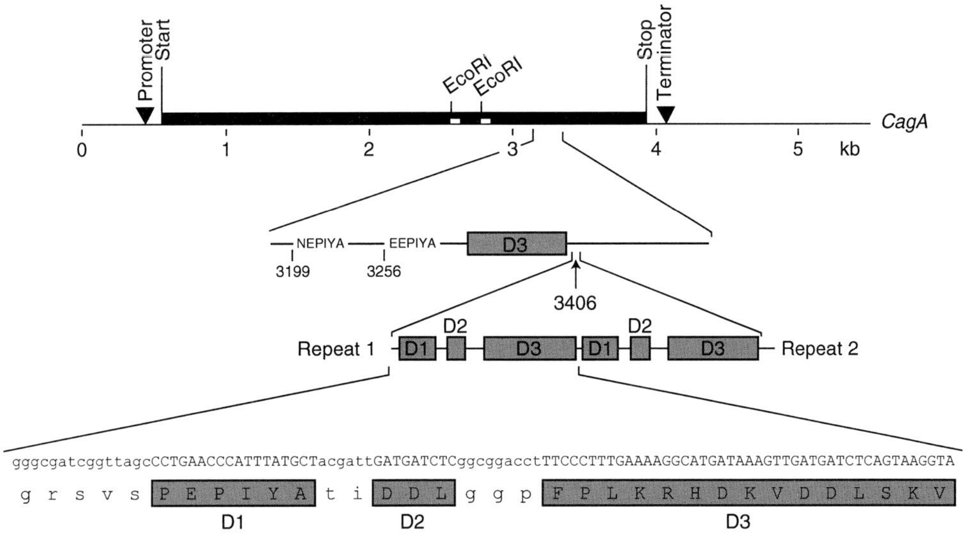

Figure 1. Repeats present in CagA are a site for tyrosine phosphorylation. Within the *cag*A gene, duplication of sequence motifs in the 5′-end region assembled into a repeat at nucleotide position 3406. The anatomy of a repeat is based on three domains: D1, D2, and D3. Repeats are 102 bp causing an increment in size of the protein of 4 kDa. The EPIYA motifs are shown.

THE TYPE IV SECRETION SYSTEM MACHINERY: FROM THE INNER CORE OF CONJUGATION SYSTEMS TO THE INNER CORE OF *cag* TFSS

Bordetella pertussis, Legionella pneumophila, and *Bartonella henselae* are members of a large bacterial family that share versions of the TFSS. New structural data were not specific for the *H. pylori* TFSS so that the latter might be different. However, there are good reasons to examine this large structure. It spans the periplasmic space and forms a chamber where other components are assembled. This is the minimal set of VirB subunits that could be useful for most, if not all, TFSSs. Our understanding of the conjugation system was developed with studies of the F plasmid of *Escherichia coli.* Conjugation, a contact-dependent process, bears a strong mechanistic resemblance to the type III secretion systems (TTSSs) and TFSSs. Pathogenic bacteria of humans and plants have coopted conjugation systems to export virulence factors to eukaryotic host cells. Thanks to the efforts of Christie et al. (Christie 2009; Sarkar et al. 2013) and Waksman et al. (Chandran et al. 2009; Fronzes et al. 2009a,b; Rivera-Calzada et al. 2013) we are now capable of describing the inner mechanics of the core complex (CC) and the effector delivery subsystem.

CC

The VirB7, VirB9, and VirB10 proteins assemble into a 1.07 megadalton (MDa) CC spanning the inner and outer membranes. This core consists of 14 copies of each of the proteins around which all other components assemble. The main body structure has two polar rings or "restrictions" that intercept the outer membrane (cap) and the inner membrane (base), respectively. The wall of the inner layer consisted of the amino-terminal domain of VirB10 and VirB9 homologs and is anchored in the inner membrane, resembling a cup that is opened at the base by a 55 Å diameter hole. The wall of the outer layer revealed a double-walled structure composed by the carboxy-terminal domain of the VirB10 homolog partially overlapping with the first half of the inner wall of the cap made by the VirB7 and VirB9 homologs. VirB7 and VirB9 contain signal peptides that target them to the periplasm. VirB7 is a small lipoprotein that interacts with and stabilizes VirB9 (Fig. 2).

The Energizers

The cytoplasmic ATPases VirD4, VirB11, and VirB4 are essential for secretion and for powering the secretory engine. VirD4 belong to the substrate receptor because they recruit substrates for secretion through the translocation channel. VirD4 are tethered to the inner membrane by an amino-terminal membrane anchor sequence that is required for assembly into hexameric structures. VirB11 is a typical traffic ATPase in a dynamic and regulated equilibrium between the cytoplasm and the membrane localization. Electron microscopy visualization of VirB11 homologs showed that they organize into hexameric rings. VirB4 forms homodimers, but the dimer might represent an intermediate state for the assembly of higher-order complexes. All of them are likely to form very large complexes of ATPases involved in energy supply for the transport of substrates along the secretory engine.

VirB7, VirB9, and VirB10 form a stable CC that spans the cell envelope and aggregate spontaneously. VirB6 and VirB8 stabilize the ATPase complex—VirB4, VirB11, and VirD4—to dock

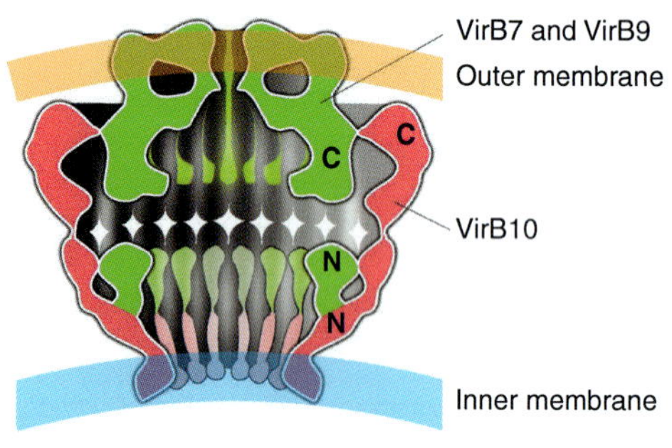

Figure 2. An artistic impression of the core complex as reported by Waksman et al. is shown. It spans the periplasmic space and is composed of a semispherical cup formed by monomers of the VirB10 protein and by a cap of assembled VirB7 and VirB9 elements. The dark areas are the inner capsular spaces.

the translocation channel at the inner membrane. The VirB2 and VirB5 pilus subunits are involved in the latter stage of pilus biogenesis.

CagA AND HOST-CELL DIFFERENTIATION

CagA Induction of Abnormal Apical and Basolateral Markers: Loss of Polarity

The reconstruction of the evolutionary path of a bacterial species should allow us to infer its population structure, the diversification of the virulence determinants that are responsible for discontinuity in clinicopathological manifestations. During lifelong infection CagA persistence increases the risk of long-term complications for the host, such as the development of peptic ulcers and gastric adenocarcinoma (Blaser and Atherton 2004; Cover and Blaser 2009; Correa and Piazuelo 2012). CagA induced abnormal localization of the gp135 apical marker to the basolateral membrane in MDCK cells. This loss of polarity occurred rapidly, because gp135 associated with the basolateral membrane is detected before changes in cell morphology are usually seen. At later stages, CagA-expressing cells elongated and freely migrated underneath neighboring cells, and gp135 localized throughout the basolateral membrane and concentrated at the trailing pole. The localization of basolateral membrane proteins is also altered in CagA-expressing cells. E-cadherin is usually absent from the apical surface of epithelia and exposed if cell polarity is lost or if the fence function of the junctions, which controls the diffusion of integral membrane proteins, fails. Antibodies to the extracellular domain of E-cadherin added to the apical side of nonpermeabilized epithelial monolayers do not stain the apical surface and cannot reach basolateral epitopes (Gumbiner and Simons 1987; Shore and Nelson 1991). However, these antibodies stained the apical surface of CagA-expressing cells but not neighboring control cells. The paracellular barrier of the tight junctions is disrupted in CagA-expressing cells, because these antibodies applied to the apex of nonpermeable cells penetrated the paracellular space to stain basolateral E-cadherin (Bagnoli et al. 2005).

CagA-Dependent Loss of Cell Adhesion

CagA-expressing cells extend pseudopodial processes between adjacent cells and also move between and under neighboring cells in a manner consistent with the loss of cell–cell adhesion. By tracking the movement of individual CagA-expressing cells and surrounding control cells CagA-expressing cells translocated almost four times further from their points of origin indicating a loss of cell adhesion and movement independent of their neighbors. The acquisition of migratory behavior in the context of loss of apicobasal polarity and the severing of intercellular junctions is reminiscent of a differentiation process known as epithelial-to-mesenchymal transition (EMT) (Birchmeier et al. 1996). This morphogenetic developmental program is observed at specialized sites during embryogenesis but is also thought to play an important part in the loss of epithelial characteristics that occurs during tumorigenic progression (Thiery 2002). Epithelial cells that undergo this mesenchymal transition are able to invade the extracellular matrix by expressing matrix metalloproteinases (MMPs) and degrading their underlying basement membrane (Yin et al. 2010).

CagA and Matrix Invasion

MDCK monolayers on filters coated with a basement membrane-like substrate containing type IV collagen and laminin indicated that CagA-expressing cells generated basal pseudopodia that invaded the matrix and penetrated into the underlying filter pores 40 times more than control cells. When GM6001, a synthetic peptide inhibitor of MMPs (Boghaert et al. 1994) is added to the matrix a four-times reduction was observed, suggesting that the process is MMPs dependent and supporting the notion that CagA induces invasive behavior in epithelial cells (Kitadai et al. 2003; El-Etr et al. 2004). Different domains of the CagA protein were expressed in polarized MDCK monolayers. The expression of the carboxy-terminal domain containing the EPIYA motifs were found tyrosine phosphorylated; cells elongated, extending basal pseudopodia that advance and retract between and below adjacent cells. The carboxyl terminus of CagA was sufficient, however, to induce MMP activity because its expression induced the formation of pseudopodia that were able to penetrate basement membranes. The amino terminus of the CagA protein, lacking the EPIYA motifs, was found sufficient to target CagA to the junctions and the plasma membrane but did not cause detectable redistribution of junctional proteins, nor did it disrupt cell polarity. Furthermore, it did not induce morphological changes, cell migration, or invasive behavior. Together, these findings show that the signaling triggered by the EPIYA motifs of CagA is necessary to cause cellular elongation and pseudopodial activity but additional conformational motifs are necessary.

In summary, CagA, a cancer-associated effector protein of *H. pylori*, is sufficient to disrupt the mechanisms that maintain normal epithelial differentiation, including cell adhesion, cell polarity, and the inhibition of migration.

CagA AND CELL INITIATION

CagA Associates with the ASPP2 Tumor Suppressor

CagA is one of the most important factors that link infection with *H. pylori* to the development of gastric cancer. Mongolian gerbils challenged with $CagA^+$ *H. pylori* strains develop gastric dysplasia and adenocarcinoma 12 wk postinfection (Franco et al. 2008). CagA has been suggested to interact with the RTK c-Met (Churin et al. 2003), the tyrosine phosphatase SHP-2 (Higashi et al. 2002), and the adaptor protein Grb2 (Mimuro et al. 2002). CagA has no clear amino acid homology with eukaryotic proteins involved in signaling pathways, but has structural similarities to proteins that can act on regulators of differentiation. For example, CagA share some of the functional properties of Gab proteins, docking molecules that serve as scaffolds to mediate the activation of multiple signals downstream from RTK (Hatakeyama 2003; Higashi et al. 2004). CagA and Gab proteins have been implicated in the activation of the Ras/MAPK (Meyer-ter-Vehn et al. 2000) and PI3-kinase pathways (Churin et al. 2003), both of which have been linked to EMT (Bakin et al. 2000; Grände et al. 2002). Earlier reports have linked injection or exogenous expression of CagA to the activation of several oncogenic pathways. For example, destabilization of the E-cadherin/β-catenin complex by CagA induces abnormal activation of the wingless/int (WNT)/β-catenin pathway (Franco et al. 2005; Botham et al. 2008; Neal et al. 2013). The role of the individual domains of CagA was investigated using an in vivo biotinylation approach combined with affinity purification to identify interacting proteins. LC-MS/MS analysis of a 150 kDa polypeptide unambiguously identified ASPP2, which interacted with full-length CagA and its amino terminus but not with its carboxyl terminus (Buti et al. 2011). The interaction between ASPP2 and p53 results in proteasomal degradation of p53 and consequently, inhibition of the apoptotic response of the host cell. CagA thus hijacks ASPP2 and alters its activity in a manner that promotes cell survival and favors transformation.

CagA Delocalizes ASPP2

ASPP2 is a member of the ASPP family, which includes the proapoptotic ASPP1 and the antiapoptotic iASPP (Bergamaschi et al. 2003). Both ASPP1 and ASPP2 are known activators of the tumor suppressor p53 in response to DNA damage or oncogenic stimuli (Samuels-Lev et al. 2001). The carboxyl terminus of ASPP2 transiently interacts with the DNA-binding domain of p53, and after this interaction, p53 induces the expression of genes involved in apoptosis. ASPP2 also binds other factors with roles in apoptosis (Bcl-2 and Yes-associated protein) (Naumovski and Cleary 1994; Yang et al. 1999; Espanel and Sudol 2001) and control of cell growth (p65/RelA, a subunit of NF-κB, APC tumor suppressor, and protein phosphatase-1) (Helps et al. 1995; Nakagawa et al. 2000). AGS cells infected with Hp and lysed in 0.5% Triton X-100 indicated that ASPP2 was delocalized to a Triton-resistant fraction, which contains cell remnants, nuclei, Triton-resistant membranes, and bacteria. Coexpression

 Cite this article as *Cold Spring Harb Perspect Med* doi: 10.1101/cshperspect.a010355

of Cherry-ASPP2 with GFP-CagA shows strong colocalization, and CagA recruits ASPP2 to a region close to the plasma membrane. Although there is no clear consensus as to the exact localization of ASPP2 (Sun et al. 2008; Uhlmann-Schiffler et al. 2009), a recent study described the targeting of ASPP2 to tight junctions (Sottocornola et al. 2010). Because full-length CagA is also found close to the plasma membrane and junctional complexes, CagA likely recruits ASPP2 to form a complex near the plasma membrane.

ASPP2 and p53

Injection of CagA facilitates the interaction between p53 and ASPP2. Treatment of cells with Doxorubicin (Dox), a DNA damaging agent that activates p53, induced the association of ASPP2 with p53 as well as apoptosis. Association between CagA and ASPP2 was detected already 90 min postinfection, increasing as time progressed. In contrast, the association between ASPP2 and cytoplasmic p53 was observed not earlier than 3 h and reached a peak at 7 h. This suggested that ASPP2 recruits the cytoplasmic pool of p53 after ASPP2 engaged by CagA (Buti et al. 2011).

Proteosomal Degradation of p53

p53 is a transcription factor that regulates expression on a variety of target genes. It is rapidly degraded by the proteasome but strongly stabilized after DNA damage or cellular stress. Because ASPP2 recruits cytoplasmic p53 during *H. pylori* infection, the transcriptional activity of p53 might be altered during translocation of CagA. Several lines of research suggested that CagA-mediated inhibition of p53 expression is ASPP2 dependent and CagA-induced degradation of p53 is mediated by the proteasome. Under normal conditions, the tumor suppressor function of the ASPP2-p53 pathway is exerted mainly through induction of the apoptotic response. However, in *H. pylori* infected cells, CagA promotes the interaction between p53 and ASPP2, leading to an enhanced degradation of p53, and therefore inhibition of its transcriptional activity. CagA inhibits apoptosis by binding the tumor suppressor ASPP2 and through this interaction p53 is degraded and its apoptotic function inhibited (Mimuro et al. 2007; Buti et al. 2011).

CagA Antiapoptotic Effect

Several viral oncogenes (e.g., T-antigen E1B-55Kd and HPV E6 and E7) inactivate the tumor suppressor activity of the host cell by targeting and degrading p53 when the infection is established (Sarnow et al. 1982; Schutzbank et al. 1982; Werness et al. 1990). In addition, *H. pylori* infection is enhanced when the apoptotic response of gastric epithelial cells is impaired. CagA injection induces the up-regulation of the antiapoptotic protein MCL-1, and consequently, the infected cells become more resistant to normal cell turnover in the stomach. CagA-mediated degradation of p53 results in resistance to the apoptotic response in an ASPP2-dependent manner. Of note, RUNX3 is also sent to the proteasome after CagA interaction, suggesting a link between CagA and the proteasomal machinery of the infected cell (Tsang et al. 2010).

ASPP2 Interactions

ASPP2 is best known for its role as tumor suppressor, and as such, it enhances the apoptotic response of the cell. The amino terminus of ASPP2 is required to enhance the transcriptional activity of p53 (Gorina and Pavletich 1996). Cytosolic DDA3 binds ASPP2 and prevents activation of p53 without affecting the ASPP2-p53 interaction (Sun et al. 2008). CagA may dislocate p53 through ASPP2, thereby altering its function. Indeed, cytoplasmic ASPP1 functions as oncogene, whereas nuclear ASPP1 acts as tumor suppressor (Vigneron et al. 2010). The function of ASPP2 depends on its localization, and CagA may interfere with its localization. The function of ASPP2 is not restricted to activation of p53 but includes regulation of cell–cell adhesion and polarity. *Drosophila* dASPP localizes at adherens junctions and regulates the activity of carboxy-terminal kinase (dCsk).

Loss of function of dASPP increases cell spreading and apoptosis. In mammalian cells, ASPP2 associates with Par-3 (Sottocornola et al. 2010), a complex crucial for the formation and localization of the apical-junctional complex. ASPP2-depleted cells are defective in the formation of tight junctions and acquire a migratory phenotype (Langton et al. 2007, 2009; Cong et al. 2010). Thus, ASPP2 acts as a tumor suppressor by enhancing the transcriptional activity of p53 (Bunz et al. 1998; Wei et al. 2010); then it may act as a regulator of cell–cell adhesion in a p53-independent manner. Interactors of CagA identified by LC-MS/MS not only reveal the presence of ASPP2 but also Par-3 (Fig. 3) (Buti et al. 2011). Several articles have also identified abnormal p53 after exposure to *H. pylori*, including truncated forms that are promoting increased survival of the host cells (Wei et al. 2010, 2012). MALT lymphomas can be distinguished from other lymphomas because they occur in various extranodal locations, primarily the stomach, where they are preceded by *H. pylori* infection (Vicente-Dueñas et al. 2012). Malt lymphomas are associated with chromosomal translocations producing the oncogene MALT-1. Mouse animal model of the disease has been generated and progression to activated B-cell diffuse large-cell lymphoma (ADC-DLBCL) is associated with loss of p53, mimicking the human form of the disease (Vicente-Dueñas et al. 2012). Loss of p53 is a crucial factor for inducing a totipotent phenotype and generating pluripotent stem cells (Marión et al. 2009; Kawamura et al. 2009). The association between *H. pylori* and diseases might

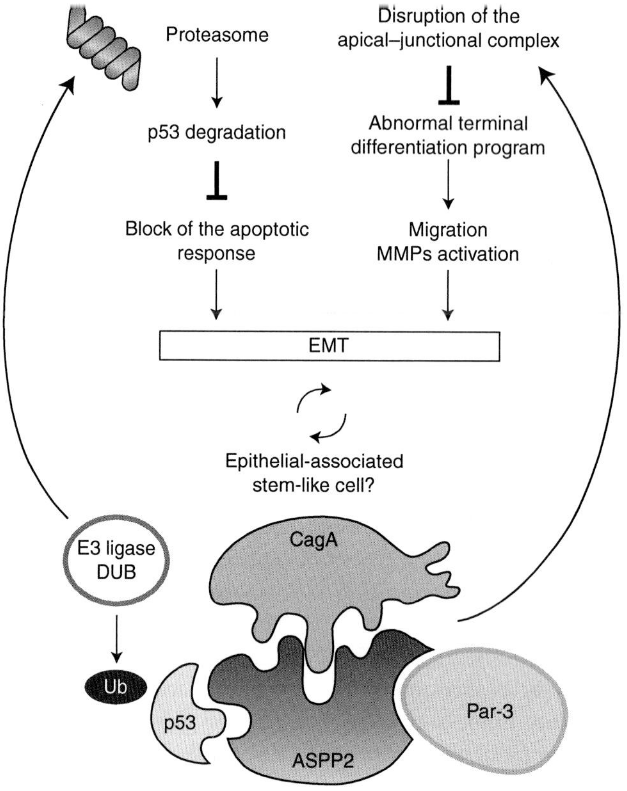

Figure 3. Schematic model of the effects induced by the CagA-ASPP2 interaction. CagA binds ASPP2, and this complex is responsible for proteasome-dependent p53 degradation. Cells harboring CagA are unable to cause apoptosis. In addition, they acquire the ability to migrate, invade, and they lack a terminal differentiation program (EMT).

hinge on the effect that controls the epigenetic status of the cells. Not only does CagA cause the functional ablation of the ASPP2/p53 tumor suppressor pathways and the altered terminal differentiation program (EMT), but may predispose to initiation/promotion before the further stage of tumorigenic progression.

CONCLUDING REMARKS

The acquisition of the *cag* PAI has targeted the mechanisms of virulence of *H. pylori* at the intracellular level.

CagA, the major effector of virulence is conveyed within the host cell by a dedicated TFSS. The contact-dependent secretion system relies on a large complex of ATPases able to translocate CagA molecules at a sustained peak rate.

Tyrosine-phosphorylated CagA is responsible for acting at the level of cellular junctions. The clinical lesions (ulceration) associated with *H. pylori* infection should be seen as resulting from loss of contact among cells.

H. pylori has infected man before the human migration and has established a status of quasispecies. Through a deep-sequencing approach migrations can be traced at an unprecedented level and factors affecting persistence, tissue specificity, and disease outcomes can be identified in the host.

At the epithelial level the CagA protein causes altered terminal differentiation, preneoplastic lesions like ETM, and abnormal cell migration.

Degradation of p53 is a consequence of the recruitment and misregulation of ASPP2 by CagA. We suggest that altered localization and/or posttranslational modification of ASPP2 might be responsible for decay in tumor suppressor function resulting in aberrant complex with p53.

ACKNOWLEDGMENTS

We thank C. Montecucco, Fabio Bagnoli, Manuel Amieva, Hidde Ploegh, Stanley Falkow, Jörg Hacker, Derya Unutmaz, and Rino Rappuoli for helpful discussions and critical reading of the manuscript and Riccardo Beltrami, Claudio Donati, and Duccio Medini for granting access to an unpublished set of data. We acknowledge the National Center for Biotechnology Information, EBI, EMBL, and The Broad Institute at MIT (Cambridge, MA) for granting access to their genomic software, and Dr. N. Figura (University of Siena, Italy) for *H. pylori* clinical isolates. We gratefully acknowledge G. Corsi for the artwork and Barbara Acosta (Cold Spring Harbor Laboratory Press) for excellent editorial assistance. A.C. thanks the "external reviewer" for previous comments on the manuscript.

REFERENCES

Achtman M, Azuma T, Berg DE, Ito Y, Morelli G, Pan ZJ, Suerbaum S, Thompson SA, Van der Ende A, Van Doorn LJ. 1999. Recombination and clonal groupings within *Helicobacter pylori* from different geographical regions. *Mol Microbiol* **32:** 459–470.

Akopyants NS, Clifton SW, Kersulyte D, Crabtree JE, Youree BE, Reece CA, Bukanov NO, Drazek ES, Roe BA, Berg DE. 1998. Analyses of the cag pathogenicity island of *Helicobacter pylori. Mol Microbiol* **28:** 37–53.

Amieva MR, Vogelmann R, Covacci A, Tompkins LS, Nelson WJ, Falkow S. 2003. Disruption of the epithelial apical-junctional complex by *Helicobacter pylori* CagA. *Science* **300:** 1430–1434.

Asahi M, Azuma T, Ito S, Ito Y, Suto H, Nagai Y, Tsubokawa M, Tohyama Y, Maeda S, Omata M, et al. 2000. *Helicobacter pylori* Caga protein can be tyrosine phosphorylated in gastric epithelial cells. *J Exp Med* **191:** 593–602.

Backert S, Ziska E, Brinkmann V, Zimny-Arndt U, Fauconnier A, Jungblut PR, Naumann M, Meyer TF. 2000. Translocation of the *Helicobacter pylori* CagA protein in gastric epithelial cells by a type IV secretion apparatus. *Cell Microbiol* **2:** 155–164.

Bagnoli F, Buti L, Tompkins L, Covacci A, Amieva MR. 2005. *Helicobacter pylori* CagA induces a transition from polarized to invasive phenotypes in MDCK cells. *Proc Natl Acad Sci* **102:** 16339–16344.

Bakin AV, Tomlinson AK, Bhowmick NA, Moses HL, Arteaga CL. 2000. Phosphatidylinositol 3-kinase function is required for transforming growth factor β-mediated epithelial to mesenchymal transition and cell migration. *J Biol Chem* **275:** 36803–36810.

Bergamaschi D, Samuels Y, O'Neil NJ, Trigiante G, Crook T, Hsieh J-K, O'Connor DJ, Zhong S, Campargue I, Tomlinson ML, et al. 2003. iASPP oncoprotein is a key inhibitor of p53 conserved from worm to human. *Nat Genet* **33:** 162–167.

Birchmeier C, Birchmeier W, Brand-Saberi B. 1996. Epithelial-mesenchymal transitions in cancer progression. *Acta Anatomica* **156:** 217–226.

Blaser MJ, Atherton JC. 2004. *Helicobacter pylori* persistence: Biology and disease. *J Clin Invest* **113:** 321–333.

Blum G, Ott M, Lischewski A, Ritter A, Imrich H, Tschäpe H, Hacker J. 1994. Excision of large DNA regions termed pathogenicity islands from tRNA-specific loci in the chromosome of an *Escherichia coli* wild-type pathogen. *Infect Immun* **62:** 606–614.

Boghaert ER, Chan SK, Zimmer C, Grobelny D, Galardy RE, Vanaman TC, Zimmer SG. 1994. Inhibition of collagenolytic activity relates to quantitative reduction of invasion in vitro in a c-Ha-ras transfected glial cell line. *J Neurooncol* **21:** 141–150.

Botham CM, Wandler AM, Guillemin K. 2008. A transgenic *Drosophila* model demonstrates that the *Helicobacter pylori* CagA protein functions as a eukaryotic Gab adaptor. *PLoS Pathog* **4:** e1000064.

Bunz F, Dutriaux A, Lengauer C, Waldman T, Zhou S, Brown JP, Sedivy JM, Kinzler KW, Vogelstein B. 1998. Requirement for p53 and p21 to sustain G2 arrest after DNA damage. *Science* **282:** 1497–1501.

Buti L, Spooner E, Van der Veen AG, Rappuoli R, Covacci A, Ploegh HL. 2011. *Helicobacter pylori* cytotoxin-associated gene A (CagA) subverts the apoptosis-stimulating protein of p53 (ASPP2) tumor suppressor pathway of the host. *Proc Natl Acad Sci* **108:** 9238–9243.

Campbell S, Fraser A, Holliss B, Schmid J, O'Toole PW. 1997. Evidence for ethnic tropism of *Helicobacter pylori*. *Infect Immun* **65:** 3708–3712.

Censini S, Lange C, Xiang Z, Crabtree JE, Ghiara P, Borodovsky M, Rappuoli R, Covacci A. 1996. cag, a pathogenicity island of *Helicobacter pylori*, encodes type I-specific and disease-associated virulence factors. *Proc Natl Acad Sci* **93:** 14648–14653.

Chandran V, Fronzes R, Duquerroy S, Cronin N, Navaza J, Waksman G. 2009. Structure of the outer membrane complex of a type IV secretion system. *Nature* **462:** 1011–1015.

Christie PJ. 2009. Structural biology: Translocation chamber's secrets. *Nature* **462:** 992–994.

Churin Y, Al-Ghoul L, Kepp O, Meyer TF, Birchmeier W, Naumann M. 2003. *Helicobacter pylori* CagA protein targets the c-Met receptor and enhances the motogenic response. *J Cell Biol* **161:** 249–255.

Cong W, Hirose T, Harita Y, Yamashita A, Mizuno K, Hirano H, Ohno S. 2010. ASPP2 regulates epithelial cell polarity through the PAR complex. *Curr Biol* **20:** 1408–1414.

Correa P, Piazuelo MB. 2012. The gastric precancerous cascade. *J Digest Dis* **13:** 2–9.

Covacci A, Rappuoli R. 2000. Tyrosine-phosphorylated bacterial proteins: Trojan horses for the host cell. *J Exp Med* **191:** 587–592.

Covacci A, Censini S, Bugnoli M, Petracca R, Burroni D, Macchia G, Massone A, Papini E, Xiang Z, Figura N. 1993. Molecular characterization of the 128-kDa immunodominant antigen of *Helicobacter pylori* associated with cytotoxicity and duodenal ulcer. *Proc Natl Acad Sci* **90:** 5791–5795.

Covacci A, Telford JL, Del Giudice G, Parsonnet J, Rappuoli R. 1999. *Helicobacter pylori* virulence and genetic geography. *Science* **284:** 1328–1333.

Cover TL, Blaser MJ. 2009. *Helicobacter pylori* in health and disease. *Gastroenterology* **136:** 1863–1873.

El-Etr SH, Mueller A, Tompkins LS, Falkow S, Merrell DS. 2004. Phosphorylation-independent effects of CagA during interaction between *Helicobacter pylori* and T84 polarized monolayers. *J Infect Dis* **190:** 1516–1523.

Espanel X, Sudol M. 2001. Yes-associated protein and p53-binding protein-2 interact through their WW and SH3 domains. *J Biol Chem* **276:** 14514–14523.

Franco AT, Israel DA, Washington MK, Krishna U, Fox JG, Rogers AB, Neish AS, Collier-Hyams L, Perez-Perez GI, Hatakeyama M, et al. 2005. Activation of β-catenin by carcinogenic *Helicobacter pylori*. *Proc Natl Acad Sci* **102:** 10646–10651.

Franco AT, Johnston E, Krishna U, Yamaoka Y, Israel DA, Nagy TA, Wroblewski LE, Piazuelo MB, Correa P, Peek RM. 2008. Regulation of gastric carcinogenesis by *Helicobacter pylori* virulence factors. *Cancer Res* **68:** 379–387.

Fronzes R, Christie PJ, Waksman G. 2009a. The structural biology of type IV secretion systems. *Nat Rev Microbiol* **7:** 703–714.

Fronzes R, Schäfer E, Wang L, Saibil HR, Orlova EV, Waksman G. 2009b. Structure of a type IV secretion system core complex. *Science* **323:** 266–268.

Gorina S, Pavletich NP. 1996. Structure of the p53 tumor suppressor bound to the Ankyrin and SH3 domains of 53BP2. *Science* **274:** 1001–1005.

Grände M, Franzen Å, Karlsson J-O, Ericson LE, Heldin N-E, Nilsson M. 2002. Transforming growth factor-β and epidermal growth factor synergistically stimulate epithelial to mesenchymal transition (EMT) through a MEK-dependent mechanism in primary cultured pig thyrocytes. *J Cell Sci* **115** : 4227–4236.

Gumbiner B, Simons K. 1987. The role of uvomorulin in the formation of epithelial occluding junctions. *Ciba Found Symp* **125:** 168–186.

Hatakeyama M. 2003. *Helicobacter pylori* CagA—A potential bacterial oncoprotein that functionally mimics the mammalian Gab family of adaptor proteins. *Microb Infect* **5:** 143–150.

Hayashi T, Senda M, Morohashi H, Higashi H, Horio M, Kashiba Y, Nagase L, Sasaya D, Shimizu T, Venugopalan N, et al. 2012. Tertiary structure-function analysis reveals the pathogenic signaling potentiation mechanism of *Helicobacter pylori* oncogenic effector CagA. *Cell Host Microbe* **12:** 20–33.

Hazell SL, Lee A, Brady L, Hennessy W. 1986. *Campylobacter pyloridis* and gastritis: Association with intercellular spaces and adaptation to an environment of mucus as important factors in colonization of the gastric epithelium. *J Infect Dis* **153:** 658–663.

Helps NR, Barker HM, Elledge SJ, Cohen PT. 1995. Protein phosphatase 1 interacts with p53BP2, a protein which binds to the tumour suppressor p53. *FEBS Lett* **377:** 295–300.

Higashi H, Tsutsumi R, Muto S, Sugiyama T, Azuma T, Asaka M, Hatakeyama M. 2002. SHP-2 tyrosine phosphatase as an intracellular target of *Helicobacter pylori* CagA protein. *Science* **295:** 683–686.

Higashi H, Nakaya A, Tsutsumi R, Yokoyama K, Fujii Y, Ishikawa S, Higuchi M, Takahashi A, Kurashima Y, Teishikata Y, et al. 2004. *Helicobacter pylori* CagA induces Ras-independent morphogenetic response through SHP-

2 recruitment and activation. *J Biol Chem* **279:** 17205–17216.

Kaplan-Türköz B, Jiménez-Soto LF, Dian C, Ertl C, Remaut H, Louche A, Tosi T, Haas R, Terradot L. 2012. Structural insights into *Helicobacter pylori* oncoprotein CagA interaction with β1 integrin. *Proc Natl Acad Sci* **109:** 14640–14645.

Kawamura T, Suzuki J, Wang YV, Menendez S, Morera LB, Raya A, Wahl GM, Izpisúa Belmonte JC. 2009. Linking the p53 tumour suppressor pathway to somatic cell reprogramming. *Nature* **460:** 1140–1144.

Kitadai Y, Sasaki A, Ito M, Tanaka S, Oue N, Yasui W, Aihara M, Imagawa K, Haruma K, Chayama K. 2003. *Helicobacter pylori* infection influences expression of genes related to angiogenesis and invasion in human gastric carcinoma cells. *Biochem Biophys Res Commun* **311:** 809–814.

Langton PF, Colombani J, Aerne BL, Tapon N. 2007. *Drosophila* ASPP regulates C-terminal Src kinase activity. *Dev Cell* **13:** 773–782.

Langton PF, Colombani J, Chan EHY, Wepf A, Gstaiger M, Tapon N. 2009. The dASPP-dRASSF8 complex regulates cell-cell adhesion during *Drosophila* retinal morphogenesis. *Curr Biol* **19:** 1969–1978.

Marión RM, Strati K, Li H, Murga M, Blanco R, Ortega S, Fernandez-Capetillo O, Serrano M, Blasco MA. 2009. A p53-mediated DNA damage response limits reprogramming to ensure iPS cell genomic integrity. *Nature* **460:** 1149–1153.

Meyer-ter-Vehn T, Covacci A, Kist M, Pahl HL. 2000. *Helicobacter pylori* activates mitogen-activated protein kinase cascades and induces expression of the proto-oncogenes c-fos and c-jun. *J Biol Chem* **275:** 16064–16072.

Mimuro H, Suzuki T, Tanaka J, Asahi M, Haas R, Sasakawa C. 2002. Grb2 is a key mediator of *Helicobacter pylori* CagA protein activities. *Mol Cell* **10:** 745–755.

Mimuro H, Suzuki T, Nagai S, Rieder G, Suzuki M, Nagai T, Fujita Y, Nagamatsu K, Ishijima N, Koyasu S, et al. 2007. *Helicobacter pylori* dampens gut epithelial self-renewal by inhibiting apoptosis, a bacterial strategy to enhance colonization of the stomach. *Cell Host Microbe* **2:** 250–263.

Moodley Y, Linz B, Yamaoka Y, Windsor HM, Breurec S, Wu J-Y, Maady A, Bernhöft S, Thiberge J-M, Phuanukoonnon S, et al. 2009. The peopling of the Pacific from a bacterial perspective. *Science* **323:** 527–530.

Moodley Y, Linz B, Bond RP, Nieuwoudt M, Soodyall H, Schlebusch CM, Bernhöft S, Hale J, Suerbaum S, Mugisha L, et al. 2012. Age of the association between *Helicobacter pylori* and man. *PLoS Patho* **8:** e1002693.

Nakagawa H, Koyama K, Murata Y, Morito M, Akiyama T, Nakamura Y. 2000. APCL, a central nervous system-specific homologue of adenomatous polyposis coli tumor suppressor, binds to p53-binding protein 2 and translocates it to the perinucleus. *Cancer Res* **60:** 101–105.

Naumovski L, Cleary M. 1994. Bcl2 inhibits apoptosis associated with terminal differentiation of HL-60 myeloid leukemia cells. *Blood* **83:** 2261–2267.

Neal JT, Peterson TS, Kent ML, Guillemin K. 2013. *H. pylori* virulence factor CagA increases intestinal cell proliferation by Wnt pathway activation in a transgenic zebrafish model. *Disease Model Mech* **6:** 802–810.

Nguyen LT, Uchida T, Murakami K, Fujioka T, Moriyama M. 2008. *Helicobacter pylori* virulence and the diversity of gastric cancer in Asia. *J Med Microbiol* **57:** 1445–1453.

Odenbreit S, Püls J, Sedlmaier B, Gerland E, Fischer W, Haas R. 2000. Translocation of *Helicobacter pylori* CagA into gastric epithelial cells by type IV secretion. *Science* **287:** 1497–1500.

Rivera-Calzada A, Fronzes R, Savva CG, Chandran V, Lian PW, Laeremans T, Pardon E, Steyaert J, Remaut H, Waksman G, et al. 2013. Structure of a bacterial type IV secretion core complex at subnanometre resolution. *EMBO J* **32:** 1195–1204.

Rohde M, Püls J, Buhrdorf R, Fischer W, Haas R. 2003. A novel sheathed surface organelle of the *Helicobacter pylori* cag type IV secretion system. *Mol Microbiol* **49:** 219–234.

Samuels-Lev Y, O'Connor DJ, Bergamaschi D, Trigiante G, Hsieh JK, Zhong S, Campargue I, Naumovski L, Crook T, Lu X. 2001. ASPP proteins specifically stimulate the apoptotic function of p53. *Mol Cell* **8:** 781–794.

Sarkar MK, Husnain SI, Jakubowski SJ, Christie PJ. 2013. Isolation of bacterial type IV machine subassemblies. *Method Mol Biol* **966:** 187–204.

Sarnow P, Ho YS, Williams J, Levine AJ. 1982. Adenovirus E1b-58kd tumor antigen and SV40 large tumor antigen are physically associated with the same 54 kd cellular protein in transformed cells. *Cell* **28:** 387–394.

Schutzbank T, Robinson R, Oren M, Levine AJ. 1982. SV40 large tumor antigen can regulate some cellular transcripts in a positive fashion. *Cell* **30:** 481–490.

Segal ED, Cha J, Lo J, Falkow S, Tompkins LS. 1999. Altered states: Involvement of phosphorylated CagA in the induction of host cellular growth changes by *Helicobacter pylori*. *Proc Natl Acad Sci* **96:** 14559–14564.

Shore EM, Nelson WJ. 1991. Biosynthesis of the cell adhesion molecule uvomorulin (E-cadherin) in Madin-Darby canine kidney epithelial cells. *J Biol Chem* **266:** 19672–19680.

Sottocornola R, Royer C, Vives V, Tordella L, Zhong S, Wang Y, Ratnayaka I, Shipman M, Cheung A, Gaston-Massuet C, et al. 2010. ASPP2 binds Par-3 and controls the polarity and proliferation of neural progenitors during CNS development. *Dev Cell* **19:** 126–137.

Steer HW. 1984. Surface morphology of the gastroduodenal mucosa in duodenal ulceration. *Gut* **25:** 1203–1210.

Stein M, Rappuoli R, Covacci A. 2000. Tyrosine phosphorylation of the *Helicobacter pylori* CagA antigen after cag-driven host cell translocation. *Proc Natl Acad Sci* **97:** 1263–1268.

Stein M, Bagnoli F, Halenbeck R, Rappuoli R, Fantl WJ, Covacci A. 2002. c-Src/Lyn kinases activate *Helicobacter pylori* CagA through tyrosine phosphorylation of the EPIYA motifs. *Mol Microbiol* **43:** 971–980.

Sun W-T, Hsieh P-C, Chiang M-L, Wang M-C, Wang F-F. 2008. p53 target DDA3 binds ASPP2 and inhibits its stimulation on p53-mediated BAX activation. *Biochem Biophys Res Commun* **376:** 395–398.

Thiery JP. 2002. Epithelial-mesenchymal transitions in tumour progression. *Nat Rev Cancer* **2:** 442–454.

Tsang YH, Lamb A, Romero-Gallo J, Huang B, Ito K, Peek RM, Ito Y, Chen LF. 2010. *Helicobacter pylori* CagA targets gastric tumor suppressor RUNX3 for proteasome-mediated degradation. *Oncogene* **29:** 5643–5650.

Tummuru MK, Cover TL, Blaser MJ. 1993. Cloning and expression of a high-molecular-mass major antigen of *Helicobacter pylori*: Evidence of linkage to cytotoxin production. *Infect Immun* **61:** 1799–1809.

Uhlmann-Schiffler H, Kiermayer S, Stahl H. 2009. The DEAD box protein Ddx42p modulates the function of ASPP2, a stimulator of apoptosis. *Oncogene* **28:** 2065–2073.

Vicente-Dueñas C, Fontán L, Gonzalez-Herrero I, Romero-Camarero I, Segura V, Aznar MA, Alonso-Escudero E, Campos-Sanchez E, Ruiz-Roca L, Barajas-Diego M, et al. 2012. Expression of MALT1 oncogene in hematopoietic stem/progenitor cells recapitulates the pathogenesis of human lymphoma in mice. *Proc Natl Acad Sci* **109:** 10534–10539.

Vigneron AM, Ludwig RL, Vousden KH. 2010. Cytoplasmic ASPP1 inhibits apoptosis through the control of YAP. *Genes Dev* **24:** 2430–2439.

Wei J, Nagy TA, Vilgelm A, Zaika E, Ogden SR, Romero-Gallo J, Piazuelo MB, Correa P, Washington MK, El-Rifai W, et al. 2010. Regulation of p53 tumor suppressor by *Helicobacter pylori* in gastric epithelial cells. *Gastroenterology* **139:** 1333–1343.

Wei J, Noto J, Zaika E, Romero-Gallo J, Correa P, El-Rifai W, Peek RM, Zaika A. 2012. Pathogenic bacterium *Helicobacter pylori* alters the expression profile of p53 protein isoforms and p53 response to cellular stresses. *Proc Natl Acad Sci* **109:** E2543–E2550.

Werness BA, Levine AJ, Howley PM. 1990. Association of human papillomavirus types 16 and 18 E6 proteins with p53. *Science* **248:** 76–79.

Xiang Z, Censini S, Bayeli P, Telford J, Figura N, Rappuoli R, Covacci A. 1995. Analysis of expression of CagA and VacA virulence factors in 43 strains of *Helicobacter pylori* reveals that clinical isolates can be divided into two major types and that CagA is not necessary for expression of the vacuolating cytotoxin. *Infect Immun* **63:** 94–98.

Yang JP, Hori M, Takahashi N, Kawabe T, Kato H, Okamoto T. 1999. NF-κB subunit p65 binds to 53BP2 and inhibits cell death induced by 53BP2. *Oncogene* **18:** 5177–5186.

Yin Y, Grabowska AM, Clarke PA, Whelband E, Robinson K, Argent RH, Tobias A, Kumari R, Atherton JC, Watson SA. 2010. *Helicobacter pylori* potentiates epithelial:mesenchymal transition in gastric cancer: Links to soluble HB-EGF, gastrin and matrix metalloproteinase-7. *Gut* **59:** 1037–1045.

Helicobacter and *Salmonella* Persistent Infection Strategies

Denise M. Monack

Department of Microbiology and Immunology, Stanford University School of Medicine, Stanford, California 94305

Correspondence: dmonack@stanford.edu

Some host-adapted bacterial pathogens are capable of causing persistent infections in humans. For example, *Helicobacter pylori* inhabits the human gastric mucosa and persistence can be lifelong. *Salmonella enterica* serovar Typhi causes systemic infections that involve colonization of the reticuloendothelial system and some individuals become lifelong carriers. In this review, I compare and contrast the different lifestyles of *Helicobacter* and *Salmonella* within the host and the strategies they have evolved to persist in mammalian hosts. Persistently infected carriers serve as the reservoirs for these pathogens, and the carrier state is an essential feature that is required for survival of the bacteria within a restricted host population. Therefore, investigating the chronic carrier state should provide insight into bacterial survival strategies, as well as new therapeutic approaches for treatments.

On infection with a pathogenic microorganism, the host usually responds by activating the innate and adaptive immune responses. If the host survives the initial battle, the adaptive immune response usually clears the foreign invader. However, some pathogens have evolved the capacity to survive the initial robust immune response and persist. The persistent phase of infection usually involves a complex balance of protective immunity and immunopathology. The interactions between the host and pathogen are very complex and likely reflect the coevolution and fine tuning of bacterial virulence mechanisms and host immune responses.

Persistent colonization with the human-specific bacterial pathogens *Helicobacter pylori* and *Salmonella enterica* serovar Typhi (*S.* Typhi) are usually not clinically apparent. However, even in the absence of clinical symptoms, infection poses some risk to the host. For example, *H. pylori* induce gastritis with varying degrees of severity. In addition, individuals who are chronically infected with S. Typhi have an increased risk of developing hepatobiliary cancer. Although a deeper understanding of the mechanisms of pathology and disease caused by chronic *Helicobacter* and *Salmonella* infections is important, this work focuses on the mechanisms of persistent colonization and transmission.

PERSISTENT *H. PYLORI* INFECTIONS

H. pylori infection is an important example of a persistent bacterial pathogen that is usually acquired in early childhood and lasts for a lifetime. The majority of those infected (80%–90%) will

Cite this article as *Cold Spring Harb Perspect Med* doi: 10.1101/cshperspect.a010348

carry and transmit *H. pylori* without any symptoms of disease (Hunt 1996; Amieva and El-Omar 2008). In some ways, *H. pylori* behave like commensal bacteria and not pathogens. However, this bacterium has evolved to successfully colonize the hostile environment of the human stomach in the face of a constant innate and adaptive immune response. Although most infected people do not develop disease, three main gastric phenotypes have been identified. The most common, a "benign gastritis" phenotype, is seen in asymptomatic subjects, who on the whole do not develop serious gastrointestinal disease. The second, a "duodenal ulcer" phenotype, accounts for up to 15% of infected subjects. The third, a "gastric cancer" phenotype, is more serious and affects ~1% of infected subjects as a result of a chronic inflammation induced by the infection and increases the risk of gastric cancer (Hunt 1996; Amieva and El-Omar 2008). Most basic research focuses on the diseases that are caused by *H. pylori*, the host factors for disease, and the bacterial virulence determinants. Here, we focus on the bacterial and host factors that contribute to the ability of this pathogen to persist.

Chemotaxis Is Required for Persistence

H. pylori colonize the harsh environment of the human stomach. To survive their journey through the acidic stomach, *H. pylori* generate large quantities of cytosolic and cell surface-associated urease. Urease is an enzyme that breaks down urea to generate ammonia and carbon dioxide and transiently buffers the acidic environment (Merrell et al. 2003; Pflock et al. 2006). Rather than persisting in the lumen of the stomach, *H. pylori* have evolved mechanisms to reach and colonize a very narrow anatomical niche (25–30 µm) near the surface of the epithelial cells (Schreiber et al. 2004). The ability to reach this niche is dependent on sensing pH gradients and chemotaxis away from low pH (Ottemann and Lowenthal 2002; Schreiber et al. 2004, 2005; Croxen et al. 2006). Mutant *H. pylori* that are defective for sensing low pH and chemotaxis are defective in their ability to colonize the stomachs of mice (Howitt et al. 2011; Lertsethtakarn et al. 2011). For example, CheW, CheA, CheY, and ChePep *H. pylori* mutants are defective for colonizing the mouse stomach when in competition with wild-type *H. pylori* (Terry et al. 2005; Williams et al. 2007; Howitt et al. 2011). Recent results with the ChePep-deficient *H. pylori* mutant in a mouse model suggest that chemotaxis is essential for colonizing the glands of the antrum of stomachs (Fig. 1). Both wild-type and ΔChePep strains colonize the mucus layer overlying the stomach surface. In contrast, the ΔChePep mutant is not found in the glandular region (Howitt et al. 2011). Taken together, these results indicate that chemotaxis is important in allowing the bacteria to colonize a specialized niche and in locating or persisting within the mid-glands. This tropism for the mid-glandular zone is intriguing because the gastric progenitor cells are known to reside in this region and a direct association with *H. pylori* and gastric progenitor cells could be related to the increased gastric cancer associated with *H. pylori* infection (Uemura et al. 2001; Qiao et al. 2007).

Life on the Surface of Gastric Epithelial Cells

Once the bacteria reach the microenvironment of the epithelial lining, they survive as two major populations: one that is free-swimming in the mucus gel and a second population (~20%) found directly adhered to the epithelial surface via multiple adhesins (Hessey et al. 1990; Ilver et al. 1998; Mahdavi et al. 2002). The more virulent strains of *H. pylori* have contact-dependent mechanisms to interact with and modify epithelial cells, including a type IV secretion system that injects the virulence factor CagA into host cells (Segal et al. 1999; Asahi et al. 2000; Backert et al. 2000; Odenbreit et al. 2000; Stein et al. 2000). CagA has multiple effects on epithelial cells, including the ability to modify apical junctions and perturb cell polarity (Amieva et al. 2003; Bagnoli et al. 2005; Murata-Kamiya et al. 2007; Saadat et al. 2007; Zeaiter et al. 2008). CagA affects multiple host cell signaling pathways, which are the subject of many reviews. However, the functions of CagA that benefit the bacteria have only recently been described

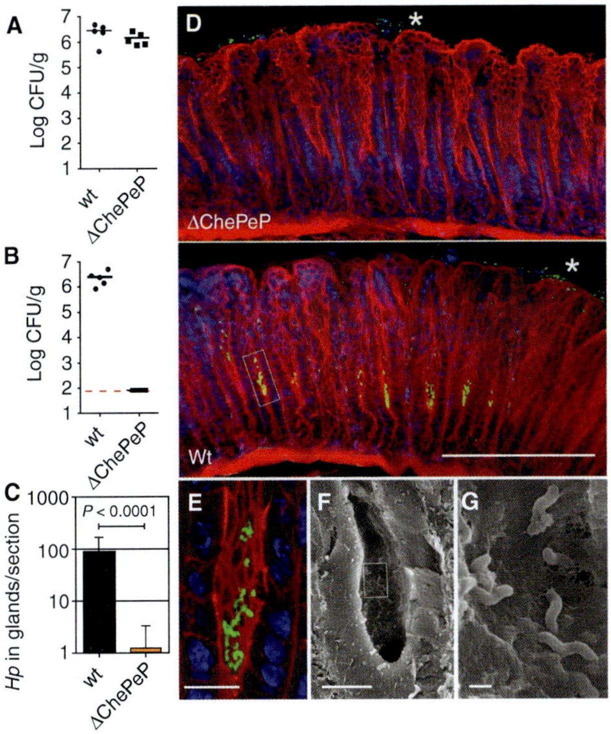

Figure 1. Chemotaxis is essential for *H. pylori* colonization of the antral gastric glands and colonization of the mouse stomach. (*A*) Colony-forming units (CFU) of *H. pylori* in the stomachs of mice colonized with either the wild-type or ΔChePep mutant for 2 weeks. Each marker represents an individual mouse. (*B*) CFU counts from the mice coinfected with both the wild-type and ΔChePep mutant in a 1:1 ratio for 2 weeks. The dashed line indicates the limit of detection. $P < 0.0001$. (*C*) Volumetric analysis of bacteria colonizing the antral glands calculated from 100-µm-thick sections imaged by three-dimensional (3D) confocal microscopy from single infections of either the wild-type or ΔChePep mutant. (*D*) 3D confocal microscopy of murine stomachs infected with either wild-type or ΔChePep mutant. F-actin is stained with phalloidin (red) and nuclei (blue), and *H. pylori* cells (green) are immunolabeled. Asterisks indicate *H. pylori* cells in the surface mucus of the stomach, whereas a box highlights bacterial colonies in mid-glands. (*E*) Magnified view of the area boxed in panel *D*. (*F*) Scanning electron microscopy (SEM) of wild-type infected gland. (*G*) Magnified view of area boxed in panel *F*. Scale bars, 100 µm (*D*), 10 µm (*E*) and (*F*), and 2 µm (*G*). *Hp*, *Helicobacter pylori*; wt, wild type.

(Hatakeyama 2004; Peek 2005; Suzuki et al. 2006; Handa et al. 2007; Hatakeyama 2008; Backert et al. 2010, 2011; Murata-Kamiya 2011). Using a polarized epithelium model system, CagA was found to play an important role in enabling *H. pylori* colonization of the epithelium (Tan et al. 2009). This occurs via a local perturbation of epithelial integrity and enables *H. pylori* to grow as microcolonies adhering to the host cell surface even in conditions that do not support growth of free-swimming bacteria (Fig. 2) (Tan et al. 2009). *H. pylori* alter the internalization, intracellular transport, and polarity of the transferrin/transferrin receptor iron uptake system by a mechanism that depends on CagA and another major virulence factor, VacA (Tan et al. 2011). CagA is important in promoting iron acquisition in vivo. This was shown in a Mongolian gerbil model by comparing wild-type and CagA-deficient strains for their ability to colonize gerbils. Although wild-type and CagA-deficient strains of *H. pylori* colonize iron-replete gerbils to the same extent, the *cagA* deletion mutants are markedly impaired in colonizing iron-deficient gerbils (Tan et al. 2011). These findings suggest that *H. pylori*

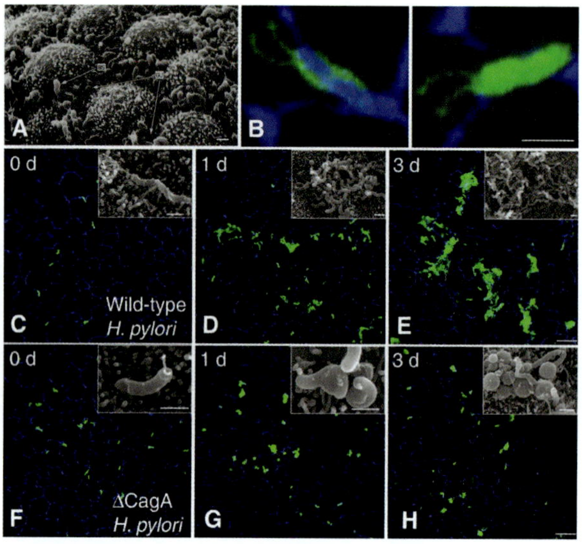

Figure 2. Formation of *H. pylori* microcolonies on the surface of epithelial cells requires CagA. (*A*) SEM of the gastric epithelial cell surface from a prepyloric biopsy from a patient with duodenal ulceration. The dome-shaped surfaces are gastric epithelial cells. Bacteria are located in the junctions or gutters between host cells (Steer 1984). (*B–D*) 3D confocal images of *H. pylori* colonizing the cell surface of polarized MDCK cells in the Transwell system described by Tan et al. (2009). View from below the junctions (*left*) and above the junctions (*right*). Bacteria are visualized with anti-*H. pylori* antibodies (green) and cell junctions are stained blue (anti-ZO-1). (*C*) Wild-type *H. pylori* is tightly adhering to junctions between epithelial cells. Scale bar, 1 μm. (*C–E*) Low-magnification 3D confocal images of wild-type *H. pylori* forming microcolonies on epithelial cells. Scale bar, 10 μm. *Insets* show replicating wild type as visualized by SEM. Scale bar, 1 μm. (*F–H*) Low-magnification 3D confocal images of *ΔcagA H. pylori* colonizing the surface of epithelial cells. Scale bar, 10 μm. *Insets* show CagA mutant colony morphology as visualized by SEM. Scale bar, 1 μm.

have evolved multiple mechanisms to manipulate the epithelium to acquire micronutrients (e.g., iron) from host cells and to grow on the cell surface.

Immune Evasion Strategies

The immunomodulatory effects of VacA and CagA and some of the *H. pylori* adhesins have been reviewed elsewhere (Monack et al. 2004b; Backert and Clyne 2011; Backert et al. 2011; Murata-Kamiya 2011; Tegtmeyer et al. 2011). In addition, a number of *H. pylori* factors have evolved to reduce recognition by the innate immune system. For example, *H. pylori* make flagellin molecules that are not recognized by TLR5 (Gewirtz et al. 2004). The mutations in *H. pylori* flagellin lead to evasion map to the TLR5 recognition domain, amino acids 89–96 of the amino-terminal D1 domain of flagellin (Andersen-Nissen et al. 2005). Mutations in these residues destroy motility in *Salmonella*. However, *H. pylori* and other α and ε Proteobacteria possess compensatory amino acid changes in other regions of the flagellin molecule that then allow them to evade TLR5 recognition and retain motility, a phenotype that is crucial for virulence (Andersen-Nissen et al. 2005).

In addition, *H. pylori* evade signaling through TLR4, which recognizes lipopolysaccharide (LPS). *H. pylori* make an LPS that is 1000 times less pyrogenic and 500-fold less toxic than other Gram-negative enteric bacteria, such as *Salmonella* (Moran 2007). This evasion is thought to be due to underphosphorylation, underacylation, and substitution of long-chain fatty acids in the lipid A moiety of *H. pylori* (Moran 2007). In addition, modifications in the *H. pylori* LPS may act as molecular mimics of human glycans to avoid immune recognition.

Cite this article as *Cold Spring Harb Perspect Med* doi: 10.1101/cshperspect.a010348

H. pylori possess several enzymes that may modify LPS with carbohydrate groups resembling human Lewis blood group antigens (Moran 2008).

Transmission

Although there is a very large population of humans throughout the world that is colonized with *H. pylori*, relatively little is known about how it is transmitted. Epidemiologic studies of *H. pylori* transmission show that the majority of infections tend to occur within families through close person-to-person contact (Falush et al. 2003; Linz et al. 2007). Most transmission occurs in childhood, and maternal-to-child and sibling–sibling transmission seem most likely (McCallion et al. 1996; Bardhan 1997; Kivi et al. 2005; Weyermann et al. 2006). Although viable *H. pylori* have not been isolated from many sources, such as sewage and water sources, *H. pylori* DNA is present, suggesting that they are a source of infective *H. pylori* (Brown 2000). However, it seems more likely that transmission occurs in situations in which gastric content can be transferred quickly from person to person. For example, gastric-oral transmission is suggested in association with gastroenteritis with vomiting (Brown 2000). Fecal–oral transmission may also be possible. Thus far, the contribution of specific *H. pylori* factors in transmission have not been studied or identified.

PERSISTENT *SALMONELLA* INFECTIONS

Salmonella enterica is a pathogenic bacterial species that is an important cause of disease in humans ranging from gastroenteritis to systemic infections (Monack et al. 2004b). Host-adapted *Salmonella* serovars disseminate from the gastrointestinal tract and colonize systemic sites. For example, *S. enterica* serovar Typhi (*S.* Typhi) causes human typhoid fever, whereas *S. enterica* serovar Typhimurium (*S.* Typhimurium) has a broad host range, causing disease in a variety of animals. Strains of *S.* Typhimurium cause a typhoid-like disease in mice and usually cause a self-limiting gastroenteritis in healthy human adults. However, *S.* Typhimurium can cause systemic infections in humans (Kariuki et al. 2006; Gordon et al. 2008; Dhanoa and Fatt 2009; Sigauque et al. 2009; Yen et al. 2009). Indeed, recent cases of invasive and recurrent infections in Malawi, Mozambique, Malaysia, and Taiwan were caused by nontyphoidal Salmonella (NTS), which were largely comprised of multidrug-resistant *S.* Typhimurium strains (Kariuki et al. 2006; Gordon et al. 2008; Dhanoa and Fatt 2009; Sigauque et al. 2009; Yen et al. 2009).

S. Typhi and *S. enterica* serovar Paratyphi (*S.* Paratyphi) are important human pathogens of immense concern to the public health (Parry et al. 2002). They are endemic in regions of the world where drinking water quality and sewage-treatment facilities are poor (Parry et al. 2002; Monack et al. 2004b). Chronically infected humans are the reservoirs for the spread of infection and disease.

Salmonella enters the host through the gastrointestinal tract and translocates by multiple mechanisms to deeper tissues (Kohbata et al. 1986; Jones et al. 1994; Vazquez-Torres et al. 1999). In order for the infection to extend beyond the intestinal mucosa, *Salmonella* must survive and replicate in macrophages, a privileged niche that allows *Salmonella* to elude the adaptive immune response (Jones and Falkow 1996; Cirillo et al. 1998; Deiwick et al. 1998; Hensel et al. 1998; Cheminay et al. 2005; Tobar et al. 2006; Halici et al. 2008; Haraga et al. 2008). A significant percentage (1%–6%) of typhoid patients become chronic carriers of *S.* Typhi, as do many people who have never had a clinical history of typhoid fever (Levine et al. 1982; Monack et al. 2004b). These individuals serve as a reservoir of infections as the bacteria are periodically shed and transmitted to new hosts (Vogelsang and Boe 1948).

What is the immune state of human typhoid carriers? Optimal operation of IL-22- and IFN-γ-dependent immunity and T-cell function appear to be essential for the eradication of *Salmonella* from reservoirs within the reticuloendothelial system (Mastroeni and Menager 2003; Dougan et al. 2011). However, the human carrier state could be associated with an incapacity to develop an effective immune response

(Thompson et al. 2009), although further studies of typhoid patients are needed to confirm this observation. Chronic carriers likely have genetic or other more transient or stochastic differences (e.g., disruption of indigenous intestinal microbiota) that predispose them to an immune state that allows for the establishment of this "equilibrated" state. This response can be efficient enough to survive the acute peak of infection, but too weak to allow eradication of the bacteria from the host.

Mechanisms of *S.* Typhi Evasion of Innate Immunity and Persistence in the Gallbladder

There are several strategies that *S.* Typhi use to evade detection by the host innate immune system (Tsolis et al. 2008). Genome sequencing revealed a unique region in the *S.* Typhi genome-designated *Salmonella* pathogenicity island 7 (SPI-7) (Baker and Dougan 2007). SPI-7 encodes functions for the production and export of the Vi-capsular polysaccharide antigen. The Vi capsule is expressed during infections of humans with *S.* Typhi and appears to be important for pathogenesis in humans and cultured human colonic epithelial cells (Tran et al. 2010). The Vi-antigen plays a role in evading detection of *S.* Typhi by Toll-like receptor 4, perhaps by "masking" detection of lipopolysaccharide (Wilson et al. 2008). In addition, they have shown that TviA, a SPI-7-encoded regulatory protein that controls Vi expression, flagellar motility, and the invasion-associated type 3 secretion system (T3SS) on SPI-1, is essential for the appropriate timing of virulence factor expression in the gastrointestinal tract (Wilson et al. 2008; Winter et al. 2009). Indeed, their findings suggest that TviA-mediated repression of flagellin expression helps *S.* Typhi avoid detection by host TLR5 (Winter et al. 2008).

Chronic infections with *S.* Typhi are classically associated with long-term excretion of bacteria and localization in the gallbladder (Young et al. 2002). Although humans who carry *Salmonella* chronically often have biliary tract disease, this condition is not an absolute requirement for development of the carrier state (Dinbar et al. 1969; Levine et al. 1982; Monack et al. 2004b). There is a strong correlation with the presence of gallstones and conversion to the chronic carrier state in a study conducted in typhoid-endemic Mexico City (Crawford et al. 2010; Gonzalez-Escobedo et al. 2011). The bile, a lipid-rich, detergent-like digestive secretion with antimicrobial properties present in the gallbladder induces the production of an exopolysaccharide matrix O-antigen that facilitates *S.* Typhi biofilm formation on human gallstones (Crawford et al. 2008; Hall-Stoodley and Stoodley 2009). Gallstone biofilms may promote carriage of this bacterial pathogen in the gallbladder and may lead to reseeding of the intestine, fecal shedding, and transmission to a new host. Indeed, mice fed a lithogenic diet developed cholesterol gallstones that supported biofilm formation during persistent *S.* Typhimurium infection and, as a result, showed enhanced fecal shedding and enhanced colonization of the gallbladder and bile (Crawford et al. 2010).

S. TYPHIMURIUM PERSISTENCE IN MICE

To study the basic aspects of host-pathogen interactions during the carrier state, models of long-term chronic *S.* Typhimurium infection in mice have been established. The genetic background of the mouse strain plays a pivotal role in determining *Salmonella* persistence (Hormaeche 1979). Nramp1 is a genetic locus originally identified as being a critical factor in host defense against intracellular pathogens including *Leishmania*, *Mycobacteria*, and *Salmonella* (Vidal et al. 1993). The *Nramp1* gene codes for an ion transporter and is expressed primarily in macrophages and dendritic cells (DC) (Vidal et al. 1993; Govoni et al. 1995). Commonly used mouse strains such as C57BL/6 carry two-point mutations resulting in increased susceptibility to intracellular pathogens. As a result, chronic infection models in C57BL/6 mice require attenuated *Salmonella* strains. Virulent *S.* Typhimurium strains cause a persistent infection in $Nramp1^{wt/wt}$ mice or their F1 hybrids (Monack et al. 2004a; Johanns et al. 2010). Oral inoculation of up to 10^8 CFU results in a persistent infection in which bacteria can be found within macrophages in the mesenteric lymph nodes up

 Cite this article as *Cold Spring Harb Perspect Med* doi: 10.1101/cshperspect.a010348

to 1 year postinoculation (Fig. 3) (Monack et al. 2004a).

S. Typhimurium persist within macrophages in the reticuloendothelial system. The metabolic state of these persisting bacteria is not known. However, results of a recent study conducted in bone marrow-derived macrophages and splenocytes from susceptible mice indicate that many intracellular bacteria do not replicate, but appear to enter a dormant-like state (Helaine et al. 2010). In addition, *S.* Typhimurium survival and replication in hemophagocytic macrophages may help establish persistent infection (Nix et al. 2007).

The *Salmonella* factors that contribute to virulence have been extensively studied and reviewed elsewhere (Haraga et al. 2008; Ibarra and Steele-Mortimer 2009; McGhie et al. 2009; Valdez et al. 2009). However, it is unclear which of these or other bacterial factors are required specifically for persistent *Salmonella* infection. Much effort has been concentrated on the secreted effector proteins that depend on the type III secretion systems encoded by *Salmonella* pathogenicity islands 1 and 2 (SPI1 and SPI2, respectively). Yet, the extent to which these pathogenicity islands actually contribute to persistent infection remains uncertain. Clearly, a variety of *Salmonella* factors that are independent of SPI1 and SPI2 also play significant roles in persistent infection. Therefore, I will discuss both SPI1- and SPI2-dependent and -independent factors that have been reported to play a role during the persistent phase of *Salmonella* infection. Because much of the work investigating persistent *Salmonella* infection has used *S.* Typhimurium, this section of the review will concentrate on the bacterial factors from this serovar.

SPI1 and SPI2

The pathogenicity islands SPI1 and SPI2 encode two distinct type III secretion systems. Each translocates a specific group of bacterial effector proteins into host cells. SPI1-dependent translocation of bacterial invasion factors into host epithelial cells enables *S.* Typhimurium to penetrate the small intestines and Peyer's patches (Galan 2001), accomplishing an initial step of *S.* Typhimurium-dissemination from the gastrointestinal tract to host systemic sites (including the MLN, spleen, liver, and bone marrow). To search for *Salmonella* factors that contribute to later stages of infection (persistent infection), a negative-selection screen was undertaken using a transposon-mutagenized library of *S.* Typhimurium and the *Nramp1*^{r} wild-type mouse strain 129X1/SvJ (Lawley et al. 2006). Although the mice were inoculated intraperitoneally instead of orogastrically (a route that bypasses the significant bottleneck that occurs during dis-

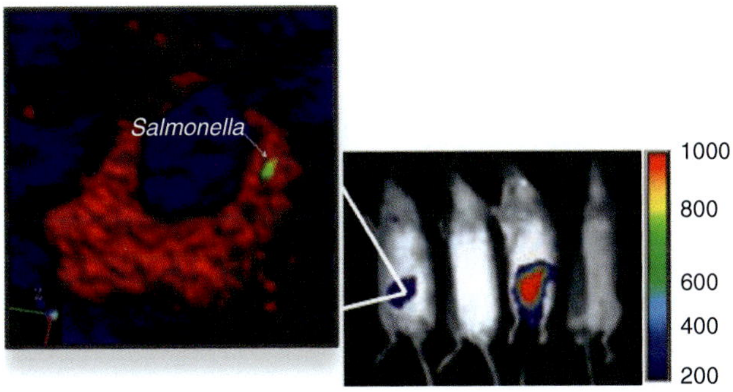

Figure 3. *S.* Typhimurium persist in macrophages within the mesenteric lymph nodes (MLN) of 129Sv mice. 129Sv mice were inoculated by oral administration of the bioluminescent labeled wild-type *S.* Typhimurium strain, SL1344*hha*::Tn5*lux* at a dose of 10^8 CFU and monitored for more than 80 days (Monack et al. 2004a). Light intensity is represented by a color scale in counts. *S.* Typhimurium (labeled with anti-*S.* Typhimurium antibodies in green) persist in the MLN and are found in macrophages (labeled with antimacrophage antibody, MOMA-2 in red). Nuclei are stained with a DNA dye, ToTo3, in blue.

semination from the gastrointestinal tract to systemic sites [Meynell 1957; Meynell and Stocker 1957; Mecsas et al. 2001]), the screen revealed that SPI1 was necessary to sustain a persistent infection for at least 1 month post-infection (Lawley et al. 2006). The SPI1 genes confirmed to be required for persistent infection in this study included the invasion and translocation effectors SipB, SipC, and SipD; however, it is possible that additional SPI1 effectors also could contribute to persistent infection. In light of the fact that *S.* Typhimurium is extruded from dying cells in the epithelia throughout an infection (Knodler et al. 2010), it is likely that the ability of *S.* Typhimurium to continuously reinvade epithelial tissues is necessary to sustain a persistent infection. In support of this notion, SPI1 is also required for a persistently infected mouse to transmit *S.* Typhimurium to naïve cage-mates (Lawley et al. 2008).

S. Typhimurium Modulates Host Cell Migration: A Trait that Mediates Persistent Infection

In addition to identifying SPI-1, this same negative-selection screen in mice identified many SPI-2 genes (Lawley et al. 2006). Although most SPI-2 genes were required for initial colonization of the spleen, the SPI-2 effector SseI did not emerge from the screen until 2 weeks post-infection, indicating that SseI plays a role in long-term carriage. SseI is a secreted effector that is expressed by intracellular *Salmonella* and translocated across the vacuolar membrane into the host cell cytosol via the SPI-2-encoded T3SS (Miao and Miller 2000). In a subsequent study, it was shown that SseI is required for maintaining a long-term systemic infection in mice by modulating normal cell migration of primary macrophages and DC (McLaughlin et al. 2009). The ability to inhibit migration requires the host factor IQ-motif-containing GTPase-activating protein 1 (IQGAP1), an important regulator of host cell migration. SseI binds directly to IQGAP1 and colocalizes with this factor at the cell periphery. Furthermore, *S.* Typhimurium inhibits DC, which are potent antigen-presenting cells that are vital to activating T-cell migration toward CCL19. Although the exact mechanism by which SseI modulates migration is still not clear, it seems to play a role in subverting the onset of adaptive immunity as spleens from mice infected with the *sseI* mutant contained more DC and $CD4^+$ T-lymphocytes compared to mice infected with wild-type *S.* Typhimurium (McLaughlin et al. 2009). It should be noted that *S.* Typhi and NTS isolates do not contain *sseI*. In addition, there are *sseI*-independent mechanisms by which *S.* Typhimurium modulates DC migration in vivo (McLaughlin et al. 2009). Future studies are needed to identify additional novel "migration modulation" factors in *Salmonella.*

In addition, it is important to note that *sseI* is a pseudogene in *S.* Typhi. The molecular bases for *S.* Typhi's host restriction and unique pathogenic attributes are just beginning to be understood, but they are believed to be the result of a combination of genome degradation and the acquisition of new genetic information (Retamal et al. 2010; Sabbagh et al. 2010; Trombert et al. 2011; Spano and Galan 2012; Song et al. 2013). For example, the proteolytic targeting of a host GTPase, Rab29, by an effector protein, GtgE, that is present in *S.* Typhimurium and not *S.* Typhi distinguishes the intracellular compartments of human-adapted Typhi and the broad-host *S.* Typhimurium (Spano and Galan 2012). Several functional studies indicate that *S.* Typhi pseudogenes contribute to its adaptation to humans (Retamal et al. 2010; Trombert et al. 2011). Perhaps the role of sseI is similar.

The Role of $Foxp3^+$ Regulatory T cells in *Salmonella* Persistence

The pathological damage that results from continued macrophage activation will, at some stage of the infection, outweigh the immediate risk that is posed by the residual, persisting *Salmonella*, and the immune response likely turns itself off, allowing bacterial persistence. Regulatory T cells (Tregs) have been implicated in playing a critical role in sustaining this balance in many infections. The role of Tregs during *Salmonella* persistence was recently investigated in a mouse model of persistence (Johanns et al. 2010). The role of Tregs during chronic infection with *Leishmania major*, viruses, *Mycobacterium tuberculosis*, *Schistosoma mansoni*, and *Plasmodium berghei* has been characterized (Long et al. 2003; Suvas et al. 2003; Hesse et al. 2004; Hisaeda et al. 2004; Suvas et al. 2004). Interestingly, these studies suggest that Treg-mediated immune suppression can provide both detrimental and protective roles in host defense against these pathogens (Uzonna et al. 2001; Suvas et al. 2003; Hesse et al. 2004; Mendez et al. 2004). Johanns et al. show that early after infection with *S.* Typhimurium, when the bacterial burden is progressively increasing, the activation of protective immune components is delayed, and this coincides with increased Treg cell suppressive potency (Johanns et al. 2010). In contrast, later during infection, when reductions in bacterial burden occur, protective immune components are highly activated and Treg cell suppressive potency is reduced (Johanns et al. 2010). They also found that when Tregs were ablated early after infection, when their suppressive potency is increased, there was an accelerated bacterial eradication. This study suggests that Tregs control the early stages of setting up a persistent *Salmonella* infection.

Salmonella Factors Required for Intestinal Persistence

Sustained colonization of the gastrointestinal tract is another important aspect of persistent *Salmonella* infection because it is necessary for live bacteria to be passed in the feces and eventually transmitted to other potential hosts (Lawley et al. 2008). Using the CBA mouse strain (*Nramp*^{r}, *S.* Typhimurium-resistant), several *Salmonella* adherence factors were shown to be important for intestinal persistence of *S.* Typhimurium. The fibronectin-binding factors, ShdA and MisL, contribute significantly to persistence in the gastrointestinal tract and fecal shedding of live bacteria (Kingsley et al. 2000, 2002a,b, 2004; Dorsey et al. 2005). In addition, several fimbrial operons (encoding outer surface structures with possible adhesive functions), including *lpf*, *bcf*, *stb*, *stc*, *std*, and *sth*, are also required for long-term intestinal carriage and fecal shedding (Weening et al. 2005). In the *Salmonella* persistence screen described above, several genes from these operons were negatively selected, including *stcC*, *stcD*, *sthA*, *bcfD*, *stbD*, and *stdA*, indicating a link between intestinal persistence and systemic persistence (Lawley et al. 2006). For example, the antivirulence modulator ZirTS is expressed only by *S.* Typhimurium colonizing the gastrointestinal tract (as well as in the feces of persistently infected mice), and such expression negatively impacts systemic colonization in *Nramp1*^{r} mice (Gal-Mor et al. 2008). How ZirTS expression affects systemic persistence requires further investigation.

S. Typhimurium factors that protect against host-derived antimicrobial peptides also contribute to persistent *Salmonella* infection. For example, it was recently shown that several *Salmonella* factors are required for *S.* Typhimurium carriage in the intestines of *Caenorhabditis elegans*, including SPI-2, pSLT (virulence plasmid), PhoPQ (a sensor/kinase system that senses antimicrobial peptides (Bader et al. 2005), acidic pH (Bearson et al. 1998), and changes in metal cation concentration (Groisman et al. 1997; Alegado and Tan 2008). The PhoPQ regulon encompasses genes that protect against antimicrobial peptides such as polymixin B, C18G, and CRAMP, such as *pgtE* and *mig-14* (Guina et al. 2000; Brodsky et al. 2005; Alegado and Tan 2008). PgtE, a modifier of the bacterial outer surface membrane, was also identified in one of the *Salmonella* persistence screens performed by Lawley et al. (2006) Mig-14, along with several other *Salmonella* factors providing resistance toward antimicrobial peptides (VirK, RcsC, and YdeI) also contribute to persistent *Salmonella* infection (Detweiler et al. 2003; Erickson and Detweiler 2006). These results show that *Salmonella* resistance toward antimicrobial peptides of the host immune response is critical for *Salmonella* to maintain persistent infection in systemic sites.

Transmission

Host-to-host transmission of a pathogen ensures its successful propagation and maintenance within a host population. Although this is an essential stage of a pathogen's life cycle, very little is known at a molecular level about this aspect of infection (Lipsitch and Moxon 1997; Ebert and Bull 2003). Transmission is a complex process involving components of both the pathogen and the host (Fig. 4). For example, *Salmonella* must contend with the indigenous microbiota, the host innate immune system, and the establishment of a replicative niche. *Salmonella* must also facilitate exit and spread to other hosts and finally, successfully colonize and multiply within a new host, thus repeating the cycle.

Understanding the host factors that affect S. Typhimurium colonization of the gut is very important for controlling persistence and transmission. Some host factors have been identified (Mastroeni and Grant 2011). For example, B cells and innate secretory antibodies contribute to control the spread of *Salmonella* beyond the gut (Mittrucker et al. 2000). $IgM^{-/-}$ mice have lower LD_{50} after oral infection and $Pigr^{-/-}$, which are unable to bind and actively transport dimeric IgA and pentameric IgM to mucosae, are more susceptible (Wijburg et al. 2006). In addition, the indigenous microbiota plays an important role in protecting the gut from colonization by pathogens by multiple mechanisms (Stecher and Hardt 2011). On the other hand, *Salmonella* has evolved multiple ways to survive in the gut. For example, S. Typhimurium requires intestinal inflammation to grow in the gut by using its virulence factors to invade the intestinal epithelium and survive in mucosal macrophages (Stecher et al. 2007; Barman et al. 2008; Lawley et al. 2008). In addition, Winter et al. has shown recently that reactive oxygen species generated during inflammation react with endogenous, luminal sulphur compounds (thiosulphate) to form a new respiratory electron acceptor, tetrathionate (Winter et al. 2010). The genes encoded by the *ttrSR ttrBCA* gene cluster confer the ability to use tetrathionate as an electron acceptor, thus providing a growth advantage for S. Typhimurium over the competing microbiota in the lumen of the inflamed gut (Bäumler and Fang 2013). In addition, fimbrial and autotransporter adhesins mediate persistence in the gut (Kingsley et al. 2000, 2002a,b, 2004; Humphries et al. 2001; Wagner and Hensel 2011) and possibly other tissues (Lawley et al. 2006).

Once *Salmonella* has colonized the gut, what are the factors that govern transmission? Using a natural model of persistently infected mice, it was found that a subset of the infected mice (~30%), termed supershedders, shed high levels (>10^8 CFU/g) of S. Typhimurium in their feces and, as a result, rapidly transmit infection (Lawley et al. 2008). Although most of the infected mice show signs of intestinal inflammation, only supershedder mice develop colitis. It was shown that development of the supershedder phenotype depends on the virulence determinants SPI-1 and -2, and it is characterized by mucosal invasion and, importantly, high luminal abundance of S. Typhimurium in the colon. In addition, the treatment of mice with antibiotics that alter the indigenous intestinal microbiota rapidly induces the supershedder phenotype and leads to rapid transmission to naïve hosts. Importantly, a single dose of an antibiotic given to chronically infected mice (125 days postinfection) reactivated the supershedder phenotype in mice that were not shedding detectable levels of S. Typhimurium, demonstrating that the intestinal microbiota has a role in controlling persistent disease (Lawley et al. 2008). Although, this model has established laboratory conditions that ensure that animals exposed to S. Typhimurium become supershedders, it is still not understood what circumstances dictate whether animals with identical genes and exposed to the identical inoculum become persistently infected and shed at low or moderate levels or become supershedders with significant pathology. Future studies are required to determine whether specific conditions lead to supershedders or whether it is a stochastic event. What determines the creation of a murine version of "Typhoid Mary?"

As mentioned previously, a theory to explain typhoid carriage proposes that bacteria originating from the systemic sites seed the intestinal tract through the gallbladder. Perhaps the bacteria that seed the intestinal tract from the gallbladder are "primed" to express crucial metabolic pathways and transmission factors that help the pathogen outcompete the indigenous

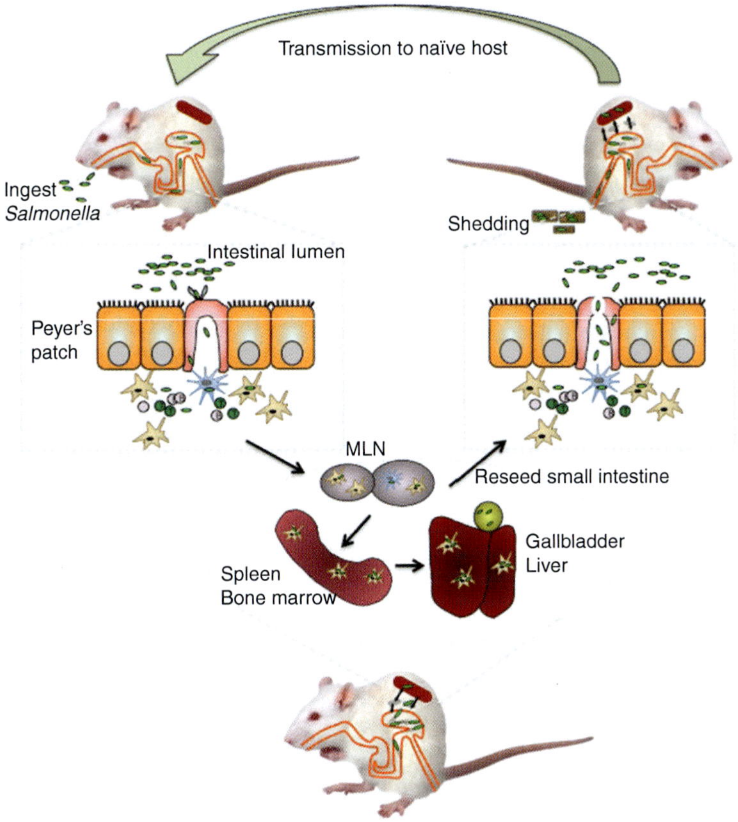

Figure 4. Life cycle of *Salmonella* in mammalian hosts. Bacteria are ingested in contaminated food or water and enter the gastrointestinal tract. *Salmonella* enters the Peyer's patches of the intestinal tract mucosal surface by invading M cells—specialized epithelial cells that take up and transcytose luminal antigens for uptake by phagocytic immune cells. This is followed by inflammation and phagocytosis of bacteria by neutrophils, macrophages, and DC. In systemic salmonellosis, such as typhoid fever, *Salmonella* may target specific types of host cells, such as DC and macrophages that favor dissemination through the lymphatics and bloodstream to the MLNs and deeper tissues. This then leads to transport to the spleen, bone marrow, liver, and gallbladder. Bacteria can persist in the MLNs, bone marrow, and gallbladder for life, and periodic reseeding of the mucosal surface via the bile ducts and/or the lymphatics occur, and shedding can take place from the mucosal surface. Some mice that are infected orally with *S.* Typhimurium develop a supershedder phenotype ($\sim$30%). Supershedders shed high levels ($>10^8$ CFU/g) of *S.* Typhimurium in their feces, and as a result, rapidly transmit infection. Some of the factors that influence the supershedder phenotype include: *Salmonella* virulence factors, gut inflammation, indigenous intestinal microbiota, and antibiotic treatment. Factors that influence efficient transmission could include survival in the environment (Foster 1995; Lee et al. 1995; Lin et al. 1995; Spector 1998; Scher et al. 2005; White et al. 2006, 2008). For example, the *Salmonella* rdar (red, dry, and rough) morphotype, a multicellular phenotype characterized by fimbria- and cellulose-mediated colony pattern formation, has been linked to survival in nutrient-limited environments and may enhance *Salmonella* survival outside the host, thereby aiding in transmission between hosts (White et al. 2006, 2008). In addition, there could be additional transmission-specific factors or conditions that aid in efficient transmission, similar to what has been described for *Vibrio cholera* (Merrell et al. 2002). Finally, factors within the naïve host that influence efficient transmission include survival in the acidic conditions of the stomach (Foster 1995), alterations in the indigenous intestinal microbiota that could be induced by alterations in diet (Sonnenburg et al. 2010), immune state (Dhanoa and Fatt 2009), or recent antibiotic treatment (Riley et al. 1984).

intestinal microbiota even in the absence of antibiotic treatment. Certainly, the finding that antibiotic treatment induces the supershedder phenotype has relevance to understanding how antibiotic therapy is a risk factor for the development of salmonellosis (Riley et al. 1984), gastroenteritis associated with antibiotic-resistant *S.* Typhimurium (Glynn et al. 2004), and increased transmission of antibiotic-resistant *S. enterica* in poultry (Bauer-Garland et al. 2006). These findings indicate that alterations in the intestinal microbiota caused by antibiotic use may induce fecal shedding and transmission of enteric pathogens.

MECHANISMS OF INTESTINAL MICROBIOTA CONTROLLING *SALMONELLA* COLONIZATION

The indigenous intestinal microbiota plays an important role in maintaining health (Sekirov et al. 2010). It provides digestive functions, modulates host metabolism, and stimulates development of lymphatic tissue and the mucosal immune system (Rakoff-Nahoum et al. 2004; Mazmanian et al. 2005; Cash et al. 2006). Importantly, it can efficiently limit infection of the gut by pathogenic microbes (Sonnenburg et al. 2010; Salzman 2011). Some of the microbiota-mediated mechanisms of protection include (1) blocking growth of invading pathogen by a mechanism referred to as colonization resistance, (2) priming the host's innate and adaptive immune defenses, and (3) helping eliminate the pathogen from the gut lumen at the end of infection. The role of colonization resistance during *Salmonella* infection has recently been reviewed and could include direct inhibition, nutrient depletion, and stimulation of immune defenses (Stecher and Hardt 2011). In addition to colonization resistance, it has been shown in a model of *Salmonella* diarrhea that the microbiota mediates *S.* Typhimurium clearance from the gut lumen (Endt et al. 2010). It was very nicely shown that the *Salmonella*-elicited secretory IgA (sIgA) prevents disease when the animal is infected with *S.* Typhimurium for a second time. However, sIgA was dispensable for pathogen clearance from the gut. Instead, clearance was mediated by the microbiota, which was necessary and sufficient for terminating long-term fecal shedding. These results have implications for developing diet- or microbiota-based therapies for curing *Salmonella* infections in humans (Sonnenburg et al. 2006, 2010; Bolam and Sonnenburg 2011).

CONCLUDING REMARKS

Bacterial persistence and host-to-host transmission are key phases of a pathogen's life cycle and represent a window when there can be intervention to reduce and control disease. Although we describe some possible mechanisms of *Helicobacter* and *Salmonella* persistence, we actually know very little about how these pathogens survive for long periods of time in the mammalian host in the presence of immunosurveillance. In addition, our current understanding of disease transmission comes largely from retrospective epidemiological analysis and mathematical modeling of infectious disease transmission within natural populations. Future applications of genome-based techniques, including high throughput sequencing analysis of libraries of bacterial mutants, bacterial and host gene expression profiling, and system-wide microbiome analyses, as well as metabolomics, will allow further investigation of the fundamental genetics and immune responses of bacterial persistence and transmission. We believe that an increased understanding of the molecular mechanisms that control host–pathogen interactions could provide new targets to disable persistence and transmission of pathogens with preventative and/or therapeutic interventions.

ACKNOWLEDGMENTS

I thank Sara Fisher for critical reading of this review. I am very grateful to Dr. Manuel Amieva for providing insight, helpful suggestions, and figures.

REFERENCES

**Reference is also in this collection.*

Alegado RA, Tan MW. 2008. Resistance to antimicrobial peptides contributes to persistence of *Salmonella typhi-*

murium in the *C. elegans* intestine. *Cell Microbiol* **10:** 1259–1273.

Amieva MR, El-Omar EM. 2008. Host-bacterial interactions in *Helicobacter pylori* infection. *Gastroenterology* **134:** 306–323.

Amieva MR, Vogelmann R, Covacci A, Tompkins LS, Nelson WJ, Falkow S. 2003. Disruption of the epithelial apical-junctional complex by *Helicobacter pylori* CagA. *Science* **300:** 1430–1434.

Andersen-Nissen E, Smith KD, Strobe KL, Barrett SL, Cookson BT, Logan SM, Aderem A. 2005. Evasion of Toll-like receptor 5 by flagellated bacteria. *Proc Natl Acad Sci* **102:** 9247–9252.

Asahi M, Azuma T, Ito S, Ito Y, Suto H, Nagai Y, Tsubokawa M, Tohyama Y, Maeda S, Omata M, et al. 2000. *Helicobacter pylori* CagA protein can be tyrosine phosphorylated in gastric epithelial cells. *J Exp Med* **191:** 593–602.

Backert S, Clyne M. 2011. Pathogenesis of *Helicobacter pylori* infection. *Helicobacter* **16:** 19–25.

Backert S, Ziska E, Brinkmann V, Zimny-Arndt U, Fauconnier A, Jungblut PR, Naumann M, Meyer TF. 2000. Translocation of the *Helicobacter pylori* CagA protein in gastric epithelial cells by a type IV secretion apparatus. *Cell Microbiol* **2:** 155–164.

Backert S, Tegtmeyer N, Selbach M. 2010. The versatility of *Helicobacter pylori* CagA effector protein functions: The master key hypothesis. *Helicobacter* **15:** 163–176.

Backert S, Clyne M, Tegtmeyer N. 2011. Molecular mechanisms of gastric epithelial cell adhesion and injection of CagA by *Helicobacter pylori*. *Cell Commun Signal* **9:** 28.

Bader MW, Sanowar S, Daley ME, Schneider AR, Cho U, Xu W, Klevit RE, Le Moual H, Miller SI. 2005. Recognition of antimicrobial peptides by a bacterial sensor kinase. *Cell* **122:** 461–472.

Bagnoli F, Buti L, Tompkins L, Covacci A, Amieva MR. 2005. *Helicobacter pylori* CagA induces a transition from polarized to invasive phenotypes in MDCK cells. *Proc Natl Acad Sci* **102:** 16339–16344.

Baker S, Dougan G. 2007. The genome of *Salmonella enterica* serovar Typhi. *Clin Infect Dis* **45:** S29–S33.

Bardhan PK. 1997. Epidemiological features of *Helicobacter pylori* infection in developing countries. *Clin Infect Dis* **25:** 973–978.

Barman M, Unold D, Shifley K, Amir E, Hung K, Bos N, Salzman N. 2008. Enteric salmonellosis disrupts the microbial ecology of the murine gastrointestinal tract. *Infect Immun* **76:** 907–915.

Bauer-Garland J, Frye JG, Gray JT, Berrang ME, Harrison MA, Fedorka-Cray PJ. 2006. Transmission of *Salmonella enterica* serotype Typhimurium in poultry with and without antimicrobial selective pressure. *J Appl Microbiol* **101:** 1301–1308.

* Bäumler A, Fang FC. 2013. Host specificity of bacterial pathogens. *Cold Spring Harb Perspect Med* doi: 10.1101/cshperspect.a010041.

Bearson BL, Wilson L, Foster JW. 1998. A low pH-inducible, PhoPQ-dependent acid tolerance response protects *Salmonella typhimurium* against inorganic acid stress. *J Bacteriol* **180:** 2409–2417.

Bolam DN, Sonnenburg JL. 2011. Mechanistic insight into polysaccharide use within the intestinal microbiota. *Gut Microbes* **2:** 86–90.

Brodsky IE, Ghori N, Falkow S, Monack D. 2005. Mig-14 is an inner membrane-associated protein that promotes *Salmonella* typhimurium resistance to CRAMP, survival within activated macrophages and persistent infection. *Mol Microbiol* **55:** 954–972.

Brown LM. 2000. *Helicobacter pylori*: Epidemiology and routes of transmission. *Epidemiol Rev* **22:** 283–297.

Cash HL, Whitham CV, Behrendt CL, Hooper LV. 2006. Symbiotic bacteria direct expression of an intestinal bactericidal lectin. *Science* **313:** 1126–1130.

Cheminay C, Mohlenbrink A, Hensel M. 2005. Intracellular *Salmonella* inhibit antigen presentation by dendritic cells. *J Immunol* **174:** 2892–2899.

Cirillo DM, Valdivia RH, Monack DM, Falkow S. 1998. Macrophage-dependent induction of the *Salmonella* pathogenicity island 2 type III secretion system and its role in intracellular survival. *Mol Microbiol* **30:** 175–188.

Crawford RW, Gibson DL, Kay WW, Gunn JS. 2008. Identification of a bile-induced exopolysaccharide required for *Salmonella* biofilm formation on gallstone surfaces. *Infect Immun* **76:** 5341–5349.

Crawford RW, Rosales-Reyes R, Ramirez-Aguilar Mde L, Chapa-Azuela O, Alpuche-Aranda C, Gunn JS. 2010. Gallstones play a significant role in *Salmonella* spp. gallbladder colonization and carriage. *Proc Natl Acad Sci* **107:** 4353–4358.

Croxen MA, Sisson G, Melano R, Hoffman PS. 2006. The *Helicobacter pylori* chemotaxis receptor TlpB (HP0103) is required for pH taxis and for colonization of the gastric mucosa. *J Bacteriol* **188:** 2656–2665.

Deiwick J, Nikolaus T, Shea JE, Gleeson C, Holden DW, Hensel M. 1998. Mutations in *Salmonella* pathogenicity island 2 (SPI2) genes affecting transcription of SPI1 genes and resistance to antimicrobial agents. *J Bacteriol* **180:** 4775–4780.

Detweiler CS, Monack DM, Brodsky IE, Mathew H, Falkow S. 2003. virK, somA and rcsC are important for systemic *Salmonella enterica* serovar Typhimurium infection and cationic peptide resistance. *Mol Microbiol* **48:** 385–400.

Dhanoa A, Fatt QK. 2009. Non-typhoidal *Salmonella* bacteraemia: Epidemiology, clinical characteristics and its association with severe immunosuppression. *Ann Clin Microbiol Antimicrob* **8:** 15.

Dinbar A, Altmann G, Tulcinsky DB. 1969. The treatment of chronic biliary *salmonella* carriers. *Am J Med* **47:** 236–242.

Dorsey CW, Laarakker MC, Humphries AD, Weening EH, Bäumler AJ. 2005. *Salmonella enterica serotype Typhimurium* MisL is an intestinal colonization factor that binds fibronectin. *Mol Microbiol* **57:** 196–211.

Dougan G, John V, Palmer S, Mastroeni P. 2011. Immunity to salmonellosis. *Immunol Rev* **240:** 196–210.

Ebert D, Bull JJ. 2003. Challenging the trade-off model for the evolution of virulence: Is virulence management feasible? *Trends Microbiol* **11:** 15–20.

Endt K, Stecher B, Chaffron S, Slack E, Tchitchek N, Benecke A, Van Maele L, Sirard JC, Mueller AJ, Heikenwalder M, et al. 2010. The microbiota mediates pathogen clearance from the gut lumen after non-typhoidal *Salmonella* diarrhea. *PLoS Pathog* **6:** e1001097.

Erickson KD, Detweiler CS. 2006. The Rcs phosphorelay system is specific to enteric pathogens/commensals and activates ydeI, a gene important for persistent *Salmonella* infection of mice. *Mol Microbiol* **62:** 883–894.

Falush D, Wirth T, Linz B, Pritchard JK, Stephens M, Kidd M, Blaser MJ, Graham DY, Vacher S, Perez-Perez GI, et al. 2003. Traces of human migrations in *Helicobacter pylori* populations. *Science* **299:** 1582–1585.

Foster JW. 1995. Low pH adaptation and the acid tolerance response of *Salmonella typhimurium*. *Crit Rev Microbiol* **21:** 215–237.

Galan JE. 2001. *Salmonella* interactions with host cells: Type III secretion at work. *Annu Rev Cell Dev Biol* **17:** 53–86.

Gal-Mor O, Gibson DL, Baluta D, Vallance BA, Finlay BB. 2008. A novel secretion pathway of *Salmonella enterica* acts as an antivirulence modulator during salmonellosis. *PLoS Pathog* **4:** e1000036.

Gewirtz AT, Yu Y, Krishna US, Israel DA, Lyons SL, Peek RM Jr. 2004. *Helicobacter pylori* flagellin evades toll-like receptor 5-mediated innate immunity. *J Infect Dis* **189:** 1914–1920.

Glynn MK, Reddy V, Hutwagner L, Rabatsky-Ehr T, Shiferaw B, Vugia DJ, Segler S, Bender J, Barrett TJ, Angulo FJ. 2004. Prior antimicrobial agent use increases the risk of sporadic infections with multidrug-resistant *Salmonella enterica* serotype Typhimurium: A FoodNet case-control study, 1996–1997. *Clin Infect Dis* **38:** S227–S236.

Gonzalez-Escobedo G, Marshall JM, Gunn JS. 2011. Chronic and acute infection of the gall bladder by *Salmonella Typhi*: Understanding the carrier state. *Nat Rev Microbiol* **9:** 9–14.

Gordon MA, Graham SM, Walsh AL, Wilson L, Phiri A, Molyneux E, Zijlstra EE, Heyderman RS, Hart CA, Molyneux ME. 2008. Epidemics of invasive *Salmonella enterica* serovar enteritidis and *S. enterica* serovar Typhimurium infection associated with multidrug resistance among adults and children in Malawi. *Clin Infect Dis* **46:** 963–969.

Govoni G, Vidal S, Cellier M, Lepage P, Malo D, Gros P. 1995. Genomic structure, promoter sequence, and induction of expression of the mouse *Nramp1* gene in macrophages. *Genomics* **27:** 9–19.

Groisman EA, Kayser J, Soncini FC. 1997. Regulation of polymyxin resistance and adaptation to low-$Mg2^+$ environments. *J Bacteriol* **179:** 7040–7045.

Guina T, Yi EC, Wang H, Hackett M, Miller SI. 2000. A PhoP-regulated outer membrane protease of *Salmonella enterica* serovar Typhimurium promotes resistance to α-helical antimicrobial peptides. *J Bacteriol* **182:** 4077–4086.

Halici S, Zenk SF, Jantsch J, Hensel M. 2008. Functional analysis of the *Salmonella* pathogenicity island 2-mediated inhibition of antigen presentation in dendritic cells. *Infect Immun* **76:** 4924–4933.

Hall-Stoodley L, Stoodley P. 2009. Evolving concepts in biofilm infections. *Cell Microbiol* **11:** 1034–1043.

Handa O, Naito Y, Yoshikawa T. 2007. CagA protein of *Helicobacter pylori*: A hijacker of gastric epithelial cell signaling. *Biochem Pharmacol* **73:** 1697–1702.

Haraga A, Ohlson MB, Miller SI. 2008. Salmonellae interplay with host cells. *Nat Rev Microbiol* **6:** 53–66.

Hatakeyama M. 2004. Oncogenic mechanisms of the *Helicobacter pylori* CagA protein. *Nat Rev Cancer* **4:** 688–694.

Hatakeyama M. 2008. Linking epithelial polarity and carcinogenesis by multitasking *Helicobacter pylori* virulence factor CagA. *Oncogene* **27:** 7047–7054.

Helaine S, Thompson JA, Watson KG, Liu M, Boyle C, Holden DW. 2010. Dynamics of intracellular bacterial replication at the single cell level. *Proc Natl Acad Sci* **107:** 3746–3751.

Hensel M, Shea JE, Waterman SR, Mundy R, Nikolaus T, Banks G, Vazquez-Torres A, Gleeson C, Fang FC, Holden DW. 1998. Genes encoding putative effector proteins of the type III secretion system of *Salmonella* pathogenicity island 2 are required for bacterial virulence and proliferation in macrophages. *Mol Microbiol* **30:** 163–174.

Hesse M, Piccirillo CA, Belkaid Y, Prufer J, Mentink-Kane M, Leusink M, Cheever AW, Shevach EM, Wynn TA. 2004. The pathogenesis of schistosomiasis is controlled by cooperating IL-10-producing innate effector and regulatory T cells. *J Immunol* **172:** 3157–3166.

Hessey SJ, Spencer J, Wyatt JI, Sobala G, Rathbone BJ, Axon AT, Dixon MF. 1990. Bacterial adhesion and disease activity in *Helicobacter* associated chronic gastritis. *Gut* **31.** 134–138.

Hisaeda H, Maekawa Y, Iwakawa D, Okada H, Himeno K, Kishihara K, Tsukumo S, Yasutomo K. 2004. Escape of malaria parasites from host immunity requires $CD4^+$ $CD25^+$ regulatory T cells. *Nat Med* **10:** 29–30.

Hormaeche CE. 1979. Natural resistance to *Salmonella typhimurium* in different inbred mouse strains. *Immunology* **37:** 311–318.

Howitt MR, Lee JY, Lertsethtakarn P, Vogelmann R, Joubert LM, Ottemann KM, Amieva MR. 2011. ChePep controls *Helicobacter pylori* infection of the gastric glands and chemotaxis in the *Epsilonproteobacteria*. *MBio* **2:** e00098-11.

Humphries AD, Townsend SM, Kingsley RA, Nicholson TL, Tsolis RM, Bäumler AJ. 2001. Role of fimbriae as antigens and intestinal colonization factors of *Salmonella* serovars. *FEMS Microbiol Lett* **201:** 121–125.

Hunt RH. 1996. The role of *Helicobacter pylori* in pathogenesis: The spectrum of clinical outcomes. *Scand J Gastroenterol Suppl* **220:** 3–9.

Ibarra JA, Steele-Mortimer O. 2009. *Salmonella*—The ultimate insider. *Salmonella* virulence factors that modulate intracellular survival. *Cell Microbiol* **11:** 1579–1586.

Ilver D, Arnqvist A, Ogren J, Frick IM, Kersulyte D, Incecik ET, Berg DE, Covacci A, Engstrand L, Boren T. 1998. *Helicobacter pylori* adhesin binding fucosylated histo-blood group antigens revealed by retagging. *Science* **279:** 373–377.

Johanns TM, Ertelt JM, Rowe JH, Way SS. 2010. Regulatory T cell suppressive potency dictates the balance between bacterial proliferation and clearance during persistent *Salmonella* infection. *PLoS Pathog* **6:** e1001043.

Jones BD, Falkow S. 1996. Salmonellosis: Host immune responses and bacterial virulence determinants. *Annu Rev Immunol* **14:** 533–561.

Jones BD, Ghori N, Falkow S. 1994. *Salmonella typhimurium* initiates murine infection by penetrating and destroying the specialized epithelial M cells of the Peyer's patches. *J Exp Med* **180:** 15–23.

Kariuki S, Revathi G, Kariuki N, Kiiru J, Mwituria J, Muyodi J, Githinji JW, Kagendo D, Munyalo A, Hart CA. 2006. Invasive multidrug-resistant non-typhoidal *Salmonella* infections in Africa: Zoonotic or anthroponotic transmission? *J Med Microbiol* **55:** 585–591.

Kingsley RA, van Amsterdam K, Kramer N, Bäumler AJ. 2000. The shdA gene is restricted to serotypes of *Salmonella enterica* subspecies I and contributes to efficient and prolonged fecal shedding. *Infect Immun* **68:** 2720–2727.

Kingsley RA, Santos RL, Keestra AM, Adams LG, Bäumler AJ. 2002a. *Salmonella enterica* serotype Typhimurium ShdA is an outer membrane fibronectin-binding protein that is expressed in the intestine. *Mol Microbiol* **43:** 895–905.

Kingsley RA, Weening EH, Keestra AM, Bäumler AJ. 2002b. Population heterogeneity of *Salmonella enterica* serotype Typhimurium resulting from phase variation of the lpf operon in vitro and in vivo. *J Bacteriol* **184:** 2352–2359.

Kingsley RA, Abi Ghanem D, Puebla-Osorio N, Keestra AM, Berghman L, Bäumler AJ. 2004. Fibronectin binding to the *Salmonella enterica serotype Typhimurium* ShdA autotransporter protein is inhibited by a monoclonal antibody recognizing the A3 repeat. *J Bacteriol* **186:** 4931–4939.

Kivi M, Johansson AL, Reilly M, Tindberg Y. 2005. *Helicobacter pylori* status in family members as risk factors for infection in children. *Epidemiol Infect* **133:** 645–652.

Knodler LA, Vallance BA, Celli J, Winfree S, Hansen B, Montero M, Steele-Mortimer O. 2010. Dissemination of invasive *Salmonella* via bacterial-induced extrusion of mucosal epithelia. *Proc Natl Acad Sci* **107:** 17733–17738.

Kohbata S, Yokoyama H, Yabuuchi E. 1986. Cytopathogenic effect of *Salmonella typhi* GIFU 10007 on M cells of murine ileal Peyer's patches in ligated ileal loops: An ultrastructural study. *Microbiol Immunol* **30:** 1225–1237.

Lawley TD, Chan K, Thompson LJ, Kim CC, Govoni GR, Monack DM. 2006. Genome-wide screen for *Salmonella* genes required for long-term systemic infection of the mouse. *PLoS Pathog* **2:** e11.

Lawley TD, Bouley DM, Hoy YE, Gerke C, Relman DA, Monack DM. 2008. Host transmission of *Salmonella enterica* serovar Typhimurium is controlled by virulence factors and indigenous intestinal microbiota. *Infect Immun* **76:** 403–416.

Lee IS, Lin J, Hall HK, Bearson B, Foster JW. 1995. The stationary-phase sigma factor sigma S (RpoS) is required for a sustained acid tolerance response in virulent *Salmonella typhimurium*. *Mol Microbiol* **17:** 155–167.

Lertsethtakarn P, Ottemann KM, Hendrixson DR. 2011. Motility and chemotaxis in *Campylobacter* and *Helicobacter*. *Annu Rev Microbiol* **65:** 389–410.

Levine MM, Black RE, Lanata C. 1982. Precise estimation of the numbers of chronic carriers of *Salmonella typhi* in Santiago, Chile, an endemic area. *J Infect Dis* **146:** 724–726.

Lin J, Lee IS, Frey J, Slonczewski JL, Foster JW. 1995. Comparative analysis of extreme acid survival in *Salmonella typhimurium*, *Shigella flexneri*, and *Escherichia coli*. *J Bacteriol* **177:** 4097–4104.

Linz B, Balloux F, Moodley Y, Manica A, Liu H, Roumagnac P, Falush D, Stamer C, Prugnolle F, van der Merwe SW, et al. 2007. An African origin for the intimate association between humans and *Helicobacter pylori*. *Nature* **445:** 915–918.

Lipsitch M, Moxon ER. 1997. Virulence and transmissibility of pathogens: What is the relationship? *Trends Microbiol* **5:** 31–37.

Long TT, Nakazawa S, Onizuka S, Huaman MC, Kanbara H. 2003. Influence of $CD4^+CD25^+$ T cells on Plasmodium berghei NK65 infection in BALB/c mice. *Int J Parasitol* **33:** 175–183.

Mahdavi J, Sonden B, Hurtig M, Olfat FO, Forsberg L, Roche N, Angstrom J, Larsson T, Teneberg S, Karlsson KA, et al. 2002. *Helicobacter pylori* SabA adhesin in persistent infection and chronic inflammation. *Science* **297:** 573–578.

Mastroeni P, Grant AJ. 2011. Spread of *Salmonella enterica* in the body during systemic infection: Unravelling host and pathogen determinants. *Expert Rev Mol Med* **13:** e12.

Mastroeni P, Menager N. 2003. Development of acquired immunity to *Salmonella*. *J Med Microbiol* **52:** 453–459.

Mazmanian SK, Liu CH, Tzianabos AO, Kasper DL. 2005. An immunomodulatory molecule of symbiotic bacteria directs maturation of the host immune system. *Cell* **122:** 107–118.

McCallion WA, Murray LJ, Bailie AG, Dalzell AM, O'Reilly DP, Bamford KB. 1996. *Helicobacter pylori* infection in children: Relation with current household living conditions. *Gut* **39:** 18–21.

McGhie EJ, Brawn LC, Hume PJ, Humphreys D, Koronakis V. 2009. Salmonella takes control: Effector-driven manipulation of the host. *Curr Opin Microbiol* **12:** 117–124.

McLaughlin LM, Govoni GR, Gerke C, Gopinath S, Peng K, Laidlaw G, Chien YH, Jeong HW, Li Z, Brown MD, et al. 2009. The *Salmonella* SPI2 effector SseI mediates long-term systemic infection by modulating host cell migration. *PLoS Pathog* **5:** e1000671.

Mecsas J, Bilis I, Falkow S. 2001. Identification of attenuated Yersinia pseudotuberculosis strains and characterization of an orogastric infection in BALB/c mice on day 5 postinfection by signature-tagged mutagenesis. *Infect Immun* **69:** 2779–2787.

Mendez S, Reckling SK, Piccirillo CA, Sacks D, Belkaid Y. 2004. Role for $CD4^+$ $CD25^+$ regulatory T cells in reactivation of persistent leishmaniasis and control of concomitant immunity. *J Exp Med* **200:** 201–210.

Merrell DS, Butler SM, Qadri F, Dolganov NA, Alam A, Cohen MB, Calderwood SB, Schoolnik GK, Camilli A. 2002. Host-induced epidemic spread of the cholera bacterium. *Nature* **417:** 642–645.

Merrell DS, Goodrich ML, Otto G, Tompkins LS, Falkow S. 2003. pH-regulated gene expression of the gastric pathogen *Helicobacter pylori*. *Infect Immun* **71:** 3529–3539.

Meynell GG. 1957. The applicability of the hypothesis of independent action to fatal infections in mice given *Salmonella typhimurium* by mouth. *J Gen Microbiol* **16:** 396–404.

Meynell GG, Stocker BA. 1957. Some hypotheses on the aetiology of fatal infections in partially resistant hosts and their application to mice challenged with *Salmonella paratyphi-B* or *Salmonella typhimurium* by intraperitoneal injection. *J Gen Microbiol* **16:** 38–58.

Miao EA, Miller SI. 2000. A conserved amino acid sequence directing intracellular type III secretion by *Salmonella typhimurium*. *Proc Natl Acad Sci* **97:** 7539–7544.

Mittrucker HW, Raupach B, Kohler A, Kaufmann SH. 2000. Cutting edge: Role of B lymphocytes in protective immunity against *Salmonella typhimurium* infection. *J Immunol* **164:** 1648–1652.

Monack DM, Bouley DM, Falkow S. 2004a. *Salmonella typhimurium* persists within macrophages in the mesenteric lymph nodes of chronically infected *Nramp1*$^{+/+}$ mice and can be reactivated by IFN-γ neutralization. *J Exp Med* **199:** 231–241.

Monack DM, Mueller A, Falkow S. 2004b. Persistent bacterial infections: The interface of the pathogen and the host immune system. *Nat Rev Microbiol* **2:** 747–765.

Moran AP. 2007. Lipopolysaccharide in bacterial chronic infection: Insights from *Helicobacter pylori* lipopolysaccharide and lipid A. *Int J Med Microbiol* **297:** 307–319.

Moran AP. 2008. Relevance of fucosylation and Lewis antigen expression in the bacterial gastroduodenal pathogen *Helicobacter pylori*. *Carbohydr Res* **343:** 1952–1965.

Murata-Kamiya N. 2011. Pathophysiological functions of the CagA oncoprotein during infection by *Helicobacter pylori*. *Microbes Infect* **13:** 799–807.

Murata-Kamiya N, Kurashima Y, Teishikata Y, Yamahashi Y, Saito Y, Higashi H, Aburatani H, Akiyama T, Peek RM Jr, Azuma T, et al. 2007. *Helicobacter pylori* CagA interacts with E-cadherin and deregulates the β-catenin signal that promotes intestinal transdifferentiation in gastric epithelial cells. *Oncogene* **26:** 4617–4626.

Nix RN, Altschuler SE, Henson PM, Detweiler CS. 2007. Hemophagocytic macrophages harbor *Salmonella enterica* during persistent infection. *PLoS Pathog* **3:** e193.

Odenbreit S, Puls J, Sedlmaier B, Gerland E, Fischer W, Haas R. 2000. Translocation of *Helicobacter pylori* CagA into gastric epithelial cells by type IV secretion. *Science* **287:** 1497–1500.

Ottemann KM, Lowenthal AC. 2002. *Helicobacter pylori* uses motility for initial colonization and to attain robust infection. *Infect Immun* **70:** 1984–1990.

Parry CM, Hien TT, Dougan G, White NJ, Farrar JJ. 2002. Typhoid fever. *N Engl J Med* **347:** 1770–1782.

Peek RM Jr. 2005. Orchestration of aberrant epithelial signaling by *Helicobacter pylori* CagA. *Sci STKE* **2005:** pe14.

Pflock M, Kennard S, Finsterer N, Beier D. 2006. Acid-responsive gene regulation in the human pathogen *Helicobacter pylori*. *J Biotechnol* **126:** 52–60.

Qiao XT, Ziel JW, McKimpson W, Madison BB, Todisco A, Merchant JL, Samuelson LC, Gumucio DL. 2007. Prospective identification of a multilineage progenitor in murine stomach epithelium. *Gastroenterology* **133:** 1989–1998.

Rakoff-Nahoum S, Paglino J, Eslami-Varzaneh F, Edberg S, Medzhitov R. 2004. Recognition of commensal microflora by toll-like receptors is required for intestinal homeostasis. *Cell* **118:** 229–241.

Retamal P, Castillo-Ruiz M, Villagra NA, Morgado J, Mora GC. 2010. Modified intracellular-associated phenotypes in a recombinant *Salmonella* Typhi expressing *S.* Typhimurium SPI-3 sequences. *PloS ONE* **5:** e9394.

Riley LW, Cohen ML, Seals JE, Blaser MJ, Birkness KA, Hargrett NT, Martin SM, Feldman RA. 1984. Importance of host factors in human salmonellosis caused by multiresistant strains of *Salmonella*. *J Infect Dis* **149:** 878–883.

Saadat I, Higashi H, Obuse C, Umeda M, Murata-Kamiya N, Saito Y, Lu H, Ohnishi N, Azuma T, Suzuki A, et al. 2007. *Helicobacter pylori* CagA targets PAR1/MARK kinase to disrupt epithelial cell polarity. *Nature* **447:** 330–333.

Sabbagh SC, Forest CG, Lepage C, Leclerc JM, Daigle F. 2010. So similar, yet so different: Uncovering distinctive features in the genomes of *Salmonella enterica* serovars Typhimurium and Typhi. *FEMS Microbiol Lett* **305:** 1–13.

Salzman NH. 2011. Microbiota-immune system interaction: An uneasy alliance. *Curr Opin Microbiol* **14:** 99–105.

Scher K, Romling U, Yaron S. 2005. Effect of heat, acidification, and chlorination on *Salmonella enterica* serovar typhimurium cells in a biofilm formed at the air-liquid interface. *Appl Environ Microbiol* **71:** 1163–1168.

Schreiber S, Konradt M, Groll C, Scheid P, Hanauer G, Werling HO, Josenhans C, Suerbaum S. 2004. The spatial orientation of *Helicobacter pylori* in the gastric mucus. *Proc Natl Acad Sci* **101:** 5024–5029.

Schreiber S, Bucker R, Groll C, Azevedo-Vethacke M, Garten D, Scheid P, Friedrich S, Gatermann S, Josenhans C, Suerbaum S. 2005. Rapid loss of motility of *Helicobacter pylori* in the gastric lumen in vivo. *Infect Immun* **73:** 1584–1589.

Segal ED, Cha J, Lo J, Falkow S, Tompkins LS. 1999. Altered states: Involvement of phosphorylated CagA in the induction of host cellular growth changes by *Helicobacter pylori*. *Proc Natl Acad Sci* **96:** 14559–14564.

Sekirov I, Russell SL, Antunes LC, Finlay BB. 2010. Gut microbiota in health and disease. *Physiol Rev* **90:** 859–904.

Sigauque B, Roca A, Mandomando I, Morais L, Quinto L, Sacarlal J, Macete E, Nhamposa T, Machevo S, Aide P, et al. 2009. Community-acquired bacteremia among children admitted to a rural hospital in Mozambique. *Pediatr Infect Dis J* **28:** 108–113.

Song J, Gao X, Galan JE. 2013. Structure and function of the *Salmonella* Typhi chimaeric A_2B_5 typhoid toxin. *Nature* **499:** 350–354.

Sonnenburg JL, Chen CT, Gordon JI. 2006. Genomic and metabolic studies of the impact of probiotics on a model gut symbiont and host. *PLoS Biol* **4:** e413.

Sonnenburg ED, Zheng H, Joglekar P, Higginbottom SK, Firbank SJ, Bolam DN, Sonnenburg JL. 2010. Specificity of polysaccharide use in intestinal bacteroides species determines diet-induced microbiota alterations. *Cell* **141:** 1241–1252.

Spano S, Galan JE. 2012. A Rab32-dependent pathway contributes to *Salmonella* Typhi host restriction. *Science* **338:** 960–963.

Spector MP. 1998. The starvation-stress response (SSR) of *Salmonella*. *Adv Microb Physiol* **40:** 233–279.

Stecher B, Hardt WD. 2011. Mechanisms controlling pathogen colonization of the gut. *Curr Opin Microbiol* **14:** 82–91.

Stecher B, Robbiani R, Walker AW, Westendorf AM, Barthel M, Kremer M, Chaffron S, Macpherson AJ, Buer J, Parkhill J, et al. 2007. *Salmonella enterica* serovar Typhimurium exploits inflammation to compete with the intestinal microbiota. *PLoS Biol* **5:** 2177–2189.

Steer HW. 1984. Surface morphology of the gastroduodenal mucosa in duodenal ulceration. *Gut* **25:** 1203–1210.

Stein M, Rappuoli R, Covacci A. 2000. Tyrosine phosphorylation of the *Helicobacter pylori* CagA antigen after cag-driven host cell translocation. *Proc Natl Acad Sci* **97:** 1263–1268.

Suvas S, Kumaraguru U, Pack CD, Lee S, Rouse BT. 2003. $CD4^+CD25^+$ T cells regulate virus-specific primary and memory $CD8^+$ T cell responses. *J Exp Med* **198:** 889–901.

Suvas S, Azkur AK, Kim BS, Kumaraguru U, Rouse BT. 2004. $CD4^+CD25^+$ regulatory T cells control the severity of viral immunoinflammatory lesions. *J Immunol* **172:** 4123–4132.

Suzuki T, Matsuo K, Sawaki A, Ito H, Hirose K, Wakai K, Sato S, Nakamura T, Yamao K, Ueda R, et al. 2006. Systematic review and meta-analysis: Importance of CagA status for successful eradication of *Helicobacter pylori* infection. *Aliment Pharmacol Ther* **24:** 273–280.

Tan S, Tompkins LS, Amieva MR. 2009. *Helicobacter pylori* usurps cell polarity to turn the cell surface into a replicative niche. *PLoS Pathog* **5:** e1000407.

Tan S, Noto JM, Romero-Gallo J, Peek RM Jr, Amieva MR. 2011. *Helicobacter pylori* perturbs iron trafficking in the epithelium to grow on the cell surface. *PLoS Pathog* **7:** e1002050.

Tegtmeyer N, Wessler S, Backert S. 2011. Role of the cag-pathogenicity island encoded type IV secretion system in *Helicobacter pylori* pathogenesis. *FEBS J* **278:** 1190–1202.

Terry K, Williams SM, Connolly L, Ottemann KM. 2005. Chemotaxis plays multiple roles during *Helicobacter pylori* animal infection. *Infect Immun* **73:** 803–811.

Thompson LJ, Dunstan SJ, Dolecek C, Perkins T, House D, Dougan G, Nguyen TH, Tran TP, Doan CD, Le TP, et al. 2009. Transcriptional response in the peripheral blood of patients infected with *Salmonella enterica* serovar Typhi. *Proc Natl Acad Sci* **106:** 22433–22438.

Tobar JA, Carreno LJ, Bueno SM, Gonzalez PA, Mora JE, Quezada SA, Kalergis AM. 2006. Virulent *Salmonella enterica* serovar Typhimurium evades adaptive immunity by preventing dendritic cells from activating T cells. *Infect Immun* **74:** 6438–6448.

Tran QT, Gomez G, Khare S, Lawhon SD, Raffatellu M, Bäumler AJ, Ajithdoss D, Dhavala S, Adams LG. 2010. The *Salmonella enterica* serotype Typhi Vi capsular antigen is expressed after the bacterium enters the ileal mucosa. *Infect Immun* **78:** 527–535.

Trombert AN, Rodas PI, Mora GC. 2011. Reduced invasion to human epithelial cell lines of *Salmonella enterica* serovar Typhi carrying *S.* Typhimurium sopD2. *FEMS Microbiol Lett* **322:** 150–156.

Tsolis RM, Young GM, Solnick JV, Bäumler AJ. 2008. From bench to bedside: Stealth of enteroinvasive pathogens. *Nat Rev Microbiol* **6:** 883–892.

Uemura N, Okamoto S, Yamamoto S, Matsumura N, Yamaguchi S, Yamakido M, Taniyama K, Sasaki N, Schlemper RJ. 2001. *Helicobacter pylori* infection and the development of gastric cancer. *N Engl J Med* **345:** 784–789.

Uzonna JE, Wei G, Yurkowski D, Bretscher P. 2001. Immune elimination of Leishmania major in mice: Implications for immune memory, vaccination, and reactivation disease. *J Immunol* **167:** 6967–6974.

Valdez Y, Ferreira RB, Finlay BB. 2009. Molecular mechanisms of *Salmonella* virulence and host resistance. *Curr Top Microbiol Immunol* **337:** 93–127.

Vazquez-Torres A, Jones-Carson J, Bäumler AJ, Falkow S, Valdivia R, Brown W, Le M, Berggren R, Parks WT, Fang FC. 1999. Extraintestinal dissemination of *Salmonella* by CD18-expressing phagocytes. *Nature* **401:** 804–808.

Vidal SM, Malo D, Vogan K, Skamene E, Gros P. 1993. Natural resistance to infection with intracellular parasites: Isolation of a candidate for Bcg. *Cell* **73:** 469–485.

Vogelsang TM, Boe J. 1948. Temporary and chronic carriers of *Salmonella* Typhi and *Salmonella* Paratyphi *B. J Hyg (Lond)* **46:** 252–261.

Wagner C, Hensel M. 2011. Adhesive mechanisms of *Salmonella enterica*. *Adv Exp Med Biol* **715:** 17–34.

Weening EH, Barker JD, Laarakker MC, Humphries AD, Tsolis RM, Bäumler AJ. 2005. The *Salmonella enterica* serotype Typhimurium *lpf, bcf, stb, stc, std,* and *sth* fimbrial operons are required for intestinal persistence in mice. *Infect Immun* **73:** 3358–3366.

Weyermann M, Adler G, Brenner H, Rothenbacher D. 2006. The mother as source of *Helicobacter pylori* infection. *Epidemiology* **17:** 332–334.

White AP, Gibson DL, Kim W, Kay WW, Surette MG. 2006. Thin aggregative fimbriae and cellulose enhance long-term survival and persistence of *Salmonella. J Bacteriol* **188:** 3219–3227.

White AP, Gibson DL, Grassl GA, Kay WW, Finlay BB, Vallance BA, Surette MG. 2008. Aggregation via the red, dry, and rough morphotype is not a virulence adaptation in *Salmonella enterica* serovar Typhimurium. *Infect Immun* **76:** 1048–1058.

Wijburg OL, Uren TK, Simpfendorfer K, Johansen FE, Brandtzaeg P, Strugnell RA. 2006. Innate secretory antibodies protect against natural *Salmonella* Typhimurium infection. *J Exp Med* **203:** 21–26.

Williams SM, Chen YT, Andermann TM, Carter JE, McGee DJ, Ottemann KM. 2007. *Helicobacter pylori* chemotaxis modulates inflammation and bacterium-gastric epithelium interactions in infected mice. *Infect Immun* **75:** 3747–3757.

Wilson RP, Raffatellu M, Chessa D, Winter SE, Tukel C, Bäumler AJ. 2008. The Vi-capsule prevents Toll-like receptor 4 recognition of *Salmonella. Cell Microbiol* **10:** 876–890.

Winter SE, Raffatellu M, Wilson RP, Rüssmann H, Bäumler AJ. 2008. The *Salmonella enterica* serotype Typhi regulator TviA reduces interleukin-8 production in intestinal epithelial cells by repressing flagellin secretion. *Cell Microbiol* **10:** 247–261.

Winter SE, Winter MG, Thiennimitr P, Gerriets VA, Nuccio SP, Rüssmann H, Bäumler AJ. 2009. The TviA auxiliary protein renders the *Salmonella enterica* serotype Typhi RcsB regulon responsive to changes in osmolarity. *Mol Microbiol* **74:** 175–193.

Winter SE, Thiennimitr P, Winter MG, Butler BP, Huseby DL, Crawford RW, Russell JM, Bevins CL, Adams LG, Tsolis RM, et al. 2010. Gut inflammation provides a respiratory electron acceptor for *Salmonella. Nature* **467:** 426–429.

Yen YF, Wang FD, Chiou CS, Chen YY, Lin ML, Chen TL, Liu CY. 2009. Prognostic factors and clinical features of non-typhoid *Salmonella* bacteremia in adults. *J Chin Med Assoc* **72:** 408–413.

Young D, Hussell T, Dougan G. 2002. Chronic bacterial infections: Living with unwanted guests. *Nat Immunol* **3:** 1026–1032.

Zeaiter Z, Huynh HQ, Kanyo R, Stein M. 2008. CagA of *Helicobacter pylori* alters the expression and cellular distribution of host proteins involved in cell signaling. *FEMS Microbiol Lett* **288:** 227–234.

Entry of *Listeria monocytogenes* in Mammalian Epithelial Cells: An Updated View

Javier Pizarro-Cerdá[1,2,3], Andreas Kühbacher[1,2,3], and Pascale Cossart[1,2,3]

[1]Institut Pasteur, Unité des Interactions Bactéries Cellules, Paris F-75015, France

[2]INSERM, U604, Paris F-75015, France

[3]INRA, USC2020, Paris F-75015, France

Correspondence: javier.pizarro-cerda@pasteur.fr; pascale.cossart@pasteur.fr

Listeria monocytogenes is a bacterial pathogen that promotes its internalization into host epithelial cells. Interaction between the bacterial surface molecules InlA and InlB and their cellular receptors E-cadherin and Met, respectively, triggers the recruitment of endocytic effectors, the subversion of the phosphoinositide metabolism, and the remodeling of the actin cytoskeleton that lead to bacterial engulfment. Additional bacterial surface and secreted virulence factors also contribute to entry, albeit to a lesser extent. Here we review the increasing number of signaling effectors that are reported as being subverted by *L. monocytogenes* during invasion of cultured cell lines. We also update the current knowledge of the early steps of in vivo cellular infection, which, as shown recently, challenges previous concepts generated from in vitro data.

Listeria monocytogenes is a saprophytic gram-positive bacterium responsible for human listeriosis, a food-borne disease characterized in immunocompromised individuals by severe septicemia and meningoencephalitis; in pregnant women, placental infection can lead to meningoencephalitis of the newborn and abortion (Allerberger and Wagner 2010). The potential of *L. monocytogenes* to cause disease correlates with its capacity to survive within macrophages, to invade nonphagocytic cells and replicate therein (Fig. 1), and also to cross the intestinal, the blood–brain, and the fetoplacental barriers (Lecuit 2005; Cossart 2011). The proteins involved in cellular invasion and tissue tropism are encoded by the *inlAB* gene locus (Gaillard et al. 1991). The factors required for intracellular survival are encoded by the major virulence locus of *L. monocytogenes*. They include the secreted listeriolysin O (LLO) and two phospholipases involved in the disruption of phagosomal membranes and bacterial escape to the cytoplasm (Gaillard et al. 1987; Mengaud et al. 1987; Vázquez-Boland et al. 1992), the surface protein ActA that mediates the polymerization of cytoplasmic actin (Domann et al. 1992; Kocks et al. 1992) and favors cell-to-cell spread (Tilney and Portnoy 1989) as well as PrfA, the transcriptional activator of bacterial virulence genes (Leimeister-Wächter et al. 1990; Mengaud et al. 1991). Comparative genomic approaches on the pathogenic *L. monocytogenes* and the nonpathogenic

Cite this article as *Cold Spring Harb Perspect Med* doi: 10.1101/cshperspect.a010009

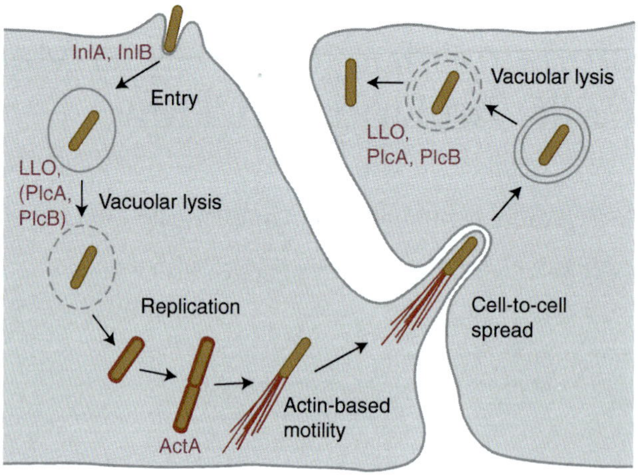

Figure 1. Intracellular cell cycle of *L. monocytogenes. L. monocytogenes* binds to epithelial host cells and promotes its own uptake in a process mediated by the two bacterial surface proteins InlA and InlB. The secreted pore-forming toxin LLO (together with the bacterial phospholipases PlcA and PlcB, depending on the cell type) promotes vacuolar rupture and bacterial escape to the cytoplasm, where *L. monocytogenes* can replicate efficiently. Surface expression of ActA allows intracellular bacteria to polymerize host cell actin and to generate actin comet tails that propel *L. monocytogenes* through the cytoplasm and through membrane protrusions into neighboring cells. There, bacteria localize in a double membrane vacuole, which can be lysed by LLO, PlcA, and PlcB to start a new infection cycle.

Listeria innocua has led in recent years to the identification of many additional bacterial factors required for infection (Camejo et al. 2011; Cossart 2011). In this article, we will summarize our current understanding of bacterial effectors and cellular signaling cascades that are subverted by *L. monocytogenes* in order to invade mammalian epithelial cells.

InlA AND InlB: TWO MAJOR VIRULENCE FACTORS INVOLVED IN CELL INVASION

The two major invasion proteins of *L. monocytogenes*, InlA and InlB, belong to the so-called internalin family. Indeed, *L. monocytogenes* genome encodes 27 proteins now known as internalins, based on the presence of tandemly arranged leucine-rich repeats (LRRs) in their amino-terminal domain, which in several cases have been shown to be involved in protein–protein interactions with host-cell ligands (Marino et al. 2000; Cabanes et al. 2002; Bierne et al. 2007). Internalins are classified in three different groups according to their association to the *L. monocytogenes* surface through their carboxy-terminal domain: 21 internalins present LPXTG motifs that direct covalent protein anchoring to the cell wall peptidoglycan; this is the case for InlA, the prototype member of the family responsible for internalization—hence the name internalin—within a specific subset of host epithelial cells (Gaillard et al. 1991). InlB has been implicated in vitro in the invasion of a broad range of host cells. It is the only internalin characterized by repeats starting with the dipeptide GW (known as GW motifs) that promote loose interactions with the cell wall lipoteichoic acid (Dramsi et al. 1995; Braun et al. 1997; Jonquières et al. 1999). Four internalins lack anchoring motifs and are secreted: Among them, InlC interferes with NF-κB signaling and also perturbs cell–cell contacts intracellularly, promoting cell-to-cell spread (Rajabian et al. 2009; Gouin et al. 2010). So far, InlA and InlB are the only internalins that have been directly implicated in *L. monocytogenes* internalization within host cells; here we detail the signaling pathways activated by InlA and InlB (Fig. 2) as well as how other bacterial proteins potentiate invasion.

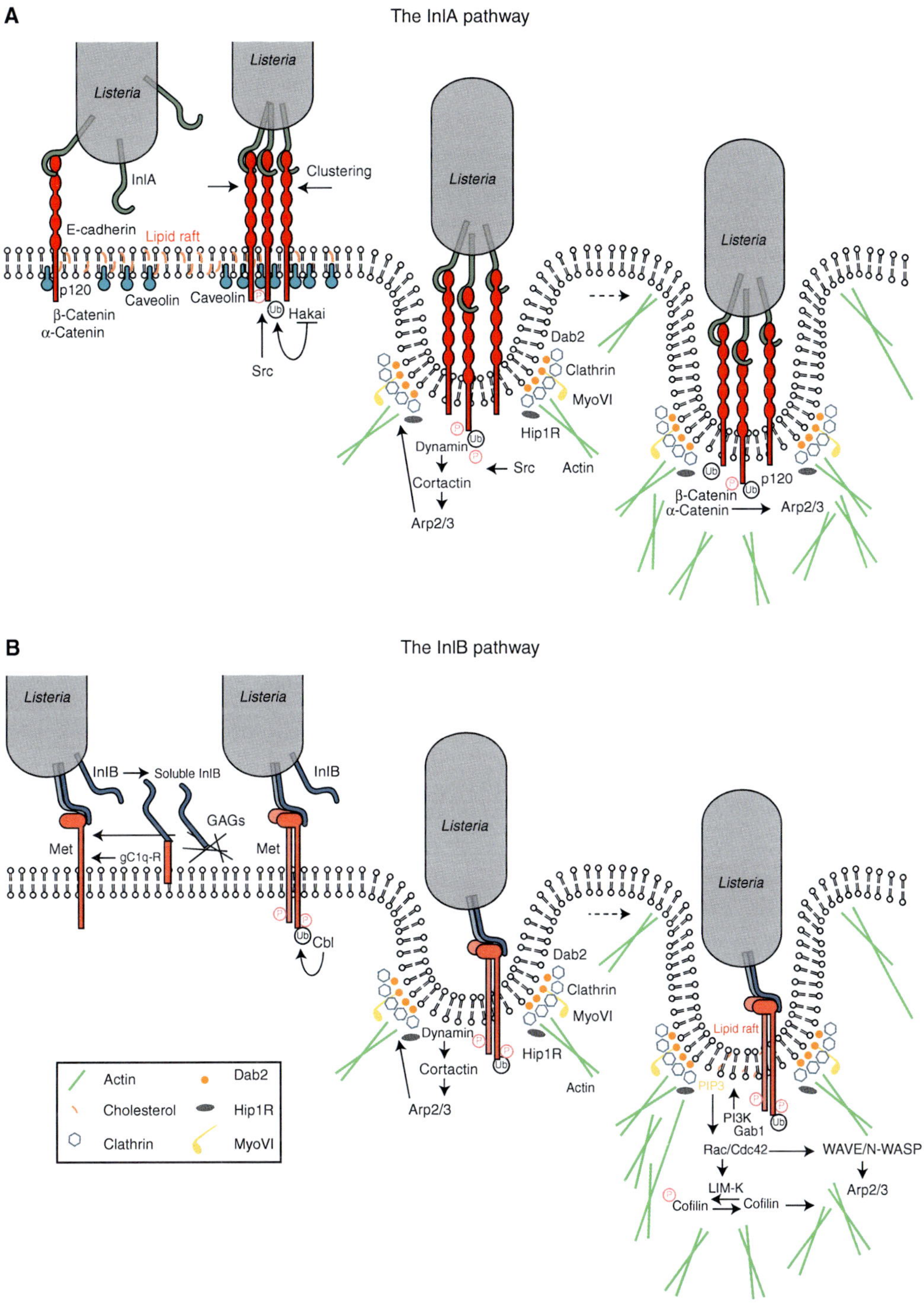

Figure 2. Signaling cascades activated via the InlA- and InlB-invasion pathways. Interaction of InlA and/or InlB with their respective host-cell surface receptors E-cadherin and Met induces ubiquitination of the receptors by the ubiquitin ligases Hakai in the case of E-cadherin or Cbl in the case of Met and subsequent recruitment of the clathrin endocytosis machinery (Dab2, clathrin, dynamin, Hip1R, MyoVI), which provides an initial platform for actin cytoskeleton polymerization. (*See following page for legend.*)

InlA Interacts with the Adhesion Molecule E-Cadherin to Promote Invasion in Specific Cellular Subpopulations

E-cadherin is an adhesion molecule involved in the formation of adherens junctions at the intestinal barrier, the blood–brain barrier, and the placenta, which has been identified as the cellular receptor for InlA (Mengaud et al. 1996). In mammalian tissues, polarized epithelial cells are characterized by the presence of several adhesion structures including tight and adherens junctions, which confer to a given organ impermeability to the external environment. Homophilic interactions between the extracellular domains of E-cadherin molecules in adjacent epithelial cells are required for the initial formation of adherens junctions; these structures are further stabilized by dynamic interactions of the E-cadherin cytoplasmic domain with elements of the clathrin-mediated endocytosis machinery and the actin cytoskeleton. As discussed below, *L. monocytogenes* subverts these cellular mechanisms to promote bacterial internalization. Interaction between the InlA LRRs and E-cadherin is sufficient to promote bacterial adhesion and entry into host cells: Indeed, LRRs expression on the surface of the noninvasive species *L. innocua* or on latex beads allows particle internalization in Caco-2 cells (Lecuit et al. 1997).

Remarkably, the InlA/E-cadherin interaction is species specific: A proline at position 16 of human or guinea pig E-cadherin is critical for its interaction with InlA, whereas a glutamic acid at the same position in mouse or rat E-cadherin precludes InlA recognition. Elegant studies have shown that expression of a murine E-cadherin displaying a proline instead of a glutamic acid at position 16 allows recognition of this molecule by InlA in vitro and in vivo (Lecuit et al. 1999, 2001). The crystal structure of the InlA LRRs in association with the first extracellular domain of the human E-cadherin has illustrated the molecular basis of this specificity: The proline 16 is accommodated in a hydrophobic pocket created by a deletion of one amino acid in the InlA LRR6, which shortens this repeat to 21 residues (instead of 22 observed in all other repeats); a bulky glutamic acid at this position does not fit the InlA LRRs hydrophobic pocket and prevents E-cadherin binding (Schubert et al. 2002). Owing to the specific distribution of E-cadherin in certain tissues, InlA plays a critical role in vivo in the bacterial traversal of the intestinal barrier and the tropism for the placenta (see below). Moreover, owing to its species-specific interaction with E-cadherin, InlA appears as the most important effector determining the susceptibility of a host to a *L. monocytogenes* infection.

InlB Interacts with the Hepatocyte Growth Factor Receptor Met to Promote Internalization in a Wide Variety of Mammalian Cells

InlB, which is encoded by the second gene of the *inlAB* locus (Gaillard et al. 1991), allows *L. monocytogenes* internalization in a broad range of cell lines including HeLa, HEp-2, HepG2, TIB73, and Vero cells (Dramsi et al. 1995; Lingnau et al. 1995; Braun et al. 1998, 1999; Greiffenberg et al. 1998). Several host-cell molecules have been identified as potential receptors for InlB; among them, the hepatocyte growth factor (HGF) receptor Met plays the most critical role for bacterial internalization (Shen et al. 2000). Met is a ubiquitous receptor tyrosine kinase that controls cell migration and growth during embryogenesis, as well as invasion and metastasis in cancer cells (Trusolino and Comoglio 2002). The mature protein is a disulphide-linked heterodimer formed by a short extracellular α chain

Figure 2. (*Continued*) Downstream from E-cadherin, this first actin polymerization wave is activated by Src and cortactin, which promote recruitment of the Arp2/3 complex; association of β/α-catenins to the bacterial entry sites favors dynamic interactions between the E-cadherin cytoplasmic tail and the actin cytoskeleton. In the case of Met, actin polymerization can be first coordinated by dynamin and cortactin upstream of the Arp2/3 complex, and subsequently by a signaling cascade downstream from the type IA PI 3-kinase, which involves the small GTPases Rac1 and Cdc42, abi1, WAVE, and N-WASP (depending on the cell type); LIM-K and cofilin play a critical role in the depolymerization of actin to allow completion of the bacterial internalization process.

and a longer β chain that comprises an ectodomain, a transmembrane domain, and a cytoplasmic tail where a kinase domain and phosphorylatable motifs are located (Birchmeier et al. 2003). InlB functionally mimics HGF, the natural Met ligand, through the binding of its LRRs (Braun et al. 1999; Marino et al. 1999; Niemann et al. 2007). As in the case of the InlA/E-cadherin interaction, InlB interaction with Met is species specific: InlB interacts with human and mouse Met, but does not recognize the guinea-pig or rabbit receptor (Khelef et al. 2006). In susceptible species, InlB is reported to dimerize on binding to Met (Ferraris et al. 2010) leading to the receptor dimerization, the autophosphorylation of the protein-docking motifs, and the recruitment of protein effectors that induce signals to the actin cytoskeleton and the endocytosis machinery (see below). CD44v6, a member of the CD44 family of type I transmembrane glycoproteins, has been reported to potentiate Met activation by HGF and also by InlB (Orian-Rousseau et al. 2002; Ponta et al. 2003) but conflicting results exist concerning its implication during *L. monocytogenes* invasion (Jung et al. 2009; Dortet et al. 2010). The "B repeat" of InlB, often referred to as a spacer domain between the LRRs and GW modules, has also been reported to participate in the activation of Met and to endow InlB with stronger biological activities than those elicited by HGF (i.e., activation of a Ras-MAP kinase pathway) (Copp et al. 2003). However, structural studies indicate that it does not interact with Met, leading to the proposal that the B repeat may interact with a nonidentified cellular receptor (Ebbes et al. 2011).

Although the amino-terminal LLR domain of InlB is critical for activating Met, the three carboxy-terminal GW modules are required to stabilize the InlB/Met interaction. As mentioned previously, the GW modules were originally described as mediating loose interactions between InlB and the lipoteichoic acids of the *L. monocytogenes* cell wall (Braun et al. 1997; Jonquières et al. 1999) and interestingly, soluble heparin can detach InlB from the bacterial cell wall (Jonquières et al. 2001). Heparin and glycosaminoglycans (GAGs), which are highly charged extracellular matrix polysaccharides, favor the surface clustering of growth factors including HGF, potentiating their activity by stabilizing their interaction with their receptors (Zhang 2010). Purified InlB was shown to bind to purified GAGs through its GW domains and invasion of Chinese hamster ovary (CHO) cells is potentiated 10-fold when GAGs are produced at the cell surface (Jonquières et al. 2001). These results suggest that the GW modules of InlB present the dual function of attaching InlB to the bacterial cell wall and of enhancing *L. monocytogenes* entry triggered by the LRRs domain, reinforcing the similarities between the mechanisms of cell activation exerted by InlB and HGF (Jonquières et al. 2001; Banerjee et al. 2004). The GW domains of InlB have been reported to bind another ubiquitous cellular molecule, the receptor of the globular part of the complement component C1q (gC1q-R) (Braun et al. 2000; Marino et al. 2002). Intriguingly, gC1q-R is a glycosylphosphatidylinositol (GPI)-anchored protein that does not present a cytoplasmic domain, suggesting that signaling on interaction with InlB is transduced through another cofactor. It has been reported that interaction between gC1q-R and InlB GW domains antagonizes rather than enhances InlB signaling (Marino et al. 2002; Banerjee et al. 2004). Therefore, the precise role of gC1q-R in *L. monocytogenes* infection remains to be clarified.

InlB expression by *L. innocua* does not induce bacterial adhesion to Vero cells (Jonquières et al. 1999) suggesting that InlB does not function as an adhesin but instead as a potent signaling invasin that mediates *L. monocytogenes* internalization through interactions with multiple host-cell ligands. The potential synergy of the InlA- and the InlB-invasion pathways has not been fully explored yet but it has been recently found that InlB plays, together with InlA, a critical role in the colonization of the placenta in vivo (Disson et al. 2008) (see below).

The Clathrin-Mediated Endocytosis Machinery Is Involved in the Early Steps of Entry

The clathrin-mediated endocytosis machinery is involved in the cell invasion by *L. monocyto-*

genes (Veiga and Cossart 2005; Veiga et al. 2007; Bonazzi et al. 2011). For both the InlA and the InlB pathways, the initial events in the internalization process are the posttranslational modifications of the two main *Listeria* receptors E-cadherin and Met: Indeed, E-cadherin is phosphorylated by Src and then ubiquitinated by Hakai (Bonazzi et al. 2008), whereas Met autophosphorylation leads to the recruitment of Cbl, which in turn ubiquitinates Met (Veiga and Cossart 2005). Recent studies have shown that the clathrin adaptor Dab2 is recruited to the *L. monocytogenes* receptors followed by the clathrin heavy and light chains; interestingly, the clathrin heavy chain is phosphorylated and this event is critical for the bacterial internalization process (Bonazzi et al. 2011). The protein Hip1R, which interacts with the clathrin light chain and also with the actin cytoskeleton, is then recruited; in addition, myosin VI can bind Hip1R and interact with actin filaments: Its capacity to move toward the minus end of the actin filaments fits well with a function in pulling the bacteria inside the cell (Bonazzi et al. 2011). Small interfering RNA (siRNA) experiments have shown that other members of the CME machinery play a critical role, in particular, dynamin (Veiga and Cossart 2005), which in turn can recruit cortactin, an activator of Arp2/3, a main actin nucleator; a first wave of actin polymerization could thus be triggered by cortactin. Structural studies reveal that clathrin-coated pits are present at the *L. monocytogenes* entry foci and are probably the sites of early actin rearrangements (Bonazzi et al. 2011). Interestingly, in JEG-3 epithelial cells, clathrin-independent mechanisms involving caveolin also mediate bacterial internalization downstream from the InlA-invasion pathway (Bonazzi et al. 2008).

Cytoskeletal and Membrane Rearrangements during Invasion Are Regulated by Actin-Binding Proteins, Small GTPases, PI 3-Kinase, and Phosphoinositides

As shown by numerous studies, following the first wave of actin polymerization described above or concomitantly, a second wave of actin rearrangements takes place during the *L. monocytogenes* invasion process, involving a PI 3-kinase, small GTPases of the Rho family, and several actin-binding proteins. The use of the PI 3-kinase inhibitors wortmannin and LY294002 showed in early studies that type IA PI 3-kinase activity is necessary for *L. monocytogenes* infection of Caco-2 and Vero cells via the InlA- and InlB-invasion pathways, respectively (Ireton et al. 1996). In Caco-2 cells, however, no increase in PI(3,4)P_2 or PI(3,4,5)P_3 levels was detected on interaction with *L. monocytogenes*, because in these cells PI3 kinase is constitutively and highly expressed. On the contrary, 10- and 20-fold increases on PI(3,4)P_2 or PI(3,4,5)P_3 levels are detected on infection of Vero cells with *L. monocytogenes* via the InlB-invasion pathway (Ireton et al. 1996). In these cells, activation of type IA PI 3-kinase is promoted by the protein adaptors Gab1, Cbl, and Shc, which are phosphorylated and recruited to the Met cytoplasmic tail, translocating with them the p85 subunit to the *L. monocytogenes* invasion foci (Ireton et al. 1999; Shen et al. 2000). Gab1 can also be translocated to the plasma membrane through its pleckstrin homology domain, which binds PI(3,4,5)P_3 (Basar et al. 2005) and indirectly boosts type IA PI 3-kinase activity via recruitment of the adaptor protein CrkII (Sun et al. 2005; Dokainish et al. 2007). PI(3,4,5)P_3 produced by the type IA PI 3-kinase plays a critical role in the recruitment of molecules that modulate actin cytoskeleton dynamics (see below) to the plasma membrane and a recent study has shown that the 5′-phosphatase OCRL restricts bacterial entry by precisely reducing the levels of PI(4,5)P_2 and PI(3,4,5)P_3 at *L. monocytogenes* entry foci (Kühbacher et al. 2012).

Actin polymerization following the InlB/Met interaction differs among cell lines: In Vero cells, plasma membrane lipid rafts promote the clustering of PI(3,4,5)P_3 and favor the translocation of Rac1 to the plasma membrane (Seveau 2004; Seveau et al. 2007); together with the Wiskott-Aldrich syndrome protein (WASP)-related protein complex WAVE2, Rac1 promotes actin polymerization by the Arp2/3 complex (Bierne 2001; Bierne et al. 2005). In HeLa cells, Rac1 and Cdc42 are required for Arp2/3 activation downstream from $PI(3,4,5)P_3$, together with N-WASP, WAVE1, and WAVE2 (Bierne et al. 2005); in this cell line, $PI(3,4,5)P_3$ is also recognized by ARAP2, a protein that functions as a GAP for the small GTPase Arf6; ARAP2 promotes actin rearrangements during *L. monocytogenes* invasion in part by antagonizing the activity of Arf6 (Gavicherla et al. 2010). In murine fibroblasts, Cdc42 participates in the activation of Rac1 via PI 3-kinase-dependent and -independent mechanisms (Bosse et al. 2007). A recent systems-biology study has identified nine genes encoding known upstream regulators or downstream effectors of the type IA PI 3-kinase required for invasion of target cells by *L. monocytogenes* including the small GTPase Rab5 and several regulators of Arf or Rac1 small GTPases (Jiwani et al. 2011). PI4P, a phosphoinositide usually present at the Golgi and produced by type II PI 4-kinases, is detected at bacterial entry sites and is critical for invasion in LoVo and HeLa cells. PI4P does not seem to function as a $PI(3,4,5)P_3$ precursor but rather acts as an independent effector fostering bacterial invasion downstream from the tetraspanin CD81 (Pizarro-Cerdá et al. 2007; Tham et al. 2010).

Concerning the InlA/E-cadherin invasion pathway, inhibition of Src kinase with the pharmacological agent PP1 in L2071hEcad cells showed that phosphorylation of cortactin by Src contributes to the recruitment of the Arp2/3 complex to bacterial invasion foci and initiates actin polymerization (Sousa et al. 2007). Among the small GTPases of the Rho family known to control actin dynamics, only Rac1 (but not Cdc42 or RhoA) appeared to be required for *L. monocytogenes* entry in Caco-2 cells (Sousa et al. 2007). The interactions between E-cadherin and the actin cytoskeleton are highly complex (Yamada et al. 2005). In both adherens junction formation and internalization of *Listeria*, they involve α and β catenins (Lecuit et al. 2000). A two-hybrid screen using α-catenin as bait identified the autosomal recessive hypercholesterolemia GTPase-activating protein 10 (ARHGAP10) as a molecule required for α-catenin recruitment and actin cytoskeleton rearrangements during *L. monocytogenes* invasion (Sousa et al. 2005a). However, ARHGAP10 promotes GTP hydrolysis by Cdc42 and RhoA but not Rac1; because Cdc42 and RhoA are not involved in InlA mediated entry, ARHGAP10 may function as an adaptor protein rather than GAP favoring interactions between α-catenin and still nonidentified molecular partners during bacterial internalization. The unconventional Myosin VIIa has also been identified as a binding partner of the transmembrane molecule vezatin and E-cadherin in the InlA-invasion pathway and could participate, together with actin, in membrane rearrangements required for bacterial entry (Sousa et al. 2004).

Elongation of actin filaments and depolymerization of actin are tightly controlled events during *L. monocytogenes* internalization. Sequestration of the actin-binding proteins Ena/VASP to the mitochondria (by the expression of mitochondrial targeting motifs) as well as siRNA inactivation of VASP inhibit the entry of bacteria or InlB-coated beads in Vero cells, demonstrating a role for VASP in entry (Bierne et al. 2005). In addition, overexpression of the LIM kinase, which phosphorylates and inactivates the actin depolymerizing factor cofilin, induces the accumulation of actin filaments beneath bacteria in Vero and REF-52 cells, preventing internalization and showing that actin depolymerization is critical during the late stages of the entry process (Bierne 2001). Interestingly, phosphorylated cofilin also drives phosphatidylcholine cleavage by host-cell phospholipase D1 (PLD1) and perturbation of the cellular levels of PLD1 blocks *L. monocytogenes* entry in Vero and HepG2 cells (Han et al. 2011).

Among the many other cellular actin interacting partners, septins represent a group of small GTPases known to polymerize into oligomeric protein complexes and filaments (Surka et al. 2002) that regulate cytokinesis, vesicle transport, and intracellular compartmentalization in a wide range of organisms (Beites et al. 2005; Estey et al. 2011). A member of the septin family, SEPT9, was initially identified as being recruited together with actin to the entry site of InlB-coated beads in LoVo cells (Pizarro-

Cerdá et al. 2002). Yeast two-hybrid screens and immunoprecipitation experiments revealed the interaction of SEPT9 with SEPT2, SEPT7, and SEPT11 in resting or InlB-stimulated HeLa cells (Mostowy et al. 2009a,b) and atomic force microscopy has shown that SEPT2 and SEPT9 contribute to the anchorage of Met to the actin cytoskeleton (Mostowy et al. 2011). Inactivation of septins by siRNA highlighted different contributions of SEPT2 and SEPT11 to *L. monocytogenes* invasion: SEPT2 knockdown severely modifies the actin cytoskeleton and inhibits bacterial entry (Mostowy et al. 2009b), whereas SEPT11 knockdown increases host-cell infection without modifying the actin cytoskeleton or affecting the Met signaling pathway, suggesting that SEPT11 rather restricts the InlB-mediated cell invasion (Mostowy and Cossart 2009).

In summary, the multiple interactions between the clathrin-mediated endocytosis machinery, the various phosphoinositides, and the actin cytoskeleton during *L. monocytogenes* cellular invasion highlight a very complex process whose dynamics vary with the nature of the infected cell. Genome-wide siRNA screens in progress should help in deciphering the orchestration of the molecular events that are key to cellular infection, and will address, in particular, whether internal membrane components are shuttled to the plasma membrane during entry.

Role of InlA and InlB in the Traversal of the Host Barriers In Vivo

Different lines of evidence show a critical contribution of InlA during *L. monocytogenes* infection in vivo. Indeed, epidemiological studies (Olier et al. 2003; Jacquet et al. 2004) show that strains in which InlA is truncated and secreted are frequently found in asymptomatic carriers or in food products, whereas full-length InlA is present far more frequently in clinical strains, suggesting that expression of an intact InlA molecule correlates with full bacterial virulence. When E-cadherin was identified as the receptor for InlA (Mengaud et al. 1996), a role for this internalin in the traversal of the intestinal barrier was proposed because E-cadherin is an adhesion molecule localized at the basolateral surface of several cell types of the small intestine villi, including enterocytes (absorptive cells) and goblet (mucus-secreting) cells. Studies in guinea pigs, which express an E-cadherin recognized by InlA, as well as in transgenic mice expressing the human E-cadherin under the control of the *iFABP* intestinal promotor, have confirmed that the presence of InlA is required for the traversal of the small intestine barrier in vivo through enterocyte invasion (Lecuit et al. 2001). Of note, in enterocytes E-cadherin is mainly present at adherens junctions where molecules are hidden below tight junctions and therefore are not normally exposed to the lumen of the intestine (Sousa et al. 2005b). Using rabbit ligated ileal loops and Madin-Darby canine kidney (MCDK) cells as a polarized epithelium model, it has been proposed that *L. monocytogenes* does not disrupt the enterocyte tight junctions to gain access to E-cadherin, but instead interacts with exposed E-cadherin at sites where apoptotic cells are expelled and detached from the epithelium by extrusion, suggesting that *L. monocytogenes* exploits the dynamic nature of epithelial renewal and junctional remodeling in small intestinal villi to breach this barrier (Fig. 3) (Pentecost et al. 2006). Of note, infection of wild-type mice intragastrically with a *L. monocytogenes* strain expressing a modified InlA that binds murine E-cadherin suggested that InlB also participates in bacterial internalization at small intestine villi; experiments in MDCK cells suggest that InlB does not act as an adhesin but rather accelerates bacterial internalization by promoting endocytosis of junctional components (Pentecost et al. 2010). It has thus been proposed that InlA provides specificity of adhesion to exposed E-cadherin at the tip of small intestine villi and InlB activates Met to accelerate junction endocytosis and bacterial invasion of enterocytes.

A recent study in *iFABP* transgenic mice has identified several discrete locations where E-cadherin is luminally accessible, including extruding enterocytes at the tip of villi as described above, but also villus epithelial folds and particularly junctions between mucus-secreting goblet cells, which are recognized as the preferred

Table 1. Cell lines used for the study of the InlA- and InlB-invasion pathways

Cell line	Species	Cell type	Receptors
Caco-2	Human	Colorectal adenocarcinoma	E-cadherin and Met
CHO	Hamster	Ovary	Met
GPC16	Guinea pig	Epithelial X	E-cadherin and Met
HeLa	Human	Cervix carcinoma	Met
HEp-2	Human	Larynx carcinoma	Met
HepG2	Human	Hepatocarcinoma	Met
JEG-3	Human	Trophoblasts	E-cadherin and Met
L2071hEcad	Mouse	Fibroblasts	E-cadherin and Met
LoVo	Human	Colorectal carcinoma	E-cadherin and Met
REF-52	Rat	Embryonic fibroblasts	Met
TIB73	Mouse	Embryonic hepatocyte	Met
Vero	African green monkey	Kidney	Met

Studies of the InlA and InlB pathways have been performed in different human and mammalian cell lines. Although some of them express both *L. monocytogenes* receptors, E-cadherin and Met, allowing the bacterium to enter by both pathways in concert or individually if either InlA or InlB is deleted, other cell lines do not express E-cadherin and can therefore only be used for studies concerning the InlB-invasion pathway.

sites for *L. monocytogenes* translocation at the small intestine (Nikitas et al. 2011). The InlA-dependent bacterial translocation through goblet cells is extremely rapid and surprisingly LLO and ActA independent; pharmacological and microscopical analyses reveal that *L. monocytogenes* does not escape from its internalization vacuole in goblet cells and rather takes advantage of a transcytosis mechanism that allows fast traversal of the small intestinal barrier (Fig. 3) (Nikitas et al. 2011). Although the small intestine is a major site for *L. monocytogenes* invasion, experiments in gerbils, which are natural hosts for *L. monocytogenes*, and in humanized transgenic mice that express E-cadherin in all physiological tissues, show that bacterial InlA-dependent translocation can also be observed at the caecum and the colon of the large intestine (Disson et al. 2008). Studies in rat ligated ileal loops from the lower part of the small intestine suggest furthermore that Peyer's patches, which are lymphoid nodules lined by antigen-sampling microfold (M) cells, may represent *L. monocytogenes* InlA- and InlB-independent invasion sites at the intestinal level (Pron et al. 1998; Corr et al. 2006).

As mentioned above, epidemiological data indicate that full-length InlA is predominant in clinically isolated *L. monocytogenes* strains, and this is particularly striking in pregnancy-related cases, in which 100% of isolated strains display an intact InlA molecule (Jacquet et al. 2004). In the placenta, E-cadherin is particularly enriched at the continuous multinucleated syncytiotrophoblast layer, which contributes to the barrier function of this organ, and studies in placental explants from human fetoplacental listeriosis cases as well as in primary cell cultures and placental cell lines indicate that interaction between InlA and the syncytiotrophoblast E-cadherin is critical for the traversal of the fetoplacental barrier (Lecuit et al. 2004). During the first trimester of pregnancy, the syncytiotrophoblast overlies a cytotrophoblast layer, which also participates in the placental barrier, and analyses of first trimester human placental cultures suggest that cytotrophoblasts can also be targeted by *L. monocytogenes* via InlA/E-cadherin interactions or cell-to-cell spread from infected phagocytes (Robbins et al. 2010; Zeldovich et al. 2011). The conjugated action of InlA and InlB to fetoplacental listeriosis was originally debated: Although it was shown in human mature placental explants that InlB participated together with InlA to syncytiotrophoblast invasion (Lecuit et al. 2004), in the guinea-pig model no role was shown for InlA despite its permissiveness to the InlA pathway (Bakardjiev et al. 2006), and in the mouse model no role was observed for InlB despite its permissiveness to the InlB pathway (Le Monnier et al. 2007). The different organization of the placenta in these animal models could partially account for differences reported in each of these studies. However, these apparently contradictory observations were subsequently explained when it was found in gerbils and knock-in mice ubiquitously expressing humanized E-cadherin, that placental infection by *L. monocytogenes* takes place only when both InlA- and InlB-invasion pathways are functional as in humans (Disson et al. 2008). It is not the case in the guinea pig in which Met is not a receptor for InlB (Khelef et al. 2006), and it is also not the case in the mouse in which E-cadherin is not a receptor for InlA (Lecuit et al. 1999).

Modulation of InlA- and InlB-Dependent Invasion Pathways by Other Internalins

To date, among the 27 known *L. monocytogenes* internalins, InlA and InlB are the major virulence factors implicated in cell invasion. Several studies suggest that other members of the internalin family may modulate *L. monocytogenes* invasion (Chatterjee et al. 2006) by behaving as adhesins, by indirectly affecting the surface exposure of InlA and/or InlB, or by directly interacting with putative novel cellular receptors. For example, expression of the LPXTG-anchored InlJ by *L. innocua* leads to bacterial adhesion (but no invasion) to human JEG-3 and HT29 polarized epithelial cells (Sabet et al. 2008), but its potential cell ligand has not been identified yet. The gene cluster *inlGHE* is required for full invasion of Caco-2 cells by the *L. monocytogenes* EGDe strain and it has been hypothesized that its expression may indirectly modulate the bacterial cell wall organization, therefore affecting InlA presentation at the *L. monocytogenes* surface (Bergmann et al. 2002).

A particular case has been reported for InlF: Under standard cell culture conditions, no role has been observed for this LPXTG internalin during invasion of Caco-2 and HepG2 cells in vitro or during Balb/c mice infection in vivo (Dramsi et al. 1997). However, in human HeLa and HEp-2 cells, pharmacological inhibition of the RhoA/Rho kinase signaling pathway using the drug Y27632 increases bacterial adhesion in an InlF-dependent manner (Kirchner and Higgins 2008). Moreover, in murine TIB75 hepatocytes and L2 fibroblasts, Y27632 treatment promotes InlF-dependent cell invasion by *L. monocytogenes* or by *L. innocua*-expressing InlF (but not InlA or InlB); Balb/c mice exposure to the HA-1077 drug, which also inhibits Rho kinases, increases the InlF-dependent virulence of *L. monocytogenes* in vivo (Kirchner and Higgins 2008). The mechanisms by which the RhoA/Rho kinase signaling pathway affects infection are unknown: Because this cascade enhances the assembly of actin-myosin structures, it was hypothesized that perturbation of actin-myosin interactions may lead to the cell surface exposure of a nonidentified InlF host-cell receptor (Carragher et al. 2006).

For several internalins including InlI, no role has been attributed to date (Sabet et al. 2005). Interestingly, although InlC2 and InlD in the strain EGD do not participate in the invasion of Caco-2 or S180 cells in vitro and do not contribute to bacterial virulence in BALB/c mice in vivo (Dramsi et al. 1997), these internalins are the major targets of the rabbit humoral response to *L. monocytogenes*, together with InlA (Yu et al. 2007). For other members of the internalin family, different roles in bacterial intracellular stages have been identified, as is the case for InlC, which perturbs NF-κB signaling (Gouin et al. 2010) and cell-to-cell spread (Rajabian et al. 2009), or InlK, which allows escape of autophagy (Dortet et al. 2011). There are 14 remaining internalins that have not been characterized yet.

ROLE OF NONINTERNALIN MOLECULES IN BACTERIAL ADHESION AND INVASION

Several molecules not belonging to the internalin family have been reported to contribute to the *L. monocytogenes* adhesion and invasion of host cells using in vitro systems. This is the case of LLO, a major virulence factor belonging to the family of cholesterol-dependent cytotoxins first identified as a molecule allowing phagosomal maturation arrest and phagosomal lysis that leads to bacterial escape to the cytoplasmic space

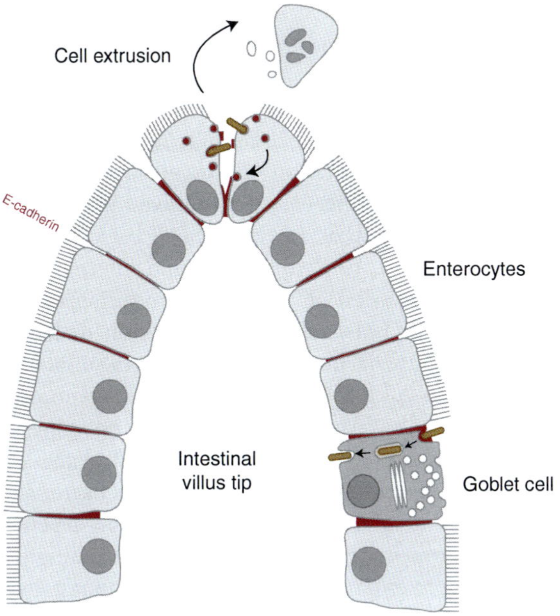

Figure 3. Traversal of the intestinal barrier by the InlA-E-cadherin interaction. The first contact between *L. monocytogenes* and host target cells takes place at the intestinal barrier level. E-cadherin, which is normally present at the basolateral face of enterocytes and therefore not exposed to the intestinal lumen, can be accessible to the *L. monocytogenes* invasion protein InlA at sites of apoptotic cell extrusion at the villus tip. However, the main route for bacterial translocation across the intestinal barrier is through mucus-secreting goblet cells via a rapid transcytosis mechanism that does not require intracellular bacterial escape from its internalizing vacuole.

(Gaillard et al. 1987; Dramsi and Cossart 2002; Henry et al. 2006). LLO secreted by extracellular bacteria has been shown to modulate a broad range of cellular activities including histone posttranscripional modifications (Hamon et al. 2007) as well as changes in the SUMOylation profile of host cells (Ribet et al. 2010) and several studies indicate that extracellular LLO also potentiates *L. monocytogenes* cell invasion in InlA/InlB-dependent and -independent ways. Indeed, transient increases in cellular calcium from the extracellular milieu (and not from intracellular reservoirs) induced by extracellular LLO increase the InlB-dependent bacterial invasion of HEp-2 cells (Dramsi and Cossart 2003). Transient mitochondrial fragmentation, mitochondrial membrane potential loss, and drop in respiration and cellular ATP levels also correlate with the LLO-induced calcium fluxes and increased bacterial entry in HeLa cells, suggesting that the bioenergetic state of resting cells represents a barrier to *L. monocytogenes* invasion and the LLO-induced metabolic cell reprogramming promotes efficient bacterial internalization (Stavru et al. 2011); the molecular mechanisms that link this reprogramming with phagocytosis are currently unknown. A direct role for LLO in promoting bacterial adhesion to epithelial cells has been proposed (Krawczyk-Balska and Bielecki 2005), and a recent study shows that LLO is sufficient to induce *L. monocytogenes* entry in HepG2 and HeLa cells in the absence of InA or InlB signaling (Vadia et al. 2011). Indeed, *L. innocua*-expressing LLO or latex beads coated with LLO are internalized in a cholesterol-, dynamin-, tyrosine kinase-, and actin-dependent but clathrin-independent manner; interestingly, complete pore formation is critical because a LLO variant that binds to host cells but only assembles into a prepore complex fails to promote invasion. Although LLO seems therefore sufficient to trigger bacterial invasion, it is proposed that, by stimulating actin polymerization and endocytosis, this cholesterol-dependent cytotoxin cooperates with InlA and InlB to potentiate efficient bacterial infection in a broad range of host cells (Vadia et al. 2011).

ActA is another major *L. monocytogenes* virulence factor that has been referred to play a role in bacterial invasion of host cells. ActA was first described as a protein promoting actin polymerization in the intracellular environment and bacterial cell-to-cell spread (Domann et al. 1992; Kocks et al. 1992). Heterologous expression of ActA in the nonpathogenic species *L. innocua* allows bacterial internalization in epithelial polarized and nonpolarized cells (Caco-2, MCDK, HeLa, and Vero) but not in COS-1 fibroblasts, suggesting that ActA is sufficient to directly activate an invasion pathway specific for epithelial cells (Suárez et al. 2001). The amino-terminal region of ActA is similar to the domain of the *Plasmodium falciparum* circumsporozoite protein involved in heparate sulfate recognition and hepatocyte binding (Pancake et al. 1992; Coppi et al. 2007) and it has been shown that the presence of heparan sulfate at the surface of CHO cells is required for the entry of *L. monocytogenes* in an ActA-dependent manner (Alvarez-Dominguez et al. 1997). The in vivo conditions in which the ActA invasion-associated function would be critical for virulence remain to be characterized.

Several cell membrane enzymes involved in the modification of the *L. monocytogenes* cell wall architecture have been also shown to affect bacterial adhesion and/or invasion in vitro as well as virulence in vivo. For example, the lipoteichoic acid modifiers GtcA (Promadej et al. 1999) and DltA (Abachin et al. 2002), the lipoprotein LpeA (Réglier-Poupet et al. 2003), the prolipoprotein diacylglyceryl transferase Lgt (Machata et al. 2008), and the protein MprF involved in the modification of the phospholipid lysylphosphatidylglycerol (Thedieck et al. 2006) play roles in *L. monocytogenes* adhesion or invasion in cell lines by probably modulating the electronegativity of the bacterial surface charge and/or altering the surface expression of adhesins/invasins. A particular class of cell membrane enzymes is the autolysins, which are involved in the digestion of the cell wall peptidoglycan: Among the six putative autolysins encoded by the *L. monocytogenes* genome, three of them (Ami, Auto, and IspC) characterized by the presence of carboxy-terminal GW repeats have been implicated in bacterial attachment to host cells or invasion. Ami, a *N*-acetylmuramoyl-L-alanine amidase (Braun et al. 1997) is required for adhesion but not for invasion of host cells in an *ΔinlA*, *ΔinlB*, or *ΔinlAB* background only (Milohanic et al. 2000), and interestingly, the eight GW repeats of Ami are sufficient to promote *L. monocytogenes* adhesion to host cells (Milohanic et al. 2001). Auto, the only *L. monocytogenes* autolysin absent from the genome of *L. innocua*, is required for invasion (not adhesion) to several mammalian cell lines (Cabanes et al. 2004), whereas IspC is required for bacterial adhesion and/or invasion in a cell line-dependent manner (Wang and Lin 2008). Intriguingly, the *Δami*, *Δauto*, and *ΔispC* strains do not display major morphological abnormalities (ActA surface expression is only affected in the *ΔispC* mutant) suggesting that these molecules probably play a minor role as autolysins in vitro; their importance is highlighted by the reduced virulence of the mutant strains in the mouse model in vivo. The potential cellular ligands for these autolysins or the precise mechanisms governing their implication in adhesion and invasion have not been identified to date.

Other bacterial molecules involved in *L. monocytogenes* adhesion and/or invasion include the LPXTG protein Vip (Cabanes et al. 2005), the adhesins Lap (Jagadeesan et al. 2011), and LapB (Reis et al. 2010) as well as the fibronectin-binding protein A (FbpA) required also as an escort protein for LLO and InlB (Dramsi et al. 2004).

CONCLUDING REMARKS

L. monocytogenes is a fascinating bacterial pathogen able to survive in a saprophytic environment and to induce disease in mammalian hosts. Adhesion to host cells and invasion of the intracellular space are critical steps for the traversal of host barriers leading to organ infection and bacterial spread in the organism; not surprisingly, many bacterial effectors have been shown to participate in every stage of the infectious process. We are starting to understand the global principles that govern the regulation of the expression of these virulence factors. The challenge for future studies is to understand how the combinatorial function of the bacterial and cellular factors specifically contributes to the pathophysiology of listeriosis in vivo.

ACKNOWLEDGMENTS

Research in P. Cossart's laboratory is supported by the Pasteur Institute, ERC advanced grant 233348, the Institut National de la Santé et de la Recherche Médicale, the Institut National de la Recherche Agronomique, the Agence Nationale de la Recherche, the Louis-Jeantet Foundation, and the Fondation Le Roch Les Mousquetaires. A.K. is a recipient of a scholarship from the Pasteur-Paris University International Doctoral Program/Institut Carnot Maladies Infectieuses.

REFERENCES

Abachin E, Poyart C, Pellegrini E, Milohanic E, Fiedler F, Berche P, Trieu-Cuot P. 2002. Formation of D-alanyl-lipoteichoic acid is required for adhesion and virulence of *Listeria monocytogenes*. *Mol Microbiol* **43:** 1–14.

Allerberger F, Wagner M. 2010. Listeriosis: A resurgent foodborne infection. *Clin Microbiol Infect* **16:** 16–23.

Alvarez-Dominguez C, Carrasco-Marín E, Leyva-Cobián F. 1993. Role of complement component C1q in phagocytosis of *Listeria monocytogenes* by murine macrophage-like cell lines. *Infect Immun* **61:** 3664–3672.

Alvarez-Dominguez C, Vázquez-Boland JA, Carrasco-Marín E, López-Mato P, Leyva-Cobián F. 1997. Host cell heparan sulfate proteoglycans mediate attachment and entry of *Listeria monocytogenes*, and the listerial surface protein ActA is involved in heparan sulfate receptor recognition. *Infect Immun* **65:** 78–88.

Bakardjiev AI, Theriot JA, Portnoy DA. 2006. *Listeria monocytogenes* traffics from maternal organs to the placenta and back. *PLoS Pathog* **2:** e66.

Banerjee M, Copp J, Vuga D, Marino M, Chapman T, Van Der Geer P, Ghosh P. 2004. GW domains of the *Listeria monocytogenes* invasion protein InlB are required for potentiation of Met activation. *Mol Microbiol* **52:** 257–271.

Basar T, Shen Y, Ireton K. 2005. Redundant roles for Met docking site tyrosines and the Gab1 pleckstrin homology domain in InlB-mediated entry of *Listeria monocytogenes*. *Infect Immun* **73:** 2061–2074.

Beites CL, Campbell KA, Trimble WS. 2005. The septin Sept5/CDCrel-1 competes with α-SNAP for binding to the SNARE complex. *Biochem J* **385:** 347–353.

Bergmann B, Raffelsbauer D, Kuhn M, Goetz M, Hom S, Goebel W. 2002. InlA- but not InlB-mediated internalization of *Listeria monocytogenes* by non-phagocytic mammalian cells needs the support of other internalins. *Mol Microbiol* **43:** 557–570.

Bierne H. 2001. A role for cofilin and LIM kinase in *Listeria*-induced phagocytosis. *J Cell Biol* **155:** 101–112.

Bierne H, Miki H, Innocenti M, Scita G, Gertler FB, Takenawa T, Cossart P. 2005. WASP-related proteins, Abi1 and Ena/VASP are required for *Listeria* invasion induced by the Met receptor. *J Cell Sci* **118:** 1537–1547.

Bierne H, Sabet C, Personnic N, Cossart P. 2007. Internalins: A complex family of leucine-rich repeat-containing proteins in *Listeria monocytogenes*. *Microbes Infect* **9:** 1156–1166.

Birchmeier C, Birchmeier W, Gherardi E, Vande Woude GF. 2003. Met, metastasis, motility and more. *Nat Rev Mol Cell Biol* **4:** 915–925.

Bonazzi M, Veiga E, Pizarro-Cerdá J, Cossart P. 2008. Successive post-translational modifications of E-cadherin are required for InlA-mediated internalization of *Listeria monocytogenes*. *Cell Microbiol* **10:** 2208–2222.

Bonazzi M, Vasudevan L, Mallet A, Sachse M, Sartori A, Prevost MC, Roberts A, Taner SB, Wilbur JD, Brodsky FM, et al. 2011. Clathrin phosphorylation is required for actin recruitment at sites of bacterial adhesion and internalization. *J Cell Biol* **195:** 525–536.

Bosse T, Ehinger J, Czuchra A, Benesch S, Steffen A, Wu X, Schloen K, Niemann HH, Scita G, Stradal TEB, et al. 2007. Cdc42 and phosphoinositide 3-kinase drive Rac-mediated actin polymerization downstream of c-Met in distinct and common pathways. *Mol Cell Biol* **27:** 6615–6628.

Braun L, Dramsi S, Dehoux P, Bierne H, Lindahl G, Cossart P. 1997. InlB: An invasion protein of *Listeria monocytogenes* with a novel type of surface association. *Mol Microbiol* **25:** 285–294.

Braun L, Ohayon H, Cossart P. 1998. The InIB protein of *Listeria monocytogenes* is sufficient to promote entry into mammalian cells. *Mol Microbiol* **27:** 1077–1087.

Braun L, Nato F, Payrastre B, Mazié JC, Cossart P. 1999. The 213-amino-acid leucine-rich repeat region of the *Listeria monocytogenes* InlB protein is sufficient for entry into mammalian cells, stimulation of PI 3-kinase and membrane ruffling. *Mol Microbiol* **34:** 10–23.

Braun L, Ghebrehiwet B, Cossart P. 2000. gC1q-R/p32, a C1q-binding protein, is a receptor for the InlB invasion protein of *Listeria monocytogenes*. *EMBO J* **19:** 1458–1466.

Cabanes D, Dehoux P, Dussurget O, Frangeul L, Cossart P. 2002. Surface proteins and the pathogenic potential of *Listeria monocytogenes*. *Trends Microbiol* **10:** 238–245.

Cabanes D, Sousa S, Cebriá A, Lecuit M, García-Del Portillo F, Cossart P. 2005. Gp96 is a receptor for a novel *Listeria monocytogenes* virulence factor, Vip, a surface protein. *EMBO J* **24:** 2827–2838.

Camejo A, Carvalho F, Reis O, Leitão E, Sousa S, Cabanes D. 2011. The arsenal of virulence factors deployed by *Listeria monocytogenes* to promote its cell infection cycle. *Virulence* **2:** 379–394.

Carragher NO, Walker SM, Scott Carragher LA, Harris F, Sawyer TK, Brunton VG, Ozanne BW, Frame MC. 2006. Calpain 2 and Src dependence distinguishes mesenchymal and amoeboid modes of tumer cell invasion: A link to integrin function. *Oncogene* **25:** 5726–5740.

Chatterjee SS, Otten S, Hain T, Lingnau A, Carl UD, Wehland J, Domann E, Chakraborty T. 2006. Invasiveness is a variable and heterogeneous phenotype in *Listeria monocytogenes* serotype strains. *Int J Med Microbiol* **296:** 277–286.

Copp J, Marino M, Banerjee M, Ghosh P, Van Der Geer P. 2003. Multiple regions of internalin B contribute to its ability to turn on the Ras-mitogen-activated protein kinase pathway. *J Biol Chem* **278:** 7783–7789.

Coppi A, Tewari R, Bishop JR, Bennett BL, Lawrence R, Esko JD, Billker O, Sinnis P. 2007. Heparan sulfate proteoglycans provide a signal to plasmodium sporozoites to stop migrating and productively invade host cells. *Cell Host Microbe* **2:** 316–327.

Corr S, Hill C, Gahan CGM. 2006. An in vitro cell-culture model demonstrates internalin- and hemolysin-independent translocation of *Listeria monocytogenes* across M cells. *Microb Pathogenesis* **41:** 241–250.

Cossart P. 2011. Illuminating the landscape of host-pathogen interactions with the bacterium *Listeria monocytogenes*. *Proc Natl Acad Sci* **108:** 19848–19491.

Cossart P, Toledo-Arana A. 2008. *Listeria monocytogenes*, a unique model in infection biology: An overview. *Microbes Infect* **10:** 1041–1050.

Disson O, Grayo S, Huillet E, Nikitas G, Langa-Vives F, Dussurget O, Ragon M, Le Monnier A, Babinet C, Cossart P, et al. 2008. Conjugated action of two species-specific invasion proteins for fetoplacental listeriosis. *Nature* **455:** 1114–1118.

Dokainish H, Gavicherla B, Shen Y, Ireton K. 2007. The carboxyl-terminal SH3 domain of the mammalian adaptor CrkII promotes internalization of *Listeria monocytogenes* through activation of host phosphoinositide 3-kinase. *Cell Microbiol* **9:** 2497–2516.

Domann E, Wehland J, Rohde M, Pistor S, Hartl M, Goebel W, Leimeister-Wächter M, Wuenscher M, Chakraborty T. 1992. A novel bacterial virulence gene in *Listeria monocytogenes* required for host cell microfilament interaction with homology to the proline-rich region of vinculin. *EMBO J* **11:** 1981–1990.

Dortet L, Veiga E, Bonazzi M, Cossart P. 2010. CD44-Independent activation of the Met signaling pathway by HGF and InlB. *Microbes Infect* **12:** 919–927.

Dortet L, Mostowy S, Louaka AS, Gouin E, Nahori M-A, Wiemer EAC, Dussurget O, Cossart P. 2011. Recruitment of the major vault protein by InlK: A *Listeria monocytogenes* strategy to avoid autophagy. *PLoS Pathog* **7:** e1002168.

Dramsi S, Cossart P. 2002. Listeriolysin O: A genuine cytolysin optimized for an intracellular parasite. *J Cell Biol* **156:** 943–946.

Dramsi S, Cossart P. 2003. Listeriolysin O-mediated calcium influx potentiates entry of *Listeria monocytogenes* into the human Hep-2 epithelial cell line. *Infect Immun* **71:** 3614–3618.

Dramsi S, Biswas I, Maguin E, Braun L, Mastroeni P, Cossart P. 1995. Entry of *Listeria monocytogenes* into hepatocytes requires expression of inIB, a surface protein of the internalin multigene family. *Mol Microbiol* **16:** 251–261.

Dramsi S, Dehoux P, Lebrun M, Goossens PL, Cossart P. 1997. Identification of four new members of the internalin multigene family of *Listeria monocytogenes* EGD. *Infect Immun* **65:** 1615–1625.

Dramsi S, Bourdichon F, Cabanes D, Lecuit M, Fsihi H, Cossart P. 2004. FbpA, a novel multifunctional *Listeria monocytogenes* virulence factor. *Mol Microbiol* **53:** 639–649.

Drevets DA, Bronze MS. 2008. *Listeria monocytogenes*: Epidemiology, human disease, and mechanisms of brain invasion. *FEMS Immunol Med Mic* **53:** 151–165.

Ebbes M, Bleymüller WM, Cernescu M, Nölker R, Brutschy B, Niemann HH. 2011. Fold and function of the InlB B-repeat. *J Biol Chem* **286:** 15496–15506.

Estey MP, Kim MS, Trimble WS. 2011. Septins. *Curr Biol* **21:** R384–R387.

Ferraris DM, Gherardi E, Di Y, Heinz DW, Niemann HH. 2010. Ligand-mediated dimerization of the Met receptor tyrosine kinase by the bacterial invasion protein InlB. *J Mol Biol* **395:** 522–532.

Gaillard JL, Berche P, Mounier J, Richard S, Sansonetti P. 1987. In vitro model of penetration and intracellular growth of *Listeria monocytogenes* in the human enterocyte-like cell line Caco-2. *Infect Immun* **55:** 2822–2829.

Gaillard JL, Berche P, Frehel C, Gouin E, Cossart P. 1991. Entry of *L. monocytogenes* into cells is mediated by internalin, a repeat protein reminiscent of surface antigens from gram-positive cocci. *Cell* **65:** 1127–1141.

Gaillard JL, Jaubert F, Berche P. 1996. The inlAB locus mediates the entry of *Listeria monocytogenes* into hepatocytes in vivo. *J Exp Med* **183:** 359–369.

Gavicherla B, Ritchey L, Gianfelice A, Kolokoltsov AA, Davey RA, Ireton K. 2010. Critical role for the host GTPase-activating protein ARAP2 in InlB-mediated entry of *Listeria monocytogenes*. *Infect Immun* **78:** 4532–4541.

Gouin E, Adib-Conquy M, Balestrino D, Nahori M-A, Villiers V, Colland F, Dramsi S, Dussurget O, Cossart P. 2010. The *Listeria monocytogenes* InlC protein interferes with innate immune responses by targeting the IκB kinase subunit IKKα. *Proc Natl Acad Sci* **107:** 17333–17338.

Gregory SH, Sagnimeni AJ, Wing EJ. 1996. Expression of the inlAB operon by *Listeria monocytogenes* is not required for entry into hepatic cells in vivo. *Infect Immun* **64:** 3983–3986.

Greiffenberg L, Goebel W, Kim KS, Weiglein I, Bubert A, Engelbrecht F, Stins M, Kuhn M. 1998. Interaction of *Listeria monocytogenes* with human brain microvascular endothelial cells: InlB-dependent invasion, long-term intracellular growth, and spread from macrophages to endothelial cells. *Infect Immun* **66:** 5260–5267.

Hamon MA, Batsché E, Régnault B, Tham TN, Seveau S, Muchardt C, Cossart P. 2007. Histone modifications induced by a family of bacterial toxins. *Proc Natl Acad Sci* **104:** 13467–13472.

Han X, Yu R, Ji L, Zhen D, Tao S, Li S, Sun Y, Huang L, Feng Z, Li X, et al. 2011. InlB-Mediated *Listeria monocytogenes* internalization requires a balanced phospholipase D activity maintained through phosphocofilin. *Mol Microbiol* **81:** 860–880.

Henry R, Shaughnessy L, Loessner MJ, Alberti-Segui C, Higgins DE, Swanson JA. 2006. Cytolysin-dependent delay of vacuole maturation in macrophages infected with *Listeria monocytogenes*. *Cell Microbiol* **8:** 107–119.

Ireton K, Payrastre B, Chap H, Ogawa W, Sakaue H, Kasuga M, Cossart P. 1996. A role for phosphoinositide 3-kinase in bacterial invasion. *Science* **274:** 780–782.

Ireton K, Payrastre B, Cossart P. 1999. The *Listeria monocytogenes* protein InlB is an agonist of mammalian phosphoinositide 3-kinase. *J Biol Chem* **274:** 17025–17032.

Jacquet C, Doumith M, Gordon JI, Martin PMV, Cossart P, Lecuit M. 2004. A molecular marker for evaluating the pathogenic potential of foodborne *Listeria monocytogenes*. *J Infect Dis* **189:** 2094–2100.

Jagadeesan B, Fleishman Littlejohn AE, Amalaradjou MAR, Singh AK, Mishra KK, La D, Kihara D, Bhunia AK. 2011. N-terminal Gly(224)-Gly(411) domain in *Listeria* adhesion protein interacts with host receptor Hsp60. *PLoS ONE* **6:** e20694.

Jiwani S, Wang Y, Dowd GC, Gianfelice A, Pichestapong P, Gavicherla B, Vanbennekom N, Ireton K. 2011. Identification of components of the host type 1A phosphoinositide 3-kinase pathway that promote internalization of *Listeria monocytogenes*. *Infect Immun* **80:** 1252–1266.

Jonquières R, Bierne H, Mengaud J, Cossart P. 1998. The inlA gene of *Listeria monocytogenes* LO28 harbors a nonsense mutation resulting in release of internalin. *Infect Immun* **66:** 3420–3422.

Jonquières R, Bierne H, Fiedler F, Gounon P, Cossart P. 1999. Interaction between the protein InlB of *Listeria monocytogenes* and lipoteichoic acid: A novel mechanism of protein association at the surface of gram-positive bacteria. *Mol Microbiol* **34:** 902–914.

Jonquières R, Pizarro-Cerda J, Cossart P. 2001. Synergy between the N- and C-terminal domains of InlB for efficient invasion of non-phagocytic cells by *Listeria monocytogenes*. *Mol Microbiol* **42:** 955–965.

Jung C, Matzke A, Niemann HH, Schwerk C, Tenenbaum T, Orian-Rousseau V. 2009. Involvement of CD44v6 in InlB-dependent Listeria invasion. *Mol Microbiol* **72:** 1196–1207.

Khelef N, Lecuit M, Bierne H, Cossart P. 2006. Species specificity of the *Listeria monocytogenes* InlB protein. *Cell Microbiol* **8:** 457–470.

Kirchner M, Higgins DE. 2008. Inhibition of ROCK activity allows InlF-mediated invasion and increased virulence of *Listeria monocytogenes*. *Mol Microbiol* **68:** 749–767.

Kocks C, Gouin E, Tabouret M, Berche P, Ohayon H, Cossart P. 1992. *L. monocytogenes*-induced actin assembly requires the actA gene product, a surface protein. *Cell* **68:** 521–531.

Krawczyk-Balska A, Bielecki J. 2005. *Listeria monocytogenes* listeriolysin O and phosphatidylinositol-specific phospholipase C affect adherence to epithelial cells. *Can J Microbiol* **51:** 745–751.

Kühbacher A, Dambournet D, Echard A, Cossart P, Pizarro-Cerda J. 2012. Phosphatidylinositol 5-phosphatase oculocerebrorenal syndrome of Lowe protein (OCRL) controls actin dynamics during early steps of *Listeria monocytogenes* infection. *J Biol Chem* **287:** 13128–13136.

Lecuit M. 2005. Understanding how *Listeria monocytogenes* targets and crosses host barriers. *Clin Microbiol Infect* **11:** 430–436.

Lecuit M, Ohayon H, Braun L, Mengaud J, Cossart P. 1997. Internalin of *Listeria monocytogenes* with an intact leucine-rich repeat region is sufficient to promote internalization. *Infect Immun* **65:** 5309–5319.

Lecuit M, Dramsi S, Gottardi C, Fedor-Chaiken M, Gumbiner B, Cossart P. 1999. A single amino acid in E-cadherin responsible for host specificity towards the human pathogen *Listeria monocytogenes*. *EMBO J* **18:** 3956–3963.

Lecuit M, Hurme R, Pizarro-Cerda J, Ohayon H, Geiger B, Cossart P. 2000. A role for α-and β-catenins in bacterial uptake. *Proc Natl Acad Sci* **97:** 10008–10013.

Lecuit M, Vandormael-Pournin S, Lefort J, Huerre M, Gounon P, Dupuy C, Babinet C, Cossart P. 2001. A transgenic model for listeriosis: Role of internalin in crossing the intestinal barrier. *Science* **292:** 1722–1725.

Lecuit M, Nelson DM, Smith SD, Khun H, Huerre M, Vacher-Lavenu M-C, Gordon JI, Cossart P. 2004. Targeting and crossing of the human maternofetal barrier by *Listeria monocytogenes*: Role of internalin interaction with trophoblast E-cadherin. *Proc Natl Acad Sci* **101:** 6152–6157.

Leimeister-Wächter M, Haffner C, Domann E, Goebel W, Chakraborty T. 1990. Identification of a gene that positively regulates expression of listeriolysin, the major virulence factor of *Listeria monocytogenes*. *Proc Natl Acad Sci* **87:** 8336–8340.

Le Monnier A, Autret N, Join-Lambert OF, Jaubert F, Charbit A, Berche P, Kayal S. 2007. ActA is required for crossing of the fetoplacental barrier by *Listeria monocytogenes*. *Infect Immun* **75:** 950–957.

Lindén SK, Bierne H, Sabet C, Png CW, Florin TH, McGuckin MA, Cossart P. 2008. *Listeria monocytogenes* internalins bind to the human intestinal mucin MUC2. *Arch Microbiol* **190:** 101–104.

Lingnau A, Domann E, Hudel M, Bock M, Nichterlein T, Wehland J, Chakraborty T. 1995. Expression of the *Listeria monocytogenes* EGD inlA and inlB genes, whose products mediate bacterial entry into tissue culture cell lines, by PrfA-dependent and -independent mechanisms. *Infect Immun* **63:** 3896–3903.

Machata S, Tchatalbachev S, Mohamed W, Jänsch L, Hain T, Chakraborty T. 2008. Lipoproteins of *Listeria monocytogenes* are critical for virulence and TLR2-mediated immune activation. *J Immunol* **181:** 2028–2035.

Madarame H, Seuberlich T, Abril C, Zurbriggen A, Vandevelde M, Oevermann A. 2011. The distribution of E-cadherin expression in listeric rhombencephalitis of ruminants indicates its involvement in *Listeria monocytogenes* neuroinvasion. *Neuropath Appl Neuro* **37:** 753–767.

Marino M, Braun L, Cossart P, Ghosh P. 1999. Structure of the lnlB leucine-rich repeats, a domain that triggers host cell invasion by the bacterial pathogen *L. monocytogenes*. *Mol Cell* **4:** 1063–1072.

Marino M, Braun L, Cossart P, Ghosh P. 2000. A framework for interpreting the leucine-rich repeats of the *Listeria* internalins. *Proc Natl Acad Sci* **97:** 8784–8788.

Marino M, Banerjee M, Jonquières R, Cossart P, Ghosh P. 2002. GW domains of the *Listeria monocytogenes* invasion protein InlB are SH3-like and mediate binding to host ligands. *EMBO J* **21:** 5623–5634.

Mengaud J, Chenevert J, Geoffroy C, Gaillard JL, Cossart P. 1987. Identification of the structural gene encoding the SH-activated hemolysin of *Listeria monocytogenes*: Listeriolysin O is homologous to streptolysin O and pneumolysin. *Infect Immun* **55:** 3225–3227.

Mengaud J, Dramsi S, Gouin E, Vazquez-Boland JA, Milon G, Cossart P. 1991. Pleiotropic control of *Listeria monocytogenes* virulence factors by a gene that is autoregulated. *Mol Microbiol* **5:** 2273–2283.

Mengaud J, Ohayon H, Gounon P, Mege R-M, Cossart P. 1996. E-cadherin is the receptor for internalin, a surface protein required for entry of *L. monocytogenes* into epithelial cells. *Cell* **84:** 923–932.

Milohanic E, Pron B, Berche P, Gaillard JL. 2000. Identification of new loci involved in adhesion of *Listeria monocytogenes* to eukaryotic cells. European Listeria Genome Consortium. *Microbiology* **146:** 731–739.

Milohanic E, Jonquières R, Cossart P, Berche P, Gaillard JL. 2001. The autolysin Ami contributes to the adhesion of *Listeria monocytogenes* to eukaryotic cells via its cell wall anchor. *Mol Microbiol* **39:** 1212–1224.

Mostowy S, Cossart P. 2009. From pathogenesis to cell biology and back. *Cell Host Microbe* **5:** 510–513.

Mostowy S, Danckaert A, Tham TN, Machu C, Guadagnini S, Pizarro-Cerdá J, Cossart P. 2009a. Septin 11 restricts InlB-mediated invasion by *Listeria*. *J Biol Chem* **284:** 11613–11621.

Mostowy S, Tham TN, Danckaert A, Guadagnini S, Boisson-Dupuis S, Pizarro-Cerdá J, Cossart P. 2009b. Septins regulate bacterial entry into host cells. *PLoS ONE* **4:** e4196.

Mostowy S, Janel S, Forestier C, Roduit C, Kasas S, Pizarro-Cerdá J, Cossart P, Lafont F. 2011. A role for septins in the interaction between the *Listeria monocytogenes* invasion protein InlB and the Met receptor. *Biophys J* **100:** 1949–1959.

Niemann HH, Jäger V, Butler PJG, van den Heuvel J, Schmidt S, Ferraris D, Gherardi E, Heinz DW. 2007. Structure of the human receptor tyrosine kinase met in complex with the *Listeria* invasion protein InlB. *Cell* **130:** 235–246.

Nikitas G, Deschamps C, Disson O, Niault T, Cossart P, Lecuit M. 2011. Transcytosis of *Listeria monocytogenes* across the intestinal barrier upon specific targeting of goblet cell accessible E-cadherin. *J Exp Med* **208:** 2263–2277.

Olier M, Pierre F, Rousseaux S, Lemaître J-P, Rousset A, Piveteau P, Guzzo J. 2003. Expression of truncated Internalin A is involved in impaired internalization of some *Listeria monocytogenes* isolates carried asymptomatically by humans. *Infect Immun* **71:** 1217–1224.

Orian-Rousseau V, Chen L, Sleeman JP, Herrlich P, Ponta H. 2002. CD44 is required for two consecutive steps in HGF/c-Met signaling. *Genes Dev* **16:** 3074–3086.

Pancake SJ, Holt GD, Mellouk S, Hoffman SL. 1992. Malaria sporozoites and circumsporozoite proteins bind specifically to sulfated glycoconjugates. *J Cell Biol* **117:** 1351–1357.

Pentecost M, Otto G, Theriot JA, Amieva MR. 2006. *Listeria monocytogenes* invades the epithelial junctions at sites of cell extrusion. *PLoS Pathog* **2:** e3.

Pentecost M, Kumaran J, Ghosh P, Amieva MR. 2010. *Listeria monocytogenes* internalin B activates junctional endocytosis to accelerate intestinal invasion. *PLoS Pathog* **6:** e1000900.

Peschard P, Fournier TM, Lamorte L, Naujokas MA, Band H, Langdon WY, Park M. 2001. Mutation of the c-Cbl TKB domain binding site on the Met receptor tyrosine kinase converts it into a transforming protein. *Mol Cell* **8:** 995–1004.

Pizarro-Cerdá J, Jonquières R, Gouin E, Vandekerckhove J, Garin J, Cossart P. 2002. Distinct protein patterns associated with *Listeria monocytogenes* InlA- or InlB-phagosomes. *Cell Microbiol* **4:** 101–115.

Pizarro-Cerdá J, Payrastre B, Wang Y-J, Veiga E, Yin HL, Cossart P. 2007. Type II phosphatidylinositol 4-kinases promote *Listeria monocytogenes* entry into target cells. *Cell Microbiol* **9:** 2381–2390.

Pizarro-Cerdá J, Bonazzi M, Cossart P. 2010. Clathrin-mediated endocytosis: What works for small, also works for big. *Bioessays* **32:** 496–504.

Ponta H, Sherman L, Herrlich PA. 2003. CD44: From adhesion molecules to signalling regulators. *Nat Rev Mol Cell Biol* **4:** 33–45.

Promadej N, Fiedler F, Cossart P, Dramsi S, Kathariou S. 1999. Cell wall teichoic acid glycosylation in *Listeria monocytogenes* serotype 4b requires gtcA, a novel, serogroup-specific gene. *J Bacteriol* **181:** 418–425.

Pron B, Boumaila C, Jaubert F, Sarnacki S, Monnet JP, Berche P, Gaillard JL. 1998. Comprehensive study of the intestinal stage of listeriosis in a rat ligated ileal loop system. *Infect Immun* **66:** 747–755.

Rajabian T, Gavicherla B, Heisig M, Müller-Altrock S, Goebel W, Gray-Owen SD, Ireton K. 2009. The bacterial virulence factor InlC perturbs apical cell junctions and promotes cell-to-cell spread of *Listeria*. *Nat Cell Biol* **11:** 1212–1218.

Réglier-Poupet H, Pellegrini E, Charbit A, Berche P. 2003. Identification of LpeA, a PsaA-like membrane protein that promotes cell entry by *Listeria monocytogenes*. *Infect Immun* **71:** 474–482.

Reis O, Sousa S, Camejo A, Villiers V, Gouin E, Cossart P, Cabanes D. 2010. LapB, a novel *Listeria monocytogenes* LPXTG surface adhesin, required for entry into eukaryotic cells. *J Infect Dis* **202:** 551–562.

Ribet D, Hamon M, Gouin E, Nahori M-A, Impens F, Neyret-Kahn H, Gevaert K, Vandekerckhove J, Dejean A, Cossart P. 2010. *Listeria monocytogenes* impairs SUMOylation for efficient infection. *Nature* **464:** 1192–1195.

Robbins JR, Skrzypczynska KM, Zeldovich VB, Kapidzic M, Bakardjiev AI. 2010. Placental syncytiotrophoblast constitutes a major barrier to vertical transmission of *Listeria monocytogenes*. *PLoS Pathog* **6:** e1000732.

Sabet C, Lecuit M, Cabanes D, Cossart P, Bierne H. 2005. LPXTG protein InlJ, a newly identified internalin involved in *Listeria monocytogenes* virulence. *Infect Immun* **73:** 6912–6922.

Sabet C, Toledo-Arana A, Personnic N, Lecuit M, Dubrac S, Poupel O, Gouin E, Nahori M-A, Cossart P, Bierne H. 2008. The *Listeria monocytogenes* virulence factor InlJ is specifically expressed in vivo and behaves as an adhesin. *Infect Immun* **76:** 1368–1378.

Schubert WD, Urbanke C, Ziehm T, Beier V, Machner MP, Domann E, Wehland J, Chakraborty T, Heinz DW. 2002. Structure of internalin, a major invasion protein of *Listeria monocytogenes*, in complex with its human receptor E-cadherin. *Cell* **111:** 825–836.

Seveau S. 2004. Role of lipid rafts in E-cadherin- and HGF-R/Met-mediated entry of *Listeria monocytogenes* into host cells. *J Cell Biol* **166:** 743–753.

Seveau S, Tham TN, Payrastre B, Hoppe AD, Swanson JA, Cossart P. 2007. A FRET analysis to unravel the role of cholesterol in Rac1 and PI 3-kinase activation in the InlB/Met signalling pathway. *Cell Microbiol* **9:** 790–803.

Shen Y, Naujokas M, Park M, Ireton K. 2000. InlB-dependent internalization of *Listeria* is mediated by the Met receptor tyrosine kinase. *Cell* **103:** 501–510.

Sousa S, Cabanes D, El-Amraoui A, Petit C, Lecuit M, Cossart P. 2004. Unconventional myosin VIIa and vezatin, two proteins crucial for *Listeria* entry into epithelial cells. *J Cell Sci* **117:** 2121–2130.

Sousa S, Cabanes D, Archambaud C, Colland F, Lemichez E, Popoff M, Boisson-Dupuis S, Gouin E, Lecuit M, Legrain P, et al. 2005a. ARHGAP10 is necessary for α-catenin recruitment at adherens junctions and for *Listeria* invasion. *Nat Cell Biol* **7:** 954–960.

Sousa S, Lecuit M, Cossart P. 2005b. Microbial strategies to target, cross or disrupt epithelia. *Curr Opin Cell Biol* **17:** 489–498.

Sousa S, Cabanes D, Bougnères L, Lecuit M, Sansonetti P, Tran-Van-Nhieu G, Cossart P. 2007. Src, cortactin and Arp2/3 complex are required for E-cadherin-mediated internalization of *Listeria* into cells. *Cell Microbiol* **9:** 2629–2643.

Stavru F, Bouillaud F, Sartori A, Ricquier D, Cossart P. 2011. *Listeria monocytogenes* transiently alters mitochondrial dynamics during infection. *Proc Natl Acad Sci* **108:** 3612–3617.

Suárez M, González-Zorn B, Vega Y, Chico-Calero I, Vázquez-Boland JA. 2001. A role for ActA in epithelial cell invasion by *Listeria monocytogenes*. *Cell Microbiol* **3:** 853–864.

Sun H, Shen Y, Dokainish H, Holgado-Madruga M, Wong A, Ireton K. 2005. Host adaptor proteins Gab1 and CrkII promote InlB-dependent entry of *Listeria monocytogenes*. *Cell Microbiol* **7:** 443–457.

Surka MC, Tsang CW, Trimble WS. 2002. The mammalian septin MSF localizes with microtubules and is required for completion of cytokinesis. *Mol Biol Cell* **13:** 3532–3545.

Tham TN, Gouin E, Rubinstein E, Boucheix C, Cossart P, Pizarro-Cerdá J. 2010. Tetraspanin CD81 is required for *Listeria monocytogenes* invasion. *Infect Immun* **78:** 204–209.

Thedieck K, Hain T, Mohamed W, Tindall BJ, Nimtz M, Chakraborty T, Wehland J, Jänsch L. 2006. The MprF protein is required for lysinylation of phospholipids in listerial membranes and confers resistance to cationic antimicrobial peptides (CAMPs) on *Listeria monocytogenes*. *Mol Microbiol* **62:** 1325–1339.

Tilney LG, Portnoy DA. 1989. Actin filaments and the growth, movement, and spread of the intracellular bacterial parasite, *Listeria monocytogenes*. *J Cell Biol* **109:** 1597–1608.

Toledo-Arana A, Dussurget O, Nikitas G, Sesto N, Guet-Revillet H, Balestrino D, Loh E, Gripenland J, Tiensuu T, Vaitkevicius K, et al. 2009. The *Listeria* transcriptional landscape from saprophytism to virulence. *Nature* **459:** 950–956.

Trusolino L, Comoglio PM. 2002. Scatter-factor and semaphorin receptors: Cell signalling for invasive growth. *Nat Rev Cancer* **2:** 289–300.

Vadia S, Arnett E, Haghighat A-C, Wilson-Kubalek EM, Tweten RK, Seveau S. 2011. The pore-forming toxin Listeriolysin O mediates a novel entry pathway of *L. monocytogenes* into human hepatocytes. *PLoS Pathog* **7:** e1002356.

Vázquez-Boland JA, Kocks C, Dramsi S, Ohayon H, Geoffroy C, Mengaud J, Cossart P. 1992. Nucleotide sequence of the lecithinase operon of *Listeria monocytogenes* and possible role of lecithinase in cell-to-cell spread. *Infect Immun* **60:** 219–230.

Veiga E, Cossart P. 2005. *Listeria* hijacks the clathrin-dependent endocytic machinery to invade mammalian cells. *Nat Cell Biol* **7:** 894–900.

Veiga E, Guttman JA, Bonazzi M, Boucrot E, Toledo-Arana A, Lin AE, Enninga J, Pizarro-Cerdá J, Finlay BB, Kirchhausen T, et al. 2007. Invasive and adherent bacterial pathogens co-Opt host clathrin for infection. *Cell Host Microbe* **2:** 340–351.

Wang L, Lin M. 2008. A novel cell wall-anchored peptidoglycan hydrolase (autolysin), IspC, essential for *Listeria monocytogenes* virulence: Genetic and proteomic analysis. *Microbiology* **154:** 1900–1913.

Yu WL, Dan H, Lin M. 2007. Novel protein targets of the humoral immune response to *Listeria monocytogenes* infection in rabbits. *J Med Microbiol* **56:** 888–895.

Zeldovich VB, Robbins JR, Kapidzic M, Lauer P, Bakardjiev AI. 2011. Invasive extravillous trophoblasts restrict intracellular growth and spread of *Listeria monocytogenes*. *PLoS Pathog* **7:** e1002005.

Yamada S, Pokutta S, Drees F, Weiss WI, Nelson WJ. 2005. Deconstructing the cadherin-catenin-actin complex. *Cell* **123:** 889–901.

Zhang L. 2010. Glycosaminoglycan (GAG) biosynthesis and GAG-binding proteins. *Prog Mol Biol Transl Sci* **93:** 1–17.

Pathogenesis of Meningococcemia

Mathieu Coureuil[1,2], Olivier Join-Lambert[1,2,3], Hervé Lécuyer[1,2,3], Sandrine Bourdoulous[2,4,5], Stefano Marullo[2,4,5], and Xavier Nassif[1,2,3]

[1]INSERM, Unité U1002, Paris, France

[2]Université Paris Descartes, Sorbonne Paris Cité, Faculté de Médecine, Paris, France

[3]Assistance Publique–Hôpitaux de Paris, Hôpital Necker Enfants Malades, Paris, France

[4]Inserm, U1016 Institut Cochin, Paris, France

[5]CNRS, UMR8104 Paris, France

Correspondence: xavier.nassif@inserm.fr

Neisseria meningitidis is responsible for two major diseases: cerebrospinal meningitis and/or septicemia. The latter can lead to a purpura fulminans, an often-fatal condition owing to the associated septic shock. These two clinical aspects of the meningococcal infection are consequences of a tight interaction of meningococci with host endothelial cells. This interaction, mediated by the type IV pili, is responsible for the formation of microcolonies on the apical surface of the cells. This interaction is followed by the activation of signaling pathways in the host cells leading to the formation of a microbiological synapse. A low level of bacteremia is likely to favor the colonization of brain vessels, leading to bacterial meningitis, whereas the colonization of a large number of vessels by a high number of bacteria is responsible for one of the most severe forms of septic shock observed.

Neisseria meningitidis (meningococcus) is a Gram-negative coccus restricted to humans, which is responsible for two major diseases: cerebrospinal meningitis and/or septicemia. The latter can lead to a purpura fulminans, an often-fatal condition owing to the associated septic shock. Paradoxically, *N. meningitidis* is a common inhabitant of the human nasopharynx, and as such is a normal, saprophytic organism that is transmitted from person to person by direct contact. Only in a small proportion of colonized subjects does the bacteria invade the bloodstream where they are responsible for septicemia and/or meningitis, after crossing of the blood–brain barrier.

Bacterial meningitis is the leading cause of central nervous system (CNS) infection. The blood–brain barrier (BBB) protects the CNS from most bacteria that may have reached the bloodstream. Most of the few types of bacteria, which can cross BBB to invade the meninges, are extracellular pathogens: *Escherichia coli* K1 and *Streptococcus agalactiae* (Group B Streptococcus) in the newborn, *N. meningitidis*, *Haemophilus influenzae* type b, and *Streptococcus pneumoniae* in children and adults (Pong and Bradley 1999; Huang et al. 2000; Van de Beek et al. 2006). Once inside the cerebrospinal fluid (CSF), bacterial multiplication is thought to be uncontrolled, owing to the local deficiency

Cite this article as *Cold Spring Harb Perspect Med* doi: 10.1101/cshperspect.a012393

in complement and immunoglobulins, and despite the influx of polymorphonuclear leukocytes induced by the local inflammatory response. The small number of bacterial species capable of invading the meninges suggests that specific virulence factors are required for bacteria to enter the subarachnoidal space. Among the above-mentioned extracellular bacteria, *N. meningitidis* is the pathogen that once in the bloodstream, is able to invade the meninges the most effectively. It has been estimated that 63% of the cases of bacteremia owing to *N. meningitidis* are associated with meningitis (InVS 2009).

The other specific clinical feature of meningococcal infection is the ability of the bacteria to multiply rapidly in the bloodstream and to be responsible for one of the most severe forms of septic shock observed, associating extensive thrombosis, vascular leakage, and cardiovascular failure (i.e., the purpura fulminans). Purpura fulminans usually occur at an early stage of the bloodstream infection, before an inflammatory response can be observed in the CSF. The epidemic nature of *N. meningitidis* infections and the high mortality rate of purpura fulminans are responsible for the fear of meningococcal infections in communities.

This article will concentrate on the pathogenesis of these two clinical aspects of the meningococcal infection.

THE MENINGOCOCCAL INFECTION

As mentioned above, *N. meningitidis* is a frequent asymptomatic colonizer of the human nasopharynx, and only a very small proportion of infections proceed to a sustained bacteremia. The mechanisms responsible for nasopharyngeal colonization and crossing of the nasopharyngeal mucosa remain mostly unexplained and will not be approached in this review. The reasons why disease occurs in some individuals and not in others remain unclear, but human genetic polymorphism is likely to be important in determining the outcome, particularly regarding the risk of developing purpura fulminans (Brouwer et al. 2009). In addition, all meningococci do not have the same pathogenic potential. Indeed, analysis of results from multilocus sequence typing (MLST) has shown the existence of distinct phylogenetic groups (clonal complexes), some of which are more likely to be isolated from patients with disease than from asymptomatic carriers (Maiden et al. 1998). These are the so-called "hypervirulent" or "hyperinvasive" lineages. Recently, the presence of a prophage has been shown to be responsible for a large proportion of invasiveness of strains belonging to hyperinvasive lineages (Bille et al. 2005, 2008). This element inserted into the bacterial chromosome can be induced to produce a filamentous phage.

N. meningitidis interacts only with human cells and there is no animal model of meningococcal sepsis. In some circumstances mice and infant rats have been used to assess the ability of the bacteria to survive in the extracellular fluids (Sun et al. 2000), but these models are unable to assess the consequences of the interaction with endothelial cells (Fig. 1).

Consequently, most hypotheses regarding the pathogenesis of meningococcal infections are derived from postmortem studies of patients who died from meningococcal meningitis or purpura fulminans or from biopsies of skin purpuric lesions (Pron et al. 1997; Harrison et al. 2002). In the brain, data obtained from a patient who died of fulminant meningoccocemia at the time bacteria were invading the CSF revealed that *N. meningitidis* adheres onto endothelial cells and forms small colonies on the endoluminal surface of the microvasculature. Bacteria are also found inside cells and in intercellular spaces (Pron et al. 1997). These data suggest that *N. meningitidis* crosses the blood–brain barrier through a direct interaction between bacteria and endothelial cells. In peripheral purpuric lesions that are commonly observed during meningococcal bacteremia, retraction of endothelial cells with capillary disruption can be observed, as well as hemorrhages, adhesion of leukocytes, and formation of small thrombi. Altogether, in vivo observations thus suggest that the pathogenesis of meningococcal infection relies on the ability of the bacteria (1) to survive and/or grow in the bloodstream, and (2) to interact and to signal to endothelial cells throughout the body.

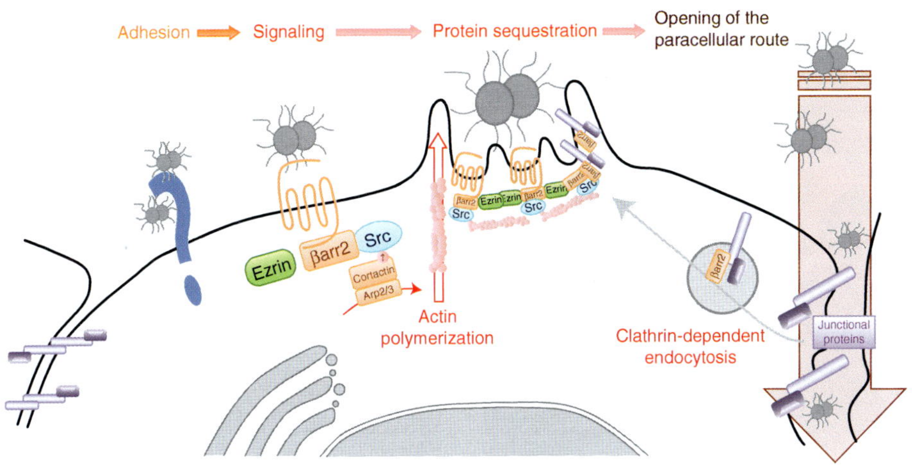

Figure 1. Interaction of *Neisseria meningitidis* with endothelial cells. *N. meningitidis* adhere to brain microvascular endothelial cells through an interaction between type IV pili and an unknown adhesion receptor (Nassif et al. 1994). Following initial bacterial adhesion, type IV pili mediate the recruitment and the activation of the β2-adrenoceptor thus leading to the organization of a specific cytoplasmic molecular complex, referred to as cortical plaques. The formation of cortical plaques results (1) from the local accumulation of ezrin and ezrin-binding proteins, and (2) from the accumulation of β-arrestins and β-arrestin-binding molecules such as Src, which induce active actin polymerization, and that of the junctional proteins p120-catenin and VE-cadherin (Hoffmann et al. 2001; Coureuil et al. 2009, 2010). Consequently, this promotes the opening of the cell–cell junctions and allows the transmigration of bacteria through the endothelium.

From these in vivo data, it can be speculated that the clinical forms of meningococcal disease mostly reflect the level of bacteremia (Fig. 2). When a low or moderate number of meningococci circulate within in the bloodstream, the bacteria interacting with peripheral capillaries cause only few localized purpuric lesions, whereas the interaction with brain endothelial cells is sufficient to lead to meningeal invasion. In that case, meningitis is the prominent clinical sign of the disease. In contrast, in the case of high bacteremia, peripheral endothelial cells are probably massively colonized by meningococci, leading to a significant increase of vascular permeability possibly associated with extensive thrombosis and purpura. In this case, bacteria also invade the meningeal spaces, but clinical presentation of disease is a septic shock associated with purpuric lesions, purpura fulminans.

THE BLOOD PHASE

Meningeal invasion by *N. meningitidis* is directly correlated with bacteremia. The bacteremia is believed to favor meningeal invasion by directly increasing the likelihood of the interaction between bacteria and the components of the blood–CSF barrier. In addition, the bacteremia is critical in the case of sepsis, as mortality is directly related to levels of bacteremia (Brandtzaeg et al. 1995). Therefore, the bacterial attributes involved in growth and/or survival in the extracellular fluids play an essential role in meningeal invasion by *N. meningitidis* and/or the sepsis outcome.

The lack of a relevant animal model has greatly hampered the identification of the bacterial factors important for bacteria survival, multiplication, or adaptation in the bloodstream. Studies aimed at identifying these factors were based on the screen of transposon mutagenesis libraries using either an infant rat model (Sun et al. 2000) or survival in complemented serum (Geoffroy et al. 2003) as readout. More recently, a transcriptomic analysis of bacteria grown in human whole blood has been performed (Echenique-Rivera et al. 2011). These studies pointed out the major role of the virulence factors

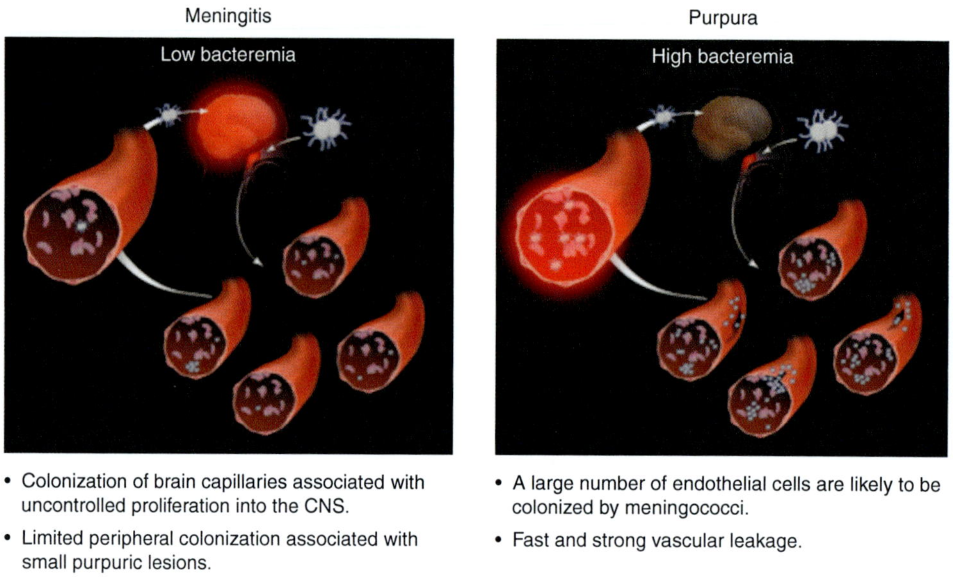

Figure 2. The clinical features of meningococcemia are dependent on bacterial load. A low level of bacteremia is likely to be associated with limited vascular colonization and few purpuric lesions. Some colonies in the brain capillaries may breach the blood–brain barrier and proliferate into the central nervous system (CNS). Thus, few colonies are sufficient to induce meningitis. On the other hand, high bacteremia is associated with an important colonization of the peripheral blood vessels that results in a fast and strong vascular leakage.

commonly observed in most extracellular pathogens. In addition to iron chelation systems, which are essential for pathogenic bacteria to obtain the necessary ferric iron in vivo (Stojiljkovic et al. 1995), the polysaccharidic capsule and the lipooligosaccharide (LOS) participate in the prevention of bacterial killing by the host innate immune system effectors, such as complement. A new virulence factor, the factor-H-binding protein (Fhbp), was recently identified (Seib et al. 2008). It consists of 28 kDa surface-exposed lipoprotein that is capable of inducing bactericidal antibodies (Masignani et al. 2003; Fletcher et al. 2004). Binding of human factor H to Fhbp in the bloodstream of infected individuals permits *N. meningitidis* to effectively evade host immune responses by preventing complement-mediated bacterial cell lysis. Based on sequence diversity, Fhbp proteins are classified into three variants and two subfamilies (Masignani et al. 2003; Fletcher et al. 2004). More recently, the modular protein structure has been described for Fhbp whereby a combination of five distinct variable segments defines six unique modular groups (Beernink and Granoff 2009). This protein is expressed by all *N. meningitidis* strains studied to date, although the level of expression varies between strains (high, intermediate, or low expressers). Antibodies directed against Fhbp are bactericidal and this protein is currently one of the best vaccine candidates (Beernink et al. 2006).

Much less is known about the contribution of other actors of the innate immune system, such as polymorphonuclear neutrophils (PMNs), in the protection against meningococcal disease. These phagocytic cells elaborate multiple mechanisms effective at killing microorganisms, such as the production of antimicrobial peptides, degradative enzymes, and reactive oxygen species derived from the oxidative burst. Recently, it has been shown that a mechanism by which *N. meningitidis* escape the killing by PMNs to survive in the bloodstream is owing to its ability to uptake available L-glutamate and to convert it to glutathione; by maintaining intracellular redox potential, this key molecule protects the bacterium from reactive oxygen species, such as hydrogen peroxide, that are produced by the oxidative burst of PMNs (Tala et al. 2011).

It should be pointed out that meningococcal infections are very often discovered at the stage of cerebrospinal meningitis, thus suggesting that the bloodstream phase can be totally asymptomatic until the bacteria cross the blood–brain barrier and develop an inflammation in the CSF. This is an apparent discrepancy with the fact that, in some patients, bacteria in the bloodstream lead to a purpura fulminans. As for all Gram-negative bacteria, the lipopolysaccharide is the major inducer of the proinflammatory response via activation of the Toll-like receptor 4. This response is believed to be beneficial to the host when induced at low levels. However, it can become detrimental for the infected host when excessive stimulation occurs (Beutler and Rietschel 2003; Palsson-McDermott and O'Neill 2004; Russell 2006). Meningococcal sepsis is generally seen as the prototypical endotoxin-mediated disease. The chemical composition of the lipid A determines how well it is recognized by TLR4. *N. meningitidis* has been reported to produce lipid A with six acyl chains, the optimal number for TLR4 recognition. Recent data have shown that in a significant percentage of invasive meningococcal diseases, the lipid A was penta-acylated. This modification was the consequence of a mutation in the *lpX1* gene. The corresponding strains were dramatically less efficient in inducing the production of cytokines. Furthermore, patients infected by these strains presented significantly less frequent rash and platelet counts were higher, consistent with reduced cytokine induction and less activation of tissue-factor-mediated coagulopathy (Fransen et al. 2009). Because underacylation of lipid A, caused by mutations in the *lpxl1* gene, helps bacteria evade the innate immune system, this might explain why the blood phase could remain asymptomatic. However, such a mutation was observed in only 9% of the isolates obtained in meningococcal invasive diseases. It is therefore likely that *N. meningitidis* also has other means to reduce the activation of the host innate immune system.

THE INTERACTION OF *N. MENINGITIDIS* WITH ENDOTHELIAL CELLS

The tropism of *N. meningitidis* for microvasculature endothelial cells is a major pathophysiological issue in meningococcal infections. This property is likely to be associated with two clinical features of meningococcal infections: the crossing of the BBB and the formation of peripheral purpuric lesions, which can lead to a purpura fulminans. Endothelial permeability is tightly regulated to maintain endothelial integrity and to prevent exposure of the highly thrombogenic subendothelial blood vessel matrix (Danese et al. 2007). Transendothelial crossing involves both transcellular and paracellular routes. Transcellular passage requires either cell fenestration or the establishment of complex systems of vesiculo–vacuolar organelles (Nourshargh et al. 2010); it is regulated by signaling pathways involved in endocytosis and vesicular trafficking. The paracellular path is based on the coordinated opening and closure of endothelial cell–cell junctions, and is regulated via the proteins that are involved in the organization of these junctions. Endothelial cells possess specialized junctional regions such as adherens (AJs) and tight junctions (TJs). They express cell-type-specific transmembrane adhesion proteins, such as VE-cadherin at AJs, and claudin-5 at TJs. Composition and morphology of junctional structures vary along the vascular tree, as a consequence of different permeability requirements. Whereas AJs are relatively conserved in all types of vessels, TJs are lost or poorly organized where dynamic and rapid exchanges between blood and tissue are required, as in postcapillary venules. On the other hand, they are extremely well developed where permeability is strictly controlled, as required in the brain microvasculature (Dejana et al. 2009). Brain microvascular endothelial cells are essential components of the BBB. The close interaction between microvascular endothelial cells and other components of the neurovascular unit (astrocytes, pericytes, neurons, and basement membrane) ensure the differentiation of endothelial cells into brain endothelial cells and the proper function of the CNS.

Transport across the BBB is strictly limited through both physical (TJs) and metabolic barriers (enzymes, diverse transport systems) (Wolburg and Lippoldt 2002). A functional polarity exists between the luminal and basal membrane surfaces of brain microvascular endothelial cells. During bacteremia, the BBB also prevents the bacteria from invading the brain.

In vivo, blood flow generates mechanical forces that vary depending on the type and diameter of the vessels, and that could prevent bacterial interaction with the endothelial cells. The ability of *N. meningitidis* to bind to endothelial cells in the presence of shear stress mimicking the bloodstream was recently investigated (Mairey et al. 2006). These data revealed that, after initial attachment, bacteria have the ability to resist high blood velocities, to multiply, and to form microcolonies onto the apical surface of the endothelial cells. This resistance to shear stress and the ability to grow at the luminal surface of endothelial cells in the presence of blood flow highlight the efficacy of the interaction between *N. meningitidis* and the host cells.

Various bacterial surface components have been described allowing the interaction of *N. meningitidis* with human cells. These are type IV pili (Tfp) and other factors such as Opa or Opc proteins (for review, see Carbonnelle et al. 2009; Virji 2009). However, in capsulated bacteria, type IV pili are the main bacterial factors capable of promoting adhesion, because nonpiliated capsulated bacteria are unable to adhere to any cell type. Early work performed with piliated capsulated meningococci has shown that meningococcal interaction with human cells can be divided in two steps. The first step allows the adhesion of single diplococci in a rather inefficient manner. The second step corresponds to the bacterial division onto the apical surface of the cells. The high number of bacteria that interact with cells is a consequence of bacterial division of the few meningococci that have initially succeeded to adhere. Type IV pili are required for both steps. They promote the initial interaction of diplococci with the endothelial cells, and then generate bacteria–bacteria interactions that lead to the spreading of the bacteria on the apical surface of the cells.

Type IV Pili

Pilus Biogenesis

Type IV pili are polymeric filaments found on many Gram-negative bacteria (Wolfgang et al. 2000). These structures correspond to the multimeric assembly of a pilin subunit. Functional Tfp are dynamic structures. Pilin subunits are constantly assembled into fibers from a platform in the inner membrane. The fiber is then extruded through the outer membrane via the secretin PilQ (Wolfgang et al. 2000). A remarkable property of Tfp is their ability to retract into the bacterium from which they originate, via the action of the force-generating ATPase PilT (Merz et al. 2000; Skerker and Berg 2001). Retraction is a consequence of the disassembly of pilin subunits that are then stored in the cytoplasmic membrane (Morand et al. 2004). Tfp retraction is essential for bacterial motility (twitching motility), competence for DNA transformation, and pilus-associated signaling to host cells (Lee et al. 2005).

The major neisserial pilus subunit, the pilin, is encoded by the *pilE* gene that is subject to antigenic variation following recombination with silent loci (*pilS*). Pilin is synthesized as a preprotein. A short leader sequence (5–6 residues) is cleaved by the prepilin peptidase PilD (Lory and Strom 1997). PilD also methylates the amino-terminal phenylalanine of the mature protein product. Pilins are packed through internal hydrophobic interactions between conserved amino-terminal α helices, leaving hypervariable carboxy-terminal globular regions exposed (Parge et al. 1995). PilE undergoes several posttranslational modifications of serine residues including glycosylation at position 63, whereas at position 68, the residue has been reported to be modified by phosphate, phosphoethanolamine, or phosphorylcholine addition (Forest et al. 1999; Hegge et al. 2004). Concerning the glycosylation, the structure of the sugar is different depending on the strain (Virji 1997; Chamot-Rooke et al. 2007).

In addition to PilE, pili also contain the less abundant proteins PilX, PilV, and ComP. These are called minor pilins (Winther-Larsen et al. 2001; Helaine et al. 2005; Mikaty et al. 2009;

Brown et al. 2010) as they structurally resemble PilE and are likely to be assembled within the filaments in a similar way (Craig and Li 2008). Importantly, each minor pilin modulates Tfp-linked properties (Brown et al. 2010). PilX is crucial for the formation of bacterial aggregates and indirectly controls adhesion to human cells by promoting bacteria–bacteria interactions (Helaine et al. 2005, 2007); ComP is essential for competence for DNA transformation (Wolfgang et al. 1999; Aas et al. 2002; Brown et al. 2010), whereas PilV affects several pilus-linked properties such as signaling to endothelial cells (see below) (Brown et al. 2010).

The PilC proteins play a crucial but still enigmatic role (Jonsson et al. 1991; Nassif et al. 1994; Rudel et al. 1995a,b; Wolfgang et al. 2000). Two alleles were originally discovered (Jonsson et al. 1991). Expression of both variants is subject to phase variation as a result of frameshift in homopolymeric "G" tracts located in the open reading frames (Jonsson et al. 1991). PilC-null strains show impaired pilus expression and lack the ability for transformation competence. In *N. meningitidis*, only PilC1 is required for adhesion. PilC2, which is expressed independently of PilC1, fails to promote adhesion despite identical functions in pilus expression and transformation competence (Nassif et al. 1994; Morand et al. 2001). Abolition of *pilT* in a PilC-null background restores piliation, confirming the hypothesis that PilC acts as an antagonist of PilT by preventing PilT-mediated retraction (Wolfgang et al. 1998; Morand et al. 2004).

Mechanism of Pilus-Mediated Interaction with Host Cells

The molecular mechanism responsible for the first step of the adhesive process, the initial attachment of individual diplococci to the cells, is still not fully understood. One report suggested that the PilC1 protein may carry a cell-binding domain (Rudel et al. 1995b); this hypothesis was based on inhibition of adhesion using purified PilC molecules. However, nonadhesive nonpiliated isolates of a serogroup B strain with high PilC expression and piliated adhesive isolates with barely detectable PilC expression have been described (Virji et al. 1995). In addition, another $PilC^{+}/PilE^{-}$ strain in which PilC is localized in the outer membrane, is unable to interact with eukaryotic cells (X Nassif, unpubl.), thus raising doubts on the role of PilC as an adhesin. It is therefore likely that the binding domain on the type IV pili remain to be identified. The CD46 receptor was proposed as being the adhesion receptor for both *Neisseria gonorrheae* and *N. meningitidis* type IV pili. However, this finding was not confirmed by subsequent studies (Kallstrom et al. 1997; Kirchner et al. 2005). The Laminin receptor was also described as a potential receptor for *N. meningitidis*. Two bacterial ligands for this receptor have been reported: the PilQ secretin and the PorA protein (Orihuela et al. 2009). However, considering that PilQ is expressed in nonpiliated mutants, and that nonpiliated noncapsulated strains are unable to interact with endothelial cells, its role remains to be determined. The I-domain-containing integrins were described to be essential for *N. gonorrheae* adhesion to primary urethral epithelial cells (Edwards and Apicella 2005). However, similar data have not been reported for *N. meningitidis* and endothelial cells.

Following the initial adhesion of a few diplococci, the formation of microcolonies on the host cells is owing to bacteria–bacteria interactions. As mentioned above, the minor pilin PilX is essential to promote interbacterial interactions, and *pilX* mutants, which are unable to form aggregates, are also unable to multiply at the cell surface. Bacterial spreading onto the host cells relies also on the ability of bacteria to retract their pili. Following the interactions between bacteria and target cells, in some systems, pili have been shown to retract and eventually adherent meningococci appear nonpiliated (Pujol et al. 1999; Merz et al. 2000). Measurements using optical tweezers showed that retraction of a single Tfp generates forces up to 110 pN in a transient manner for each fiber. Bundles of Tfp, which result from the association of 8 to 10 pili, act as coordinated retractable units. Bundles can generate retraction forces in the nanonewton range (Biais et al. 2008). The successive extension, binding, and retraction of Tfp enable bacteria to move by twitching motility and spread on the apical surface of the host cells. In addition, it has been recently shown that the addition of phosphoglycerol molecules on pilins, which is increased during bacterial cell interaction, reduces type IV pili-dependent contacts between bacteria and enhances the ability of the bacteria to detach from the adherent aggregates and to disseminate (Chamot-Rooke et al. 2011).

The Consequence of the Type IV Pilus-Mediated Interaction with Endothelial Cells

Type IV pilus-mediated adhesion allows the activation of a signaling receptor at the site of bacterial cell interaction, which mediates the formation of a specific membrane domain designated the "cortical plaque." This membrane domain is enriched in structural proteins such as the ERM proteins (ezrin-radixin-moesin family) that anchor the actin network to the site of bacterial adhesion (Merz et al. 1999). The concomitant activation of Rho GTPases and Src tyrosine kinase promotes cortactin-dependent actin polymerization (Coureuil et al. 2009). In addition to the accumulation of the structural proteins, the cortical plaque is enriched in adhesion receptors or membrane receptors of receptor tyrosine kinase and G-protein-coupled receptor (GPCR) families (Hoffmann et al. 2001; Doulet et al. 2006; personal observation).

The interaction domain between the bacterial colony and the host cell can be compared to a synapse because (1) it is composed of adhesion receptors and signaling receptors (that are mostly immunoglobulin domain-containing receptors and GPCRs, respectively), and (2) it transmits signals. One particularity of this "bacterial synapse" is that adhesion receptor and signaling receptor are not internalized but sequestrated underneath the meningococcal colonies (Doulet et al. 2006; personal observation). The "bacterial synapse" is maintained by a cortical cytoskeletal protein network highly enriched in members of the ERM protein family, which are anchored to the plasma membrane by their PIP2-binding domain and control the organization of the cortical actin cytoskeleton through their carboxy-terminal F-actin binding sites (Eugene et al. 2002; Lambotin et al. 2005; Doulet et al. 2006). ERM proteins also bind the cytoplasmic domain of several ERM-binding transmembrane receptors such as CD44 and ICAM-1 through their carboxy-terminal domain (Merz and So 1997; Merz et al. 1999; Eugene et al. 2002). As a consequence of the recruitment of ERM proteins, many receptors known to be involved in leukocyte adhesion are sequestered underneath bacterial colonies (Doulet et al. 2006). During leukocyte adhesion, these components that form the "endothelial docking structures" or "transmigratory cups" are essential to promote firm adhesion and extravasation of leukocytes through the endothelium. Transmigratory cups result from the dynamic redistribution of ICAM-1, VCAM-1, E-selectin, and CD44 at the endothelial-leukocyte contact area, accompanied by the recruitment of activated ERM proteins, and leads to cortical actin polymerization (Carman and Springer 2004; Shaw et al. 2004; Barreiro et al. 2005). Because the same set of endothelial proteins is present within the membrane protrusions induced by *N. meningitidis* and within the docking structures promoted by leukocyte adhesion, it was suggested that *N. meningitidis* highjack the leukocyte adhesion pathway (Doulet et al. 2006).

The β2 Adrenergic Receptor Is the Signaling Receptor

The initial cellular receptor responsible for pilus-mediated adhesion remains unknown. However, the β2-adrenergic receptor was recently shown to be recruited at the site of bacterial cell interaction following this initial event and was described as an important signaling receptor for *N. meningitidis* (Coureuil et al. 2010; Lecuyer et al. 2012). The β2-adrenergic receptor is a GPCR that signals via the heterotrimeric Gαs protein and β-arrestins. This receptor is also known for its implication in vascular homeostasis and disease. The expression of both the β2-adrenergic receptor and β-arrestins is sufficient

 Cite this article as *Cold Spring Harb Perspect Med* doi: 10.1101/cshperspect.a012393

to promote an *N. meningitides*-induced cell response in an incompetent cell line, such as HEK293 cells (Coureuil et al. 2010). It has been shown that PilE and PilV, the major pilin subunit and a minor pilin subunit, respectively, directly interact with the extracellular amino-terminal domain of the β2-adrenergic receptor to transmit the signal (Coureuil et al. 2010). This interaction is believed to modify the conformation of the receptor, resulting in the activation of β-arrestin-mediated signaling without activating the heterotrimeric Gαs protein and the downstream adenyl cyclase/cAMP pathway, a property referred to as biased activation (Coureuil et al. 2010). β-arrestins are scaffolding proteins involved in many cellular processes such as receptor internalization and actin polymerization (DeWire et al. 2007). Following the activation of the β2-adrenergic receptor by *N. meningitidis* and the accumulation of β-arrestins, the signal leads to the formation of a "raftlike" membrane domain enriched in cholesterol and PIP2, in which several transmembrane receptors and structural proteins are sequestered, thus leading to the subsequent formation of the cortical plaque (Doulet et al. 2006; Mikaty et al. 2009). It is likely that accumulated β-arrestins underneath bacteria play a major role in the sequestration of these signaling molecules. Nevertheless, inhibition of the β2-adrenergic-mediated signaling does not affect bacterial adhesion (Coureuil et al. 2010), thus confirming that adhesion and signaling are two independent events, owing to the recognition of two separate receptors.

In summary, following the initial adhesion of the bacteria to a yet-unknown receptor, the β2-adrenergic receptor is recruited and activated by components of the type IV pili, thus inducing the formation of the cortical plaque.

Consequences of N. meningitidis-Induced Signaling on Endothelial Cells

The formation of the cortical plaque is accompanied by a robust local stimulation of actin polymerization leading to the elongation of membrane protrusions around bacteria (Merz and So 1997; Eugene et al. 2002). Interestingly, the formation of such protrusions was also observed in vivo on brain endothelial cells by transmission electron microscopy analysis of brain sections from a child who died of fulminant meningoccemia (Pujol et al. 1997). These cellular projections have been shown to be required to allow the microcolonies to stand up to the shear stress of the bloodstream (Mairey et al. 2006). Actin polymerization relies on the activation of the small GTPases of the Rho family (Eugene et al. 2002; Lambotin et al. 2005). In addition, it has been shown that proper actin polymerization in these membrane protrusions relies on cortactin phosphorylation. Cortactin (or cortical actin-binding protein) is a perinuclear cytoplasmic protein that is involved in the reorganization of the cell cortical actin cytoskeleton. It appears that cortactin recruitment and phosphorylation is finely tuned by *N. meningitidis*:

1. The recruitment of cortactin at the site of *N. meningitidis* adhesion is controlled by the Cdc42-Par6/PKCζ pathway (Coureuil et al. 2009).

2. The activation of cortactin through phosphorylation is controlled by the tyrosine kinase Src that is sequestered in the cortical plaque and activated by direct interaction with β-arrestins (Luttrell et al. 1999; Coureuil et al. 2010; Lecuyer et al. 2012).

3. The ErbB2 tyrosine kinase receptor also regulates Src activity and the subsequent cortactin phosphorylation (Hoffmann et al. 2001). The ErbB2 tyrosine kinase receptor belongs to the family of epidermal growth factor (EGF) receptors. The interaction of *N. meningitidis* with human endothelial cells leads to ErbB2 activation in the cortical plaque most likely via homodimerization. This is an example of secondary signaling activated by the accumulation of a cellular receptor at the site of bacterial adhesion that contributes to the formation of the cortical plaque.

Another consequence of the formation of the cortical plaque is the opening of the inter-endothelial junctions allowing the transmigration of bacteria through a paracellular route. Indeed, it has been shown that *N. meningitidis* recruits junctional proteins underneath the colony into the cortical plaque. This recruitment is owing to the activation of the Cdc42-Par3/Par6/PKCζ pathway that is usually involved in the formation of adherens and TJs at cell–cell contact. Here, the ectopic activation of the polarity complex Par3/Par6/PKCζ leads to abnormal recruitment of junctional proteins that are sequestrated underneath bacterial colonies through their interaction with β-arrestins. As a consequence, these molecules are depleted at the intercellular junctions causing endothelium leakage. Adhesion of *N. meningitidis* lately promotes the cleavage of occludin (a component of the tight junction) by the metalloproteinase MMP-8 (Schubert-Unkmeir et al. 2010), further altering the intercellular junctions.

In summary, two important consequences for meningococcal pathogenesis are associated with type IV pilus-mediated adhesion: (1) the ability to be protected from the shear stress by the formation of a microvilli-like structure on the apical surface of endothelial cells, which is a prerequisite to the formation of adhesive bacterial microcolonies, and (2) the retraction of the opening of a paracellular route between endothelial cells, thus locally jeopardizing the integrity of the capillaries.

CONSEQUENCES OF *N. meningitidis* INTERACTION WITH ENDOTHELIAL CELLS FOR MENINGOCOCCAL PATHOGENESIS

As described above, *N. meningitidis* is an extracellular pathogen that once disseminated into the bloodstream, has two specificities: (1) its ability to interact freely with the BBB to reach the subarachnoidal space, and (2) in some circumstances to be responsible for a severe shock with extensive purpura (i.e., the purpura fulminans).

The Crossing of the BBB

Following its interaction of bacteria with endothelial cells, there are at least four strategies for a pathogen to cross a monolayer of brain endothelial cell monolayers: (1) transcellular transport by passive or adhesion-induced transcytosis, (2) paracellular passage through open TJs, (3) disruption of the endothelial barrier owing to a direct cytotoxic effect, and (4) leukocyte-facilitated transport by infected phagocytes.

Extracellular pathogens, such as *N. meningitidis*, do not use leukocytes as vehicles to cross the blood–CSF barrier. A breakdown of the blood–CNS barrier owing to apoptosis or bacterial cytotoxity is unlikely, because tissue lesions, such as hemorrhages in the subarachnoidal space, are uncommon during bacterial meningitis. Therefore, the entry of blood-borne pathogen meningococcus most probably respects the architecture of the blood–CNS barrier (Koedel et al. 2002). Accordingly, adhesion of *N. meningitidis* to endothelial cells can induce an intracellular signaling pathway leading to disruption of intercellular TJs or, alternatively, bacteria may induce their own transcytosis through the cell monolayer. In vitro, transcytosis of bacteria through human brain endothelial cells has been shown for *S. pneumoniae* (Ring et al. 1998), *E. coli* (Prasadarao et al. 1999; Stins et al. 1999, 2001), *S. agalactiae* (Nizet et al. 1997), and through human umbilical vein endothelial cells for *H. influenzae* (Virji et al. 1991, 1992). Although *N. meningitidis* can readily be internalized in vitro within vacuoles in human brain microvascular endothelial cells (Nikulin et al. 2006), the fact that *N. meningitidis* can open gaps in a monolayer of brain endothelial cells as a consequence of the delocalization of junctional components suggests that *N. meningitidis* cross the BBB using the paracellular route.

The Purpura Fulminans

In purpuric lesions, endothelial cell retraction is observed at the site of bacterial endothelial cells interaction, associated with a loss of the integrity of the capillary resulting in an increase in permeability, hemorrhages, leukocyte aggregation and formation of thrombi hemorrhages, and adhesion of leukocytes (Dupin et al. 2012). As the signaling events induced by adherent microcolonies of *N. meningitidis* are responsible for disjunction of endothelial cells in vitro, it is most likely that these lesions are also owing to initiation of similar signaling pathways in

 Cite this article as *Cold Spring Harb Perspect Med* doi: 10.1101/cshperspect.a012393

vivo. The colonization of a large number of vessels by a high number of bacteria as those seen in purpura fulminans and the consequences of the corresponding signaling may explain the extensive purpuric lesions and thrombosis as well as the loss of vascular integrity responsible for the severity of the shock.

Recent exciting findings have considerably expanded our understanding of the cellular events involved in meningococcal interaction with endothelial cells, thus leading to the establishment of a hypothesis regarding the pathogenesis of meningeal invasion and of the purpura. However, despite recent advances in our understanding of these molecular mechanisms, much remains to be discovered about the complex molecular networks involved. Among the major issues is the identification of the receptor for meningococcal adhesion, which would constitute a significant breakthrough in the field, and the identification of animal models that would allow one to confirm the in vitro observations.

ACKNOWLEDGMENTS

The laboratory of X.N. is supported by INSERM, Université Paris Descartes, a grant from "La Fondation pour la Recherche Médicale," and ANR Grant No. ANR-AAP-2009-06-26. M.C. is supported by an ANR Grant No. ANR-11-JSV3-0002 and a grant "Mairie de Paris–Programme EMERGENCE."

REFERENCES

Aas FE, Wolfgang M, Frye S, Dunham S, Lovold C, Koomey M. 2002. Competence for natural transformation in *Neisseria gonorrhoeae*: Components of DNA binding and uptake linked to type IV pilus expression. *Mol Microbiol* **46:** 749–760.

Barreiro O, Yanez-Mo M, Sala-Valdes M, Gutierrez-Lopez MD, Ovalle S, Higginbottom A, Monk PN, Cabanas C, Sanchez-Madrid F. 2005. Endothelial tetraspanin microdomains regulate leukocyte firm adhesion during extravasation. *Blood* **105:** 2852–2861.

Beernink PT, Granoff DM. 2009. The modular architecture of meningococcal factor H-binding protein. *Microbiology* **155:** 2873–2883.

Beernink PT, Leipus A, Granoff DM. 2006. Rapid genetic grouping of factor h-binding protein (genome-derived neisserial antigen 1870), a promising group B meningococcal vaccine candidate. *Clin Vaccine Immunol* **13:** 758–763.

Beutler B, Rietschel ET. 2003. Innate immune sensing and its roots: The story of endotoxin. *Nat Rev Immunol* **3:** 169–176.

Biais N, Ladoux B, Higashi D, So M, Sheetz M. 2008. Cooperative retraction of bundled type IV pili enables nanonewton force generation. *PLoS Biol* **6:** e87.

Bille E, Zahar JR, Perrin A, Morelle S, Kriz P, Jolley KA, Maiden MC, Dervin C, Nassif X, Tinsley CR. 2005. A chromosomally integrated bacteriophage in invasive meningococci. *J Exp Med* **201:** 1905–1913.

Bille E, Ure R, Gray SJ, Kaczmarski EB, McCarthy ND, Nassif X, Maiden MC, Tinsley CR. 2008. Association of a bacteriophage with meningococcal disease in young adults. *PloS ONE* **3:** e3885.

Brandtzaeg P, Ovstebo R, Kierulf P. 1995. Bacteremia and compartmentalization of LPS in meningococcal disease. *Prog Clin Biol Res* **392:** 219–233.

Brouwer MC, de Gans J, Heckenberg SGB, Zwinderman AH, van der Poll T, van de Beek D. 2009. Host genetic susceptibility to pneumococcal and meningococcal disease: A systematic review and meta-analysis. *Lancet Infect Dis* **9:** 31–44.

Brown DR, Helaine S, Carbonnelle E, Pelicic V. 2010. Systematic functional analysis reveals that a set of seven genes is involved in fine-tuning of the multiple functions mediated by type IV Pili in *Neisseria meningitidis. Infect Immun* **78:** 3053–3063.

Carbonnelle E, Hill DJ, Morand P, Griffiths NJ, Bourdoulous S, Murillo I, Nassif X, Virji M. 2009. Meningococcal interactions with the host. *Vaccine* **27** (Suppl 2): B78–B89.

Carman CV, Springer TA. 2004. A transmigratory cup in leukocyte diapedesis both through individual vascular endothelial cells and between them. *J Cell Biol* **167:** 377–388.

Chamot-Rooke J, Rousseau B, Lanternier F, Mikaty G, Mairey E, Malosse C, Bouchoux G, Pelicic V, Camoin L, Nassif X, et al. 2007. Alternative *Neisseria spp.* type IV pilin glycosylation with a glyceramido acetamido trideoxyhexose residue. *Proc Natl Acad Sci* **104:** 14783–14788.

Chamot-Rooke J, Mikaty G, Malosse C, Soyer M, Dumont A, Gault J, Imhaus AF, Martin P, Trellet M, Clary G, et al. 2011. Posttranslational modification of pili upon cell contact triggers *N. meningitidis* dissemination. *Science* **331:** 778–782.

Coureuil M, Mikaty G, Miller F, Lecuyer H, Bernard C, Bourdoulous S, Dumenil G, Mege RM, Weksler BB, Romero IA, et al. 2009. Meningococcal type IV Pili recruit the polarity complex to cross the brain endothelium. *Science* **25:** 83–87.

Coureuil M, Lecuyer H, Scott MG, Boularan C, Enslen H, Soyer M, Mikaty G, Bourdoulous S, Nassif X, Marullo S. 2010. Meningococcus hijacks a β2-adrenoceptor/β-arrestin pathway to cross brain microvasculature endothelium. *Cell* **143:** 1149–1160.

Craig L, Li J. 2008. Type IV pili: Paradoxes in form and function. *Curr Opin Struc Biol* **18:** 267–277.

Danese S, Dejana E, Fiocchi C. 2007. Immune regulation by microvascular endothelial cells: Directing innate and adaptive immunity, coagulation, and inflammation. *J Immunol* **178:** 6017–6022.

Dejana E, Tournier-Lasserve E, Weinstein BM. 2009. The control of vascular integrity by endothelial cell junctions: Molecular basis and pathological implications. *Dev Cell* **16:** 209–221.

DeWire SM, Ahn S, Lefkowitz RJ, Shenoy SK. 2007. β-arrestins and cell signaling. *Ann Rev Physiol* **69:** 483–510.

Doulet N, Donnadieu E, Laran-Chich MP, Niedergang F, Nassif X, Couraud PO, Bourdoulous S. 2006. *Neisseria meningitidis* infection of human endothelial cells interferes with leukocyte transmigration by preventing the formation of endothelial docking structures. *J Cell Biol* **173:** 627–637.

Dupin N, Lecuyer H, Carlotti A, Poyart C, Coureuil M, Chanal J, Schmitt A, Vacher-Lavenu MC, Taha MK, Nassif X, et al. 2012. Chronic meningoccemia involves perivascular invasion by *N. meningitidis* through the remodelling of endothelial barriers. *Clin Infect Dis* **54:** 1162–1165.

Echenique-Rivera H, Muzzi A, Del Tordello E, Seib KL, Francois P, Rappuoli R, Pizza M, Serruto D. 2011. Transcriptome analysis of *Neisseria meningitidis* in human whole blood and mutagenesis studies identify virulence factors involved in blood survival. *PLoS Pathog* **7:** e1002027.

Edwards JL, Apicella MA. 2005. I-domain-containing integrins serve as pilus receptors for *Neisseria gonorrhoeae* adherence to human epithelial cells. *Cell Microbiol* **7:** 1197–1211.

Eugene E, Hoffmann I, Pujol C, Couraud PO, Bourdoulous S, Nassif X. 2002. Microvilli-like structures are associated with the internalization of virulent capsulated *Neisseria meningitidis* into vascular endothelial cells. *J Cell Sci* **115:** 1231–1241.

Fletcher LD, Bernfield L, Barniak V, Farley JE, Howell A, Knauf M, Ooi P, Smith RP, Weise P, Wetherell M, et al. 2004. Vaccine potential of the *Neisseria meningitidis* 2086 lipoprotein. *Infect Immun* **72:** 2088–2100.

Forest KT, Dunham SA, Koomey M, Tainer JA. 1999. Crystallographic structure reveals phosphorylated pilin from *Neisseria*: Phosphoserine sites modify type IV pilus surface chemistry and fibre morphology. *Mol Microbiol* **31:** 743–752.

Fransen F, Heckenberg SG, Hamstra HJ, Feller M, Boog CJ, van Putten JP, van de Beek D, van der Ende A, van der Ley P. 2009. Naturally occurring lipid A mutants in *Neisseria meningitidis* from patients with invasive meningococcal disease are associated with reduced coagulopathy. *PLoS Pathog* **5:** e1000396.

Geoffroy MC, Floquet S, Metais A, Nassif X, Pelicic V. 2003. Large-scale analysis of the meningococcus genome by gene disruption: Resistance to complement-mediated lysis. *Genome Res* **13:** 391–398.

Harrison OB, Robertson BD, Faust SN, Jepson MA, Goldin RD, Levin M, Heyderman RS. 2002. Analysis of pathogen-host cell interactions in purpura fulminans: Expression of capsule, type IV pili, and PorA by *Neisseria meningitidis* in vivo. *Infect Immun* **70:** 5193–5201.

Hegge FT, Hitchen PG, Aas FE, Kristiansen H, Lovold C, Egge-Jacobsen W, Panico M, Leong WY, Bull V, Virji M, et al. 2004. Unique modifications with phosphocholine and phosphoethanolamine define alternate antigenic forms of *Neisseria gonorrhoeae* type IV pili. *Proc Natl Acad Sci* **101:** 10798–10803.

Helaine S, Carbonnelle E, Prouvensier L, Beretti JL, Nassif X, Pelicic V. 2005. PilX, a pilus-associated protein essential for bacterial aggregation, is a key to pilus-facilitated attachment of *Neisseria meningitidis* to human cells. *Mol Microbiol* **55:** 65–77.

Helaine S, Dyer DH, Nassif X, Pelicic V, Forest KT. 2007. 3D structure/function analysis of PilX reveals how minor pilins can modulate the virulence properties of type IV pili. *Proc Natl Acad Sci* **104:** 15888–15893.

Hoffmann I, Eugene E, Nassif X, Couraud PO, Bourdoulous S. 2001. Activation of ErbB2 receptor tyrosine kinase supports invasion of endothelial cells by *Neisseria meningitidis*. *J Cell Biol* **155:** 133–143.

Huang SH, Stins MF, Kim KS. 2000. Bacterial penetration across the blood–brain barrier during the development of neonatal meningitis. *Microbes Infect* **2:** 1237–1244.

In VS. 2009. Réseau EPIBAC. Surveillance des infections invasives à *Haemophilus influenzae, Listeria monocytogenes, Neisseria meningitidis, Streptococcus pneumoniae, Streptococcus agalactiae* (B) et *Streptococcus pyogenes* (A) en France métropolitaine. Institut National de Veille Sanitaire (InVs), Paris.

Jonsson AB, Nyberg G, Normark S. 1991. Phase variation of gonococcal pili by frameshift mutation in pilC, a novel gene for pilus assembly. *EMBO J* **10:** 477–488.

Kallstrom H, Liszewski MK, Atkinson JP, Jonsson AB. 1997. Membrane cofactor protein (MCP or CD46) is a cellular pilus receptor for pathogenic *Neisseria*. *Mol Microbiol* **25:** 639–647.

Kirchner M, Heuer D, Meyer TF. 2005. CD46-independent binding of neisserial type IV pili and the major pilus adhesin, PilC, to human epithelial cells. *Infect Immun* **73:** 3072–3082.

Koedel U, Scheld WM, Pfister HW. 2002. Pathogenesis and pathophysiology of pneumococcal meningitis. *Lancet Infect Dis* **2:** 721–736.

Lambotin M, Hoffmann I, Laran-Chich MP, Nassif X, Couraud PO, Bourdoulous S. 2005. Invasion of endothelial cells by *Neisseria meningitidis* requires cortactin recruitment by a phosphoinositide-3-kinase/Rac1 signalling pathway triggered by the lipo-oligosaccharide. *J Cell Sci* **118:** 3805–3816.

Lecuyer H, Nassif X, Coureuil M. 2012. Two strikingly different signaling pathways are induced by Meningococcal type IV pili on endothelial and epithelial cells. *Infect Immun* **80:** 175–186.

Lee SW, Higashi DL, Snyder A, Merz AJ, Potter L, So M. 2005. PilT is required for PI(3,4,5)P3-mediated crosstalk between *Neisseria gonorrhoeae* and epithelial cells. *Cell Microbiol* **7:** 1271–1284.

Lory S, Strom MS. 1997. Structure-function relationship of type-IV prepilin peptidase of *Pseudomonas aeruginosa*—A review. *Gene* **192:** 117–121.

Luttrell LM, Ferguson SS, Daaka Y, Miller WE, Maudsley S, Della Rocca GJ, Lin F, Kawakatsu H, Owada K, Luttrell DK, et al. 1999. β-arrestin-dependent formation of β2 adrenergic receptor-Src protein kinase complexes. *Science* **283:** 655–661.

Maiden MC, Bygraves JA, Feil E, Morelli G, Russell JE, Urwin R, Zhang Q, Zhou J, Zurth K, Caugant DA, et al. 1998. Multilocus sequence typing: A portable approach to the identification of clones within populations of pathogenic microorganisms. *Proc Natl Acad Sci* **95:** 3140–3145.

Mairey E, Genovesio A, Donnadieu E, Bernard C, Jaubert F, Pinard E, Seylaz J, Olivo-Marin JC, Nassif X, Dumenil G. 2006. Cerebral microcirculation shear stress levels determine *Neisseria meningitidis* attachment sites along the blood–brain barrier. *J Exp Med* **203:** 1939–1950.

Masignani V, Comanducci M, Giuliani MM, Bambini S, Adu-Bobie J, Arico B, Brunelli B, Pieri A, Santini L, Savino S, et al. 2003. Vaccination against *Neisseria meningitidis* using three variants of the lipoprotein GNA1870. *J Exp Med* **197:** 789–799.

Merz AJ, So M. 1997. Attachment of piliated, Opa- and Opc-gonococci and meningococci to epithelial cells elicits cortical actin rearrangements and clustering of tyrosine-phosphorylated proteins. *Infect Immun* **65:** 4341–4349.

Merz AJ, Enns CA, So M. 1999. Type IV pili of pathogenic *Neisseriae* elicit cortical plaque formation in epithelial cells. *Mol Microbiol* **32:** 1316–1332.

Merz AJ, So M, Sheetz MP. 2000. Pilus retraction powers bacterial twitching motility. *Nature* **407:** 98–102.

Mikaty G, Soyer M, Mairey E, Henry N, Dyer D, Forest KT, Morand P, Guadagnini S, Prevost MC, Nassif X, et al. 2009. Extracellular bacterial pathogen induces host cell surface reorganization to resist shear stress. *PLoS Pathog* **5:** e1000314.

Morand PC, Tattevin P, Eugene E, Beretti JL, Nassif X. 2001. The adhesive property of the type IV pilus-associated component PilC1 of pathogenic *Neisseria* is supported by the conformational structure of the N-terminal part of the molecule. *Mol Microbiol* **40:** 846–856.

Morand PC, Bille E, Morelle S, Eugene E, Beretti JL, Wolfgang M, Meyer TF, Koomey M, Nassif X. 2004. Type IV pilus retraction in pathogenic *Neisseria* is regulated by the PilC proteins. *EMBO J* **23:** 2009–2017.

Nassif X, Beretti JL, Lowy J, Stenberg P, O'Gaora P, Pfeifer J, Normark S, So M. 1994. Roles of pilin and PilC in adhesion of *Neisseria meningitidis* to human epithelial and endothelial cells. *Proc Natl Acad Sci* **91:** 3769–3773.

Nikulin J, Panzner U, Frosch M, Schubert-Unkmeir A. 2006. Intracellular survival and replication of *Neisseria meningitidis* in human brain microvascular endothelial cells. *Int J Med Microbiol* **296:** 553–558.

Nizet V, Kim KS, Stins M, Jonas M, Chi EY, Nguyen D, Rubens CE. 1997. Invasion of brain microvascular endothelial cells by group B streptococci. *Infect Immun* **65:** 5074–5081.

Nourshargh S, Hordijk PL, Sixt M. 2010. Breaching multiple barriers: Leukocyte motility through venular walls and the interstitium. *Nat Rev* **11:** 366–378.

Orihuela CJ, Mahdavi J, Thornton J, Mann B, Wooldridge KG, Abouseada N, Oldfield NJ, Self T, Ala'Aldeen DA, Tuomanen EI. 2009. Laminin receptor initiates bacterial contact with the blood brain barrier in experimental meningitis models. *J Clin Invest* **119:** 1638–1646.

Palsson-McDermott EM, O'Neill LA. 2004. Signal transduction by the lipopolysaccharide receptor, Toll-like receptor-4. *Immunology* **113:** 153–162.

Parge HE, Forest KT, Hickey MJ, Christensen DA, Getzoff ED, Tainer JA. 1995. Structure of the fibre-forming protein pilin at 2.6 A resolution. *Nature* **378:** 32–38.

Pong A, Bradley JS. 1999. Bacterial meningitis and the newborn infant. *Infect Dis Clin North Am* **13:** 711–733, viii.

Prasadarao NV, Wass CA, Stins MF, Kim KS. 1999. Outer membrane protein A-promoted actin condensation of brain microvascular endothelial cells is required for *Escherichia coli* invasion. *Infect Immun* **67:** 5775–5783.

Pron B, Taha MK, Rambaud C, Fournet JC, Pattey N, Monnet JP, Musilek M, Beretti JL, Nassif X. 1997. Interaction of *Neisseria meningitidis* with the components of the blood–brain barrier correlates with an increased expression of PilC. *J Infect Dis* **176:** 1285–1292.

Pujol C, Eugene E, de Saint Martin L, Nassif X. 1997. Interaction of *Neisseria meningitidis* with a polarized monolayer of epithelial cells. *Infect Immun* **65:** 4836–4842.

Pujol C, Eugene E, Marceau M, Nassif X. 1999. The meningococcal PilT protein is required for induction of intimate attachment to epithelial cells following pilus-mediated adhesion. *Proc Natl Acad Sci* **96:** 4017–4022.

Ring A, Weiser JN, Tuomanen EI. 1998. Pneumococcal trafficking across the blood-brain barrier. Molecular analysis of a novel bidirectional pathway. *J Clin Invest* **102:** 347–360.

Rudel T, Facius D, Barten R, Scheuerpflug I, Nonnenmacher E, Meyer TF. 1995a. Role of pili and the phase-variable PilC protein in natural competence for transformation of *Neisseria gonorrhoeae*. *Proc Natl Acad Sci* **92:** 7986–7990.

Rudel T, Scheurerpflug I, Meyer TF. 1995b. *Neisseria* PilC protein identified as type-4 pilus tip-located adhesin. *Nature* **373:** 357–359.

Russell JA. 2006. Management of sepsis. *N Engl J Med* **355:** 1699–1713.

Schubert-Unkmeir A, Konrad C, Slanina H, Czapek F, Hebling S, Frosch M. 2010. Neisseria meningitidis induces brain microvascular endothelial cell detachment from the matrix and cleavage of occludin: A role for MMP-8. *PLoS Pathog* **6:** e1000874.

Seib KL, Serruto D, Oriente F, Delany I, Adu-Bobie J, Veggi D, Arico B, Rappuoli R, Pizza M. 2008. Factor H-binding protein is important for meningococcal survival in human whole blood and serum, and in the presence of the antimicrobial peptide LL-37. *Infect Immun* **77:** 292–299.

Shaw SK, Ma S, Kim MB, Rao RM, Hartman CU, Froio RM, Yang L, Jones T, Liu Y, Nusrat A, et al. 2004. Coordinated redistribution of leukocyte LFA-1 and endothelial cell ICAM-1 accompany neutrophil transmigration. *J Exp Med* **200:** 1571–1580.

Skerker JM, Berg HC. 2001. Direct observation of extension and retraction of type IV pili. *Proc Natl Acad Sci* **98:** 6901–6904.

Stins MF, Nemani PV, Wass C, Kim KS. 1999. *Escherichia coli* binding to and invasion of brain microvascular endothelial cells derived from humans and rats of different ages. *Infect Immun* **67:** 5522–5525.

Stins MF, Badger J, Sik Kim K. 2001. Bacterial invasion and transcytosis in transfected human brain microvascular endothelial cells. *Microb Pathog* **30:** 19–28.

Stojiljkovic I, Hwa V, de Saint Martin L, O'Gaora P, Nassif X, Heffron F, So M. 1995. The *Neisseria meningitidis* haemoglobin receptor: Its role in iron utilization and virulence. *Mol Microbiol* **15:** 531–541.

Sun YH, Bakshi S, Chalmers R, Tang CM. 2000. Functional genomics of *Neisseria meningitidis* pathogenesis. *Nat Med* **6:** 1269–1273.

Tala A, Monaco C, Nagorska K, Exley RM, Corbett A, Zychlinsky A, Alifano P, Tang CM. 2011. Glutamate utilization promotes meningococcal survival in vivo through avoidance of the neutrophil oxidative burst. *Mol Microbiol* **81:** 1330–1342.

Van de Beek D, de Gans J, Tunkel AR, Wijdicks EF. 2006. Community-acquired bacterial meningitis in adults. *N Engl J Med* **354:** 44–53.

Virji M. 1997. Post-translational modifications of meningococcal pili. Identification of common substituents: Glycans and α-glycerophosphate—A review. *Gene* **192:** 141–147.

Virji M. 2009. Pathogenic neisseriae: Surface modulation, pathogenesis and infection control. *Nat Rev Microbiol* **7:** 274–286.

Virji M, Kayhty H, Ferguson DJ, Alexandrescu C, Moxon ER. 1991. Interactions of *Haemophilus influenzae* with cultured human endothelial cells. *Microb Pathog* **10:** 231–245.

Virji M, Kayhty H, Ferguson DJ, Alexandrescu C, Moxon ER. 1992. Interactions of *Haemophilus influenzae* with human endothelial cells in vitro. *J Infect Dis* **165** (Suppl 1)**:** S115–S116.

Virji M, Makepeace K, Peak I, Payne G, Saunders JR, Ferguson DJ, Moxon ER. 1995. Functional implications of the expression of PilC proteins in meningococci. *Mol Microbiol* **16:** 1087–1097.

Winther-Larsen HC, Hegge FT, Wolfgang M, Hayes SF, van Putten JP, Koomey M. 2001. *Neisseria gonorrhoeae* PilV, a type IV pilus-associated protein essential to human epithelial cell adherence. *Proc Natl Acad Sci* **98:** 15276–15281.

Wolburg H, Lippoldt A. 2002. Tight junctions of the blood-brain barrier: Development, composition and regulation. *Vascul Pharmacol* **38:** 323–337.

Wolfgang M, Lauer P, Park HS, Brossay L, Hebert J, Koomey M. 1998. PilT mutations lead to simultaneous defects in competence for natural transformation and twitching motility in piliated *Neisseria gonorrhoeae*. *Mol Microbiol* **29:** 321–330.

Wolfgang M, van Putten JP, Hayes SF, Koomey M. 1999. The comP locus of *Neisseria gonorrhoeae* encodes a type IV prepilin that is dispensable for pilus biogenesis but essential for natural transformation. *Mol Microbiol* **31:** 1345–1357.

Wolfgang M, van Putten JP, Hayes SF, Dorward D, Koomey M. 2000. Components and dynamics of fiber formation define a ubiquitous biogenesis pathway for bacterial pili. *EMBO J* **19:** 6408–6418.

The Pneumococcus: Epidemiology, Microbiology, and Pathogenesis

Birgitta Henriques-Normark[1] and Elaine I. Tuomanen[2]

[1]Department of Microbiology, Tumor, and Cell Biology, Karolinska Institutet, SE-171 77 Stockholm, Sweden

[2]Department of Infectious Diseases, St. Jude Children's Research Hospital, Memphis, Tennessee 39105

Correspondence: elaine.tuomanen@stjude.org

The pneumococcus is the classic Gram-positive extracellular pathogen. The medical burden of diseases it causes is amongst the greatest in the world. Intense study for more than 100 years has yielded an understanding of fundamental aspects of its physiology, pathogenesis, and immunity. Efforts to control infection have led to the deployment of polysaccharide vaccines and an understanding of antibiotic resistance. The inflammatory response to pneumococci, one of the most potent in medicine, has revealed the double-edged sword of clearance of infection but at a cost of damage to host cells. In virtually every aspect of the infectious process, the pneumococcus has set the rules of the Gram-positive pathogenesis game.

Streptococcus pneumoniae (the pneumococcus) is the classic example of a highly invasive, Gram-positive, extracellular bacterial pathogen. It is a major cause of morbidity and mortality globally causing more deaths than any other infectious disease. At highest risk are the smallest children and the elderly with ~1 million children below 5 years of age dying yearly (Centers for Disease Control 2008). Pneumococcal diseases range from mild respiratory tract mucosal infections such as otitis media and sinusitis to more severe diseases such as pneumonia, septicemia, and meningitis. It is estimated in the United States that about 5% of those that get pneumococcal pneumonia die from the disease. The corresponding figure for septicemia is 20% and for meningitis, lethality is as high as 30% (Tomasz 1997). Although pneumococcus can cause lethal diseases, it is more commonly a quiescent colonizer of the upper respiratory tract where up to 60% of small children may carry pneumococci in the nose asymptomatically (Henriques-Normark et al. 2003; Nunes and Sa-Leao 2005).

Medically, the three major bacterial pathogens of children have historically been the pneumococcus, *Haemophilus*, and meningococcus. All three are naturally transformable and undergo autolysis in the stationary phase. They share the same progression of disease from the nasopharynx to blood and brain and use phosphorylcholine on their surfaces to bind to the same series of receptors in the pathway of innate invasion (Thornton et al. 2010). This uniquely shared pattern of disease and microbial biology makes understanding this group of enormous importance. Whereas vaccination has succeeded in dramatically reducing disease caused by

Cite this article as *Cold Spring Harb Perspect Med* doi: 10.1101/cshperspect.a010215

Haemophilus and meningococcus, pneumococcus remains a poorly controlled outlier. The huge number of pneumococcal serotypes (compared to the small number for the other two pathogens) remains the main challenge in developing strategies to control this global pathogen.

EPIDEMIOLOGY

Pneumococcal infections are thought to spread from person to person via droplets/aerosols and nasopharyngeal colonization is a prerequisite for pneumococcal disease. The carriage rate peaks around 2–3 years of age and diminishes thereafter to <10% in the adult population. However, adults with small children at home may have a higher carriage rate. The bacteria enter the nasal cavity and attach to the nasopharyngeal epithelial cells and may then either stay as a colonizer or spread further to other organs, such as the ears, sinuses, or via bronchi down to the lungs and then potentially penetrate the mucosal barrier to enter the bloodstream and/or cross the blood–brain barrier to cause meningitis (Fig. 1).

Tracking the global or local spread of pneumococci is commonly done by serotyping of the capsular polysaccharide. So far, at least 93 different capsular structures/serotypes have been described. Certain serotypes are most commonly found in young children, such as types 6B, 9V, 14, 19F, and 23F. A comparison of invasive disease isolates and carriage isolates from the same region during the same time period suggests that virulence differs depending on the capsular serotype (Brueggemann and Peto 2004; Sandgren et al. 2004, 2005). Some serotypes, such as 1 and 7F, are more prominent in invasive disease (they have a high invasive disease potential) whereas other types are mainly involved in carriage. However, only rare serotypes cause only carriage. Patients infected with serotypes with a

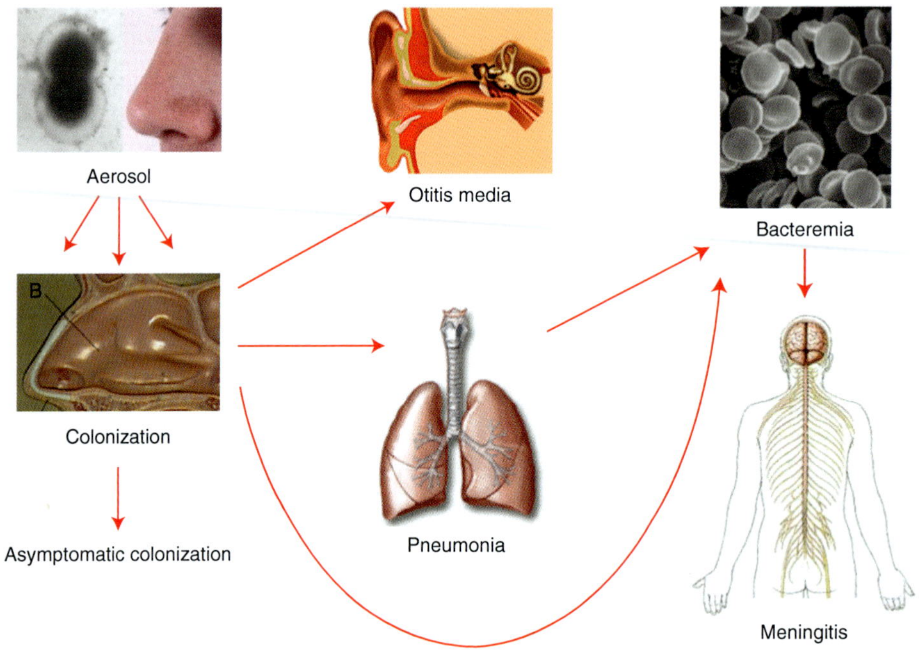

Figure 1. Progression of pneumococcal disease. Spread by aerosol, pneumococci are harbored in the nasopharynx. This interaction most commonly leads to clearance and serotype-specific immunity. Progression to otitis media is very common in children. Invasive disease involves spread to the lungs and bloodstream. The most serious development is meningitis. This series of invasive steps is a pattern shared by the three major bacterial pathogens of children (pneumococcus, *Haemophilus*, and meningococcus) and is driven by a very basic interplay between innate invasion and innate immunity.

low invasive disease potential more often had an underlying disease before the pneumococcal infection as compared to patients infected with serotypes with a high invasive disease potential (Sjöström et al. 2006). However, the highest mortality rates were found in patients infected with serotypes of a lower invasive disease potential, such as types 3, 6B, and 19F (Sandgren et al. 2004).

Since pneumococci naturally take up DNA from other bacteria and potentially switch capsular serotype, other methods than serotyping need to be used to study genetic relationships between clinical pneumococcal isolates, i.e., clonal types. Such molecular techniques include PFGE (pulsed field gel electrophoresis), whereby bacterial chromosomal DNA is cleaved using restriction enzymes and then put on a pulsed field gel separating the bands depending on their charge and size. Another sequenced based method is MLST (multilocus sequence typing) in which seven housekeeping genes are sequenced and the strains are assigned a sequence type (ST) depending on allelic variations. PFGE is more discriminatory than MLST and can be used also to study outbreaks. Furthermore, genomic comparisons using whole genome-based microarrays have shown that MLST can be used as a good method to estimate genome content (Dagerhamn et al. 2008).

Pneumococcal serotypes causing invasive disease have been shown to differ between geographic areas and the distribution is also dependent on the time period studied. This is especially important because current pneumococcal vaccines are based on a limited number of capsular polysaccharides. For decades, a 23-valent vaccine consisting of only polysaccharides was used in the elderly and immunocompromised individuals. As this vaccine is not immunogenic in the smallest children, newer vaccines are now used in which proteins are coupled to the polysaccharides to get a more T-cell-dependent immune response. For technical reasons and because of high costs, only a limited number of polysaccharides have been included so far in the conjugate vaccines (7, 10, and recently 13 serotypes). The serotypes have been chosen because they are the dominant serotypes causing invasive disease in the United States, a choice that means they are not well matched to other geographic areas with different serotype distribution. A review by Lin compiling data from studies performed in the Asia–Pacific region shows that the potential coverage rate for the seven-valent conjugate vaccine ranges from >70% to <50% (Lin 2010). The seven-valent vaccine was introduced into the childhood vaccination program in the United States in year 2000. Since then a tremendous decrease has been observed in vaccine-related invasive pneumococcal disease in children (Whitney et al. 2003). Also, a reduction in carriage of vaccine types has been noticed as well as a herd immunity effect in the adult population. However, an increase of invasive disease caused by nonvaccine types and clonal expansion of nonvaccine clones of certain serotypes has now been observed. It is worrisome that pneumococci may circumvent the vaccine by capsular switch events and that the immune pressure by the vaccine may select for minor serotypes not included in the vaccines. Therefore, pneumococcal molecular epidemiology must be followed thoroughly after vaccine introduction to detect possible adverse effects of vaccination, such as an increase of nonvaccine-type diseases.

For decades, the standard treatment of pneumococcal infections has been penicillin. However, a widespread increase of pneumococci resistant to most antibiotics (except for vancomycin) has been witnessed. Data from the European networks EARSS (www.rivm.nl/earss) on the rates of resistance to certain antibiotics among invasive disease isolates in Europe indicate an increasing trend of resistance to both penicillin and erythromycin. Multiresistant clonal types have emerged even in countries with a low frequency of resistance, such as in Sweden. Molecular typing of antibiotic resistant isolates shows that some pneumococcal clonal types are more prone to be spread globally. One such successfully spread clonal type is ST156, carrying resistance to penicillin and trimethoprim sulfamethoxazole, originated in Spain and spread to other countries facilitated by a pathogenicity islet encoding a pilus-like structure on the surface of pneumococci, shown to be important for colonization in a mouse model (Barocchi et al. 2006; Sjöström et al. 2007). The introduction of pneumococcal vaccines into the childhood vaccination programs in many countries is also expected to affect resistance rates. However, the experiences differ somewhat because some areas observe a decreased rate of resistance after vaccination whereas others do not see the same effect because they also have an increase of nonvaccine types. In the United States an expansion of serotype 19A was observed with resistance to many antibiotic drugs, creating treatment problems for common infections.

MICROBIOLOGY

The Surface

The pneumococcal surface is covered by a polysaccharide capsule that overlays the cell wall comprised of peptidoglycan and teichoic acid. Although the peptidoglycan has the classical Gram-positive structure of *N*-acetylglucosamine, *N*-acetylmuramic acid, and a lysine-containing stem peptide, the teichoic acid is unusual in containing a ribitol phosphate backbone and covalently attached phosphorylcholine (PCho) (Bean and Tomasz 1977). A nutritional requirement for choline is unique to *S. pneumoniae*. PCho serves as a docking station for noncovalent linkage of many pneumococcal proteins to the bacterial surface (Rosenow et al. 1997). In addition, the LPxTG motif is targeted by sortases that covalently link proteins to the cell wall (Löfling et al. 2011).

The surface composition of the bacteria varies between two distinct states: transparent and opaque colony types (Weiser et al. 1994). The precise mechanism of this phase variation is unclear but its impact is to aid in evading host defenses, such as C-reactive protein (CRP), which targets PCho, and leukocytes in blood while promoting attachment to host cells in the nasopharynx. The transparent phenotype dominates in the nasopharynx and expresses less capsule and less of the choline binding protein PspA, but more of the choline binding proteins CbpA and LytA autolysin (see below). In contrast, the opaque phase predominates in the blood and is characterized by increased capsular polysaccharide and PspA and less CbpA. Opaque bacteria also produce a greater biofilm and are more invasive for lungs and brain (Trappetti et al. 2011).

Capsule

The pneumococcal capsule has played a major role in the history of many medical discoveries and basic science tenets. In 1928, Griffith reported genetic exchange in bacteria when nonencapsulated, avirulent derivatives were transformed to a virulent, encapsulated phenotype following co-infection in mice (Griffith 1928). Avery, MacLeod, and McCarty then determined that Griffith's "transforming principle" was DNA and hence the discovery of the genetic material (Avery et al. 1944). The polysaccharide capsules are antiphagocytic and sterically hinder the access of leukocytes to complement fixed on the underlying cell wall. The capsule locus, transcribed as a single operon, shows a similar organization in all strains with genes encoding a specific capsule type flanked by genes common to all types.

Pili

Pili are multimeric filamentous surface structures composed of subunit proteins with LPxTG motifs recognized by sortases and attached to the cell wall. Two pathogenicity islets encoding pili, PI-1 and PI-2, are involved in adhesion (Barocchi et al. 2006; Bagnoli et al. 2008). PI-1, encoded by the *rlrA* accessory region, has been shown to influence colonization, virulence, and the inflammatory response in mouse challenge models. This pilus is composed of three covalently attached structural proteins in which RrgB is the major stalk protein, and RrgC and RrgA are ancillary proteins decorating the pilus shaft and tip. RrgA is the major adhesion and its three-dimensional structure has recently been solved (Nelson et al. 2007; Izoré et al. 2010). The role of the three pilus associated sortases in the assembly of the pilus has also been explored in which both Sortase B and Sortase C have been suggested to act as pilus subunit polymerases, whereas Sortase D is required for wild-type focal presentation of the pilus at the cell surface (Fälker et al. 2008).

Transformation

Pneumococci are naturally transformable, a state permissive for DNA uptake that arises for an entire population of pneumococci at one time in a highly regulated manner. The first description of quorum sensing is attributed to the induction of pneumococcal transformation (Tomasz 1965). The competence stimulating peptides (CSP1 and 2) are released and act externally on the two component system ComD/E. Binding of the cognate CSP induces the ComD histidine kinase to phosphorylate ComE, the response regulator, activating the *comAB* and *comCDE* operons resulting in a wave of competence that proceeds throughout the pneumococcal culture. This surge of early competence gene expression is followed by a second wave of expression of 16 late competence genes found in eight operons bearing the common combox. Many genes in this second wave are required for DNA binding and uptake.

Surface Proteins

Surface-exposed pneumococcal proteins are marked by one of three sequence motifs: the LPxTG cell wall anchor, a choline-binding domain, or a lipoprotein domain. The LPxTG bridge covalently anchors a class of more than a dozen degradative enzymes and pilus proteins to the cell wall. The choline binding protein (Cbp) family is unique to *S. pneumoniae* and the ~12 members share a 20 amino acid choline-binding domain repeated 3–9 times at the carboxyl terminus (Rosenow et al. 1997). Key proteins belong in this class. The LytA autolysin hydrolyses peptidoglycan and is critical for cell division. Pneumococcal surface protein A (PspA) inhibits complement deposition and activation, binds lactoferrin, and interferes with uptake into phagocytes. Cbp's A, D, E, and G play roles in adherence. CbpA is an adhesin that presents a specific motif for binding to the polymeric immunoglobulin receptor on the nasopharyngeal mucosa and a second motif for binding the laminin receptor at the blood–brain barrier (Hammerschmidt et al. 1997; Orihuela et al. 2009). Both interactions enable bacterial translocation across their respective barriers. CbpA also binds complement C3 and factor H.

Genome

The complete annotated 2.16 Mbps genome of *S. pneumoniae* encoding 2236 predicted proteins was first published in 2001 (Tettelin et al. 2001). Sequence analysis (www.sanger.ac.uk/resources/downloads/bacteria/streptococcus-pneumoniae.html or www.biomedcentral.com/1471-2180/11/25) indicates a nutritionally fastidious, lactic acid fermenting, facultative anaerobe with a G + C content of 40%. As pneumococcus lacks the TCA cycle, it depends on fermentation of carbohydrates for energy and can catabolize 14 different sugars. One-third of the predicted coding regions have no known function and a third of the genome varies sufficiently between strains to be considered to encode noncore functions. With the ability to naturally take up DNA and undergo transformation, the genome is exceptionally highly variable. Up to 10% of genes show altered sequences between isolates and 3%–5% of the genome shows insertion sequences and truncations. The number of DNA repeat regions is at the top of all available genomes and likely contributes to plasticity. Next generation sequencing has shown that pneumococci carried in the nasopharynx progressively accumulate recombination events indicative of the sequential transfer of DNA in vivo (Hiller et al. 2010). Sequencing of hundreds of globally dispersed strains of a single serotype 23F lineage showed large-scale recombination events varying in size between 3 bp and 72,038 bp (Croucher et al. 2011). Important major changes resulted in the frequent spread of antibiotic resistance and a shift in capsular serotype from 23F to 19A, indicating escape under vaccine pressure.

Large-scale identification of *S. pneumoniae* virulence determinants at the genomic level has been attempted. Signature-tagged mutagenesis (STM) (Lau et al. 2001; Hava and Camilli 2002) and Tn-Seq (van Opijnen et al. 2009) have used libraries of mutants to screen for survival in animal models. Gene expression in vivo and in various tissues has been performed at the transcriptome level using microarrays, deep RNA sequencing, and tiling array studies (Sorek and Cossart 2010). The genomic content of invasive versus noninvasive strains or clonotypes has suggested that the swapping of cassettes of "regions of diversity" changes infection dynamics leading to organ specific virulence (Orihuela et al. 2004). These methods have also revealed complex regulatory pathways, including at least 13 two-component systems (Paterson et al. 2006) and many sRNAs that act in the absence of Hfq (Mann et al. 2012).

COLONIZATION BIOLOGY

The ecological niche for pneumococci is the nasopharynx of small children. Carriage rates decrease with age and duration of carriage is longer in younger (~43 d) than older children (~25 d) (Högberg et al. 2007). When the bacteria enter the nasal cavity, the negatively charged capsule protects the bacteria from being trapped in the mucous and the bacteria can reach the epithelial surface. At the surface, transparent variants express a thinner capsule and better adherence during initial colonization (Weiser et al. 1994).

Many surface proteins affect nasopharyngeal colonization. A prominent example is CbpA (also called PspC or SpsA) that binds to the polymeric immunoglobulin receptor (pIgR) and promotes adherence and uptake of bacteria into nasopharyngeal cells (Hammerschmidt et al. 1997; Zhang et al. 2000; Braathen et al. 2006). Other adhesins include CbpG (Mann et al. 2006), CbpD (Gosink et al. 2000), and LPxTG proteins such as the pilus proteins and IgA protease (Weiser et al. 2003). The formation of a biofilm is promoted by PsrP which mediates bacterial aggregation and intraspecies adherence (Sanchez et al. 2010).

Enzymes, including NanA (neuraminidase), BgaA (β-glucosidase), and SrtH (β-*N*-glucosaminidase), affect colonization by cleaving terminal sugars from human glycoconjugates, thereby potentially exposing receptors for adherence. NanA also promotes biofilm formation and the development of otitis media (Tong et al. 2000; Munoz-Elias et al. 2008). Hyaluronidase helps to spread bacteria in the tissue and components like PavA and enolase interact with extracellular matrix proteins such as fibronectin and plasminogen (Bergmann et al. 2001; Kolberg et al. 2006). The serine protease HtrA also strongly influences nasopharyngeal colonization (Sebert et al. 2002).

An experimental colonization study in humans showed that carriage induces both mucosal and systemic immunoglobulin production and that the response is strain and type specific (McCool and Weiser 2002). Prevention of colonization involves Th17-cell adaptive responses (Lu et al. 2008) whereas progression against invasive disease is mediated by antibody (Cohen et al. 2011). Mouse studies indicate that clearance requires bacterial phagocytosis by CCR2-bearing monocyte/macrophages. Recruitment of these cells and production of anticapsular antibodies depend on the induction of CCL2 from macrophages that involves LysM fragmentation of peptidoglycan, uptake of the fragments, and sensing by Nod2 (Davis et al. 2011). Thus, peptidoglycan hydrolases such as LysM, LytB (glucosaminidase), LytC (lyzosyme), and Pce or CbpE (esterase) may affect colonization by programming immune responses (Bergmann and Hammerschmidt 2006).

An important aspect of the epidemiology of pneumococcal colonization is the common occurrence of co-infections. It is widely recognized that mortality during influenza outbreaks is due in large part to secondary pneumococcal pneumonia. Colonization with pneumococci concurrently with influenza leads to a synergistic type-1 interferon response with decreased clearance of bacteria from the nasopharynx and increased pneumonia (Nakamura et al. 2011).

HOST–BACTERIA INTERACTIONS IN THE COURSE OF INVASIVE DISEASE

Pneumonia: Common Invasion Strategy

As bacteria enter the lower respiratory tract, the initiation of pneumonia requires escape from mucous defenses and bacterial migration into the alveolus. Neuraminidase cleaves mucin and the absence of NanA attenuates pneumonia (Berry and Paton 2000). NanA also uncovers glycoconjugates, such as *N*-acetylgalactosamine β1–3 galactose, on host cells that provide sites for loose bacterial attachment. Interference with this step by neuraminidase inhibitors, such as oseltamivir, blocks the synergy in pneumonia between influenza and pneumococcus.

Firm adherence of bacteria to the alveolar epithelium and subsequent replication and initiation of host damage responses lead to the classic evolution of lobar pneumonia. The interaction of secreted products and cell surface-anchored bacterial components with the alveolar epithelium and innate immune defenses drives this process. The pore-forming toxin pneumolysin and hydrogen peroxide released in copious amounts by the bacteria disrupt the alveolar epithelium and edema fluid accumulates in the alveolar space (Canvin et al. 1995). Lipoteichoic acids, lipoproteins, and an array of proteins associated with the bacterial surface further amplify interactions with the host. During this early host response, bacteria gain firm adherence to the alveolar epithelium using several protein adhesins, including the histidine triad family of lipoproteins, LPxTG proteins, choline-binding proteins, and pili (Löfling et al. 2011). Proteins anchored by the LPxTG motif that bind the extracellular matrix include PavB (contains the highly repetitive SSURE domain), PfbA, PclA, and PsrP, an enormous serine rich protein (Jensch et al. 2010). Pili, encoded by PI-1 or PI-2, bind bacteria to respiratory epithelial cells (Barocchi et al. 2006; Bagnoli et al. 2008). The Cbp family of proteins noncovalently attached to the phosphorylcholine determinant on the pneumococcal teichoic acid has a broad range of functions from proteases to cell wall-modifying enzymes to complement interactors to adhesins (Rosenow et al. 1997). CbpA in particular is important in binding pneumococci to host cells and PspA interacts with many innate host defense components.

Many respiratory pathogens have converged on a common strategy to advance mucosal disease to bacteremia. This process is termed "innate invasion" and is counteracted by the very early participants induced in innate immunity (Gould and Weiser 2002; Thornton et al. 2010) (Fig. 2). The key host receptor for innate invasion is the platelet-activating factor receptor (PAFr) (Cundell et al. 1995). Many respiratory pathogens decorate their surfaces with phosphorylcholine (PCho), an example of molecular mimickry as PCho is the ligand critical to the activity of the chemokine PAF. Binding of PCho to PAFr initiates uptake of the bacteria into a vacuole that is then trafficked across the host cell where bacteria can be released to seed the bloodstream (Radin et al. 2005). Mice lacking PAFr and mice treated with PAFr antagonists fail to progress from pneumonia to sepsis. The fundamental PCho-PAFr step in pathogenesis is countered by the innate acute phase response element C-reactive protein (CRP), which binds PCho and competitively inhibits the PAFr/PCho-mediated bacterial invasion (Gould and Weiser 2002). Similarly, surfactant, which is rich in PCho, and an antibody to phosphorylcholine are competitive inhibitors of bacterial ligand binding to PAFr and CRP in the lungs (Briles et al. 1981). In this setup, innate invasion is counteracted by early response elements in innate immunity (Fig. 2).

Control of bacterial multiplication in the lung requires innate immune recognition of the pneumococccal pathogen-associated molecular patterns (PAMPs) that initiate recruitment of neutrophils and the accumulation of hemorrhagic debris as the hallmark of pulmonary consolidation. Pneumococcal peptidoglycan components induce signaling by TLR2 and LPS binding protein (Yoshimura et al. 1999; Weber et al. 2003), PCho-bearing teichoic acid binds to PAFr (Cundell et al. 1995), and small intracellular peptidoglycan fragments are recognized by Nod2 (Girardin et al. 2003; Moreira et al. 2008 (Fig. 3). Pneumolysin is detected by TLR4 (Malley et al. 2003) and DNA is sensed by TLR9 (Albiger et al. 2007). The importance of the inflammatory response through the TLR pathways to activation of NF-κB is highlighted by the altered disease susceptibilities in individuals with polymorphisms in this pathway (Brouwer et al. 2009). SNPs in *NFKBIA*,

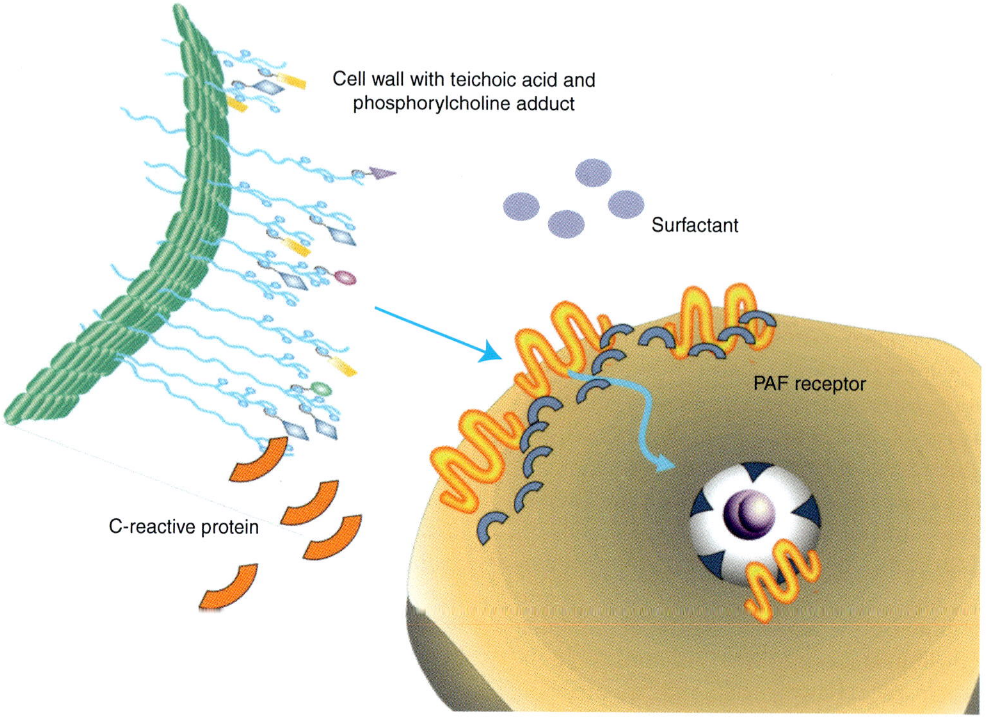

Figure 2. Innate invasion versus innate immunity. The majority of respiratory pathogens have phosphorylcholine (PCho) on their surfaces. PCho is on the cell wall teichoic acid of pneumococci (pictured as blue circles on cell wall surface extensions). PCho, by mimicking the host chemokine PAF, binds to the host cell platelet-activating factor receptor (PAFr) and its scaffold of β arrestins, leading to the uptake of bacteria into a vesicle and transmigration across the epithelial or endothelial barrier. This simple invasion step is countered by innate immune elements. C-reactive protein binds to PCho on the bacteria and surfactant contains abundant PCho, both of which inhibit bacterial contact with the host cell.

NFKBIE, and *TIRAP* that lead to decreased inflammatory signaling show a protective effect, whereas impaired function of IRAK4 or NEMO increases susceptibility to pneumococcal infection. NF-κB signaling leads to the release of cytokines, particularly IL-6 and IL-1ß, which then induces neutrophil recruitment and macrophage activation. The lung is unique in enabling neutrophil recruitment specifically in response to pneumococcus in both a CD-11/CD-18 dependent and independent manner (Doerschuk et al. 2000). Although required for clearance of bacteria, the intense inflammation induced by neutrophils contributes to lung damage; neutropenia (although not a complete absence of neutrophils) may attenuate damage and improve outcome (Marks et al. 2007).

Pneumococcus secretes a potent cytotoxin, pneumolysin, which is a member of the cholesterol-dependent, pore-forming cytotoxin family (Boulnois et al. 1991; Rossjohn et al. 1997). It plays an important role in the respiratory tract, not only by lysing host cells but also by inhibiting the mucociliary beat of respiratory cells and separating epithelial cell tight junctions (Berry et al. 1989; Kadioglu et al. 2002). It similarly contributes to the pathogenesis of otitis media by cytotoxic effects on the hair cells of the cochlea in the middle ear (Sato et al. 1996). Pneumolysin activates the classical complement pathway (Brown et al. 2002) and may be recognized by TLR4, thereby evoking an inflammatory response (Malley et al. 2003). Pneumolysin has been shown to influence bacterial uptake into human dendritic cells, to inhibit cellular

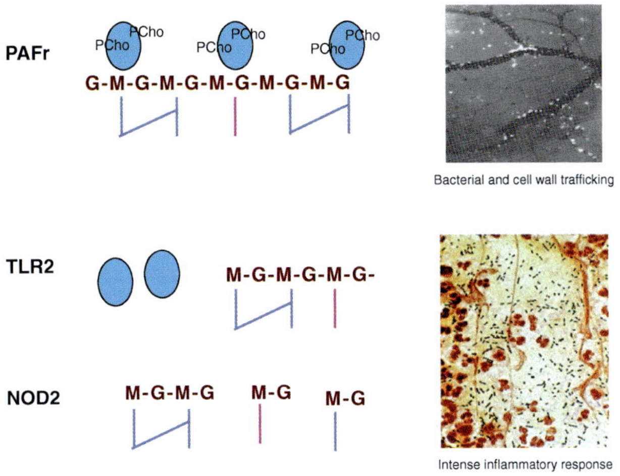

Figure 3. Pneumococcal cell wall PAMPs in disease progression. The cell wall of pneumococci is a major determinant of the course of disease. The interaction of PCho on the cell wall teichoic acid binds to PAFr and enables bacterial invasion of cells and transmigration across barriers. It also enables the cell wall itself to traffic to all organs (*inset top*: view through the cranial window of fluorescein-tagged cell wall pieces migrating from vasculature to brain). In the tissues, cell wall subcomponents generate intense inflammation by interacting with TLR2 and Nod2. Signaling through these receptors activates NF-κB pathways, cytokine, and chemokine production, leading to intense neutrophil recruitment (*inset bottom*: Gram stain of pneumococcus [blue] and neutrophils [red] in the lung).

activation and cytokine secretion, and to induce a caspase-dependent apoptosis (Braun et al. 2002). A nonfunctional pneumolysin, such as in some serotype 1 isolates, also influences inflammation induced by the interaction of bacteria and dendritic cells and might affect the development of diseases such as empyema (Littmann et al. 2009).

Bacteremia

The most important element in the development of high titer bacteremia is the pneumococcal polysaccharide capsule (Austrian 1981; Briles et al. 1992). Capsule is the key antiphagocytic element and antibody to capsule is opsonic and highly protective. Each of the >90 serotypes has a structurally unique polysaccharide and these vary in their antiphagocytic efficacy. Steric hindrance and negative charge determine the ability of a capsular type to inhibit surface deposition of complement, CRP, mannose-binding proteins, and antibodies that can bridge to receptors on the phagocytes (Briles et al. 1992). The amount of capsule is also important in that it is a phase variable trait and unencapsulated strains are essentially avirulent.

In addition to capsule, bacterial surface components interact with the complement cascade to impede complement deposition (Brown et al. 2002). PspA inhibits classical complement deposition and CbpA binds C3 and Factor H. Scavenger receptors, such as SR-A, macrophage receptor with collagenous structure (MARCO), and LOX-1, target the polyanionic teichoic acid and play a role in the clearance of bacteria. These interactions bridge bacteria to complement components or cellular phagocytic machinery.

Meningitis: A Common Entry Path

With high titer bacteremia, pneumococci come in contact with blood–brain barrier vascular endothelial cells. Crossing the tight junctions of this specialized endothelium enables entry into the cerebrospinal fluid and the brain parenchyma. Using pneumococcus as a probe, a common entry portal for meningeal pathogens was found for this site. Bacterial adhesins for *Strep-*

tococcus pneumoniae, *Haemophilus influenzae*, and *Neisseria meningitidis* share a common epitope that permits binding to the endothelial 37/67-kDa laminin receptor (LR) (Orihuela et al. 2009). This receptor is also targeted by many neurotropic viruses, implicating a common mechanism for the development of meningitis in general. In the case of pneumococci, the bacterial adhesin CbpA binds LR by a surface exposed loop distinct from that binding to the polymeric immunoglobulin receptor in the nasopharynx. Recently NanA was suggested to also promote pneumococcal adherence to brain endothelial cells through its laminin G-like lectin domain (Uchiyama et al. 2009). Once binding occurs to the vascular wall, the bacteria resist blood flow and activate underlying endothelial cells, specifically increasing the amount of PAFr on the endothelial surface (Fillon et al. 2006). PAFr then engages phosphorylcholine on the cell wall and translocates the bacteria across the endothelium intracellularly allowing passage from blood into brain. Mice lacking PAFr or bacteria deficient in CbpA do not develop meningitis even at very high titer bacteremia. Antibodies to CbpA are protective against meningitis across several species of bacteria.

Meningitis as a Model of Inflammatory Damage

Bacterial multiplication in the CSF space is fulminant as there are no innate defenses in this compartment. Thus, brain damage can be clearly tracked over time as a function of the interplay between bacterial components and the host response. The bacterial components responsible for the signs and symptoms of disease have been mapped by fractionating the pneumococcus, individually instilling pieces into the CSF space, and following inflammation and injury (Tuomanen et al. 1985). The hallmarks of meningitis are recruitment of leukocytes, blood–brain barrier permeability as measured by increased CSF protein, and vascular dysregulation, leading to increased intracranial pressure. This symptom complex can be recapitulated by instilling only components of the pneumococcal cell wall or toxins in the absence of living bacteria. The significance of this finding is twofold: First, cell wall and toxin interactions that drive inflammation are key to injury and second, antibiotics cause lysis and the release of cell wall components which, in the first few doses, increase CSF inflammation. This provides the rationale for mitigating damage to host cells by decreasing the inflammatory response during antibiotic therapy (Tuomanen et al. 1987). Blocking leukocyte invasion and subsequent inflammation decreases brain edema, intracranial pressure, blood flow alterations, and neuronal damage (Tuomanen et al. 1989; Weber et al. 1995). Although the complete absence of TLR2 or MyD88 in mice leads to higher mortality with bacterial overgrowth (Koedel 2009), down-modulation of downstream signals has been shown to be beneficial, including targeting MAP kinases, NF-κB (Koedel et al. 2010), and pro- and anti inflammatory cytokines (Saukkonen et al. 1990) as well as chemokines. In animal models, survival and burden of sequelae is improved by using anti-inflammatory agents together with antibiotics (Braun and Tuomanen 1999), including dexamethasone (Odio et al. 1991), antileukocyte agents (antiselectins or anti-CD18 integrin (Tuomanen et al. 1989), antimatrix metalloproteases, or inhibitors of nitric oxide.

The details of the host–pathogen interaction in the CSF have provided an excellent example of the overall picture of host response elements in Gram-positive disease. During the inflammatory response, neuronal toxicity is observed in two waves (Fig. 4). Direct toxin-mediated neuronal death occurs early in sepsis mediated by the release and penetration of the blood–brain barrier by pneumolysin and hydrogen peroxide (Braun et al. 2002; Weber and Tuomanen 2007). Approximately half of neuronal death is toxin mediated and is caspase independent. Later, during the host response to cell wall debris, indirect damage causes apoptosis of neurons and glia as a bystander effect (Mitchell et al. 2004). As leukocytes and serum components appear in the CSF over hours, bacterial peptidoglycan and teichoic acid bind to TLR2 that then leads to activation NF-κB (Yoshimura et al. 1999; Orihuela et al. 2006). Small-

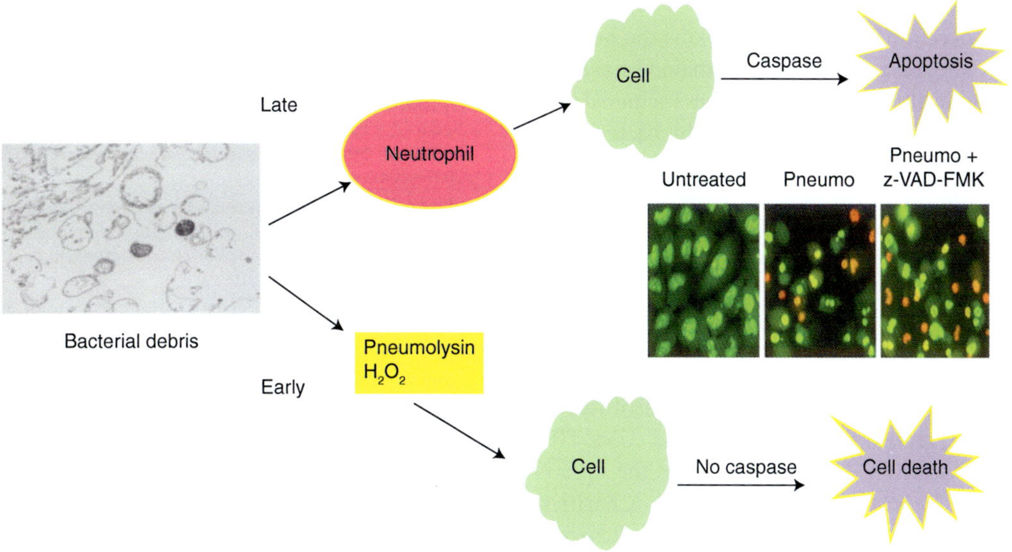

Figure 4. Host cell damage by pneumococcal components. During the inflammatory response, host cells are subjected to bacterial and host-derived toxic stimuli. Early in the course of disease, the bacteria produce the potent pore-forming cytotoxin pneumolysin and copious amounts of hydrogen peroxide, both of which kill host cells independent of caspases. Later in the course of the inflammatory response, when cell wall debris has induced neutrophil recruitment, host cells undergo caspase-dependent apoptosis. The caspase inhibitor z-VAD-FMK attenuates only half of the damage in infected tissues (*inset*: orange, dead cells; green, live cells).

er cell wall-derived muramyl peptides bind to NOD proteins. The accumulation of leukocytes is hastened by activation of the chemokine/cytokine cascade, the complement cascade, and the coagulation cascade. The damage caused by the host inflammatory response is greatest in the hippocampus where neuronal apoptosis is driven by caspase-dependent pathways (Braun et al. 1999). Oxidative damage from host-derived free radicals amplifies tissue damage (Braun et al. 2002; Hoffmann et al. 2006).

CONCLUDING REMARKS

The pneumococcus is the classic Gram-positive extracellular bacterial pathogen. It universally colonizes the nasopharynx and is the major cause of pneumonia, sepsis, and meningitis, causing more morbidity and mortality worldwide than any other infection. Its microbial physiology has revealed the discovery of DNA, the structure and assembly of Gram-positive peptidoglycan, the processes of quorum sensing, autolysis and transformation, and the first Gram-positive genome sequence. It led to the deployment of polysaccharide vaccines and an understanding of antibiotic resistance. Study of the pneumococcus mapped out the basic interplay between innate invasion and innate immunity for the major infections of childhood. Phosphorylcholine, present on nearly all respiratory pathogens, interacts with PAFr for invasion of the respiratory tract, and is counteracted by CRP and surfactant. Meningeal pathogens recognize the blood–brain barrier by interactions with laminin receptor on the vascular endothelium. Gram-positive teichoic acids and peptidoglycan are recognized as PAMPs by TLR2 and Nod2, DNA by TLR9. Pneumolysin is a pore-forming, cholesterol-dependent cytolysin. The inflammatory response to pneumococci, one of the most potent in medicine, underlies both protection and host cell damage and this understanding has lead to the clinical use of anti-inflammatory agents to improve the outcome of disease. In virtually every aspect of the infectious process, the pneumococcus has set the rules of the Gram-positive bacterial pathogenesis game. Yet more details of why a pneumococcal invasion ultimately kills the host are still needed and will make a major impact on world health. Current challenges include developing a universal vaccine that the bacteria will not be able to circumvent by transformation and improving the outcome of infection by balancing inflammation with bacterial death during antibiotic therapy.

ACKNOWLEDGMENTS

E.I.T. is supported by National Institutes of Health grants R01 AI-27913 and U54 HL-070590, the PATH Foundation, and the American Lebanese Syrian Associated Charities. B.H.N. is supported by the Swedish Research Council, Torsten and Ragnar Söderbergs Foundation, ALF bidrag from Stockholm City Council, EU project Pneumopath with the 7th Framework Program, and VINNOVA.

REFERENCES

Albiger B, Dahlberg S, Sandgren A, Wartha F, Beiter K, Katsuragi H, Akira S, Normark S, Henriques-Normark B. 2007. Toll-like receptor 9 acts at an early stage in host defence against pneumococcal infection. *Cell Microbiol* **9:** 633–644.

Austrian R. 1981. Pneumococcus: The first one hundred years. *Rev Infect Dis* **3:** 183–189.

Avery OT, Macleod CM, McCarty M. 1944. Studies on the chemical nature of the substance inducing transformation of pneumococcal types: Induction of transformation by a desoxyribonucleic acid fraction isolated from pneumococcus type III. *J Exp Med* **79:** 137–158.

Bagnoli F, Moschioni M, Donati C, Dimitrovska V, Ferlenghi I, Facciotti C, Muzzi A, Giusti F, Emolo C, Sinisi A, et al. 2008. A second pilus type in *Streptococcus pneumoniae* is prevalent in emerging serotypes and mediates adhesion to host cells. *J Bacteriol* **190:** 5480–5492.

Barocchi M, Ries J, Zogaj X, Albiger B, Kanth A, Dahlberg S, Moschioni M, Masignani V, Hultenby K, von Euler A, et al. 2006. A pneumococcal pilus influences virulence and host inflammatory responses. *Proc Natl Acad Sci* **103:** 2857–2862.

Bean B, Tomasz A. 1977. Choline metabolism in pneumococci. *J Bacteriol* **130:** 571–574.

Bergmann S, Hammerschmidt S. 2006. Versatility of pneumococcal surface proteins. *Microbiology* **152:** 295–303.

Bergmann S, Rohde M, Chhatwal GS, Hammerschmidt S. 2001. α-Enolase of *Streptococcus pneumoniae* is a plasmin(ogen)-binding protein displayed on the bacterial cell surface. *Mol Microbiol* **40:** 1273–1287.

Berry AM, Paton JC. 2000. Additive attenuation of virulence of *Streptococcus pneumoniae* by mutation of the genes encoding pneumolysin and other putative pneumococcal virulence proteins. *Infect Immun* **68:** 133–140.

Berry AM, Yother J, Briles DE, Hansman D, Paton JC. 1989. Reduced virulence of a defined pneumolysin-negative mutant of *Streptococcus pneumoniae*. *Infect Immun* **57:** 2037–2042.

Boulnois GJ, Paton JC, Mitchell TJ, Andrew PW. 1991. Structure and function of pneumolysin, the multifunctional, thiol-activated toxin of *Streptococcus pneumoniae*. *Mol Microbiol* **5:** 2611–2616.

Braathen R, Sandvik A, Berntzen G, Hammerschmidt S, Fleckenstein B, Sandlie I, Brandtzaeg P, Johansen F, Lauvrak V. 2006. Identification of a polymeric Ig receptor binding phage-displayed peptide that exploits epithelial transcytosis without dimeric IgA competition. *J Biol Chem* **281:** 7075–7081.

Braun JS, Tuomanen EI. 1999. Molecular mechanisms of brain damage in bacterial meningitis. *Adv Pediatr Infect Dis* **14:** 49–71.

Braun JS, Novak R, Herzog KH, Bodner SM, Cleveland JL, Tuomanen EI. 1999. Neuroprotection by a caspase inhibitor in acute bacterial meningitis. *Nat Med* **5:** 298–302.

Braun JS, Sublett JE, Freyer D, Mitchell TJ, Cleveland JL, Tuomanen EI, Weber JR. 2002. Pneumococcal pneumolysin and H_2O_2 mediate brain cell apoptosis during meningitis. *J Clin Invest* **109:** 19–27.

Briles DE, Nahm M, Schroer K, Davie J, Baker P, Kearney J, Barletta R. 1981. Antiphosphocholine antibodies found in normal mouse serum are protective against intravenous infection with type 3 *Streptococcus pneumoniae*. *J Exp Med* **153:** 694–705.

Briles DE, Crain MJ, Gray BM, Forman C, Yother J. 1992. Strong association between capsular type and virulence for mice among human isolates of *Streptococcus pneumoniae*. *Infect Immun* **60:** 111–116.

Brouwer M, de Gans J, Heckenberg S, Zwinderman A, van der Poll T, van de Beek D. 2009. Host genetic susceptibility to pneumococcal and meningococcal disease: A systematic review and meta-analysis. *Lancet Infect Dis* **9:** 31–44.

Brown JS, Hussell T, Gilliland SM, Holden DW, Paton JC, Ehrenstein MR, Walport MJ, Botto M. 2002. The classical pathway is the dominant complement pathway required for innate immunity to *Streptococcus pneumoniae* infection in mice. *Proc Natl Acad Sci* **99:** 16969–16974.

Brueggemann A, Peto T. 2004. Temporal and geographic stability of the serogroup-specific invasive disease potential of *Streptococcus pneumoniae* in children. *Infect Dis* **190:** 1203–1211.

Canvin JR, Marvin AP, Sivakumaran M, Paton JC, Boulnois GJ, Andrew PW, Mitchell TJ. 1995. The role of pneumolysin and autolysin in the pathology of pneumonia and septicemia in mice infected with a type 2 pneumococcus. *J Infect Dis* **172:** 119–123.

Centers for Disease Control and Prevention. 2008. Progress in introduction of pneumococcal conjugate vaccine worldwide. *MMWR Morb Mortal Wkly Rep* **57:** 1148–1151.

Cohen JM, Khandavilli S, Camberlein E, Hyams C, Baxendale HE, Brown JS. 2011. Protective contributions against invasive *Streptococcus pneumoniae* pneumonia of antibody and Th17-cell responses to nasopharyngeal colonisation. *PLoS ONE* **6:** e25558.

Croucher N, Harris S, Fraser C, Quail M, Burton J, van der Linden M, McGee L, von Gottberg A, Song J, Ko K, et al. 2011. Rapid pneumococcal evolution in response to clinical interventions. *Science* **331:** 430–434.

Cundell DR, Gerard NP, Gerard C, Idanpaan-Heikkila I, Tuomanen EI. 1995. *Streptococcus pneumoniae* anchor to activated human cells by the receptor for platelet-activating factor. *Nature* **377:** 435–438.

Dagerhamn J, Blomberg C, Browall S, Sjöström K, Morfeldt E, Henriques-Normark B. 2008. Determination of accessory gene patterns predicts the same relatedness among strains of *Streptococcus pneumoniae* as sequencing of housekeeping genes does and represents a novel approach in molecular epidemiology. *J Clin Microbiol* **46:** 863–868.

Davis K, Nakamura S, Weiser JN. 2011. Nod2 sensing of lysozyme-digested peptidoglycan promotes macrophage recruitment and clearance of *S. pneumoniae* colonization in mice. *J Clin Invest* **121:** 3666–3676.

Doerschuk C, Tasaka S, Wang Q. 2000. CD11/CD18-dependent and -independent neutrophil emigration in the lungs: How do neutrophils know which route to take? *Am J Respir Cell Mol Biol* **23:** 133–136.

Fälker S, Nelson AL, Morfeldt E, Jonas K, Hultenby K, Ries J, Melefors O, Normark S, Henriques-Normark B. 2008. Sortase mediate assembly and surface topology of adhesive pneumococcal pili. *Mol Microbiol* **70:** 595–607.

Fillon S, Soulis K, Rajasekaran S, Benedict-Hamilton H, Radin JN, Orihuela CJ, El Kasmi KC, Murti G, Kaushal D, Gaber MW, et al. 2006. Platelet-activating factor receptor and innate immunity: Uptake of Gram-positive bacterial cell wall into host cells and cell-specific pathophysiology. *J Immunol* **177:** 6182–6191.

Girardin S, Boneca I, Viala J, Chamaillard M, Labigne A, Thomas G, Philpott D, Sansonetti P. 2003. Nod2 is a general sensor of peptidoglycan through muramyl dipeptide (MDP) detection. *J Biol Chem* **8:** 8869–8872.

Gosink KK, Mann ER, Guglielmo C, Tuomanen EI, Masure H. 2000. Role of novel choline binding proteins in virulence of *Streptococcus pneumoniae. Infect Immun* **68:** 5690–5695.

Gould JM, Weiser JN. 2002. The inhibitory effect of C-reactive protein on bacterial phosphorylcholine platelet-activating factor receptor-mediated adherence is blocked by surfactant. *J Infect Dis* **186:** 361–371.

Griffith F. 1928. The significance of pneumococcal types. *J Hyg (Lond)* **27:** 113–159.

Hammerschmidt S, Talay SR, Brandtzaeg P, Chhatwal GS. 1997. SpsA, a novel pneumococcal surface protein with specific binding to secretory immunoglobulin A and secretory component. *Mol Microbiol* **25:** 1113–1124.

Hava DL, Camilli A. 2002. Large-scale identification of serotype 4 *Streptococcus pneumoniae* virulence factors. *Mol Microbiol* **45:** 1389–1406.

Hiller NLAA, Powell E, Martin DP, Eutsey R, Earl J, Janto B, Boissy RJ, Hogg J, Barbadora K, Sampath R, et al. 2010. Generation of genic diversity among *Streptococcus pneumoniae* strains via horizontal gene transfer during a chronic polyclonal pediatric infection. *PLoS Pathog* **6:** e1001108.

Hoffmann O, Zweigner J, Smith SH, Freyer D, Mahrhofer C, Dagand E, Tuomanen EI, Weber JR. 2006. Interplay of pneumococcal hydrogen peroxide and host-derived nitric oxide. *Infect Immun* **74:** 5058–5066.

Högberg L, Geli P, Ringberg H, Melander E, Lipsitch M, Ekdahl K. 2007. Age- and serogroup-related differences in observed durations of nasopharyngeal carriage of penicillin-resistant pneumococci. *J Clin Microbiol* **45:** 948–952.

Izoré T, Contreras-Martel C, El Mortaji L, Manzano C, Terrasse R, Vernet T, Di Guilmi AM, Dessen A. 2010. Structural basis of host cell recognition by the pilus adhesin of *Streptococcus pneumoniae. Structure* **18:** 106–115.

Jensch I, Gámez G, Rothe M, Ebert S, Fulde M, Somplatzki D, Bergmann S, Petruschka L, Rohde M, Nau R, et al. 2010. PavB is a surface-exposed adhesin of *Streptococcus pneumoniae* contributing to nasopharyngeal colonization and airways infections. *Mol Microbiol* **77:** 22–43.

Kadioglu A, Taylor S, Iannelli F, Pozzi G, Mitchell TJ, Andrew PW. 2002. Upper and lower respiratory tract infection by *Streptococcus pneumoniae* is affected by pneumolysin deficiency and differences in capsule type. *Infect Immun* **70:** 2886–2890.

Koedel U. 2009. Toll-like receptors in bacterial meningitis. *Curr Top Microbiol Immunol* **336:** 15–40.

Koedel U, Klein M, Pfister H. 2010. Modulation of brain injury as a target of adjunctive therapy in bacterial meningitis. *Curr Infect Dis Rep* **12:** 266–273.

Kolberg J, Aase A, Bergmann S, Herstad T, Rødal G, Frank R, Rohde M, Hammerschmidt S. 2006. *Streptococcus pneumoniae* enolase is important for plasminogen binding despite low abundance of enolase protein on the bacterial cell surface. *Microbiology* **152:** 1307–1317.

Lau GW, Haataja S, Lonetto M, Kensit SE, Marra A, Bryant AP, McDevitt D, Morrison DA, Holden DW. 2001. A functional genomic analysis of type 3 *Streptococcus pneumoniae* virulence. *Mol Microbiol* **40:** 555–571.

Lin C. 2010. Summary of invasive pneumococcal disease burden among children. *Vaccine* **28:** 7589–7605.

Littmann M, Albiger B, Frentzen A, Normark S, Henriques-Normark B, Plant L. 2009. Pneumolysin expression allows evasion of human dendritic cell responses to *Streptococcus pneumoniae. EMBO Mol Med* **1:** 211–222.

Löfling J, Vimberg V, Battig P, Henriques-Normark B. 2011. Cellular interactions by LPxTG-anchored pneumococcal adhesins and their streptococcal homologues. *Cell Microbiol* **13:** 186–197.

Lu Y, Gross J, Bogaert D, Finn A, Bagrade L, Zhang Q, Kolls J, Srivastava A, Lundgren A, Forte S, et al. 2008. Interleukin-17A mediates acquired immunity to pneumococcal colonization. *PLoS Pathog* **4:** e1000159.

Malley R, Henneke P, Morse SC, Cieslewicz MJ, Lipsitch M, Thompson CM, Kurt-Jones E, Paton JC, Wessels MR, Golenbock DT. 2003. Recognition of pneumolysin by Toll-like receptor 4 confers resistance to pneumococcal infection. *Proc Natl Acad Sci* **100:** 1966–1971.

Mann B, Orihuela C, Antikainen J, Gao G, Sublett J, Korhonen TET. 2006. Multifunctional role of choline binding protein G in pneumococcal pathogenesis. *Infect Immun* **74:** 821–829.

Mann B, van Opijnen T, Wang J, Obert C, Wang YD, Carter R, McGoldrick DJ, Ridout G, Camilli A, Tuomanen EI, et al. 2012. Control of virulence by small RNAs in *Streptococcus pneumoniae. PLoS Pathog* **8:** e1002788.

Marks M, Burns T, Abadi M, Seyoum B, Thornton J, Tuomanen E, Pirofski LA. 2007. Influence of neutropenia on the course of serotype 8 pneumococcal pneumonia in mice. *Infect Immun* **75:** 1586–1597.

McCool, Weiser JN. 2002. The immune response to pneumococcal proteins during experimental human carriage. *J Exp Med* **195:** 359–365.

Mitchell L, Smith SH, Braun JS, Herzog KH, Weber JR, Tuomanen EI. 2004. Dual phases of apoptosis in pneumococcal meningitis. *J Infect Dis* **190:** 2039–2046.

Moreira LO, El Kasmi KC, Smith AM, Finkelstein D, Fillon S, Kim YG, Nunez G, Tuomanen E, Murray PJ. 2008. The TLR2-MyD88-NOD2-RIPK2 signalling axis regulates a balanced pro-inflammatory and IL-10-mediated anti-inflammatory cytokine response to Gram-positive cell walls. *Cell Microbiol* **10:** 2067–2077.

Munoz-Elias E, Marcano J, Camilli A. 2008. Isolation of *Streptococcus pneumoniae* biofilm mutants and their characterization during nasopharyngeal colonization. *Infect Immun* **76:** 5049–5061.

Nakamura S, Davis KM, Weiser JN. 2011. Synergistic stimulation of type 1 interferons during influenza virus coinfection promotes *Streptococcus pneumoniae* colonization in mice. *J Clin Invest* **121:** 3657–3665.

Nelson AL, Ries J, Bagnoli F, Dahlberg S, Fälker S, Rounioja S, Tschöp J, Morfeldt E, Ferlenghi I, Hilleringmann M, et al. 2007. RrgA is a pilus-associated adhesin in *Streptococcus pneumoniae. Mol Microbiol* **66:** 329–40.

Nunes T, Sa-Leao R. 2005. Trends in drug resistance, serotypes, and molecular types of *Streptococcus pneumoniae* colonizing preschool-age children attending day care centers in Lisbon, Portugal: A summary of 4 years of annual surveillance. *J Clin Microbiol* **43:** 1285–1293.

Odio C, Faingezicht I, Paris M, Nassar M, Baltodano A, Rogers J, Sáez-Llorens X, Olsen K, McCracken GJ. 1991. The beneficial effects of early dexamethasone administration in infants and children with bacterial meningitis. *N Engl J Med* **324:** 1525–1531.

Orihuela CJ, Gao G, Francis KP, Yu J, Tuomanen EI. 2004. Tissue-specific contributions of pneumococcal virulence factors to pathogenesis. *J Infect Dis* **190:** 1661–1669.

Orihuela CJ, Fillon S, Smith-Sielicki SH, El Kasmi KC, Gao G, Soulis K, Patil A, Murray PJ, Tuomanen EI. 2006. Cell wall-mediated neuronal damage in early sepsis. *Infect Immun* **74:** 3783–3789.

Orihuela CJ, Mahdavi J, Thornton J, Mann B, Wooldridge KG, Abouseada N, Oldfield NJ, Self T, Ala'Aldeen DA, Tuomanen EI. 2009. Laminin receptor initiates bacterial contact with the blood brain barrier in experimental meningitis models. *J Clin Invest* **119:** 1638–1646.

Paterson GK, Blue CE, Mitchell TJ. 2006. Role of two-component systems in the virulence of *Streptococcus pneumoniae. J Med Microbiol* **55:** 355–363.

Radin JN, Orihuela CJ, Murti G, Guglielmo C, Murray PJ, Tuomanen EI. 2005. β-Arrestin 1 participates in platelet-activating factor receptor-mediated endocytosis of *Streptococcus pneumoniae. Infect Immun* **73:** 7827–7835.

Rosenow C, Ryan P, Weiser JN, Johnson A, Fontan P, Ortqvist A, Masure HR. 1997. Contribution of novel choline-binding proteins to adherence, colonization and immunogenicity of *Streptococcus pneumoniae. Mol Microbiol* **25:** 819–829.

Rossjohn J, Feil SC, McKinstry WJ, Tweten RK, Parker MW. 1997. Structure of a cholesterol-binding, thiol-activated cytolysin and a model of its membrane form. *Cell* **89:** 685–692.

Sanchez CJ, Shivshankar P, Stol K, Trakhtenbroit S, Sullam PM, Sauer K, Hermans PW, Orihuela CJ. 2010. The pneumococcal serine-rich repeat protein is an intra-species bacterial adhesin that promotes bacterial aggregation in vivo and in biofilms. *PLoS Pathog* **6:** e1001044.

Sandgren A, Sjöström K, Olsson-Liljequist B, Christensson B, Samuelsson A, Kronvall G, Henriques Normark B. 2004. Effect of clonal and serotype-specific properties on the invasive capacity of *Streptococcus pneumoniae. J Infect Dis* **189:** 785–796.

Sandgren A, Albiger B, Orihuela C, Tuomanen E, Normark S, Henriques-Normark B. 2005. Virulence in mice of pneumococcal clonal types with known invasive disease potential in man. *J Infect Dis* **192:** 791–800.

Sato K, Quartey MK, Liebeler CL, Le CT, Giebink GS. 1996. Roles of autolysin and pneumolysin in middle ear inflammation caused by a type 3 *Streptococcus pneumoniae* strain in the chinchilla otitis media model. *Infect Immun* **64:** 1140–1145.

Saukkonen K, Sande S, Cioffe C, Wolpe S, Sherry B, Cerami A, Tuomanen E. 1990. The role of cytokines in the generation of inflammation and tissue damage in experimental Gram-positive meningitis. *J Exp Med* **171:** 439–448.

Sebert M, Palmer L, Rosenberg M, Weiser J. 2002. Microarray-based identification of htrA, a *Streptococcus pneumoniae* gene that is regulated by the CiaRH two-component system and contributes to nasopharyngeal colonization. *Infect Immun* **70:** 4059–4067.

Sjöström K, Spindler C, Ortqvist A, Kalin M, Sandgren A, Kuhlmann-Berenzon S, Henriques-Normark B. 2006. Clonal and capsular types decide whether pneumococci will act as a primary or opportunistic pathogen. *Clin Infect Dis* **42:** 451–459.

Sjöström K, Blomberg C, Fernebro J, Dagerhamn J, Morfeldt E, Barocchi M, Andersson M, Browall S, Henriques F, Rappuoli R, et al. 2007. Clonal success of piliated penicillin non-susceptible pneumococci. *Proc Natl Acad Sci* **104:** 12907–12912.

Sorek R, Cossart P. 2010. Prokaryotic transcriptomics: A new view on regulation, physiology and pathogenicity. *Nat Rev Genet* **11:** 9–16.

Tettelin H, Nelson KE, Paulsen IT, Eisen JA, Read TD, Peterson S, Heidelberg J, DeBoy RT, Haft DH, Dodson RJ, et al. 2001. Complete genome sequence of a virulent isolate of *Streptococcus pneumoniae. Science* **293:** 498–506.

Thornton JA, Durick-Eder K, Tuomanen EI. 2010. Pneumococcal pathogenesis: "Innate invasion" yet organ-specific damage. *J Mol Med (Berl)* **88:** 103–107.

Tomasz A. 1965. Control of the competent state in pneumococcus by a hormone-like cell product: An example for a new type of regulatory mechanism in bacteria. *Nature* **208:** 155–159.

Tomasz A. 1997. Antibiotic resistance in *Streptococcus pneumoniae*. *Clin Infect Dis* **24:** S85–S88.

Tong HH, Blue LE, James MA, DeMaria TF. 2000. Evaluation of the virulence of a *Streptococcus pneumoniae* neuraminidase-deficient mutant in nasopharyngeal colonization and development of otitis media in the chinchilla model. *Infect Immun* **68:** 921–924.

Trappetti C, Ogunniyi A, Oggioni M, Paton J. 2011. Extracellular matrix formation enhances the ability of *Streptococcus pneumoniae* to cause invasive disease. *PLoS ONE* **6:** e19844.

Tuomanen E, Liu H, Hengstler B, Zak O, Tomasz A. 1985. The induction of meningeal inflammation by components of the pneumococcal cell wall. *J Infect Dis* **151:** 859–868.

Tuomanen E, Hengstler B, Rich R, Bray MA, Zak O, Tomasz A. 1987. Nonsteroidal anti-inflammatory agents in the therapy for experimental pneumococcal meningitis. *J Infect Dis* **155:** 985–990.

Tuomanen EI, Saukkonen K, Sande S, Cioffe C, Wright SD. 1989. Reduction of inflammation, tissue damage, and mortality in bacterial meningitis in rabbits treated with monoclonal antibodies against adhesion-promoting receptors of leukocytes. *J Exp Med* **170:** 959–969.

Uchiyama S, Carlin A, Khosravi A, Weiman S, Banerjee A, Quach D, Hightower G, Mitchell T, Doran K, Nizet V. 2009. The surface-anchored NanA protein promotes pneumococcal brain endothelial cell invasion. *J Exp Med* **206:** 1845–1852.

van Opijnen T, Bodi KL, Camilli A. 2009. Tn-seq: high-throughput parallel sequencing for fitness and genetic interaction studies in microorganisms. *Nat Methods* **6:** 767–772.

Weber JR, Tuomanen EI. 2007. Cellular damage in bacterial meningitis: An interplay of bacterial and host driven toxicity. *J Neuroimmunol* **184:** 45–52.

Weber J, Angstwurm K, Bürger W, Einhäupl K, Dirnagl U. 1995. Anti ICAM-1 (CD 54) monoclonal antibody reduces inflammatory changes in experimental bacterial meningitis. *J Neuroimmunol* **63:** 63–68.

Weber JR, Freyer D, Alexander C, Schroder NW, Reiss A, Kuster C, Pfeil D, Tuomanen EI, Schumann RR. 2003. Recognition of pneumococcal peptidoglycan: An expanded, pivotal role for LPS binding protein. *Immunity* **19:** 269–279.

Weiser JN, Austrian R, Sreenivasan PK, Masure HR. 1994. Phase variation in pneumococcal opacity: Relationship between colonial morphology and nasopharyngeal colonization. *Infect Immun* **62:** 2582–2589.

Weiser JN, Bae D, Fasching C, Scamurra RW, Ratner AJ, Janoff EN. 2003. Antibody-enhanced pneumococcal adherence requires IgA1 protease. *Proc Natl Acad Sci* **100:** 4215–4220.

Whitney CG, Farley MM, Hadler J, Harrison LH, Bennett NM, Lynfield R, Reingold A, Cieslak PR, Pilishvili T, Jackson D, et al. 2003. Decline in invasive pneumococcal disease after the introduction of protein-polysaccharide conjugate vaccine. *N Engl J Med* **348:** 1737–1746.

Yoshimura A, Lien E, Ingalls RR, Tuomanen E, Dziarski R, Golenbock D. 1999. Cutting edge: Recognition of Gram-positive bacterial cell wall components by the innate immune system occurs via Toll-like receptor 2. *J Immunol* **163:** 1–5.

Zhang JR, Mostov KE, Lamm ME, Nanno M, Shimida S, Ohwaki M, Tuomanen E. 2000. The polymeric immunoglobulin receptor translocates pneumococci across human nasopharyngeal epithelial cells. *Cell* **102:** 827–837.

The Inside Story of *Shigella* Invasion of Intestinal Epithelial Cells

Nathalie Carayol[1,2,3,4] and Guy Tran Van Nhieu[1,2,3,4]

[1]Equipe Communication Intercellulaire et Infections Microbiennes, Centre de Recherche Interdisciplinaire en Biologie (CIRB), Collège de France, Paris, France

[2]Institut National de la Santé et de la Recherche Médicale (INSERM), U1050 Paris, France

[3]Centre National de la Recherche Scientifique (CNRS), UMR 7241 Paris, France

[4]MEMOLIFE Laboratory of Excellence and Paris Science Lettre, Research University, 75005 Paris, France

Correspondence: guy.tran-van-nhieu@college-de-france.fr

As opposed to other invasive pathogens that reside into host cells in a parasitic mode, *Shigella*, the causative agent of bacillary dysentery, invades the colonic mucosa but does not penetrate further to survive into deeper tissues. Instead, *Shigella* invades, replicates, and disseminates within the colonic mucosa. Bacterial invasion and spreading in intestinal epithelium lead to the elicitation of inflammatory responses responsible for the tissue destruction and shedding in the environment for further infection of other hosts. In this article, we highlight specific features of the *Shigella* arsenal of virulence determinants injected by a type III secretion apparatus (T3SA) that point to the targeting of intestinal epithelial cells as a discrete route of invasion during the initial event of the infectious process.

Shigella spp. are Gram-negative enteroinvasive bacteria responsible for diarrheal diseases. *Shigella dysenteria* and *Shigella flexneri* are responsible for bacillary dysentery, associated with epidemics and the endemic form of the disease, respectively, mostly in developing countries in children under the age of 5 with an estimated 100 million cases yearly. There is no known reservoir of these enteropathogenic bacteria, and transmission has been mostly reported from human-to-human through feco–oral contamination. On ingestion, *Shigella* invades the colonic mucosa in which it elicits an intense inflammatory reaction responsible for tissue destruction. In its most severe forms, *Shigella* infections are associated with the dysenteric syndrome with the emission of bloody, mucopurulent stools, and eventually death. Other intestinal complications include renal failure associated with the hematouremic syndrome and bacterial expression of the Shiga toxin, as well as rheumatoid arthritis.

PART I. THE INFECTIOUS STRATEGY OF *SHIGELLA*

Shigella, a Pathogenic *Escherichia coli* Adapted to Intracellular Lifestyle

Although related to enteroinvasive *E. coli* responsible for similar symptoms, phylogenetic analysis indicates that *Shigella* has evolved from a commensal *E. coli*, through horizontal gene

Cite this article as *Cold Spring Harb Perspect Med* doi: 10.1101/cshperspect.a016717

transfer and genomic mutations in a pathoadaptative process. Hence, as opposed to *E. coli* commensals, *Shigella* presents mutations in genes encoding enzymes, such as spermidine acetylase, lysine decarboxylase, and arginine acetyltransferase, which regulates the polyamine contents in bacteria shown detrimental to the infection process (Bliven and Maurelli 2012; Prosseda et al. 2012). Also, *Shigella* does not express any known adhesin or curlin, does not form biofilm, and is nonmotile because of the lack of flagellum expression. Instead, *Shigella* expresses IcsA, a surface protein allowing intracellular motility, required for bacterial spreading in intestinal epithelial cells (Bliven and Maurelli 2012). Importantly, *Shigella* carries a large virulence plasmid encoding the majority of the virulence determinants, including a T3SA allowing the injection of bacterial effectors into host cells to divert cellular processes.

Environmental Cues for the Regulation of Virulence Genes

How *Shigella* persists in the environment is not known. There are, however, environmental conditions that turn on virulence genes, allowing bacteria to invade and to replicate within the colonic mucosa. Under low temperature and low osmolarity conditions, corresponding to environmental conditions, the H-NS repressor inhibits the expression of the VirF master transcriptional regulator, and thereby the expression of virulence genes (Dorman et al. 2001). On ingestion by the host, the temperature shift leads to the activation of VirF, which induces the expression of another transcriptional regulator VirB (Dorman et al. 2001). VirB directly controls the synthesis of virulence genes, including that of the type III secretion system (T3SS). The acidic pH conditions encountered in the stomach activate the RpoS-dependent expression of periplasmic proteins allowing acid resistance, as well as the CpxA/CpxR two-component system, which, together with the temperature and osmolarity, control the expression of VirF/VirB (Marteyn et al. 2012). The conditions encountered in the colon trigger the expression of genes involved in anaerobic energy metabolism, modification of the LPS and outer membrane layers, resistance to oxidative stress, as well as that of various transcriptional regulators, including FNR. FNR was shown to regulate the synthesis of the *T3SA* genes, resulting in secretion apparatus with elongated needles, presumably primed but repressed for injection of effectors. At the close vicinity of colonic intestinal epithelial cells, the microaerophilic environment leads to FNR repression and activation of secretion to promote bacterial invasion (Marteyn et al. 2010). In the intracellular environment, carbon source and iron limitation may also regulate the expression of VirF/VirB-dependent genes.

The expression of T3SA substrates is subjected to another level of regulation dependent on the secretion activity. Among the 25 identified T3S effectors, 16 are constitutively expressed at the basal state (Fig. 1) (Parsot 2009). These include the Ipa proteins implicated in invasion, which on cell contact, are injected to induce cytoskeletal reorganization. Thirteen effectors, however, are up-regulated following induction of secretion under intracellular growth conditions. Transcriptional control of these genes is performed by MxiE. MxiE senses the secretory state of the T3SA through its binding to IpgC, a chaperone of IpaB and IpaC, as well as to the T3SA substrate OspD1 (Parsot et al. 2005). On activation of the T3SA, secretion of IpaB and IpaC leads to the release of free IpgC in the bacterial cytoplasm (Mavris et al. 2002). Secretion of OspD1 that sequesters MxiE at basal state, leads to increased levels of free MxiE, which presumably in association with IpgC, up-regulate the transcription of a second wave of type III secretion (T3S) effectors. As will be developed further, the function of these second wave effectors for those characterized, is consistent with a strategy of a "discrete" intracellular replication, where the bacteria prevents inflammation and alert signals.

Shigella Invasion of the Colonic Epithelium via Multiple Routes

Shigella invasion of the colonic mucosa is associated with an intense inflammatory reaction,

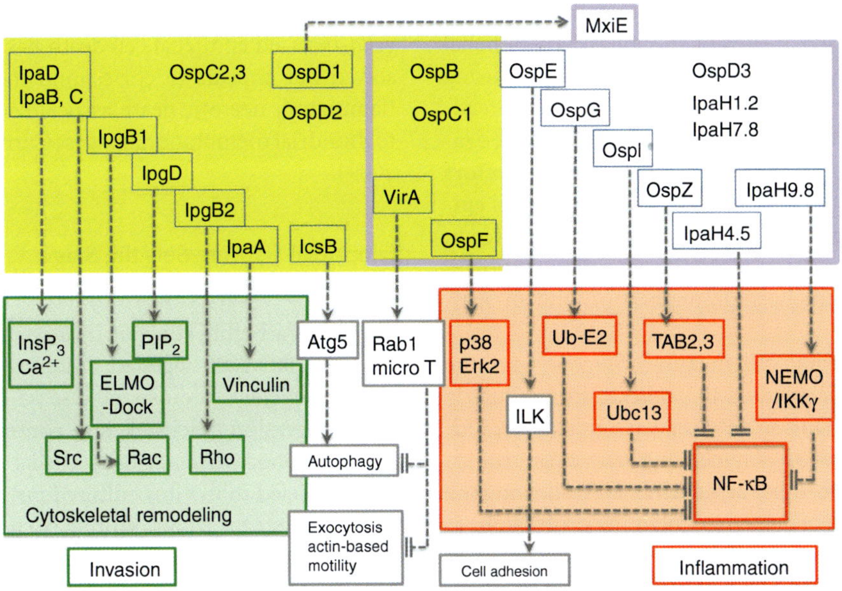

Figure 1. The control of cell responses by *Shigella* T3S effectors. *Shigella* T3S effectors are shown in yellow and purple boxes. The T3S effectors boxed in blue represent those for which a function has been assigned. T3S effectors that are constitutively expressed (yellow box) or up-regulated following cell-contact induction of secretion (boxed in purple) and activation by MxiE. On cell contact, IpaB and IpaC insert into the host-cell membrane to form the "translocon," required for the induction of $InsP_3$-dependent Ca^{2+} responses. The targets of T3S effectors involved in *Shigella* invasion are boxed in green. IpaC, through its carboxy-terminal domain induces the recruitment and activation of the Src kinase. IpgB1 and IpgD, together with IpaC, participate in actin polymerization and membrane ruffle formation by targeting ELMO/Dock and hydrolyzing PIP_2, respectively. IpgB2 that activates Rho, and IpaA that binds to vinculin, further reorganize the actin cytoskeleton to promote invasion. IcsB prevents autophagic responses by inhibiting binding of Atg5 to IcsA. The targets of T3S effectors that down-regulate inflammation are boxed in red. VirA prevents autophagy and IL-8 secretion through its GAP activity toward Rab1. VirA also inhibits microtubule polymerization to favor actin-based motility. Among the T3S effectors up-regulated by MxiE, OspF prevents the activation of the p38 and Erk2 MAP kinases via its phospho-threonine lyase activity. OspE reinforces cell adhesion by targeting ILK. The targets of OspG, OspI, OspZ, IpaH4.5, and IpaH9.8 that down-regulate NF-κB activation are indicated.

and experiments performed in rabbit ileal loops indicate that the inhibition of inflammation through the perfusion of IL-1 Ra, an antagonist of IL-1 receptor-mediated signaling, prevents not only tissular destruction but also bacterial dissemination in the intestinal mucosa (Phalipon and Sansonetti 2007). This has led to speculation that inflammation is triggered by the bacteria for efficient tissue colonization. Consistently, on ingestion by macrophage, *Shigella* was shown to induce cell death in a so-called "pyroptotic" process, involving the cleavage of caspase-1 and the release of proinflammatory IL-1β and IL-18 (Zychlinsky et al. 1992; Sansonetti et al. 2000). Using an *icsA* mutant deficient for cell-to-cell spreading, it was observed that *Shigella* invades at the levels of Peyer's patches, through M-cells of the follicle-associated epithelium (Sansonetti and Phalipon 1999). Via this route of entry, *Shigella*-induced pyroptotic death of macrophage leads to the recruitment of polymorphonuclear cells and contributes to the destabilization of the epithelial layer, thus permitting de novo invasion of intestinal epithelial cells through a basolateral route. The rabbit ileal loop infection model, however, does not allow the study of primary events during *Shigella* infection. Although relatively inefficient in vitro, *Shigella* invasion can occur at the apical surface of polarized intestinal epithelial cells

(Fig. 2). This infection route, which only involves discrete sites of the intestinal epithelial layer, may have been overlooked in studies using an *icsA* mutant deficient in cell-to-cell spreading. In recent years, the functional characterization of various T3S second wave effectors suggests that discrete invasion of intestinal epithelial cells by *Shigella* represents a route, in which the down-regulation of inflammation is required for bacterial replication and dissemination.

Following cell invasion, *Shigella* lyses the phagocytic vacuole to replicate intracellularly (Fig. 3). During intracellular replication, *Shigella* moves by polymerizing actin at one bacterial pole, forming actin comet tails, allowing the formation of bacteria-containing protrusions at the cell plasma membrane that invade adjacent cells. Following lysis of the donor and recipient cell membranes, bacteria reinitiate intracellular replication to disseminate into the epithelium. Bacterial intracellular replication occurs at a doubling time estimated at 10–15 min, emphasizing the fitness of the bacterium to the intracellular environment. Infected cells loaded with bacteria die a few hours following infection. The precise mechanism implicated in *Shigella*-induced epithelial cell death has not been solved but appears to correspond to a proinflammatory necrotic death associated with mitochondrial membrane damages (Carneiro et al. 2009).

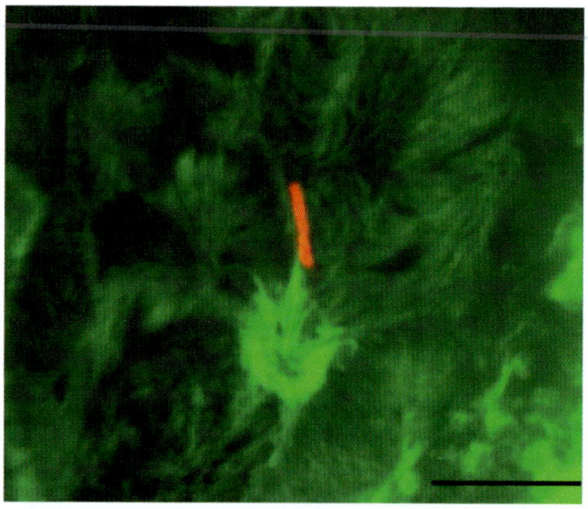

Figure 2. Filopodial capture and *Shigella* invasion at the apical side of polarized Caco-2 cells. Shown is a *Z* projection of confocal fluorescence images of *Shigella*-induced actin foci at the apical side of polarized Caco-2 cells. Red, anti-LPS immunostaining; green, Alexa488-phalloidin staining. *Z* sections were acquired via spinning-disk confocal microscopy. Scale bar, 5 mm.

Filopodial Capture Sets the Stage for Bacterial Invasion

Shigella is a highly efficient invasive pathogen, because 100–1000 ingested bacteria were estimated sufficient to cause the disease. The high in vivo invasion efficiency is in contrast to the relatively poor efficiency of epithelial cell invasion modeled in in vitro cultured polarized intestinal cells. One factor limiting bacterial invasion is the absence of constitutive cell-binding activity. This lack of cell-binding activity probably stresses the bacterial fitness to the intracellular lifestyle, and perhaps underlines a strategy of invasion associated with a lack of exposure at the cell surfaces during the initial stages of the infectious process.

Although *Shigella* does not express adhesin that would allow constitutive cell-binding activity, during the initial steps of cell infection, only a minute fraction of bacteria interacts with epithelial cells to trigger their internalization. This peculiar characteristic is linked to a sampling of invasive bacteria by finger-like cell extensions protruding from the cell surface termed filopodia, corresponding to sensory organelles probing the cell environment to establish adhesion structures (Fig. 2) (Romero et al. 2011). The interaction between one bacterial pole and the filopodial tip triggers the retraction of bacteria-bound extensions toward the cell body in which invasion occurs. Bacterial capture by filopodia is dependent on the T3SA and appears to result from direct binding of the T3SA tip complex to unidentified receptors at the filopodial tip. Measurements based on optical tweezers indicate that a discrete numbers of T3SA are sufficient to promote filopodial retraction, with a strength that is comparable to that induced by several thousands of integrin ligands (Romero et al. 2012). There is no evidence for active type III secretion during these initial events to pro-

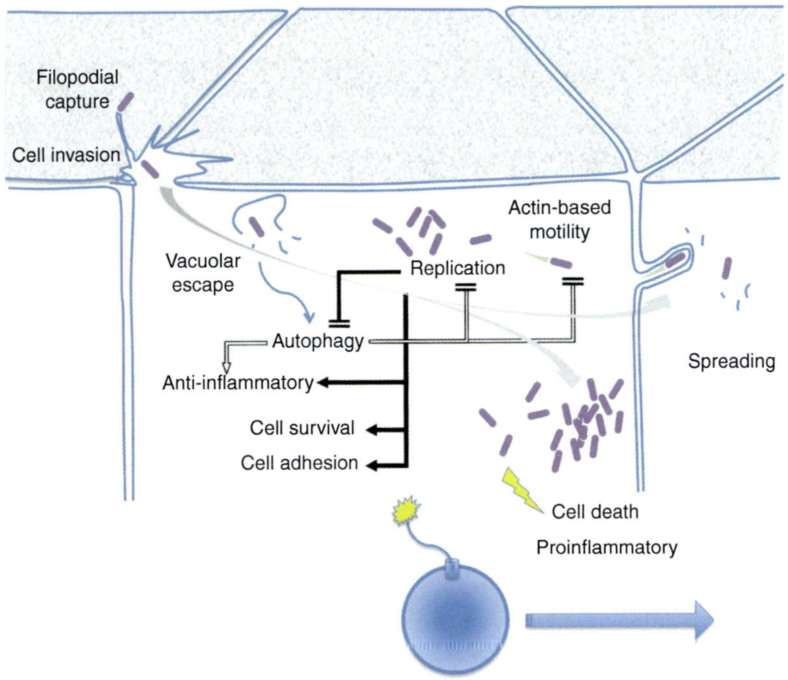

Figure 3. *Shigella* invasion, intracellular replication, and spreading in epithelial cells. Filopodial capture initiates the invasion process, accompanied by localized membrane ruffles at the epithelial cell apical surface. Following invasion, bacterial escape from the vacuole triggers autophagic responses that dampen inflammatory signals and limit bacterial replication and actin-based motility. During the early stages of bacterial intracellular replication, T3S effectors counter autophagy, down-regulate inflammatory responses, promote cell survival, and reinforce cell adhesion (solid arrows) to preserve the epithelial cell integrity (gray arrow), to allow bacterial spreading. At late stages of intracellular replication, the increase in cytosolic Ca^{2+} concentration and mitochondrial damage lead to necrotic cell death and the emission of proinflammatory signals (yellow). *Shigella* invasion of epithelial cells can be seen as the ignition of a time bomb, in which bacterial spreading must occur before necrotic death of the primary infected cell.

mote retraction. The discrete number of T3SA sufficient to promote retraction probably reflects elementary events at the filopodial tip regulating the dynamics of actin assembly on the few tens of filaments in the filopodial shaft. Bacterial-induced filopodial retraction results from inhibition of actin polymerization at the filopodial tip, and from the action of the treadmilling of actin filaments in the cell cortex that pulls the embedded filopodial actin shaft inward. Retraction is controlled by the MAP kinase Erk that also controls the dynamics of bacterial capture by filopodia. Filopodial capture is increased during *Shigella* invasion, leading to the internalization of multiple bacteria at a single invasion site (Romero et al. 2011).

Vacuolar Lysis

Following internalization, *Shigella* rapidly lyses the phagocytic vacuole. Time-lapse fluorescence microscopy using markers of the vacuolar membrane or using a FRET sensor to detect bacterial access to the cytosol, indicate that vacuolar lysis occurs in the continuation of the invasion process, within minutes following bacterial-cell contact (Ehsani et al. 2012). Studies involving the complementation of mutants with orthologs from *Salmonella typhimurium*, an enteroinvasive bacterium that does not lyse, but replicates within an intracellular vacuole, point to a role for translocon components in vacuolar rupture (Hermant et al. 1995), but the precise mechanism remains undefined. As for invasion, efficient vacuolar rupture probably results from the concerted action of T3S effectors.

Autophagy and Inflammation

On vacuolar rupture, membrane remnants act as danger signals that are controlled by the autophagic responses. Proteins contained in membrane remnants are polyubiquinated and recruit autophagic markers such as LC3, as well as the Nalp3, Asc, and Ipaf inflammasome markers (Dupont et al. 2009). By targeting the inflammasome markers to degradation, autophagy acts as a rheostat controlling the levels of danger signals emitted by bacteria invading the cell cytosol (Fig. 3). Thus, during early phases of invasion, discrete events of vacuolar rupture controlled by the autophagic pathway may have a limited impact on inflammatory signals, in particular because T3S effectors also negatively regulate these signals (Ashida et al. 2011).

At later stages, this balance may switch to a proinflammatory context, when the accumulation of invading bacteria and vacuolar membrane remnants overwhelm the autophagy machinery. Bacteria intracellular replication leads to cytosolic accumulation of PAMPs (pathogen associated molecular patterns), such as muramyl dipeptide, which activates the NLR inflammasome. Increase in cytosolic calcium concentration leads to the activation of protease, such as calpain, and mitochondrial damage leads to the death of infected cells via a necrotic pathway (Fig. 3) (Carneiro et al. 2009; Bergounioux et al. 2012).

As for many pathogens that use actin to move intracellularly, such as *Listeria* or *Shigella*, activation of autophagy may control intracellular replication. For *Shigella*, autophagy is associated with rings of septins, described as molecular scaffolds implicated in cell division (Mostowy et al. 2010). Septins form "cages" entrapping *Shigella*, preventing intracellular motility and targeting bacteria to autophagy. To counter autophagy, *Shigella* secretes the IcsB T3S effector that prevents Atg5 recognition of IcsA and bacterial degradation (Figs. 1 and 3) (Ogawa et al. 2005).

Cell-to-Cell Spreading

In vitro experiments, using reconstituted motility assays with purified proteins, show that IcsA is necessary and sufficient for *Shigella* motility. IcsA binds to N-WASP (neuronal Wiskott–Aldrich syndrome protein), which in turn, activates the Arp2/3 actin nucleator complex (Egile et al. 1999). In cells, IcsA-mediated bacterial motility is also controlled by tyrosine kinases, such as Btk or Abl/Arg that phosphorylate N-WASP and allow its recruitment at the bacterial surface (Burton et al. 2005; Dragoi et al. 2012). Intracellular *Shigella* also recruits the TOCA-1 protein, necessary to relieve the autoinhibition of N-WASP (Leung et al. 2008). Actin-based motility is favored by the action of the T3S effector VirA that prevents the assembly of microtubules representing potential physical obstacles (Yoshida et al. 2006).

Although actin-based motility allows *Shigella* to probe its intracellular environment, it is not sufficient for bacterial spreading. Cadherin-based junctions are required for *Shigella* cell-to-cell spread, presumably by providing the machinery necessary for protrusion formation by the donor cell, and its endocytosis by the recipient cell in a process requiring the activity of the T3SA and myosin II (Sansonetti et al. 1994; Schuch et al. 1999; Rathman et al. 2000). Recently, *Shigella* was observed to disseminate preferentially at multijunctions, corresponding to sites within the epithelium in which several cells intersect (Fukumatsu et al. 2012). Tricellulin, a marker of these multijunctions, has been involved in *Shigella* spreading, by participating in a noncanonical clathrin-dependent endocytic pathway (Fukumatsu et al. 2012).

Preserving the Epithelium Integrity to Allow Dissemination

As mentioned earlier, increased loads of bacteria linked to intracellular replication and multiple invasion events will lead infected epithelial cells to undergo necrotic death. The death of infected cells, however, may be counterproductive for the infectious process if occurring too early. Through the action of injected T3S effectors, *Shigella* preserves epithelial cells' integrity by delaying cell detachment, thus increasing possibility to disseminate.

IpgD is a T3S effector that hydrolyzes phosphatidylinositol (PI) (4, 5) biphosphate (PIP_2) to generate PI5P. PI5P activates a cell-survival pathway dependent on the PI3 kinase and the Akt kinase (Pendaries et al. 2006). Another T3S effector, OspE, binds to ILK (integrin-like kinase) to stabilize integrin-dependent attachment of the cell to the basal matrix (Fig. 1) (Kim et al. 2009). Reinforcement of cell adhesion to the matrix may represent an important factor during infection, by preventing their clearing in the intestinal lumen (Kim et al. 2009).

Shigella also induces calcium signals during the early phase of epithelial cell invasion, leading to the opening of connexin hemichannels at the plasma membrane (Tran Van Nhieu et al. 2003). ATP, released in the extracellular medium through the opening of hemichannels, signals in a paracrine manner to neighboring cells to enhance bacterial spreading, as well as invasion by stimulating Erk-dependent filopodial capture (Clair et al. 2008; Romero et al. 2011).

Silencing Inflammation and Danger Signals

Some constitutively expressed T3S effectors are up-regulated following activation of the T3SA and *Shigella* invasion (Parsot 2009). For these, it is possible that their activity is required early following the invasion process, as well as at later stages, during intracellular replication and spreading. Modulation of T3S effectors activity can occur consequently to endoproteolytic cleavage by Caspase 3 (Srikanth et al. 2010). T3S effectors and translocon components can have multiple activities. For example, in addition to its role in translocation, IpaB participates in the pyroptotic death observed in macrophages, and dissolution of the Golgi observed during epithelial cell invasion (Fig. 1) (Mounier et al. 2012). The disruption of the Golgi apparatus may impair processes such as exocytosis, involved in membrane receptor signaling, and the mounting of inflammatory responses. Along these lines, VirA shows GAP activity toward Rab1 and prevents the targeting of bacteria to autophagy, as well as the secretion of the proinflammatory cytokine IL-8 (Fig. 1) (Dong et al. 2012). Six T3S effectors have been reported to directly or indirectly inhibit the activation of the proinflammatory transcription factor NF-κ. OspF removes phosphothreonine residues on the p38 and Erk2 MAP kinases, preventing their activation (Arbibe et al. 2007; Li et al. 2007); OspG prevents the ubiquitinylation of IκBα and therefore its degradation, by targeting ubiquitinylated ubiquitin-conjugating enzymes (Kim et al. 2005); OspI deamidates Ubc13 to prevent its ubiquitin ligase activity and TRAF-6 signaling (Sanada et al. 2012); OspZ is related to the EPEC NleE T3S effector, that prevents NF-κ translocation through the cystein methylation of ubiquitin chain-sensing protein (Newton 2010; Zhang et al. 2011); the IpaH proteins are E3 ubiquitin ligases. IpaH4.5 ubiquitinates the p65 subunit of NF-κB (Wang et al. 2013); IpaH9.8 promotes the polyubiquitination of NEMO/IκB kinase and its degradation (Fig. 1) (Ashida et al. 2010).

PART II. MOLECULAR INSIGHTS INTO *SHIGELLA* INVASION OF EPITHELIAL CELLS

The *Shigella* Mxi-Spa T3SA

T3SAs are expressed by many of Gram-negative bacterial pathogens to mammals, fish, and plants (Mueller et al. 2008). These secretion systems are repressed under basal state conditions, and are activated on cell contact to allow the delivery of effectors, directly from the bacterial cytosol to the cell cytoplasm (Cossart and Sansonetti 2004; Münter et al. 2006; Matteï et al. 2011). This inducible feature of the secretory activity may be an important reason for the wide representation of T3SAs in bacterial pathogens, because it allows the masking of virulence determinants to host defense systems and a *cis*-acting activity of injected effectors. T3SAs present high levels of synteny, conserved structure, and function. These apparatus are related to flagellar systems and many of the general T3SS features have been inferred from studies on the *Shigella* T3SS.

The *Shigella* Mxi-Spa T3SA is composed of 21 different proteins forming a basal body embedded in the bacterial inner and outer membranes prolonged by a needle capped by a so-called "tip complex" (Blocker et al. 2008; Mueller et al. 2008). As for flagella, the basal body consists of a series of rings formed by the oligomerization of different T3SA components. Structural evidence indicates that at basal state, the tip complex corresponds to a homopentamer of IpaD (Epler et al. 2012). On binding to deoxycholate, which may correspond to the bile salts enriched environment encountered by *Shigella* during transit in the intestinal lumen, IpaD switches conformation to allow the tip presentation of IpaB (Dickenson et al. 2011). On contact with host cells, IpaC is secreted and together with IpaB insert into the host-cell plasma membrane to form the "translocon," endowed with pore-forming activity and required for the translocation of injected effectors into the host cell cytosol.

How the tip complex senses host-cell membrane to trigger secretion remains an open question. Host-cell sensing has been proposed to involve interaction between IpaB and cell-surface receptors, such as hyaluronic acid receptor CD44. Interestingly, IpaD thought to connect IpaB and the MxiH-needle, also appears to control the various stages of secretion between early and late effectors on cell contact (Schiavolin et al. 2013). The precise mechanism controlling the switch from nonactive to active secretion following tip contact is not known. The change of configuration of the tip complex proteins associated with interaction with a cell-surface receptor may lead to the emission of signals, involving MxiC and perhaps a tip-to-base transduction of signals through change of configuration of MxiH (Blocker et al. 2008; Botteaux et al. 2009; Fujii et al. 2012).

The IpaB and IpaC Translocon Components and Tyrosine Kinase Signaling

Shigella invasion of epithelial cells is associated with the formation of actin-rich membrane ruffles at the site of bacterial contact that engulf the bacterium in a macropinocytic process. This invasion process is triggered by the cell-contact-dependent injection of type-III effectors acting in concert to reorganize the actin cytoskeleton (Figs. 1 and 2).

Invasion of epithelial cells was shown to depend on cholesterol-rich microdomains, presumably because they represent signaling platforms readily mobilized that are proficient for bacterial-induced cytoskeletal reorganization. Targeting of microdomains may occur through direct preferential binding of the IpaB translocon components to cholesterol, or through IpaB interaction with cell-surface receptors such as CD44, that triggers its mobilization to microdomains (Lafont et al. 2002). On cell contact, IpaC is secreted and together with IpaB form the translocon within the host-cell plasma membrane, required for the injection of T3S effectors. In addition to its role in translocation, IpaC participates in the early stages of the invasion process by inducing actin polymerization. IpaC-mediated actin polymerization involves the recruitment and activation of the Src tyrosine kinase, which allows further amplification of actin polymerization and ruffle formation (Fig. 1) (Mounier et al. 2009). During *Shigella* invasion, Src mediates the phosphorylation of cortactin, an F-actin binding protein that can promote Arp2/3 complex-dependent actin polymerization. Tyrosyl-phosphorylated cortactin is then recruited at the plasma membrane, via interaction with the Crk adaptor to amplify actin polymerization and membrane ruffle formation (Bougnères et al. 2004). Tyrosine kinase signaling may also further allow the amplification of actin polymerization at invasion sites by acting upstream of the Rac GTPase. In addition to its role in the recruitment of cortactin, Crk was shown to be a target for Abl/Arg tyrosine kinases that are activated and required during *Shigella* invasion (Burton et al. 2003). Because in other systems Crk has been implicated in the recruitment and activation of Rac, the Abl/Arg kinases may maintain and amplify Rac activation at entry sites (Vuori and Ruoslahti 1995). Interestingly, Abl/Arg kinases are also activated downstream from Src kinases (Plattner et al. 1999). Thus, tyrosine kinases via Crk may feed in a loop that amplifies signals and actin polymerization leading to ruffle formation at *Shigella* invasion sites.

The Role of IpgB1 and IpgD in *Shigella*-Induced Actin Polymerization

Among the T3S injected effectors, IpgB1 has also been implicated in *Shigella* invasion of epithelial cells. IpgB1 belongs to the so-called "WXXXE" type-III effector family, formerly proposed to act as mimics of RhoGTPases, but for which there are structural and enzymatic activity evidence that this family rather corresponds to GEFs (guanosine exchange factor) for small GTPases (Fig. 1) (Alto et al. 2006; Bulgin et al. 2010; Orchard and Alto 2012). The picture is blurred by evidence that the same WXXXE effectors recruit GEFs for the GTPase. For example, IpgB1 first described as a mimic for Rac, may also recruit the ELMO-Dock180 complex with GEF activity toward Rac (Handa et al. 2007). Rac activation leads to actin polymerization mediated by the activation of the Arp2/3 nucleator complex. Although it is clear that Rac-dependent actin polymerization occurs in the absence of IpgB1, IpgB1-mediated activation increases ruffle formation at entry sites (Mounier et al. 2009).

IpgD also participates in the induction of membrane ruffling at *Shigella* invasion sites, although its precise mode of action is unclear. PIP_2-binding proteins are involved in the anchoring of cortical actin to the plasma membrane. Through PIP_2 hydrolysis, IpgD may help free cortical actin to favor de novo polymerization at invasion sites (Fig. 1) (Niebuhr et al. 2000). It is possible, however, that IpgD participates through different means in *Shigella* invasion. The *Salmonella* ortholog SopB/ SigD, also hydrolyzes PIP_2. SopB, however, was also shown to act as an inositolpolyphosphate (InsP) phosphatase leading to the production of $InsP_4$, shown to stimulate Rac-dependent membrane ruffling (Zhou et al. 2001). More recently, SopB has been involved in the activation of the Abi/Wave complex through the stimulation of PIP_3 production and the activation of the ARNO GEF. SopB activity may synergize with that of the SopE GEF for Rac to produce actin polymerization, or alternatively act in a SopE-independent pathway to induce bacterial invasion in a myosin-II-dependent process (Hänisch et al. 2011; Humphreys et al. 2012).

The Role of IpgB2 and IpaA in *Shigella*-Induced Cytoskeletal Reorganization

Although Rac and actin polymerization are required for ruffle formation and the initiation of *Shigella* invasion, Rho activation appears to be required for the recruitment of components such as ezrin, involved in later stages of the invasion process (Duménil et al. 2000). Other lines of evidence implicate Rho at later stages of *Shigella* invasion through the action of two injected T3S effectors, IpaA and IpgB2 (Fig. 1) (Alto et al. 2006; Demali et al. 2006; Klink et al. 2010). IpaA binds to the focal adhesion protein vinculin via its carboxy-terminal domain that mimics talin (Izard et al. 2006). Talin is also a component of adhesion structures that binds to the cytosolic domain of the integrin β1 subunit and activates vinculin. Following activation, vinculin unfolds and its carboxy-terminal domain associates with actin filaments. The talin–vinculin interaction is used by cells to reinforce cytoskeletal anchorage to integrin receptors (del Rio et al. 2009; Parsons et al. 2010). In the absence of classical receptor–ligand interaction explaining how *Shigella* recruits focal adhesion proteins, it is possible that signaling occurs through injected IpaA. Consistent with its vinculin-binding activity, IpaA allows the anchorage of bacteria within membrane ruffle (Tran Van Nhieu et al. 1997). Surprisingly, however, not only does IpaA promote this bacterial anchorage activity, but IpaA also permits the depolymerization of actin filaments during the completion of the invasion process (Bourdet-Sicard et al. 1999). The mechanism underlying IpaA-mediated actin depolymerization is still unclear. The amino-terminal part of IpaA has been proposed to induce the activation of Rho by an undefined mechanism which, together with IpaA-mediated inhibition of talin association with β1 integrin, leads to the disappearance of actin stress fibers (Demali et al. 2006). IpaA in association with vinculin was also shown to regulate the dynamics of actin polymerization/depolymerization by partially capping the barbed end of actin filaments (Ramarao et al. 2007).

IpgB2, first described as a Rho mimic, may also activate Rho through its GEF activity, or through the activation of GEF-H1 (Alto et al. 2006; Fukazawa et al. 2008). A role for IpgB2 in *Shigella* invasion is observed in polarized but not in nonpolarized cells and only in combination with IpgB1 (Hachani et al. 2008). Because polarized cells differ from nonpolarized cells by the establishment of cell–cell junctions, it is tempting to speculate that the combined action of IpgB1 and IpgB2 serves to regulate Rac- and Rho-dependent junctional structures such as cadherin-based junctions (Harris and Tepass 2012).

$InsP_3$-Dependent Local Calcium Signals during *Shigella* Invasion

How precisely initial events leading to bacterial-induced actin polymerization during *Shigella* invasion are triggered is not known. There is evidence indicating that this early signaling is directly linked to insertion of the translocon in the plasma membrane. First, although inactivation of the above-mentioned injected T3S effectors only lead to a partial decrease of bacterial invasion, mutations in the translocon component IpaC that does not prevent effector translocation totally abolishing bacterial-induced actin polymerization and *Shigella* uptake (Mounier et al. 2009). Interestingly, *Shigella* induces calcium responses localized at the site of bacterial invasion that, unlike bacterial-induced global calcium responses, are required for actin polymerization at entry sites (Tran Van Nhieu et al. 2013). These calcium responses can occur in the absence of extracellular calcium and are elicited by $InsP_3$-mediated release of intracellular pools, in a process implicating PLC-β1 and PLC-δ1. As for actin polymerization, *Shigella*-induced $InsP_3$-mediated signaling does not appear to depend on injected T3S effectors, and is totally abolished in mutants for the translocon component IpaC (Tran Van Nhieu et al. 2013). As described for pore-forming toxins, insertion of the *Shigella* translocon in the plasma membrane may lead to the destabilization of lipid bilayers leading to the activation of PLCs, that hydrolyze PIP_2 to induce $InsP_3$-mediated signaling (García-Sáez et al. 2011). Future works will be required to investigate the relationship between this $Insp_3$-mediated signaling, tyrosine kinase activation, and stimulation of Rac-dependent actin polymerization during *Shigella* invasion.

CONCLUDING REMARKS

Shigella can be considered as a primitive, poorly evolved pathogen because it invades the colonic mucosa, without persisting in deeper tissues, and elicits an intense inflammation and tissue destruction that may kill its host. We propose a scheme where *Shigella* invasion occurs at discrete sites of the intestinal epithelium, timing its intracellular replication to ensure spreading before the death of the primary infected cell. This discrete route of invasion plants the seeds leading to the massive inflammation and tissue destruction observed at the late stages of infection.

In this article, we have focused on intestinal epithelial cells as the spreading ground for this pathogen. Clearly, this is a reductionist view, as the role of other players such as macrophages and M cells, Paneth cells secreting defensins, or the enteric nervous system, shown to be important in the control of *Shigella* infection has not been developed. While compared with other pathogens, the invasion and destruction of the colonic mucosa could appear as a reflection of a primitive adaptation to its host, the extent of sophistication developed by *Shigella* through the action of multiple effectors dedicated to the timely subversion of host-cell processes keeps revealing facets of the complexity of the infectious process.

ACKNOWLEDGMENTS

The authors thank the Collège de France, the Institut National de la Santé et de la Recherche Médical, the Centre National de la Recherche Scientifique, and l'Agence Nationale de la Recherche for funding. N.C. is a Marie Curie IRG grant recipient.

REFERENCES

Alto NM, Shao F, Lazar CS, Brost RL, Chua G, Mattoo S, McMahon SA, Ghosh P, Hughes TR, Boone C, et al. 2006. Identification of a bacterial type III effector family with G protein mimicry functions. *Cell* **124:** 133–145.

Arbibe L, Kim DW, Batsche E, Pedron T, Mateescu B, Muchardt C, Parsot C, Sansonetti PJ. 2007. An injected bacterial effector targets chromatin access for transcription factor NF-κB to alter transcription of host genes involved in immune responses. *Nat Immunol* **8:** 47–56.

Ashida H, Kim M, Schmidt-Supprian M, Ma A, Ogawa M, Sasakawa C. 2010. A bacterial E3 ubiquitin ligase IpaH9.8 targets NEMO/IKKγ to dampen the host NF-κB-mediated inflammatory response. *Nat Cell Biol* **12:** 66–73.

Ashida H, Ogawa M, Kim M, Suzuki S, Sanada T, Punginelli C, Mimuro H, Sasakawa C. 2011. *Shigella* deploys multiple countermeasures against host innate immune responses. *Curr Opin Microbiol* **14:** 16–23.

Bergounioux J, Elisee R, Prunier AL, Donnadieu F, Sperandio B, Sansonetti P, Arbibe L. 2012. Calpain activation by the *Shigella flexneri* effector VirA regulates key steps in the formation and life of the bacterium's epithelial niche. *Cell Host Microbe* **11:** 240–252.

Bliven KA, Maurelli AT. 2012. Antivirulence genes: Insights into pathogen evolution through gene loss. *Infect Immun* **80:** 4061–4070.

Blocker AJ, Deane JE, Veenendaal AK, Roversi P, Hodgkinson JL, Johnson S, Lea SM. 2008. What's the point of the type III secretion system needle? *Proc Natl Acad Sci* **105:** 6507–6513.

Botteaux A, Sory MP, Biskri L, Parsot C, Allaoui A. 2009. MxiC is secreted by and controls the substrate specificity of the *Shigella flexneri* type III secretion apparatus. *Mol Microbiol* **71:** 449–460.

Bougnères L, Girardin SE, Weed SA, Karginov AV, Olivo-Marin JC, Parsons JT, Sansonetti PJ, Tran Van Nhieu G. 2004. Cortactin and Crk cooperate to trigger actin polymerization during *Shigella* invasion of epithelial cells. *J Cell Biol* **166:** 225–235.

Bourdet-Sicard R, Rüdiger M, Jockusch BM, Gounon P, Sansonetti PJ, Tran Van Nhieu G. 1999. Binding of the *Shigella* protein IpaA to vinculin induces F-actin depolymerization. *EMBO J* **18:** 5853–5862.

Bulgin R, Raymond B, Garnett JA, Frankel G, Crepin VF, Berger CN, Arbeloa A. 2010. Bacterial guanine nucleotide exchange factors SopE-like and WxxxE effectors. *Infect Immun* **78:** 1417–1425.

Burton EA, Plattner R, Pendergast AM. 2003. Abl tyrosine kinases are required for infection by *Shigella flexneri*. *EMBO J* **22:** 5471–5479.

Burton EA, Oliver TN, Pendergast AM. 2005. Abl kinases regulate actin comet tail elongation via an N-WASP-dependent pathway. *Mol Cell Biol* **25:** 8834–8843.

Carneiro LA, Travassos LH, Soares F, Tattoli I, Magalhaes JG, Bozza MT, Plotkowski MC, Sansonetti PJ, Molkentin JD, Philpott DJ, et al. 2009. *Shigella* induces mitochondrial dysfunction and cell death in nonmyleoid cells. *Cell Host Microbe* **5:** 123–136.

Clair C, Combettes L, Pierre F, Sansonetti P, Tran Van Nhieu G. 2008. Extracellular-loop peptide antibodies reveal a predominant hemichannel organization of connexins in polarized intestinal cells. *Exp Cell Res* **314:** 1250–1265.

Cossart P, Sansonetti PJ. 2004. Bacterial invasion: The paradigms of enteroinvasive pathogens. *Science* **304:** 242–248.

del Rio A, Perez-Jimenez R, Liu R, Roca-Cusachs P, Fernandez JM, Sheetz MP. 2009. Stretching single talin rod molecules activates vinculin binding. *Science* **323:** 638–641.

Demali KA, Jue AL, Burridge K. 2006. IpaA targets β1 integrins and ρ to promote actin cytoskeleton rearrangements necessary for *Shigella* entry. *J Biol Chem* **281:** 39534–39541.

Dickenson NE, Zhang L, Epler CR, Adam PR, Picking WL, Picking WD. 2011. Conformational changes in IpaD from *Shigella flexneri* upon binding bile salts provide insight into the second step of type III secretion. *Biochemistry* **50:** 172–180.

Dong N, Zhu Y, Lu Q, Hu L, Zheng Y, Shao F. 2012. Structurally distinct bacterial TBC-like GAPs link Arf GTPase to Rab1 inactivation to counteract host defenses. *Cell* **150:** 1029–1041.

Dorman CJ, McKenna S, Beloin C. 2001. Regulation of virulence gene expression in *Shigella flexneri*, a facultative intracellular pathogen. *Int J Med Microbiol* **291:** 89–96.

Dragoi AM, Talman AM, Agaisse H. 2012. Bruton's tyrosine kinase regulates *Shigella flexneri* dissemination in HT-29 intestinal cells. *Infect Immun* **81:** 598–607.

Duménil G, Sansonetti P, Tran Van Nhieu G. 2000. Src tyrosine kinase activity down-regulates ρ-dependent responses during *Shigella* entry into epithelial cells and stress fibre formation. *J Cell Sci* **113:** 71–80.

Dupont N, Lacas-Gervais S, Bertout J, Paz I, Freche B, Tran Van Nhieu G, van der Goot FG, Sansonetti PJ, Lafont F. 2009. *Shigella* phagocytic vacuolar membrane remnants participate in the cellular response to pathogen invasion and are regulated by autophagy. *Cell Host Microbe* **6:** 137–149.

Egile C, Loisel TP, Laurent V, Li R, Pantaloni D, Sansonetti PJ, Carlier MF. 1999. Activation of the CDC42 effector N-WASP by the *Shigella flexneri* IcsA protein promotes actin nucleation by Arp2/3 complex and bacterial actin-based motility. *J Cell Biol* **146:** 1319–1332.

Ehsani S, Santos JC, Rodrigues CD, Henriques R, Audry L, Zimmer C, Sansonetti P, Tran Van Nhieu G, Enninga J. 2012. Hierarchies of host factor dynamics at the entry site of *Shigella flexneri* during host cell invasion. *Infect Immun* **80:** 2548–2557.

Epler CR, Dickenson NE, Bullitt E, Picking WL. 2012. Ultrastructural analysis of IpaD at the tip of the nascent MxiH type III secretion apparatus of *Shigella flexneri*. *J Mol Biol* **420:** 29–39.

Fujii T, Cheung M, Blanco A, Kato T, Blocker AJ, Namba K. 2012. Structure of a type III secretion needle at 7-A resolution provides insights into its assembly and signaling mechanisms. *Proc Natl Acad Sci* **109:** 4461–4466.

Fukazawa A, Alonso C, Kurachi K, Gupta S, Lesser CF, McCormick BA, Reinecker HC. 2008. GEF-H1 mediated control of NOD1 dependent NF-κB activation by *Shigella* effectors. *PLoS Pathog* **4:** e1000228.

Fukumatsu M, Ogawa M, Arakawa S, Suzuki M, Nakayama K, Shimizu S, Kim M, Mimuro H, Sasakawa C. 2012.

Shigella targets epithelial tricellular junctions and uses a noncanonical clathrin-dependent endocytic pathway to spread between cells. *Cell Host Microbe* **11:** 325–336.

García-Sáez AJ, Buschhorn SB, Keller H, Anderluh G, Simons K, Schwille P. 2011. Oligomerization and pore formation by equinatoxin II inhibit endocytosis and lead to plasma membrane reorganization. *J Biol Chem* **286:** 37768–37777.

Hachani A, Biskri L, Rossi G, Marty A, Ménard R, Sansonetti P, Parsot C, Tran Van Nhieu G, Bernardini ML, Allaoui A. 2008. IpgB1 and IpgB2, two homologous effectors secreted via the Mxi-Spa type III secretion apparatus, cooperate to mediate polarized cell invasion and inflammatory potential of *Shigella flexenri*. *Microbes Infect* **10:** 260–268.

Handa Y, Suzuki M, Ohya K, Iwai H, Ishijima N, Koleske AJ, Fukui Y, Sasakawa C. 2007. *Shigella* IpgB1 promotes bacterial entry through the ELMO-Dock180 machinery. *Nat Cell Biol* **9:** 121–128.

Hänisch J, Kölm R, Wozniczka M, Bumann D, Rottner K, Stradal TE. 2011. Activation of a RhoA/myosin II-dependent but Arp2/3 complex-independent pathway facilitates *Salmonella* invasion. *Cell Host Microbe* **9:** 273–285.

Harris TJ, Tepass U. 2012. Adherens junctions: From molecules to morphogenesis. *Nat Rev Mol Cell Biol* **11:** 502–514.

Hermant D, Ménard R, Arricau N, Parsot C, Popoff MY. 1995. Functional conservation of the *Salmonella* and *Shigella* effectors of entry into epithelial cells. *Mol Microbiol* **17:** 781–789.

Humphreys D, Davidson A, Hume PJ, Koronakis V. 2012. *Salmonella* virulence effector SopE and Host GEF ARNO cooperate to recruit and activate WAVE to trigger bacterial invasion. *Cell Host Microbe* **11:** 129–139.

Izard T, Tran Van Nhieu G, Bois PR. 2006. *Shigella* applies molecular mimicry to subvert vinculin and invade host cells. *J Cell Biol* **175:** 465–475.

Kim DW, Lenzen G, Page AL, Legrain P, Sansonetti PJ, Parsot C. 2005. The *Shigella flexneri* effector OspG interferes with innate immune responses by targeting ubiquitin-conjugating enzymes. *Proc Natl Acad Sci* **102:** 14046–14051.

Kim M, Ogawa M, Fujita Y, Yoshikawa Y, Nagai T, Koyama T, Nagai S, Lange A, Fässler R, Sasakawa C. 2009. Bacteria hijack integrin-linked kinase to stabilize focal adhesions and block cell detachment. *Nature* **459:** 578–582.

Klink BU, Barden S, Heidler TV, Borchers C, Ladwein M, Stradal TE, Rottner K, Heinz DW. 2010. Structure of *Shigella* IpgB2 in complex with human ρA: Implications for the mechanism of bacterial guanine nucleotide exchange factor mimicry. *J Biol Chem* **285:** 17197–17208.

Lafont F, Tran Van Nhieu G, Hanada K, Sansonetti P, van der Goot FG. 2002. Initial steps of *Shigella* infection depend on the cholesterol/sphingolipid raft-mediated CD44-IpaB interaction. *EMBO J* **21:** 4449–4457.

Leung Y, Ally S, Goldberg MB. 2008. Bacterial actin assembly requires toca-1 to relieve N-wasp autoinhibition. *Cell Host Microbe* **3:** 39–47.

Li H, Xu H, Zhou Y, Zhang J, Long C, Li S, Chen S, Zhou JM, Shao F. 2007. The phosphothreonine lyase activity of a bacterial type III effector family. *Science* **315:** 1000–1003.

Marteyn B, West NP, Browning DF, Cole JA, Shaw JG, Palm F, Mounier J, Prévost MC, Sansonetti P, Tang CM. 2010. Modulation of *Shigella* virulence in response to available oxygen in vivo. *Nature* **465:** 355–358.

Marteyn B, Gazi A, Sansonetti P. 2012. *Shigella*: A model of virulence regulation in vivo. *Gut Microbes* **3:** 104–120.

Matteï PJ, Faudry E, Job V, Izoré T, Attree I, Dessen A. 2011. Membrane targeting and pore formation by the type III secretion system translocon. *FEBS J* **278:** 414–426.

Mavris M, Sansonetti PJ, Parsot C. 2002. Identification of the *cis*-acting site involved in activation of promoters regulated by activity of the type III secretion apparatus in *Shigella flexneri*. *J Bacteriol* **184:** 6751–6759.

Mostowy S, Bonazzi M, Hamon MA, Tham TN, Mallet A, Lelek M, Gouin E, Demangel C, Brosch R, Zimmer C, et al. 2010. Entrapment of intracytosolic bacteria by septin cage-like structures. *Cell Host Microbe* **8:** 433–444.

Mounier J, Popoff MR, Enninga J, Frame MC, Sansonetti PJ, Tran Van Nhieu G. 2009. The IpaC carboxy terminal effector domain mediates Src-dependent actin polymerization during *Shigella* invasion of epithelial cells. *PLoS Pathog* **5:** e1000271.

Mounier J, Boncompain G, Senerovic L, Lagache T, Chrétien F, Perez F, Kolbe M, Olivo-Marin JC, Sansonetti PJ, Sauvonnet N. 2012. *Shigella* effector IpaB-induced cholesterol relocation disrupts the Golgi complex and recycling network to inhibit host cell secretion. *Cell Host Microbe* **12:** 381–389.

Mueller CA, Broz P, Cornelis GR. 2008. The type III secretion system tip complex and translocon. *Mol Microbiol* **68:** 1085–1095.

Münter S, Way M, Frischknecht F. 2006. Signaling during pathogen infection. *Sci STKE* **2006:** pre5.

Newton HJ, Pearson JS, Badea L, Kelly M, Lucas M, Holloway G, Wagstaff KM, Dunstone MA, Sloan J, Whisstock JC, et al. 2010. The type III effectors NleE and NleB from enteropathogenic *E. coli* and OspZ from *Shigella* block nuclear translocation of NF-κB p65. *PLoS Pathog* **6:** e1000898.

Niebuhr K, Jouihri N, Allaoui A, Gounon P, Sansonetti PJ, Parsot C. 2000. IpgD, a protein secreted by the type III secretion machinery of *Shigella flexneri*, is chaperoned by IpgE and implicated in entry focus formation. *Mol Microbiol* **38:** 8–19.

Ogawa M, Yoshimori T, Suzuki T, Sagara H, Mizushima N, Sasakawa C. 2005. Escape of intracellular *Shigella* from autophagy. *Science* **307:** 727–731.

Orchard RC, Alto NM. 2012. Mimicking GEFs: A common theme for bacterial pathogens. *Cell Microbiol* **14:** 10–18.

Parsons JT, Horwitz AR, Schwartz MA. 2010. Cell adhesion: Integrating cytoskeletal dynamics and cellular tension. *Nat Rev Mol Cell Biol* **11:** 633–643.

Parsot C. 2009. *Shigella* type III secretion effectors: How, where, when, for what purposes? *Curr Opin Microbiol* **12:** 110–116.

Parsot C, Ageron E, Penno C, Mavris M, Jamoussi K, d'Hauteville H, Sansonetti P, Demers B. 2005. A secreted anti-activator, OspD1, and its chaperone, Spa15, are involved in the control of transcription by the type III secretion apparatus activity in *Shigella flexneri*. *Mol Microbiol* **56:** 1627–1635.

Pendaries C, Tronchère H, Arbibe L, Mounier J, Gozani O, Cantley L, Fry MJ, Gaits-Iacovoni F, Sansonetti PJ, Payrastre B. 2006. PtdIns5P activates the host cell PI3-kinase/Akt pathway during *Shigella flexneri* infection. *EMBO J* **25:** 1024–1034.

Phalipon A, Sansonetti PJ. 2007. *Shigella's* ways of manipulating the host intestinal innate and adaptive immune system: A tool box for survival? *Immunol Cell Biol* **85:** 119–129.

Plattner R, Kadlec L, DeMali KA, Kazlauskas A, Pendergast AM. 1999. c-Abl is activated by growth factors and Src family kinases and has a role in the cellular response to PDGF. *Genes Dev* **13:** 2400–2411.

Prosseda G, Di Martino ML, Campilongo R, Fioravanti R, Micheli G, Casalino M, Colonna B. 2012. Shedding of genes that interfere with the pathogenic lifestyle: The *Shigella* model. *Res Microbiol* **163:** 399–406.

Ramarao N, Le Clainche C, Izard T, Bourdet-Sicard R, Ageron E, Sansonetti PJ, Carlier MF, Tran Van Nhieu G. 2007. Capping of actin filaments by vinculin activated by the *Shigella* IpaA carboxyl-terminal domain. *FEBS Lett* **581:** 853–857.

Rathman M, de Lanerolle P, Ohayon H, Gounon P, Sansonetti P. 2000. Myosin light chain kinase plays an essential role in *S. flexneri* dissemination. *J Cell Sci* **113:** 3375–3386.

Romero S, Grompone G, Carayol N, Mounier J, Guadagnini S, Prevost MC, Sansonetti PJ, Tran Van Nhieu G. 2011. ATP-mediated Erk1/2 activation stimulates bacterial capture by filopodia, which precedes *Shigella* invasion of epithelial cells. *Cell Host Microbe* **9:** 508–519.

Romero S, Quatela A, Bornschlög T, Guadagnini S, Bassereau P, Tran Van Nhieu G. 2012. Filopodium retraction is controlled by adhesion to its tip. *J Cell Sci* **125:** 4999–5004.

Sanada T, Kim M, Mimuro H, Suzuki M, Ogawa M, Oyama A, Ashida H, Kobayashi T, Koyama T, Nagai S, et al. 2012. The *Shigella flexneri* effector OspI deamidates UBC13 to dampen the inflammatory response. *Nature* **483:** 623–626.

Sansonetti PJ, Phalipon A. 1999. M cells as ports of entry for enteroinvasive pathogens: Mechanisms of interaction, consequences for the disease process. *Semin Immunol* **11:** 193–203.

Sansonetti PJ, Mounier J, Prévost MC, Mège RM. 1994. Cadherin expression is required for the spread of *Shigella flexneri* between epithelial cells. *Cell* **76:** 829–839.

Sansonetti PJ, Phalipon A, Arondel J, Thirumalai K, Banerjee S, Akira S, Takeda K, Zychlinsky A. 2000. Caspase-1 activation of IL-1β and IL-18 are essential for *Shigella flexneri*-induced inflammation. *Immunity* **12:** 581–590.

Schiavolin L, Meghraoui A, Cherradi Y, Biskri L, Botteaux A, Allaoui A. 2013. Functional insights into the *Shigella* type III needle tip IpaD in secretion control and cell contact. *Mol Microbiol* **88:** 268–282.

Schuch R, Sandlin RC, Maurelli AT. 1999. A system for identifying post-invasion functions of invasion genes: Requirements for the Mxi-Spa type III secretion pathway of *Shigella flexneri* in intercellular dissemination. *Mol Microbiol* **34:** 675–689.

Srikanth CV, Wall DM, Maldonado-Contreras A, Shi HN, Zhou D, Demma Z, Mumy KL, McCormick BA. 2010. *Salmonella* pathogenesis and processing of secreted effectors by caspase-3. *Science* **330:** 390–393.

Tran Van Nhieu G, Ben-Ze'ev A, Sansonetti PJ. 1997. Modulation of bacterial entry into epithelial cells by association between vinculin and the *Shigella* IpaA invasin. *EMBO J* **16:** 2717–2729.

Tran Van Nhieu G, Clair C, Bruzzone R, Mesnil M, Sansonetti P, Combettes L. 2003. Connexin-dependent intercellular communication increases invasion and dissemination of *Shigella* in epithelial cells. *Nat Cell Biol* **5:** 720–726.

Tran Van Nhieu G, Kai Liu B, Zhang J, Pierre F, Prigent S, Sansonetti P, Erneux C, Kuk Kim J, Suh PG, Dupont G, et al. 2013. Actin-based confinement of calcium responses during *Shigella* invasion. *Nat Commun* **4:** 1567.

Vuori K, Ruoslahti E. 1995. Tyrosine phosphorylation of p130Cas and cortactin accompanies integrin-mediated cell adhesion to extracellular matrix. *J Biol Chem* **270:** 22259–22262.

Wang F, Jiang Z, Li Y, He X, Zhao J, Yang X, Zhu L, Yin Z, Li X, Wang X, et al. 2013. *Shigella flexneri* T3SS effector IpaH4.5 modulates the host inflammatory response via interaction with NF-κB p65 protein. *Cell Microbiol* **15:** 474–485.

Yoshida S, Handa Y, Suzuki T, Ogawa M, Suzuki M, Tamai A, Abe A, Katayama E, Sasakawa C. 2006. Microtubule-severing activity of *Shigella* is pivotal for intercellular spreading. *Science* **314:** 985–989.

Zhang L, Ding X, Cui J, Xu H, Chen J, Gong YN, Hu L, Zhou Y, Ge J, Lu Q, et al. 2011. Cysteine methylation disrupts ubiquitin-chain sensing in NF-κB activation. *Nature* **481:** 204–208.

Zhou D, Chen LM, Hernandez L, Shears SB, Galán JE. 2001. A *Salmonella* inositol polyphosphatase acts in conjunction with other bacterial effectors to promote host cell actin cytoskeleton rearrangements and bacterial internalization. *Mol Microbiol* **39:** 248–259.

Zychlinsky A, Prevost MC, Sansonetti PJ. 1992. *Shigella flexneri* induces apoptosis in infected macrophages. *Nature* **358:** 167–169.

RNA-Mediated Regulation in Pathogenic Bacteria

Isabelle Caldelari[1], Yanjie Chao[2], Pascale Romby[1], and Jörg Vogel[2]

[1]Architecture et Réactivité de l'ARN, Université de Strasbourg, CNRS, IBMC, F-67084 Strasbourg, France

[2]Institute for Molecular Infection Biology (IMIB) and Zentrum für Infektionsforschung (ZINF), University of Würzburg, 97080 Würzburg, Germany

Correspondence: joerg.vogel@uni-wuerzburg.de

Pathogenic bacteria possess intricate regulatory networks that temporally control the production of virulence factors, and enable the bacteria to survive and proliferate after host infection. Regulatory RNAs are now recognized as important components of these networks, and their study may not only identify new approaches to combat infectious diseases but also reveal new general control mechanisms involved in bacterial gene expression. In this review, we illustrate the diversity of regulatory RNAs in bacterial pathogens, their mechanism of action, and how they can be integrated into the regulatory circuits that govern virulence-factor production.

Successful host infection by pathogenic bacteria largely depends on the coordinated expression of a multitude of virulence factors and genes involved in the infection process. Over the last decade, much has been learned about the complexities of bacterial gene expression and it is now established that bacterial genes are regulated at many different levels, including and beyond transcriptional control at the DNA level. One class of macromolecules that have seen tremendous progress with respect to their regulatory scope and mechanisms is RNA. Regulatory RNAs are now recognized as important players in many physiological and adaptive responses in pathogenic bacteria (Gripenland et al. 2010; Papenfort and Vogel 2010). Some of them have been identified as previous missing links in the regulatory pathways that allow bacteria to sense population density, to modulate and modify cell-surface properties, to fine-tune their metabolism during cell growth, and to regulate virulence gene expression. Most of these findings followed the pioneering discoveries of the unexpected large numbers of many small noncoding RNAs (sRNAs) and riboswitches in nonpathogenic model bacteria (Argaman et al. 2001; Wassarman et al. 2001; Mandal et al. 2003), which prompted many microbiologists to also systematically search for sRNAs in pathogenic bacteria. More recently, new technologies such as high-throughput RNA sequencing (RNA-seq) and high-density microarrays have facilitated the genome-wide detection of expressed RNAs in various pathogens and nonpathogens (Papen-

Cite this article as *Cold Spring Harb Perspect Med* doi: 10.1101/cshperspect.a010298

fort and Vogel 2010; Romby and Charpentier 2010; Papenfort et al. 2013a). Unlike 10 years ago when transposon insertions outside protein-coding regions were not deemed worth pursuing in virulence-factor screens, there is now a fast-growing list of RNA genes within intergenic regions, which seem essential for bacterial virulence. For instance, a recent study combining RNA-seq, transposon mutagenesis, and targeted deletions in *Streptococcus pneumoniae* identified 89 sRNAs of which a large proportion play important roles in the progression of the infection as well as in tissue tropism (Mann et al. 2012). Overall, the study of regulatory RNAs in bacteria has become a fast-growing field, which is reflected by the many reviews of their general functions and mechanisms in Gram-negative and Gram-positive bacteria (Storz et al. 2011; Vogel and Luisi 2011).

In this article, we aim to illustrate the fascinating diversity of sRNA action on their cellular targets and the importance of the dynamics in regulation using selected examples from pathogenic bacteria. These examples include independently transcribed (or unique) regulatory RNAs, that either act on other RNA molecules, or target the activity of cellular proteins; and regulatory RNAs that are intricate parts of the messenger RNAs (mRNAs) they regulate. For a more detailed classification of these riboregulators and further work in nonpathogenic organisms, we refer the reader to a number of general reviews of bacterial sRNAs (Gottesman and Storz 2011; Storz et al. 2011) and an excellent book dedicated to the subject (Marchfelder and Hess 2011).

CIS-REGULATORY REGIONS OF mRNAS

Genome-wide transcriptomic analyses performed in several pathogens, including *Helicobacter pylori* and *Listeria monocytogenes*, revealed that the average size of the 5′ untranslated region (5′ UTR) of bacterial mRNAs is short, around 20–40 nucleotides long (Toledo-Arana et al. 2009; Sharma et al. 2010; Wurtzel et al. 2012), which most likely reflects an optimization for ribosome recruitment and efficient translation. However, highly regulated pathways such as growth control, cell division, virulence, stress responses, and metabolism involve many genes with longer 5′ UTRs. These often contain regulatory RNA elements that function as direct sensors of physical (thermosensor) or metabolic cues (Breaker 2009; Narberhaus 2010; Ramesh and Winkler 2010), many of which are well known to alter the expression of virulence factors (Somerville and Proctor 2009).

A widespread *cis*-acting RNA element in mRNAs is the metabolite-sensing riboswitch (Breaker 2011). These RNA sequences have the ability to bind metabolites with strict specificity, to typically cause a conformational change in the mRNA and alter the expression of the downstream coding sequence at either the levels of transcriptional elongation or translational initiation (Dambach and Winkler 2009). Typically, the riboswitch feedback regulates the associated genes, which are typically involved in the uptake or metabolism of the sensed metabolite (Breaker 2011; Serganov and Nudler 2013). According to sequence and structure conservation (Barrick and Breaker 2007; Yao et al. 2007), riboswitches seem underrepresented in Gram-negative pathogens but may regulate almost 2% of the genes in Gram-positive bacteria such as *Staphylococcus aureus* (Caldelari et al. 2011) and *L. monocytogenes* (Toledo-Arana et al. 2009). Common riboswitches respond to the intracellular concentration of diverse ligands including *S*-adenosylmethionine (SAM), thiamine pyrophosphate (TPP), flavin mononucleotide (FMN), lysine, glycine, guanine, 7-aminomethyl-7-deazaguanine (preQ1), cobalamin, Mg^{2+}, adenosine-triphosphate (ATP), or glucosamine-6-phosphate (GlcN6P). Riboswitches that sense signaling molecules such as the secondary messenger cyclic diguanosine monophosphate (c-di-GMP) have been described in *Clostridium difficile* and *Vibrio cholera* (Sudarsan et al. 2008). This dinucleotide is known to regulate a wide variety of functions including the transition from a motile to a biofilm lifestyle, as well as virulence gene expression. As an example of high regulatory complexity, the synthesis of a putative virulence factor of *C. difficile* is subject to control by an allosteric group I intron, whose self-splicing (and therefore, formation of an intact mRNA) is regulated by a c-di-GMP-sensing riboswitch (Lee et al. 2010). In *Salmonella typhimurium*, a leader mRNA with tandem attenuators senses the intracellular concentration of ATP and of proline in an independent but additive manner (Lee and Groisman 2012a,b). This mRNA encodes MgtC, a protein that is highly produced within macrophages and required for virulence. Most recently, an aminoglycoside-binding riboswitch that is widely distributed among antibiotic-resistant bacterial pathogens has been discovered. This riboswitch is present in the leader RNA of genes and activates the production of enzymes that confer resistance to aminoglycoside antibiotics when these are present in the environment (Jia et al. 2013).

Riboswitches represent promising targets for novel antibacterial compounds that avoid current mechanisms of resistance (Breaker 2009; Mulhbacher et al. 2010b). First, they are generally absent in the eukaryotic host. Second, they are amenable to structural analysis, which has provided atomic-resolution structures for disparate riboswitch classes (Serganov 2010). Nonmetabolizable agonists were recently designed based on the crystal structure of the purine-sensing riboswitch (Mulhbacher et al. 2010a); a pyrimidine derivative compound, PC1, binds the guanine riboswitch and constitutively switches off the essential *guaA* gene-encoding GMP synthase. PC1 shows bactericidal activity against *S. aureus* and reduces infection in mice. Importantly, this potential drug has a narrow spectrum activity as it targets exclusively bacteria containing the purine riboswitch, which should reduce selective pressure for resistance in nontargeted bacteria; indeed, no resistant bacteria arose during successive PC1 treatments (Mulhbacher et al. 2010a). Other studies revealed the bacterial toxicity of lysine and thiamine pyrophosphate analogs, which is a result of binding to their respective riboswitches (Sudarsan et al. 2005; Blount et al. 2006). Moreover, a natural analog of flavin, roseoflavin, is able to recognize the FMN riboswitch (Lee et al. 2009). Roseoflavin is synthesized by *Streptomyces davawensis*, its only known producer, which itself is resistant to this natural antibiotic thanks to a highly specialized FMN riboswitch that discriminates FMN from roseoflavin (Pedrolli et al. 2012). Roseoflavin affects growth and infectivity of *L. monocytogenes* through the constitutive repression of a riboflavin transporter, which is controlled by an FMN-sensing riboswitch; in parallel, roseoflavin stimulates virulence gene expression independently of the FMN-sensing riboswitch (Mansjö and Johansson 2011).

Another major class of *cis*-acting regulatory RNA elements in mRNAs responds to temperature. These RNA thermometers are usually characterized by irregular stem-loop structures (ROSE, FourU motifs) that sequester the Shine-Dalgarno (SD) region of the associated mRNA (Klinkert and Narberhaus 2009; Kortmann and Narberhaus 2012). At low temperature (below 30°C), these hairpins are typically stable and, hence, the ribosome binding site (RBS) is not accessible for translation. Increasing temperature melts the stem-loop structure to promote translation. At least two major transcriptional activators of virulence genes, LcrF in *Yersinia pestis* (Hoe and Goguen 1993; Bohme et al. 2012) and PrfA in *L. monocytogenes* (Johansson et al. 2002), are known to be regulated by thermosensors. LcrF synthesis is regulated by two hierarchical thermosensitive regulators in *Yersinia pseudotuberculosis*. At moderate temperatures, the *lcrF* gene is transcriptionally repressed by the thermosensitive modulator protein YmoA, whereas a FourU thermosensor in the 5′ UTR represses mRNA translation. At the host temperature, the SD is liberated for translation initiation (Bohme et al. 2012). Of note, mutations that both stabilize and destabilize the RNA thermometer attenuate the virulence of *Y. pseudotuberculosis*. Therefore, RNA thermosensors critically contribute to adjusting the concentration of transcriptional activators to promote optimal infection programs.

Last but not least, intracellular pH has been identified as a regulatory cue for *cis*-regulation within bacterial mRNAs. In *Salmonella*, the *alx* leader of the mRNA of a putative transporter acts as a pH-responsive element. At high pH, the progression of RNA polymerase is affected allowing the refolding of the mRNA to promote its translation (Nechooshtan et al. 2009).

GENERAL PROPERTIES OF REGULATORY RNAs AND THEIR ASSOCIATED PROTEIN MACHINERY

A staggering number of sRNAs that are bona fide regulators of gene expression have been identified in bacterial pathogens (Felden et al. 2011; Gottesman and Storz 2011; Storz et al. 2011). One group includes sRNAs, which modulate the activities of proteins (Marzi and Romby 2012; Romeo et al. 2012). As outlined further below, these sRNAs possess specific sequence or structural elements that mimic the natural substrates of regulatory proteins, and indirectly impact gene regulation at a global level. The second and the largest group of sRNAs so far act by base-pairing with cellular RNAs, primarily mRNAs. These sRNAs are classified based on genomic localization relative to their mRNA targets. In the first case, the antisense (asRNA) and its target RNA are transcribed in an overlapping fashion from the opposite strands of a DNA region, which usually entails considerable base complementarity. Short and long *cis*-encoded asRNAs regulate a variety of functions such as virulence, toxins, motility, and biofilm formation (Toledo-Arana et al. 2009; Beaume et al. 2010; Sharma et al. 2010; Lasa et al. 2011; Lioliou et al. 2012; Wurtzel et al. 2012). In the second case, the sRNA is encoded distant to its target(s), and typically shows only partial complementarity, which can be as short as half a dozen nucleotides. These sRNAs tend to regulate multiple rather than single target mRNAs and act in concert with transcriptional regulatory proteins or two-component systems to respond to environmental changes, envelope stress (Gogol et al. 2011), alter nutrient conditions (Beisel and Storz 2011), control quorum sensing (Sonnleitner et al. 2009; Felden et al. 2011; Rutherford et al. 2011), or express virulence genes (Chao and Vogel 2010; Sonnleitner and Haas 2011; Mellin and Cossart 2012).

It has emerged that most of the *trans*-acting sRNAs in Gram-negative bacteria functionally require and interact with the RNA chaperone protein Hfq (Vogel and Luisi 2011). Coimmunoprecipitation of cellular RNAs with Hfq followed by RNA-seq and other approaches have provided a genome-wide picture of the abundance of sRNAs in the model pathogen *S. typhimurium*, hinting at a large posttranscriptional network wherein 100–200 sRNAs together with Hfq may control >25% of all mRNAs (Padalon-Brauch et al. 2008; Sittka et al. 2008, 2009; Ansong et al. 2009; Chao et al. 2012; Kröger et al. 2012). Many of these mRNAs encode virulence factors, or proteins involved in biofilm formation and motility (Fig. 1). Hfq is also important for the fitness and virulence of many other Gram-negative bacteria and an increasing number of Gram-positive pathogens (Chao and Vogel 2010). There has been considerable progress in understanding how Hfq facilitates sRNA-dependent regulation (i.e., by protecting the sRNAs against degradation, helping them anneal to their mRNA targets, modifying mRNA structure for better accessibility, and recruiting important nucleases such as RNase E) (Vogel and Luisi 2011). The situation is less straightforward in Gram-positive bacteria, in which several important pathogens including *Streptococci* lack a recognizable *hfq* gene, whereas others encode Hfq but express it only in a subset of strains within the same species (Romby and Charpentier 2010). In *L. monocytogenes*, however, Hfq is required in at least one case of sRNA-mediated mRNA regulation (Nielsen et al. 2011). The various levels of dispensability or requirement of Hfq in several pathogens may result from specific features of Hfq or sRNAs and their RNA duplexes that still need to be characterized (Jousselin et al. 2009; Horstmann et al. 2012; Romilly et al. 2012).

There is currently very limited information on other RNA-binding proteins that might be involved in sRNA regulation, but new RNA-binding proteins keep emerging through serendipitous findings. For example in *S. aureus*, the pleiotropic transcriptional regulatory protein SarA was unexpectedly identified as an RNA-binding protein, which altered mRNA turnover (Roberts et al. 2006; Morrison et al. 2012). Whether SarA affects mRNA degradation through sRNAs is still an open question. Moreover, several ribonucleases have been implicated in *S. aureus* virulence (reviewed in Romilly et al. 2012). Among these enzymes, RNase

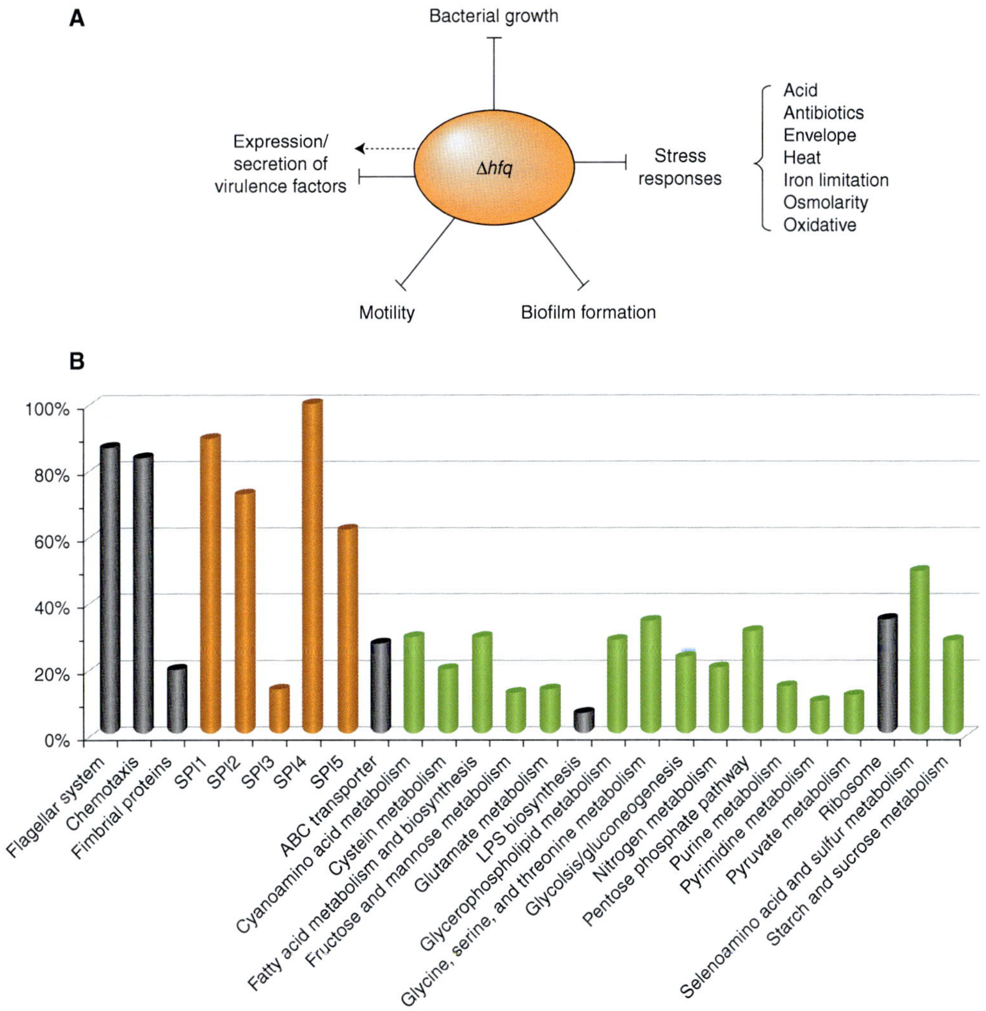

Figure 1. Prominent functions of Hfq in bacterial pathogens. (*A*) Summary of Hfq protein functions as inferred from the phenotypes of *hfq* mutants in Gram-negative bacteria (reviewed in Chao and Vogel 2010). In most bacteria examined, inactivation of the *hfq* gene causes a growth defect, a decrease in motility and biofilm formation, and an inability to cope with environmental stresses. Expression and/or secretion of the type III secretion system can be affected in either a negative or positive manner. (*B*) Gene ontology analysis of disregulated genes in a *Salmonella hfq* mutant strain (Sittka et al. 2008). Bars show the percentage of Hfq-dependent genes in each pathway. Orange bars denote the five major virulence regions of *S. typhimurium*; green bars denote metabolic functions.

III has been clearly identified as a major partner in antisense regulation (Lasa et al. 2011; Lioliou et al. 2012). These latter global analyses using RNA deep sequencing have produced compelling evidence that RNase III processes overlapping 5′ UTRs of divergently transcribed genes, and facilitates RNA quality control of pervasive transcription. In *Sinorhizobium meliloti*, the conserved and ubiquitous metalloprotein SMc01113, ortholog to *E. coli* YbeY, affects the accumulation of sRNAs and their mRNA targets in a way similar to Hfq (Pandey et al. 2011). This protein, which is conserved in many pathogenic bacteria, shares structural similarities with the MID domain of eukaryotic AGO proteins. These preliminary analyses suggest that bacteria may use a protein similar to a core component of the corresponding eukaryotic RNA machinery to achieve sRNA-mediated regulation. A recent work shows that *E. coli* YbeY is a single strand-specific endoribonuclease, which plays key roles in 70S ribosome quality control and 16S rRNA maturation (Jacob et al. 2013). More generally, we expect a surge in studies in the future that use genetic screens and/or purification of ribonucleoprotein particles followed by high-throughput RNA sequencing to identify the protein cofactors involved in regulating the activity of sRNAs in pathogens.

NOVEL REGULATORY MECHANISMS OF *TRANS*-ACTING sRNAS

Pathogenic bacteria do not necessarily use different strategies to regulate gene expression by sRNAs compared to nonpathogenic species. Nonetheless, new mechanisms have been revealed through studies of pathogens. For example, the prototypical mechanism of *trans*-acting sRNAs is to sequester the RBS of target mRNAs by base-pairing to the SD sequence or start codon. However, recent work in *Salmonella* discovered alternative repression mechanisms by MicC sRNA and RNase E in the coding sequence of *ompD* porin mRNA (Pfeiffer et al. 2009; Bandyra et al. 2012) or by Rho-dependent transcription termination within the *chiP* porin gene by ChiX sRNA (Bossi et al. 2012). Note that work on ChiX also revealed an indirect positive regulation wherein this sRNA is trapped by a pseudotarget mRNA, causing a derepression of its actual target (Figueroa-Bossi et al. 2009).

The prototypical mechanism of target activation was originally discovered in a pathogen (i.e., where RNAIII of *S. aureus* induces hemolysin α synthesis by resolving an inhibitory secondary structure around the RBS of this mRNA through competitive binding) (Fig. 2A). However, various other modes of positive regulation have also been recently discovered in pathogens. For example, the VR-RNA of *Clostridium perfringens* interacts with the 5′ part of a hairpin structure that sequesters the SD sequence of collagenase mRNA, promoting a specific cleavage and stabilization of this virulence-factor-encoding transcript (Obana et al. 2010). In the human pathogen group A *Streptococcus* (GAS), the FasX sRNA activates the synthesis of the secreted virulence-factor streptokinase SKA, which aids the dissemination of the pathogen by converting the host plasminogen into the fibrin-degrading protease plasmin (Ramirez-Pena et al. 2010). FasX binds to the *ska* mRNA within its 5′ end at a UCCC sequence motif, 30 nucleotides upstream of the RBS. This interaction enhances both the stability of the transcript and protein synthesis. Of note, Gram-positive bacteria possess 5′-to-3′ exoribonuclease activity (Condon and Bechhofer 2011), which might explain why blocking the 5′ end of *ska* mRNA

Figure 2. Dual functions of peptide-coding small RNAs. (*A*) Quorum sensing and RNAIII-controlled gene expression. *S. aureus* produces an autoinducing peptide that is sensed by a histidine kinase (AgrC). Sensing of the autoinducing peptide by AgrC leads to phosphorylation of the response regulator AgrA, which, in turn, is a transcriptional activator of the bifunctional RNAIII. RNAIII encodes the *hld* gene (coding for δ-hemolysin), and posttranscriptionally regulates several target mRNAs. Whereas *spa*, *coa*, *rot*, SA1000, and SA2353 mRNAs are repressed by RNAIII (Boisset et al. 2007), the *hla* mRNA is activated by this bifunctional RNA (Morfeldt et al. 1995). Red and blue arrows denote sRNA-mediated posttranscriptional activation and repression, respectively. (*B*) Glucose-phosphate stress and SgrS modulates virulence gene expression. The accumulation of glucose phosphate in the cytoplasm is sensed by the transcription factor SgrR, which, in turn, activates transcription of the SgrS RNA. The 5′ region of SgrS encodes a small peptide SgrT (43 amino acids), whereas the 3′ region contains a seed-pairing domain that targets several mRNAs with the help of the RNA chaperone Hfq. In addition to targeting the conserved sugar transporters *ptsG* and *manXYZ*, SgrS also represses the mRNA of the T3SS secreted virulence effector SopD in *Salmonella* (Papenfort et al. 2012). (*C*) PhrS sRNA in *Pseudomonas*. Binding of the PhrS sRNA induces the translation of the quorum-sensing regulator PqsR translational activation of an upstream ORF (uof), which is coupled to PqsR (Sonnleitner et al. 2011). Expression of PhrS is induced by ANR, an oxygen-responsive regulator, at high-cell density populations when oxygen becomes limiting.

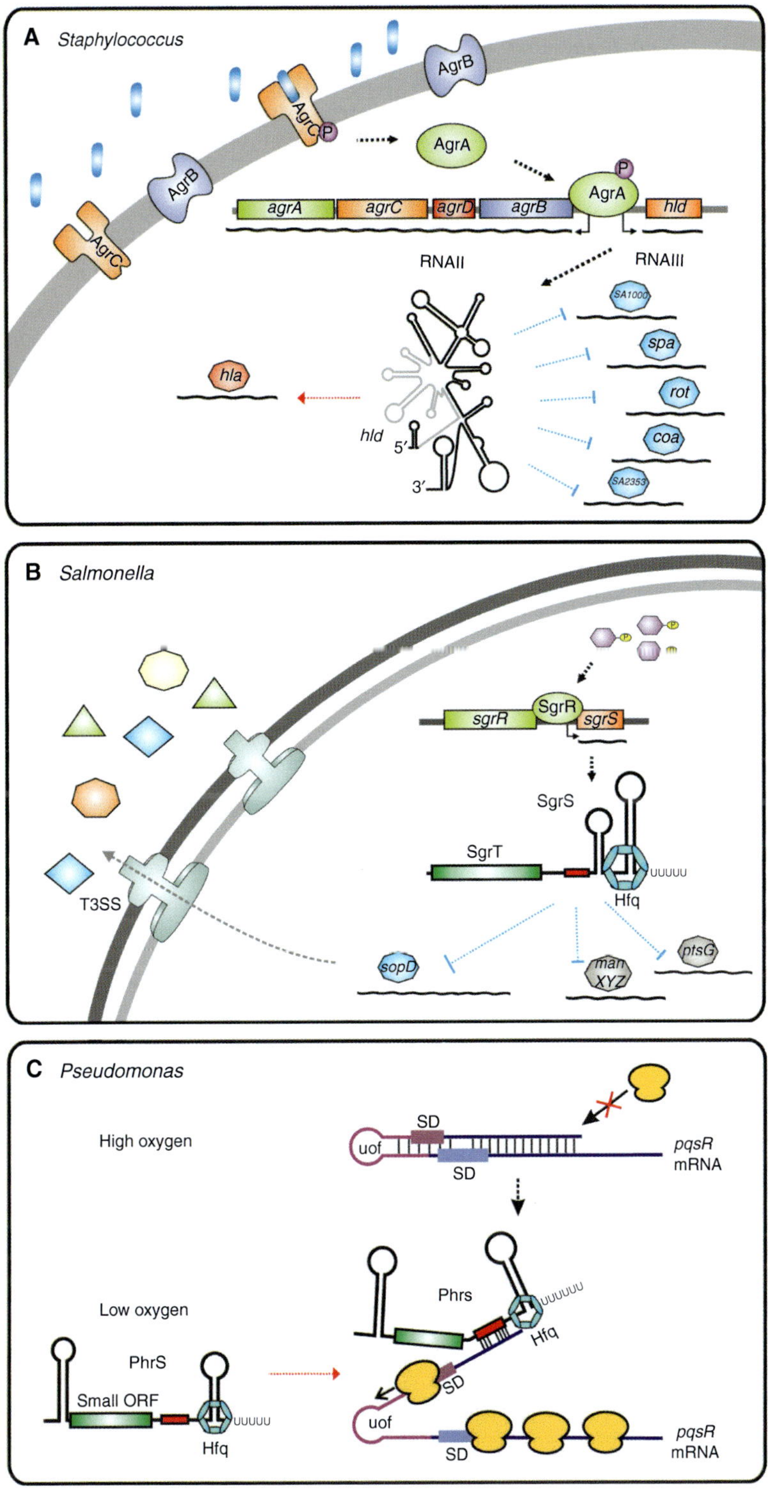

Figure 2. (*See legend on facing page.*)

increases mRNA stability. Finally, work on the PhrS RNA in *Pseudomonas aeruginosa* (Fig. 2C) has revealed a mode of indirect target activation through promoting ribosome binding to a small upstream open reading frame (uof) whose translation is coupled to *pqsR* (Sonnleitner et al. 2011).

In the following section, several representative examples will illustrate the variety of sRNA-dependent mechanisms and functions that contribute to pathogenesis.

SMALL RNAs ESTABLISH A GRADIENT OF RESPONSES IN QUORUM SENSING

In *Vibrio* species, the control of quorum sensing (QS) by 4–5 paralogous sRNAs (Qrr1–5) is a well-studied process (Ng and Bassler 2009). At low cell density, the transcriptional regulator LuxO is continuously phosphorylated and in conjunction with RNA polymerase-bound σ^{54} drives the transcription of the Qrr sRNAs that base-pair to and down-regulate the *hapR* mRNA (Fig. 3) (Lenz et al. 2005; Bardill et al. 2011). Active HapR inhibits the expression of virulence genes and type III secretion, whereas it induces the production of proteases and several other genes. In contrast, at high cell density, LuxO phosphorylation is inhibited and HapR is expressed (Fig. 3). Interestingly, Qrr4 also represses the *luxO* mRNA, which effectively creates a posttranscriptional feedback loop that facilitates precise gene regulation during the transition from the exponential to the stationary phase of growth (Tu et al. 2010). The Qrr sRNAs also directly activate the expression of AphA (Fig. 3), which is a transcriptional repressor of the *hapR* gene and the dominant QS regulator at low cell density conditions (Rutherford et al. 2011; Shao and Bassler 2012). The regulatory architecture linking the Qrr sRNAs, HapR, and AphA allows maximal AphA levels at low cell

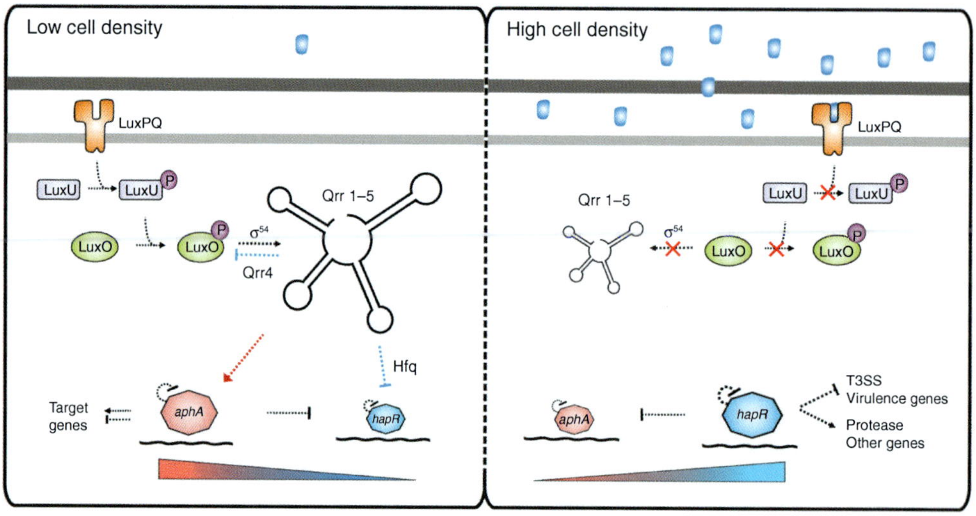

Figure 3. Several sRNAs regulate quorum sensing in *Vibrio cholerae*. At low cell density, the histidine kinase-receptor LuxPQ phosphorylates LuxU, which, in turn, phosphorylates LuxO (LuxO-P). LuxO-P, together with the RNAP-σ^{54} holoenzyme, activates the transcription of 5 sRNAs, Qrr1-5 (Rutherford et al. 2011). These sRNAs together with Hfq repress the translation of *hapR* mRNA and activate the translation of *ahpA* mRNA through an anti-antisense mechanism. Qrr4 also represses the translation of *luxO* mRNA, thus creating a negative-feedback regulation. At high cell density and high concentrations of extracellular autoinducer-2, the LuxPQ receptor acts as a phosphatase and dephosphorylates LuxO. In consequence, the synthesis of Qrr sRNAs is arrested and the synthesis of HapR is derepressed, and a gradient of higher concentrations of HapR (in blue) over AphA (in pink) is established to provide a large range of responses. In addition, AphA and HapR autoregulate their own synthesis at the transcriptional level. Red and blue arrows denote sRNA-mediated posttranscriptional activation and repression, respectively.

density, whereas HapR production is strongest at high cell density. Hence, the Qrr sRNAs act as the switch determining which transcription factor is expressed, and cause reciprocal gradients of AphA and HapR concentrations to establish the gene expression patterns at low and high cell density (Fig. 3) (Rutherford et al. 2011). Mechanistically, two conserved regions in the Qrr4 of *V. cholerae* were identified to base-pair with mRNA targets of this sRNA, *aphA* and *hapR*, and are used to discriminate between them (Shao and Bassler 2012). In this context, sRNAs are at the heart of a dynamic system to precisely control the expression of quorum-sensing-controlled genes by balancing the levels of cell density master regulators during growth.

SMALL RNAs ARE REGULATORS OF VIRULENCE GENES AND SIGNALING PATHWAYS

The two most recent additions to the list of *Vibrio* sRNA regulators affecting pathogenicity are TarA and TarB sRNAs, both of which are controlled by the master virulence regulator ToxT. TarA negatively regulates the major *Vibrio* glucose transporter *ptsG* (Richard et al. 2010), whereas TarB down-regulates the secreted colonization factor TcpF (Bradley et al. 2011). TarB also decreases the expression of another transcription regulator, VspR, belonging to the *Vibrio* seventh pandemic island-1 (VSP-1) (Davies et al. 2012). VspR directly represses a gene that encodes a new family of dinucleotide cyclases. Cyclic nucleotides play a role in biofilm formation, flagellum biosynthesis, DNA integrity, and cell membrane stress. In *V. cholerae*, the signal molecule seems to be predominantly cAMP-GMP, whose accumulation represses chemotaxis, and as an indirect effect enhances intestinal colonization by the bacterium (Davies et al. 2012). This study shows for the first time a link between sRNA-dependent regulation and intracellular signal molecules as cyclic dinucleotides, which modulate various cellular activities including pathogenesis and the transition between sessility and motility. Another sRNA, VrrA, of *V. cholerae* modulates the infection of the host intestinal tract and contributes to bacterial fitness in response to stressful environments. This sRNA represses the translation of *ompA* mRNA and causes a significant reduction of TcpA, a toxin coregulated pilus (Song et al. 2008).

In *Yersinia pestis*, the formation of biofilms is dependent on exopolysaccharide synthesis, which is controlled by the intracellular yields of c-di-GMP. Hfq was recently shown to activate the expression of the c-di-GMP phosphodiesterase HmsP at the transcriptional level while it also induces rapid degradation of *hmsT* mRNA encoding the diguanylate cyclase via the 5′ UTR (Bellows et al. 2012). Whether these regulatory events are mediated through the binding of sRNAs is unknown. However, dysregulation of c-di-GMP turnover and severe attenuation of pathogenesis in the *hfq* mutant suggest that sRNAs are involved in the c-di-GMP-mediated response to environmental conditions regulating virulence.

DUAL-FUNCTION RNAs ENCODING BOTH A PROTEIN AND A RIBOREGULATOR

The prime example of a bifunctional sRNA is the well-characterized staphylococcal RNAIII, which is the main intracellular effector of QS (Novick 2003). The noncoding parts of RNAIII control gene expression at the posttranscriptional level via an antisense mechanism (Fig. 2A). The 5′ UTR of RNAIII binds to the leader region of *hla* mRNA to facilitate ribosome recruitment, whereas the large 3′ UTR primarily acts as a repressor domain. The 3′ UTR contains three redundant hairpin structures with conserved C-rich sequences located in their apical loops. This motif is often used to bind G-rich sequences in mRNAs located primarily in the RBS (Huntzinger et al. 2005; Boisset et al. 2007; Chevalier et al. 2010). Although the topologies of the RNAIII-mRNA duplexes are different, they all efficiently prevent ribosome binding and recruit the double-strand-specific RNase III, which initiates the rapid degradation of the repressed mRNAs. The repressed mRNAs encode adhesin factors (protein A, coagulase, SA1000) and the transcriptional repressor of toxins, Rot (Geisinger et al. 2006; Boisset et al. 2007). Thus, the temporal regulation of virulence factors favors the transition of the bacterial population from a defensive mode (colonization) toward an offensive mode (dissemination). Besides the regulatory functions, RNAIII contains the hemolysin δ ORF (*hld*), which produces a phenol-soluble modulin. Both hemolysin δ peptide and the riboregulatory part of RNAIII play roles in virulence (Novick and Geisinger 2008). Surprisingly, the synthesis of hemolysin δ was found delayed by 1 h after the transcription of RNAIII and this delay was abolished when the repressor domain located at the 3′ end of RNAIII was deleted (Balaban and Novick 1995). Whether the regulatory activities of RNAIII might influence *hld* translation awaits further experimental works.

SgrS is a well-conserved bifunctional RNA in enteric bacteria and is transcribed in response to glucose-phosphate stress to prevent the inhibition of cell growth. It acts by inhibiting translation followed by degradation of mRNAs encoding sugar transporters (Fig. 2B). In the particular case of the *manXYZ* operon in *E. coli*, SgrS binds to two specific sites in the *manX* coding region and in the *manX-manY* intergenic region. Pairing at both sites is required for efficient mRNA degradation, but pairing at one site only is sufficient to inhibit translation of the respective genes (Rice et al. 2012). SgrS also contains a small ORF, SgrT, which prevents glucose uptake by the transporters (Wadler and Vanderpool 2007). In *Salmonella*, the SgrS sRNA uses its conserved seed-pairing domain in its 3′ end to repress both the major sugar uptake proteins (as in *E. coli*) and the *Salmonella*-specific secreted effector protein SopD (Fig. 2B). Strikingly, the almost identical SopD2 protein is not regulated by SgrS, owing to a single hydrogen bond difference in the duplexes formed by SgrS with the *sopD* or *sopD2* mRNAs (Papenfort et al. 2012). SgrS is up-regulated under different stress conditions, and its genomic deletion attenuates *Salmonella* virulence in mice (Santiviago et al. 2009).

A third example of a bifunctional RNA from a pathogen is PhrS of *P. aeruginosa* (Fig. 2C). PhrS is an activator of one of the key quorum-sensing regulators, PqsR, in this organism, and achieves activation by an interesting mechanism. It activates by base-pairing a short upstream ORF to which the *pqsR* gene is translationally coupled, and so increases protein synthesis of PqsR, too. The PhrS RNA also encodes a conserved protein whose expression has been confirmed but whose cellular function remains to be elucidated (Sonnleitner et al. 2011).

TRANS-ACTING sRNAs MADE FROM THE UTRs OF mRNAs

Work in several pathogenic bacteria has shown that 5′ UTRs and 3′ UTRs of mRNAs can accumulate as stable RNA species, suggesting that they might function as riboregulators in *trans* (e.g., Toledo-Arana et al. 2009; Beaume et al. 2010; Sharma et al. 2010). This concept was originally termed "parallel transcriptional output" following observations of short distinct RNA species that accumulated from UTRs and riboswitch regions in nonpathogenic *E. coli* (Vogel et al. 2003).

Recent work in *L. monocytogenes* has produced proof of this concept, demonstrating that *cis*-acting riboswitches located in the 5′ UTR of mRNAs can also regulate gene expression in *trans*. The *L. monocytogenes* genome encodes for seven putative SAM riboswitches, named SAM riboswitch element (SreA-G). In low-nutrient conditions, SreA is transcribed together with its downstream genes. In contrast, under rich medium conditions, SreA is produced as an sRNA owing to premature transcriptional arrest, and binds to the *prfA* mRNA 80 nucleotides upstream of the SD sequence to repress translation (Fig. 4A). Because the *prfA* mRNA encodes a transcriptional activator of virulence factors, the data establishes a link between metabolism and pathogenesis. The regulation of PrfA synthesis is unique because the *prfA* mRNA also contains a thermosensor allowing the coordinated expression of virulence factors at the permissive temperature of the host. Its 5′ UTR folds into a stable hairpin structure sequestering the SD at 30°C (Johansson et al. 2002). At 37°C, this structure is destabilized allowing PrfA expression and consequently the synthesis of virulence factors. The *trans*-acting regulatory effect of SreA only occurs at 37°C

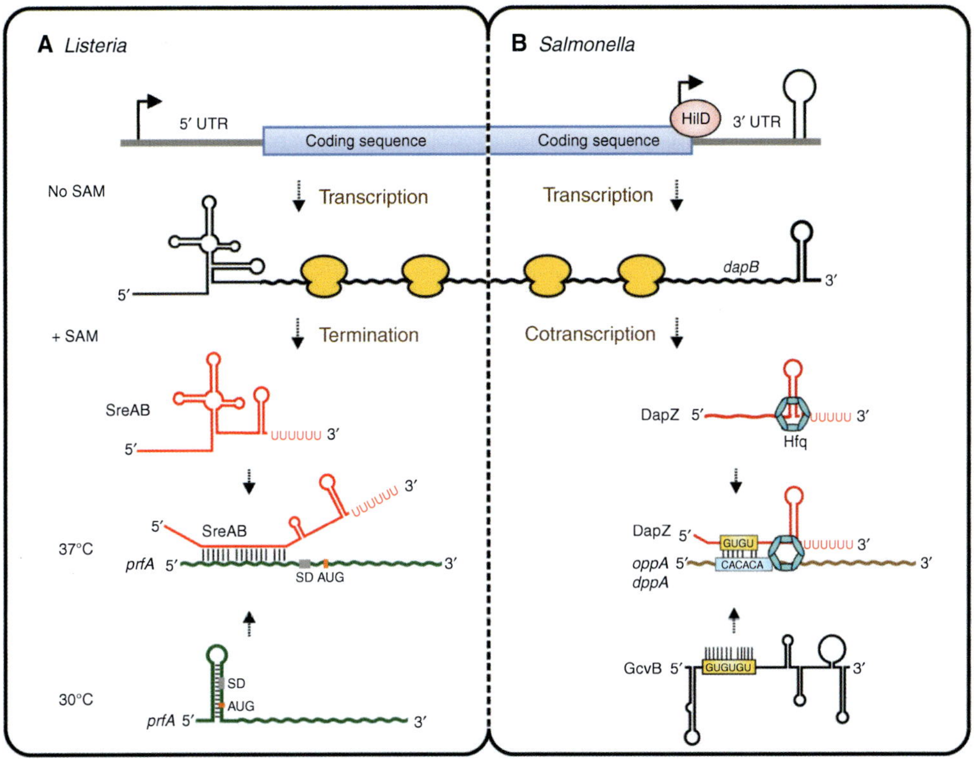

Figure 4. UTR-derived *trans*-acting small RNAs. (*A*) 5′ UTR-derived sRNA in *L. monocytogenes*. During the exponential phase of growth (i.e., under rich nutrient conditions), *S*-adenosylmethionine (SAM) binds to SreA, the terminator structure (in red) is formed, and transcription is prematurely halted (Loh et al. 2009). At 37°C, the *prfA* thermosensor (in green) melts to expose the SD for PrfA synthesis and consequently the virulence factors are produced. Under these conditions, the SreA-terminated transcript represses *prfA* mRNA translation by binding to the 5′ UTR 80 bp upstream of the SD and turning off its expression by an unknown mechanism. (*B*) 3′ UTR-derived sRNA in *S. typhimurium*. The synthesis of DapZ (in red) is induced by HilD, the master regulator of *Salmonella* invasion genes. DapZ carries a short unpaired G/U-rich region that initiates binding to the conserved C/A-rich regions in *dppA* and *oppA* mRNAs, which encode the major ABC transporters (Chao et al. 2012). The same target sites are recognized by a similar G/U-rich seed region of GcvB sRNA, the master sRNA regulator of amino acid transporters.

when the thermosensor is melted rendering the complementary sequences to SreA available for binding. This novel mechanism of riboswitch functioning in *trans* suggests that revisiting known *cis*-acting RNAs in different growth conditions may provide new insights into RNA regulation (Loh et al. 2009). Indeed, other riboswitches such as the ones sensing riboflavin or thiamine have also been known to produce sRNAs (Vogel et al. 2003).

There is now evidence that 3′ UTRs also produce RNA species that act in *trans*. A recent study analyzed Hfq-associated RNA species from multiple stages of growth in *S. typhimurium*, and revealed a large number of new sRNAs that originated from mRNA loci (Chao et al. 2012). The Hfq association strongly suggests that these new sRNAs are functional, rather than being passive degradation products. One of them, called DapZ, coincides with the 3′ region of the biosynthetic gene *dapB*. It is transcribed from a promoter just upstream of the *dapB* stop codon, but terminates at the same stem-loop structure as the longer, protein-coding *dapB* mRNA. Importantly, in *Salmonella*, this gene-internal *dapZ* promoter evolved to be activated by the regulator of invasion genes, HilD, rendering DapZ sRNA production part of a *Salmonella*-specific virulence program (Fig. 4B). DapZ uses a short G/U-rich seed sequence to repress the *dppA* and *oppA* mRNAs encoding major ABC transporters (Chao et al. 2012). Thus, this Hfq-dependent sRNA from an mRNA 3′ UTR is functionally homologous to GcvB, an sRNA that regulates many amino acid transporters (Sharma et al. 2011). There are many other *Salmonella* mRNAs with potential double output, producing both a protein and an Hfq-dependent sRNA from their 3′ UTRs under virulence conditions (Chao et al. 2012).

THE FUNCTIONAL IMPACT OF LONG AND SHORT FULLY COMPLEMENTARY asRNAS

Transcriptomic analyses in several pathogenic bacteria have revealed a high occurrence of antisense transcription throughout their genomes. Recent studies have reported different roles of such antisense transcripts (Fig. 5). An unusually long antisense RNA of 1.2 kb (lasRNA), AmgR, was identified in *Salmonella* (Lee and Groisman 2010). AmgR is transcribed from a promoter located in the *mgtC-mgtB* intergenic region of the polycistronic *mgtCBR* mRNA, which encodes an inner membrane Mg^{2+} transporter, and a 30-amino-acid-long peptide involved in the proteolysis of MgtC, respectively (Fig. 5A). The MgtC protein is necessary for survival within macrophages, virulence in mice, and growth at low Mg^{2+} concentrations (Lee and Groisman 2010). Expression of AmgR induces rapid degradation of the *mgtC* and *mgtB* mRNAs in an RNase-E-dependent manner that does not require Hfq. Intriguingly, the transcription of both *mgtCBR* and *amgR* is activated by the PhoP transcription factor but the promoter of *amgR* is weaker. Surprisingly, inactivation of the chromosomal *amgR* promoter enhances *Salmonella* virulence most likely owing to enhanced levels of MgtC and MgtB. Thus, AmgR functions as a temporal regulator to alter MgtC and MgtB levels after the onset of PhoP-inducing conditions, which is critical for *Salmonella* virulence.

Another type of lasRNA was identified in pathogenic and nonpathogenic *Listeria* strains. They are rather large in size (up to 6.5 kb), overlap one ORF, and serve as the 5′ UTR of a neighboring encoded ORF. For one of these lasRNAs, it was experimentally shown that its proximal part negatively regulates the expression of the gene transcribed on the opposite strand and that its distal part acts as a polycistronic mRNA for the downstream operon (Fig. 5B). Interestingly, the expression of one of these dual-functional lasRNAs leads to an activation of a permease and results in a simultaneous silencing of the associated efflux pump. This particular type of genomic locus, encoding an unusually long asRNA that spans divergent genes or operons with related or opposing functions, has been termed an "excludon" (Toledo-Arana et al. 2009; Wurtzel et al. 2012; Sesto et al. 2013). Overlapping transcripts from divergently transcribed genes may also generate a positive regulation. In *S. aureus*, several divergent mRNAs contain overlapping 5′ UTRs, which are processed by RNase III (Lioliou et al. 2012). This processing generates functional mRNAs but with shorter 5′ UTRs. Whether this processing has a positive effect on translation has yet to be analyzed.

Another study using RNA deep sequencing to analyze short RNA fractions of *S. aureus* has revealed a large collection of 22-nucleotide RNA fragments generated by RNase III digestion of sense/antisense transcripts from all over the chromosome (Lasa et al. 2011). More than 75% of the mRNAs were subjected to specific RNase III processing as the result of antisense regulation. Deletion of RNase III significantly reduces the amount of short RNA fragments and concomitantly leads to the accumulation of low levels of antisense transcripts. These data are indicative of genome-wide antisense transcription, which is hidden owing to RNase III processing of sense/antisense transcripts. Interestingly, the involvement of RNase III in this novel posttranscriptional process appears to be restricted to Gram-positive bacteria (Lasa et al. 2011).

A particular mechanism of antisense regulation has been recently proposed for the staphylococcal type I toxin–antitoxin SprA1/SprA1-AS (Sayed et al. 2011). SprA1 encodes a peptide,

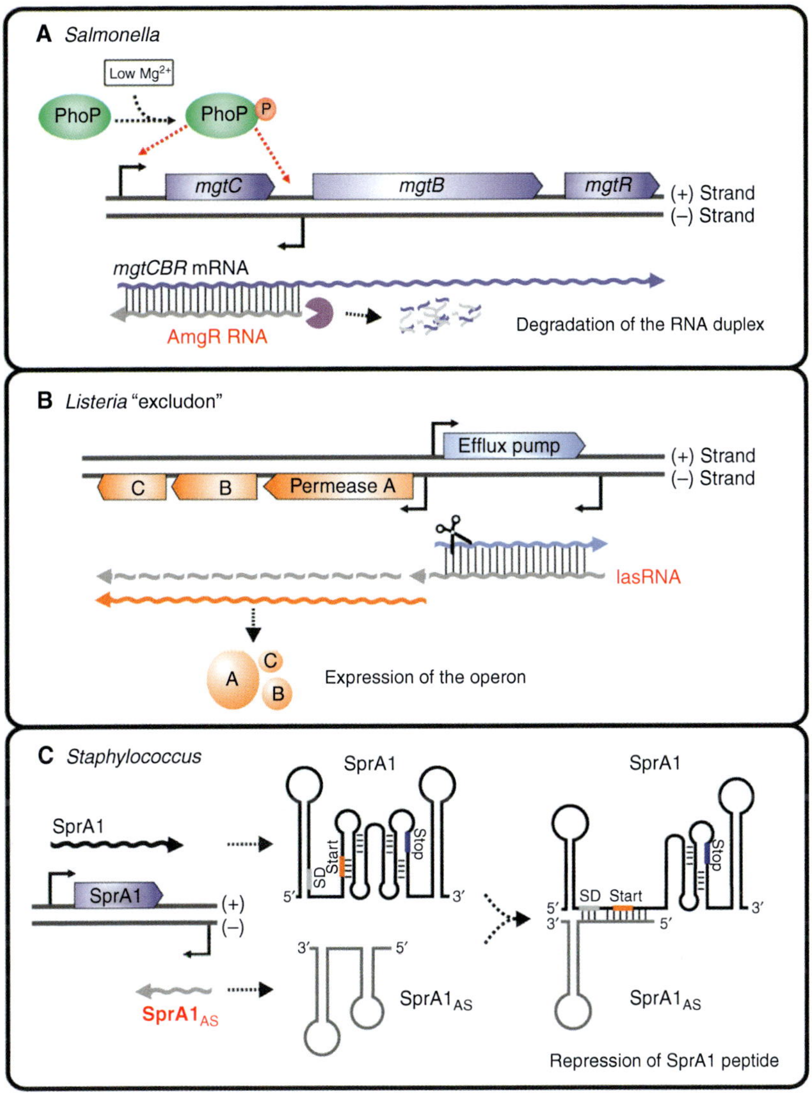

Figure 5. Novel regulations by long and short antisense RNAs. (*A*) AmgR is a 1.2-kb *cis*-encoded RNA that is expressed convergently to the *mgtC* ORF in *Salmonella* (Lee and Groisman 2010). When Mg^{2+} concentrations are low, the PhoPQ two-component system activates expression of both the AmgR and *mgtC* transcripts. The interaction of these two RNAs results in RNase-E-dependent degradation of the RNA duplex. (*B*) Unusually long antisense RNAs have been predicted upstream of many operons in *Listeria* (Wurtzel et al. 2012). Expression of these lasRNAs may activate the downstream operons while inhibiting the expression of genes located antisense to them, which usually have opposite functions to the operon. (*C*) $SprA1_{AS}$ is a short, ~60-nucleotide-long antisense RNA of a toxin–antitoxin (TA) system in *S. aureus*. It overlaps with the 3′ end of the dual-functional SprA1 sRNA. $SprA1_{AS}$ was shown to use its 5′ end to base-pair with SprA1 sRNA through imperfect pairing, and to repress the translation of the SprA1-encoded small cytolytic peptide (Sayed et al. 2011).

which has lytic activity toward human erythrocytes and displays antimicrobial activity against Gram-positive and Gram-negative bacteria. Surprisingly, structure probing combined with mutational analysis shows that the short SprA1-AS binds to the RBS of SprA1 through imperfect pairings to prevent translation of the peptide. Unexpectedly, the interacting region does not involve the 3′ end of SprA1-AS, which is fully complementary to the 3′ end of SprA1. Instead, the active region is located in its 5′ part that is partially complementary to the RBS of SprA1 (Fig. 5C). The fully complementary regions of both RNAs corresponded to stable Rho-independent terminators, which are probably unfavorable to promote the rapid formation of stable intermolecular pairings. Hence, the 5′ unpaired region of SprA1-AS may be more appropriate to promote fast binding with the connecting loop L2 of the pseudoknot of SprA1, a mechanism often used by sRNAs from Gram-negative bacteria (Papenfort et al. 2010).

SMALL RNAs AS DECOYS TO REGULATE THE REGULATORY ACTIVITIES OF PROTEINS

As mentioned earlier, a limited number of sRNAs modulate the activity of posttranscriptional regulators (Fig. 6). Among them, Crc and CsrA/RsmA are RNA-binding proteins discovered in *Pseudomonas* species (Babitzke et al. 2009; Rojo 2010; Romeo et al. 2012). They share functional resemblances, but no amino acids sequence and structure similarities. Both proteins are involved in gene regulation of several processes like secondary metabolism, carbon storage, stress response, and virulence, and repress translation of target mRNAs. For example, Crc binds with high affinity to an unpaired AANAANAA sequence (in which N is C = U > A) located upstream of or downstream from the AUG initiation codon of mRNAs to repress their translation (Moreno et al. 2009; Sonnleitner et al. 2009). Furthermore, they are themselves controlled by sRNA-mediated regulation (Fig. 6). Indeed, CsrZ sRNA from *P. aeruginosa* (Sonnleitner et al. 2009) and CsrY and CrsZ in *Pseudomonas putida* (Moreno et al. 2012) contain several Crc recognition motifs AANAANAA. CsrZ and CsrY act as antagonists and sequester Crc allowing translation of the target mRNAs (Fig. 6B). Similarly, the activity of RsmA is antagonized by two sRNAs, RsmY and RsmZ, whose expression is exclusively regulated by the GacS/GacA two-component system, the master regulator of virulence in *P. aeruginosa* (Fig. 6A). The GacA-response regulator transduces external regulatory signals and binds exclusively to two chromosomal loci to activate the expression of RsmY and RsmZ (Brencic et al. 2009). Both RsmY and RsmZ contain multiple GGA motifs to sequester RsmA (Heurlier et al. 2004; Kay et al. 2005). RsmA itself binds to the conserved GGA motifs located in the 5′ UTR of target mRNAs, resulting in translational repression. In summary, RNA mimicry regulates many processes that can directly alter the function of nucleic acid-binding proteins (Marzi and Romby 2012).

Finally, 6S RNA is a (structurally) well-conserved sRNA in all bacteria. It interacts with the housekeeping form of RNA polymerase in complex with σ^{70} and inhibits expression from certain promoters when the bacteria shift from the exponential to the stationary phase of growth (Wassarman 2007). The sequestration of σ^{70} by 6S RNA down-regulates σ^{70}-dependent transcription thereby facilitating transcription from stationary phase σ-factor-dependent promoters. The involvement of 6S RNA in pathogenesis was first shown in *Legionella pneumophila*, which is a Gram-negative opportunistic human pathogen that infects and multiplies in a broad range of phagocytic protozoa and mammalian phagocytes. The 6S RNA was shown to positively regulate the expression of genes encoding type IVB secretion system effectors, stress response genes, as well as many genes involved in the acquisition of nutrients (Faucher et al. 2010). Deletion of 6S RNA significantly reduces *L. pneumophila* intracellular multiplication in both protist and mammalian host cells.

GENERAL CONCLUSION AND PERSPECTIVES

The analysis of sRNAs in bacterial pathogens is undergoing a paradigm shift because deep se-

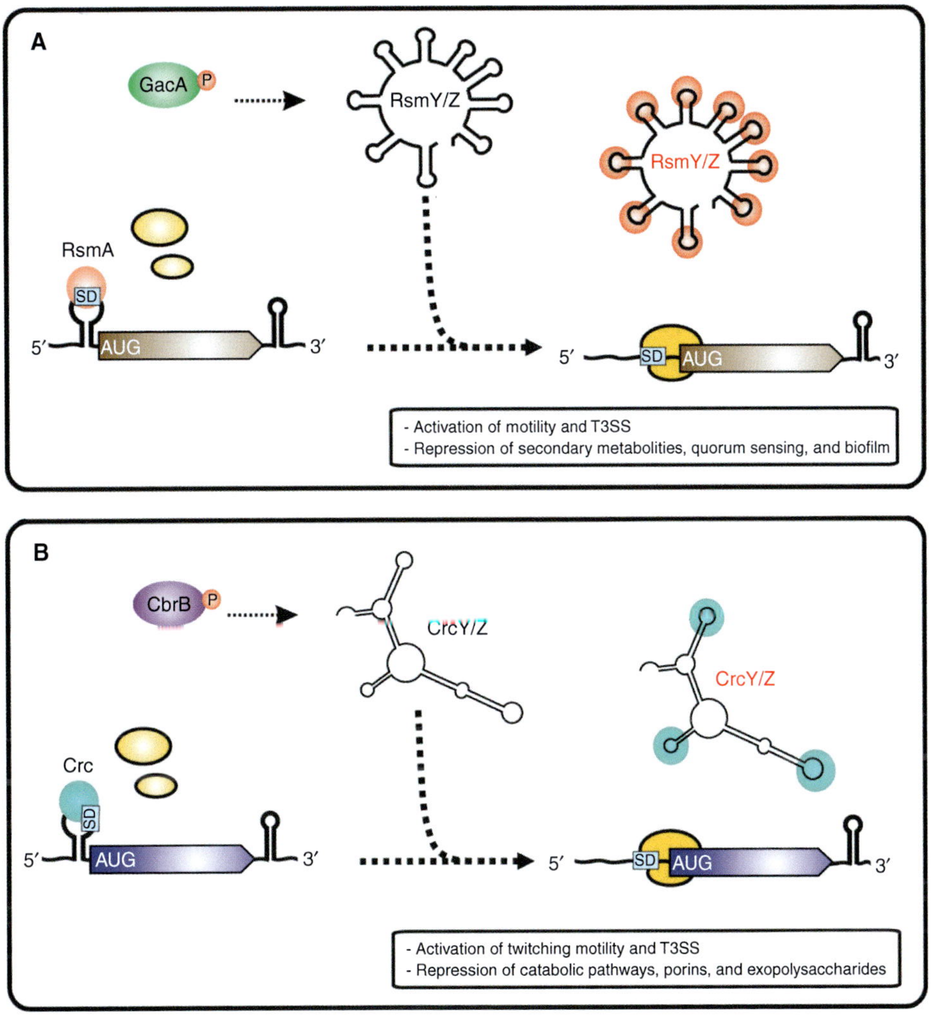

Figure 6. Small RNAs directly regulate protein activity in *Pseudomonas*. (*A*) RsmA is an RNA-binding protein that modulates gene expression by antagonizing translational initiation (Lapouge et al. 2008). The RsmY/Z RNAs contain multiple RsmA binding sites and counteract RsmA activity via a titration mechanism in several different plant and human pathogens. Similar small RNAs (e.g., CsrB) target the homologous CsrA protein in enterobacteria (Romeo et al. 2012). (*B*) Crc is an RNA-binding protein that modulates expression by antagonizing translational initiation (Moreno et al. 2009). CrcY/Z RNAs carry up to six Crc-binding sites and counteract Crc activity via a titration mechanism.

quencing can rapidly discover and report expression profiles of, in principle, all RNAs of an organism. Recently, deep sequencing performed on several major pathogens such as *S. aureus* (Beaume et al. 2010; Bohn et al. 2010; Lasa et al. 2011), *H. pylori* (Sharma et al. 2010), *L. monocytogenes* (Oliver et al. 2009; Wurtzel et al. 2012), *S. pneumoniae* (Mann et al. 2012), or *V. cholerae* (Mandlik et al. 2011) has revealed hundreds of previously undetected transcripts in both sense and antisense orientations. Numerous short transcripts derived from 5′ or 3′ UTRs, or internal regions of mRNAs have been detected, and some of them were recently shown to behave as independent modules (Loh et al. 2009; Chao et al. 2012). Novel mechanisms have been found in representative pathogens of Gram-positive bacteria in which sRNAs bind at the 5′ end of the target mRNA to protect the mRNAs against degradation, which is a mechanism that has not yet been found in Gram-negative bacteria. It is now time to explore other avenues such as sRNA-mediated modulation of chromosome structure, sRNA-dependent activation of virulence factors, sRNAs that affect the mobility of pathogenicity islands, the expression of host genes, or cell-to-cell communication. Efforts to combine experimental and computational approaches will advance our knowledge of sRNA functions and of their regulons. The importance of the sRNA and mRNA structures, the identification of the initial contacts, the possibility to form multiple binding sites and noncanonical base pairs, the stoichiometry of sRNA–mRNA complexes are criteria that need to be taken into account to improve the search for primary targets.

Most studies have provided a static picture of sRNA-dependent regulation in a population of cells. However, identical genotype and growth environments are insufficient to ensure that individual cells within clonal microbial cultures will show the same phenotype (Gefen and Balaban 2009). This heterogeneity confers the ability of a population to persist in response to external changes and to exploit new niches. Biofilm formation represents such heterogeneity of cells within an isogenic bacterial cell population. A recent study has shown that the major biofilm regulator CsgD in *Salmonella* is expressed in a bistable manner during the biofilm development (Grantcharova et al. 2010). Interestingly, ArcZ sRNA repressed the translation of the *csgD* mRNA and contributed to the transition between sessility and biofilm formation (Papenfort et al. 2009; Monteiro et al. 2012). Whether ArcZ contributes to the bistable expression of CsgD remains to be studied.

Only a few of the known sRNAs have been functionally characterized in vivo (Mann et al. 2012). A novel approach, TRADIS (i.e., transposon directed insertion-site sequencing), has been developed to simultaneously assay every gene of *S. typhimurium* for essentiality (Langridge et al. 2009). Therefore, one could obtain, in the near future, the list of sRNAs that are essential under specific conditions, or for pathogenesis of a given bacterium. The clinical significance of *S. aureus* sRNAs has been recently shown in human colonization and infection (Song et al. 2012). The expression levels of five sRNAs from *S. aureus* were shown to be highly variable in abscesses and in sputum samples of cystic fibrosis patients suggesting specific host responses. Conversely, they were highly uniform in the nasal carrier samples, which probably reflects the commensalism of *S. aureus* in this niche. This study also revealed that the expression of sRNAs in cells grown in rich medium culture does not uniformly mimic the in vivo conditions (Song et al. 2012). Therefore, the host response or the human microbiome may influence the expression pattern of sRNAs or vice versa. This suggests that a Dual RNA-seq approach on the host and pathogen will provide a more comprehensive view of the gene expression changes occurring in both the pathogen and the host (Westermann et al. 2012). Another interesting issue would be to search for secreted bacterial sRNAs, which could act as signaling molecules or perturb the host response. A recent study has shown that efficient cytosolic immune sensing requires the release of RNA and DNA from live *Listeria* to the cytoplasm of infected cells (Abdullah et al. 2012). These secreted bacterial RNA/DNA were recognized by the cytosolic sensors to trigger interferon β production and to cause inflammasome activation.

Clearly we are only beginning to fully understand the roles of sRNAs in bacterial pathogenesis and persistence. Continued characterization of the functions, structures, and mechanism of action of individual sRNAs and their machineries will help us to fully understand the regulatory circuits that enable the bacteria to survive within the host and to cause disease.

Recent observations suggest that the Crc protein is devoid of RNA-binding activity and might act together with Hfq (Milojevic et al. 2013). From the genetic data, Crc counteracts the action of the small ncRNA CrcZ/Y, but experimental work is further required to reveal whether Crc is acting indirectly through Hfq or not. Regarding the recently reported aminoglycoside-binding riboswitch, Roth and Breaker

(2013) have debated the physiological role of this RNA element with respect to its potential antibiotic sensing. There have been recent new findings related to the small RNA activities illustrated in this article. SgrS RNA (Fig. 2B) was revealed to function as an activator in addition to its known repressor functions (Papenfort et al. 2013b). The Qrr sRNAs (Fig. 3) have been found to control an unexpected large suite of new mRNA targets (Shao et al. 2013).

ACKNOWLEDGMENTS

We thank all the members of our respective laboratories for stimulating discussions, and Stan Gorski for help with the manuscript. This work is supported by the Federal Ministry of Education, Science, Research and Technology (BMBF NGFN, grant 01GS0806 to J.V.), the German Research Council Priority Programmes SPP1258 (grant Vo857/4-2 to J.V.) and TRR34 (project B04 to J.V.), the "Centre National de la Recherche Scientifique" (CNRS to P.R.), the "Agence Nationale pour la Recherche" (ANR10-Pathogenomics-ARMSA to P.R.), and the labex Net RNA (ANR-10-LABX-36 to P.R.).

REFERENCES

Abdullah Z, Schlee M, Roth S, Mraheil MA, Barchet W, Bottcher J, Hain T, Geiger S, Hayakawa Y, Fritz JH, et al. 2012. RIG-I detects infection with live *Listeria* by sensing secreted bacterial nucleic acids. *EMBO J* **31:** 4153–4164.

Ansong C, Yoon H, Porwollik S, Mottaz-Brewer H, Petritis BO, Jaitly N, Adkins JN, McClelland M, Heffron F, Smith RD. 2009. Global systems-level analysis of Hfq and SmpB deletion mutants in *Salmonella*: Implications for virulence and global protein translation. *PLoS ONE* **4:** e4809.

Argaman L, Hershberg R, Vogel J, Bejerano G, Wagner EG, Margalit H, Altuvia S. 2001. Novel small RNA-encoding genes in the intergenic regions of *Escherichia coli*. *Curr Biol* **11:** 941–950.

Babitzke P, Baker CS, Romeo T. 2009. Regulation of translation initiation by RNA binding proteins. *Annu Rev Microbiol* **63:** 27–44.

Balaban N, Novick RP. 1995. Translation of RNAIII, the *Staphylococcus aureus* agr regulatory RNA molecule, can be activated by a 3′ end deletion. *FEMS Microbiol Lett* **133:** 155–161.

Bandyra KJ, Said N, Pfeiffer V, Gorna MW, Vogel J, Luisi BF. 2012. The seed region of a small RNA drives the controlled destruction of the target mRNA by the endoribonuclease RNase E. *Mol Cell* **47:** 943–953.

Bardill JP, Zhao X, Hammer BK. 2011. The *Vibrio cholerae* quorum sensing response is mediated by Hfq-dependent sRNA/mRNA base pairing interactions. *Mol Microbiol* **80:** 1381–1394.

Barrick JE, Breaker RR. 2007. The distributions, mechanisms, and structures of metabolite-binding riboswitches. *Genome Biol* **8:** R239.

Beaume M, Hernandez D, Farinelli L, Deluen C, Linder P, Gaspin C, Romby P, Schrenzel J, Francois P. 2010. Cartography of methicillin-resistant *S. aureus* transcripts: Detection, orientation and temporal expression during growth phase and stress conditions. *PLoS ONE* **5:** e10725.

Beisel CL, Storz G. 2011. Discriminating tastes: Physiological contributions of the Hfq-binding small RNA Spot 42 to catabolite repression. *RNA Biol* **8:** 766–770.

Bellows LE, Koestler BJ, Karaba SM, Waters CM, Lathem WW. 2012. Hfq-dependent, co-ordinate control of cyclic diguanylate synthesis and catabolism in the plague pathogen *Yersinia pestis*. *Mol Microbiol* **86:** 661–674.

Blount KF, Wang JX, Lim J, Sudarsan N, Breaker RR. 2006. Antibacterial lysine analogs that target lysine riboswitches. *Nat Chem Biol* **3:** 44–49.

Bohme K, Steinmann R, Kortmann J, Seekircher S, Heroven AK, Berger E, Pisano F, Thiermann T, Wolf-Watz H, Narberhaus F, et al. 2012. Concerted actions of a thermo-labile regulator and a unique intergenic RNA thermosensor control *Yersinia* virulence. *PLoS Pathog* **8:** e1002518.

Bohn C, Rigoulay C, Chabelskaya S, Sharma CM, Marchais A, Skorski P, Borezee-Durant E, Barbet R, Jacquet E, Jacq A, et al. 2010. Experimental discovery of small RNAs in *Staphylococcus aureus* reveals a riboregulator of central metabolism. *Nucleic Acids Res* **38:** 6620–6636.

Boisset S, Geissmann T, Huntzinger E, Fechter P, Bendridi N, Possedko M, Chevalier C, Helfer AC, Benito Y, Jacquier A, et al. 2007. *Staphylococcus aureus* RNAIII coordinately represses the synthesis of virulence factors and the transcription regulator Rot by an antisense mechanism. *Genes Dev* **21:** 1353–1366.

Bossi L, Schwartz A, Guillemardet B, Boudvillain M, Figueroa-Bossi N. 2012. A role for Rho-dependent polarity in gene regulation by a noncoding small RNA. *Genes Dev* **26:** 1864–1873.

Bradley ES, Bodi K, Ismail AM, Camilli A. 2011. A genome-wide approach to discovery of small RNAs involved in regulation of virulence in *Vibrio cholerae*. *PLoS Pathog* **7:** e1002126.

Breaker RR. 2009. Riboswitches: From ancient gene-control systems to modern drug targets. *Future Microbiol* **4:** 771–773.

Breaker RR. 2011. Prospects for riboswitch discovery and analysis. *Mol Cell* **43:** 867–879.

Brencic A, McFarland KA, McManus HR, Castang S, Mogno I, Dove SL, Lory S. 2009. The GacS/GacA signal transduction system of *Pseudomonas aeruginosa* acts exclusively through its control over the transcription of the RsmY and RsmZ regulatory small RNAs. *Mol Microbiol* **73:** 434–445.

Caldelari I, Fechter P, Lioliou E, Romilly C, Chevalier C, Gaspin C, Romby P. 2011. A current overview of regulatory RNAs in *Staphylococcus aureus*. In *Regulatory RNAs in prokaryotes* (ed. Marchfelder A, Hess W), pp. 51–75. Wiley, New York.

Chao Y, Vogel J. 2010. The role of Hfq in bacterial pathogens. *Curr Opin Microbiol* **13:** 24–33.

Chao Y, Papenfort K, Reinhardt R, Sharma CM, Vogel J. 2012. An atlas of Hfq-bound transcripts reveals 3′ UTRs as a genomic reservoir of regulatory small RNAs. *EMBO J* **31:** 4005–4019.

Chevalier C, Boisset S, Romilly C, Masquida B, Fechter P, Geissmann T, Vandenesch F, Romby P. 2010. *Staphylococcus aureus* RNAIII binds to two distant regions of *coa* mRNA to arrest translation and promote mRNA degradation. *PLoS Pathog* **6:** e1000809.

Condon C, Bechhofer DH. 2011. Regulated RNA stability in the Gram positives. *Curr Opin Microbiol* **14:** 148–154.

Dambach MD, Winkler WC. 2009. Expanding roles for metabolite-sensing regulatory RNAs. *Curr Opin Microbiol* **12:** 161–169.

Davies BW, Bogard RW, Young TS, Mekalanos JJ. 2012. Coordinated regulation of accessory genetic elements produces cyclic di-nucleotides for *V. cholerae* virulence. *Cell* **149:** 358–370.

Faucher SP, Friedlander G, Livny J, Margalit H, Shuman HA. 2010. *Legionella pneumophila* 6S RNA optimizes intracellular multiplication. *Proc Natl Acad Sci* **107:** 7533–7538.

Felden B, Vandenesch F, Bouloc P, Romby P. 2011. The *Staphylococcus aureus* RNome and its commitment to virulence. *PLoS Pathog* **7:** e1002006.

Figueroa-Bossi N, Valentini M, Malleret L, Fiorini F, Bossi L. 2009. Caught at its own game: Regulatory small RNA inactivated by an inducible transcript mimicking its target. *Genes Dev* **23:** 2004–2015.

Gefen O, Balaban NQ. 2009. The importance of being persistent: Heterogeneity of bacterial populations under antibiotic stress. *FEMS Microbiol Rev* **33:** 704–717.

Geisinger E, Adhikari RP, Jin R, Ross HF, Novick RP. 2006. Inhibition of rot translation by RNAIII, a key feature of *agr* function. *Mol Microbiol* **61:** 1038–1048.

Gogol EB, Rhodius VA, Papenfort K, Vogel J, Gross CA. 2011. Small RNAs endow a transcriptional activator with essential repressor functions for single-tier control of a global stress regulon. *Proc Natl Acad Sci* **108:** 12875–12880.

Gottesman S, Storz G. 2011. Bacterial small RNA regulators: Versatile roles and rapidly evolving variations. *Cold Spring Harb Perspect Biol* **3:** a003798.

Grantcharova N, Peters V, Monteiro C, Zakikhany K, Romling U. 2010. Bistable expression of CsgD in biofilm development of *Salmonella enterica serovar typhimurium*. *J Bacteriol* **192:** 456–466.

Gripenland J, Netterling S, Loh E, Tiensuu T, Toledo-Arana A, Johansson J. 2010. RNAs: Regulators of bacterial virulence. *Nat Rev Microbiol* **8:** 857–866.

Heurlier K, Williams F, Heeb S, Dormond C, Pessi G, Singer D, Camara M, Williams P, Haas D. 2004. Positive control of swarming, rhamnolipid synthesis, and lipase production by the posttranscriptional RsmA/RsmZ system in *Pseudomonas aeruginosa* PAO1. *J Bacteriol* **186:** 2936–2945.

Hoe NP, Goguen JD. 1993. Temperature sensing in *Yersinia pestis*: Translation of the LcrF activator protein is thermally regulated. *J Bacteriol* **175:** 7901–7909.

Horstmann N, Orans J, Valentin-Hansen P, Shelburne SAR, Brennan RG. 2012. Structural mechanism of *Staphylococcus aureus* Hfq binding to an RNA A-tract. *Nucleic Acids Res* **40:** 11023–11035.

Huntzinger E, Boisset S, Saveanu C, Benito Y, Geissmann A, Namane G, Lina J, Etienne B, Ehresmann C, Ehresmann, et al. 2005. *Staphylococcus aureus* RNAIII and the endoribonuclease III coordinately regulate spa gene expression. *EMBO J* **24:** 824–835.

Jacob AI, Köhrer C, Davies BW, RajBhandary UL, Walker GC. 2013. Conserved bacterial RNase YbeY plays key roles in 70S ribosome quality control and 16S rRNA maturation. *Mol Cell* **49:** 427–438.

Jia X, Zhang J, Sun W, He W, Jiang H, Chen D, Murchie AIL. 2013. Riboswitch control of aminoglycoside antibiotic resistance. *Cell* **152:** 68–81.

Johansson J, Mandin P, Renzoni A, Chiaruttini C, Springer M, Cossart P. 2002. An RNA thermosensor controls expression of virulence genes in *Listeria monocytogenes*. *Cell* **110.** 551–561.

Jousselin A, Metzinger L, Felden B. 2009. On the facultative requirement of the bacterial RNA chaperone, Hfq. *Trends Microbiol* **17:** 399–405.

Kay E, Dubuis C, Haas D. 2005. Three small RNAs jointly ensure secondary metabolism and biocontrol in *Pseudomonas fluorescens* CHA0. *Proc Natl Acad Sci* **102:** 17136–17141.

Klinkert B, Narberhaus F. 2009. Microbial thermosensors. *Cell Mol Life Sci* **66:** 2661–2676.

Kortmann J, Narberhaus F. 2012. Bacterial RNA thermometers: Molecular zippers and switches. *Nat Rev Microbiol* **10:** 255–265.

Kröger C, Dillon SC, Cameron AD, Papenfort K, Sivasankaran SK, Hokamp K, Chao Y, Sittka A, Hebrard M, Handler K, et al. 2012. The transcriptional landscape and small RNAs of *Salmonella enterica serovar Typhimurium*. *Proc Natl Acad Sci* **109:** E1277–E1286.

Langridge GC, Phan MD, Turner DJ, Perkins TT, Parts L, Haase J, Charles I, Maskell DJ, Peters SE, Dougan G, et al. 2009. Simultaneous assay of every *Salmonella Typhi* gene using one million transposon mutants. *Genome Res* **19:** 2308–2316.

Lapouge K, Schubert M, Allain FH, Haas D. 2008. Gac/Rsm signal transduction pathway of γ-proteobacteria: From RNA recognition to regulation of social behaviour. *Mol Microbiol* **67:** 241–253.

Lasa I, Toledo-Arana A, Dobin A, Villanueva M, de Los Mozos IR, Vergara-Irigaray M, Segura V, Fagegaltier D, Penades JR, Valle J, et al. 2011. Genome-wide antisense transcription drives mRNA processing in bacteria. *Proc Natl Acad Sci* **108:** 20172–20177.

Lee EJ, Groisman EA. 2010. An antisense RNA that governs the expression kinetics of a multifunctional virulence gene. *Mol Microbiol* **76:** 1020–1033.

Lee EJ, Groisman EA. 2012a. Control of a *Salmonella* virulence locus by an ATP-sensing leader messenger RNA. *Nature* **486:** 271–275.

Lee EJ, Groisman EA. 2012b. Tandem attenuators control expression of the *Salmonella* mgtCBR virulence operon. *Mol Microbiol* **86:** 212–224.

Lee ER, Blount KF, Breaker RR. 2009. Roseoflavin is a natural antibacterial compound that binds to FMN riboswitches and regulates gene expression. *RNA Biol* **6:** 187–194.

Lee ER, Baker JL, Weinberg Z, Sudarsan N, Breaker RR. 2010. An allosteric self-splicing ribozyme triggered by a bacterial second messenger. *Science* **329:** 845–848.

Lenz DH, Miller MB, Zhu J, Kulkarni RV, Bassler BL. 2005. CsrA and three redundant small RNAs regulate quorum sensing in *Vibrio cholerae*. *Mol Microbiol* **58:** 1186–1202.

Lioliou E, Sharma CM, Caldelari I, Helfer AC, Fechter P, Vandenesch F, Vogel J, Romby P. 2012. Global regulatory functions of the *Staphylococcus aureus* endoribonuclease iii in gene expression. *PLoS Genet* **8:** e1002782.

Loh E, Dussurget O, Gripenland J, Vaitkevicius K, Tiensuu T, Mandin P, Repoila F, Buchrieser C, Cossart P, Johansson J. 2009. A *trans*-acting riboswitch controls expression of the virulence regulator PrfA in *Listeria monocytogenes*. *Cell* **139:** 770–779.

Mandal M, Boese B, Barrick JE, Winkler WC, Breaker RR. 2003. Riboswitches control fundamental biochemical pathways in *Bacillus subtilis* and other bacteria. *Cell* **113:** 577–586.

Mandlik A, Livny J, Robins WP, Ritchie JM, Mekalanos JJ, Waldor MK. 2011. RNA-Seq-based monitoring of infection-linked changes in *Vibrio cholerae* gene expression. *Cell Host Microbe* **10:**165–174.

Mann B, van Opijnen T, Wang J, Obert C, Wang YD, Carter R, McGoldrick DJ, Ridout G, Camilli A, Tuomanen EI, et al. 2012. Control of virulence by small RNAs in *Streptococcus pneumoniae*. *PLoS Pathogen* **8:** e1002788.

Mansjö M, Johansson J. 2011. The riboflavin analog roseoflavin targets an FMN-riboswitch and blocks *Listeria monocytogenes* growth, but also stimulates virulence gene-expression and infection. *RNA Biol* **8:** 674–680.

Marchfelder A, Hess W. 2011. *Regulatory RNAs in prokaryotes* (ed. Marchfelder A, Hess W), pp. 1–253. Springer, New York.

Marzi S, Romby P. 2012. RNA mimicry, a decoy for regulatory proteins. *Mol Microbiol* **83:** 1–6.

Mellin JR, Cossart P. 2012. The non-coding RNA world of the bacterial pathogen *Listeria monocytogenes*. *RNA Biol* **9:** 372–378.

Milojevic T, Grishkovskaya I, Sonntleitner E, Djinovic-Carugo K, Blasi U. 2013. The *Pseudomonas aeruginosa* catabolite repression control protein Crc is devoid of RNA binding activity. *PloS ONE* **23:** e64609.

Monteiro C, Papenfort K, Hentrich K, Ahmad I, Le Guyon S, Reimann R, Grantcharova N, Romling U. 2012. Hfq and Hfq-dependent small RNAs are major contributors to multicellular development in *Salmonella enterica serovar Typhimurium*. *RNA Biol* **9:** 489–502.

Moreno R, Marzi S, Romby P, Rojo F. 2009. The Crc global regulator binds to an unpaired A-rich motif at the *Pseudomonas putida alkS* mRNA coding sequence and inhibits its translation initiation. *Nucleic Acids Res* **37:** 7678–7690.

Moreno R, Fonseca P, Rojo F. 2012. Two small RNAs, CrcY and CrcZ, act in concert to sequester the Crc global regulator in *Pseudomonas putida*, modulating catabolite repression. *Mol Microbiol* **83:** 24–40.

Morfeldt E, Taylor D, von Gabain A, Arvidson S. 1995. Activation of α-toxin translation in *Staphylococcus aureus* by the *trans*-encoded antisense RNA, RNAIII. *EMBO J* **14:** 4569–4577.

Morrison JM, Anderson KL, Beenken KE, Smeltzer MS, Dunman PM. 2012. The staphylococcal accessory regulator, SarA, is an RNA-binding protein that modulates the mRNA turnover properties of late-exponential and stationary phase *Staphylococcus aureus* cells. *Front Cell Infect Microbiol* **2:** 1–11.

Mulhbacher J, Brouillette E, Allard M, Fortier LC, Malouin F, Lafontaine DA. 2010a. Novel riboswitch ligand analogs as selective inhibitors of guanine-related metabolic pathways. *PLoS Pathog* **6:** e1000865.

Mulhbacher J, St-Pierre P, Lafontaine DA. 2010b. Therapeutic applications of ribozymes and riboswitches. *Curr Opin Pharmacol* **10:** 551–556.

Narberhaus F. 2010. Translational control of bacterial heat shock and virulence genes by temperature-sensing mRNAs. *RNA Biol* **7:** 84–89.

Nechooshtan G, Elgrably-Weiss M, Sheaffer A, Westhof E, Altuvia S. 2009. A pH-responsive riboregulator. *Genes Dev* **23:** 2650–2662.

Ng WL, Bassler BL. 2009. Bacterial quorum-sensing network architectures. *Annu Rev Genet* **43:** 197–222.

Nielsen JS, Larsen MH, Lillebaek EM, Bergholz TM, Christiansen MH, Boor KJ, Wiedmann M, Kallipolitis BH. 2011. A small RNA controls expression of the chitinase ChiA in *Listeria monocytogenes*. *PLoS ONE* **6:** e19019.

Novick RP. 2003. Autoinduction and signal transduction in the regulation of staphylococcal virulence. *Mol Microbiol* **48:** 1429–1449.

Novick RP, Geisinger E. 2008. Quorum sensing in Staphylococci. *Annu Rev Genet* **42:** 541–564.

Obana N, Shirahama Y, Abe K, Nakamura K. 2010. Stabilization of *Clostridium perfringens* collagenase mRNA by VR-RNA-dependent cleavage in 5′ leader sequence. *Mol Microbiol* **77:** 1416–1428.

Oliver HF, Orsi RH, Ponnala L, Keich U, Wang W, Sun Q, Cartinhour SW, Filiatrault MJ, Wiedmann M, Boor KJ. 2009. *BMC Genomics* **10:** 641–663.

Padalon-Brauch G, Hershberg R, Elgrably-Weiss M, Baruch K, Rosenshine I, Margalit H, Altuvia S. 2008. Small RNAs encoded within genetic islands of *Salmonella typhimurium* show host-induced expression and role in virulence. *Nucleic Acids Res* **36:** 1913–1927.

Pandey SP, Minesinger BK, Kumar J, Walker GC. 2011. A highly conserved protein of unknown function in *Sinorhizobium meliloti* affects sRNA regulation similar to Hfq. *Nucleic Acids Res* **39:** 4691–4708.

Papenfort K, Vogel J. 2010. Regulatory RNA in bacterial pathogens. *Cell Host Microbe* **8:** 116–127.

Papenfort K, Said N, Welsink T, Lucchini S, Hinton JC, Vogel J. 2009. Specific and pleiotropic patterns of mRNA regulation by ArcZ, a conserved, Hfq-dependent small RNA. *Mol Microbiol* **74:** 139–158.

Papenfort K, Bouvier M, Mika F, Sharma CM, Vogel J. 2010. Evidence for an autonomous 5′ target recognition domain in an Hfq-associated small RNA. *Proc Natl Acad Sci* **107:** 20435–20440.

Papenfort K, Podkaminski D, Hinton JC, Vogel J. 2012. The ancestral SgrS RNA discriminates horizontally acquired *Salmonella* mRNAs through a single G-U wobble pair. *Proc Natl Acad Sci* **109:** E757–E764.

Papenfort K, Corcoran C, Gupta SK, Miyakoshi M, Heidrich N, Chao Y, Frölich KS, Sharma CM, Ziebuhr W, Böhm A, et al. 2013a. Regulatory mechanisms of special significance: Role of small RNAs in virulence regulation. In *Regulation of bacterial virulence* (ed. Vasil M, Darwin A), pp. 492–527. ASM, Washington, DC.

Papenfort K, Sun Y, Miyakoshi M, Vanderpool CK, Vogel J. 2013b. Small RNA-mediated activation of sugar phosphatase mRNA regulates glucose homeostasis. *Cell* **153:** 426–437.

Pedrolli DB, Matern A, Wang J, Ester M, Siedler K, Breaker R, Mack M. 2012. A highly specialized flavin mononucleotide riboswitch responds differently to similar ligands and confers roseoflavin resistance to *Streptomyces davawensis*. *Nucleic Acids Res* **40:** 8662–8673.

Pfeiffer V, Papenfort K, Lucchini S, Hinton JC, Vogel J. 2009. Coding sequence targeting by MicC RNA reveals bacterial mRNA silencing downstream of translational initiation. *Nat Struct Mol Biol* **16:** 840–846.

Ramesh A, Winkler WC. 2010. Magnesium-sensing riboswitches in bacteria. *RNA Biol* **7:** 77–83.

Ramirez-Pena E, Trevino J, Liu Z, Perez N, Sumby P. 2010. The group A *Streptococcus* small regulatory RNA FasX enhances streptokinase activity by increasing the stability of the *ska* mRNA transcript. *Mol Microbiol* **78:** 1332–1347.

Rice JB, Balasubramanian D, Vanderpool CK. 2012. Small RNA binding-site multiplicity involved in translational regulation of a polycistronic mRNA. *Proc Natl Acad Sci* **109:** E2691–E2698.

Richard AL, Withey JH, Beyhan S, Yildiz F, DiRita VJ. 2010. The *Vibrio cholerae* virulence regulatory cascade controls glucose uptake through activation of TarA, a small regulatory RNA. *Mol Microbiol* **78:** 1171–1181.

Roberts C, Anderson KL, Murphy E, Projan SJ, Mounts W, Hurlburt B, Smeltzer M, Overbeek R, Disz T, Dunman PM. 2006. Characterizing the effect of the *Staphylococcus aureus* virulence factor regulator, SarA, on log-phase mRNA half-lives. *J Bacteriol* **188:** 2593–2603.

Rojo F. 2010. Carbon catabolite repression in Pseudomonas: Optimizing metabolic versatility and interactions with the environment. *FEMS Microbiol Rev* **34:** 658–684.

Romby P, Charpentier E. 2010. An overview of RNAs with regulatory functions in Gram-positive bacteria. *Cell Mol Life Sci* **67:** 217–237.

Romeo T, Vakulskas CA, Babitzke P. 2012. Post-transcriptional regulation on a global scale: Form and function of Csr/Rsm systems. *Environ Microbiol* **15:** 313–324.

Romilly C, Caldelari I, Parmentier D, Lioliou E, Romby P, Fechter P. 2012. Current knowledge on regulatory RNAs and their machineries in *Staphylococcus aureus*. *RNA Biol* **9:** 402–413.

Roth A, Breaker RR. 2013. Integron attI1 sites, not riboswitches, associate with antibiotic resistance genes. *Cell* **153:** 1417–1418.

Rutherford ST, van Kessel JC, Shao Y, Bassler BL. 2011. AphA and LuxR/HapR reciprocally control quorum sensing in vibrios. *Genes Dev* **25:** 397–408.

Santiviago CA, Reynolds MM, Porwollik S, Choi SH, Long F, Andrews-Polymenis HL, McClelland M. 2009. Analysis of pools of targeted *Salmonella* deletion mutants identifies novel genes affecting fitness during competitive infection in mice. *PLoS Pathog* **5:** e1000477.

Sayed N, Jousselin A, Felden B. 2011. A *cis*-antisense RNA acts in *trans* in *Staphylococcus aureus* to control translation of a human cytolytic peptide. *Nat Struct Mol Biol* **19:** 105–112.

Serganov A. 2010. Determination of riboswitch structures: Light at the end of the tunnel? *RNA Biol* **7:** 98–103.

Serganov A, Nudler E. 2013. A decade of riboswitches. *Cell* **152:** 17–24.

Sesto N, Wurtzel O, Archambaud C, Sorek R, Cossart P. 2013. The excludon: A new concept in bacterial antisense RNA-mediated gene regulation. *Nat Rev Microbiol* **11:** 75–82.

Shao Y, Bassler BL. 2012. Quorum-sensing non-coding small RNAs use unique pairing regions to differentially control mRNA targets. *Mol Microbiol* **83:** 599–611.

Shao Y, Feng L, Rutherford ST, Papenfort K, Bassler BL. 2013. Functional determinants of the quorum-sensing non-coding RNAs and their roles in target regulation. *EMBO J* **32:** 2158–2171.

Sharma CM, Hoffmann S, Darfeuille F, Reignier J, Findeiss S, Sittka A, Chabas S, Reiche K, Hackermüller J, Reinhardt R, et al. 2010. The primary transcriptome of the major human pathogen *Helicobacter pylori*. *Nature* **464:** 250–255.

Sharma CM, Papenfort K, Pernitzsch SR, Mollenkopf HJ, Hinton JC, Vogel J. 2011. Pervasive post-transcriptional control of genes involved in amino acid metabolism by the Hfq-dependent GcvB small RNA. *Mol Microbiol* **81:** 1144–1165.

Sittka A, Lucchini S, Papenfort K, Sharma CM, Rolle K, Binnewies TT, Hinton JC, Vogel J. 2008. Deep sequencing analysis of small noncoding RNA and mRNA targets of the global post-transcriptional regulator, Hfq. *PLoS Genet* **4:** e1000163.

Sittka A, Sharma CM, Rolle K, Vogel J. 2009. Deep sequencing of *Salmonella* RNA associated with heterologous Hfq proteins *in vivo* reveals small RNAs as a major target class and identifies RNA processing phenotypes. *RNA Biol* **6:** 266–275.

Somerville GA, Proctor RA. 2009. At the crossroads of bacterial metabolism and virulence factor synthesis in staphylococci. *Microbiol Mol Biol Rev* **73:** 233–248.

Song T, Mika F, Lindmark B, Liu Z, Schild S, Bishop A, Zhu J, Camilli A, Johansson J, Vogel J, et al. 2008. A new *Vibrio cholerae* sRNA modulates colonization and affects release of outer membrane vesicles. *Mol Microbiol* **70:** 100–111.

Song J, Lays C, Vandenesch F, Benito Y, Bes M, Chu Y, Lina G, Romby P, Geissmann T, Boisset S. 2012. The expression of small regulatory RNAs in clinical samples reflects the different life styles of *Staphylococcus aureus* in colonization vs. infection. *PLoS ONE* **7:** e37294.

Sonnleitner E, Haas D. 2011. Small RNAs as regulators of primary and secondary metabolism in *Pseudomonas* species. *Appl Microbiol Biotechnol* **91:** 63–79.

Sonnleitner E, Abdou L, Haas D. 2009. Small RNA as global regulator of carbon catabolite repression in *Pseudomonas aeruginosa*. *Proc Natl Acad Sci* **106:** 21866–21871.

Sonnleitner E, Gonzalez N, Sorger-Domenigg T, Heeb S, Richter AS, Backofen R, Williams P, Huttenhofer A, Haas D, Blasi U. 2011. The small RNA PhrS stimulates synthesis of the *Pseudomonas aeruginosa* quinolone signal. *Mol Microbiol* **80:** 868–885.

Storz G, Vogel J, Wassarman KM. 2011. Regulation by small RNAs in bacteria: Expanding frontiers. *Mol Cell* **43:** 880–891.

Sudarsan N, Cohen-Chalamish S, Nakamura S, Emilsson GM, Breaker RR. 2005. Thiamine pyrophosphate riboswitches are targets for the antimicrobial compound pyrithiamine. *Chem Biol* **12:** 1325–1335.

Sudarsan N, Lee ER, Weinberg Z, Moy RH, Kim JN, Link KH, Breaker RR. 2008. Riboswitches in eubacteria sense the second messenger cyclic di-GMP. *Science* **321:** 411–413.

Toledo-Arana A, Dussurget O, Nikitas G, Sesto N, Guet-Revillet H, Balestrino D, Loh E, Gripenland J, Tiensuu T, Vaitkevicius K, et al. 2009. The *Listeria* transcriptional landscape from saprophytism to virulence. *Nature* **459:** 950–956.

Tu KC, Long T, Svenningsen SL, Wingreen NS, Bassler BL. 2010. Negative feedback loops involving small regulatory RNAs precisely control the *Vibrio harveyi* quorum-sensing response. *Mol Cell* **37:** 567–579.

Vogel J, Luisi BF. 2011. Hfq and its constellation of RNA. *Nat Rev Microbiol* **9:** 578–589.

Vogel J, Bartels V, Tang TH, Churakov G, Slagter-Jäger JG, Hüttenhofer A, Wagner EGH. 2003. RNomics in *Escherichia coli* detects new sRNA species and indicates parallel transcriptional output in bacteria. *Nucleic Acids Res* **31:** 6435–6443.

Wadler CS, Vanderpool CK. 2007. A dual function for a bacterial small RNA: SgrS performs base pairing-dependent regulation and encodes a functional polypeptide. *Proc Natl Acad Sci* **104:** 20454–20459.

Wassarman KM. 2007. 6S RNA: A regulator of transcription. *Mol Microbiol* **65:** 1425–1431.

Wassarman KM, Repoila F, Rosenow C, Storz G, Gottesman S. 2001. Identification of novel small RNAs using comparative genomics and microarrays. *Genes Dev* **15:** 1637–1651.

Westermann AJ, Gorski SA, Vogel J. 2012. Dual RNA-seq of pathogen and host. *Nat Rev Microbiol* **10:** 618–630.

Wurtzel O, Sesto N, Mellin JR, Karunker I, Edelheit S, Becavin C, Archambaud C, Cossart P, Sorek R. 2012. Comparative transcriptomics of pathogenic and non-pathogenic *Listeria* species. *Mol Syst Biol* **8:** 583.

Yao Z, Barrick J, Weinberg Z, Neph S, Breaker R, Tompa M, Ruzzo WL. 2007. A computational pipeline for high-throughput discovery of *cis*-regulatory noncoding RNA in prokaryotes. *PLoS Comput Biol* **3:** e126.

Bacterial Quorum Sensing: Its Role in Virulence and Possibilities for Its Control

Steven T. Rutherford[1] and Bonnie L. Bassler[1,2]

[1]Department of Molecular Biology, Princeton University, Princeton, New Jersey 08544

[2]Howard Hughes Medical Institute, Chevy Chase, Maryland 20815

Correspondence: bbassler@princeton.edu

Quorum sensing is a process of cell–cell communication that allows bacteria to share information about cell density and adjust gene expression accordingly. This process enables bacteria to express energetically expensive processes as a collective only when the impact of those processes on the environment or on a host will be maximized. Among the many traits controlled by quorum sensing is the expression of virulence factors by pathogenic bacteria. Here we review the quorum-sensing circuits of *Staphylococcus aureus, Bacillus cereus, Pseudomonas aeruginosa,* and *Vibrio cholerae.* We outline these canonical quorum-sensing mechanisms and how each uniquely controls virulence factor production. Additionally, we examine recent efforts to inhibit quorum sensing in these pathogens with the goal of designing novel antimicrobial therapeutics.

Quorum sensing (QS) is a bacterial cell–cell communication process that involves the production, detection, and response to extracellular signaling molecules called autoinducers (AIs). AIs accumulate in the environment as the bacterial population density increases, and bacteria monitor this information to track changes in their cell numbers and collectively alter gene expression. QS controls genes that direct activities that are beneficial when performed by groups of bacteria acting in synchrony. Processes controlled by QS include bioluminescence, sporulation, competence, antibiotic production, biofilm formation, and virulence factor secretion (reviewed in Novick and Geisinger 2008; Ng and Bassler 2009; Williams and Camara 2009).

Despite differences in regulatory components and molecular mechanisms, all known QS systems depend on three basic principles. First, the members of the community produce AIs, which are the signaling molecules. At low cell density (LCD), AIs diffuse away, and, therefore, are present at concentrations below the threshold required for detection. At high cell density (HCD), the cumulative production of AIs leads to a local high concentration, enabling detection and response (Kaplan and Greenberg 1985). Second, AIs are detected by receptors that exist in the cytoplasm or in the membrane. Third, in addition to activating expression of genes necessary for cooperative behaviors, detection of AIs results in activation of AI production (Novick et al. 1995; Seed et al. 1995). This

Cite this article as *Cold Spring Harb Perspect Med* doi: 10.1101/cshperspect.a012427

feed-forward autoinduction loop presumably promotes synchrony in the population.

Gram-positive and Gram-negative bacteria use different types of QS systems (Fig. 1 shows the four paradigmatic QS wiring diagrams). Gram-positive bacteria use peptides, called autoinducing peptides (AIPs), as signaling molecules. Once produced in the cell, AIPs are processed and secreted. When the extracellular concentration of the AIP is high, which occurs at HCD, it binds to a cognate membrane-bound two-component histidine kinase receptor. Usually, binding activates the receptor's kinase activity, it autophosphorylates, and passes phosphate to a cognate cytoplasmic response regulator. The phosphorylated response regulator activates transcription of the genes in the QS regulon (Fig. 1A). In some cases of Gram-positive bacterial QS, AIPs are transported back into the cell cytoplasm where they interact with transcription factors to modulate the transcription factor's activity and, in turn, modulate gene expression changes (Fig. 1B).

Gram-negative bacteria communicate using small molecules as AIs. These are either acyl-homoserine lactones (AHLs) or other molecules whose production depends on *S*-adenosylmethionine (SAM) as a substrate (Wei et al. 2011). AIs are produced in the cell and freely diffuse across the inner and outer membranes. When the concentration of AIs is sufficiently high, which occurs at HCD, they bind cytoplasmic receptors that are transcription factors. The AI-bound receptors regulate expression of the genes in the QS regulon (Fig. 1C). In some cases of Gram-negative bacterial QS, AIs are detected by two-component histidine kinase receptors that function analogously to those described in the preceding paragraph for Gram-positive QS bacteria (Fig. 1D).

Dozens of clinically-relevant bacteria use QS to regulate the collective production of virulence factors. Here, we highlight the QS systems of four human pathogens that exemplify the diversity of QS systems. First, we outline how *Staphylococcus aureus* uses the paradigmatic Agr system to regulate adhesion and production of virulence factors. We also discuss the PlcR/PapR system that controls virulence factor production in the Gram-positive bacterium *Bacillus cereus*.

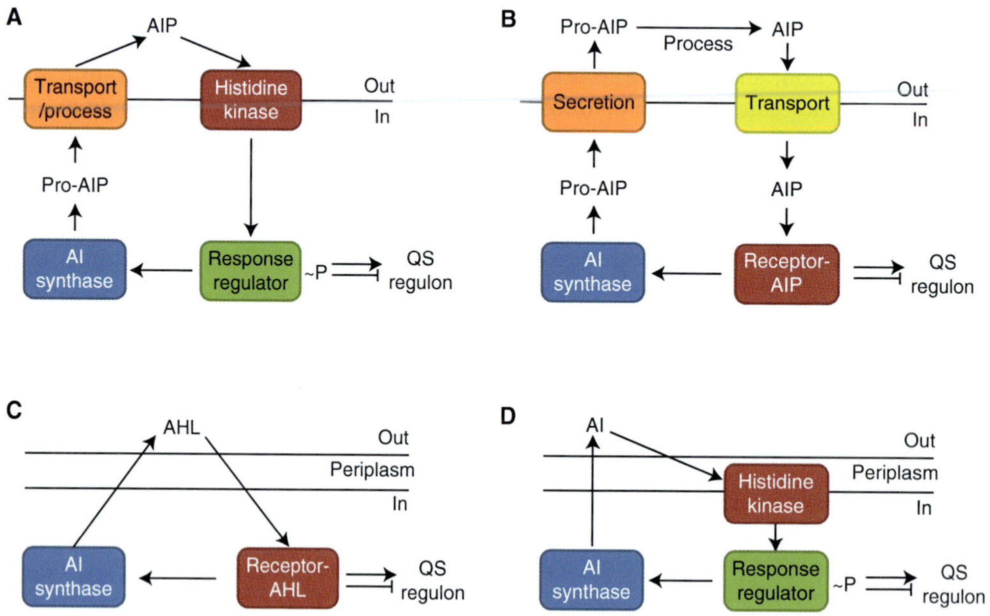

Figure 1. Canonical bacterial quorum-sensing (QS) circuits. Autoinducing peptide (AIP) QS in Gram-positive bacteria by (*A*) two-component signaling, or (*B*) an AIP-binding transcription factor. Small molecule QS in Gram-negative bacteria by (*C*) a LuxI/LuxR-type system, or (*D*) two-component signaling.

Next, we describe the canonical Gram-negative LuxI/LuxR QS circuit and how *Pseudomonas aeruginosa* exploits two such circuits arranged in tandem to control virulence factor production and biofilm formation. Finally, we discuss QS in *Vibrio cholerae*, a Gram-negative bacterium that uses two parallel two-component QS systems to control virulence factor production and biofilm formation. Importantly, in all of these cases, the QS circuits are tailored to promote the specific disease. *P. aeruginosa* and *S. aureus* cause persistent diseases while *V. cholerae* and *B. cereus* cause acute infections. We also outline efforts to develop inhibitors of these QS systems to be deployed as novel antimicrobials.

QS CONTROL OF VIRULENCE FACTORS IN GRAM-POSITIVE BACTERIA

Two-Component QS in Gram-Positive Bacteria

QS in Gram-positive bacteria relies on principles common to all QS circuits: production, detection, and response to AIs. In many Gram-positive bacteria, the AIs are oligopeptide AIPs that are detected by membrane-bound two-component signal transduction systems (Fig. 1A) (Havarstein et al. 1995; Ji et al. 1995; Solomon et al. 1996).

The AIPs are encoded as precursors (pro-AIPs) and are diverse in sequence and structure (Havarstein et al. 1995; Otto et al. 1998; Lazazzera 2001; Nakayama et al. 2001; Kalkum et al. 2003; Okada et al. 2005; Thoendel et al. 2011). Because the cell membrane is impermeable to peptides, specialized transporters are required to secrete AIPs. The AIP transporters also process the pro-AIPs. The final processed AIPs range in size from 5 to 17 amino acids, can be posttranslationally modified, and can be linear or cyclized (Magnuson et al. 1994; Havarstein et al. 1995; Mayville et al. 1999; Okada et al. 2005; Bouillaut et al. 2008). Extracellular AIPs are detected via membrane-bound two-component sensor kinases (Hoch and Silhavy 1995; Inouye and Dutta 2003; Simon et al. 2007). The sensor kinases autophosphorylate at conserved histidines when bound by the AIP. The phosphoryl group is passed from the histidine to a conserved aspartate on a cognate cytoplasmic response-regulator protein, and the phosphorylated response regulator controls expression of QS-target genes. In these Gram-positive QS circuits, the pro-AIP, transporter, histidine kinase receptor, and response regulator are typically encoded in an operon (Ji et al. 1995; Peterson et al. 2000). Expression of this operon is activated by the phosphorylated response regulator, resulting in an autoinducing feed-forward loop that synchronizes the QS response.

Some examples of Gram-positive QS behaviors are competence in *Streptococcus pneumonia* and *Bacillus subtilis* and sporulation in *B. subtilis* (Kleerebezem et al. 1997). QS controls virulence factor production in Gram-positive human pathogens including *S. aureus*, *Listeria monocytogenes*, *Enterococcus faecalis*, and *Clostridium perfringens* (Autret et al. 2003; Podbielski and Kreikemeyer 2004; Ohtani et al. 2009; Riedel et al. 2009; Thoendel et al. 2011). The most well-studied system in this group of pathogens is the *S. aureus* Agr system (reviewed extensively in Thoendel et al. 2011).

S. aureus Quorum Sensing

S. aureus is found among the normal human skin flora. If the epithelial barrier is compromised, *S. aureus* can cause minor skin infections. These infections can lead to pneumonia, bacteremia, and sepsis (Lowy 1998; Massey et al. 2006). *S. aureus* is the leading cause of hospital-related infections in the United States. Its ability to cause disease depends on expression of an array of adhesion molecules, toxins, and compounds that affect the immune system. QS regulates expression of genes encoding these virulence factors.

S. aureus uses a canonical Gram-positive two-component QS system encoded by the *agr* locus (Fig. 2). The P2 promoter drives expression of a transcript (RNAII), which encodes the four components of the QS system (Novick et al. 1995). *agrD* encodes the pro-AIP, which is processed to the final AIP and secreted by the *trans*-membrane transporter protein AgrB (Ji et al.

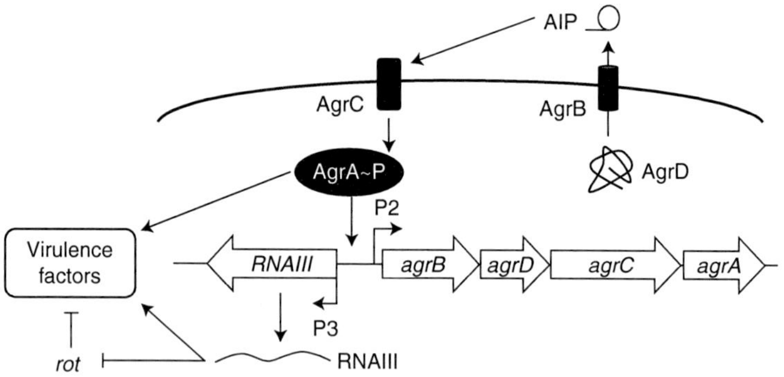

Figure 2. *S. aureus* Agr QS circuit. The autoinducing peptide (AIP) is synthesized as a precursor from *agrD*. The AIP transporter AgrB processes the precursor to the mature AIP and transports it out of the cell. AIPs are detected by a two-component signal transduction pathway. AgrC is the membrane-bound histidine kinase and AgrA is the response regulator. Phosphorylated AgrA activates the P2 and P3 promoters encoding the *agr* operon (called RNAII) and the RNAIII regulatory RNA, respectively. RNAIII posttranscriptionally activates virulence factor production and represses expression of *rot*, the repressor of toxins, leading to further derepression of virulence factors.

1995; Saenz et al. 2000; Thoendel and Horswill 2009, 2010; Thoendel et al. 2011). Processing involves truncating the 45-47 residue pro-AIP to a 7-9 residues peptide, coupled with cyclization of a five-membered peptide ring via a thiolactone bond between a central cysteine residue and the carboxyl terminus (Table 1). When the AIP accumulates, it binds the membrane-bound histidine kinase AgrC, which autophosphorylates at a conserved histidine and transfers the phosphate group to an aspartate on the response regulator AgrA (Lina et al. 1998). Phosphorylated AgrA binds upstream of the P2 promoter to autoinduce the *agr* operon (Novick et al. 1995).

In addition to activating the P2 promoter, phosphorylated AgrA activates the divergently encoded P3 promoter. The P3 promoter controls expression of RNAIII (Novick et al. 1993). The 5′ region of RNAIII harbors the *hld* gene, which encodes the virulence factor δ-hemolysin (Janzon and Arvidson 1990). A more prominent role for RNAIII is as a regulatory RNA (Novick et al. 1993). RNAIII has the dual-function of activating production of α-toxin and repressing expression of *rot*, fibronectin binding proteins A and B, protein A, coagulase, and other surface proteins (Morfeldt et al. 1995; Dunman et al. 2001). Repression of *rot*, which encodes a *r*epressor *o*f *t*oxins, leads to de-repression of additional toxins, proteases, lipases, enterotoxins, superantigens, and urease (Said-Salim et al. 2003; Geisinger et al. 2006). The net result of this QS regulatory cascade is down-regulation of surface virulence factors (such as protein A), and up-regulation of secreted virulence factors (such as α-toxin). Most of the effects of QS on regulation of virulence in *S. aureus* are mediated through direct and indirect regulation by RNAIII, however, phosphorylated AgrA also directly activates at least two additional virulence genes encoding phenol-soluble modulines (Queck et al. 2008).

Another key component of the *S. aureus* virulence program is biofilm development. In *S. aureus*, the *agr* system inhibits biofilm formation (Vuong et al. 2000; Boles and Horswill 2008). One interpretation of this finding is that establishing a biofilm community at LCD allows *S. aureus* time to grow to HCD and, at that point, it is optimally poised to secrete virulence factors (Yarwood et al. 2004; Boles and Horswill 2008). To facilitate its dispersal, *S. aureus* terminates biofilm production and decreases surface proteins and adhesions at HCD. This behavior is analogous to the strategy used by *V. cholerae* (see below).

Numerous additional regulatory factors converge at the P2 and P3 promoters, which add

 Cite this article as *Cold Spring Harb Perspect Med* doi: 10.1101/cshperspect.a012427

Table 1. *S. aureus* AIPs and inhibitors

	AIP	Truncations	Substitutions	Hybrids
AIP-I	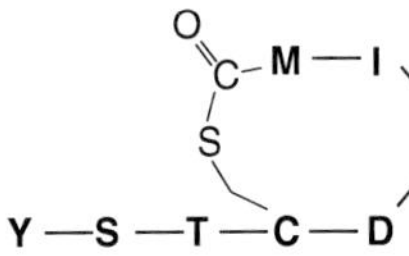	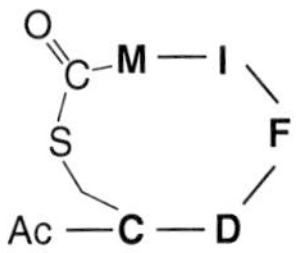 (11)	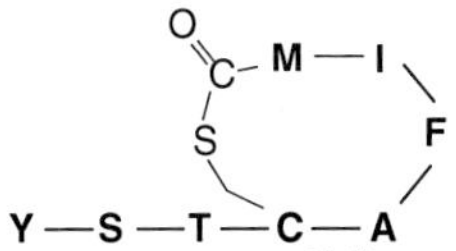 (14)	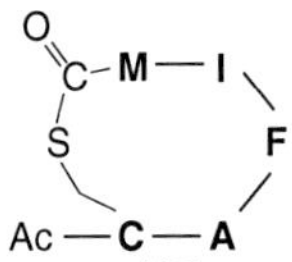 (15)
AIP-II	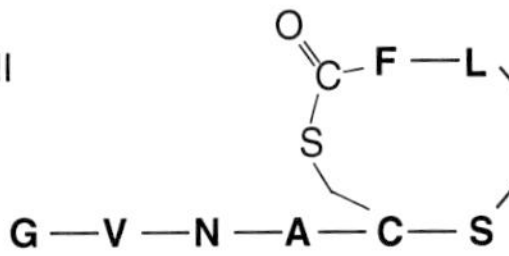	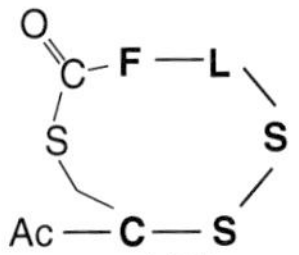 (12)		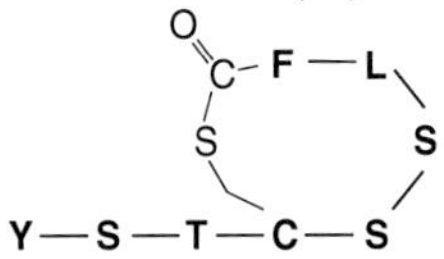 (16)
AIP-III	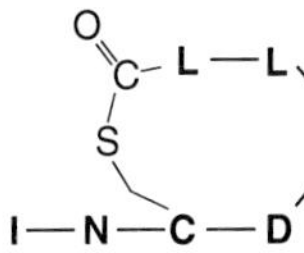			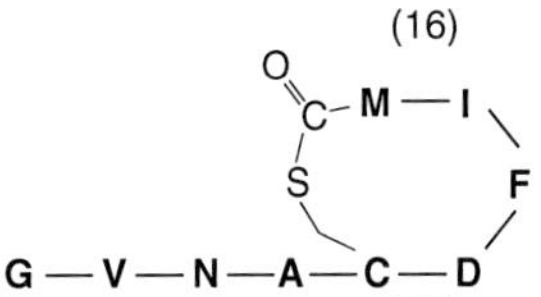 (17)
AIP-IV	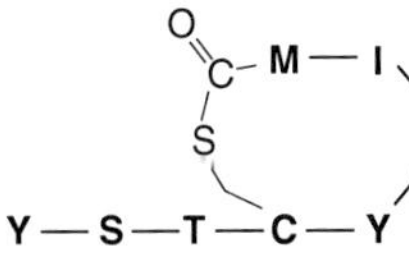	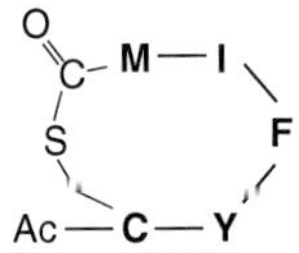 (13)		

Inhibitors 11, 13, 15, 16, and 17 are from Lyon et al. (2002); inhibitor 12 is from Lyon et al. (2000); and inhibitor 14 is from McDowell et al. (2001).

layers of regulation to the *S. aureus* QS circuitry, presumably ensuring that the QS program is expressed under the most ideal situations. There is evidence for regulation by ArlRS, CcpA, CodY, Rsr, σ^B, SarA, SarR, SarT, SarU, SarX, SarZ, Rsr, and SrrA/SrrB (Cheung and Projan 1994; Schmidt et al. 2001; Yarwood et al. 2001; Manna and Cheung 2003, 2006a,b; Liang et al. 2005; Seidl et al. 2006; Majerczyk et al. 2008; Lauderdale et al. 2009; Tamber and Cheung 2009; Tamber et al. 2010). One of the interesting themes common to several of these accessory *agr* regulators is that they provide *S. aureus* the ability to respond to extracellular environmental signals, in addition to AIPs, using the QS circuitry. For example, following extracellular stress, the alternative sigma factor σ^B interacts with core RNAP to direct transcription of surface proteins and pigment production, and inhibit expression of secreted toxins and proteases. RNAIII levels also increase when the σ^B gene is deleted (Lauderdale et al. 2009). No consensus sequence for σ^B exists at the P2 or P3 promoters, so it is likely that σ^B regulates some as-yet-unknown regulator of *agr*. σ^B regulation presumably ensures that *S. aureus* does not undergo QS, which is energetically costly, under conditions when the bacteria must dedicate resources to alleviating stress. Another example involves CodY, which responds to isoleucine limitation in *S. aureus*. Deletion of *codY* increases RNAII and RNAIII expression (Majerczyk et al. 2008; Pohl et al. 2009). CodY binding sites are located in the *agrC* gene (Majerczyk et al. 2010). Thus, CodY allows *S. aureus* to delay its QS response until it exists in a nutrient-rich environment. Finally, the SrrA/SrrB two-component system, which regulates gene expression in response to anaerobic environments, also controls the P2 and P3 promoter region via SrrB binding (Yarwood et al. 2001). Overexpression of SrrA/SrrB decreases virulence, likely due, at least in part, to inhibition of *agr* expression. Again, this regulation is thought to ensure that QS is not undertaken in a less than ideal environment.

A final interesting aspect of the *S. aureus* QS system is cross-competition among AIP specificity types. Hypervariability exists in the *agrD* gene and a portion of the *agrB* gene (Dufour et al. 2002). This variability leads to production of one of four different types of *S. aureus* AIPs depending on the strain (Table 1). The identity of the AIP allows typing of *S. aureus* into four specificity groups (I–IV) (Ji et al. 1997; Jarraud et al. 2000). A corresponding hypervariability exists in the portion of the *agrC* gene encoding the sensing domain of the AIP receptor (Dufour et al. 2002). Thus, each specific AIP is detected by a coevolved cognate AgrC sensor. The presence of a noncognate AIP results in inhibition of QS, highlighting the specific nature of the AIP-receptor interactions. Specificity is proposed to be determined sterically and binding of the incorrect AIP-type can lead to stabilization of an inhibitory confirmation of AgrC, thus halting cell–cell signaling (Mayville et al. 1999; Lyon et al. 2002; Geisinger et al. 2009). The biological consequence of this mechanism is that the first strain to establish its QS cascade in the host is the one that causes the infection (Fleming et al. 2006).

S. aureus QS as a Therapeutic Target

The prevalence of *S. aureus* infections coupled with the emergence of methicillin-resistant *S. aureus* (MRSA) highlight the importance of studying and controlling this pathogen (Brumfitt and Hamilton-Miller 1989). The rationale for targeting the *S. aureus* QS system is that *agr* mutants are defective in virulence. The principal target considered to date is the receptor AgrC. Because the AIPs naturally cross-inhibit, analyses of their variations coupled with synthetic chemistry have provided a mechanistic understanding of AIP antagonism. By varying the length (Table 2, peptides 11–13) or amino acid sequence (Table 2, peptides 14) of the AIP or by combining substitutions and truncations to make hybrid AIPs (Table 2, peptides 15–17), universal inhibitors capable of out-competing all four AIP types have been designed (Lyon et al. 2000, 2002; McDowell et al. 2001).

The AIPs themselves have also been targeted for inhibition by inactivation through interaction with specific antibodies. The AIPs are poor antigens, however, monoclonal antibodies against AIPs coupled to carriers have been raised and are promising (Park et al. 2007). For example, screening of a library of monoclonal antibodies identified one antibody that reduced RNAIII and toxin production. This antibody acts as a highly-specific QS quencher.

One of the concerns with targeting the *S. aureus* QS system is the trade-off between virulence factor production and biofilm formation (Otto 2004). Inhibition of *agr* causes *S. aureus* to become more adherent because of increased biofilm formation. This facet of the circuit can lead to persistent colonization by *S. aureus*, which is of utmost concern for medically-implanted devices. Although there are reports that deletion of *agr* does not enhance biofilm formation under all conditions (Shenkman et al. 2001), targeting *agr* will likely become one part of a combination therapy for combating *S. aureus* and MRSA.

QS by Peptide-Binding Transcription Factors in Gram-Positive Bacteria

Some Gram-positive bacteria use a QS system in which AIPs, following release, are imported back into the cell and detected by cytoplasmic transcription factors (Fig. 1B). In these systems, the pro-AIP is secreted and processed into the mature AIP by extracellular proteases. Following transport back into the cell, the AIP binds to and alters the activity of a transcription factor (Slamti and Lereclus 2002).

Some examples of Gram-positive systems in which AIPs are imported include sporulation, competence, and enzyme production in *B. subtilis* (Pottathil and Lazazzera 2003) and plasmid transfer in *E. faecalis* (Dunny 2007). Additionally, AIP-bound transcription factors control virulence factor production in the *Bacillus cereus* group (Slamti and Lereclus 2002; Bouillaut et al. 2008) and we describe this system here (see review by Rocha-Estrada et al. 2010).

B. cereus Quorum Sensing

The *B. cereus* group of Gram-positive bacteria consists of several closely-related bacteria relevant to human health including *B. cereus*, *B.*

anthracis, and *B. thuringiensis*. *B. cereus* causes both intestinal and nonintestinal infections in humans and is most commonly associated with food poisoning (Bottone 2010). Its ability to cause acute diarrheal disease is caused by the production and secretion of a variety of hemolysins, phospholipases, and toxins (Bottone 2010).

QS in *B. cereus* requires the transcription factor PlcR, which controls expression of most *B. cereus* virulence factors following binding to the intracellular AIP derived from the PapR protein (Fig. 3) (Slamti and Lereclus 2002). *papR* is encoded 70 basepairs downstream of *plcR*. PapR is 48 amino acids long and contains an amino-terminal signal peptide that targets it for the secretory pathway (Okstad et al. 1999). Once outside the cell, the PapR pro-AIP is processed by the secreted neutral protease B (NprB) to form the active AIP (Pomerantsev et al. 2009). *nprB* is encoded divergently from *plcR* and *nprB* expression is activated by AIP-bound PlcR (Okstad et al. 1999). NprB cleaves the pro-AIP PapR into peptides of 5, 7, 8, and 11 amino acids in length, all of which are derived from the carboxyl terminus of full-length PapR (Pomerantsev et al. 2009). Only the pentapetide and heptapeptide activate PlcR activity, however; the heptapeptide causes maximal activation and is more prevalent in vivo (Bouillaut et al. 2008; Pomerantsev et al. 2009).

Similar to the *S. aureus* Agr system (see above), there is sequence diversity in the PapR AIPs that enables classification of members of the *B. cereus* group into four pherotypes. Specificity for the PlcR receptor stems from the identities of the first and fifth amino acids in the pentapetides: LPFE(F/Y), VP(F/Y)E(F/Y), MPFEF, and LPFEH (Slamti and Lereclus 2005). There is low cross-reactivity with heterologous PlcRs, indicating that the AIP and its receptor (PlcR) coevolved (Slamti and Lereclus 2005). Another interesting aspect of the pherotypes is that some contain different species, and can therefore communicate across species, whereas in other cases, different isolates of the same species fall into different subgroups and are prevented from communicating with one another (Slamti and Lereclus 2005; Rocha-Estrada et al. 2010).

The processed PapR AIP is imported back into the cell by the oligopeptide permease system (Opp) (Fig. 3) (Gominet et al. 2001). Once inside the cell, the AIP binds to the transcription factor PlcR, and this causes conformational changes in the DNA-binding domain of PlcR, facilitates PlcR oligomerization, DNA binding, and regulation of transcription (Declerck et al. 2007).

When PlcR interacts with the PapR AIP and oligomerizes, it binds to "PlcR boxes" to regulate transcription of target genes (Agaisse et al. 1999). PlcR controls expression of 45 genes, many of which encode extracellular proteins including several enterotoxins, hemolysins, phos-

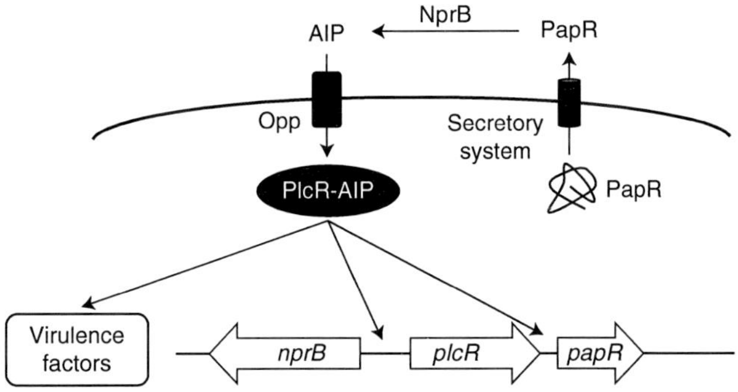

Figure 3. *B. cereus* QS circuit. The pro-AIP PapR is secreted and is then processed to the mature heptapeptide AIP by the extracellular protease NprB. The mature AIP is transported back into the cell by Opp. Intracellular AIP binds to the transcription factor PlcR and activates it. The PlcR–AIP complex regulates virulence factor production and activates expression of *papR*.

pholipases, and proteases (Lereclus et al. 1996; Gohar et al. 2002, 2008). As is the case in all QS systems, AIP-bound PlcR also feedback activates expression of *papR* (Lereclus et al. 1996). In addition to its role in regulating extracellular virulence factors, AIP-bound PlcR has other regulatory targets. These include two-component systems, transport systems, and GGDEF-containing proteins (Gohar et al. 2008). Finally, deletion of *plcR* results in increased biofilm production (Hsueh et al. 2006). However, despite its important role in regulating virulence factor production, disruption of *plcR* does not fully eliminate virulence (Lereclus et al. 2000; Callegan et al. 2003) because several additional systems feed into the *B. cereus* QS circuit and virulence factor regulation. These additional sensory inputs include sporulation through SpoOA ~ P, nutritional state through CodY, motility through FlhA, and other two-component systems (Lereclus et al. 2000; Bouillaut et al. 2005; Brillard et al. 2008; Frenzel et al. 2012).

To date, no anti-PlcR/PapR QS molecules or screens for inhibitors of QS in *B. cereus* have been reported. Analogous to the *S. aureus* anti-QS strategies described above, the *B. cereus* AIPs could certainly be investigated for development into QS inhibitors. The NprB protease, the Opp transporter, and the PlcR receptor are other obvious candidates for exploration of whether or not they are vulnerable to small molecule inhibition.

QS CONTROL OF VIRULENCE IN GRAM-NEGATIVE BACTERIA

LuxI/LuxR Quorum Sensing

Gram-negative bacteria typically use LuxI/LuxR-type QS systems homologous to the first described QS system from the bioluminescent marine symbiotic bacterium *Vibrio fischeri* (Fig. 1C) (Ruby 1996; Hastings and Greenberg 1999). In these systems, the LuxI homolog is an AI synthase that catalyzes a reaction between SAM and an acyl carrier protein (ACP) to produce a freely diffusible acyl homoserine lactone (AHL) AI (Engebrecht and Silverman 1984; More et al. 1996; Schaefer et al. 1996; Ng and Bassler 2009). At high concentrations, AHL AIs bind to cognate cytoplasmic LuxR-like transcription factors. When not bound by AI, LuxR-type proteins are rapidly degraded, presumably to prevent bacteria from "short-circuiting" their QS systems. AI binding stabilizes the LuxR-type proteins, allowing them to fold, bind DNA, and activate transcription of target genes (Engebrecht et al. 1983; Engebrecht and Silverman 1984; Stevens et al. 1994; Zhu and Winans 1999, 2001). Typically, AHL-bound LuxR-type proteins also activate *luxI* expression, forming a feed-forward autoinduction loop that floods the vicinity with AI (Engebrecht et al. 1983; Fuqua and Winans 1994; Seed et al. 1995).

LuxI/LuxR homologs have been identified in more than 100 Gram-negative bacterial species (Case et al. 2008). AHLs produced by different bacteria possess different length side chains and side-chain decorations. Acyl chains ranging from C4 to C18 have been identified with modifications such as carbonyl and hydroxy moieties at the C3 position (Fuqua et al. 2001; Ng and Bassler 2009). This chemical diversity promotes intraspecies-specific bacterial cell–cell communication in that interactions with partner LuxR proteins are usually highly specific. The substrate binding pockets of the various LuxI homologs have correspondingly different sizes and shapes, ensuring accommodation of only a particular acyl-ACP for synthesis of a particular AI (Watson et al. 2002; Gould et al. 2004). Likewise, the AI-detecting LuxR homologs possess unique binding pockets that specifically accommodate particular AHL ligands (Vannini et al. 2002; Zhang et al. 2002; Yao et al. 2006; Bottomley et al. 2007; Chen et al. 2011).

Numerous Gram-negative pathogens control virulence factor production using LuxI/LuxR-type QS circuits. Some examples are LasI/LasR and RhlI/RhlR in *P. aeruginosa*, SmaI/SmaR in *Serratia marcescens*, VjbR (an orphan LuxR homolog) in *Brucella melitensis*, and CviI/CviR in *Chromobacterium violaceum* (Passador et al. 1993; Brint and Ohman 1995; McClean et al. 1997; Thomson et al. 2000; Weeks et al. 2010). We describe the *P. aeruginosa* systems as the canonical example here (reviewed in detail in Williams and Camara 2009).

P. aeruginosa QS

P. aeruginosa is a ubiquitous Gram-negative bacterium that can cause both acute and chronic infections in humans (Chugani and Greenberg 2007; Mena and Gerba 2009). *P. aeruginosa* infection generally depends on the host having a compromised immune system. Typically, *P. aeruginosa* infections are found in the lungs of people with cystic fibrosis (Lyczak et al. 2002; Zemanick et al. 2011). This infection leads to declined pulmonary function and increased mortality. *P. aeruginosa* also causes acute infections in people with compromised epithelial barriers; a common trait among patients with severe burns or tracheal intubation and mechanical ventilation (Bielecki et al. 2008). In both scenarios, *P. aeruginosa* uses QS to collectively produce a suite of virulence factors that contribute to its disease-causing ability.

P. aeruginosa harbors three QS systems: two LuxI/LuxR-type QS circuits that function in series to control expression of virulence factors as well as a third, non-LuxI/LuxR-type system called the *Pseudomonas* quinolone signal (PQS) system (Fig. 4). In the first circuit, the LuxI homolog LasI synthesizes 3-oxo-C12-homoserine lactone (3OC12HSL) (Table 2) (More et al. 1996; Val and Cronan 1998; Parsek et al. 1999; Gould et al. 2004; Bottomley et al. 2007). At HCD, this AI is detected by the cytoplasmic LuxR homolog LasR. The LasR–3OC12HSL complex activates transcription of target genes including those encoding virulence factors such as elastase, proteases, and exotoxin A (Gambello and Iglewski 1991; Gambello et al. 1993; Schuster et al. 2003, 2004). One of the LasR–3OC12HSL targets is *lasI*, which establishes an autoinducing feed-forward loop (Seed et al. 1995). Another target of regulation by LasR–3OC12HSL is a second *luxI* homolog called *rhlI* (Latifi et al. 1996; Pesci et al. 1997). RhlI synthesizes a second AI, butanoyl homoserine lactone (C4HSL) (Table 2) (Ochsner et al. 1994; Pearson et al. 1995). At high concentrations, this AI binds to RhlR, a second LuxR homolog. RhlR–C4HSL activates target genes, including those encoding elastase, proteases, pyocyanin, and siderophores (Schuster et al. 2003; Schuster and Greenberg 2007). Among

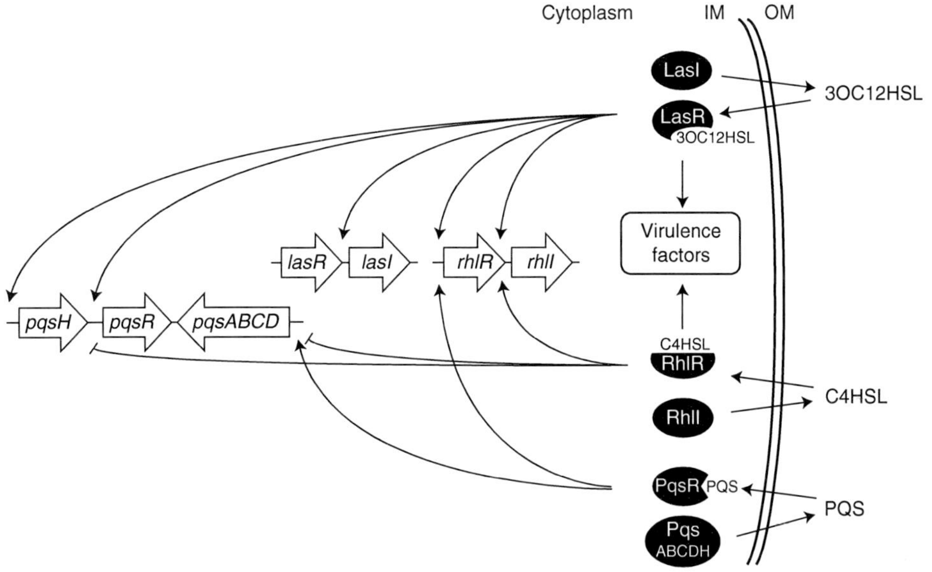

Figure 4. *P. aeruginosa* QS circuits. The three AI synthases, LasI, RhlI, and PqsABCDH, produce the AIs 3OC12HSL, C4HSL, and PQS, respectively. The AIs are detected by the cytoplasmic transcription factors LasR, RhlR, and PqsR, respectively. Each transcription factor regulates expression of its corresponding AI synthase as well as additional targets as indicated by the arrows.

Table 2. *P. aeruginosa* AIs and inhibitors

Receptor/target	Autoinducer	Antagonists	
LasR	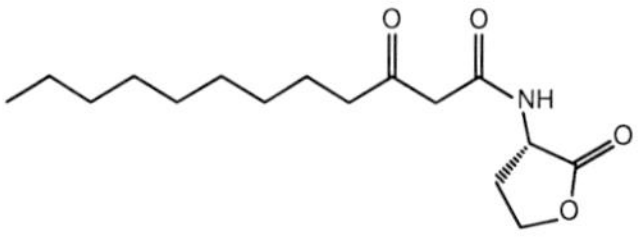	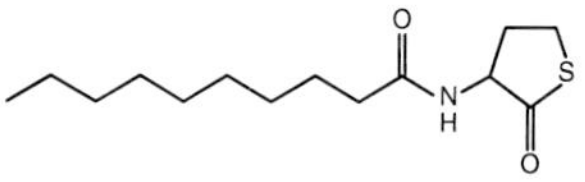	(1)
		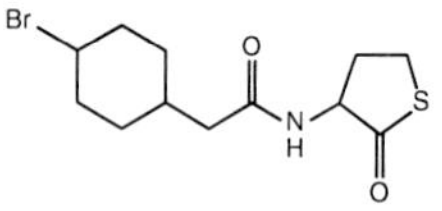	(2)
		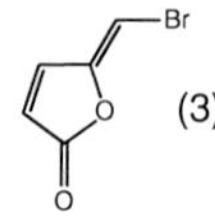 (3) 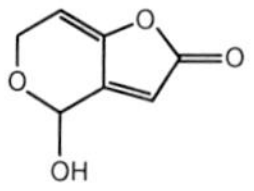	(4)
		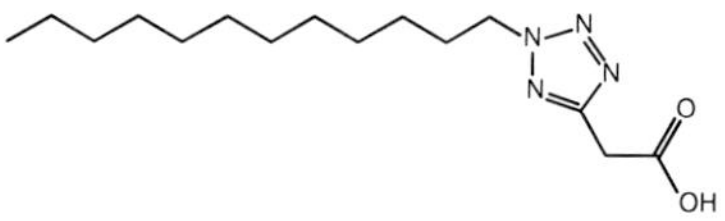	(5)
		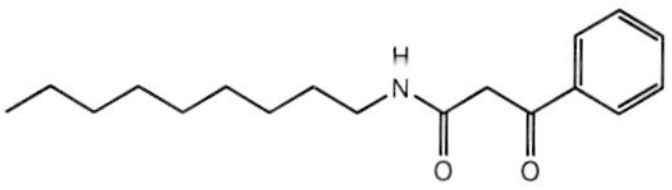	(6)
		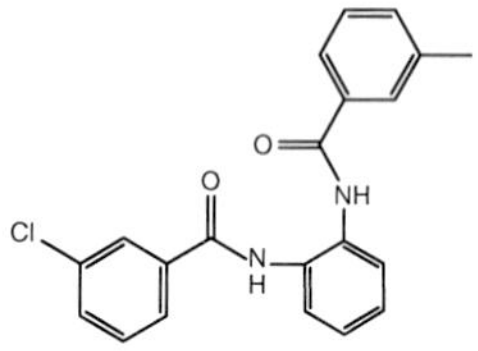	(7)
RhlR	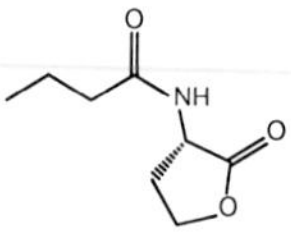		(8)
		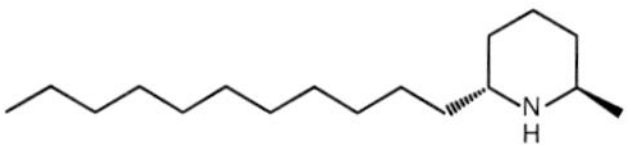	(9)
PqsR	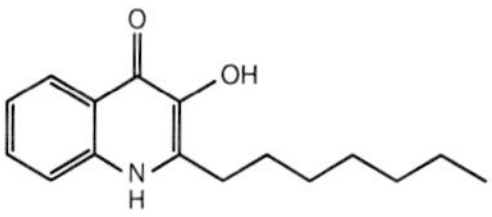		
QscR		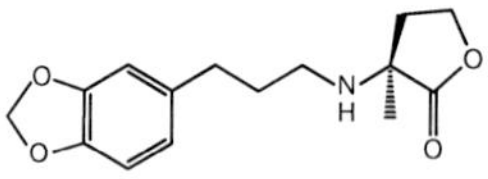	(10)

The three AIs (and their cognate receptors) are 3OC12HSL (LasR), C4HSL (RhlR), and PQS (PqsR). Inhibitors 1 and 2 are modified from 3OC12HSL by McInnis and Blackwell (2011); inhibitor 3 is a furanone described in Hentzer et al. (2002) and Wu et al. (2004); inhibitor 4 is patulin described in Rasmussen et al. (2005); inhibitors 5 and 6 are from a chemical library described in Muh et al. (2006b); inhibitor 7 is a triphenyl derivative from Muh et al. (2006a); inhibitor 8 is C10-acyl-cyclopentylamine from Ishida et al. (2007); inhibitor 9 is solenopsin from Park et al. (2008); and compound 10 is propanoyl homoserine lactone from Mattmann et al. (2008).

its targets is *rhlI*, leading to autoinduction of this second QS circuit.

The crystal structure of the ligand binding domain of LasR bound to 3OC12HSL has been solved, revealing interactions between the receptor protein and the AI (Bottomley et al. 2007). Structures of other LuxR-type QS regulators, TraR, SdiR, and CviR, have also been solved enabling comparative analysis (Vannini et al. 2002; Zhang et al. 2002; Yao et al. 2006; Chen et al. 2011). In all cases, the common lactone head group of each AHL is bound by conserved residues in the binding pocket, whereas the residues forming the pockets for the different acyl tails are diverse. Comparing the LasR binding pocket for 3OC12HSL with the TraR binding pocket for 3OC8HSL shows different sizes and hydrophobic characteristics in the channels that accommodate the different acyl chains. These structures coupled with biochemical data indicate that binding of ligand to the receptor stabilizes folding of the hydrophobic core of the protein. Once folded properly, LasR (and presumably other LuxR-type receptors) can dimerize, bind DNA, and activate transcription (Zhu and Winans 2001; Kiratisin et al. 2002; Schuster et al. 2004; Urbanowski et al. 2004).

P. aeruginosa uses an additional non-LuxI/LuxR QS system to control virulence factor gene expression. PQS, 2-heptyl-3-hydroxi-4-quinolone, is produced by PqsA, PqsB, PqsC, PqsD, and PqsH and is detected by the regulator PqsR (also called MvfR) (Fig. 4 and Table 2). Expression of *pqsH* and *pqsR* is activated by LasR–3OC12HSL, whereas RhlR–C4HSL represses *pqsABCD* and *pqsR* (Gallagher et al. 2002; Deziel et al. 2004; Xiao et al. 2006b). PqsR–PQS autoinduces PQS synthesis and further activates *rhlI* and *rhlR* expression (Xiao et al. 2006a; Diggle et al. 2007). Thus, the PQS circuit is intimately tied to the LasI/LasR and RhlI/RhlR QS systems and, therefore, also influences virulence factor production.

P. aeruginosa QS-activated virulence factors include elastase, proteases, pyocyanin, lectin, swarming motility, rhamnolipids, and toxins. Although distinct regulons for LasR–3OC12HSL and RhlR–C4HSL have been reported, it is apparent that there is extensive overlap between the regulons (Whiteley et al. 1999; Schuster et al. 2003, 2004; Wagner et al. 2004; Schuster and Greenberg 2007). Thus, most genes originally thought to be targeted only by LasR or only by RhlR can in fact be activated by either one. For example, a Δ*lasR* mutant, which is defective for *rhlI* induction, expresses virulence factors originally reported to be LasR-dependent (Dekimpe and Deziel 2009). Apparently, low level *rhlI* and *rhlR* expression promotes the eventual accumulation of C4HSL and autoinduction of the RhlI/RhlR system. Accordingly, activation of virulence factors is delayed in the Δ*lasR* strain. These findings are of medical relevance because many clinical isolates of *P. aeruginosa* possess mutations in *lasR* (Smith et al. 2006).

Another QS-controlled activity in *P. aeruginosa* is biofilm formation. Although regulation of biofilm formation in *P. aeruginosa* largely depends on additional environmental signals, QS regulation of rhamnolipids, swarming motility, and siderophores also contribute to *P. aeruginosa* biofilm formation (Ochsner et al. 1994; Deziel et al. 2003; De Kievit 2008; Patriquin et al. 2008; Rahman et al. 2010).

Up to 10% of the *P. aeruginosa* genome is controlled by QS (Schuster and Greenberg 2006), thus it is not surprising that there are additional levels of regulation impinging on the QS circuits. Vfr, AlgR, PhoR/B, RpoN, RpoS, DksA, RelA, GidA, QscR, RsmY, RsmZ, RsmA, PrrF, PhrS, VqsM, VqsR, and RsaL have all been reported to feed information into the *P. aeruginosa* QS circuit (Albus et al. 1997; de Kievit et al. 1999; Branny et al. 2001; Pessi et al. 2001; van Delden et al. 2001; Heurlier et al. 2003; Ledgham et al. 2003; Medina et al. 2003; Hogardt et al. 2004; Juhas et al. 2004; Dong et al. 2005; Jensen et al. 2006; Morici et al. 2007; Schuster and Greenberg 2007; Gupta et al. 2009; Sonnleitner et al. 2010). These accessory regulators of *P. aeruginosa* QS presumably fine-tune the network so that virulence factor production occurs with optimal precision. Indeed, when AIs are added to exponentially growing *P. aeruginosa*, they do not induce expression of the entire QS regulon (Schuster and Greenberg 2007). Only when cells are exposed to stimuli associated with stationary phase as well as AIs is the complete set of QS targets launched. Although there are several fascinating regulatory events, here we highlight only two posttranscriptional regulation mechanisms.

QscR

QscR is an orphan LuxR homolog that does not have a partner LuxI homolog. QscR can, however, bind to the AI produced by LasI, 3OC12HSL (Lequette et al. 2006; Lintz et al. 2011; Oinuma and Greenberg 2011). Additionally, it forms mixed dimers with LasR and RhlR, rendering them inactive (Ledgham et al. 2003). Thus, QscR likely prevents aberrant QS responses before the cells reaching "a quorum."

sRNAs

Small regulatory RNAs (sRNAs) in bacteria act by basepairing with mRNA targets to repress or activate gene expression (reviewed in Gottesman 2004; Papenfort and Vogel 2010; Storz et al. 2011). In *P. aeruginosa*, the sRNAs PrrF1, PrrF2, and PhrS regulate the PQS system in response to limiting iron (PrrF1 and PrrF2) or limiting oxygen (PhrS) (Wilderman et al. 2004; Sonnleitner et al. 2010; Sonnleitner and Haas 2011). PhrS stimulates translation of an open reading frame in the 5′ end of the *pqsR* mRNA and this action activates *pqsR* translation by altering the secondary structure surrounding the *pqrR* RBS (Sonnleitner et al. 2010). PrrF1 and PrrF2 repress expression of genes encoding enzymes that degrade a precursor of PQS (anthranilate), thus allowing increased PQS under low iron conditions (Oglesby et al. 2008; Oglesby-Sherrouse and Vasil 2010; Sonnleitner and Haas 2011). Two other sRNAs involved in *P. aeruginosa* QS regulation are RsmY and RsmZ, sRNAs that are activated by the GacS/GacA two-component system in response to limiting magnesium and other unknown signals (Kay et al. 2006; Mulcahy and Lewenza 2011; O'Callaghan et al. 2011). RsmY and RsmZ repress the transcription factor RsmA, which, in turn, represses production of both 3OC12HSL and C4HSL (Pessi et al. 2001; Kay et al. 2006).

P. aeruginosa QS as a Therapeutic Target

The *P. aeruginosa* QS circuits make attractive targets for novel antimicrobials because QS controls virulence factor production and no homologs to known QS components exist in humans. This is especially critical in the treatment of persistent infections in cystic fibrosis patients given the resistance of many *P. aeruginosa* isolates to available antibiotics. Small molecule inhibitors of *P. aeruginosa* QS have been extensively reviewed (Mattmann and Blackwell 2010). Here we highlight a few recent advances.

Because LasR sits at the top of the *P. aeruginosa* QS cascade, identifying LasR inhibitors has been a major focus. Competitive inhibitors have been reported that contain modifications to the native 3OC12HSL ligand (Geske et al. 2005, 2007, 2008; Mattmann and Blackwell 2010; McInnis and Blackwell 2011). In this realm, alterations to both the head group and the acyl tail have led to molecules that out-compete 3OC12HSL for binding to LasR (Table 2, compounds 1 and 2). Importantly, some of these modifications, such as substitution of the lactone ring for a thiolactone ring (McInnis and Blackwell 2011), are useful because of their increased stability under biological conditions. Second, natural products have been isolated that inhibit QS by antagonizing LasR (Hentzer et al. 2002; Wu et al. 2004; Rasmussen et al. 2005; Ren et al. 2005; Kim et al. 2008). These inhibitors include furanones and patulin (Table 2, compounds 3 and 4, respectively), which have been further modified to increase their efficacy. Finally, high-throughput screens of small molecule libraries have revealed additional scaffolds for the design of inhibitors (Muh et al. 2006a,b; Borlee et al. 2010) (for example, Table 2, compounds 5–7). A potential complication in targeting LasR is that, as mentioned, some clinical *P. aeruginosa* isolates possess defective LasR proteins (Smith et al. 2006). Nonetheless, establishment of the initial infection is known to be LasI/LasR-dependent, suggesting that, at a minimum, LasR inhibitors could be used as prophylactics.

Other factors involved in *P. aeruginosa* QS have also been targeted for drug discovery. For

 Cite this article as *Cold Spring Harb Perspect Med* doi: 10.1101/cshperspect.a012427

example, inhibitors of LasR have been successfully modified to act as competitive inhibitors of RhlR and in some cases these molecules have proven to be potent inhibitors of both LasR and RhlR (Table 2, compound 8) (Smith et al. 2003; Glansdorp et al. 2004; Ishida et al. 2007). Natural products have been isolated that inhibit RhlR (Table 2, compound 9) (Park et al. 2008).

A final approach is to target regulators that affect both the LasI/LasR and RhlI/RhlR QS systems. For example, a small molecule library was screened for agonists and antagonists of QscR, which, as described, influences both the LasI/LasR and RhlI/RhlR systems (Lee et al. 2006; Mattmann et al. 2008, 2011; Amara et al. 2009; Liu et al. 2010). By agonizing QscR, it could be possible to diminish the overall QS response and thus prevent or delay expression of virulence factors. Interestingly, some non-natural AHLs that target QscR also inhibit LasR, raising the intriguing possibility of a compound that can act broadly to target all of the *P. aeruginosa* QS systems (Table 2, compound 10) (Mattmann and Blackwell 2010).

V. cholerae Quorum Sensing

V. cholerae is the causative agent of the disease cholera (Faruque et al. 1998). This diarrheal disease is endemic in underdeveloped regions and has epidemic and pandemic potential, especially when clean water supplies are compromised. The principal symptom of the disease, profuse watery diarrhea, can lead to dehydration and death if not properly treated. This symptom is caused by an enterotoxin called cholera toxin, expression of which is controlled by QS (Jobling and Holmes 1997; Zhu et al. 2002).

V. cholerae, a Gram-negative bacterium, produces and responds to two AIs using two parallel QS circuits. One AI is (S)-3-hyroxytridecan-4-one (CAI-1), which is synthesized by the CqsA enzyme using SAM and decanoyl-coenzyme A as substrates (Table 3) (Higgins et al. 2007; Ng et al. 2011; Wei et al. 2011). Homologs of CqsA have been identified in all vibrio species consistent with a proposed role for CAI-1 as an intergenus communication molecule, allowing communication among other vibrios harboring *cqsA* (Miller et al. 2002; Henke and Bassler 2004; Higgins et al. 2007). The second AI, AI-2, is synthesized by LuxS. LuxS converts the SAM cycle intermediate *S*-ribosylhomocysteine to 4,5-dihydroxy-2,3-penanedionine (DPD) and homocysteine. DPD spontaneously converts into AI-2 (Table 3) (Surette et al. 1999; Schauder et al. 2001; Chen et al. 2002). Homologs of *luxS* exist in hundreds of Gram-negative and Gram-positive bacteria, consistent with a role for AI-2 in interspecies communication that allows other *luxS*-encoding bacteria in a particular environment to contribute to the overall cell-density information (Federle and Bassler 2003; Xavier and Bassler 2005). Thus, using two different AIs presumably allows *V. cholerae* to detect both the number of other vibrios and the total number of bacteria in the environment.

V. cholerae detects CAI-1 and AI-2 using two parallel membrane-bound two-component receptors (Fig. 5). CqsS detects CAI-1 and LuxPQ detects AI-2 (Bassler et al. 1994; Neiditch et al. 2005, 2006; Ng et al. 2010, 2011). In the absence of ligand, at LCD, both of these receptors function as kinases. Upon autophosphorylation, CqsS and LuxPQ transfer phosphate to the shared response regulator LuxO via the phosphotransfer protein LuxU (Bassler et al. 1994; Freeman and Bassler 1999a,b; Lilley and Bassler 2000). Phosphorylated LuxO activates expression of genes encoding four homologous quorum-regulatory sRNAs, called Qrr1-4 (Lenz et al. 2004). Transcription of the *qrr* sRNA genes is directed by the alternative sigma factor σ^{54}. To date, the four *qrr* promoters and the *luxO* promoter itself are the only known targets for regulation by LuxO.

At LCD, the Qrr sRNAs are expressed and regulate target mRNAs by base-pairing. The principal targets of the Qrr RNAs are the mRNAs encoding the master QS regulators HapR and AphA (Lenz et al. 2004; Rutherford et al. 2011). However, the Qrr sRNAs have opposite effects on these two mRNA targets. The Qrr sRNAs base-pair with the *hapR* mRNA overlapping the RBS, prevent ribosome binding, and facilitate degradation of *hapR* mRNA (Lenz et al. 2004; Tu and Bassler 2007). The Qrr sRNAs activate production of *aphA* by binding to the

Table 3. *V. cholerae* AIs and inhibitors

Receptor/target	Autoinducer	Antagonists
CqsS	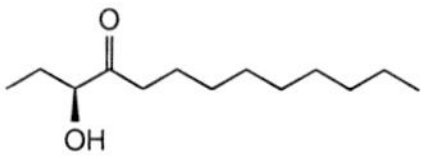	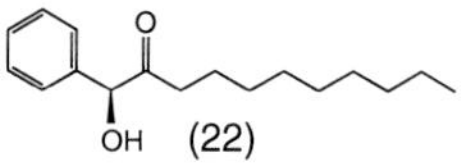 (22) 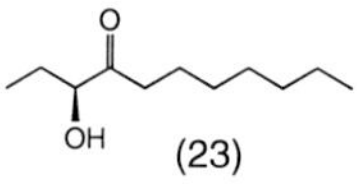 (23)
LuxPQ	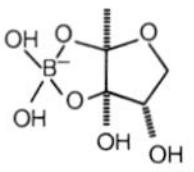	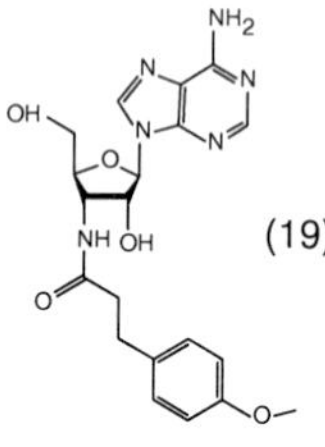 (19)
MTAN		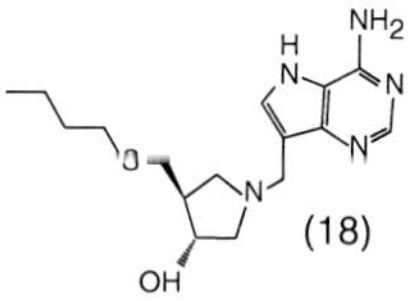 (18)
HapR		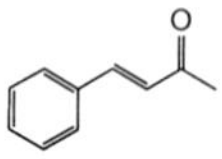 (20) 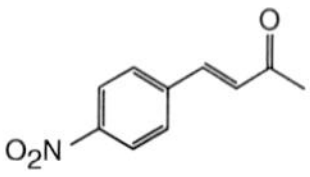 (21)
LuxO		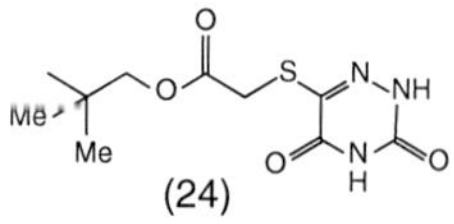 (24)

The autoinducers are CAI-1 and AI-2. Inhibitor 18 is a MTAN transition state analog from Schramm et al. (2008); inhibitor 19 is a modified nucleoside from Brackman et al. (2009); inhibitors 20 and 21 are cinnamaldehydes described in Niu et al. (2006); Brackman et al. (2008, 2011); compounds 22 (phenyl-CAI-1) and 23 (C8-CAI-1) are from Ng et al. (2010); and compound 24 a 6-thio-5-azauracil derivative from Ng et al. (2012).

untranslated region of the *aphA* mRNA and inducing an alternative secondary structure, which reveals the RBS to allow translation (Rutherford et al. 2011; Shao and Bassler 2012). Thus, at LCD, the Qrr sRNAs prevent *hapR* expression and activate *aphA* expression so that HapR levels are low and AphA levels are high.

When CAI-1 and AI-2 accumulate at HCD, they bind CqsS and LuxPQ, respectively, which switches the receptors to phosphatases (Neiditch et al. 2006; Ng et al. 2010). It is not known how AI-2 impinges on in the LuxPQ kinase/phosphatase reaction but CAI-1 functions by inhibiting CqsS autophosphorylation (Wei et al. 2012). CAI-1 has no effect on the phosphotransfer step or the CqsS phosphatase activity (Wei et al. 2012). Once bound by AIs, LuxPQ and CqsS de-phosphorylate LuxO, leading to cessation of *qrr* expression, and the freeing of the *hapR* mRNA for translation. In the absence of Qrr sRNAs, in contrast, the *aphA* mRNA assumes a conformation refractory to

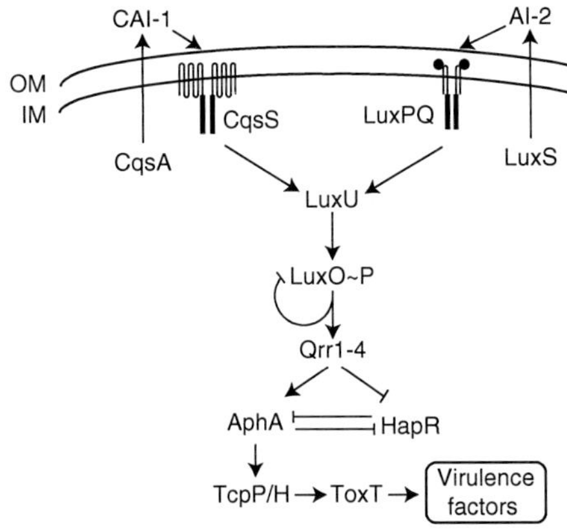

Figure 5. *V. cholerae* QS circuit. *V. cholerae* detects CAI-1 (produced by CqsA) and AI-2 (produced by LuxS) via two two-component histidine kinases, CqsS and LuxPQ, respectively. In the absence of AIs at LCD, the receptors function as kinases resulting in phosphorylation of LuxO, via LuxU, and activation of *qrr* gene expression. The Qrr sRNAs activate *aphA* and repress *hapR* expression. AphA represses *hapR* and activates *tcpPH*. TcpPH activates *toxT* expression and ToxT activates expression of the major virulence factors. At HCD, AIs bind the receptors switching them to phosphatases resulting in dephosphorylation of LuxO, cessation of *qrr* gene expression, and increased *hapR* expression. HapR represses *aphA* to shut down virulence factor production.

ribosome binding. The two master QS regulators are thus produced in reciprocal gradients with AphA highest at LCD and HapR highest at HCD. This dichotomy is reinforced by mutual repression whereby AphA represses *hapR* expression and HapR represses *aphA* expression (Kovacikova and Skorupski 2002; Lin et al. 2007; Pompeani et al. 2008; Rutherford et al. 2011).

Interestingly, *V. cholerae* expresses its suite of virulence factors at LCD (Jobling and Holmes 1997; Zhu et al. 2002). At LCD, AphA, a winged-helix transcription factor, works together with a LysR-type transcription factor, AphB, to activate transcription of *tcpPH* (Kovacikova and Skorupski 1999, 2001; Skorupski and Taylor 1999). TcpP is a *trans*-membrane DNA binding protein, the function of which is enhanced by TcpH (Carroll et al. 1997; Beck et al. 2004; Matson et al. 2007). TcpPH activates expression of *toxT*. ToxT is a transcription factor that activates expression of the genes encoding cholera toxin and the toxin co-regulated pilus, which are the major *V. cholerae* virulence factors (Fig. 5) (Higgins et al. 1992; Withey and DiRita 2006; Matson et al. 2007). Expression of *toxT* is also activated by ToxRS, which is homologous to TcpPH (Peterson and Mekalanos 1988; Skorupski and Taylor 1997; Matson et al. 2007).

Along with other regulators, QS also controls biofilm formation in *V. cholerae* (Hammer and Bassler 2003; Zhu and Mekalanos 2003). Biofilm formation is activated at LCD and repressed at HCD. HapR, expressed at HCD, plays a major role in this aspect of the program. First, HapR directly represses genes encoding components of the biofilm factory (Yildiz et al. 2004). Second, HapR represses expression of two transcription factors that activate biofilm formation genes, VpsR and VpsT (Beyhan et al. 2007; Waters et al. 2008; Shikuma et al. 2009; Srivastava et al. 2011). Third, HapR represses expression of genes required for the synthesis of the second messenger cyclic-di-GMP (Lim et al. 2006, 2007; Waters et al. 2008). Cyclic-di-GMP is sensed through the transcription factor VpsT to activate biofilm formation (Krasteva et al. 2010). This LCD biofilm lifestyle allows *V. cholerae* to remain attached to host tissue when virulence factors are expressed. At HCD, QS repression of biofilms and virulence factor production facilitates dispersal of *V. cholerae* back into the environment. This strategy allows *V. cholerae* to maximally compete for nutrients in the host at LCD and maximize its ability to exit and spread to other hosts once it reaches HCD (Nadell and Bassler 2011). Because a high infective dose is required, the host serves as the incubator, and it appears that when the titer is high enough to ensure infection of a new host, the population is dispersed.

V. cholerae QS Components as Targets for Therapeutics

The *V. cholerae* QS circuit has several possible targets for the discovery or design of novel antimicrobial agents. Some interesting approaches have targeted pathways involved in the synthesis of the AIs. In the SAM cycle, MTAN, which is found only in bacteria, is a hydrolase that turns over methylthioadenosine to maintain the homeostatic SAM pool. This step is critical for the synthesis of both CAI-1 and AI-2 (Schauder et al. 2001; Wei et al. 2011). A synthetic inhibitor of MTAN blocks QS in *V. cholerae* without affecting growth (Table 3, compound 18) (Schramm et al. 2008). This MTAN inhibitor blocks AI-2 production in enterohemorrhagic *Escherichia coli* as well. In a screen for nucleoside analogs able to disrupt AI-2 QS, another molecule was identified that disrupts AI-2-based QS without affecting growth in *V. harveyi*, a close relative of *V. cholerae* (Table 3, compound 19) (Brackman et al. 2009). This compound likely targets the signaling pathway for the AI-2 receptor, LuxPQ, which is present in numerous vibrio species including *V. cholerae* (Fig. 5).

Cinnamaldehyde is a natural product that inhibits QS in vibrios (Niu et al. 2006). Cinnamaldehyde and derivatives are proposed to target HapR and its homologs (Table 3, compounds 20 and 21), thus inhibiting virulence factor production and biofilm formation without dramatically affecting growth (Brackman et al. 2008, 2011). These compounds increase survival of model host nematodes infected with different vibrios leading to the interesting possibility that HapR can be regulated by a cell permeable ligand.

Several synthetic ligands of the receptor CqsS have been identified. Using these molecules has been instrumental in determining the mechanism of signal recognition and discrimination in *V. cholerae* QS. For example, CqsS does not respond to CAI-1 molecules with bulky head groups (Table 3, compound 22) or shortened tail groups (Table 3, compound 23) (Ng et al. 2010). Indeed, such molecules are CqsS antagonists. Owing to the fact that *V. cholerae* expresses virulence factors at LCD in the absence of AIs, CqsS represents an interesting target because compounds that prematurely trigger QS (i.e., potent CqsS agonists) could prevent virulence factor expression.

Finally, in addition to the receptors and the transcription factor HapR, intermediate components in the QS circuit are targets for therapeutics. For example, a high-throughput chemical screen led to the identification of a LuxO antagonist (Ng et al. 2012). The molecule and potent analogs are all 6-thio-5-azauracil derivatives (Table 3, compound 24, for example). This family of molecules uncompetitively bind the preformed LuxO–ATP complex and inhibit ATP hydrolysis, which prevents transcription activation. When inhibited, LuxO is not able to activate expression of the *qrr* sRNA genes, thus locking *V. cholerae* into HCD mode and preventing expression of virulence factors (Fig. 5). Because LuxO exists in all vibrio QS circuits, this family of molecules has broad spectrum antivirulence activity in marine vibiros (Ng et al. 2012).

CONCLUDING REMARKS

QS is a vital regulatory mechanism used by many bacteria to control collective traits that allow bacteria to exploit particular niches. For example, QS enables access of symbionts to nutrient-rich environments in hosts (Ruby 1996). Bacterial populations use QS to control biofilm formation, which provides members of the population superior access nutrients and thus enables them to out-compete non-biofilm-producing neighbors (Nadell and Bassler 2011). Finally, as discussed in the above examples, bacteria that make their living by exploiting eukaryotic hosts have coupled production of the virulence factors necessary for a pathogenic lifestyle to their ability to detect changes in cell population density.

Four trends emerge when considering the examples described above. First, it is clear that QS gene regulation is the result of regulatory networks layered onto the basic AI production and detection apparatuses. Because of the global genetic programs controlled by QS, these additional regulatory mechanisms likely ensure that altered gene expression occurs only under precisely defined conditions. Alternatively, given that QS regulates global gene expression patterns, perhaps other regulatory systems have co-opted QS systems to control additional products in response to particular environmental conditions. Second, it is interesting that in the cases described here, as well as in other cases, QS is linked to biofilm production. Although the mechanism and timing of production differ among these species, in each case, QS plays a role in dictating when this process occurs. Members of biofilm communities most certainly benefit from the ability to communicate and, likewise, bacterial communication is probably more robust when the cells are in the intimate associations provided by biofilms. Third, there appears to be a prominent role for regulation involving sRNAs. QS controls expression of hundreds of genes and a commitment to a lifestyle that could be detrimental to the individual. One of the general benefits of regulation by sRNAs is the rapidity with which regulation can occur because translation and protein folding are not required to produce the regulator. Thus, it is possible that QS systems rely heavily on sRNAs to quickly adjust global gene expression programs as bacteria transition between disparate niches. Finally, small molecules analogous to those involved in QS are increasingly becoming recognized as key modulators of pathogenic behaviors and important players in biofilm formation in bacteria. Thus, strategies aimed at manipulating small molecule control of bacterial behaviors are now viewed as especially promising. Presumably, therapies that affect bacterial behavior will not be as prone to resistance as are the targets of traditional antibiotics that result in outright killing of bacteria or inhibition of their growth. Thus, therapeutics that interfere with small molecule-controlled pathways could have longer functional shelf lives than second and third generation antibiotics.

ACKNOWLEDGMENTS

We thank members of the Bassler Laboratory for insightful discussion and especially Dr. Wai-Leung Ng, Dr. Julia van Kessel, and Ms. Colleen O'Loughlin for helpful suggestions. S.T.R. is supported by an NRSA postdoctoral fellowship F32AI085922 and work in the Bassler Laboratory is supported by the Howard Hughes Medical Institute, National Institutes of Health (NIH) grant 5R01GM065859, NIH grant 5R01AI054442, and National Science Foundation (NSF) grant MCB-0343821.

REFERENCES

Agaisse H, Gominet M, Okstad OA, Kolsto AB, Lereclus D. 1999. PlcR is a pleiotropic regulator of extracellular virulence factor gene expression in *Bacillus thuringiensis*. *Mol Microbiol* **32:** 1043–1053.

Albus AM, Pesci EC, Runyen-Janecky LJ, West SE, Iglewski BH. 1997. Vfr controls quorum sensing in *Pseudomonas aeruginosa*. *J Bacteriol* **179:** 3928–3935.

Amara N, Mashiach R, Amar D, Krief P, Spieser SA, Bottomley MJ, Aharoni A, Meijler MM. 2009. Covalent inhibition of bacterial quorum sensing. *J Am Chem Soc* **131:** 10610–10619.

Autret N, Raynaud C, Dubail I, Berche P, Charbit A. 2003. Identification of the *agr* locus of *Listeria monocytogenes*: Role in bacterial virulence. *Infect Immun* **71:** 4463–4471.

Bassler BL, Wright M, Silverman MR. 1994. Multiple signalling systems controlling expression of luminescence in *Vibrio harveyi*: Sequence and function of genes encoding a second sensory pathway. *Mol Microbiol* **13:** 273–286.

Beck NA, Krukonis ES, DiRita VJ. 2004. TcpH influences virulence gene expression in *Vibrio cholerae* by inhibiting degradation of the transcription activator TcpP. *J Bacteriol* **186:** 8309–8316.

Beyhan S, Bilecen K, Salama SR, Casper-Lindley C, Yildiz FH. 2007. Regulation of rugosity and biofilm formation in *Vibrio cholerae*: Comparison of VpsT and VpsR regulons and epistasis analysis of vpsT, vpsR, and hapR. *J Bacteriol* **189:** 388–402.

Bielecki P, Glik J, Kawecki M, Martins dos Santos VA. 2008. Towards understanding *Pseudomonas aeruginosa* burn wound infections by profiling gene expression. *Biotechnol Lett* **30:** 777–790.

Boles BR, Horswill AR. 2008. Agr-mediated dispersal of *Staphylococcus aureus* biofilms. *PLoS Pathog* **4:** e1000052.

Borlee BR, Geske GD, Blackwell HE, Handelsman J. 2010. Identification of synthetic inducers and inhibitors of the quorum-sensing regulator LasR in *Pseudomonas aeruginosa* by high-throughput screening. *Appl Environ Microbiol* **76:** 8255–8258.

Bottomley MJ, Muraglia E, Bazzo R, Carfi A. 2007. Molecular insights into quorum sensing in the human pathogen *Pseudomonas aeruginosa* from the structure of the virulence regulator LasR bound to its autoinducer. *J Biol Chem* **282:** 13592–13600.

Bottone EJ. 2010. *Bacillus cereus*, a volatile human pathogen. *Clin Microbiol Rev* **23:** 382–398.

Bouillaut L, Ramarao N, Buisson C, Gilois N, Gohar M, Lereclus D, Nielsen-Leroux C. 2005. FlhA influences *Bacillus thuringiensis* PlcR-regulated gene transcription, protein production, and virulence. *Appl Environ Microbiol* **71:** 8903–8910.

Bouillaut L, Perchat S, Arold S, Zorrilla S, Slamti L, Henry C, Gohar M, Declerck N, Lereclus D. 2008. Molecular basis for group-specific activation of the virulence regulator PlcR by PapR heptapeptides. *Nucleic Acids Res* **36:** 3791–3801.

Brackman G, Defoirdt T, Miyamoto C, Bossier P, Van Calenbergh S, Nelis H, Coenye T. 2008. Cinnamaldehyde and cinnamaldehyde derivatives reduce virulence in Vibrio spp. by decreasing the DNA-binding activity of the quorum sensing response regulator LuxR. *BMC Microbiol* **8:** 149.

Brackman G, Celen S, Baruah K, Bossier P, Van Calenbergh S, Nelis HJ, Coenye T. 2009. AI-2 quorum-sensing inhibitors affect the starvation response and reduce virulence in several Vibrio species, most likely by interfering with LuxPQ. *Microbiology* **155:** 4114–4122.

Brackman G, Celen S, Hillaert U, Van Calenbergh S, Cos P, Maes L, Nelis HJ, Coenye T. 2011. Structure-activity relationship of cinnamaldehyde analogs as inhibitors of AI-2 based quorum sensing and their effect on virulence of Vibrio spp. *PLoS ONE* **6:** e16084.

Branny P, Pearson JP, Pesci EC, Kohler T, Iglewski BH, Van Delden C. 2001. Inhibition of quorum sensing by a *Pseudomonas aeruginosa dksA* homologue. *J Bacteriol* **183:** 1531–1539.

Brillard J, Susanna K, Michaud C, Dargaignaratz C, Gohar M, Nielsen-Leroux C, Ramarao N, Kolsto AB, Nguyen-the C, Lereclus D, et al. 2008. The YvfTU two-component system is involved in *plcR* expression in *Bacillus cereus*. *BMC Microbiol* **8:** 183.

Brint JM, Ohman DE. 1995. Synthesis of multiple exoproducts in *Pseudomonas aeruginosa* is under the control of RhlR-RhlI, another set of regulators in strain PAO1 with homology to the autoinducer-responsive LuxR-LuxI family. *J Bacteriol* **177:** 7155–7163.

Brumfitt W, Hamilton-Miller J. 1989. Methicillin-resistant *Staphylococcus aureus. N Engl J Med* **320:** 1188–1196.

Callegan MC, Kane ST, Cochran DC, Gilmore MS, Gominet M, Lereclus D. 2003. Relationship of *plcR*-regulated factors to *Bacillus endophthalmitis* virulence. *Infect Immun* **71:** 3116–3124.

Carroll PA, Tashima KT, Rogers MB, DiRita VJ, Calderwood SB. 1997. Phase variation in *tcpH* modulates expression of the ToxR regulon in *Vibrio cholerae. Mol Microbiol* **25:** 1099–1111.

Case RJ, Labbate M, Kjelleberg S. 2008. AHL-driven quorum-sensing circuits: their frequency and function among the Proteobacteria. *Isme J* **2:** 345–349.

Chen X, Schauder S, Potier N, Van Dorsselaer A, Pelczer I, Bassler BL, Hughson FM. 2002. Structural identification of a bacterial quorum-sensing signal containing boron. *Nature* **415:** 545–549.

Chen G, Swem LR, Swem DL, Stauff DL, O'Loughlin CT, Jeffrey PD, Bassler BL, Hughson FM. 2011. A strategy for antagonizing quorum sensing. *Mol Cell* **42:** 199–209.

Cheung AL, Projan SJ. 1994. Cloning and sequencing of sarA of *Staphylococcus aureus*, a gene required for the expression of *agr. J Bacteriol* **176:** 4168–4172.

Chugani S, Greenberg EP. 2007. The influence of human respiratory epithelia on *Pseudomonas aeruginosa* gene expression. *Microb Pathog* **42:** 29–35.

Declerck N, Bouillaut L, Chaix D, Rugani N, Slamti L, Hoh F, Lereclus D, Arold ST. 2007. Structure of PlcR: Insights into virulence regulation and evolution of quorum sensing in Gram-positive bacteria. *Proc Natl Acad Sci* **104:** 18490–18495.

De Kievit TR. 2008. Quorum sensing in *Pseudomonas aeruginosa* biofilms. *Environ Microbiol* **11:** 279–288.

de Kievit T, Seed PC, Nezezon J, Passador L, Iglewski BH. 1999. RsaL, a novel repressor of virulence gene expression in *Pseudomonas aeruginosa. J Bacteriol* **181:** 2175–2184.

Dekimpe V, Deziel E. 2009. Revisiting the quorum-sensing hierarchy in *Pseudomonas aeruginosa*: The transcriptional regulator RhlR regulates LasR-specific factors. *Microbiology* **155:** 712–723.

Deziel E, Lepine F, Milot S, Villemur R. 2003. rhlA is required for the production of a novel biosurfactant promoting swarming motility in *Pseudomonas aeruginosa*: 3-(3-hydroxyalkanoyloxy)alkanoic acids (HAAs), the precursors of rhamnolipids. *Microbiology* **149:** 2005–2013.

Deziel E, Lepine F, Milot S, He J, Mindrinos MN, Tompkins RG, Rahme LG. 2004. Analysis of *Pseudomonas aeruginosa* 4-hydroxy-2-alkylquinolines (HAQs) reveals a role for 4-hydroxy-2-heptylquinoline in cell-to-cell communication. *Proc Natl Acad Sci* **101:** 1339–1344.

Diggle SP, Matthijs S, Wright VJ, Fletcher MP, Chhabra SR, Lamont IL, Kong X, Hider RC, Cornelis P, Camara M, et al. 2007. The *Pseudomonas aeruginosa* 4-quinolone signal molecules HHQ and PQS play multifunctional roles in quorum sensing and iron entrapment. *Chem Biol* **14:** 87–96.

Dong YH, Zhang XF, Xu JL, Tan AT, Zhang LH. 2005. VqsM, a novel AraC-type global regulator of quorum-sensing signalling and virulence in *Pseudomonas aeruginosa. Mol Microbiol* **58:** 552–564.

Dufour P, Jarraud S, Vandenesch F, Greenland T, Novick RP, Bes M, Etienne J, Lina G. 2002. High genetic variability of the *agr* locus in Staphylococcus species. *J Bacteriol* **184:** 1180–1186.

Dunman PM, Murphy E, Haney S, Palacios D, Tucker-Kellogg G, Wu S, Brown EL, Zagursky RJ, Shlaes D, Projan SJ. 2001. Transcription profiling-based identification of *Staphylococcus aureus* genes regulated by the *agr* and/or *sarA* loci. *J Bacteriol* **183:** 7341–7353.

Dunny GM. 2007. The peptide pheromone-inducible conjugation system of *Enterococcus faecalis* plasmid pCF10: Cell-cell signalling, gene transfer, complexity and evolution. *Philos Trans R Soc Lond B Biol Sci* **362:** 1185–1193.

Engebrecht J, Silverman M. 1984. Identification of genes and gene products necessary for bacterial bioluminescence. *Proc Natl Acad Sci* **81:** 4154–4158.

Engebrecht J, Nealson K, Silverman M. 1983. Bacterial bioluminescence: Isolation and genetic analysis of functions from *Vibrio fischeri. Cell* **32:** 773–781.

Faruque SM, Albert MJ, Mekalanos JJ. 1998. Epidemiology, genetics, and ecology of toxigenic *Vibrio cholerae. Microbiol Mol Biol Rev* **62:** 1301–1314.

Federle MJ, Bassler BL. 2003. Interspecies communication in bacteria. *J Clin Invest* **112:** 1291–1299.

Fleming V, Feil E, Sewell AK, Day N, Buckling A, Massey RC. 2006. Agr interference between clinical *Staphylococcus aureus* strains in an insect model of virulence. *J Bacteriol* **188:** 7686–7688.

Freeman JA, Bassler BL. 1999a. A genetic analysis of the function of LuxO, a two-component response regulator involved in quorum sensing in *Vibrio harveyi. Mol Microbiol* **31:** 665–677.

Freeman JA, Bassler BL. 1999b. Sequence and function of LuxU: A two-component phosphorelay protein that regulates quorum sensing in *Vibrio harveyi*. *J Bacteriol* **181:** 899–906.

Frenzel E, Doll V, Pauthner M, Lucking G, Scherer S, Ehling-Schulz M. 2012. CodY orchestrates the expression of virulence determinants in emetic *Bacillus cereus* by impacting key regulatory circuits. *Mol Microbiol* **85:** 67–88.

Fuqua WC, Winans SC. 1994. A LuxR-LuxI type regulatory system activates Agrobacterium Ti plasmid conjugal transfer in the presence of a plant tumor metabolite. *J Bacteriol* **176:** 2796–2806.

Fuqua C, Parsek MR, Greenberg EP. 2001. Regulation of gene expression by cell-to-cell communication: Acyl-homoserine lactone quorum sensing. *Annu Rev Genet* **35:** 439–468.

Gallagher LA, McKnight SL, Kuznetsova MS, Pesci EC, Manoil C. 2002. Functions required for extracellular quinolone signaling by *Pseudomonas aeruginosa*. *J Bacteriol* **184:** 6472–6480.

Gambello MJ, Iglewski BH. 1991. Cloning and characterization of the *Pseudomonas aeruginosa lasR* gene, a transcriptional activator of elastase expression. *J Bacteriol* **173:** 3000–3009.

Gambello MJ, Kaye S, Iglewski BH. 1993. LasR of *Pseudomonas aeruginosa* is a transcriptional activator of the alkaline protease gene (*apr*) and an enhancer of exotoxin A expression. *Infect Immun* **61:** 1180–1184.

Geisinger E, Adhikari RP, Jin R, Ross HF, Novick RP. 2006. Inhibition of *rot* translation by RNAIII, a key feature of *agr* function. *Mol Microbiol* **61:** 1038–1048.

Geisinger E, Muir TW, Novick RP. 2009. agr receptor mutants reveal distinct modes of inhibition by staphylococcal autoinducing peptides. *Proc Natl Acad Sci* **106:** 1216–1221.

Geske GD, Wezeman RJ, Siegel AP, Blackwell HE. 2005. Small molecule inhibitors of bacterial quorum sensing and biofilm formation. *J Am Chem Soc* **127:** 12762–12763.

Geske GD, O'Neill JC, Miller DM, Mattmann ME, Blackwell HE. 2007. Modulation of bacterial quorum sensing with synthetic ligands: Systematic evaluation of *N*-acylated homoserine lactones in multiple species and new insights into their mechanisms of action. *J Am Chem Soc* **129:** 13613–13625.

Geske GD, Mattmann ME, Blackwell HE. 2008. Evaluation of a focused library of *N*-aryl L-homoserine lactones reveals a new set of potent quorum sensing modulators. *Bioorg Med Chem Lett* **18:** 5978–5981.

Glansdorp FG, Thomas GL, Lee JK, Dutton JM, Salmond GP, Welch M, Spring DR. 2004. Synthesis and stability of small molecule probes for *Pseudomonas aeruginosa* quorum sensing modulation. *Org Biomol Chem* **2:** 3329–3336.

Gohar M, Okstad OA, Gilois N, Sanchis V, Kolsto AB, Lereclus D. 2002. Two-dimensional electrophoresis analysis of the extracellular proteome of *Bacillus cereus* reveals the importance of the PlcR regulon. *Proteomics* **2:** 784–791.

Gohar M, Faegri K, Perchat S, Ravnum S, Okstad OA, Gominet M, Kolsto AB, Lereclus D. 2008. The PlcR virulence regulon of *Bacillus cereus*. *PLoS ONE* **3:** e2793.

Gominet M, Slamti L, Gilois N, Rose M, Lereclus D. 2001. Oligopeptide permease is required for expression of the *Bacillus thuringiensis plcR* regulon and for virulence. *Mol Microbiol* **40:** 963–975.

Gottesman S. 2004. The small RNA regulators of *Escherichia coli*: Roles and mechanisms. *Annu Rev Microbiol* **58:** 303–328.

Gould TA, Schweizer HP, Churchill ME. 2004. Structure of the *Pseudomonas aeruginosa* acyl-homoserinelactone synthase LasI. *Mol Microbiol* **53:** 1135–1146.

Gupta R, Gobble TR, Schuster M. 2009. GidA posttranscriptionally regulates *rhl* quorum sensing in *Pseudomonas aeruginosa*. *J Bacteriol* **191:** 5785–5792.

Hammer BK, Bassler BL. 2003. Quorum sensing controls biofilm formation in *Vibrio cholerae*. *Mol Microbiol* **50:** 101–104.

Hastings JW, Greenberg EP. 1999. Quorum sensing: The explanation of a curious phenomenon reveals a common characteristic of bacteria. *J Bacteriol* **181:** 2667–2668.

Havarstein LS, Coomaraswamy G, Morrison DA. 1995. An unmodified heptadecapeptide pheromone induces competence for genetic transformation in *Streptococcus pneumoniae*. *Proc Natl Acad Sci* **92:** 11140–11144.

Henke JM, Bassler BL. 2004. Three parallel quorum-sensing systems regulate gene expression in *Vibrio harveyi*. *J Bacteriol* **186:** 6902–6914.

Hentzer M, Riedel K, Rasmussen TB, Heydorn A, Andersen JB, Parsek MR, Rice SA, Eberl L, Molin S, Hoiby N, et al. 2002. Inhibition of quorum sensing in *Pseudomonas aeruginosa* biofilm bacteria by a halogenated furanone compound. *Microbiology* **148:** 87–102.

Heurlier K, Denervaud V, Pessi G, Reimmann C, Haas D. 2003. Negative control of quorum sensing by RpoN (sigma54) in *Pseudomonas aeruginosa* PAO1. *J Bacteriol* **185:** 2227–2235.

Higgins DE, Nazareno E, DiRita VJ. 1992. The virulence gene activator ToxT from *Vibrio cholerae* is a member of the AraC family of transcriptional activators. *J Bacteriol* **174:** 6974–6980.

Higgins DA, Pomianek ME, Kraml CM, Taylor RK, Semmelhack MF, Bassler BL. 2007. The major *Vibrio cholerae* autoinducer and its role in virulence factor production. *Nature* **450:** 883–886.

Hoch JA, Silhavy TJ. 1995. *Two-component signal transduction*. ASM Press, Washington, DC.

Hogardt M, Roeder M, Schreff AM, Eberl L, Heesemann J. 2004. Expression of *Pseudomonas aeruginosa exoS* is controlled by quorum sensing and RpoS. *Microbiology* **150:** 843–851.

Hsueh YH, Somers EB, Lereclus D, Wong AC. 2006. Biofilm formation by *Bacillus cereus* is influenced by PlcR, a pleiotropic regulator. *Appl Environ Microbiol* **72:** 5089–5092.

Inouye M, Dutta R. 2003. *Histidine kinases in signal transduction*. Academic, London.

Ishida T, Ikeda T, Takiguchi N, Kuroda A, Ohtake H, Kato J. 2007. Inhibition of quorum sensing in *Pseudomonas aeruginosa* by *N*-acyl cyclopentylamides. *Appl Environ Microbiol* **73:** 3183–3188.

Janzon L, Arvidson S. 1990. The role of the δ-lysin gene (*hld*) in the regulation of virulence genes by the accessory gene regulator (*agr*) in *Staphylococcus aureus*. *EMBO J* **9:** 1391–1399.

Jarraud S, Lyon GJ, Figueiredo AM, Gerard L, Vandenesch F, Etienne J, Muir TW, Novick RP. 2000. Exfoliatin-producing strains define a fourth *agr* specificity group in *Staphylococcus aureus*. *J Bacteriol* **182:** 6517–6522.

Jensen V, Lons D, Zaoui C, Bredenbruch F, Meissner A, Dieterich G, Munch R, Haussler S. 2006. RhlR expression in *Pseudomonas aeruginosa* is modulated by the Pseudomonas quinolone signal via PhoB-dependent and -independent pathways. *J Bacteriol* **188:** 8601–8606.

Ji G, Beavis R, Novick RP. 1997. Bacterial interference caused by autoinducing peptide variants. *Science* **276:** 2027–2030.

Ji G, Beavis RC, Novick RP. 1995. Cell density control of staphylococcal virulence mediated by an octapeptide pheromone. *Proc Natl Acad Sci* **92:** 12055–12059.

Jobling MG, Holmes RK. 1997. Characterization of hapR, a positive regulator of the *Vibrio cholerae* HA/protease gene *hap*, and its identification as a functional homologue of the *Vibrio harveyi luxR* gene. *Mol Microbiol* **26:** 1023–1034.

Juhas M, Wiehlmann L, Huber B, Jordan D, Lauber J, Salunkhe P, Limpert AS, von Gotz F, Steinmetz I, Eberl L, et al. 2004. Global regulation of quorum sensing and virulence by VqsR in *Pseudomonas aeruginosa*. *Microbiology* **150:** 831–841.

Kalkum M, Lyon GJ, Chait BT. 2003. Detection of secreted peptides by using hypothesis-driven multistage mass spectrometry. *Proc Natl Acad Sci* **100:** 2795–2800.

Kaplan HB, Greenberg EP. 1985. Diffusion of autoinducer is involved in regulation of the *Vibrio fischeri* luminescence system. *J Bacteriol* **163:** 1210–1214.

Kay E, Humair B, Denervaud V, Riedel K, Spahr S, Eberl L, Valverde C, Haas D. 2006. Two GacA-dependent small RNAs modulate the quorum-sensing response in *Pseudomonas aeruginosa*. *J Bacteriol* **188:** 6026–6033.

Kim C, Kim J, Park HY, Park HJ, Lee JH, Kim CK, Yoon J. 2008. Furanone derivatives as quorum-sensing antagonists of *Pseudomonas aeruginosa*. *Appl Microbiol Biotechnol* **80:** 37–47.

Kiratisin P, Tucker KD, Passador L. 2002. LasR, a transcriptional activator of *Pseudomonas aeruginosa* virulence genes, functions as a multimer. *J Bacteriol* **184:** 4912–4919.

Kleerebezem M, Quadri LE, Kuipers OP, de Vos WM. 1997. Quorum sensing by peptide pheromones and two-component signal-transduction systems in Gram-positive bacteria. *Mol Microbiol* **24:** 895–904.

Kovacikova G, Skorupski K. 1999. A *Vibrio cholerae* LysR homolog, AphB, cooperates with AphA at the *tcpPH* promoter to activate expression of the ToxR virulence cascade. *J Bacteriol* **181:** 4250–4256.

Kovacikova G, Skorupski K. 2001. Overlapping binding sites for the virulence gene regulators AphA, AphB and cAMP-CRP at the *Vibrio cholerae tcpPH* promoter. *Mol Microbiol* **41:** 393–407.

Kovacikova G, Skorupski K. 2002. Regulation of virulence gene expression in *Vibrio cholerae* by quorum sensing: HapR functions at the *aphA* promoter. *Mol Microbiol* **46:** 1135–1147.

Krasteva PV, Fong JC, Shikuma NJ, Beyhan S, Navarro MV, Yildiz FH, Sondermann H. 2010. *Vibrio cholerae* VpsT regulates matrix production and motility by directly sensing cyclic di-GMP. *Science* **327:** 866–868.

Latifi A, Foglino M, Tanaka K, Williams P, Lazdunski A. 1996. A hierarchical quorum-sensing cascade in *Pseudomonas aeruginosa* links the transcriptional activators LasR and RhIR (VsmR) to expression of the stationary-phase sigma factor RpoS. *Mol Microbiol* **21:** 1137–1146.

Lauderdale KJ, Boles BR, Cheung AL, Horswill AR. 2009. Interconnections between Sigma B, agr, and proteolytic activity in *Staphylococcus aureus* biofilm maturation. *Infect Immun* **77:** 1623–1635.

Lazazzera BA. 2001. The intracellular function of extracellular signaling peptides. *Peptides* **22:** 1519–1527.

Ledgham F, Ventre I, Soscia C, Foglino M, Sturgis JN, Lazdunski A. 2003. Interactions of the quorum sensing regulator QscR: interaction with itself and the other regulators of *Pseudomonas aeruginosa* LasR and RhlR. *Mol Microbiol* **48:** 199–210.

Lee JH, Lequette Y, Greenberg EP. 2006. Activity of purified QscR, a *Pseudomonas aeruginosa* orphan quorum-sensing transcription factor. *Mol Microbiol* **59:** 602–609.

Lenz DH, Mok KC, Lilley BN, Kulkarni RV, Wingreen NS, Bassler BL. 2004. The small RNA chaperone Hfq and multiple small RNAs control quorum sensing in *Vibrio harveyi* and *Vibrio cholerae*. *Cell* **118:** 69–82.

Lequette Y, Lee JH, Ledgham F, Lazdunski A, Greenberg EP. 2006. A distinct QscR regulon in the *Pseudomonas aeruginosa* quorum-sensing circuit. *J Bacteriol* **188:** 3365–3370.

Lereclus D, Agaisse H, Gominet M, Salamitou S, Sanchis V. 1996. Identification of a *Bacillus thuringiensis* gene that positively regulates transcription of the phosphatidylinositol-specific phospholipase C gene at the onset of the stationary phase. *J Bacteriol* **178:** 2749–2756.

Lereclus D, Agaisse H, Grandvalet C, Salamitou S, Gominet M. 2000. Regulation of toxin and virulence gene transcription in *Bacillus thuringiensis*. *Int J Med Microbiol* **290:** 295–299.

Liang X, Zheng L, Landwehr C, Lunsford D, Holmes D, Ji Y. 2005. Global regulation of gene expression by ArlRS, a two-component signal transduction regulatory system of *Staphylococcus aureus*. *J Bacteriol* **187:** 5486–5492.

Lilley BN, Bassler BL. 2000. Regulation of quorum sensing in *Vibrio harveyi* by LuxO and sigma-54. *Mol Microbiol* **36:** 940–954.

Lim B, Beyhan S, Meir J, Yildiz FH. 2006. Cyclic-diGMP signal transduction systems in *Vibrio cholerae*: Modulation of rugosity and biofilm formation. *Mol Microbiol* **60:** 331–348.

Lim B, Beyhan S, Yildiz FH. 2007. Regulation of Vibrio polysaccharide synthesis and virulence factor production by CdgC, a GGDEF-EAL domain protein, in *Vibrio cholerae*. *J Bacteriol* **189:** 717–729.

Lin W, Kovacikova G, Skorupski K. 2007. The quorum sensing regulator HapR downregulates the expression of the virulence gene transcription factor AphA in *Vibrio cholerae* by antagonizing Lrp- and VpsR-mediated activation. *Mol Microbiol* **64:** 953–967.

Lina G, Jarraud S, Ji G, Greenland T, Pedraza A, Etienne J, Novick RP, Vandenesch F. 1998. Transmembrane topology and histidine protein kinase activity of AgrC, the *agr* signal receptor in *Staphylococcus aureus*. *Mol Microbiol* **28:** 655–662.

Lintz MJ, Oinuma K, Wysoczynski CL, Greenberg EP, Churchill ME. 2011. Crystal structure of QscR, a *Pseudomonas aeruginosa* quorum sensing signal receptor. *Proc Natl Acad Sci* **108:** 15763–15768.

Liu HB, Lee JH, Kim JS, Park S. 2010. Inhibitors of the *Pseudomonas aeruginosa* quorum-sensing regulator QscR. *Biotechnol Bioeng* **106:** 119–126.

Lowy FD. 1998. *Staphylococcus aureus* infections. *N Engl J Med* **339:** 520–532.

Lyczak JB, Cannon CL, Pier GB. 2002. Lung infections associated with cystic fibrosis. *Clin Microbiol Rev* **15:** 194–222.

Lyon GJ, Mayville P, Muir TW, Novick RP. 2000. Rational design of a global inhibitor of the virulence response in *Staphylococcus aureus*, based in part on localization of the site of inhibition to the receptor-histidine kinase AgrC. *Proc Natl Acad Sci* **97:** 13330–13335.

Lyon GJ, Wright JS, Muir TW, Novick RP. 2002. Key determinants of receptor activation in the *agr* autoinducing peptides of *Staphylococcus aureus*. *Biochemistry* **41:** 10095–10104.

Magnuson R, Solomon J, Grossman AD. 1994. Biochemical and genetic characterization of a competence pheromone from *B. subtilis*. *Cell* **77:** 207–216.

Majerczyk CD, Sadykov MR, Luong TT, Lee C, Somerville GA, Sonenshein AL. 2008. *Staphylococcus aureus* CodY negatively regulates virulence gene expression. *J Bacteriol* **190:** 2257–2265.

Majerczyk CD, Dunman PM, Luong TT, Lee CY, Sadykov MR, Somerville GA, Bodi K, Sonenshein AL. 2010. Direct targets of CodY in *Staphylococcus aureus*. *J Bacteriol* **192:** 2861–2877.

Manna AC, Cheung AL. 2003. sarU, a sarA homolog, is repressed by SarT and regulates virulence genes in *Staphylococcus aureus*. *Infect Immun* **71:** 343–353.

Manna AC, Cheung AL. 2006a. Expression of SarX, a negative regulator of *agr* and exoprotein synthesis, is activated by MgrA in *Staphylococcus aureus*. *J Bacteriol* **188:** 4288–4299.

Manna AC, Cheung AL. 2006b. Transcriptional regulation of the *agr* locus and the identification of DNA binding residues of the global regulatory protein SarR in *Staphylococcus aureus*. *Mol Microbiol* **60:** 1289–1301.

Massey RC, Horsburgh MJ, Lina G, Hook M, Recker M. 2006. The evolution and maintenance of virulence in *Staphylococcus aureus*: A role for host-to-host transmission? *Nat Rev Microbiol* **4:** 953–958.

Matson JS, Withey JH, DiRita VJ. 2007. Regulatory networks controlling *Vibrio cholerae* virulence gene expression. *Infect Immun* **75:** 5542–5549.

Mattmann ME, Blackwell HE. 2010. Small molecules that modulate quorum sensing and control virulence in *Pseudomonas aeruginosa*. *J Org Chem* **75:** 6737–6746.

Mattmann ME, Geske GD, Worzalla GA, Chandler JR, Sappington KJ, Greenberg EP, Blackwell HE. 2008. Synthetic ligands that activate and inhibit a quorum-sensing regulator in *Pseudomonas aeruginosa*. *Bioorg Med Chem Lett* **18:** 3072–3075.

Mattmann ME, Shipway PM, Heth NJ, Blackwell HE. 2011. Potent and selective synthetic modulators of a quorum sensing repressor in *Pseudomonas aeruginosa* identified from second-generation libraries of *N*-acylated L-homoserine lactones. *Chembiochem* **12:** 942–949.

Mayville P, Ji G, Beavis R, Yang H, Goger M, Novick RP, Muir TW. 1999. Structure-activity analysis of synthetic autoinducing thiolactone peptides from *Staphylococcus aureus* responsible for virulence. *Proc Natl Acad Sci* **96:** 1218–1223.

McClean KH, Winson MK, Fish L, Taylor A, Chhabra SR, Camara M, Daykin M, Lamb JH, Swift S, Bycroft BW, et al. 1997. Quorum sensing and *Chromobacterium violaceum*: Exploitation of violacein production and inhibition for the detection of *N*-acylhomoserine lactones. *Microbiology* **143** (Pt 12): 3703–3711.

McDowell P, Affas Z, Reynolds C, Holden MT, Wood SJ, Saint S, Cockayne A, Hill PJ, Dodd CE, Bycroft BW, et al. 2001. Structure, activity and evolution of the group I thiolactone peptide quorum-sensing system of *Staphylococcus aureus*. *Mol Microbiol* **41:** 503–512.

McInnis CE, Blackwell HE. 2011. Thiolactone modulators of quorum sensing revealed through library design and screening. *Bioorg Med Chem* **19:** 4820–4828.

Medina G, Juarez K, Diaz R, Soberon-Chavez G. 2003. Transcriptional regulation of *Pseudomonas aeruginosa* rhlR, encoding a quorum-sensing regulatory protein. *Microbiology* **149:** 3073–3081.

Mena KD, Gerba CP. 2009. Risk assessment of *Pseudomonas aeruginosa* in water. *Rev Environ Contam Toxicol* **201:** 71–115.

Miller MB, Skorupski K, Lenz DH, Taylor RK, Bassler BL. 2002. Parallel quorum sensing systems converge to regulate virulence in *Vibrio cholerae*. *Cell* **110:** 303–314.

More MI, Finger LD, Stryker JL, Fuqua C, Eberhard A, Winans SC. 1996. Enzymatic synthesis of a quorum-sensing autoinducer through use of defined substrates. *Science* **272:** 1655–1658.

Morfeldt E, Taylor D, von Gabain A, Arvidson S. 1995. Activation of α-toxin translation in *Staphylococcus aureus* by the *trans*-encoded antisense RNA RNAIII. *EMBO J* **14:** 4569–4577.

Morici LA, Carterson AJ, Wagner VE, Frisk A, Schurr JR, Honer zu Bentrup K, Hassett DJ, Iglewski BH, Sauer K, Schurr MJ. 2007. *Pseudomonas aeruginosa* AlgR represses the Rhl quorum-sensing system in a biofilm-specific manner. *J Bacteriol* **189:** 7752–7764.

Muh U, Hare BJ, Duerkop BA, Schuster M, Hanzelka BL, Heim R, Olson ER, Greenberg EP. 2006a. A structurally unrelated mimic of a *Pseudomonas aeruginosa* acyl-homoserine lactone quorum-sensing signal. *Proc Natl Acad Sci* **103:** 16948–16952.

Muh U, Schuster M, Heim R, Singh A, Olson ER, Greenberg EP. 2006b. Novel *Pseudomonas aeruginosa* quorum-sensing inhibitors identified in an ultra-high-throughput screen. *Antimicrob Agents Chemother* **50:** 3674–3679.

Mulcahy H, Lewenza S. 2011. Magnesium limitation is an environmental trigger of the *Pseudomonas aeruginosa* biofilm lifestyle. *PLoS ONE* **6:** e23307.

Nadell CD, Bassler BL. 2011. A fitness trade-off between local competition and dispersal in *Vibrio cholerae* biofilms. *Proc Natl Acad Sci* **108:** 14181–14185.

Nakayama J, Cao Y, Horii T, Sakuda S, Akkermans AD, de Vos WM, Nagasawa H. 2001. Gelatinase biosynthesis-activating pheromone: A peptide lactone that mediates a quorum sensing in *Enterococcus faecalis*. *Mol Microbiol* **41:** 145–154.

Neiditch MB, Federle MJ, Miller ST, Bassler BL, Hughson FM. 2005. Regulation of LuxPQ receptor activity by the quorum-sensing signal autoinducer-2. *Mol Cell* **18:** 507–518.

Neiditch MB, Federle MJ, Pompeani AJ, Kelly RC, Swem DL, Jeffrey PD, Bassler BL, Hughson FM. 2006. Ligand-induced asymmetry in histidine sensor kinase complex regulates quorum sensing. *Cell* **126:** 1095–1108.

Ng WL, Bassler BL. 2009. Bacterial quorum-sensing network architectures. *Annu Rev Genet* **43:** 197–222.

Ng WL, Wei Y, Perez LJ, Cong J, Long T, Koch M, Semmelhack MF, Wingreen NS, Bassler BL. 2010. Probing bacterial transmembrane histidine kinase receptor-ligand interactions with natural and synthetic molecules. *Proc Natl Acad Sci* **107:** 5575–5580.

Ng WL, Perez LJ, Wei Y, Kraml C, Semmelhack MF, Bassler BL. 2011. Signal production and detection specificity in Vibrio CqsA/CqsS quorum-sensing systems. *Mol Microbiol* **79:** 1407–1417.

Ng WL, Perez L, Cong J, Semmelhack MF, Bassler BL. 2012. Broad spectrum pro-quorum-sensing molecules as inhibitors of virulence in vibrios. *PLoS Pathog* **8:** e1002767.

Niu C, Afre S, Gilbert ES. 2006. Subinhibitory concentrations of cinnamaldehyde interfere with quorum sensing. *Lett Appl Microbiol* **43:** 489–494.

Novick RP, Geisinger E. 2008. Quorum sensing in staphylococci. *Annu Rev Genet* **42:** 541–564.

Novick RP, Ross HF, Projan SJ, Kornblum J, Kreiswirth B, Moghazeh S. 1993. Synthesis of staphylococcal virulence factors is controlled by a regulatory RNA molecule. *EMBO J* **12:** 3967–3975.

Novick RP, Projan SJ, Kornblum J, Ross HF, Ji G, Kreiswirth B, Vandenesch F, Moghazeh S. 1995. The *agr* P2 operon: An autocatalytic sensory transduction system in *Staphylococcus aureus*. *Mol Gen Genet* **248:** 446–458.

O'Callaghan J, Reen FJ, Adams C, O'Gara F. 2011. Low oxygen induces the type III secretion system in *Pseudomonas aeruginosa* via modulation of the small RNAs *rsmZ and rsmY*. *Microbiology* **157** (Pt 12): 3417–3428.

Ochsner UA, Koch AK, Fiechter A, Reiser J. 1994. Isolation and characterization of a regulatory gene affecting rhamnolipid biosurfactant synthesis in *Pseudomonas aeruginosa*. *J Bacteriol* **176:** 2044–2054.

Oglesby AG, Farrow JM 3rd, Lee JH, Tomaras AP, Greenberg EP, Pesci EC, Vasil ML. 2008. The influence of iron on *Pseudomonas aeruginosa* physiology: A regulatory link between iron and quorum sensing. *J Biol Chem* **283:** 15558–15567.

Oglesby-Sherrouse AG, Vasil ML. 2010. Characterization of a heme-regulated non-coding RNA encoded by the *prrF* locus of *Pseudomonas aeruginosa*. *PLoS ONE* **5:** e9930.

Ohtani K, Yuan Y, Hassan S, Wang R, Wang Y, Shimizu T. 2009. Virulence gene regulation by the *agr* system in *Clostridium perfringens*. *J Bacteriol* **191:** 3919–3927.

Oinuma K, Greenberg EP. 2011. Acyl-homoserine lactone binding to and stability of the orphan *Pseudomonas aeruginosa* quorum-sensing signal receptor QscR. *J Bacteriol* **193:** 421–428.

Okada M, Sato I, Cho SJ, Iwata H, Nishio T, Dubnau D, Sakagami Y. 2005. Structure of the *Bacillus subtilis* quorum-sensing peptide pheromone ComX. *Nat Chem Biol* **1:** 23–24.

Okstad OA, Gominet M, Purnelle B, Rose M, Lereclus D, Kolsto AB. 1999. Sequence analysis of three *Bacillus cereus* loci carrying PlcR-regulated genes encoding degradative enzymes and enterotoxin. *Microbiology* **145** (Pt 11): 3129–3138.

Otto M. 2004. Quorum-sensing control in Staphylococci—a target for antimicrobial drug therapy? *FEMS Microbiol Lett* **241:** 135–141.

Otto M, Sussmuth R, Jung G, Gotz F. 1998. Structure of the pheromone peptide of the *Staphylococcus epidermidis agr* system. *FEBS Lett* **424:** 89–94.

Papenfort K, Vogel J. 2010. Regulatory RNA in bacterial pathogens. *Cell Host Microbe* **8:** 116–127.

Park J, Jagasia R, Kaufmann GF, Mathison JC, Ruiz DI, Moss JA, Meijler MM, Ulevitch RJ, Janda KD. 2007. Infection control by antibody disruption of bacterial quorum sensing signaling. *Chem Biol* **14:** 1119–1127.

Park J, Kaufmann GF, Bowen JP, Arbiser JL, Janda KD. 2008. Solenopsin A, a venom alkaloid from the fire ant *Solenopsis invicta*, inhibits quorum-sensing signaling in *Pseudomonas aeruginosa*. *J Infect Dis* **198:** 1198–1201.

Parsek MR, Val DL, Hanzelka BL, Cronan JE Jr, Greenberg EP. 1999. Acyl homoserine-lactone quorum-sensing signal generation. *Proc Natl Acad Sci* **96:** 4360–4365.

Passador L, Cook JM, Gambello MJ, Rust L, Iglewski BH. 1993. Expression of *Pseudomonas aeruginosa* virulence genes requires cell-to-cell communication. *Science* **260:** 1127–1130.

Patriquin GM, Banin E, Gilmour C, Tuchman R, Greenberg EP, Poole K. 2008. Influence of quorum sensing and iron on twitching motility and biofilm formation in *Pseudomonas aeruginosa*. *J Bacteriol* **190:** 662–671.

Pearson JP, Passador L, Iglewski BH, Greenberg EP. 1995. A second *N*-acylhomoserine lactone signal produced by *Pseudomonas aeruginosa*. *Proc Natl Acad Sci* **92:** 1490–1494.

Pesci EC, Pearson JP, Seed PC, Iglewski BH. 1997. Regulation of las and rhl quorum sensing in *Pseudomonas aeruginosa*. *J Bacteriol* **179:** 3127–3132.

Pessi G, Williams F, Hindle Z, Heurlier K, Holden MT, Camara M, Haas D, Williams P. 2001. The global post-transcriptional regulator RsmA modulates production of virulence determinants and *N*-acylhomoserine lactones in *Pseudomonas aeruginosa*. *J Bacteriol* **183:** 6676–6683.

Peterson KM, Mekalanos JJ. 1988. Characterization of the *Vibrio cholerae* ToxR regulon: Identification of novel genes involved in intestinal colonization. *Infect Immun* **56:** 2822–2829.

Peterson S, Cline RT, Tettelin H, Sharov V, Morrison DA. 2000. Gene expression analysis of the *Streptococcus pneumoniae* competence regulons by use of DNA microarrays. *J Bacteriol* **182:** 6192–6202.

Podbielski A, Kreikemeyer B. 2004. Cell density-dependent regulation: Basic principles and effects on the virulence of Gram-positive cocci. *Int J Infect Dis* **8:** 81–95.

Pohl K, Francois P, Stenz L, Schlink F, Geiger T, Herbert S, Goerke C, Schrenzel J, Wolz C. 2009. CodY in *Staphylococcus aureus*: a regulatory link between metabolism and virulence gene expression. *J Bacteriol* **191:** 2953–2963.

Pomerantsev AP, Pomerantseva OM, Camp AS, Mukkamala R, Goldman S, Leppla SH. 2009. PapR peptide maturation: Role of the NprB protease in *Bacillus cereus* 569 PlcR/PapR global gene regulation. *FEMS Immunol Med Microbiol* **55:** 361–377.

Pompeani AJ, Irgon JJ, Berger MF, Bulyk ML, Wingreen NS, Bassler BL. 2008. The *Vibrio harveyi* master quorum-sensing regulator, LuxR, a TetR-type protein is both an activator and a repressor: DNA recognition and binding specificity at target promoters. *Mol Microbiol* **70:** 76–88.

Pottathil M, Lazazzera BA. 2003. The extracellular Phr peptide-Rap phosphatase signaling circuit of *Bacillus subtilis*. *Front Biosci* **8:** d32–d45.

Queck SY, Jameson-Lee M, Villaruz AE, Bach TH, Khan BA, Sturdevant DE, Ricklefs SM, Li M, Otto M. 2008. RNAIII-independent target gene control by the agr quorum-sensing system: Insight into the evolution of virulence regulation in *Staphylococcus aureus*. *Mol Cell* **32:** 150–158.

Rahman P, Dusane D, Zinjarde S, Venugopalan V, McLean R, Weber M. 2010. Quorum sensing: Implications on rhamnolipid biosurfactant production. *Biotechnol Genet Eng Rev* **27:** 159–184.

Rasmussen TB, Skindersoe ME, Bjarnsholt T, Phipps RK, Christensen KB, Jensen PO, Andersen JB, Koch B, Larsen TO, Hentzer M, et al. 2005. Identity and effects of quorum-sensing inhibitors produced by Penicillium species. *Microbiology* **151:** 1325–1340.

Ren D, Zuo R, Wood TK. 2005. Quorum-sensing antagonist (5Z)-4-bromo-5-(bromomethylene)-3-butyl-2(5H)-furanone influences siderophore biosynthesis in *Pseudomonas putida* and *Pseudomonas aeruginosa*. *Appl Microbiol Biotechnol* **66:** 689–695.

Riedel CU, Monk IR, Casey PG, Waidmann MS, Gahan CG, Hill C. 2009. AgrD-dependent quorum sensing affects biofilm formation, invasion, virulence and global gene expression profiles in *Listeria monocytogenes*. *Mol Microbiol* **71:** 1177–1189.

Rocha-Estrada J, Aceves-Diez AE, Guarneros G, de la Torre M. 2010. The RNPP family of quorum-sensing proteins in Gram-positive bacteria. *Appl Microbiol Biotechnol* **87:** 913–923.

Ruby EG. 1996. Lessons from a cooperative, bacterial-animal association: The *Vibrio fischeri-Euprymna scolopes* light organ symbiosis. *Annu Rev Microbiol* **50:** 591–624.

Rutherford ST, van Kessel JC, Shao Y, Bassler BL. 2011. AphA and LuxR/HapR reciprocally control quorum sensing in vibrios. *Genes Dev* **25:** 397–408.

Saenz HL, Augsburger V, Vuong C, Jack RW, Gotz F, Otto M. 2000. Inducible expression and cellular location of AgrB, a protein involved in the maturation of the staphylococcal quorum-sensing pheromone. *Arch Microbiol* **174:** 452–455.

Said-Salim B, Dunman PM, McAleese FM, Macapagal D, Murphy E, McNamara PJ, Arvidson S, Foster TJ, Projan SJ, Kreiswirth BN. 2003. Global regulation of *Staphylococcus aureus* genes by Rot. *J Bacteriol* **185:** 610–619.

Schaefer AL, Val DL, Hanzelka BL, Cronan JE Jr, Greenberg EP. 1996. Generation of cell-to-cell signals in quorum sensing: Acyl homoserine lactone synthase activity of a purified *Vibrio fischeri* LuxI protein. *Proc Natl Acad Sci* **93:** 9505–9509.

Schauder S, Shokat K, Surette MG, Bassler BL. 2001. The LuxS family of bacterial autoinducers: Biosynthesis of a novel quorum-sensing signal molecule. *Mol Microbiol* **41:** 463–476.

Schmidt KA, Manna AC, Gill S, Cheung AL. 2001. SarT, a repressor of α-hemolysin in *Staphylococcus aureus*. *Infect Immun* **69:** 4749–4758.

Schramm VL, Gutierrez JA, Cordovano G, Basu I, Guha C, Belbin TJ, Evans GB, Tyler PC, Furneaux RH. 2008. Transition state analogues in quorum sensing and SAM recycling. *Nucleic Acids Symp Ser (Oxf)* 75–76.

Schuster M, Greenberg EP. 2006. A network of networks: Quorum-sensing gene regulation in *Pseudomonas aeruginosa*. *Int J Med Microbiol* **296:** 73–81.

Schuster M, Greenberg EP. 2007. Early activation of quorum sensing in *Pseudomonas aeruginosa* reveals the architecture of a complex regulon. *BMC Genomics* **8:** 287.

Schuster M, Lostroh CP, Ogi T, Greenberg EP. 2003. Identification, timing, and signal specificity of *Pseudomonas aeruginosa* quorum-controlled genes: A transcriptome analysis. *J Bacteriol* **185:** 2066–2079.

Schuster M, Urbanowski ML, Greenberg EP. 2004. Promoter specificity in *Pseudomonas aeruginosa* quorum sensing revealed by DNA binding of purified LasR. *Proc Natl Acad Sci* **101:** 15833–15839.

Seed PC, Passador L, Iglewski BH. 1995. Activation of the *Pseudomonas aeruginosa lasI* gene by LasR and the Pseudomonas autoinducer PAI: An autoinduction regulatory hierarchy. *J Bacteriol* **177:** 654–659.

Seidl K, Stucki M, Ruegg M, Goerke C, Wolz C, Harris L, Berger-Bachi B, Bischoff M. 2006. *Staphylococcus aureus* CcpA affects virulence determinant production and antibiotic resistance. *Antimicrob Agents Chemother* **50:** 1183–1194.

Shao Y, Bassler BL. 2012. Quorum-sensing non-coding small RNAs use unique pairing regions to differentially control mRNA targets. *Mol Microbiol* **83:** 599–611.

Shenkman B, Rubinstein E, Cheung AL, Brill GE, Dardik R, Tamarin I, Savion N, Varon D. 2001. Adherence properties of *Staphylococcus aureus* under static and flow conditions: Roles of *agr* and *sar* loci, platelets, and plasma ligands. *Infect Immun* **69:** 4473–4478.

Shikuma NJ, Fong JC, Odell LS, Perchuk BS, Laub MT, Yildiz FH. 2009. Overexpression of VpsS, a hybrid sensor kinase, enhances biofilm formation in *Vibrio cholerae*. *J Bacteriol* **191:** 5147–5158.

Simon MI, Crane BR, Crane A. 2007. *Two-component signaling systems.* Academic, San Diego.

Skorupski K, Taylor RK. 1997. Control of the ToxR virulence regulon in *Vibrio cholerae* by environmental stimuli. *Mol Microbiol* **25:** 1003–1009.

Skorupski K, Taylor RK. 1999. A new level in the *Vibrio cholerae* ToxR virulence cascade: AphA is required for transcriptional activation of the *tcpPH* operon. *Mol Microbiol* **31:** 763–771.

Slamti L, Lereclus D. 2002. A cell–cell signaling peptide activates the PlcR virulence regulon in bacteria of the *Bacillus cereus* group. *EMBO J* **21:** 4550–4559.

Slamti L, Lereclus D. 2005. Specificity and polymorphism of the PlcR-PapR quorum-sensing system in the *Bacillus cereus* group. *J Bacteriol* **187:** 1182–1187.

Smith KM, Bu Y, Suga H. 2003. Induction and inhibition of *Pseudomonas aeruginosa* quorum sensing by synthetic autoinducer analogs. *Chem Biol* **10:** 81–89.

Smith EE, Buckley DG, Wu Z, Saenphimmachak C, Hoffman LR, D'Argenio DA, Miller SI, Ramsey BW, Speert DP, Moskowitz SM, et al. 2006. Genetic adaptation by *Pseudomonas aeruginosa* to the airways of cystic fibrosis patients. *Proc Natl Acad Sci* **103:** 8487–8492.

Solomon JM, Lazazzera BA, Grossman AD. 1996. Purification and characterization of an extracellular peptide factor that affects two different developmental pathways in *Bacillus subtilis*. *Genes Dev* **10:** 2014–2024.

Sonnleitner E, Haas D. 2011. Small RNAs as regulators of primary and secondary metabolism in Pseudomonas species. *Appl Microbiol Biotechnol* **91:** 63–79.

Sonnleitner E, Gonzalez N, Sorger-Domenigg T, Heeb S, Richter AS, Backofen R, Williams P, Huttenhofer A, Haas D, Blasi U. 2010. The small RNA PhrS stimulates synthesis of the *Pseudomonas aeruginosa* quinolone signal. *Mol Microbiol* **80:** 868–885.

Srivastava D, Harris RC, Waters CM. 2011. Integration of cyclic di-GMP and quorum sensing in the control of *vpsT* and *aphA* in *Vibrio cholerae*. *J Bacteriol* **193:** 6331–6341.

Stevens AM, Dolan KM, Greenberg EP. 1994. Synergistic binding of the *Vibrio fischeri* LuxR transcriptional activator domain and RNA polymerase to the *lux* promoter region. *Proc Natl Acad Sci* **91:** 12619–12623.

Storz G, Vogel J, Wassarman KM. 2011. Regulation by small RNAs in bacteria: Expanding frontiers. *Mol Cell* **43:** 880–891.

Surette MG, Miller MB, Bassler BL. 1999. Quorum sensing in *Escherichia coli*, *Salmonella typhimurium*, and *Vibrio harveyi*: A new family of genes responsible for autoinducer production. *Proc Natl Acad Sci* **96:** 1639–1644.

Tamber S, Cheung AL. 2009. SarZ promotes the expression of virulence factors and represses biofilm formation by modulating SarA and *agr* in *Staphylococcus aureus*. *Infect Immun* **77:** 419–428.

Tamber S, Reyes D, Donegan NP, Schwartzman JD, Cheung AL, Memmi G. 2010. The staphylococcus-specific gene *rsr* represses *agr* and virulence in *Staphylococcus aureus*. *Infect Immun* **78:** 4384–4391.

Thoendel M, Horswill AR. 2009. Identification of *Staphylococcus aureus* AgrD residues required for autoinducing peptide biosynthesis. *J Biol Chem* **284:** 21828–21838.

Thoendel M, Horswill AR. 2010. Biosynthesis of peptide signals in gram-positive bacteria. *Adv Appl Microbiol* **71:** 91–112.

Thoendel M, Kavanaugh JS, Flack CE, Horswill AR. 2011. Peptide signaling in the staphylococci. *Chem Rev* **111:** 117–151.

Thomson NR, Crow MA, McGowan SJ, Cox A, Salmond GP. 2000. Biosynthesis of carbapenem antibiotic and prodigiosin pigment in Serratia is under quorum sensing control. *Mol Microbiol* **36:** 539–556.

Tu KC, Bassler BL. 2007. Multiple small RNAs act additively to integrate sensory information and control quorum sensing in *Vibrio harveyi*. *Genes Dev* **21:** 221–233.

Urbanowski ML, Lostroh CP, Greenberg EP. 2004. Reversible acyl-homoserine lactone binding to purified *Vibrio fischeri* LuxR protein. *J Bacteriol* **186:** 631–637.

Val DL, Cronan JE Jr. 1998. In vivo evidence that *S*-adenosylmethionine and fatty acid synthesis intermediates are the substrates for the LuxI family of autoinducer synthases. *J Bacteriol* **180:** 2644–2651.

van Delden C, Comte R, Bally AM. 2001. Stringent response activates quorum sensing and modulates cell density-dependent gene expression in *Pseudomonas aeruginosa*. *J Bacteriol* **183:** 5376–5384.

Vannini A, Volpari C, Gargioli C, Muraglia E, Cortese R, De Francesco R, Neddermann P, Marco SD. 2002. The crystal structure of the quorum sensing protein TraR bound to its autoinducer and target DNA. *EMBO J* **21:** 4393–4401.

Vuong C, Saenz HL, Gotz F, Otto M. 2000. Impact of the agr quorum-sensing system on adherence to polystyrene in *Staphylococcus aureus*. *J Infect Dis* **182:** 1688–1693.

Wagner VE, Gillis RJ, Iglewski BH. 2004. Transcriptome analysis of quorum-sensing regulation and virulence factor expression in *Pseudomonas aeruginosa*. *Vaccine* **22** (Suppl 1): S15–S20.

Waters CM, Lu W, Rabinowitz JD, Bassler BL. 2008. Quorum sensing controls biofilm formation in *Vibrio cholerae* through modulation of cyclic di-GMP levels and repression of vpsT. *J Bacteriol* **190:** 2527–2536.

Watson WT, Minogue TD, Val DL, von Bodman SB, Churchill ME. 2002. Structural basis and specificity of acyl-homoserine lactone signal production in bacterial quorum sensing. *Mol Cell* **9:** 685–694.

Weeks JN, Galindo CL, Drake KL, Adams GL, Garner HR, Ficht TA. 2010. *Brucella melitensis* VjbR and C12-HSL regulons: Contributions of the *N*-dodecanoyl homoserine lactone signaling molecule and LuxR homologue VjbR to gene expression. *BMC Microbiol* **10:** 167.

Wei Y, Perez LJ, Ng WL, Semmelhack MF, Bassler BL. 2011. Mechanism of *Vibrio cholerae* autoinducer-1 biosynthesis. *ACS Chem Biol* **6:** 356–365.

Wei Y, Ng WL, Cong J, Bassler BL. 2012. Ligand and antagonist driven regulation of the *Vibrio cholerae* quorum-sensing receptor CqsS. *Mol Microbiol* **83:** 1095–1108.

Whiteley M, Lee KM, Greenberg EP. 1999. Identification of genes controlled by quorum sensing in *Pseudomonas aeruginosa*. *Proc Natl Acad Sci* **96:** 13904–13909.

Wilderman PJ, Sowa NA, FitzGerald DJ, FitzGerald PC, Gottesman S, Ochsner UA, Vasil ML. 2004. Identification of tandem duplicate regulatory small RNAs in *Pseudomonas aeruginosa* involved in iron homeostasis. *Proc Natl Acad Sci* **101:** 9792–9797.

Williams P, Camara M. 2009. Quorum sensing and environmental adaptation in *Pseudomonas aeruginosa*: A tale of regulatory networks and multifunctional signal molecules. *Curr Opin Microbiol* **12:** 182–191.

Withey JH, DiRita VJ. 2006. The toxbox: Specific DNA sequence requirements for activation of *Vibrio cholerae* virulence genes by ToxT. *Mol Microbiol* **59:** 1779–1789.

Wu H, Song Z, Hentzer M, Andersen JB, Molin S, Givskov M, Hoiby N. 2004. Synthetic furanones inhibit quorum-sensing and enhance bacterial clearance in *Pseudomonas aeruginosa* lung infection in mice. *J Antimicrob Chemother* **53:** 1054–1061.

Xavier KB, Bassler BL. 2005. Interference with AI-2-mediated bacterial cell–cell communication. *Nature* **437:** 750–753.

Xiao G, Deziel E, He J, Lepine F, Lesic B, Castonguay MH, Milot S, Tampakaki AP, Stachel SE, Rahme LG. 2006a. MvfR, a key *Pseudomonas aeruginosa* pathogenicity LTTR-class regulatory protein, has dual ligands. *Mol Microbiol* **62:** 1689–1699.

Xiao G, He J, Rahme LG. 2006b. Mutation analysis of the *Pseudomonas aeruginosa mvfR* and *pqsABCDE* gene promoters demonstrates complex quorum-sensing circuitry. *Microbiology* **152:** 1679–1686.

Yao Y, Martinez-Yamout MA, Dickerson TJ, Brogan AP, Wright PE, Dyson HJ. 2006. Structure of the *Escherichia coli* quorum sensing protein SdiA: Activation of the folding switch by acyl homoserine lactones. *J Mol Biol* **355:** 262–273.

Yarwood JM, McCormick JK, Schlievert PM. 2001. Identification of a novel two-component regulatory system that acts in global regulation of virulence factors of *Staphylococcus aureus*. *J Bacteriol* **183:** 1113–1123.

Yarwood JM, Bartels DJ, Volper EM, Greenberg EP. 2004. Quorum sensing in *Staphylococcus aureus* biofilms. *J Bacteriol* **186:** 1838–1850.

Yildiz FH, Liu XS, Heydorn A, Schoolnik GK. 2004. Molecular analysis of rugosity in a *Vibrio cholerae* O1 El Tor phase variant. *Mol Microbiol* **53:** 497–515.

Zemanick ET, Sagel SD, Harris JK. 2011. The airway microbiome in cystic fibrosis and implications for treatment. *Curr Opin Pediatr* **23:** 319–324.

Zhang RG, Pappas KM, Brace JL, Miller PC, Oulmassov T, Molyneaux JM, Anderson JC, Bashkin JK, Winans SC, Joachimiak A. 2002. Structure of a bacterial quorum-sensing transcription factor complexed with pheromone and DNA. *Nature* **417:** 971–974.

Zhu J, Mekalanos JJ. 2003. Quorum sensing-dependent biofilms enhance colonization in *Vibrio cholerae*. *Dev Cell* **5:** 647–656.

Zhu J, Winans SC. 1999. Autoinducer binding by the quorum-sensing regulator TraR increases affinity for target promoters in vitro and decreases TraR turnover rates in whole cells. *Proc Natl Acad Sci* **96:** 4832–4837.

Zhu J, Winans SC. 2001. The quorum-sensing transcriptional regulator TraR requires its cognate signaling ligand for protein folding, protease resistance, and dimerization. *Proc Natl Acad Sci* **98:** 1507–1512.

Zhu J, Miller MB, Vance RE, Dziejman M, Bassler BL, Mekalanos JJ. 2002. Quorum-sensing regulators control virulence gene expression in *Vibrio cholerae*. *Proc Natl Acad Sci* **99:** 3129–3134.

Mechanisms and Biological Roles of Contact-Dependent Growth Inhibition Systems

Christopher S. Hayes[1,2], Sanna Koskiniemi[1], Zachary C. Ruhe[1], Stephen J. Poole[1], and David A. Low[1,2]

[1]Department of Molecular, Cellular and Developmental Biology, University of California, Santa Barbara, Santa Barbara, California 93106-9625

[2]Biomolecular Science and Engineering Program, University of California, Santa Barbara, Santa Barbara, California 93106-9625

Correspondence: low@lifesci.ucsb.edu

Bacterial contact-dependent growth inhibition (CDI) is mediated by the CdiA/CdiB family of two-partner secretion proteins. CDI^+ cells bind to susceptible target bacteria and deliver a toxic effector domain derived from the carboxyl terminus of CdiA (CdiA-CT). More than 60 distinct CdiA-CT sequence types have been identified, and all CDI toxins characterized thus far display RNase, DNase, or pore-forming activities. CDI systems also encode CdiI immunity proteins, which specifically bind and inactivate cognate CdiA-CT toxins to prevent autoinhibition. CDI activity appears to be limited to target cells of the same species, suggesting that these systems play a role in competition between closely related bacteria. Recent work on the CDI system from uropathogenic *Escherichia coli* (UPEC 536) has revealed that its CdiA-CT toxin binds tightly to a cysteine biosynthetic enzyme (CysK) in the cytoplasm of target cells. The unanticipated complexity in the UPEC CDI pathway raises the possibility that these systems perform other functions in addition to growth inhibition. Finally, we propose that the phenomenon of CDI is more widespread than previously appreciated. Rhs (rearrangement hotspot) systems encode toxin-immunity pairs, some of which share significant sequence identity with CdiA-CT/CdiI proteins. A number of recent observations suggest that Rhs proteins mediate a distinct form of CDI.

The term "microbiome" was introduced to encompass microbes, their combined genomes, and their physiological interactions within a specific environment (Lederberg 2000). We have coevolved with our microbiome, and the microorganisms that colonize us play a significant role in determining health and resistance to disease (Dominguez-Bello and Blaser 2008; Benson et al. 2010; Ley 2010). The enormous diversity of bacterial species that reside within our intestines presents a daunting task if we wish to determine all the antagonistic and synergistic interactions they have with us and each other. Bacterial pathogenesis focuses on the relatively few species that cause adverse outcomes for the host, and has been explored to a much greater extent than bacterial ecology, which encompasses all interactions within the

Cite this article as *Cold Spring Harb Perspect Med* doi: 10.1101/cshperspect.a010025

microbiome. However, this situation is likely to change as we realize that microbial ecology is central to our understanding of bacterial disease and the design of effective interventions. For example, *Clostridium difficile* infections are often precipitated by perturbations to the microbiome resulting from antibiotic treatment (Bakken et al. 2011). Moreover, antibiotic treatment can increase susceptibility to influenza virus through alterations in the microbiome and its interplay with the host immune system (Ichinohe et al. 2011). These examples illustrate the importance of developing therapies that target specific bacterial pathogens. One way to accomplish this goal is to understand the mechanisms by which bacteria interact and compete with one another. Significant progress has been made in understanding how bacteria communicate and compete via secreted chemicals and peptides (Keller and Surette 2006; Ng and Bassler 2009). Some secreted factors control the expression of virulence factors, and therefore blocking these signaling pathways provides a strategy for specific intervention (Chen et al. 2011). In contrast, comparatively little is known about bacterial communication mediated by direct cell-to-cell contact (Hayes et al. 2010; Konovalova and Sogaard-Andersen 2011; Pathak et al. 2012). Because bacteria in diverse environments grow on solid surfaces in biofilms, contact-dependent communication could play important roles that are distinct from the communication mediated by secreted signaling factors. This review focuses on contact-dependent growth inhibition (CDI), a mechanism that we discovered in 2005 (Aoki et al. 2005) and have recently shown plays a role in bacterial competition (Aoki et al. 2010).

CDI

The phenomenon of CDI was first described in *Escherichia coli* EC93, a wild-type isolate cultured from the intestinal tract of rats (Aoki et al. 2005). *E. coli* EC93 was found to profoundly inhibit the growth of most other *E. coli* strains in coculture experiments. Although enteric bacteria often produce soluble factors that kill related bacteria, the growth inhibition mediated by *E. coli* EC93 is unique in that it requires direct cell-to-cell contact with target cells. The CDI locus in *E. coli* EC93 is comprised of a three-gene cluster (*cdiBAI*), which is sufficient to confer the CDI inhibitor phenotype when introduced into laboratory strains of *E. coli* K-12 (Fig. 1). The *cdiB* and *cdiA* genes encode members of the two-partner secretion protein family (Jacob-Dubuisson et al. 2013). CdiB is a predicted outer membrane β-barrel protein that is required for the secretion and presentation of the CdiA exoprotein onto the cell surface. Like other two-partner secretion exoproteins, CdiA contains an amino-terminal transport domain followed by a hemagglutinin repeat region that is predicted to adopt a filamentous β-helical structure (Kajava et al. 2001). The CDI growth inhibitory activity is localized to the carboxy-terminal region of CdiA (CdiA-CT), and the *cdiI* gene product neutralizes the CdiA-CT toxin to protect *E. coli* EC93 cells from autoinhibition. The CdiI immunity protein is also sufficient to protect CDI$^-$ target cells from growth inhibition mediated by the *E. coli* EC93 CdiA-CT toxin. Thus, the *E. coli* EC93 CDI system encodes both toxin and immunity functions that are presumably used for growth competition.

CDI Systems Are Modular

The proliferation of bacterial genome sequence information has revealed that CDI systems are widely distributed throughout the α-, β-, and γ-proteobacteria and are commonly found in human, plant, and arthropod pathogens. Proteobacterial species contain zero to multiple CDI loci (Fig. 1). In Table 1, CdiA proteins present in a number of pathogenic bacterial species are annotated with accession numbers. CdiA proteins from γ-proteobacteria typically share sequence identity along most of the amino-terminal hemagglutinin repeat region, but the CdiA-CT sequences diverge abruptly after a common VENN peptide motif (Aoki et al. 2010; Poole et al. 2011). CdiA-CT sequences can also vary considerably between CDI systems from different strains of the same species. The predicted CdiI immunity proteins are also variable, indicating that CdiA-CT/CdiI pairs constitute a family of polymorphic toxin-immunity

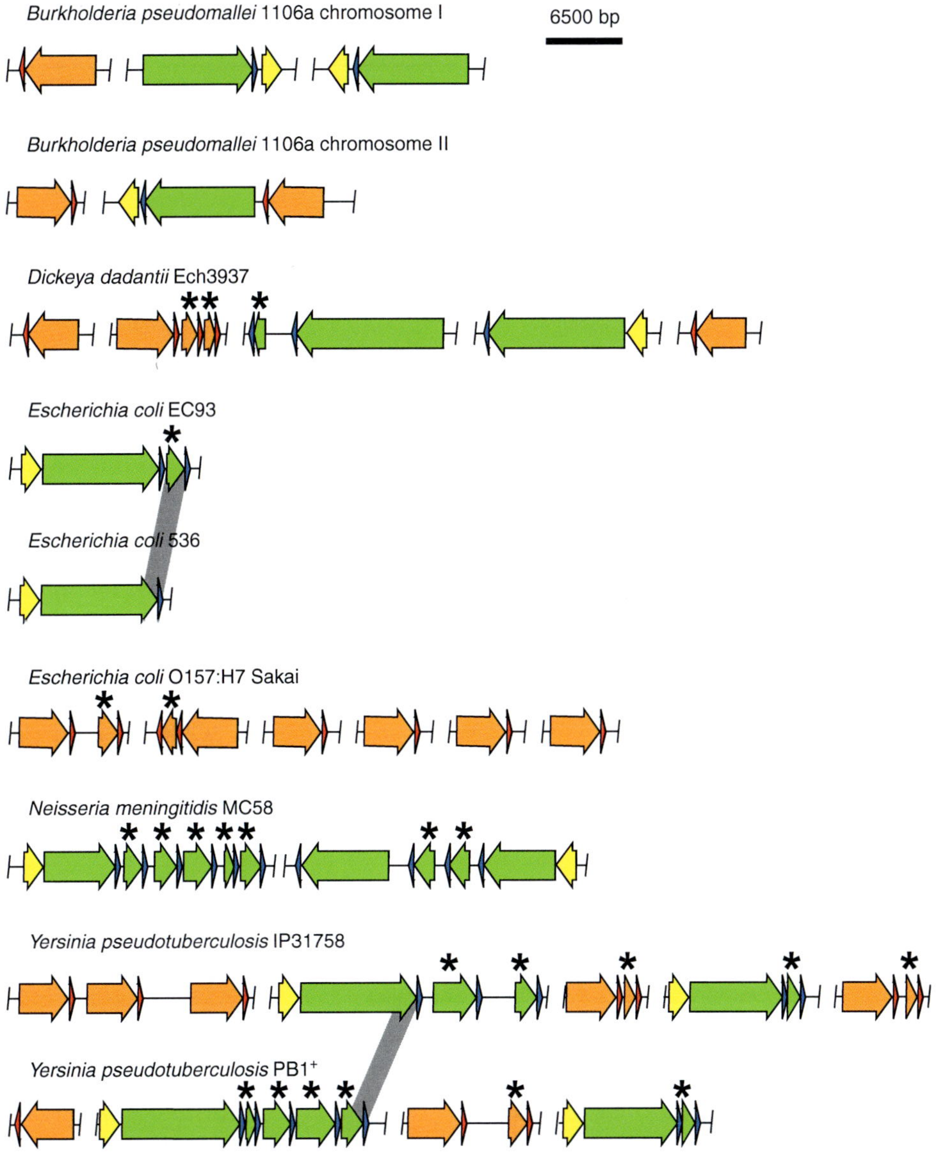

Figure 1. Organization of *cdi* and *rhs* genes in select bacterial pathogens. The *cdi* and *rhs* loci present in the chromosome(s) of the indicated bacterial pathogens are shown. Gaps between loci indicate that they are not closely linked. The order of gene regions and orientation of genes is as in the published genomic sequences, except for *E. coli* 536 and *Yersinia pseudotuberculosis* PB1$^+$, which have been reversed. Orphan *cdiA-CT* and *rhs-CT* regions are denoted by asterisks. Symbol key: *cdiA*, green arrows; *cdiB*, yellow arrows; *cdiI*, short blue arrows; *rhs*, orange arrows; and *rhsI*, short red arrows. Shaded regions show toxin-immunity modules that are found as an orphan in one strain and part of a full *CdiA* gene in another. Selected pathogens illustrate different aspects of gene organization. *Burkholderia pseudomallei* 1106a, gene order *cdiA-I-B* instead of the typical *cdiB-A-I* gene order (chromosomes I and II) and *rhs* gene linkage to a CDI module (chromosome II); *E. coli* O157:H7 Sakai, *rhs* genes present but CDI modules absent; *E. coli* EC93, *E. coli* 536, and *Neisseria meningitides* MC58, CDI modules but *rhs* genes absent; *N. meningitides* MC58, multiple CDI modules with orphan *cdiA-CT* genes; *B. pseudomallei* 1106a, *Dickeya dadantii* Ech3937, *Y. pseudotuberculosis* IP31758, and *Y. pseudotuberculosis* PB1$^+$, CDI modules together with Rhs modules; *E. coli* EC93, *E. coli* 536, *Y. pseudotuberculosis* IP31758, and *Y. pseudotuberculosis* PB1$^+$, exchange of toxin-immunity pairs between orphan *cdiA-CT* and full *cdiA-CT* genes.

Table 1. CdiA and Rhs proteins encoded in select pathogens

Species and strain	CdiA protein IDs	Rhs protein IDs
Acinetobacter baumannii str. ACICU	ACC57223 (no orphans)	ACC56430 (no orphans); ACC56574 (no orphans)
Bartonella grahamii as4aup	YP_002971801 (no orphans), YP_002972394 (no orphans), YP_002972429 (no orphans), YP_002972520 (no orphans); two additional loci lack a Pfam PF04829 domain	No *rhs* genes
Bartonella tribocorum str. CIP 105476	CAK00851 (two orphans), CAK00866 (four orphans), CAK01426 (one orphan), CAK01664 (two orphans)	No *rhs* genes
Burkholderia pseudomallei 1106a chromosome I	ABN92080 (no orphans), ABN91683 (no orphans)	ABN90290 (no orphans)
Burkholderia pseudomallei 1106a chromosome II	ABN92652 (no orphans)	ABN92960 (no orphans), ABN93053 (no orphans)
Dickeya dadantii Ech586	ACZ76770 (one orphan), ACZ78808 (three orphans)	ACZ76163 (no orphan), ACZ76342 (no orphans)
Dickeya dadantii 3937	ADM98580 (one orphan), ADM98660 (no orphans)	ADM97066 (no orphans), ADM97666 (two orphans), ADM99131 (no orphans)
Dickeya zeae Ech1591	ACT06763 (two orphans), ACT06855 (no orphans)	ACT05308 (no orphans), ACT06027 (no orphans), ACT06362 (one orphan), ACT07636 (no orphans), ACT07641 (no orphans, but ACT07636 is immediately downstream)
Edwardsiella ictaluri 93-146	ACR70209 (one orphan), ACR70465 (frameshift in *cdiA*; five orphans)	No *rhs* genes
Erwinia pyrifoliae Ep1/96	CAX56670 (three orphans)	CAX54630 (no orphans), CAX54893 (one orphan), CAX56922 (three orphans)
Escherichia coli 536	ABG72516 (no orphans)	No *rhs* genes
Escherichia coli O157:H7 str. Sakai	No *cdiA* genes	BAB33660 (one orphan), BAB34028 (one orphan), BAB34152 (no orphans), BAB35484 (no orphans), BAB37893 (no orphans), BAB38287 (no orphans)
Escherichia coli CFT073	AAN78830 (two orphans)	No *rhs* genes
Escherichia coli UT189	ABE10349 (one orphan)	No *rhs* genes
Klebsiella pneumoniae str. 342	ACI08381 (no orphans)	No *rhs* genes
Neisseria meningitidis Z2491	CAM07945 (three orphans)	No *rhs* genes
Neisseria meningitidis MC58	AAF40927 (no orphans), AAF40929 (five orphans), AAF42109 (no orphans; linked to AAF42119), AAF42119 (one orphan)	No *rhs* genes

Continued

Table 1. *Continued*

Species and strain	CdiA protein IDs	Rhs protein IDs
Pectobacterium carotovorum WPP163	ACX86275 (no orphans), ACX88282 (no orphans)	ACX86200 (seven orphans), ACX86474 (no orphans), ACX87399 (no orphans), ACX88449 (one orphan)
Photorhabdus asymbiotica subsp. asymbiotica ATCC 43949	CAQ82538 (no *cdiB*; two orphans), CAQ83255 (no orphans), CAQ83577 (five orphans)	CAQ85102 (no orphans), CAQ83059 (no orphans), CAQ85097 (no orphans)
Photorhabdus luminescens subsp. laumondii	NP_927589 (no orphans), NP_927898 (no orphans), NP_928461 (one orphan), NP_928667 (no orphans), NP_930296 (no orphans), NP_930925 (one orphan)	NP_930352 (no orphans)
Yersinia pestis bv. Antiqua str Antiqua	ABG13952 (one orphan), ABG15152 (no orphans)	ABG15350 (no orphans), ABG15382 (has an upstream orphan)
Yersinia pseudotuberculosis serotype O:3 str. YPIII	ACA67915 (one orphan)	ACA66708 (ACA66715 just downstream), ACA66715 (no orphans), ACA67067 (three orphans)

Protein accession numbers are listed. Notations in parentheses indicate whether the CDI or Rhs module contains orphan *cdiA-CT* or *rhs*-CT genes.

proteins. Together these observations suggest that CdiA proteins are modular, with the large amino-terminal β-helical "stick" region capable of delivering a variety of CdiA-CT toxins, each associated with a distinct immunity protein. To test this hypothesis, we generated *cdiA* chimeras that encode the amino-terminal region of CdiA from uropathogenic *E. coli* 536 (UPEC 536) fused at the VENN motif to heterologous CdiA-CTs from *Yersinia pestis* CO92, *E. coli* EC93, and *Dickeya dadantii* 3937. The resulting chimeric CdiA proteins are functional in CDI against *E. coli* target cells (Aoki et al. 2010). Moreover, target cells are only protected if they express the CdiI immunity protein that corresponds to the grafted CdiA-CT, consistent with the specificity of cognate CdiA-CT/CdiI interactions. This modularity suggests that *cdiA* genes may evolve rapidly by horizontal gene transfer and recombination to deploy novel toxins. This hypothesis is supported by metagenomic analysis indicating that *cdiA-CT*/*cdiI* modules are actively exchanged between CDI systems (Poole et al. 2011).

CDI Orphan Toxin-Immunity Modules

Many CDI systems contain tandem arrays of *cdiA-CT*/*cdiI* toxin-immunity gene pairs downstream from the main *cdiBAI* gene cluster (Fig. 1, green arrows with asterisks). The *cdiA-CT* sequences typically encode the conserved VENN peptide motif, but lack translation initiation signals. We have termed these gene pairs "orphan modules," because the *cdiA-CT* sequences appear to be displaced from full-length *cdiA* genes. Orphan *cdiA-CT*/*cdiI* modules from *E. coli* EC93, *D. dadantii* 3937 and *E. coli* EC869 each encode a functional CdiA-CT toxin that is specifically neutralized by the corresponding orphan CdiI immunity protein (Poole et al. 2011). Thus, orphan *cdiA-CT*/*cdiI* modules have the capacity to mediate CDI, but their deployment requires a mechanism to load orphan toxins onto full-length CdiA for export and assembly on the cell surface. There is evidence of orphan *cdiA-CT*/*cdiI* module translocation based on metagenomic analysis. For example, comparison of CDI loci from *E. coli*

EC93/*E. coli* 536 or from *Yersinia pseudotuberculosis* IP31758/*Y. pseudotuberculosis* PB1^{+} indicates that an orphan *cdiA-CT*/*cdiI* module in one strain is fused to the full-length *cdiA* gene in another strain (Fig. 1) (Poole et al. 2011). Because CdiA-CTs are interchangeable, it is likely that the resulting CdiA chimera is functional and delivers the orphan CdiA-CT to neighboring cells. We propose that orphan *cdiA-CT*/*cdiI* modules represent a reservoir of toxin diversity that confers a growth advantage to bacteria expressing them (S Koskiniemi, CS Hayes, and DA Low, unpubl.). Most orphan *cdiA-CT* fragments also encode nonvariable, CdiA "stick" sequences that are identical to the upstream full-length *cdiA* gene. These direct repeat regions could facilitate homologous recombination, but such an event would delete the *cdiI* immunity gene and render the recombinants susceptible to growth inhibition by neighboring sibling cells. Alternatively, duplication of the CDI region could occur before recombination of an orphan *cdiA-CT*/*cdiI* module onto one of the duplicated *cdiA* genes. Under this scenario, the recombinant cell would express a novel CdiA protein loaded with an orphan CdiA-CT, yet would remain immune to the parental CDI system because it retains the main *cdiI* gene. Such recombinants could conceivably have a competitive advantage if the parental cells do not express sufficient orphan CdiI protein to confer immunity to the new chimeric CdiA protein.

THE CDI PATHWAY

Initial work on the CDI pathway has focused on the system deployed by *E. coli* EC93 (Aoki et al. 2005). A model for CDI is shown in Figure 2, which depicts the events known to occur following contact between a CDI^{+} inhibitor cell and a target cell. These events are described below in detail.

Binding to Target Cells

Genetic selections for *E. coli* K-12 mutants that are resistant to the CDI system from *E. coli* isolate EC93 (CDI^{EC93}) identified the outer membrane protein BamA (also known as YaeT and Omp85) as a possible CDI receptor on target cells (Aoki et al. 2008). A single transposon insertion just upstream of the *bamA* open reading frame was found to confer a CDI^{R} phenotype, and mutants carrying this allele (termed *bamA101*) have ~10-fold less BamA on the cell surface compared with wild-type *E. coli* cells. BamA is a highly conserved, essential protein that is required for the assembly of all β-barrel proteins into the outer membrane (Bos et al. 2007; Knowles et al. 2009; Ricci and Silhavy 2012) and therefore the resistance conferred by *bamA101* could be an indirect effect. To test whether BamA itself is the CDI receptor, target cells were preincubated with α-BamA antibodies and then challenged with inhibitor cells expressing the CDI^{EC93} system. This antibody treatment protected target cells from growth inhibition, whereas treatment with preimmune serum or antibodies to the periplasmic POTRA-3 domain of BamA showed no protective effect (Aoki et al. 2008). Moreover, expression of BamA lacking the POTRA-3 domain (BamAΔP3) in the *bamA101* background restored susceptibility to CDI. BamAΔP3 is stable and localizes to the outer membrane, but is not functional in β-barrel protein assembly. Thus, the biogenesis function of BamA, which involves interaction with other members of the β-barrel assembly machine (Ricci and Silhavy 2012), is not required for CDI. BamA is also required for the stable binding of CDI^{+} inhibitor cells to target cells. Coculture of fluorescently labeled CDI^{+} inhibitor cells and target cells produces a distinct population of inhibitor-target cell aggregates that can be quantified by flow cytometry. The *bamA101* mutation reduces target cell adhesion to inhibitor cells ~10-fold, and the interaction between CDI^{+} cells and wild-type *bamA*$^{+}$ target cells can be blocked by α-BamA antisera but not antibodies to the POTRA-3 domain (Aoki et al. 2008). Taken together, these results indicate that BamA is the target cell receptor for the CDI^{EC93} system.

Growth inhibition mediated by the *E. coli* EC93 CDI system is species restricted. Although *E. coli* EC93 inhibits most other *E. coli* strains, it has no effect on the growth of related γ-proteobacteria including *Salmonella enterica* sero-

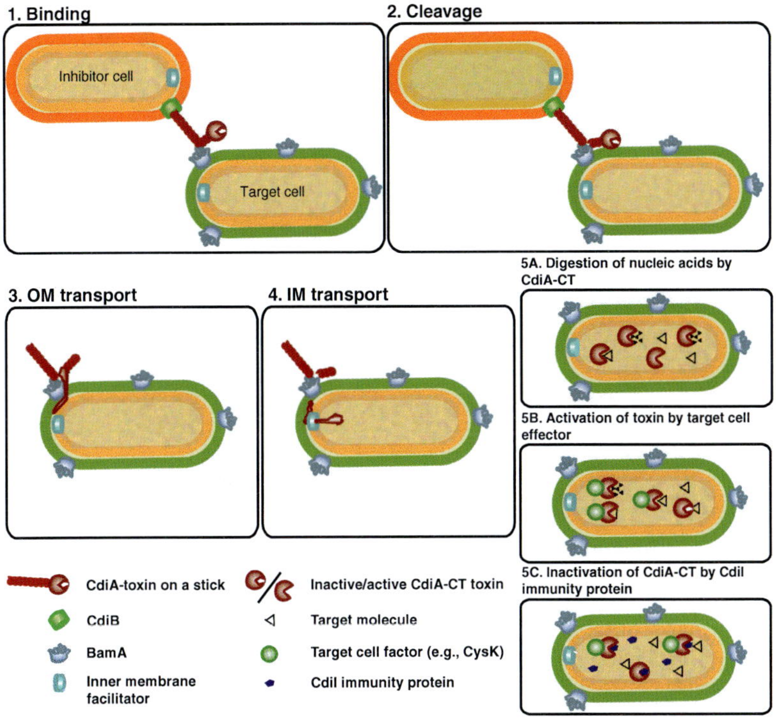

Figure 2. The CDI model. The major steps involved in CDI are depicted. CDI^{+} inhibitor cells make contact with susceptible target cells through interactions between CdiA and the BamA receptor (panel 1). The CdiA protein is then thought to undergo proteolytic cleavage (panel 2) to liberate the CdiA-CT for transport through the target cell outer membrane (panel 3). Enzymatic CdiA-CT toxins are presumably transported through the inner membrane with the assistance of integral membrane proteins (panel 4). Once delivered into the target cell cytoplasm, nuclease toxins degrade DNA or RNA to inhibit cell growth. For the UPEC 536 CDI system, the CdiA-CT toxin requires activation by the CysK "permissive factor" (panels 5A and 5B). If target cells express the cognate CdiI immunity protein (e.g., isogenic CDI^{+} siblings), then the CdiA-CT toxin is neutralized to prevent growth inhibition (panel 5C).

var Typhimurium, inhibitor-target cell aggregates, *Citrobacter freundii*, and *Proteus mirabilis* (Ruhe et al. 2013). These observations suggest that some step(s) in the CDI^{EC93} pathway is specific for *E. coli* target cells. This finding is somewhat unexpected because CDI systems are widely distributed across proteobacterial clades and appear to be horizontally transferred (Table 1). Multiple sequence alignment of BamA proteins from γ-proteobacteria show a high degree of sequence identity, but remarkably extracellular loops 4, 6, and 7 are highly variable in sequence and length (Fig. 3) (Noinaj et al. 2013). Through allelic exchange it was shown that *bamA* alleles from related species, such as *Enterobacter cloacae*, could compliment a *bamA* deletion in *E. coli* while rendering these strains resistant to CdiA^{EC93}. Additionally, binding of CdiA^{EC93} to BamA was shown to require both loop 6 and loop 7 of *E. coli* BamA. These results suggest that the CDI target cell range results from a direct interaction between CdiA and extracellular loops 6 and 7 of *E. coli* BamA. This hypothesis was tested by grafting loops 6 and 7 from *E. coli* BamA onto BamA from *Enterobacter cloacae*. *E. coli* expressing this chimeric BamA bound to and were rendered susceptible to CDI^{EC93}, supporting the hypothesis that intraspecies restriction of CDI is caused by cell receptor heterogeneity (Ruhe et al. 2013). Notably, the BamA proteins from *Burkholderia* species appear to lack extracellular loops entirely, suggesting that a different surface

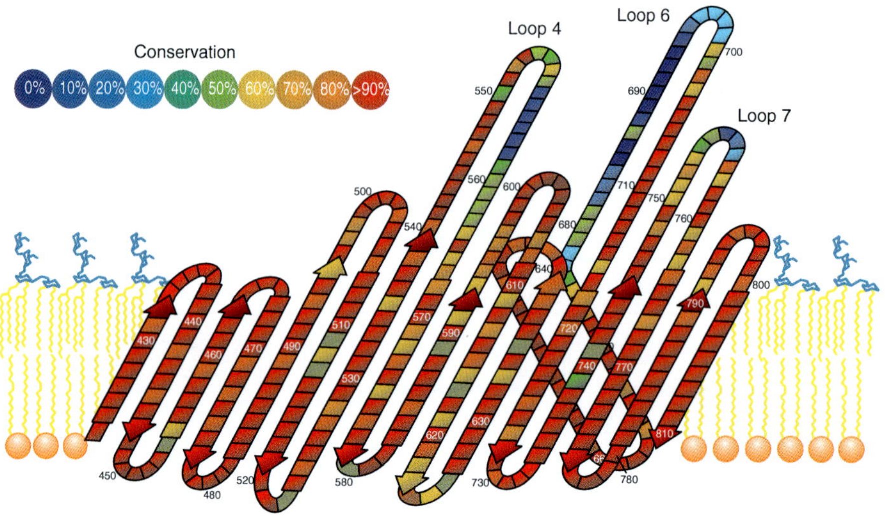

Figure 3. Conservation of BamA in the γ-proteobacteria. The predicted topology of BamA is presented with each residue color coded with a heat map indicating the degree of conservation (Ruhe et al. 2013). High and low degrees of conservation are shown as warm (red) and cool (blue) colors, respectively. Amino acid numbering is based on the preprocessed (including signal sequence) form of *E. coli* BamA. Conservation analysis was performed using the PRALINE multiple alignment tool (www.ibi.vu.nl/programs/pralinewww/) with an input of 40 different BamA sequences from γ-proteobacteria.

receptor may be exploited by the CDI systems from this group of β-proteobacteria.

CdiA-CT Toxin Activities

How do CDI systems inhibit the growth of target bacteria? Although there are approximately 60 distinct CdiA-CT/CdiI families, all characterized CDI toxins appear to either disrupt target cell membrane integrity or degrade cellular nucleic acids. During *E. coli* EC93-mediated CDI, the target cell proton gradient is dissipated with an attendant decrease in respiration and ATP levels (Aoki et al. 2009). These findings suggest a CDI-dependent membrane pore that leaks ions. Presumably, the very hydrophobic $\text{CdiI}^{\text{EC83}}$ immunity protein localizes to the inner membrane and prevents the assembly and/or gating of the membrane pore. Other CdiA-CTs are nucleases that are specific for RNA or DNA. The $\text{CdiA-CT}^{\text{UPEC536}}$ from UPEC 536 is a tRNase that cleaves in the anticodon loop of all tRNA isoacceptors, and $\text{CdiA-CT}_{\text{II}}^{\text{Dd3937}}$ from *D. dadantii* 3937 has Mg^{2+}-dependent DNase activity (Aoki et al. 2010). Both of these CdiA-CT nuclease activities are specifically blocked by the binding of cognate CdiI immunity proteins, suggesting that the nuclease activities are responsible for growth inhibition during CDI. The importance of nuclease activity has been established for the UPEC 536 CDI system. A His^{3193}Ala missense change in the carboxy-terminal region of $\text{CdiA}^{\text{UPEC536}}$ ablates tRNase activity in vitro and completely blocks UPEC 536-mediated CDI. The mutated $\text{CdiA-CT}^{\text{UPEC536}}$ domain still binds to its cognate $\text{CdiI}^{\text{UPEC536}}$ immunity protein, indicating that it retains its native tertiary structure. Moreover, the inactive variant is expressed on the inhibitor cell surface at the same level as wild-type $\text{CdiA}^{\text{UPEC536}}$, demonstrating that the His^{3193}Ala mutation does not interfere with exoprotein biogenesis. Thus, although $\text{CdiA}^{\text{UPEC536}}$ is a very large protein (331 kDa) with the potential to carry multiple toxin activities, the tRNase activity contained within the carboxy-terminal ~140 residues accounts for all growth inhibition during CDI (Diner et al. 2012). Bioinformatic analyses indicate that a number of CdiI immunity proteins belong to the SUKH superfamily, which includes eukaryotic members such as Smi1/Knr4 in yeast, which is involved in conferring resistance to a killer toxin (Zhang et al. 2011). By examining genes adjacent to those coding for SUKH family members a variety of putative nuclease and deaminase enzymes have been identified (Zhang et al. 2011).

Identification of a Permissive Factor Required for CdiA-CT Activity

Recent work has revealed unexpected complexity in the CDI pathway of UPEC 536 (Diner et al. 2012). As described above, the $CDI^{UPEC536}$ system delivers a CdiA-CT with toxic tRNase activity into target cells. However, purified $CdiA\text{-}CT^{UPEC536}$ does not cleave tRNAs in vitro unless the reactions are supplemented with *E. coli* cell lysate, suggesting that an activating factor(s) is required for nuclease activity. Biochemical fractionation of *E. coli* lysates revealed that the CysK protein is necessary and sufficient to activate the $CdiA\text{-}CT^{UPEC536}$ tRNase in vitro. CysK is a biosynthetic enzyme that catalyzes the second step of L-cysteine synthesis from L-serine, but its enzymatic activity is not required to activate $CdiA\text{-}CT^{UPEC536}$. Instead, CysK binds to the carboxy-terminal GYGI sequence of $CdiA\text{-}CT^{UPEC536}$ to form a stable, active tRNase complex. Remarkably, the $CdiA\text{-}CT^{UPEC536}$ GYGI motif mimics the GDGI peptide sequence found at the carboxyl terminus of CysE, which normally binds CysK to form the cysteine synthase complex in bacteria and plants. Thus, the $CdiA\text{-}CT^{UPEC536}$ toxin exploits an evolutionarily conserved protein–protein interaction to promote nuclease activity. CysK is also required for $CdiA\text{-}CT^{UPEC536}$ tRNase activity in vivo, and Δ*cysK* mutants are completely resistant to cell-mediated $CDI^{UPEC536}$ (Diner et al. 2012). Because target cells must express CysK to be inhibited by UPEC 536, we have designated CysK a "permissive factor." The requirement for a permissive factor is perplexing if CDI is viewed simply as a mechanism for bacterial competition, because target cells can readily acquire resistance through *cysK* mutations. Although there are likely to be fitness consequences for *cysK* mutations, these mutants are still prototrophic for cysteine owing to the presence of CysM—an isoenzyme that catalyzes the same reaction as CysK, but does not activate the $CdiA\text{-}CT^{UPEC536}$ toxin. One intriguing possibility is that CysK provides a mechanism for communication between UPEC 536 cells. There is no known exclusion mechanism in CDI and CDI^+ cells appear to actively deliver CdiA-CTs to each other (manuscript in preparation). At face value, this exchange of toxins between immune cells appears futile, but perhaps the "inactive" CdiA-CT/CdiI complexes serve an unappreciated signaling function. The $CdiA\text{-}CT^{UPEC536}$ can bind to both CysK and the $CdiI^{UPEC536}$ immunity protein simultaneously, suggesting that this ternary complex could affect cysteine metabolism or other pathways in UPEC 536 cells. We have recently found that the UPEC 536 CDI system modulates biofilm formation, although the molecular mechanism(s) underlying this phenotype has yet to be determined (Diner et al. 2012).

CDI Systems in *Burkholderia* Species

The CDI systems found in *Burkholderia*, *Cupriavidus*, and *Variovorax* species are distinct from the systems found in other proteobacteria (Aoki et al. 2010; Anderson et al. 2012; Nikolakakis et al. 2012). These CDI systems are arranged in an alternative *cdiAIB* gene order (Fig. 1) and lack the orphan *cdiA-CT/cdiI* modules present in most other bacteria. *B. pseudomallei* serves as a convenient model to study *Burkholderia* CDI diversity because several clinical and environmental isolates have been sequenced, revealing significant genome heterogeneity between different strains. Our recent analysis of 85 assembled and unassembled *B. pseudomallei* genomes shows that almost every strain contains one to three predicted CDI systems (Nikolakakis et al. 2012). These systems fall into two classes based on the sequence of the amino-terminal hemagglutinin repeat region within CdiA proteins. The *B. pseudomallei* CDI systems can be further subdivided into 10 sequence types based on the polymorphic CdiA-CT region, which is demarcated by a conserved (Q/E)LYN peptide motif instead of the VENN sequence found in most other bacteria (Nikolakakis et al. 2012). Analysis of three different *B. pseudomallei* CdiA-CT/CdiI pairs confirmed that these proteins do indeed function as toxin-immunity pairs. Each toxin inhibits cell growth by cleaving tRNA at distinct positions within the anticodon loop, T-loop, or aminoacyl acceptor stem; and each tRNase activity is specifically blocked by its cognate CdiI immunity protein (Nikolakakis et al. 2012). The *B. pseudomallei* toxins are modular and can be exchanged between different *Burkholderia* CDI systems when fused at the (Q/E)LYN sequence within CdiA (Nikolakakis et al. 2012). However, *B. pseudomallei* CdiA-CTs are not functional when grafted onto *E. coli* CdiA proteins (Aoki et al. 2010; SK Aoki and DA Low, unpubl.), indicating that CDI pathways are significantly different in these two bacteria.

Rhs Systems Are Functionally Similar to CDI

CdiA proteins share a number of characteristics with the Rhs protein family. The *rhs* genes were first identified in *E. coli* by C.W. Hill and colleagues and were named rearrangement hotspots based on their role in chromosome duplications (Lin et al. 1984; Wang et al. 1998). Rhs proteins are widely distributed throughout the eubacteria, but their function is poorly understood. Like CdiA, Rhs proteins are large, ranging from ~1500 residues in Gram-negative bacteria to over 2000 residues in some Gram-positive species. Rhs proteins also possess a central repeat region, although the characteristic YD peptide repeat of Rhs proteins is unrelated to the hemagglutinin repeats within CdiA. Moreover, Rhs proteins have variable carboxy-terminal domains that are sharply demarcated by a conserved peptide motif (PxxxxDPxGL in the Enterobacteriaceae (Jackson et al. 2009). Remarkably, some Rhs and CdiA proteins share related carboxy-terminal sequences (Poole et al. 2011), indicating that the two protein families carry similar toxin domains. Moreover, CDI and Rhs systems have a similar genetic organization with several *rhs* loci containing one or more "silent cassettes" that resemble orphan *cdiA-CT/cdiI* toxin-immunity modules (Hill et al. 1994; Poole et al. 2011). Like CDI systems, there are zero to multiple *rhs* loci in different bacterial strains (Fig. 1, Rhs = orange and RhsI = red; Table 1). Taken together, the striking parallels between CDI and Rhs suggest that the two systems are functionally analogous.

Several independent lines of experimental evidence indicate that Rhs systems encode functional toxin-immunity protein pairs. Early work from Hill and colleagues showed that overexpression of the carboxy-terminal region of *E. coli* RhsA inhibited recovery from stationary phase (Vlazny and Hill 1995). This inhibitory activity appeared to be neutralized by coexpression of the downstream open reading frame (Vlazny and Hill 1995), suggesting this gene (*yibA*) encodes an antitoxin protein that blocks RhsA activity. A more recent study has shown that RhsA inhibits protein synthesis in *E. coli*, but the precise biochemical mechanism remains unknown (Aggarwal and Lee 2011). We recently showed that the Rhs systems in *D. dadantii* 3937 encode functional toxin-immunity pairs (Poole et al. 2011). The carboxy-terminal regions of *D. dadantii* RhsA and RhsB proteins contain endonuclease NS (PF13930) and HNH endonuclease (PF01844) domains (respectively), and both proteins degrade cellular DNA when expressed in *E. coli* (Koskiniemi et al. 2013). These toxic activities are blocked specifically by small immunity proteins encoded immediately downstream from each *D. dadantii* 3937 *rhs* gene (Poole et al. 2011). These results, together with other reports indicating that Rhs proteins may be exported to the cell surface (Hill et al. 1994; McNulty et al. 2006) are consistent with the hypothesis that Rhs systems mediate a distinct form of CDI. We recently showed that RhsA and RhsB systems in the plant pathogen *D. dadantii*, and WapA, an Rhs homolog in *Bacillus subtilis*, mediate CDI and interbacterial competition (Koskiniemi et al. 2013). Previous work showed that *Pseudomonas aeruginosa* delivers an Rhs protein into eukaryotic cells where it acts as a virulence factor (Kung et al. 2012). These findings not only confirm that Rhs proteins are secreted toxins, but also define a new role for these systems in bacterial pathogenesis.

 Cite this article as *Cold Spring Harb Perspect Med* doi: 10.1101/cshperspect.a010025

CONCLUDING REMARKS

Although CDI was discovered nearly a decade ago, its importance in bacterial competition and cell-to-cell signaling is only beginning to be explored. The mechanisms of toxin delivery during CDI are not well understood, nor are the mechanisms responsible for the remarkable toxin-immunity diversity found throughout CDI systems. Insight into these fundamental features of CDI will significantly advance our understanding of interbacterial competition and contact-mediated signaling between bacterial cells. The orphan modules found in CDI and Rhs systems represent reservoirs of toxin-immunity protein diversity that can be potentially mobilized and deployed when needed, and rapidly lost under nonselective conditions. This concept is novel, but built on a foundation of research showing that gene duplications influence bacterial fitness (Reams et al. 2010) and that recombination provides a mechanism to rapidly alter the bacterial cell surface (Cahoon and Seifert 2009). Our recent data shows that Rhs constitutes a previously unrecognized CDI system (Koskiniemi et al. 2013). The function(s) of Rhs systems has been a mystery since their discovery in *E. coli* 30 years ago (Capage and Hill 1979). Rhs proteins are found throughout the eubacteria and eukaryotes (Minet and Chiquet-Ehrismann 2000), and therefore the elucidation of Rhs mechanisms and biological functions will have an impact on the study of a number of organisms.

ACKNOWLEDGMENTS

We thank Stephanie Aoki for unpublished data and members of the Hayes and Low laboratories for comments and helpful suggestions. This work was supported by grants GM078634 (to C.S.H.) and U54AI065359 (to C.S.H. and D.A.L.) from the National Institutes of Health, and grant 0642052 (to D.A.L.) from the National Science Foundation.

REFERENCES

Aggarwal K, Lee KH. 2011. Overexpression of cloned RhsA sequences perturbs the cellular translational machinery in *Escherichia coli*. *J Bacteriol* **193:** 4869–4880.

Anderson MS, Garcia EC, Cotter PA. 2012. The Burkholderia *bcpAIOB* genes define unique classes of two-partner secretion and contact dependent growth inhibition systems. *PLoS Genet* **8:** e1002877.

Aoki SK, Pamma R, Hernday AD, Bickham JE, Braaten BA, Low DA. 2005. Contact-dependent inhibition of growth in *Escherichia coli*. *Science* **309:** 1245–1248.

Aoki SK, Malinverni JC, Jacoby K, Thomas B, Pamma R, Trinh BN, Remers S, Webb J, Braaten BA, Silhavy TJ, et al. 2008. Contact-dependent growth inhibition requires the essential outer membrane protein BamA (YaeT) as the receptor and the inner membrane transport protein AcrB. *Mol Microbiol* **70:** 323–340.

Aoki SK, Webb JS, Braaten BA, Low DA. 2009. Contact-dependent growth inhibition causes reversible metabolic downregulation in *Escherichia coli*. *J Bacteriol* **191:** 1777–1786.

Aoki SK, Diner EJ, de Roodenbeke CT, Burgess BR, Poole SJ, Braaten BA, Jones AM, Webb JS, Hayes CS, Cotter PA, et al. 2010. A widespread family of polymorphic contact-dependent toxin delivery systems in bacteria. *Nature* **468:** 439–442.

Bakken JS, Borody T, Brandt LJ, Brill JV, Demarco DC, Franzos MA, Kelly C, Khoruts A, Louie T, Martinelli LP, et al. 2011. Treating *Clostridium difficile* infection with fecal microbiota transplantation. *Clin Gastroenterol Hepatol* **9:** 1044–1049.

Benson AK, Kelly SA, Legge R, Ma F, Low SJ, Kim J, Zhang M, Oh PL, Nehrenberg D, Hua K, et al. 2010. Individuality in gut microbiota composition is a complex polygenic trait shaped by multiple environmental and host genetic factors. *Proc Natl Acad Sci* **107:** 18933–18938.

Bos MP, Robert V, Tommassen J. 2007. Biogenesis of the Gram-negative bacterial outer membrane. *Ann Rev Microbiol* **61:** 191–214.

Cahoon LA, Seifert HS. 2009. An alternative DNA structure is necessary for pilin antigenic variation in *Neisseria gonorrhoeae*. *Science* **325:** 764–767.

Capage M, Hill CW. 1979. Preferential unequal recombination in the glyS region of the *Escherichia coli* chromosome. *J Mol Biol* **127:** 73–87.

Chen G, Swem LR, Swem DL, Stauff DL, O'Loughlin CT, Jeffrey PD, Bassler BL, Hughson FM. 2011. A strategy for antagonizing quorum sensing. *Mol Cell* **42:** 199–209.

Diner EJ, Beck CM, Webb JS, Low DA, Hayes CS. 2012. Identification of a target cell permissive factor required for contact-dependent growth inhibition (CDI). *Genes Dev* **26:** 515–525.

Dominguez-Bello MG, Blaser MJ. 2008. Do you have a probiotic in your future? *Microbes Infect /Institut Pasteur* **10:** 1072–1076.

Hayes CS, Aoki SK, Low DA. 2010. Bacterial contact-dependent delivery systems. *Annu Rev Genet* **44:** 71–90.

Hill CW, Sandt CH, Vlazny DA. 1994. Rhs elements of *Escherichia coli:* A family of genetic composites each encoding a large mosaic protein. *Mol Microbiol* **12:** 865–871.

Ichinohe T, Pang IK, Kumamoto Y, Peaper DR, Ho JH, Murray TS, Iwasaki A. 2011. Microbiota regulates immune defense against respiratory tract influenza A virus infection. *Proc Natl Acad Sci* **108:** 5354–5359.

Jackson AP, Thomas GH, Parkhill J, Thomson NR. 2009. Evolutionary diversification of an ancient gene family (*rhs*) through C-terminal displacement. *BMC Genomics* **10:** 584.

Jacob-Dubuisson F, Guerin J, Baelen S, Clantin B. 2013. Two-partner secretion: As simple as it sounds? *Res Microbiol* **164:** 583–595.

Kajava AV, Cheng N, Cleaver R, Kessel M, Simon MN, Willery E, Jacob-Dubuisson F, Locht C, Steven AC. 2001. β-Helix model for the filamentous haemagglutinin adhesin of *Bordetella pertussis* and related bacterial secretory proteins. *Mol Microbiol* **42:** 279–292.

Keller L, Surette MG. 2006. Communication in bacteria: An ecological and evolutionary perspective. *Nat Rev Microbiol* **4:** 249–258.

Knowles TJ, Scott-Tucker A, Overduin M, Henderson IR. 2009. Membrane protein architects: The role of the BAM complex in outer membrane protein assembly. *Nat Rev Microbiol* **7:** 206–214.

Konovalova A, Sogaard-Andersen L. 2011. Close encounters: Contact-dependent interactions in bacteria. *Mol Microbiol* **81:** 297–301.

Koskiniemi S, Lamoureux JG, Nikolakakis KC, t'Kint de Roodenbeke C, Kaplan MD, Low DA, Hayes CS. 2013. Rhs proteins from diverse bacteria mediate intercellular competition. *Proc Natl Acad Sci* **110:** 7032–7037.

Kung VL, Khare S, Stehlik C, Bacon EM, Hughes AJ, Hauser AR. 2012. An rhs gene of *Pseudomonas aeruginosa* encodes a virulence protein that activates the inflammasome. *Proc Natl Acad Sci* **109:** 1275–1280.

Lederberg J. 2000. Infectious history. *Science* **288:** 287–293.

Ley RE. 2010. Obesity and the human microbiome. *Curr Opin Gastroenterol* **26:** 5–11.

Lin RJ, Capage M, Hill CW. 1984. A repetitive DNA sequence, rhs, responsible for duplications within the *Escherichia coli* K-12 chromosome. *J Mol Biol* **177:** 1–18.

McNulty C, Thompson J, Barrett B, Lord L, Andersen C, Roberts IS. 2006. The cell surface expression of group 2 capsular polysaccharides in *Escherichia coli:* The role of KpsD, RhsA and a multi-protein complex at the pole of the cell. *Mol Microbiol* **59:** 907–922.

Minet AD, Chiquet-Ehrismann R. 2000. Phylogenetic analysis of teneurin genes and comparison to the rearrangement hot spot elements of *E. coli*. *Gene* **257:** 87–97.

Ng WL, Bassler BL. 2009. Bacterial quorum-sensing network architectures. *Annu Rev Genet* **43:** 197–222.

Nikolakakis K, Amber S, Wilbur JS, Diner EJ, Aoki SK, Poole SJ, Tuanyok A, Keim PS, Peacock S, Hayes CS, et al. 2012. The toxin/immunity network of *Burkholderia pseudomallei* contact-dependent growth inhibition (CDI) systems. *Mol Microbiol* **84:** 516–529.

Noinaj N, Kuszak AJ, Gumbart JC, Lukacik P, Chang H, Easley NC, Lithgow T, Buchanan SK. 2013. Structural insight into the biogenesis of β-barrel membrane proteins. *Nature* **501:** 385–390.

Pathak DT, Wei X, Bucuvalas A, Haft DH, Gerloff DL, Wall D. 2012. Cell contact-dependent outer membrane exchange in myxobacteria: Genetic determinants and mechanism. *PLoS Genet* **8:** e1002626.

Poole SJ, Diner EJ, Aoki SK, Braaten BA, T'Kint de Roodenbeke C, Low DA, Hayes CS. 2011. Identification of functional toxin/immunity genes linked to contact-dependent growth inhibition (CDI) and rearrangement hotspot (Rhs) systems. *PLoS Genet* **7:** e1002217.

Reams AB, Kofoid E, Savageau M, Roth JR. 2010. Duplication frequency in a population of *Salmonella enterica* rapidly approaches steady state with or without recombination. *Genetics* **184:** 1077–1094.

Ricci DP, Silhavy TJ. 2012. The Bam machine: A molecular cooper. *Biochim Biophys Acta* **1818:** 1067–1084.

Ruhe ZC, Wallace AB, Low DA, Hayes CS. 2013. Receptor polymorphism restricts contact-dependent growth inhibition to members of the same species. *mBio* **4:** e00480.

Vlazny DA, Hill CW. 1995. A stationary-phase-dependent viability block governed by two different polypeptides from the RhsA genetic element of *Escherichia coli* K-12. *J Bacteriol* **177:** 2209–2213.

Wang YD, Zhao S, Hill CW. 1998. Rhs elements comprise three subfamilies which diverged prior to acquisition by *Escherichia coli*. *J Bacteriol* **180:** 4102–4110.

Zhang D, Iyer LM, Aravind L. 2011. A novel immunity system for bacterial nucleic acid degrading toxins and its recruitment in various eukaryotic and DNA viral systems. *Nucleic Acids Res* **39:** 4532–4552.

Bacterial Biofilms: Development, Dispersal, and Therapeutic Strategies in the Dawn of the Postantibiotic Era

Maria Kostakioti[1,4,5], Maria Hadjifrangiskou[2,4], and Scott J. Hultgren[1,3]

[1]Department of Molecular Microbiology and Microbial Pathogenesis, Washington University in Saint Louis School of Medicine, St. Louis, Missouri 63110-1010

[2]Department of Pathology, Microbiology and Immunology, Vanderbilt University School of Medicine, Nashville, Tennessee 37232-2363

[3]Center for Women's Infectious Disease Research, Washington University in Saint Louis School of Medicine, St. Louis, Missouri 63110-1010

Correspondence: hultgren@borcim.wustl.edu

Biofilm formation constitutes an alternative lifestyle in which microorganisms adopt a multicellular behavior that facilitates and/or prolongs survival in diverse environmental niches. Biofilms form on biotic and abiotic surfaces both in the environment and in the healthcare setting. In hospital wards, the formation of biofilms on vents and medical equipment enables pathogens to persist as reservoirs that can readily spread to patients. Inside the host, biofilms allow pathogens to subvert innate immune defenses and are thus associated with long-term persistence. Here we provide a general review of the steps leading to biofilm formation on surfaces and within eukaryotic cells, highlighting several medically important pathogens, and discuss recent advances on novel strategies aimed at biofilm prevention and/or dissolution.

Biofilm formation enables single-cell organisms to assume a temporary multicellular lifestyle, in which "group behavior" facilitates survival in adverse environments. What was once defined as the formation of a community of microorganisms attached to a surface has come to be recognized as a complex developmental process that is multifaceted and dynamic in nature. The transition from planktonic growth to biofilm occurs in response to environmental changes, and involves multiple regulatory networks, which translate signals to concerted gene expression changes thereby mediating the spatial and temporal reorganization of the bacterial cell (Pratt and Kolter 1998; O'Toole et al. 2000; Prigent-Combaret et al. 2001; Parsek and Singh 2003; Lenz et al. 2008; Monds and O'Toole 2009). This cellular reprogramming alters the expression of surface molecules, nutrient utilization, and virulence factors and equips bacteria with an arsenal of properties that enable their survival in unfavorable

[4]These authors contributed equally to this work.

[5]Present address: Monsanto Company, 800 N. Lindbergh Blvd., Saint Louis, Missouri 63167.

Cite this article as *Cold Spring Harb Perspect Med* doi: 10.1101/cshperspect.a010306

conditions (Whiteley et al. 2001; Schembri et al. 2003; Stanley et al. 2003; Bagge et al. 2004; Beloin et al. 2004; Vuong et al. 2004; Lenz et al. 2008; Zhang and Mah 2008; Klebensberger et al. 2009).

Within the biofilm, bacteria are cocooned in a self-produced extracellular matrix, which accounts for ~90% of the biomass (Flemming and Wingender 2010). The matrix is composed of extracellular polymeric substances (EPS) that, along with carbohydrate-binding proteins (Tielker et al. 2005; Branda et al. 2006; Diggle et al. 2006), pili, flagella, other adhesive fibers (Zogaj et al. 2001; Pinkner et al. 2006; Cegelski et al. 2009), and extracellular DNA (eDNA) (Whitchurch et al. 2002; Palchevskiy and Finkel 2006; Qin et al. 2007; Yang et al. 2007; Thomas et al. 2008, 2009; Guiton et al. 2009; Vilain et al. 2009), act as a stabilizing scaffold for the three-dimensional biofilm structure (Fig. 1). In the matrix, nutrients are trapped for metabolic utilizations by the resident bacteria and water is efficiently retained through H-bond interactions with hydrophilic polysaccharides (Conrad et al. 2003; Flemming and Wingender 2010). Enzymes secreted by the bacteria modify EPS composition in response to changes in nutrient availability (Sauer et al. 2004; Gjermansen et al. 2005), thereby tailoring biofilm architecture to the specific environment (Sauer et al. 2004; Ma et al. 2009). Thus, the structural components of the matrix give rise to a highly hydrated, robust structure with high tensile strength that keeps bacteria in close proximity, enabling intimate cell-to-cell interactions and DNA exchange (Flemming and Wingender 2010; Koo et al. 2010), while protecting the biomass from desiccation, predation, oxidizing molecules, radiation, and other damaging agents (Walters et al. 2003; Jefferson et al. 2005; Mai-Prochnow et al. 2008; Flemming and Wingender 2010). The resilient nature of biofilms is also partly attributed to the presence of environmental gradients within the biomass, which give rise to community "division of labor" with subpopulations of bacteria showing differential gene expression in response to local nutrient and oxygen availability (Lewis 2005; Domka et al. 2007). Studies have shown the presence of metabolically inactive nondividing persister cells within biofilms, which are tolerant to a number of antibiotics despite the fact that they are genetically identical to the rest of the bacterial population (Lewis 2005, 2008). These are believed to be responsible for the reseeding of biofilms on cessation of antibiotic treatment in the clinical setting (Lewis 2005, 2008).

Inside the host, the matrix protects biofilm bacteria from exposure to innate immune defenses (such as opsonization and phagocytosis) and antibiotic treatments (Jesaitis et al. 2003; Walters et al. 2003; Jefferson et al. 2005; Leid et al. 2005; Cerca et al. 2006, 2007). Interbacterial interactions can promote the spread of drug-resistance markers and other virulence factors (Vuong et al. 2004). As a result, biofilm-forming pathogens persist, establishing chronic and recalcitrant infections such as upper respiratory infections (*Pseudomonas aeruginosa*) (Koch and Hoiby 1993; Govan and Deretic 1996), urinary tract infections (UTIs) (uropathogenic *Escherichia coli* [UPEC], *Klebsiella pneumoniae*) (Foxman 2010), periodontitis (mixed biofilms of *Streptococcus mutans* and other bacteria) (Kuramitsu and Wang 2011), catheter-induced and other device-associated infections (*E. coli*, *Enterococcus faecalis*, and others) (Venditti et al. 1993; Ferrieres et al. 2007; Jacobsen et al. 2008; Fey 2010). Especially in immunocompromised patients, the manifestation of infections by opportunistic biofilm-forming pathogens can be devastating, leading to severe symptoms and, in many instances, death.

Here we review the processes leading to the formation of extracellular and intracellular biofilms, highlighting several medically important pathogens. Given the prevalence and recalcitrance of biofilm-related infections, we also provide a synopsis of the most recent advances in the development of novel antibiofilm strategies.

EXTRACELLULAR BIOFILM FORMATION

Bacterial Adherence on Surfaces—What Does It Take to Stick and Stick Around?

Bacterial aggregation and subsequent biofilm maturation consists of reversible and irreversible stages and involves numerous conserved

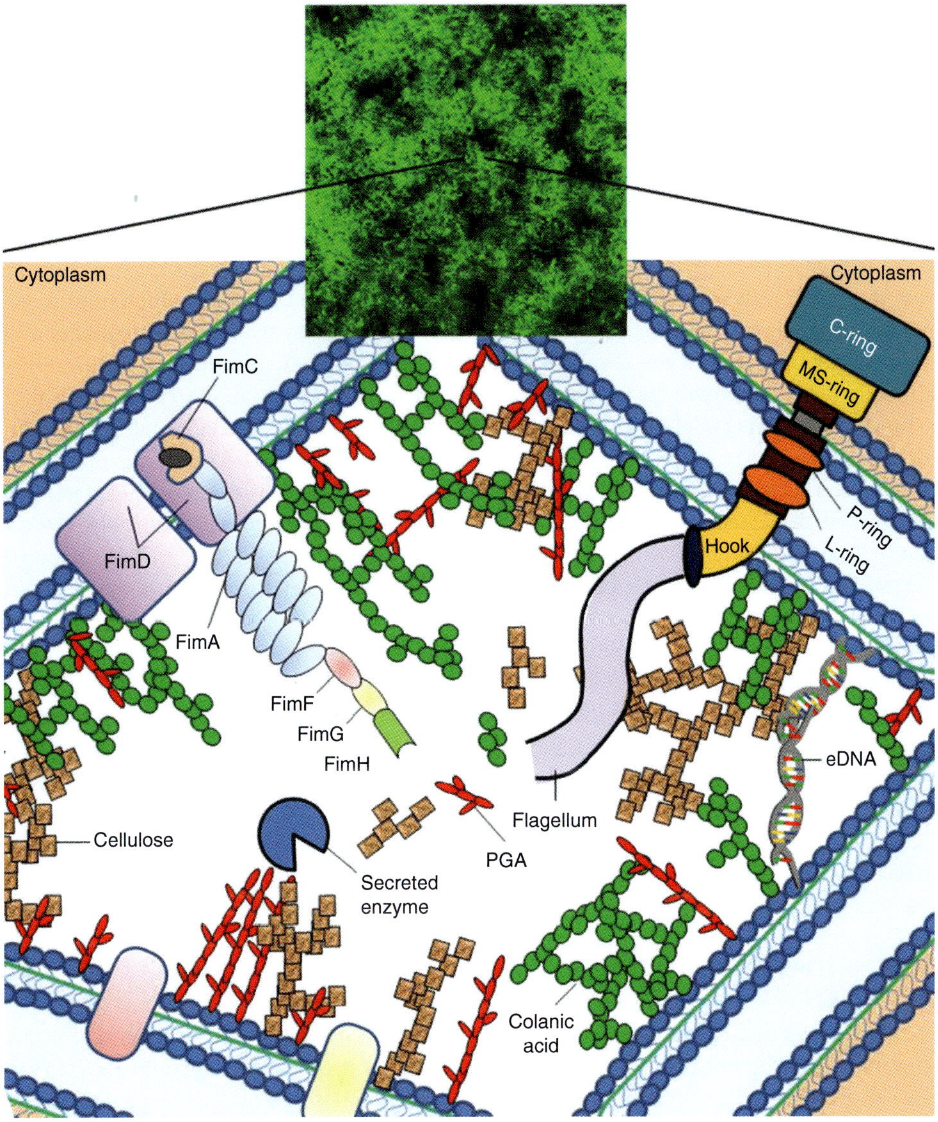

Figure 1. Schematic of the extracellular matrix composition in *E. coli*. Structural components include the EPS molecules colanic acid, cellulose, and PGA (polyglucosamine), which enable intercellular interactions, keeping bacteria in close proximity with each other. eDNA also serves as a connecting agent, as well as a nutritional source. Extracellular organelles such as flagella and CUP (chaperone usher pathway) pili enable bacterial aggregation strengthening the biofilm lattice. Secreted enzymes modify EPS components in response to environmental changes.

and/or species-specific factors. The first step involves the introduction of bacteria to a surface, a process which is at least in part stochastic, driven by Brownian motion and gravitational forces, and influenced by surrounding hydrodynamic forces (Donlan 2002; Beloin et al. 2008). Within a niche, bacteria encounter attractive or repelling forces that vary depending on nutrient levels, pH, ionic strength, and temperature. Medium properties, along with bacterial cell-surface composition affect velocity and direction toward or away from the contact surface

(Donlan 2002). Motile bacteria have a competitive advantage, utilizing flagella to overcome hydrodynamic and repulsive forces. The importance of flagellar motility for initial attachment has been documented for several pathogens, including *P. aeruginosa*, *Vibrio cholerae*, *Listeria monocytogenes*, and *E. coli* (O'Toole and Kolter 1998; Pratt and Kolter 1998; Watnick and Kolter 1999; Klausen et al. 2003a,b; Lemon et al. 2007; Toutain et al. 2007). In some bacterial species, chemotaxis also plays a role in directing attachment in response to nutrient composition; mutations in the CheR1 methyltransferase have been shown to alter the amino acid response of *P. aeruginosa* and impair attachment and biofilm maturation (Schmidt et al. 2011). Previous studies showed that chemotaxis is dispensable in *E. coli* (Pratt and Kolter 1998); however, recent investigations have revealed that disruption of the methyl-accepting chemotaxis protein II (*tar*), imparts biofilm defects in UPEC (Hadjifrangiskou et al. 2012).

Upon intercepting the surface, adherence is mediated by additional extracellular adhesive appendages and secreted adhesins. However, the decision to "stick" is not absolute; initial attachment is dynamic and reversible, during which bacteria can detach and rejoin the planktonic population if perturbed by hydrodynamic forces (sloughing bacteria off the surface), repulsive forces (Dunne 2002), or in response to nutrient availability (Banin et al. 2005; Anderson et al. 2008; Wu and Outten 2009).

Irreversible attachment is attained by bacteria that can weather shear forces and maintain a steadfast grip on the surface. UPEC and other *E. coli* pathotypes rely heavily on type 1 pili (Mulvey et al. 1998; Pratt and Kolter 1998; Martinez et al. 2000; Hung et al. 2002; Anderson et al. 2003; Beloin et al. 2008), which are multisubunit adhesive organelles assembled by the chaperone usher pathway (CUP) (Waksman and Hultgren 2009). UPEC harbor numerous CUP pili systems, which are differentially expressed and are presumed to facilitate adherence in a niche-specific manner (Welch et al. 2002; Chen et al. 2006, 2009; Hadjifrangiskou et al. 2011; Spurbeck et al. 2011). Adherence is mediated by the FimH adhesin at the tip of type 1 pili, which recognizes mannosylated moieties (Thankavel et al. 1997; Martinez et al. 2000; Zhou et al. 2001; Hung et al. 2002; Bouckaert et al. 2005; Eto et al. 2007; Nilsson et al. 2007; Wellens et al. 2008; Thumbikat et al. 2009). FimH is thought to play a critical role in UPEC pathogenesis; it mediates binding and invasion to human bladder epithelial cells, binds to human uroplakin, and is critical in a murine preclinical model of cystitis, which mimics human disease (Mulvey et al. 1998; Martinez et al. 2000; Kau et al. 2005; Bishop et al. 2007; Eto et al. 2007; Garofalo et al. 2007; Rosen et al. 2007; Wright et al. 2007; Chen et al. 2009). FimH is found under positive selection in UPEC, consistent with its role as a virulence factor in human disease (Sokurenko et al. 1994, 1995, 1998, 2004; Chen et al. 2006, 2009; Weissman et al. 2007; Wright et al. 2007) and has been said to have fulfilled Koch's postulates (Connell et al. 1996; Snyder et al. 2006). In addition to type 1 pili, curli fibers and Antigen 43 have been shown to mediate attachment and interbacterial interactions on abiotic surfaces (Henderson et al. 1997; Hasman et al. 1999; Danese et al. 2000a; Kjaergaard et al. 2000; Ulett et al. 2007; Cegelski et al. 2009). Curli also facilitates binding to the eukaryotic extracellular matrix components laminin, fibronectin, and plasminogen (Vidal et al. 1998; Cookson et al. 2002; Uhlich et al. 2006).

P. aeruginosa, an important pathogen and avid biofilm former, also uses several attachment organelles to irreversibly adhere to a surface. Besides flagella, *P. aeruginosa* uses type IV pili-mediated twitching motility to wade through the liquid interface and contact the surface, maintain adherence, and move across the attachment plane (O'Toole and Kolter 1998; Klausen et al. 2003a,b). Similar to UPEC, *P. aeruginosa* express numerous CUP fimbriae, of which CupA is involved in surface adherence and autoaggregation (Vallet et al. 2001; Klebensberger et al. 2009).

In contrast to *Pseudomonas* and UPEC, the Gram-positive *Enterococci* are nonmotile and, up until recently, were thought to possess no adhesive pili. Over the years, investigations identified a panel of enterococcal adhesins that mediate adherence to eukaryotic extracellular matrix components. Examples include SagA, Acm (*E. faecium*), and Ace (*E. faecalis*), which bind collagen (Mohamed et al. 2006), and the surface protein Esp, which has been shown to promote biofilm formation on abiotic surfaces in *esp*-expressing *E. faecalis* strains (Toledo-Arana et al. 2001). Recent studies elucidated the presence of *Enterococcal* biofilm pili (Ebp) in *E. faecalis* and showed their contribution to biofilm formation, endocarditis, and urinary tract infection (Ton-That and Schneewind 2003; Ton-That et al. 2004; Nallapareddy et al. 2006; Kemp et al. 2007; Guiton et al. 2009; Kline et al. 2010).

Biofilm Maturation—Keeping It Together

Surface contact triggers responses that lead to gene expression changes, up-regulating factors favoring sessility, such as those implicated in the formation of the extracellular matrix (Prigent-Combaret and Lejeune 1999; Otto and Silhavy 2002; Inagaki et al. 2005; Morici et al. 2007; Beloin et al. 2008; Bhomkar et al. 2010). In the case of *E. coli*, relatively little is known about matrix constituents. Cellulose was first identified as an important component of commensal *E. coli* pellicle biofilms, and was later shown to be coexpressed with curli in UPEC and gastrointestinal *E. coli* isolates (Zogaj et al. 2001, 2003; Romling 2002; Bokranz et al. 2005; Kai-Larsen et al. 2010). Curli are amyloid fibers that are critical for the formation of pellicle biofilms, as curli inhibitors (curlicides) inhibit pellicle formation and curli mutants cannot form pellicles (Cegelski et al. 2009). Additional studies showed that polyglucosamine (PGA) and colanic acid contribute to biofilm architecture (Prigent-Combaret and Lejeune 1999; Danese et al. 2000b; Prigent-Combaret et al. 2001; Wang et al. 2004; Agladze et al. 2005), with PGA being prevalent among clinical isolates, including UPEC (Cerca et al. 2007). More detailed analyses are required for a complete characterization of the extracellular matrix in pathogenic *E. coli.*

Extracellular matrix composition has been more extensively investigated in *P. aeruginosa*, and has been shown to vary depending on environmental conditions (Harmsen et al. 2010). Two primary EPS components are Pel and Psl (Friedman and Kolter 2004a,b; Jackson et al. 2004; Matsukawa and Greenberg 2004; Vasseur et al. 2005; Ma et al. 2006). Psl augments *Pseudomonas* attachment to mucin and airway epithelial cells (Ma et al. 2006), whereas increased expression of *pel* in small colony variants isolated from cystic fibrosis patients has been associated with *P. aeruginosa* persistence in lung airways (Starkey et al. 2009). Recently, Borlee and colleagues identified CdrA, a large secreted adhesin, which is expressed in the biofilm in response to high levels of the universal signal 3,5-cyclic diguanylic acid (c-di-GMP) and binds Psl, stabilizing biofilm structures (Borlee et al. 2010). Alginate, another *P. aeruginosa* EPS component, has been associated with increased resistance to antibiotic treatments and host immune defenses during chronic infection (Govan and Deretic 1996; Hatch and Schiller 1998; Hentzer et al. 2001; Leid et al. 2005). As is the case with Pel and Psl, alginate production is subject to regulation by fluctuating levels of c-di-GMP. Recent studies have shown that a surface-bound diguanylate cyclase MucR positively activates alginate synthesis, presumably through high local concentrations of c-di-GMP (Hay et al. 2009). In addition to EPS, several studies have shown that eDNA is critical for cell-to-cell connections and stabilization of *Pseudomonas* biofilms (Whitchurch et al. 2002; Yang et al. 2007). Young *Pseudomonas* biofilms are more sensitive to DNase treatment compared with mature biofilms, indicating a stabilizing role for eDNA during the initial biofilm stages when EPS components are not as abundant (Whitchurch et al. 2002). As the biofilm matures, eDNA amounts increase through lysis of a bacterial subpopulation in response to the *P. aeruginosa* quinolone signal (Pqs) quorum sensing system (Allesen-Holm et al. 2006). Allesen-Holm et al. showed that eDNA is organized in distinct patterns and localizes in the stalk portion of the mushroom-shaped biofilms (Allesen-Holm et al. 2006). This localization may act as a scaffold for the formation of the mushroom structure, as type IV pili show high eDNA binding affinity, inducing the accumulation of migrating bacteria toward areas of high eDNA concentration (Barken et al. 2008).

The contribution of eDNA to biofilm architecture has also been reported for *E. faecalis*, making it one of the few known *E. faecalis* matrix components. Thomas et al. first reported that eDNA is critical for *E. faecalis* biofilms and identified that the secreted enzymes GelE (zinc metalloprotease) and SprE (serine protease) influence biofilm formation by affecting cellular autolysis and DNA release (Thomas et al. 2008, 2009). In a separate study, Mohamed et al. reported that a mutant lacking the Atn autolysin had 30% reduction in biofilm (Mohamed et al. 2004). Guiton and colleagues later established that Atn plays a role in the temporal regulation of DNA release at specific stages during biofilm formation (Guiton et al. 2009).

Escape from the Matrix—Dispersing Mechanisms

Within the mature biofilm there is a bustling community that actively exchanges and shares products that play a pivotal role in maintaining biofilm architecture and providing a favorable living environment for the resident bacteria. However, as biofilms mature, dispersal becomes an option. Besides passive dispersal, brought about by shear stresses, bacteria have evolved ways to perceive environmental changes and gauge whether it is still beneficial to reside within the biofilm or whether it is time to resume a planktonic lifestyle. Biofilm dispersal can be the result of several cues, such as alterations in nutrient availability, oxygen fluctuations and increase of toxic products, or other stress-inducing conditions (Sauer et al. 2004; Karatan and Watnick 2009; Hong et al. 2010; Rowe et al. 2010). In UPEC, increase in extracellular iron induces biofilm dispersal (Rowe et al. 2010), whereas *P. aeruginosa* biofilms disperse in response to increased amounts of various carbon and nitrogen sources (Sauer et al. 2004; Karatan and Watnick 2009). Several sensory systems monitor the levels of small molecules, as a proxy to environmental changes, and alter gene expression accordingly, promoting dispersal (Hammer and Bassler 2003; Kaplan 2010). Among other signals, the universal c-di-GMP has been extensively implicated in the shift between sessility and motility in bacteria, including *P. aeruginosa* and *E. coli*. Typically, an increase in c-di-GMP favors sessility, whereas reduced c-di-GMP leads to up-regulation of motility (Morgan et al. 2006; Pruss et al. 2006; Barraud et al. 2009; Wood et al. 2010). Ma et al. recently reported that a c-di-GMP binding protein, BdcA, is at least partly responsible for the reduction of available c-di-GMP in biofilm communities, down-regulation of EPS, and up-regulation of swimming and swarming motility; a phenomenon that the investigators showed also occurs in *Pseudomonas* species and *Rhizobium mellioti* (Ma et al. 2011a,b).

EPS-degrading enzymes, such as alginate lyase in *P. aeruginosa*, also contribute to bacterial detachment from the matrix (Boyd and Chakrabarty 1994). In *E. coli*, the CsrA protein was shown to repress PGA synthesis, also aiding in dispersion (Wang et al. 2005). Besides down-regulating EPS, surfactant molecules are produced, reducing surface-bacterial interactions; for example, although controlled rhamnolipid production contributes to channel formation within mature *P. aeruginosa* biofilms, an increase in rhamnolipid levels aids bacterial dispersal (Boles et al. 2005; Dong et al. 2008; Harmsen et al. 2010). In addition, studies have identified flagellated subpopulations within *P. aeruginosa* biofilms, which emigrate from the biofilm, creating microcolonies with a central void (Purevdorj-Gage et al. 2005; Harmsen et al. 2010). Voids within the biofilm are also created by cell death, serving as an additional dispersal mechanism that frees resident live bacteria, as shown by studies in *P. aeruginosa* (Webb et al. 2003). Dispersing bacteria have the capacity to reinitiate the process of biofilm formation, on encountering a suitable environment.

Studies using *Bacillus subtilis* as a model organism revealed another sophisticated dispersal mechanism that may be widespread among bacteria. *B. subtilis* forms robust biofilms, which lose their integrity after 5–8 d; Kolodkin-Gal and colleagues found that biofilm disassembly is facilitated by a mixture of D-amino acids

(D-leucine, D-methionine, D-tyrosine, and D-tryptophan) that are produced during the stationary phase of growth and get incorporated into the peptide side chains of peptidoglycan in place of the terminal D-alanine (Lam et al. 2009; Kolodkin-Gal et al. 2010). This D-amino acid incorporation interferes with the anchoring of adhesive fibers on the cell surface, leading to fiber dissociation and loss of bacterial adherence, without influencing bacterial growth or expression of matrix components (Kolodkin-Gal et al. 2010). Exogenous addition of the D-amino acid mixture or the individual D-amino acids disrupted preformed biofilms of *B. subtilis* and other bacterial species (Kolodkin-Gal et al. 2010). Further studies revealed that D-amino acids work together with norspermidine, another factor produced by *B. subtilus*, to cause biofilm disassembly (Kolodkin-Gal et al. 2012). Thus, D-amino acid/norspcrimidine treatment may hold promising potential in preventing or eradicating biofilms.

THE LIFE WITHIN—INTRACELLULAR BIOFILMS

Accumulating evidence indicates that many bacterial pathogens previously considered as strictly extracellular can persist inside the host by adapting an intracellular lifestyle that involves the formation of bacterial communities with biofilm-like properties. These intracellular bacterial communities (IBCs) were first documented for UPEC, using a murine model of infection (Mulvey et al. 1998; Anderson et al. 2003; Justice et al. 2004). UPEC use type 1 pili to bind mannosylated receptors on the superficial umbrella bladder cells (Zhou et al. 2001; Hung et al. 2002; Bouckaert et al. 2005; Eto et al. 2007; Wellens et al. 2008; Thumbikat et al. 2009), triggering events that lead to bacterial internalization. Although internalized UPEC are expelled in a TLR-4-dependent process (Bishop et al. 2007), some bacteria avoid the exocytic process and escape into the host-cell cytoplasm, where they replicate into IBCs (Anderson et al. 2003; Justice et al. 2004).

IBCs progress through several developmental stages that show distinct morphological characteristics (Fig. 2) (Justice et al. 2004). During the first 6 h following bladder inoculation, UPEC divide rapidly (doubling time of ~30–35 min) resulting in small clusters of loosely associated rods (early IBCs), morphing into coccoid-shaped bacteria, with an average length of 0.7 μm that begin packing into a tight biomass. Then, between 6 and 8 h, the growth rate drops dramatically, resulting in doubling times >60 min. At this stage, bacteria are tightly packed together forming a highly organized sphere inside the cell that comprises the mature middle-stage IBC (Fig. 2). The number of IBCs can range between 3 and 700 IBCs in an infected bladder; each IBC is clonal and composed of ~10^4–10^5 bacteria (Anderson et al. 2003; Schwartz et al. 2011). IBC bacteria are surrounded by numerous fibers that emanate from the bacterial surface, resembling an extracellular matrix and encasing bacteria in individualized compartments (Anderson et al. 2004). Polysaccharides, such as the sialic acid capsule, are also present throughout the IBC and function, in part, to protect the bacteria from neutrophil attack (Anderson et al. 2010). Similar to extracellular biofilms, IBCs are heterogeneous, composed of subpopulations with different gene expression patterns (Anderson et al. 2004).

As IBCs enlarge, the bacterial mass pushes against the host-cell membrane creating a pod-like protrusion on the surface of the infected cell (Anderson et al. 2003). Eventually, UPEC at the IBC periphery detach as single rods or filaments, and flux out of the infected cell into the bladder lumen where they can reinitiate the process by binding and invading naive epithelial cells (Justice et al. 2004). The cell division inhibitor SulA has been shown to be important for filamentation and dispersal of UPEC from the biomass and, thus, establishment of next-generation IBCs (Justice et al. 2006). UPEC filaments have been shown to be a common feature in the urines of patients with UTI, but not in otherwise healthy controls (Rosen et al. 2007). Further, UPEC isolated from the urine of patients with a UTI have been shown to form IBCs when inoculated into the bladders of six different strains of mice, indicating that IBCs are important for human infection (Garofalo

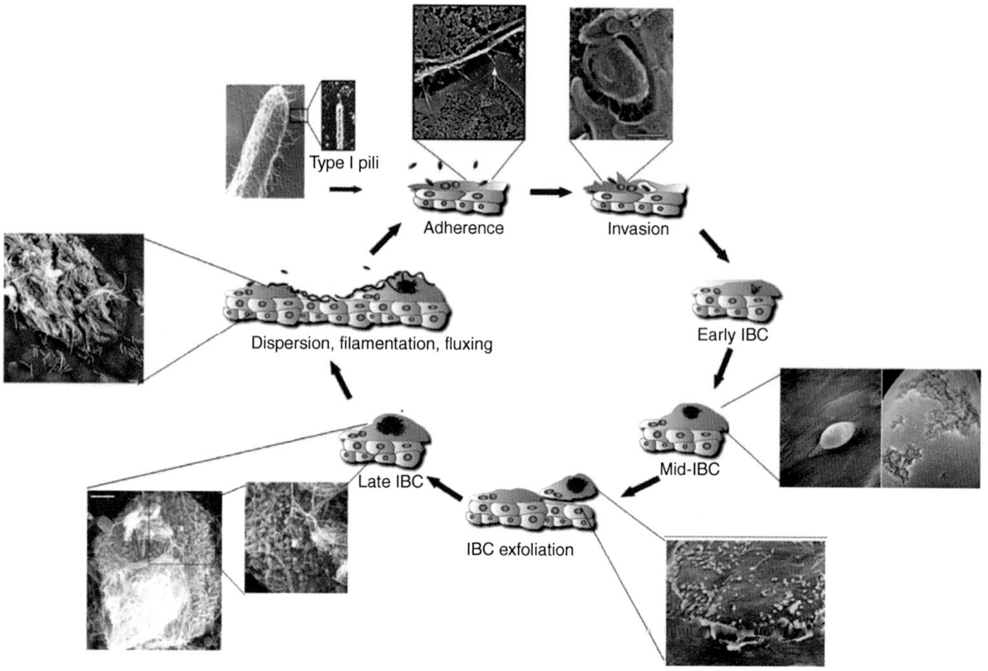

Figure 2. Schematic of the IBC developmental cascade in UPEC (uropathogenic *Escherichia coli*), accompanied by SEM (scanning electron microscopy) images depicting the distinct morphological changes from attachment and invasion to filamentation and dispersal. (SEM images from Anderson et al. 2003, Hultgren Lab.)

et al. 2007). IBC formation is restricted by severe molecular bottlenecks, and higher IBC numbers during acute infection are associated with founding the development of chronic cystitis, indicating the importance of the intracellular pathway in UTI pathogenesis (Hannan et al. 2010; Schwartz et al. 2011).

The IBC cycle is FimH dependent, as interruption of type 1 pili expression after the invasion step, disrupts normal IBC development and leads to UPEC attenuation (Wright et al. 2007). The QseBC two-component system is one of the factors influencing type 1 pili, curli expression, and IBC formation. Recent studies showed that deletion of the QseC sensor results in overactivation of the cognate response regulator QseB, which leads to virulence gene down-regulation by interfering with core metabolic processes (Kostakioti et al. 2009; Hadjifrangiskou et al. 2011). These studies also showed that the UPEC intracellular pathway requires completion of the TCA cycle (Hadjifrangiskou et al. 2011). Microarray and qPCR analyses probing the expression patterns of UPEC within IBCs revealed that iron-acquisition systems are highly up-regulated, indicating the importance of these systems for intracellular biofilm formation (Reigstad et al. 2007). Henderson et al. later showed that the same iron-acquisition systems are prevalent among UPEC isolates (Henderson et al. 2009).

Intracellular communities have also been reported for *K. pneumoniae*, which accounts for up to 5% of community-acquired UTIs, and is more prevalent in diabetic patients and in the nosocomial setting (Lye et al. 1992; Hansen et al. 1998). Similar to UPEC, type 1 pili mediate *K. pneumoniae* invasion and IBC formation, albeit with differences in pili expression kinetics and numbers of formed IBCs and filaments (Rosen et al. 2008a,b).

The ability to occupy an intracellular niche and persist inside the host by transitioning from single cell to a multicellular community is not confined to uropathogens. Using cell lines and animal models of acute lung infection, Garcia-

Medina and colleagues showed that following infection, *P. aeruginosa* can form clusters within the airway cells that matured to a podlike structure, similar in morphology to UPEC and *Klebsiella* IBCs (Garcia-Medina et al. 2005). Bacteria within the *Pseudomonas* pod structure showed regional variation in their expression patterns similar to what was reported for UPEC, which is a typical characteristic of extracellular biofilms (Garcia-Medina et al. 2005).

The ability to form intracellular biofilms may be an evolutionary adaptation that facilitates bacterial persistence to a level extending even beyond that attained by extracellular biofilms. For example, during UTI, hordes of neutrophils infiltrate the bladder migrating toward the infected superficial umbrella cells, but are unable to effectively penetrate the IBC or engulf dispersing filamentous bacteria (Justice et al. 2008; Horvath et al. 2010). The ability of IBCs to repel neutrophil penetration is lost in K1 capsule mutants (Anderson et al. 2010). Moreover, Blango and Mulvey showed that 17 different antibiotics capable of killing the virulent cystitis isolate UTI89 in vitro or in tissue culture were unable to eliminate UTI89 from bladder tissue during infection (Blango and Mulvey 2010). These findings indicate that IBC formation is a mechanism that enables rapid bacterial expansion within the host and contributes to bacterial persistence.

BIOFILM INHIBITION: TREATMENT STRATEGIES IN THE POSTANTIBIOTIC ERA

Antibiotics are currently the preferred treatment strategy for bacterial infections. Conventional antibiotics work by either preventing bacterial cell division (bacteriostatic) or killing the cell (bactericidal). Although over the years antibiotics have proven critical in eliminating bacterial pathogens, overwhelming evidence indicates that they extensively damage the host microbiota, creating an environment where opportunistic pathogens can prevail, and they increase the selective pressure toward antibiotic resistance (Dethlefsen et al. 2008; Dethlefsen and Relman 2010; Ubeda et al. 2010). Moreover, although prophylactic antibiotic administration preceding surgery is highly successful in reducing infection rate, it has little or no protective effects in surgical procedures involving implants or prostheses (Secinti et al. 2011). In most cases, the best treatment for foreign body-associated biofilm infections is to remove the infected device. However, in cases like implantable prostheses, pacemakers, and cardiac implants, device removal is difficult (Fey 2010).

Biofilm bacteria are particularly recalcitrant to antibiotic treatments not only owing to increased transmission of resistance markers within the biofilm community, but also because of diffusion limitations posed by the extracellular matrix, antibiotic inactivation by high metal ion concentration and low pH, and the presence of metabolically inactive persister cells that survive treatment (Mack et al. 2004; Lewis 2005; Costerton et al. 2007; Lewis 2008). Combined, these attributes make biofilm bacteria up to 1000-fold more tolerant and/or resistant to antibiotics than planktonic cells (Hoiby et al. 2010). Thus, the need for more effective biofilm dissolution treatments becomes imperative. Below we present some of the most recent advances in strategies designed to thwart biofilm formation by killing the bacteria or targeting different biofilm developmental stages (also summarized in Fig. 3).

Bactericidal Strategies

Phage Therapy

Phage therapy is a promising alternative to antibiotic treatments (Donlan 2009); phages are abundant and can be easily isolated from a wide range of environments, they are usually specific to narrow host ranges (thus not likely to perturb the host microbiota), and their self-replicating mode permits low dosage (Burrowes et al. 2011). Moreover, the high phage mutation rate facilitates adaptation as the corresponding bacterial hosts accumulate mutations to persist in a given environment. Phage therapy takes advantage of lytic phages that do not enter a prophage state and thus rarely contain or transfer virulence genes, although they result in rapid destruction of the bacterial cell. Many phages have been shown to encode EPS-degrading

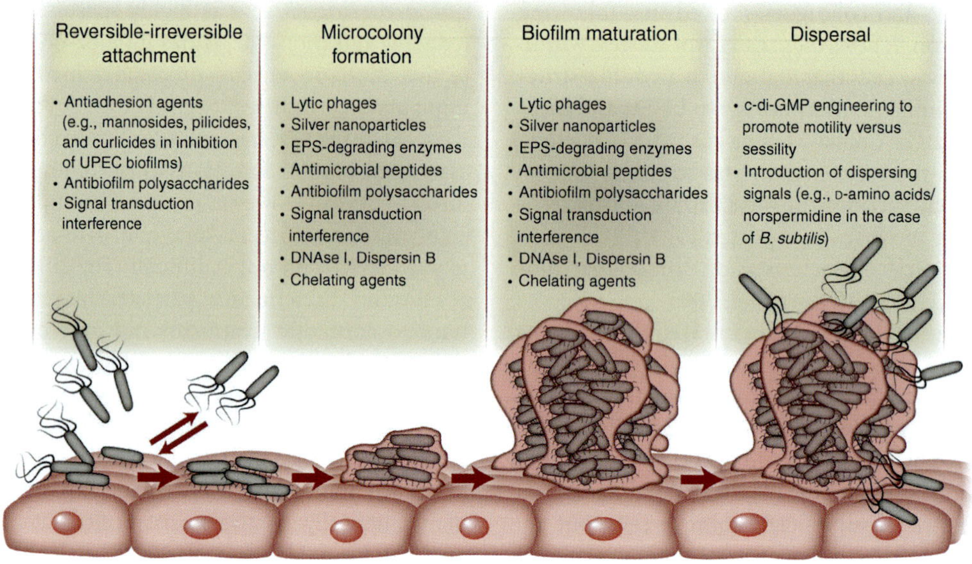

Figure 3. Schematic outlining the stages in biofilm development and listing the strategies aimed at inhibiting and/or disrupting biofilm formation at specific stages.

enzymes (Hughes et al. 1998a,b; Sutherland et al. 2004), or propagate on stationary-phase bacteria, making them more likely to persist within the biofilm (Burrowes et al. 2011).

Silver Nanoparticles

Impregnation of medical devices with antimicrobial agents has been the most commonly used approach for preventing device-associated biofilms (Fey 2010). One of the frequently used agents is silver, which has been used as an anti-infective for hundreds of years and has been extensively used to sterilize wound infections during World War I (Rupp et al. 2005; Chen and Schluesener 2008). The positively charged silver ions facilitate electrostatic attractions between the metal and the negatively charged bacterial membrane, augmenting uptake and antimicrobial activity (Kim et al. 2007). The lethality of silver for bacteria is partly owing to thiol-group reactions that inactivate enzymes (Chen and Schluesener 2008). As a result, silver treatment inhibits DNA replication, expression of ribosomal and other cellular proteins, and interferes with the bacterial electron transport chain (Bragg and Rainnie 1974; Feng et al. 2000; Yamanaka et al. 2005).

The potential toxicity of silver in humans led to its dwindling use for some time. However, the popularity of silver has been revived with the advent of nanotechnology (Chen and Schluesener 2008). Nanoparticles are typically no greater than 100 nm in size and their biocidal effectiveness is suggested to be owing to a combination of their small size and high surface-to-volume ratio, which enable intimate interactions with microbial membranes (Morones et al. 2005; Allaker 2010). Silver nanoparticles have been shown to inhibit *P. aeruginosa* and *Staphylococcus epidermidis* biofilms by >95%; studies in rabbits showed that nanoparticle silver ion-coated implants inhibited *Staphylococcus aureus* biofilm formation without causing silver accumulation in host tissues, even 28 d after impregnation (Kalishwaralal et al. 2010; Secinti et al. 2011).

Antimicrobial Peptides

Antimicrobial peptides are produced by the innate immune response system and have been proposed as attractive candidates for the development of novel types of antibiotics (Yang et al. 2002). However, their activity spectrum and mechanism of action need to be more precisely defined before they can be considered as possible therapeutic strategies (Pompilio et al. 2011). Cathelicidins constitute one of the most important classes of antimicrobial peptides. Recent work indicated that SMAP-29, BMAP-28, and BMAP-27 significantly reduced biofilm formation by multidrug-resistant (MDR) *P. aeruginosa* strains isolated from patients with CF (cystic fibrosis), and killed bacteria within preformed biofilms (Pompilio et al. 2011). This study compared the bactericidal activity of cathelicidins to that of tobramycin, the frontline antibiotic used to treat *P. aeruginosa* airway infections in CF patients. Pompilio et al. found that in contrast to tobramycin, the active cathelicidin peptides showed faster kinetics, exerting a rapid bactericidal activity regardless of the species tested. Although the extent of bacterial killing was overall higher with tobramycin, this study shows that cathelicidins may hold potential as antibiofilm agents in the case of MDR strains (Pompilio et al. 2011).

Lytic peptides are another group of antimicrobial peptides assessed for their inhibitory effects on biofilm formation. Lytic peptides bind the LPS (lipopolysaccharide) moieties of the bacterial cell membrane, disrupting membrane stability. Studies in *Staphylococcus aureus* have shown that the lytic peptide PTP-7 prevented in vitro biofilm formation and was also capable of diffusing into the deep layer of preformed biofilm, killing 99.9% of biofilm bacteria. This peptide retained activity under highly acidic environments and in the presence of excess of metals, conditions that mimic the *S. aureus* biofilm environment (Kharidia and Liang 2011).

Antiadhesion Agents

Mannosides, Pilicides, and Curlicides

Attachment constitutes the first step in virtually all types of biofilm formation, thus numerous studies have focused on ablating bacterial adherence. In UPEC, efforts have concentrated on the development of compounds that interfere with the adhesive properties or assembly of type 1 pili, because they are the prevalent means for UPEC adherence during biofilm formation in vitro and within the host. The X-ray crystal structures of the FimH adhesin bound to mannose have been used to rationally design molecules, termed mannosides, which fit the FimH mannose binding pocket and competitively inhibit FimH binding to its host receptor (Fader and Davis 1980; Schaeffer et al. 1980; Hung et al. 2002; Bouckaert et al. 2005; Sperling et al. 2006; Wellens et al. 2008; Klein et al. 2010). Recent advances in this field led to the development of monomeric biphenyl mannosides with largely enhanced potency, relative to previously reported FimH inhibitors (Han et al. 2010). These optimized mannosides prevented UPEC biofilm formation in vitro and were shown to disrupt preformed biofilms (Cusumano et al. 2011). In vivo studies indicated that mannoside administration as a prophylactic measure for UTIs interfered with UPEC adherence and invasion, reducing IBC formation and attenuating UPEC during the acute infection stages (Cusumano et al. 2011). Moreover, mannosides enhanced the antimicrobial effects of trimethoprim-sulfamethoxazole (TMP-SMZ) preventing infection by PBC-1, a UPEC isolate that was resistant to TMP-SMZ treatment in the clinical setting (Cusumano et al. 2011). Given that the intracellular niche protects UPEC from antibiotic treatment (Blango and Mulvey 2010), the conferred PBC-1 susceptibility on dual mannoside/TMP-SMZ treatment is likely owing to bacterial sequestration in the extracellular environment, where antibiotic concentrations are higher relative to the intracellular compartment (Patel and Welling 1980; Cusumano et al. 2011). Mannosides were also efficient as a therapeutic strategy for chronic UTIs, significantly reducing the bladder bacterial load of orally treated mice within 6 h (Cusumano et al. 2011). Thus, if translated to the clinic mannosides, they hold great potential for the elimination of complicated UTIs associated with antibiotic-resistant UPEC strains.

In parallel, efforts have been made to inhibit assembly of type 1 pili and other CUP pili, through the use of pilicides, which are compounds rationally designed to interfere with export of the corresponding pilin subunits. Pilicides were shown to inhibit UPEC biofilm formation in vitro by 50%, at concentrations as low as 3 µM (Pinkner et al. 2006; Berg et al. 2008; Chorell et al. 2010, 2011). Similar compounds have been shown to be effective against curli ("curlicides"), inhibiting in vitro curli biogenesis, biofilm formation, and potentiating UPEC clearance from the urinary tract (Cegelski et al. 2009).

Polysaccharides

Exopolysaccharides mediate cell-to-surface and cell-to-cell interactions that are critical for biofilm formation and stabilization. Mutants unable to synthesize or export such polysaccharides are typically deficient in adherence and biofilm formation and thus are highly sensitive to killing by antibiotics and host immune defenses (Rendueles et al. 2012). However, recent evidence indicates that some bacterial exopolysaccharides inhibit or destabilize biofilm formation by other species. For example, in the case of *P. aeruginosa*, Qin and colleagues showed that Pel and Psl-containing culture supernatants disrupted preformed *S. epidermidis* and *S. aureus* biofilms without inhibiting bacterial growth (Qin et al. 2009). Moreover, the presence of *P. aeruginosa* inhibited *S. epidermidis* biofilm formation in dual-species in vitro biofilm experiments (Pihl et al. 2010). Polysaccharides with nonbiocidal antibiofilm properties have also been isolated from cell-free biofilm extracts of several species (Rendueles et al. 2012). Their antibiofilm properties are believed to lie on their ability to (a) alter the physical characteristics of bacterial cells or abiotic surfaces; (b) act as signaling molecules that impact the gene expression patterns of susceptible bacteria; or (c) competitively inhibit multivalent carbohydrate–protein interactions, thereby interfering with adhesion. Most antibiofilm polysaccharides show a broad spectrum of biofilm inhibition, whereas some are capable of dispersing preformed biofilms. Given their nonbiocidal mode of action, as well as their biocompatibility and biodegradability, antibiofilm polysaccharides could be a promising strategy suitable for the treatment and prevention of biofilm-related infections. Furthermore, several studies suggest that antibiofilm polysaccharides could be valuable as an adjuvant, because they enhance antibiotic functions when administered together. Other potential applications could be coating surfaces of indwelling medical devices or even using antibiofilm polysaccharide-producing bacteria in probiotics to outcompete pathogens (Rendueles et al. 2012).

Signal Transduction Interference

Many studies have focused on inhibiting biofilm initiation by interfering with bacterial signaling cascades, given that two-component systems constitute a central means of intercepting and translating environmental changes (Lyon et al. 2000; Okada et al. 2007; Cegelski et al. 2008; Rasko et al. 2008; Watanabe et al. 2008; Njoroge and Sperandio 2009). Inhibition of signal transduction systems poses an attractive means of antivirulence therapy, because interference with signaling does not kill the bacteria but rather deprograms optimal gene expression and ablates virulence without applying pressure for selection of resistance. Among the systems that appear to be attractive drug target candidates are the QseBC two-component system that is common among biofilm-forming Gram-negative pathogens (Clarke et al. 2006; Bearson and Bearson 2008; Rasko et al. 2008; Kostakioti et al. 2009; Khajanchi et al. 2011; Wang et al. 2011). Previous studies have investigated the potential of inhibiting QseC kinase activity and showed efficacy in reducing enterohemorrhagic *Escherichia coli* (EHEC) virulence (Rasko et al. 2008). Studies in UPEC and EHEC have shown that deletion of QseC results in the overactivation of the QseB response regulator owing to the specific phosphatase activity of QseC required for QseB deactivation. Thus, targeting QseC phosphatase activity would be an optimized strategy to decouple normal gene expression in QseC-bearing pathogens (Kostakioti et al. 2009; Hadjifrangiskou et al. 2011).

In *E. faecalis*, studies have targeted the FsrC/FsrA TCS. FsrC/FsrA controls the expression of *fsrBDC* and *gelE-sprE*, leading to increased production of gelatinase and serine protease, both of which are required for appropriate production of eDNA (Qin et al. 2000; Thomas et al. 2008, 2009). High-throughput screening of compounds that inhibited gelatinase and the gelatinase biosynthesis-activating pheromone (GBAP), identified a peptide antibiotic, Siamycin I, as an inhibitor of GBAP signaling by FsrC/FsrA (Nakayama et al. 2007; Gotoh et al. 2010).

"Antimatrix" Agents

Besides ablating adherence, bacterial aggregation can be targeted by disrupting components of the extracellular matrix. Several investigations exploited the potential of inhibiting enzymes involved in the synthesis or modification of cell wall-associated or secreted EPS components and other matrix constituents. These studies use the direct use of naturally occurring or engineered enzymes, use bacteriophages (phage therapy) as a vehicle of enzyme delivery and expression, or take advantage of metal chelators as a means to disrupt matrix integrity.

Enzymes

N-acetyl-D-glucosamine-1-phosphate acetyltransferase (GlmU), which is involved in the biosynthesis of activated UDP-GlcNAc, an essential peptidoglycan and lipopolysaccharide (LPS) precursor in Gram-positive and Gram-negative pathogens, respectively, is among the enzymes targeted for matrix disruption (Burton et al. 2006). Burton et al. tested the effects of GlmU inhibitors, including *N*-ethyl maleimide (NEM), and the NEM analogs *N*-phenyl maleimide, *N,N'*-(1,2-phenylene)dimaleimide (oPDM), and *N*-(1-pyrenyl)maleimide (PyrM), on catheter-associated uropathogen biofilms. All NEM analogs showed antibiofilm activity against clinical isolates of *E. coli*, *P. aeruginosa*, *K. pneumoniae*, *S. epidermidis*, and *E. faecalis* (Burton et al. 2006). The same study showed that coating silicone catheters with a mixture of oPDM and the cationic polypeptide protamine sulfate (PS) enhanced the antibiofilm activity of PS against *P. aeruginosa* and *S. epidermidis*, demonstrating that this dual administration could be used as a broad-spectrum anti-infective coating for medical devices (Burton et al. 2006).

The enzymes DNase I and Dispersin B have also recently gained attention as potential antibiofilm agents, particularly against Gram-positive pathogens. The effects of DNase I lie on its ability to digest the eDNA found within the biofilm structure (Qin et al. 2007; Zhu et al. 2007; Guiton et al. 2009). DNase treatment prevented *Staphylococcus* and *Enterococcus* biofilm formation (Guiton et al. 2009; Mann et al. 2009) and dispersed preformed biofilms in vitro (Guiton et al. 2009). A recombinant form of DNase I, pulmozyme, is used in certain cases to treat patients with CF (Shak et al. 1990; Fey 2010). Dispersin B is a glycoside hydrolase produced by *Actinobacillus actinomycetemcomitans* that cleaves β 1–6 *N*-acetylglucosamine polymers (PNAG) in the bacterial peptidoglycan layer (Fey 2010). Dispersin-B treatment has been shown to be effective against *S. aureus* and *S. epidermidis* biofilms and other PNAG-containing bacteria (Izano et al. 2008; Kaplan 2010).

Engineering Dispersin B into a phage that replicates in stationary-phase cells led to complete disruption of preformed *E. coli* biofilms in vitro, an effect that was more dramatic than administration of either nonengineered phage or Dispersin B alone, likely because Dispersin B-mediated degradation of EPS allowed the phage to access the deeper layers of the biofilm structure (Lu and Collins 2007).

Chelating Agents

Metal cations, such as calcium, magnesium, and iron have been implicated in maintaining matrix integrity (Patrauchan et al. 2005; Raad et al. 2008). Consistent with this observation, chelating agents have been shown to destabilize biofilm architecture besides interfering with bacterial membrane stability (Donlan 2011). For example, sodium citrate inhibited biofilm formation by several *Staphylococci* species in vitro (Shanks et al. 2006). In addition, tetrasodium-EDTA eradicated biofilms in an in vitro biofilm model and on explanted hemodialysis catheters

(Kite et al. 2004; Percival et al. 2005), whereas disodium-EDTA, in combination with tigecyclin or gentamicin, reduced biofilm formation by *Staphylococcus* species and *P. aeruginosa* on Hickman catheter segments in vitro (Bookstaver et al. 2009). Raad et al. showed efficacy of a combination of minocycline and disodium-EDTA against biofilms in vitro or on explanted catheter tips, as well as in the treatment of catheter-related bloodstream infections in three different patient studies (Raad et al. 1997, 2003). A minocycline-EDTA solution was also successfully used to prevent indwelling implantable-port infections in children with cancer; no port infections or other adverse effects were observed in patients whose ports were flushed with the monocycline-EDTA solution, whereas 21% of the patients in the untreated control group developed infection (Chatzinikolaou et al. 2003). Moreover, reduction of catheter-related bloodstream infections was observed in hemodialysis patients after treating catheters with minocycline-EDTA (Bleyer et al. 2005; Feely et al. 2007).

Manipulating Dispersal Signals to Disassemble Biofilms

Given that planktonic cells are more susceptible to treatments, a novel treatment strategy in which a signal for biofilm dispersion is combined with administration of an antimicrobial agent for killing the dispersed organisms could be successful. As we discussed earlier, one of the signals most associated with bacterial dispersal is c-di-GMP. In a recent study, Ma et al. exploited the potential of engineering a protein that causes biofilms to disperse (Ma et al. 2011b). Using a knockout library of previously uncharacterized genes known to be influenced by impaired autoinducer-2 secretion, the investigators identified BdcA as a protein that enhances biofilm dispersal, by sequestering c-di-GMP and reducing its local concentration (Ma et al. 2011b). Subsequent analyses showed that mutating the BdcA E50 residue to V or Q, resulted in higher induction of motility by increasing the c-di-GMP binding affinity of BdcA (Ma et al. 2011b). Given that BdcA is conserved in several pathogens, this analysis provides new tools, which in combination with novel delivery strategies, such as phage therapy, could facilitate active biofilm dispersal as a therapeutic approach.

D-Amino Acids and Norspermidine

The idea of manipulating natural dispersion factors to combat biofilms has also been exploited for Gram-positive organisms. Losick and colleagues showed that exogenous addition of the D-amino acids produced by dispersing *B. subtilis* disrupted preformed biofilms and were also effective in preventing biofilm formation by *S. aureus* and *P. aeruginosa* (Kolodkin-Gal et al. 2010). It is likely that D-amino acids promote biofilm disassembly by disrupting adhesive fiber interactions (Cava et al. 2010). Indeed, studies by Hochbaum et al. showed that D-amino acids inhibit *S. aureus* biofilm formation by preventing protein localization to the cell surface (Hochbaum et al. 2011). Given that D-amino acids are produced by many bacterial species, they may provide a general strategy for biofilm disassembly (Kolodkin-Gal et al. 2010) and thus might be useful in medical and industrial antibiofilm applications. A second biofilm-disassembly molecule was recently discovered in *B. subtilus*; norspermidine works in a manner complimentary to D-amino acids by targeting the exopolysaccharide (Kolodkin-Gal et al. 2012). As is the case for D-amino acids, the biofilm-inhibiting properties of norspermidine were not limited to *B. subtilus*, but biofilm inhibition was also observed in the case of *S. aureus* and *E. coli* pellicle biofilm. Thus, norspermidine and other polyamines synthesized to bind to specific exopolysaccharides could be exploited in conjunction with D-amino acids as a novel antibiofilm approach (Kolodkin-Gal et al. 2012).

CONCLUDING REMARKS

Biofilm formation enables bacterial pathogens to colonize a wide variety of host niches and persist in harsh environments, making their eradication particularly difficult. Biofilm characteristics determine whether, to what extent, and which antimicrobial treatments may be effective. The age and composition of the biofilm are the major factors influencing the susceptibility of the resident microorganisms. As the biofilm matures, increased EPS accumulation, combined with the nutrient and oxygen gradients that affect cell metabolism and growth rates, result in reduced entry and activity of antimicrobial agents making biofilm-forming pathogens progressively more resistant to antibiotic regimens. Thus, novel strategies, designed to block a specific biofilm step without killing the bacteria, such as the use of antiadhesion agents, or using natural, bacterially produced signals to promote bacterial dispersal, are exciting avenues for exploration and ultimately the development of fast-acting, potent, and bioavailable treatment strategies.

REFERENCES

Agladze K, Wang X, Romeo T. 2005. Spatial periodicity of *Escherichia coli* K-12 biofilm microstructure initiates during a reversible, polar attachment phase of development and requires the polysaccharide adhesin PGA. *J Bacteriol* **187:** 8237–8246.

Allaker RP. 2010. The use of nanoparticles to control oral biofilm formation. *J Dent Res* **89:** 1175–1186.

Anderson GG, Palermo JJ, Schilling JD, Roth R, Heuser J, Hultgren SJ. 2003. Intracellular bacterial biofilm-like pods in urinary tract infections. *Science* **301:** 105–107.

Anderson GG, Martin SM, Hultgren SJ. 2004. Host subversion by formation of intracellular bacterial communities in the urinary tract. *Microbes Infect* **6:** 1094–1101.

Allesen-Holm M, Barken KB, Yang L, Klausen M, Webb JS, Kjelleberg S, Molin S, Givskov M, Tolker-Nielsen T. 2006. A characterization of DNA release in *Pseudomonas aeruginosa* cultures and biofilms. *Mol Microbiol* **59:** 1114–1128.

Anderson GG, Moreau-Marquis S, Stanton BA, O'Toole GA. 2008. In vitro analysis of tobramycin-treated *Pseudomonas aeruginosa* biofilms on cystic fibrosis-derived airway epithelial cells. *Infect Immun* **76:** 1423–1433.

Anderson GG, Goller CC, Justice S, Hultgren SJ, Seed PC. 2010. Polysaccharide capsule and sialic acid-mediated regulation promote biofilm-like intracellular bacterial communities during cystitis. *Infect Immun* **78:** 963–975.

Bagge N, Hentzer M, Andersen JB, Ciofu O, Givskov M, Hoiby N. 2004. Dynamics and spatial distribution of β-lactamase expression in *Pseudomonas aeruginosa* biofilms. *Antimicrob Agents Chemother* **48:** 1168–1174.

Banin E, Vasil ML, Greenberg EP. 2005. Iron and *Pseudomonas aeruginosa* biofilm formation. *Proc Natl Acad Sci* **102:** 11076–11081.

Barken KB, Pamp SJ, Yang L, Gjermansen M, Bertrand JJ, Klausen M, Givskov M, Whitchurch CB, Engel JN, Tolker-Nielsen T. 2008. Roles of type IV pili, flagellum-mediated motility and extracellular DNA in the formation of mature multicellular structures in *Pseudomonas aeruginosa* biofilms. *Environ Microbiol* **10:** 2331–2343.

Barraud N, Schleheck D, Klebensberger J, Webb JS, Hassett DJ, Rice SA, Kjelleberg S. 2009. Nitric oxide signaling in *Pseudomonas aeruginosa* biofilms mediates phosphodiesterase activity, decreased cyclic di-GMP levels, and enhanced dispersal. *J Bacteriol* **191:** 7333–7342.

Bearson BL, Bearson SM. 2008. The role of the QseC quorum-sensing sensor kinase in colonization and norepinephrine-enhanced motility of *Salmonella enterica* serovar *Typhimurium*. *Microb Pathog* **44:** 271–278.

Beloin C, Roux A, Ghigo JM. 2008. *Escherichia coli* biofilms. *Curr Top Microbiol Immunol* **322:** 249–289.

Beloin C, Valle J, Latour-Lambert P, Faure P, Kzreminski M, Balestrino D, Haagensen JA, Molin S, Prensier G, Arbeille B, et al. 2004. Global impact of mature biofilm lifestyle on *Escherichia coli* K-12 gene expression. *Mol Microbiol* **51:** 659–674.

Berg V, Das P, Chorell E, Hedenstrom M, Pinkner JS, Hultgren SJ, Almqvist F. 2008. Carboxylic acid isosteres improve the activity of ring-fused 2-pyridones that inhibit pilus biogenesis in *E. coli*. *Bioorg Med Chem Lett* **18:** 3536–3540.

Bhomkar P, Materi W, Semenchenko V, Wishart DS. 2010. Transcriptional response of *E. coli* upon FimH-mediated fimbrial adhesion. *Gene Regul Syst Bio* **4:** 1–17.

Bishop BL, Duncan MJ, Song J, Li G, Zaas D, Abraham SN. 2007. Cyclic AMP-regulated exocytosis of *Escherichia coli* from infected bladder epithelial cells. *Nat Med* **13:** 625–630.

Blango MG, Mulvey MA. 2010. Persistence of uropathogenic *Escherichia coli* in the face of multiple antibiotics. *Antimicrob Agents Chemother* **54:** 1855–1863.

Bleyer AJ, Mason L, Russell G, Raad II, Sherertz RJ. 2005. A randomized, controlled trial of a new vascular catheter flush solution (minocycline-EDTA) in temporary hemodialysis access. *Infect Control Hosp Epidemiol* **26:** 520–524.

Bokranz W, Wang X, Tschape H, Romling U. 2005. Expression of cellulose and curli fimbriae by *Escherichia coli* isolated from the gastrointestinal tract. *J Med Microbiol* **54:** 1171–1182.

Boles BR, Thoendel M, Singh PK. 2005. Rhamnolipids mediate detachment of *Pseudomonas aeruginosa* from biofilms. *Mol Microbiol* **57:** 1210–1223.

Bookstaver PB, Williamson JC, Tucker BK, Raad II, Sherertz RJ. 2009. Activity of novel antibiotic lock solutions in a model against isolates of catheter-related bloodstream infections. *Ann Pharmacother* **43:** 210–219.

Borlee BR, Goldman AD, Murakami K, Samudrala R, Wozniak DJ, Parsek MR. 2010. *Pseudomonas aeruginosa* uses a cyclic-di-GMP-regulated adhesin to reinforce the biofilm extracellular matrix. *Mol Microbiol* **75:** 827–842.

Bouckaert J, Berglund J, Schembri M, De Genst E, Cools L, Wuhrer M, Hung CS, Pinkner J, Slattegard R, Zavialov A, et al. 2005. Receptor binding studies disclose a novel class of high-affinity inhibitors of the *Escherichia coli* FimH adhesin. *Mol Microbiol* **55:** 441–455.

Boyd A, Chakrabarty AM. 1994. Role of alginate lyase in cell detachment of *Pseudomonas aeruginosa*. *Appl Environ Microbiol* **60:** 2355–2359.

Bragg PD, Rainnie DJ. 1974. The effect of silver ions on the respiratory chain of *Escherichia coli*. *Can J Microbiol* **20:** 883–889.

Branda SS, Chu F, Kearns DB, Losick R, Kolter R. 2006. A major protein component of the *Bacillus subtilis* biofilm matrix. *Mol Microbiol* **59:** 1229–1238.

Burrowes B, Harper DR, Anderson J, McConville M, Enright MC. 2011. Bacteriophage therapy: Potential uses in the control of antibiotic-resistant pathogens. *Expert Rev Anti Infect Ther* **9:** 775–785.

Burton E, Gawande PV, Yakandawala N, LoVetri K, Zhanel GG, Romeo T, Friesen AD, Madhyastha S. 2006. Antibiofilm activity of GlmU enzyme inhibitors against catheter-associated uropathogens. *Antimicrob Agents Chemother* **50:** 1835–1840.

Cava F, Lam H, de Pedro MA, Waldor MK. 2010. Emerging knowledge of regulatory roles of D-amino acids in bacteria. *Cell Mol Life Sci* **68:** 817–831.

Cegelski L, Marshall GR, Eldridge GR, Hultgren SJ. 2008. The biology and future prospects of antivirulence therapies. *Nat Rev Microbiol* **6:** 17–27.

Cegelski L, Pinkner JS, Hammer ND, Cusumano CK, Hung CS, Chorell E, Aberg V, Walker JN, Seed PC, Almqvist F, et al. 2009. Small-molecule inhibitors target *Escherichia coli* amyloid biogenesis and biofilm formation. *Nat Chem Biol* **5:** 913–919.

Cerca N, Jefferson KK, Oliveira R, Pier GB, Azeredo J. 2006. Comparative antibody-mediated phagocytosis of *Staphylococcus epidermidis* cells grown in a biofilm or in the planktonic state. *Infect Immun* **74:** 4849–4855.

Cerca N, Maira-Litran T, Jefferson KK, Grout M, Goldmann DA, Pier GB. 2007. Protection against *Escherichia coli* infection by antibody to the *Staphylococcus aureus* poly-*N*-acetylglucosamine surface polysaccharide. *Proc Natl Acad Sci* **104:** 7528–7533.

Chatzinikolaou I, Zipf TF, Hanna H, Umphrey J, Roberts WM, Sherertz R, Hachem R, Raad I. 2003. Minocycline-ethylenediaminetetraacetate lock solution for the prevention of implantable port infections in children with cancer. *Clin Infect Dis* **36:** 116–119.

Chen X, Schluesener HJ. 2008. Nanosilver: A nanoproduct in medical application. *Toxicol Lett* **176:** 1–12.

Chen SL, Hung CS, Xu J, Reigstad CS, Magrini V, Sabo A, Blasiar D, Bieri T, Meyer RR, Ozersky P, et al. 2006. Identification of genes subject to positive selection in uropathogenic strains of *Escherichia coli*: A comparative genomics approach. *Proc Natl Acad Sci* **103:** 5977–5982.

Chen SL, Hung CS, Pinkner JS, Walker JN, Cusumano CK, Li Z, Bouckaert J, Gordon JI, Hultgren SJ. 2009. Positive selection identifies an in vivo role for FimH during urinary tract infection in addition to mannose binding. *Proc Natl Acad Sci* **106:** 22439–22444.

Chorell E, Pinkner JS, Phan G, Edvinsson S, Buelens F, Remaut H, Waksman G, Hultgren SJ, Almqvist F. 2010. Design and synthesis of C-2 substituted thiazolo and dihydrothiazolo ring-fused 2-pyridones: Pilicides with increased antivirulence activity. *J Med Chem* **53:** 5690–5695.

Chorell E, Bengtsson C, Sainte-Luce Banchelin T, Das P, Uvell H, Sinha AK, Pinkner JS, Hultgren SJ, Almqvist F. 2011. Synthesis and application of a bromomethyl substituted scaffold to be used for efficient optimization of anti-virulence activity. *Eur J Med Chem* **46:** 1103–1116.

Clarke MB, Hughes DT, Zhu C, Boedeker EC, Sperandio V. 2006. The QseC sensor kinase: A bacterial adrenergic receptor. *Proc Natl Acad Sci* **103:** 10420–10425.

Connell I, Agace W, Klemm P, Schembri M, Marild S, Svanborg C. 1996. Type 1 fimbrial expression enhances *Escherichia coli* virulence for the urinary tract. *Proc Natl Acad Sci* **93:** 9827–9832.

Conrad A, Suutari MK, Keinanen MM, Cadoret A, Faure P, Mansuy-Huault L, Block JC. 2003. Fatty acids of lipid fractions in extracellular polymeric substances of activated sludge flocs. *Lipids* **38:** 1093–1105.

Cookson AL, Cooley WA, Woodward MJ. 2002. The role of type 1 and curli fimbriae of Shiga toxin-producing *Escherichia coli* in adherence to abiotic surfaces. *Int J Med Microbiol* **292:** 195–205.

Costerton JW, Montanaro L, Arciola CR. 2007. Bacterial communications in implant infections: A target for an intelligence war. *Int J Artif Organs* **30:** 757–763.

Cusumano CK, Pinkner J, Han Z, Greene SE, Ford B, Crowley JR, Henderson JP, Janetka JW, Hultgren S. 2011. Treatment and prevention of UTI with orally active mannoside FimH inhibitors. *Sci Transl Med* **3:** 109ra115.

Danese PN, Pratt LA, Dove SL, Kolter R. 2000a. The outer membrane protein, antigen 43, mediates cell-to-cell interactions within *Escherichia coli* biofilms. *Mol Microbiol* **37:** 424–432.

Danese PN, Pratt LA, Kolter R. 2000b. Exopolysaccharide production is required for development of *Escherichia coli* K-12 biofilm architecture. *J Bacteriol* **182:** 3593–3596.

Dethlefsen L, Relman DA. 2010. Incomplete recovery and individualized responses of the human distal gut microbiota to repeated antibiotic perturbation. *Proc Natl Acad Sci* **108** (Suppl 1): 4554–4561.

Dethlefsen L, Huse S, Sogin ML, Relman DA. 2008. The pervasive effects of an antibiotic on the human gut microbiota, as revealed by deep 16S rRNA sequencing. *PLoS Biol* **6:** e280.

Diggle SP, Stacey RE, Dodd C, Camara M, Williams P, Winzer K. 2006. The galactophilic lectin, LecA, contributes to biofilm development in *Pseudomonas aeruginosa*. *Environ Microbiol* **8:** 1095–1104.

Domka J, Lee J, Bansal T, Wood TK. 2007. Temporal gene-expression in *Escherichia coli* K-12 biofilms. *Environ Microbiol* **9:** 332–346.

Dong YH, Zhang XF, An SW, Xu JL, Zhang LH. 2008. A novel two-component system BqsS-BqsR modulates quorum sensing-dependent biofilm decay in *Pseudomonas aeruginosa*. *Commun Integr Biol* **1:** 88–96.

Donlan RM. 2002. Biofilms: Microbial life on surfaces. *Emerg Infect Dis* **8:** 881–890.

Donlan RM. 2009. Preventing biofilms of clinically relevant organisms using bacteriophage. *Trends Microbiol* **17:** 66–72.

Donlan RM. 2011. Biofilm elimination on intravascular catheters: Important considerations for the infectious disease practitioner. *Clin Infect Dis* **52:** 1038–1045.

Dunne WM Jr. 2002. Bacterial adhesion: Seen any good biofilms lately? *Clin Microbiol Rev* **15:** 155–166.

Eto DS, Jones TA, Sundsbak JL, Mulvey MA. 2007. Integrin-mediated host cell invasion by type 1-piliated uropathogenic *Escherichia coli*. *PLoS Pathog* **3:** e100.

Fader RC, Davis CP. 1980. Effect of piliation on *Klebsiella pneumoniae* infection in rat bladders. *Infect Immun* **30:** 554–561.

Feely T, Copley A, Bleyer AJ. 2007. Catheter lock solutions to prevent bloodstream infections in high-risk hemodialysis patients. *Am J Nephrol* **27:** 24–29.

Feng QL, Wu J, Chen GQ, Cui FZ, Kim TN, Kim JO. 2000. A mechanistic study of the antibacterial effect of silver ions on *Escherichia coli* and *Staphylococcus aureus*. *J Biomed Mater Res* **52:** 662–668.

Ferrieres L, Hancock V, Klemm P. 2007. Specific selection for virulent urinary tract infectious *Escherichia coli* strains during catheter-associated biofilm formation. *FEMS Immunol Med Microbiol* **51:** 212–219.

Fey PD. 2010. Modality of bacterial growth presents unique targets: How do we treat biofilm-mediated infections? *Curr Opin Microbiol* **13:** 610–615.

Flemming HC, Wingender J. 2010. The biofilm matrix. *Nat Rev Microbiol* **8:** 623–633.

Foxman B. 2010. The epidemiology of urinary tract infection. *Nat Rev Urol* **7:** 653–660.

Friedman L, Kolter R. 2004a. Genes involved in matrix formation in *Pseudomonas aeruginosa* PA14 biofilms. *Mol Microbiol* **51:** 675–690.

Friedman L, Kolter R. 2004b. Two genetic loci produce distinct carbohydrate-rich structural components of the *Pseudomonas aeruginosa* biofilm matrix. *J Bacteriol* **186:** 4457–4465.

Garcia-Medina R, Dunne WM, Singh PK, Brody SL. 2005. *Pseudomonas aeruginosa* acquires biofilm-like properties within airway epithelial cells. *Infect Immun* **73:** 8298–8305.

Garofalo CK, Hooton TM, Martin SM, Stamm WE, Palermo JJ, Gordon JI, Hultgren SJ. 2007. *Escherichia coli* from urine of female patients with urinary tract infections is competent for intracellular bacterial community formation. *Infect Immun* **75:** 52–60.

Gjermansen M, Ragas P, Sternberg C, Molin S, Tolker-Nielsen T. 2005. Characterization of starvation-induced dispersion in *Pseudomonas putida* biofilms. *Environ Microbiol* **7:** 894–906.

Gotoh Y, Eguchi Y, Watanabe T, Okamoto S, Doi A, Utsumi R. 2010. Two-component signal transduction as potential drug targets in pathogenic bacteria. *Curr Opin Microbiol* **13:** 232–239.

Govan JR, Deretic V. 1996. Microbial pathogenesis in cystic fibrosis: Mucoid *Pseudomonas aeruginosa* and *Burkholderia cepacia*. *Microbiol Rev* **60:** 539–574.

Guiton PS, Hung CS, Kline KA, Roth R, Kau AL, Hayes E, Heuser J, Dodson KW, Caparon MG, Hultgren SJ. 2009. Contribution of autolysin and Sortase A during *Enterococcus faecalis* DNA-dependent biofilm development. *Infect Immun* **77:** 3626–3638.

Hadjifrangiskou M, Kostakioti M, Chen SL, Henderson JP, Greene SE, Hultgren SJ. 2011. A central metabolic circuit controlled by QseC in pathogenic *Escherichia coli*. *Mol Microbiol* **80:** 1516–1529.

Hadjifrangiskou M, Gu AP, Pinkner JS, Kostakioti M, Zhang EW, Greene SE, Hultgren SJ. 2012. Transposon mutagenesis identifies uropathogenic *Escherichia coli* biofilm factors. *J Bacteriol* **194:** 6195–6205.

Hammer BK, Bassler BL. 2003. Quorum sensing controls biofilm formation in *Vibrio cholerae*. *Mol Microbiol* **50:** 101–104.

Han Z, Pinkner JS, Ford B, Obermann R, Nolan W, Wildman SA, Hobbs D, Ellenberger T, Cusumano CK, Hultgren SJ, et al. 2010. Structure-based drug design and optimization of mannoside bacterial FimH antagonists. *J Med Chem* **53:** 4779–4792.

Hannan TJ, Mysorekar IU, Hung CS, Isaacson-Schmid ML, Hultgren SJ. 2010. Early severe inflammatory responses to uropathogenic *E. coli* predispose to chronic and recurrent urinary tract infection. *PLoS Pathog* **6:** e1001042.

Hansen DS, Gottschau A, Kolmos HJ. 1998. Epidemiology of *Klebsiella bacteraemia*: A case control study using *Escherichia coli bacteraemia* as control. *J Hosp Infect* **38:** 119–132.

Harmsen M, Yang L, Pamp SJ, Tolker-Nielsen T. 2010. An update on *Pseudomonas aeruginosa* biofilm formation, tolerance, and dispersal. *FEMS Immunol Med Microbiol* **59:** 253–268.

Hasman H, Chakraborty T, Klemm P. 1999. Antigen-43-mediated autoaggregation of *Escherichia coli* is blocked by fimbriation. *J Bacteriol* **181:** 4834–4841.

Hatch RA, Schiller NL. 1998. Alginate lyase promotes diffusion of aminoglycosides through the extracellular polysaccharide of mucoid *Pseudomonas aeruginosa*. *Antimicrob Agents Chemother* **42:** 974–977.

Hay ID, Remminghorst U, Rehm BH. 2009. MucR, a novel membrane-associated regulator of alginate biosynthesis in *Pseudomonas aeruginosa*. *Appl Environ Microbiol* **75:** 1110–1120.

Henderson IR, Meehan M, Owen P. 1997. Antigen 43, a phase-variable bipartite outer membrane protein, determines colony morphology and autoaggregation in *Escherichia coli* K-12. *FEMS Microbiol Lett* **149:** 115–120.

Henderson JP, Crowley JR, Pinkner JS, Walker JN, Tsukayama P, Stamm WE, Hooton TM, Hultgren SJ. 2009. Quantitative metabolomics reveals an epigenetic blueprint for iron acquisition in uropathogenic *Escherichia coli*. *PLoS Pathog* **5:** e1000305.

Hentzer M, Teitzel GM, Balzer GJ, Heydorn A, Molin S, Givskov M, Parsek MR. 2001. Alginate overproduction affects *Pseudomonas aeruginosa* biofilm structure and function. *J Bacteriol* **183:** 5395–5401.

Hochbaum AI, Kolodkin-Gal I, Foulston L, Kolter R, Aizenberg J, Losick R. 2011. Inhibitory effects of D-amino acids on *Staphylococcus aureus* biofilm development. *J Bacteriol* **193:** 5616–5622.

Hoiby N, Bjarnsholt T, Givskov M, Molin S, Ciofu O. 2010. Antibiotic resistance of bacterial biofilms. *Int J Antimicrob Agents* **35:** 322–332.

Hong SH, Lee J, Wood TK. 2010. Engineering global regulator Hha of *Escherichia coli* to control biofilm dispersal. *Microb Biotechnol* **3:** 717–728.

Horvath DJ Jr, Li B, Casper T, Partida-Sanchez S, Hunstad DA, Hultgren SJ, Justice SS. 2010. Morphological plasticity promotes resistance to phagocyte killing of uropathogenic *Escherichia coli*. *Microbes Infect* **13:** 426–437.

Hughes KA, Sutherland IW, Clark J, Jones MV. 1998a. Bacteriophage and associated polysaccharide depolymerases—Novel tools for study of bacterial biofilms. *J Appl Microbiol* **85:** 583–590.

Hughes KA, Sutherland IW, Jones MV. 1998b. Biofilm susceptibility to bacteriophage attack: The role of phage-borne polysaccharide depolymerase. *Microbiology* **144 (Pt 11):** 3039–3047.

Hung CS, Bouckaert J, Hung D, Pinkner J, Widberg C, DeFusco A, Auguste CG, Strouse R, Langermann S, Waksman G, et al. 2002. Structural basis of tropism of *Escherichia coli* to the bladder during urinary tract infection. *Mol Microbiol* **44:** 903–915.

Inagaki S, Kuramitsu HK, Sharma A. 2005. Contact-dependent regulation of a *Tannerella forsythia* virulence factor, BspA, in biofilms. *FEMS Microbiol Lett* **249:** 291–296.

Izano EA, Amarante MA, Kher WB, Kaplan JB. 2008. Differential roles of poly-N-acetylglucosamine surface polysaccharide and extracellular DNA in *Staphylococcus aureus* and *Staphylococcus epidermidis* biofilms. *Appl Environ Microbiol* **74:** 470–476.

Jackson KD, Starkey M, Kremer S, Parsek MR, Wozniak DJ. 2004. Identification of *psl*, a locus encoding a potential exopolysaccharide that is essential for *Pseudomonas aeruginosa* PAO1 biofilm formation. *J Bacteriol* **186:** 4466–4475.

Jacobsen SM, Stickler DJ, Mobley HL, Shirtliff ME. 2008. Complicated catheter-associated urinary tract infections due to *Escherichia coli* and *Proteus mirabilis*. *Clin Microbiol Rev* **21:** 26–59.

Jefferson KK, Goldmann DA, Pier GB. 2005. Use of confocal microscopy to analyze the rate of vancomycin penetration through *Staphylococcus aureus* biofilms. *Antimicrob Agents Chemother* **49:** 2467–2473.

Jesaitis AJ, Franklin MJ, Berglund D, Sasaki M, Lord CI, Bleazard JB, Duffy JE, Beyenal H, Lewandowski Z. 2003. Compromised host defense on *Pseudomonas aeruginosa* biofilms: Characterization of neutrophil and biofilm interactions. *J Immunol* **171:** 4329–4339.

Justice SS, Hung C, Theriot JA, Fletcher DA, Anderson GG, Footer MJ, Hultgren SJ. 2004. Differentiation and developmental pathways of uropathogenic *Escherichia coli* in urinary tract pathogenesis. *Proc Natl Acad Sci* **101:** 1333–1338.

Justice SS, Hunstad DA, Seed PC, Hultgren SJ. 2006. Filamentation by *Escherichia coli* subverts innate defenses during urinary tract infection. *Proc Natl Acad Sci* **103:** 19884–19889.

Justice SS, Hunstad DA, Cegelski L, Hultgren SJ. 2008. Morphological plasticity as a bacterial survival strategy. *Nat Rev Microbiol* **6:** 162–168.

Kai-Larsen Y, Luthje P, Chromek M, Peters V, Wang X, Holm A, Kadas L, Hedlund KO, Johansson J, Chapman MR, et al. 2010. Uropathogenic *Escherichia coli* modulates immune responses and its curli fimbriae interact with the antimicrobial peptide LL-37. *PLoS Pathog* **6:** e1001010.

Kalishwaralal K, BarathManiKanth S, Pandian SR, Deepak V, Gurunathan S. 2010. Silver nanoparticles impede the biofilm formation by *Pseudomonas aeruginosa* and *Staphylococcus epidermidis*. *Colloids Surf B Biointerfaces* **79:** 340–344.

Kaplan JB. 2010. Biofilm dispersal: Mechanisms, clinical implications, and potential therapeutic uses. *J Dent Res* **89:** 205–218.

Karatan E, Watnick P. 2009. Signals, regulatory networks, and materials that build and break bacterial biofilms. *Microbiol Mol Biol Rev* **73:** 310–347.

Kau AL, Hunstad DA, Hultgren SJ. 2005. Interaction of uropathogenic *Escherichia coli* with host uroepithelium. *Curr Opin Microbiol* **8:** 54–59.

Kemp KD, Singh KV, Nallapareddy SR, Murray BE. 2007. Relative contributions of *Enterococcus faecalis* OG1RF sortase-encoding genes, srtA and bps (srtC), to biofilm formation and a murine model of urinary tract infection. *Infect Immun* **75:** 5399–5404.

Khajanchi BK, Kozlova EV, Sha J, Popov VL, Chopra AK. 2011. Two-component QseBC signaling system regulates in vitro and in vivo virulence of *Aeromonas hydrophila*. *Microbiology* **158:** 259–271.

Kharidia R, Liang JF. 2011. The activity of a small lytic peptide PTP-7 on *Staphylococcus aureus* biofilms. *J Microbiol* **49:** 663–668.

Kim JS, Kuk E, Yu KN, Kim JH, Park SJ, Lee HJ, Kim SH, Park YK, Park YH, Hwang CY, et al. 2007. Antimicrobial effects of silver nanoparticles. *Nanomedicine* **3:** 95–101.

Kite P, Eastwood K, Sugden S, Percival SL. 2004. Use of in vivo-generated biofilms from hemodialysis catheters to test the efficacy of a novel antimicrobial catheter lock for biofilm eradication in vitro. *J Clin Microbiol* **42:** 3073–3076.

Kjaergaard K, Schembri MA, Ramos C, Molin S, Klemm P. 2000. Antigen 43 facilitates formation of multispecies biofilms. *Environ Microbiol* **2:** 695–702.

Klausen M, Aaes-Jorgensen A, Molin S, Tolker-Nielsen T. 2003a. Involvement of bacterial migration in the development of complex multicellular structures in *Pseudomonas aeruginosa* biofilms. *Mol Microbiol* **50:** 61–68.

Klausen M, Heydorn A, Ragas P, Lambertsen L, Aaes-Jorgensen A, Molin S, Tolker-Nielsen T. 2003b. Biofilm formation by *Pseudomonas aeruginosa* wild type, flagella and type IV pili mutants. *Mol Microbiol* **48:** 1511–1524.

Klebensberger J, Birkenmaier A, Geffers R, Kjelleberg S, Philipp B. 2009. SiaA and SiaD are essential for inducing autoaggregation as a specific response to detergent stress in *Pseudomonas aeruginosa*. *Environ Microbiol* **11:** 3073–3086.

Klein T, Abgottspon D, Wittwer M, Rabbani S, Herold J, Jiang X, Kleeb S, Luthi C, Scharenberg M, Bezencon J, et al. 2010. FimH antagonists for the oral treatment of urinary tract infections: From design and synthesis to in vitro and in vivo evaluation. *J Med Chem* **53:** 8627–8641.

Kline KA, Dodson KW, Caparon MG, Hultgren SJ. 2010. A tale of two pili: Assembly and function of pili in bacteria. *Trends Microbiol* **18:** 224–232.

Koch C, Hoiby N. 1993. Pathogenesis of cystic fibrosis. *Lancet* **341:** 1065–1069.

Kolodkin-Gal I, Romero D, Cao S, Clardy J, Kolter R, Losick R. 2010. D-amino acids trigger biofilm disassembly. *Science* **328:** 627–629.

Kolodkin-Gal I, Cao S, Chai L, Bottcher T, Kolter R, Clardy J, Losick R. 2012. A self-produced trigger for biofilm disassembly that targets exopolysaccharide. *Cell* **149:** 684–692.

Koo H, Xiao J, Klein MI, Jeon JG. 2010. Exopolysaccharides produced by *Streptococcus mutans* glucosyltransferases modulate the establishment of microcolonies within multispecies biofilms. *J Bacteriol* **192:** 3024–3032.

Kostakioti M, Hadjifrangiskou M, Pinkner JS, Hultgren SJ. 2009. QseC-mediated dephosphorylation of QseB is required for expression of genes associated with virulence in uropathogenic *Escherichia coli*. *Mol Microbiol* **73:** 1020–1031.

Kuramitsu HK, Wang BY. 2011. The whole is greater than the sum of its parts: Dental plaque bacterial interactions can affect the virulence properties of cariogenic *Streptococcus mutans*. *Am J Dent* **24:** 153–154.

Lam H, Oh DC, Cava F, Takacs CN, Clardy J, de Pedro MA, Waldor MK. 2009. D-amino acids govern stationary phase cell wall remodeling in bacteria. *Science* **325:** 1552–1555.

Leid JG, Willson CJ, Shirtliff ME, Hassett DJ, Parsek MR, Jeffers AK. 2005. The exopolysaccharide alginate protects *Pseudomonas aeruginosa* biofilm bacteria from IFN-γ-mediated macrophage killing. *J Immunol* **175:** 7512–7518.

Lemon KP, Higgins DE, Kolter R. 2007. Flagellar motility is critical for *Listeria monocytogenes* biofilm formation. *J Bacteriol* **189:** 4418–4424.

Lenz AP, Williamson KS, Pitts B, Stewart PS, Franklin MJ. 2008. Localized gene expression in *Pseudomonas aeruginosa* biofilms. *Appl Environ Microbiol* **74:** 4463–4471.

Lewis K. 2005. Persister cells and the riddle of biofilm survival. *Biochemistry (Mosc)* **70:** 267–274.

Lewis K. 2008. Multidrug tolerance of biofilms and persister cells. *Curr Top Microbiol Immunol* **322:** 107–131.

Lu TK, Collins JJ. 2007. Dispersing biofilms with engineered enzymatic bacteriophage. *Proc Natl Acad Sci* **104:** 11197–11202.

Lye WC, Chan RK, Lee EJ, Kumarasinghe G. 1992. Urinary tract infections in patients with diabetes mellitus. *J Infect* **24:** 169–174.

Lyon GJ, Mayville P, Muir TW, Novick RP. 2000. Rational design of a global inhibitor of the virulence response in *Staphylococcus aureus*, based in part on localization of the site of inhibition to the receptor-histidine kinase, AgrC. *Proc Natl Acad Sci* **97:** 13330–13335.

Ma L, Jackson KD, Landry RM, Parsek MR, Wozniak DJ. 2006. Analysis of *Pseudomonas aeruginosa* conditional psl variants reveals roles for the psl polysaccharide in adhesion and maintaining biofilm structure postattachment. *J Bacteriol* **188:** 8213–8221.

Ma L, Conover M, Lu H, Parsek MR, Bayles K, Wozniak DJ. 2009. Assembly and development of the *Pseudomonas aeruginosa* biofilm matrix. *PLoS Pathog* **5:** e1000354.

Ma Q, Guishan Z, Wood TK. 2011a. *Escherichia coli* BdcA controls biofilm dispersal in *Pseudomonas aeruginosa* and *Rhizobium meliloti*. *BMC Res Notes* **4:** 447.

Ma Q, Yang Z, Pu M, Peti W, Wood TK. 2011b. Engineering a novel c-di-GMP-binding protein for biofilm dispersal. *Environ Microbiol* **13:** 631–642.

Mack D, Becker P, Chatterjee I, Dobinsky S, Knobloch JK, Peters G, Rohde H, Herrmann M. 2004. Mechanisms of biofilm formation in *Staphylococcus epidermidis* and *Staphylococcus aureus*: Functional molecules, regulatory circuits, and adaptive responses. *Int J Med Microbiol* **294:** 203–212.

Mai-Prochnow A, Lucas-Elio P, Egan S, Thomas T, Webb JS, Sanchez-Amat A, Kjelleberg S. 2008. Hydrogen peroxide linked to lysine oxidase activity facilitates biofilm differentiation and dispersal in several gram-negative bacteria. *J Bacteriol* **190:** 5493–5501.

Mann EE, Rice KC, Boles BR, Endres JL, Ranjit D, Chandramohan L, Tsang LH, Smeltzer MS, Horswill AR, Bayles KW. 2009. Modulation of eDNA release and degradation affects *Staphylococcus aureus* biofilm maturation. *PLoS ONE* **4:** e5822.

Martinez JJ, Mulvey MA, Schilling JD, Pinkner JS, Hultgren SJ. 2000. Type 1 pilus-mediated bacterial invasion of bladder epithelial cells. *Embo J* **19:** 2803–2812.

Matsukawa M, Greenberg EP. 2004. Putative exopolysaccharide synthesis genes influence *Pseudomonas aeruginosa* biofilm development. *J Bacteriol* **186:** 4449–4456.

Mohamed JA, Huang W, Nallapareddy SR, Teng F, Murray BE. 2004. Influence of origin of isolates, especially endocarditis isolates, and various genes on biofilm formation by *Enterococcus faecalis*. *Infect Immun* **72:** 3658–3663.

Mohamed JA, Teng F, Nallapareddy SR, Murray BE. 2006. Pleiotrophic effects of 2 *Enterococcus faecalis* sagA-like genes, *salA* and *salB*, which encode proteins that are antigenic during human infection, on biofilm formation and binding to collagen type I and fibronectin. *J Infect Dis* **193:** 231–240.

Monds RD, O'Toole GA. 2009. The developmental model of microbial biofilms: Ten years of a paradigm up for review. *Trends Microbiol* **17:** 73–87.

Morgan R, Kohn S, Hwang SH, Hassett DJ, Sauer K. 2006. BdlA, a chemotaxis regulator essential for biofilm dispersion in *Pseudomonas aeruginosa*. *J Bacteriol* **188:** 7335–7343.

Morici LA, Carterson AJ, Wagner VE, Frisk A, Schurr JR, Honer zu Bentrup K, Hassett DJ, Iglewski BH, Sauer K, Schurr MJ. 2007. *Pseudomonas aeruginosa* AlgR represses the Rhl quorum-sensing system in a biofilm-specific manner. *J Bacteriol* **189:** 7752–7764.

Morones JR, Elechiguerra JL, Camacho A, Holt K, Kouri JB, Ramirez JT, Yacaman MJ. 2005. The bactericidal effect of silver nanoparticles. *Nanotechnology* **16:** 2346–2353.

Mulvey MA, Lopez-Boado YS, Wilson CL, Roth R, Parks WC, Heuser J, Hultgren SJ. 1998. Induction and evasion of host defenses by type 1-piliated uropathogenic *Escherichia coli*. *Science* **282:** 1494–1497.

Nakayama J, Tanaka E, Kariyama R, Nagata K, Nishiguchi K, Mitsuhata R, Uemura Y, Tanokura M, Kumon H, Sonomoto K. 2007. Siamycin attenuates fsr quorum sensing mediated by a gelatinase biosynthesis-activating pheromone in *Enterococcus faecalis*. *J Bacteriol* **189:** 1358–1365.

Nallapareddy SR, Singh KV, Sillanpaa J, Garsin DA, Hook M, Erlandsen SL, Murray BE. 2006. Endocarditis and biofilm-associated pili of *Enterococcus faecalis*. *J Clin Invest* **116:** 2799–2807.

Nilsson LM, Yakovenko O, Tchesnokova V, Thomas WE, Schembri MA, Vogel V, Klemm P, Sokurenko EV. 2007. The cysteine bond in the *Escherichia coli* FimH adhesin is critical for adhesion under flow conditions. *Mol Microbiol* **65:** 1158–1169.

Njoroge J, Sperandio V. 2009. Jamming bacterial communication: New approaches for the treatment of infectious diseases. *EMBO Mol Med* **1:** 201–210.

Okada A, Gotoh Y, Watanabe T, Furuta E, Yamamoto K, Utsumi R. 2007. Targeting two-component signal transduction: A novel drug discovery system. *Methods Enzymol* **422:** 386–395.

O'Toole GA, Kolter R. 1998. Flagellar and twitching motility are necessary for *Pseudomonas aeruginosa* biofilm development. *Mol Microbiol* **30:** 295–304.

O'Toole G, Kaplan HB, Kolter R. 2000. Biofilm formation as microbial development. *Annu Rev Microbiol* **54:** 49–79.

Otto K, Silhavy TJ. 2002. Surface sensing and adhesion of *Escherichia coli* controlled by the Cpx-signaling pathway. *Proc Natl Acad Sci* **99:** 2287–2292.

Palchevskiy V, Finkel SE. 2006. *Escherichia coli* competence gene homologs are essential for competitive fitness and the use of DNA as a nutrient. *J Bacteriol* **188:** 3902–3910.

Parsek MR, Singh PK. 2003. Bacterial biofilms: An emerging link to disease pathogenesis. *Annu Rev Microbiol* **57:** 677–701.

Patel RB, Welling PG. 1980. Clinical pharmacokinetics of co-trimoxazole (trimethoprim-sulphamethoxazole). *Clin Pharmacokinet* **5:** 405–423.

Patrauchan MA, Sarkisova S, Sauer K, Franklin MJ. 2005. Calcium influences cellular and extracellular product formation during biofilm-associated growth of a marine *Pseudoalteromonas sp. Microbiology* **151:** 2885–2897.

Percival SL, Kite P, Eastwood K, Murga R, Carr J, Arduino MJ, Donlan RM. 2005. Tetrasodium EDTA as a novel central venous catheter lock solution against biofilm. *Infect Control Hosp Epidemiol* **26:** 515–519.

Pihl M, Davies JR, Chavez de Paz LE, Svensater G. 2010. Differential effects of *Pseudomonas aeruginosa* on biofilm formation by different strains of *Staphylococcus epidermidis*. *FEMS Immunol Med Microbiol* **59:** 439–446.

Pinkner JS, Remaut H, Buelens F, Miller E, Aberg V, Pemberton N, Hedenstrom M, Larsson A, Seed P, Waksman G, et al. 2006. Rationally designed small compounds inhibit pilus biogenesis in uropathogenic bacteria. *Proc Natl Acad Sci* **103:** 17897–17902.

Pompilio A, Scocchi M, Pomponio S, Guida F, Di Primio A, Fiscarelli E, Gennaro R, Di Bonaventura G. 2011. Antibacterial and anti-biofilm effects of cathelicidin peptides against pathogens isolated from cystic fibrosis patients. *Peptides* **32:** 1807–1814.

Pratt LA, Kolter R. 1998. Genetic analysis of *Escherichia coli* biofilm formation: Roles of flagella, motility, chemotaxis and type I pili. *Mol Microbiol* **30:** 285–293.

Prigent-Combaret C, Lejeune P. 1999. Monitoring gene expression in biofilms. *Methods Enzymol* **310:** 56–79.

Prigent-Combaret C, Brombacher E, Vidal O, Ambert A, Lejeune P, Landini P, Dorel C. 2001. Complex regulatory network controls initial adhesion and biofilm formation in *Escherichia coli* via regulation of the csgD gene. *J Bacteriol* **183:** 7213–7223.

Pruss BM, Besemann C, Denton A, Wolfe AJ. 2006. A complex transcription network controls the early stages of biofilm development by *Escherichia coli*. *J Bacteriol* **188:** 3731–3739.

Purevdorj-Gage B, Costerton WJ, Stoodley P. 2005. Phenotypic differentiation and seeding dispersal in non-mucoid and mucoid *Pseudomonas aeruginosa* biofilms. *Microbiology* **151:** 1569–1576.

Qin X, Singh KV, Weinstock GM, Murray BE. 2000. Effects of *Enterococcus faecalis fsr* genes on production of gelatinase and a serine protease and virulence. *Infect Immun* **68:** 2579–2586.

Qin Z, Ou Y, Yang L, Zhu Y, Tolker-Nielsen T, Molin S, Qu D. 2007. Role of autolysin-mediated DNA release in biofilm formation of *Staphylococcus epidermidis*. *Microbiology* **153:** 2083–2092.

Qin Z, Yang L, Qu D, Molin S, Tolker-Nielsen T. 2009. *Pseudomonas aeruginosa* extracellular products inhibit staphylococcal growth, and disrupt established biofilms produced by *Staphylococcus epidermidis*. *Microbiology* **155:** 2148–2156.

Raad I, Buzaid A, Rhyne J, Hachem R, Darouiche R, Safar H, Albitar M, Sherertz RJ. 1997. Minocycline and ethylenediaminetetraacetate for the prevention of recurrent vascular catheter infections. *Clin Infect Dis* **25:** 149–151.

Raad I, Chatzinikolaou I, Chaiban G, Hanna H, Hachem R, Dvorak T, Cook G, Costerton W. 2003. In vitro and ex vivo activities of minocycline and EDTA against microorganisms embedded in biofilm on catheter surfaces. *Antimicrob Agents Chemother* **47:** 3580–3585.

Raad II, Fang X, Keutgen XM, Jiang Y, Sherertz R, Hachem R. 2008. The role of chelators in preventing biofilm formation and catheter-related bloodstream infections. *Curr Opin Infect Dis* **21:** 385–392.

Rasko DA, Moreira CG, Li de R, Reading NC, Ritchie JM, Waldor MK, Williams N, Taussig R, Wei S, Roth M, et al. 2008. Targeting QseC signaling and virulence for antibiotic development. *Science* **321:** 1078–1080.

Reigstad CS, Hultgren SJ, Gordon JI. 2007. Functional genomic studies of uropathogenic *Escherichia coli* and host urothelial cells when intracellular bacterial communities are assembled. *J Biol Chem* **282:** 21259–21267.

Rendueles O, Kaplan JB, Ghigo JM. 2012. Antibiofilm polysaccharides. *Environ Microbiol* doi: 10.1111/j.1462-2920.2012.02810.x.

Romling U. 2002. Molecular biology of cellulose production in bacteria. *Res Microbiol* **153:** 205–212.

Rosen DA, Hooton TM, Stamm WE, Humphrey PA, Hultgren SJ. 2007. Detection of intracellular bacterial communities in human urinary tract infection. *PLoS Med* **4:** e329.

Rosen DA, Pinkner JS, Jones JM, Walker JN, Clegg S, Hultgren SJ. 2008a. Utilization of an intracellular bacterial community pathway in *Klebsiella pneumoniae* urinary tract infection and the effects of FimK on type 1 pilus expression. *Infect Immun* **76:** 3337–3345.

Rosen DA, Pinkner JS, Walker JN, Elam JS, Jones JM, Hultgren SJ. 2008b. Molecular variations in *Klebsiella pneumoniae* and *Escherichia coli* FimH affect function and pathogenesis in the urinary tract. *Infect Immun* **76:** 3346–3356.

Rowe MC, Withers HL, Swift S. 2010. Uropathogenic *Escherichia coli* forms biofilm aggregates under iron restriction that disperse upon the supply of iron. *FEMS Microbiol Lett* **307:** 102–109.

Rupp ME, Lisco SJ, Lipsett PA, Perl TM, Keating K, Civetta JM, Mermel LA, Lee D, Dellinger EP, Donahoe M, et al. 2005. Effect of a second-generation venous catheter impregnated with chlorhexidine and silver sulfadiazine on central catheter-related infections: A randomized, controlled trial. *Ann Intern Med* **143:** 570–580.

Sauer K, Cullen MC, Rickard AH, Zeef LA, Davies DG, Gilbert P. 2004. Characterization of nutrient-induced dispersion in *Pseudomonas aeruginosa* PAO1 biofilm. *J Bacteriol* **186:** 7312–7326.

Schaeffer AJ, Amundsen SK, Jones JM. 1980. Effect of carbohydrates on adherence of *Escherichica coli* to human urinary tract epithelial cells. *Infect Immun* **30:** 531–537.

Schembri MA, Kjaergaard K, Klemm P. 2003. Global gene expression in *Escherichia coli* biofilms. *Mol Microbiol* **48:** 253–267.

Schmidt J, Musken M, Becker T, Magnowska Z, Bertinetti D, Moller S, Zimmermann B, Herberg FW, Jansch L, Haussler S. 2011. The *Pseudomonas aeruginosa* chemotaxis methyltransferase CheR1 impacts on bacterial surface sampling. *PLoS ONE* **6:** e18184.

Schwartz DJ, Chen SL, Hultgren SJ, Seed PC. 2011. Population dynamics and niche distribution of uropathogenic *Escherichia coli* during acute and chronic urinary tract infection. *Infect Immun* **79:** 4250–4259.

Secinti KD, Ozalp H, Attar A, Sargon MF. 2011. Nanoparticle silver ion coatings inhibit biofilm formation on titanium implants. *J Clin Neurosci* **18:** 391–395.

Shak S, Capon DJ, Hellmiss R, Marsters SA, Baker CL. 1990. Recombinant human DNase I reduces the viscosity of cystic fibrosis sputum. *Proc Natl Acad Sci* **87:** 9188–9192.

Shanks RM, Sargent JL, Martinez RM, Graber ML, O'Toole GA. 2006. Catheter lock solutions influence staphylococcal biofilm formation on abiotic surfaces. *Nephrol Dial Transplant* **21:** 2247–2255.

Snyder JA, Lloyd AL, Lockatell CV, Johnson DE, Mobley HL. 2006. Role of phase variation of type 1 fimbriae in a uropathogenic *Escherichia coli* cystitis isolate during urinary tract infection. *Infect Immun* **74:** 1387–1393.

Sokurenko EV, Courtney HS, Ohman DE, Klemm P, Hasty DL. 1994. FimH family of type 1 fimbrial adhesins: Functional heterogeneity due to minor sequence variations among fimH genes. *J Bacteriol* **176:** 748–755.

Sokurenko EV, Courtney HS, Maslow J, Siitonen A, Hasty DL. 1995. Quantitative differences in adhesiveness of type 1 fimbriated *Escherichia coli* due to structural differences in fimH genes. *J Bacteriol* **177:** 3680–3686.

Sokurenko EV, Chesnokova V, Dykhuizen DE, Ofek I, Wu XR, Krogfelt KA, Struve C, Schembri MA, Hasty DL. 1998. Pathogenic adaptation of *Escherichia coli* by natural variation of the FimH adhesin. *Proc Natl Acad Sci* **95:** 8922–8926.

Sokurenko EV, Feldgarden M, Trintchina E, Weissman SJ, Avagyan S, Chattopadhyay S, Johnson JR, Dykhuizen DE. 2004. Selection footprint in the FimH adhesin shows pathoadaptive niche differentiation in *Escherichia coli*. *Mol Biol Evol* **21:** 1373–1383.

Sperling O, Fuchs A, Lindhorst TK. 2006. Evaluation of the carbohydrate recognition domain of the bacterial adhesin FimH: Design, synthesis and binding properties of mannoside ligands. *Org Biomol Chem* **4:** 3913–3922.

Spurbeck RR, Stapleton AE, Johnson JR, Walk ST, Hooton TM, Mobley HL. 2011. Fimbrial profiles predict virulence of uropathogenic *E. coli* strains: Contribution of Ygi and Yad fimbriae. *Infect Immun* **79:** 4753–4763.

Stanley NR, Britton RA, Grossman AD, Lazazzera BA. 2003. Identification of catabolite repression as a physiological regulator of biofilm formation by *Bacillus subtilis* by use of DNA microarrays. *J Bacteriol* **185:** 1951–1957.

Starkey M, Hickman JH, Ma L, Zhang N, De Long S, Hinz A, Palacios S, Manoil C, Kirisits MJ, Starner TD, et al. 2009. *Pseudomonas aeruginosa* rugose small-colony variants have adaptations that likely promote persistence in the cystic fibrosis lung. *J Bacteriol* **191:** 3492–3503.

Sutherland IW, Hughes KA, Skillman LC, Tait K. 2004. The interaction of phage and biofilms. *FEMS Microbiol Lett* **232:** 1–6.

Thankavel K, Madison B, Ikeda T, Malaviya R, Shah AH, Arumugam PM, Abraham SN. 1997. Localization of a domain in the FimH adhesin of *Escherichia coli* type 1 fimbriae capable of receptor recognition and use of a domain-specific antibody to confer protection against experimental urinary tract infection. *J Clin Invest* **100:** 1123–1136.

Thomas VC, Thurlow LR, Boyle D, Hancock LE. 2008. Regulation of autolysis-dependent extracellular DNA release by *Enterococcus faecalis* extracellular proteases influences biofilm development. *J Bacteriol* **190:** 5690–5698.

Thomas VC, Hiromasa Y, Harms N, Thurlow L, Tomich J, Hancock LE. 2009. A fratricidal mechanism is responsible for eDNA release and contributes to biofilm development of *Enterococcus faecalis*. *Mol Microbiol* **72:** 1022–1036.

Thumbikat P, Berry RE, Zhou G, Billips BK, Yaggie RE, Zaichuk T, Sun TT, Schaeffer AJ, Klumpp DJ. 2009. Bacteria-induced uroplakin signaling mediates bladder response to infection. *PLoS Pathog* **5:** e1000415.

Tielker D, Hacker S, Loris R, Strathmann M, Wingender J, Wilhelm S, Rosenau F, Jaeger KE. 2005. *Pseudomonas aeruginosa* lectin LecB is located in the outer membrane and is involved in biofilm formation. *Microbiology* **151:** 1313–1323.

Toledo-Arana A, Valle J, Solano C, Arrizubieta MJ, Cucarella C, Lamata M, Amorena B, Leiva J, Penades JR, Lasa I. 2001. The enterococcal surface protein, Esp, is involved in *Enterococcus faecalis* biofilm formation. *Appl Environ Microbiol* **67:** 4538–4545.

Ton-That H, Schneewind O. 2003. Assembly of pili on the surface of *Corynebacterium diphtheriae*. *Mol Microbiol* **50:** 1429–1438.

Ton-That H, Marraffini LA, Schneewind O. 2004. Sortases and pilin elements involved in pilus assembly of *Corynebacterium diphtheriae*. *Mol Microbiol* **53:** 251–261.

Toutain CM, Caizza NC, Zegans ME, O'Toole GA. 2007. Roles for flagellar stators in biofilm formation by *Pseudomonas aeruginosa*. *Res Microbiol* **158:** 471–477.

Ubeda C, Taur Y, Jenq RR, Equinda MJ, Son T, Samstein M, Viale A, Socci ND, van den Brink MR, Kamboj M, et al. 2010. Vancomycin-resistant *Enterococcus* domination of intestinal microbiota is enabled by antibiotic treatment in mice and precedes bloodstream invasion in humans. *J Clin Invest* **120:** 4332–4341.

Uhlich GA, Cooke PH, Solomon EB. 2006. Analyses of the red-dry-rough phenotype of an *Escherichia coli* O157:H7 strain and its role in biofilm formation and resistance to antibacterial agents. *Appl Environ Microbiol* **72:** 2564–2572.

Ulett GC, Valle J, Beloin C, Sherlock O, Ghigo JM, Schembri MA. 2007. Functional analysis of antigen 43 in uropathogenic *Escherichia coli* reveals a role in long-term persistence in the urinary tract. *Infect Immun* **75:** 3233–3244.

Vallet I, Olson JW, Lory S, Lazdunski A, Filloux A. 2001. The chaperone/usher pathways of *Pseudomonas aeruginosa*: Identification of fimbrial gene clusters (cup) and their involvement in biofilm formation. *Proc Natl Acad Sci* **98:** 6911–6916.

Vasseur P, Vallet-Gely I, Soscia C, Genin S, Filloux A. 2005. The pel genes of the *Pseudomonas aeruginosa* PAK strain are involved at early and late stages of biofilm formation. *Microbiology* **151:** 985–997.

Venditti M, Biavasco F, Varaldo PE, Macchiarelli A, De Biase L, Marino B, Serra P. 1993. Catheter-related endocarditis due to glycopeptide-resistant *Enterococcus faecalis* in a transplanted heart. *Clin Infect Dis* **17:** 524–525.

Vidal O, Longin R, Prigent-Combaret C, Dorel C, Hooreman M, Lejeune P. 1998. Isolation of an *Escherichia coli* K-12 mutant strain able to form biofilms on inert surfaces: Involvement of a new ompR allele that increases curli expression. *J Bacteriol* **180:** 2442–2449.

Vilain S, Pretorius JM, Theron J, Brozel VS. 2009. DNA as an adhesin: *Bacillus cereus* requires extracellular DNA to form biofilms. *Appl Environ Microbiol* **75:** 2861–2868.

Vuong C, Voyich JM, Fischer ER, Braughton KR, Whitney AR, DeLeo FR, Otto M. 2004. Polysaccharide intercellular adhesin (PIA) protects *Staphylococcus epidermidis* against major components of the human innate immune system. *Cell Microbiol* **6:** 269–275.

Waksman G, Hultgren SJ. 2009. Structural biology of the chaperone-usher pathway of pilus biogenesis. *Nat Rev Microbiol* **7:** 765–774.

Walters MC 3rd, Roe F, Bugnicourt A, Franklin MJ, Stewart PS. 2003. Contributions of antibiotic penetration, oxygen limitation, and low metabolic activity to tolerance of *Pseudomonas aeruginosa* biofilms to ciprofloxacin and tobramycin. *Antimicrob Agents Chemother* **47:** 317–323.

Wang X, Preston JF 3rd, Romeo T. 2004. The pgaABCD locus of *Escherichia coli* promotes the synthesis of a polysaccharide adhesin required for biofilm formation. *J Bacteriol* **186:** 2724–2734.

Wang X, Dubey AK, Suzuki K, Baker CS, Babitzke P, Romeo T. 2005. CsrA post-transcriptionally represses *pgaABCD*, responsible for synthesis of a biofilm polysaccharide adhesin of *Escherichia coli*. *Mol Microbiol* **56:** 1648–1663.

Wang X, Wang Q, Yang M, Xiao J, Liu Q, Wu H, Zhang Y. 2011. QseBC controls flagellar motility, fimbrial hemagglutination and intracellular virulence in fish pathogen *Edwardsiella tarda*. *Fish Shellfish Immunol* **30:** 944–953.

Watanabe T, Okada A, Gotoh Y, Utsumi R. 2008. Inhibitors targeting two-component signal transduction. *Adv Exp Med Biol* **631:** 229–236.

Watnick PI, Kolter R. 1999. Steps in the development of a *Vibrio cholerae* El Tor biofilm. *Mol Microbiol* **34:** 586–595.

Webb JS, Thompson LS, James S, Charlton T, Tolker-Nielsen T, Koch B, Givskov M, Kjelleberg S. 2003. Cell death in *Pseudomonas aeruginosa* biofilm development. *J Bacteriol* **185:** 4585–4592.

Weissman SJ, Beskhlebnaya V, Chesnokova V, Chattopadhyay S, Stamm WE, Hooton TM, Sokurenko EV. 2007. Differential stability and trade-off effects of pathoadaptive mutations in the *Escherichia coli* FimH adhesin. *Infect Immun* **75:** 3548–3555.

Welch RA, Burland V, Plunkett G 3rd, Redford P, Roesch P, Rasko D, Buckles EL, Liou SR, Boutin A, Hackett J, et al. 2002. Extensive mosaic structure revealed by the complete genome sequence of uropathogenic *Escherichia coli*. *Proc Natl Acad Sci* **99:** 17020–17024.

Wellens A, Garofalo C, Nguyen H, Van Gerven N, Slattegard R, Hernalsteens JP, Wyns L, Oscarson S, De Greve H, Hultgren S, et al. 2008. Intervening with urinary tract infections using anti-adhesives based on the crystal structure of the FimH-oligomannose-3 complex. *PLoS ONE* **3:** e2040.

Whitchurch CB, Tolker-Nielsen T, Ragas PC, Mattick JS. 2002. Extracellular DNA required for bacterial biofilm formation. *Science* **295:** 1487.

Whiteley M, Bangera MG, Bumgarner RE, Parsek MR, Teitzel GM, Lory S, Greenberg EP. 2001. Gene expression in *Pseudomonas aeruginosa* biofilms. *Nature* **413:** 860–864.

Wood TK, Hong SH, Ma Q. 2010. Engineering biofilm formation and dispersal. *Trends Biotechnol* **29:** 87–94.

Wright KJ, Seed PC, Hultgren SJ. 2007. Development of intracellular bacterial communities of uropathogenic *Escherichia coli* depends on type 1 pili. *Cell Microbiol* **9:** 2230–2241.

Wu Y, Outten FW. 2009. IscR controls iron-dependent biofilm formation in *Escherichia coli* by regulating type I fimbria expression. *J Bacteriol* **191:** 1248–1257.

Yamanaka M, Hara K, Kudo J. 2005. Bactericidal actions of a silver ion solution on *Escherichia coli*, studied by energy-filtering transmission electron microscopy and proteomic analysis. *Appl Environ Microbiol* **71:** 7589–7593.

Yang D, Biragyn A, Kwak LW, Oppenheim JJ. 2002. Mammalian defensins in immunity: More than just microbicidal. *Trends Immunol* **23:** 291–296.

Yang L, Barken KB, Skindersoe ME, Christensen AB, Givskov M, Tolker-Nielsen T. 2007. Effects of iron on

DNA release and biofilm development by *Pseudomonas aeruginosa*. *Microbiology* **153:** 1318–1328.

Zhang L, Mah TF. 2008. Involvement of a novel efflux system in biofilm-specific resistance to antibiotics. *J Bacteriol* **190:** 4447–4452.

Zhou G, Mo WJ, Sebbel P, Min G, Neubert TA, Glockshuber R, Wu XR, Sun TT, Kong XP. 2001. Uroplakin Ia is the urothelial receptor for uropathogenic *Escherichia coli*: Evidence from in vitro FimH binding. *J Cell Sci* **114:** 4095–4103.

Zhu Y, Weiss EC, Otto M, Fey PD, Smeltzer MS, Somerville GA. 2007. *Staphylococcus aureus* biofilm metabolism and the influence of arginine on polysaccharide intercellular adhesin synthesis, biofilm formation, and pathogenesis. *Infect Immun* **75:** 4219–4226.

Zogaj X, Nimtz M, Rohde M, Bokranz W, Romling U. 2001. The multicellular morphotypes of *Salmonella typhimurium* and *Escherichia coli* produce cellulose as the second component of the extracellular matrix. *Mol Microbiol* **39:** 1452–1463.

Zogaj X, Bokranz W, Nimtz M, Romling U. 2003. Production of cellulose and curli fimbriae by members of the family Enterobacteriaceae isolated from the human gastrointestinal tract. *Infect Immun* **71:** 4151–4158.

General Aspects and Recent Advances on Bacterial Protein Toxins

Emmanuel Lemichez[1] and Joseph T. Barbieri[2]

[1]INSERM, U1065, Centre Méditerranéen de Médecine Moléculaire, Microbial Toxins in Host-Pathogen Interactions, C3M, Université de Nice-Sophia-Antipolis, UFR Médecine, 06204 Nice, France

[2]Medical College of Wisconsin, Microbiology and Molecular Genetics, Milwaukee, Wisconsin 53151

Correspondence: lemichez@unice.fr

Bacterial pathogens produce protein toxins to influence host–pathogen interactions and tip the outcome of these encounters toward the benefit of the pathogen. Protein toxins modify host-specific targets through posttranslational modifications (PTMs) or noncovalent interactions that may inhibit or activate host cell physiology to benefit the pathogen. Recent advances have identified new PTMs and host targets for toxin action. Understanding the mechanisms of toxin action provides a basis to develop vaccines and therapies to combat bacterial pathogens and to develop new strategies to use toxin derivatives for the treatment of human disease.

During the dialogue established between bacterial pathogens and their host, secreted protein toxins play a key role in the specificity of the outcome of the infection. These toxins are fascinating objects if one considers that in some instances injection of small amounts of these purified proteins recapitulates the deadly symptoms of the infection, such as in the case of the spastic paralysis triggered by tetanus neurotoxin (TeNT). Here we will briefly introduce the diversity of bacterial protein toxins and then focus on new advances made on toxins acting on intracellular host targets.

Bacterial protein toxins can be classified into several functional groups despite their diverse structures and modes of action (Fig. 1) (for general reviews, see Alouf and Popoff 2006). These groups are:

1. A group of toxins that act directly on host cell plasma membrane receptors. By acting as host receptor agonists or antagonists, they corrupt signal transduction pathways. For instance, superantigens produced by staphylococci can bridge nonspecifically major histocompatibility complex class II at the surface of antigen-presenting cells with the T-cell receptor, leading to a nonclonal activation of immune cells and resulting in inflammatory storms (Alouf and Popoff 2006).
2. A group of toxins that disrupt membrane lipid bilayer integrity by forming pores of different size and molecular selectivity or because of their phospholipase activity (Bischofberger et al. 2009). For example, the pore-forming toxin listeriolysin O (LLO) produced by bacteria internalized into phagocytic com-

Cite this article as *Cold Spring Harb Perspect Med* doi: 10.1101/cshperspect.a013573

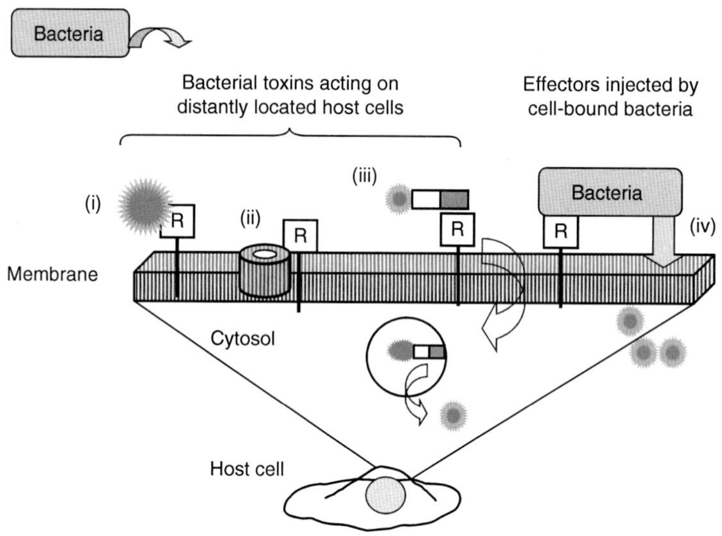

Figure 1. Classes of bacterial protein toxins. There are four major classes of bacterial toxins, including (i) toxins that bind to the surface of host cell plasma membrane receptors and modify host cell physiology by triggering intracellular signaling; (ii) toxins that bind to host cell plasma membranes and disrupt the membrane lipid bilayer through pore formation or expression of phospholipase activity; (iii) AB toxins with an A domain that possesses enzymatic activity and a B domain that binds and enters the host cell; and (iv) toxins with an enzymatic activity that is delivered into the host cell by an injection apparatus that is a component of the bacterial pathogens, which include the type III secreted cytotoxins. The first three groups of toxins often act at a site within the host that is distant from the bacterial pathogen, whereas type III secreted cytotoxins are delivered into the host cell directly by the bacterium, often paralyzing the host cell's ability to neutralize the bacterial pathogen.

partments disrupts the phagosomal membrane, releasing bacteria into the cytosol for dissemination (Cossart 2011). In addition, loss of plasma membrane integrity by pore-forming toxins corrupts cell signaling (Bischofberger et al. 2009; Cossart 2011).

3. A group of toxins that inject toxic enzymatic components into distant host cells once they have entered cells by receptor-mediated endocytosis. These sophisticated proteins can be functionally described as molecular syringes (Fig. 1). They are termed AB toxins because of the presence of a polypeptidic A domain that possesses enzymatic activity and one or more B domains that bind and enter the host cell. These toxins, such as TeNT and botulinum neurotoxins (BoNTs), are frequently endowed with a major virulent function. Therefore, these proteins are of major interest in the study of bacterial virulence mechanisms, such as transfer of virulence genes among bacteria, regulation of virulence factor expression, and the secretion apparatuses of these virulence factors. In addition, they represent valuable tools to study cell biological processes, that is: (1) receptor endocytosis and vesicular trafficking, such as the retrograde transport of Shiga toxin (Stx) from the surface to the endoplasmic reticulum (ER); (2) translocation of their enzymatic domain through intracellular membranes; and (3) the regulation and function of their cellular targets, as reviewed here (Fig. 2). Finally, it is of major interest to use these potent factors as medical tools, for instance, to kill cancer cells as well as to boost immune responses for vaccination.

4. A group of toxins that are synthesized within the bacterium and delivered directly into the host cell by an injection needle, including type III secreted cytotoxins of Gram-negative bacteria (Fig. 1) (Galan 2009). The type III delivery apparatus is derived by gene duplication from the flagellum apparatus.

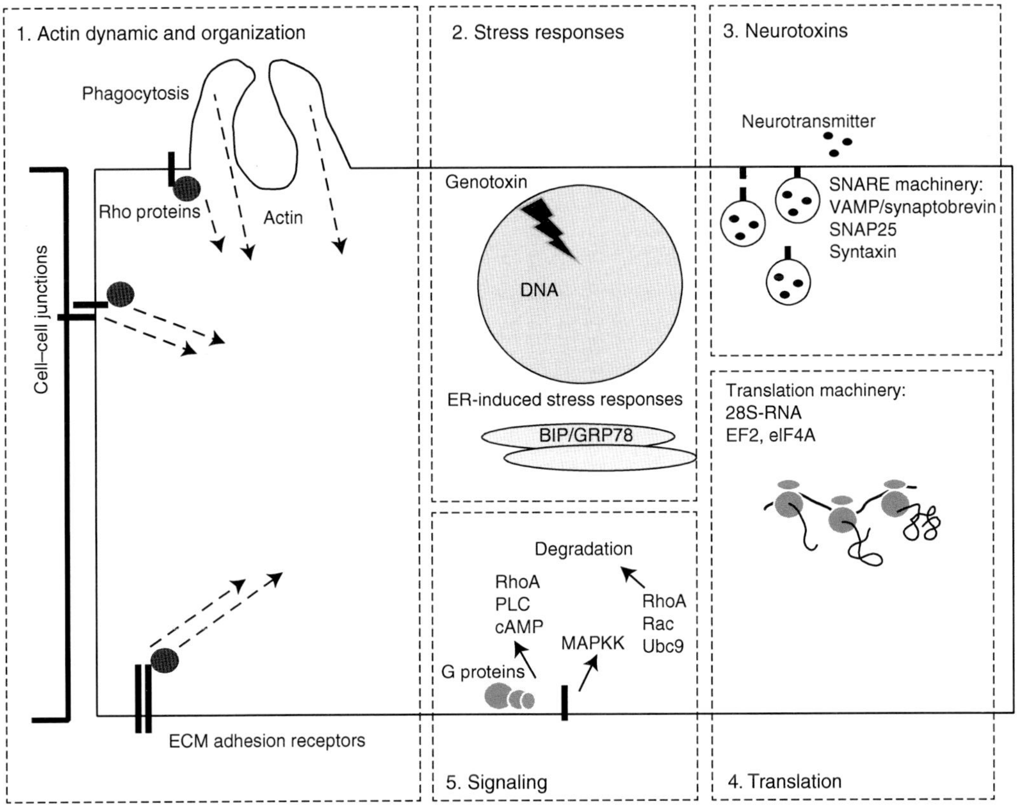

Figure 2. Cellular Achilles' heels targeted by bacterial toxins. Despite differences in provenance, structure, or enzymatic activities, potent bacterial toxins known to date target conserved key cellular factors and pathways (see Table 1 for details). (1) Several toxins directly target actin or Rho GTPases, upstream regulators of the actomyosin cytoskeleton organization and dynamic. These toxins have a major impact on cell–cell or cell–matrix adhesion, motility, and phagocytosis. (2) One group of toxins triggers stress responses, for example, by catalyzing DNA damage. (3) Neurotoxins directly cleave components of the SNARE machinery, thereby impairing exocytosis of neurotransmitters. (4) Several toxins impair cell translation by targeting ribosomal RNA or translation factors (eIF4A or EF2). (5) Other toxins corrupt major signaling pathways, notably resulting in high level of production of cAMP or a shutdown of MAP kinase signaling. They also exacerbate the destruction of signaling molecules, such as Ubc9, RhoA, and Rac1.

These cytotoxins (also called effectors) possess a catalytic activity that covalently modifies a host protein through a posttranslational modification (PTM) or through a noncovalent interaction. These cytotoxins are unique in that the bacterium delivers a large amount of molecules over a short period of time to overwhelm the physiology of the host cell. For example, *Salmonella* spp. inject SopE, a guanine nucleotide exchange factor that stimulates Rac GTPase activity to facilitate entry into pathogen-associated vacuoles in epithelial cells (Galan 2009).

TOXINS TARGETING ACTIN AND SMALL RHO GTPASES

A large number of potent bacterial toxins, notably those produced by several *Clostridium* spp., such as *Clostridium difficile*, and some pathogenic Gram-negative bacteria, such as uropathogenic strains of *Escherichia coli*, catalyze PTM of components of the cell actomyosin cytoskeleton (Table 1). The actomyosin cytoskeleton confers on host cells their shape and membrane dynamics. This function is required for cells to properly control their adhesion to

Table 1. Bacterial enzymatic toxins acting on host cell cytosolic factors

Toxin	Target	Activity	Effect	Bacteria	Reference
Toxins targeting actin cytoskeleton components					
C2	Actin R177	ADP-ribosylation	Depolymerization	*Clostridium botulinum*	Aktories et al. 1986
Iota	Actin R177	ADP-ribosylation	Depolymerization	*Clostridium perfringens*	Aktories et al. 2011
CDT	Actin R177	ADP-ribosylation	Depolymerization	*Clostridium difficile*	Aktories et al. 2011
CST	Actin R177	ADP-ribosylation	Depolymerization	*Clostridium spiroforme*	Aktories et al. 2011
VIP	Actin R177	ADP-ribosylation	Depolymerization	*Bacillus cereus*	Aktories et al. 2011
Tc (TccC3-subunit)	Actin T148	ADP-ribosylation	Actin clustering	*Photorhabdus luminescens*	Lang et al. 2010
MARTXvc (ACD domain)	Actin K50–actin E270	Actin cross-linking	Depolymerization	*Vibrio cholerae*	Aktories et al. 2011
CNF1, -2, -3	Rac1, RhoA, Cdc42 (Q61 or Q63)	Deamidase	Actin polymerization	*Escherichia coli*	Lemonnier et al. 2007
CNFy	RhoA (Q63)	Deamidase	Actin polymerization	*Yersinia pseudotuberculosis*	Hoffmann et al. 2004
DNT	Rac1, RhoA, Cdc42 (Q61 or Q63)	Transglutaminase/deamidase	Actin polymerization	*Bordetella pertussis, Bordetella parapertussis*	Lemonnier et al. 2007
C3 and C3-like	RhoA (N41)	ADP-ribosyltransferase	Actin cable depolymerization	*C. botulinum, Clostridium limosum, B. cereus, Staphylococcus aureus*	Aktories 2011
TcdA and TcdB	Rho GTPases (T35 or T37)	Glucosylation (UDP-glucose)	Actin depolymerization	*C. difficile*	Just et al. 1995
HT	Rho GTPases (T35 or T37)	Glucosylation (UDP-glucose)	Actin depolymerization	*Clostridium sordellii*	Aktories 2011
LT	Ras GTPases, Rac (T35)	Glucosylation (UDP-glucose)	Actin depolymerization	*C. sordellii*	Aktories 2011
α-Toxin	Rho GTPases (T35 or T37)	*N*-Acetyl-glucosamination (UDP-*N*-acetylglucosamine)	Actin depolymerization	*Clostridium novyi*	Aktories 2011
TpeL	Ras GTPases, Rac (T35)	*N*-Acetyl-glucosamination (UDP-*N*-acetylglucosamine)	Actin depolymerization	*C. perfrigens*	Aktories 2011
lbpA	Rho GTPases (Y32 or Y34)	Adenylylation (AMPylation)	Actin depolymerization	*Histophilus somni*	Worby et al. 2009
Tc (TccC5 subunit)	Rho GTPases (Q61 or Q63)	ADP-ribosylation	Actin polymerization	*P. luminescens*	Lang et al. 2010

Toxins corrupting ubiquitin and ubiquitin-like signaling					
LLO	Cellular membranes	Pore-forming toxin	Cell-induced degradation of Ubc9	*Listeria monocytogenes*	Ribet 2010
CNF1	RhoA	Deamidase	Smurf1-induced degradation of activated RhoA	*E. coli*	Visvikis 2010
CNF1	Rac1	Deamidase	HACE1 (XIAP, cIAP1/2)-induced degradation of activated Rac1	*E. coli*	Oberoi 2011; Torrino 2011
Toxins targeting cell translational machinery					
DT	EF2 (diphthamide-715)	ADP-ribosylation	Translation inhibition	*Corynebacterium diphtheriae*	Murphy 2011
ETA	EF2 (diphthamide-715)	ADP-ribosylation	Translation inhibition	*Pseudomonas aeruginosa*	Murphy 2011
Stx	28S ribosomal RNA (adenine base at position 4324)	*N*-Glycosidase	Translation inhibition	*Shigella dysenteriae*	Johannes and Romer 2010
Stx-like (1, 1c, 2, 2c, 2d, 2e, 2f)	28S ribosomal RNA	*N*-Glycosidase	Translation inhibition	Shigatoxigenic *E. coli* (STEC)	Johannes and Romer 2010
LT-1 (BPSL1549)	eIF4A (Q339)	Deamidase	Translation inhibition (uncouples ATPase and helicase activities)	*Burkholderia pseudomallei*	Cruz-Migoni et al. 2011
Toxins targeting SNARE machinery					
TeNT	VAMP/synaptobrevin	Zinc metalloprotease	Neurotransmission inhibition	*Clostridium tetani*	Hill et al. 2007
BoNT A, E	SNAP25	Zinc metalloprotease	Neurotransmission inhibition	*C. botulinum*	Hill et al. 2007
BoNT B, D, F, G	VAMP/synaptobrevin	Zinc metalloprotease	Neurotransmission inhibition	*C. botulinum*	Hill et al. 2007
BoNT C	SNAP25, syntaxin	Zinc metalloprotease	Neurotransmission inhibition	*C. botulinum*	Hill et al. 2007
Toxins targeting DNA and inducing endoplasmic reticulum stress responses					
EcCDT-I, II, III, IV	DNA	DNAse-I-like (B subunit)	Cell cycle arrest	*E. coli*	Guerra et al. 2011
SdCDT	DNA	DNAse-I-like (B subunit)	Cell cycle arrest	*S. dysenteriae*	Guerra et al. 2011
HdCDT	DNA	DNAse-I-like (B subunit)	Cell cycle arrest	*Haemophilus ducreyi*	Guerra et al. 2011
AaCDT	DNA	DNAse-I-like (B subunit) and PIP_3 phosphatase	Cell cycle arrest (lymphocyte apoptosis)	*Aggregatibacter actinomycetemcomitans*	Shenker et al., 2007; Guerra et al. 2011
HhCDT	DNA	DNAse-I-like (B subunit)	Cell cycle arrest	*Helicobacter hepaticus*	Guerra et al. 2011
CjCDT	DNA	DNAse-I-like (B subunit)	Cell cycle arrest	*Campylobacter jejuni*	Guerra et al. 2011
SubAB	BIP/GRP78	Serine protease	ER stress responses	STEC	Byres et al. 2008

Continued

Table 1. *Continued*

Toxin	Target	Activity	Effect	Bacteria	Reference
Toxins targeting cAMP and MAP kinase signaling components					
Ctx	$G\alpha_s$ (R201, activation)	ADP-ribosylation	cAMP induction	*V. cholerae*	Aktories et al. 2011
LT	$G\alpha_s$ (R201, activation)	ADP-ribosylation	cAMP induction	*E. coli*	Aktories et al. 2011
Ptx	$G\alpha_i$ (C351, inactivation)	ADP-ribosylation	cAMP induction	*B. pertussis*	Aktories et al. 2011
PMT	$G\alpha_q$, $G\alpha_i$, $G\alpha_{12/13}$ (Q209, Q205, or Q226, activation)	Deamidation	cAMP inhibition, RhoA activation, PLC activation	*Pasteurella multocida*	Aktories et al. 2011
ET	cAMP production	Adenylate cyclase	cAMP production	*Bacillus anthracis*	Leppla 1982
CyaA	cAMP production	Adenylate cyclase	cAMP production	*B. pertussis*	Ahuja et al. 2004
LT	MAP kinase kinases (except MEK5)	Metalloprotease	MAP kinase signaling	*B. anthracis*	Collier and Young 2003

BoNT, botulinum neurotoxin; CDT, cytolethal distending toxin; CNF, cytotoxic necrotizing factor; CST, *C. spiroforme* toxin; Ctx, cholera toxin; DNT, dermonecrotic toxin; DT, diphtheria toxin; EF2, elongation factor-2; eIF, translation initiation factor; ET, edema toxin; ETA, exotoxin A; HT, hemorrhagic toxin; LLO, listeriolysin O LT, lethal toxin; MAP, mitogen-activated protein; MARTX, multifunctional autoprocessing repeats-in-toxin; PLC, phospholipase C; PMT, *P. multocida* toxin; PTX, pertussis toxin; SNAP, synaptosomal-associated protein; SNARE, soluble *N*-ethylmaleimide-sensitive factor attachment protein receptor; Stx, Shiga toxin; Sub, subtilase toxin; TeNT, tetanus toxin; VAMP, vesicle-associated membrane protein; VIP, vegetative insect cidal protein.

the matrix substratum and intercellular monolayer barrier integrity, as well as for immune cells to undergo migration and perform phagocytosis (Jaffe and Hall 2005). This group of bacterial weapons is thus particularly efficient for disruption of host epithelium barriers, for efficient bacterial dissemination, and for freezing distantly located immune cells to prevent both their migration to the site of infection and bactericidal action (Boquet and Lemichez 2003). Actomyosin cytoskeleton disruption by direct cross-linking of ADP-ribose on actin's arginine-177 residue is an enzymatic property of several clostridial toxins (Table 1) (Aktories et al. 1986, 2011). The ADP-ribosylated actin monomers are extremely toxic, although actin is a very abundant protein in cells. Indeed, ADP-ribosylated actin monomers act dominantly once incorporated into actin filaments by blocking addition of new monomers. This produces a disassembly of actin filaments, leading to cell retraction and collapse.

The property of disassembling actomyosin cytoskeleton is shared by several large clostridial toxins that modify Rho proteins (Aktories 2011). This group of key upstream regulators of actomyosin cytoskeleton organization and dynamic belongs to the superfamilly of small Ras GTPases (Jaffe and Hall 2005). Large glucosylating toxins catalyze the addition of a sugar molecule, such as glucose from UDP-glucose, on a key threonine residue of several small GTPases (Just et al. 1995). Addition of bulky glucose groups to this key amino acid residue poisons their biochemical activity and association with downstream effectors. Glucosylating toxins share as a common target Rac, a small Rho GTPase that orchestrates cellular innate immune responses against pathogens (Bokoch 2005; Boyer et al. 2011). This effector-binding domain of Rho proteins is subjected to several other PTMs by bacterial effectors and toxins, such as ADP-ribosylation and AMPylation (Table 1) (Worby et al. 2009; Yarbrough et al. 2009; Aktories 2011).

Photorhabdus luminescens organisms establish a symbiotic relationship with some entomopathogenic nematode species, conferring on them the capacity to kill susceptible insects. Exciting new findings on Tc toxin (TccC3 and TccC5 subunits) of *P. luminescens* has shed light on two new toxin-mediated PTMs of actin and Rho GTPases (Lang et al. 2010). Hence, determination of the structure of the closely homologous Tc toxin of *Yersinia pseudotuberculosis* has revealed its organization into large multisubunit complexes (Landsberg et al. 2011). RhoA is ADP-ribosylated by TccC5 on glutamine-63 (Q61 for Rac and Cdc42) (Lang et al. 2010). This glutamine-63 residue is a hot spot of modification by bacterial toxins given its key function in catalyzing the hydrolysis of the guanosine triphosphate into guanosine diphosphate for switching GTPases to an inactive form (Flatau et al. 1997; Schmidt et al. 1997; Lemonnier et al. 2007). Thus, ADP-ribosylation of RhoA by TccC5 triggers its activation and the resulting formation of actin stress fibers (Lang et al. 2010). The TccC3 toxin subunit ADP-ribosylates actin on threonine-148 instead of arginine-177, as for other known actin-targeting toxins (Table 1). This modification blocks actin sequestration by thymosin-β4, thereby resulting in actin polymerization and aggregation of actin filament into clusters scattered through the cytosol. Intoxication of cells by both factors leads to an aggravation of actin reorganization into clusters because of the combined activation of Rho GTPases and actin polymerization.

TOXINS CORRUPTING UBIQUITIN AND UBIQUITIN-LIKE SIGNALING

An efficient way to interfere with the function of proteins is to catalyze their proteolysis. Some bacterial metalloprotease toxins such as the lethal factor of *Bacillus anthracis* and the neurotoxins of clostridia directly catalyze the endoproteolytic cleavage of key host proteins (Table 1). Recent advances also indicate that some bacterial toxins exacerbate cellular protein degradation by the ubiquitin/proteasomal system (UPS) (Munro et al. 2007). PTM of proteins by ubiquitin and ubiquitin-like molecules, such as SUMO or NEDD8, control the fate and activity of large numbers of proteins and are targeted by numerous bacterial virulence factors (Munro et al. 2007). The ubiquitylation reaction consists of the covalent attachment of ubiquitin, an 8-kDa polypeptide, to lysine residues on the target protein (Weissman 2001). This involves a cascade of transfer reactions between ubiquitin carrier proteins. Among these factors, the E3 ubiquitin ligase enzymes confer the specificity to the reaction by binding distinctively to a panel of target proteins. Additional molecules of ubiquitin can be subsequently attached to one of the seven lysines of the previously cross-linked ubiquitin molecule, leading to the formation of various types of mono-, multi-, or polyubiquitin chain assemblies, notably lysine-48 (K48) polyubiquitin chains for substrate targeting to proteasomal destruction (Dikic et al. 2009). Several toxins, by catalyzing the permanent activation of Rho proteins, in fact exacerbate the cellular UPS-mediated regulation of these GTPases (Doye et al. 2002; Visvikis et al. 2010). This represents a remarkable example of how toxins can be useful to unravel new cellular regulations and determine in this case E3 ubiquitin ligases that are responsible for Rho protein ubiquitylation. For instance, the study of cytotoxic necrotizing factor 1 (CNF1) recently revealed the critical function of HACE1 and implicated XIAP and cIAP1, E3 ubiquitin ligases, in the targeting of Rac1 to UPS (Oberoi et al. 2011; Torrino et al. 2011). The importance of UPS regulation of Rho proteins during infection remains to be fully determined, although it likely fosters endothelium invasion by bacteria (Doye et al. 2002; Torrino et al. 2011). In addition, some toxins target ubiquitin-like molecules. For instance, *Listeria* listeriolysin O acts on distant host cells to stimulate the degradation of Ubc9, a key enzyme of protein modification by SUMOylation (Ribet et al. 2010). Although bacterial toxins interfere only indirectly with UPS regulation of host cell proteins, several bacterial effectors catalyze a direct PTM of ubiquitin and ubiquitin-like molecules. Indeed, the type III secreted effectors CHBP from *Burkholderia pseudomallei* and Cif from enteropathogenic *E. coli* deamidate the ubiquitin-like protein NEDD8 (Cui et al. 2010; Taieb et al. 2011). This abolishes the activity of multimeric Cullin-RING ubiquitin ligases and impairs numerous major signaling pathways.

TOXINS TARGETING CELL TRANSLATIONAL MACHINERY

An efficient way to kill host cells is through inhibition of protein synthesis. Several pathogenic bacteria share the capacity to inhibit protein translation by PTM of translational factors, as, for example, elongation factor-2 (EF2), which is ADP-ribosylated by diphtheria toxin (DT) of *Corynebacterium diphtheriae* and exotoxin A (ETA) of *Pseudomonas aeruginosa* (Table 1) (Murphy 2011). Both DTand ETA target a PTM-modified histidine residue of EF2 called diphthamide, which is only present in eukaryotes and archaea (Zhang et al. 2010). Most recently, a report shows that the lethal toxin of *B. pseudomallei* (BPSL1549) targets the translation initiation factor 4A (eIF4A) (Cruz-Migoni et al. 2011). This toxin belongs to the growing family of deamidase virulence factors first identified for CNF1 toxin, converting a specific target glutamine into a glutamic acid residue (Flatau et al. 1997; Schmidt et al. 1997; Cui et al. 2010). Despite an absence of primary sequence homology with the CNF1 deamidase domain, BPSL1549 possesses a conservation of the amino acid residues within the catalytic site. The deamidase activity of BPSL1549 toward eIF4A glutamine-339 has an inhibitory effect on the helicase activity, thereby blocking translation (Cruz-Migoni et al. 2011).

The group of Stx and Shiga-like *N*-glycosidase toxins of *Shigella dysenteriae* and enterohemorrhagic strains of *E. coli* remove a specific adenine from 28S ribosomal RNA (Table 1) (Johannes and Romer 2010). Stx toxins play key role in the induction of hemolytic uremic syndrome and at sublethal doses induced inflammatory reactions (Sandvig et al. 1992; Johannes and Romer 2010). A remarkable property of these toxins is their ability to enter cells by triggering membrane deformation at high concentration and undergo a retrograde transport to the ER, where the catalytic domain exits to the cytosol by hijacking the ER-associated degradation machinery (Sandvig et al. 1992; Johannes and Romer 2010).

TOXINS TARGETING SNARE MACHINERY

The clostridial neurotoxins (CNTs) include TeNT and the BoNTs, which are the most toxic proteins for humans but are also commonly used therapy for many human neurological disorders (Table 1). The neurologic specificity of the CNTs lies in their ability to bind neuron-specific receptors and then cleave neuron-specific soluble *N*-ethylmaleimide-sensitive factor attachment protein receptor (SNARE) proteins. These highly specific interactions between BoNTs and synaptic vesicle proteins have been recently reviewed (Brunger et al. 2008). There are seven BoNT serotypes (A–G), and each serotype cleaves one or more proteins within the SNARE protein complex (Hill et al. 2007). SNARE protein cleavage inhibits neurotransmitter vesicle fusion to the plasma membrane and subsequent release of neurotransmitters. The unique pathologies elicited by TeNT and BoNTs are due to their intracellular trafficking in neurons. BoNTs bind gangliosides on resting motor neurons at the neuromuscular junction and enter neurons within cycling synaptic vesicles. SV2 is the protein receptor for BoNT A (Dong et al. 2006). Upon acidification of the synaptic vesicle, the catalytic domain of the BoNTs translocates into the cytoplasm to cleave SNARE proteins and inhibit neurotransmitter release at the neuromuscular junction to elicit flaccid paralysis. In contrast, TeNT binds gangliosides on the cell surface of resting motor neurons and enters endosomes that transcytose TeNT into the central nervous system, where TeNT enters inhibitory neurons to block neurotransmitter release and elicit spastic paralysis. TeNT appears to be transported via a novel neuronal compartment through specialized pH regulation (Bohnert and Schiavo 2005; Chen et al. 2009).

TOXINS TARGETING DNA AND INDUCING ER STRESS RESPONSES

In addition to playing key roles during infection, in some cases bacterial toxins may represent risk factors for cancer (Collins et al. 2011). For example, the genotoxic activity of cytolethal distending toxins (CDTs), a group of toxins produced by a large number of Gram-negative pathogenic bacteria, likely represents a risk factor for carcinogenesis (Table 1) (Lara-Tejero and Galan 2000; Guerra et al. 2011). Structural analysis of *Haemophilus ducreyi* CDT (HdCDT) shows interactions between the three globular subunits (CdtA, -B, and -C) (Nesic et al. 2004). CdtA and CdtC adopt a lectin-type conformation, with structural homologies with the cell-binding subunit of the plant toxin ricin, and with affinity to sphingomyelin. The catalytic subunit CdtB shares conserved amino acid residues with the active site of mammalian deoxyribonuclease (DNase) I (Lara-Tejero and Galan 2000) at residues that contribute to cleavage of the phosphodiester bond of DNA and within a pentapeptide motif (Elwell and Dreyfus 2000). Hence, CdtB of *Aggregatibacter actinomycetemcomitans* bears a phosphatidylinositol 3,4,5-triphosphate 5-phosphatase enzymatic activity, similar to that of PTEN or SHIP1 (Shenker et al. 2007). The relative contribution of both biochemical activities in CdtB-induced DNA damage responses, G_2M cell cycle arrest, and cell death may vary depending on cell types (for review, see Guerra et al. 2011).

TOXINS TARGETING cAMP SIGNALING COMPONENTS

Several pathogenic bacteria have evolved toxins to manipulate the cellular flux of cAMP (Table 1). This comprises Ca^{2+}/calmodulin-dependent adenylate cyclase toxins producing cAMP from ATP, such as edema toxin (ET) of *B. anthracis*, as well as a group of toxins targeting heterotrimeric G proteins, such as cholera toxin of *Vibrio cholera* and pertussis toxin of *Bordetella pertussis* (Table 1) (Ahuja et al. 2004). cAMP is a crucial mediator of cell signaling and is thus implicated in numerous cell biological and physiological functions through the activation of effector proteins such as protein kinase A and the Rap1 exchange factor EPAC. Consistently, manipulation of cAMP signaling confers on bacteria a large panel of features. For example, a recent study shed light on how ET of *B. anthracis* manipulates macrophage genetic responses (Kim et al. 2008). Production of cAMP leads to a reprogramming of macrophage transcription toward a genetic program with close similarities to anti-inflammatory G-protein-coupled receptors, for induction of inflammation resolution, a phenomenon requiring active cell signaling (Kim et al. 2008). This comprises a CREB- and syndecan-1-dependent induction of macrophage motility and vascular endothelial growth factor-triggered lymphangiogenesis. Toxins inducing cAMP flux (notably cholera toxin) are used to manipulate immune responses, a feature that is also under study for Rho-activating toxins, to develop potent immunoadjuvants for mucosal vaccination (Munro et al. 2005; Fabbri et al. 2008).

Recent advances have determined that ET and lethal toxin (LT) of *B. anthracis* directly target endothelial cell actin cytoskeleton and corrupt the proper localization of adherens junction molecules, effects that likely account for the major vascular dysfunctions resulting from anthrax infection or toxemia (Guichard et al. 2010; Lemichez et al. 2010; Maddugoda et al. 2011). At early time periods of endothelial cell infection or intoxication, ET has a dominant effect over LT, inducing the formation of transendothelial cell macroaperture tunnels ~10 μm in diameter (Maddugoda et al. 2011). Formation of these transcellular tunnels likely contributes to the rupture of endothelium barrier function and triggers edema that is particularly visible in the intestine of mice intoxicated by ET (Maddugoda et al. 2011). The tunnels are induced by a group of toxins and exoenzymes, now referred to as tunnel-forming toxins, also comprising the RhoA ADP-ribosyltranferase EDIN of *Staphylococcus aureus* (Boyer et al. 2006; Lemichez et al. 2010). LT, through its proteolytic activity toward mitogen-associated protein (MAP) kinase kinases, induces at late time periods of intoxication the formation of thick actin cables, compromising endothelial cell junction integrity and possibly the viability of some endothelial cell types (Rolando et al. 2010). Rupture of endothelial cell junctions by both toxins at late time periods of intoxication also involves the targeting of the Rab11/Sec15 exocyst components, thereby inhibiting the recycling pathways for proper localization of the vascular endothelial cadherin critical adherens junction molecule at cell–cell junctions (Guichard et al. 2010).

TOWARD DECIPHERING THE ROLE OF TOXINS DURING THE COURSE OF BACTERIAL INFECTION

Major challenges to date consist in defining how toxins affect differentiated host cell types in order to define their contribution to the various steps of the infection process. This exciting challenge is rendered possible by defining toxins' biochemical activities, as described above. New challenges include the identification of host cell toxin receptors, a major determinant of host species and cell type specificity. The importance of receptor specificity was recently presented, whereby humans do not synthesize sialic acid *N*-glycolylneuraminic acid but metabolize this monosaccharide, which is abundant in some food products, thereby becoming susceptible to the subtilase toxin (SubAB) (Byres et al. 2008). This toxin is produced by some strains of the shigatoxigenic group of *E. coli* (STEC) associated with severe forms of hemolytic uremic syndrome.

Determination of the toxin's mode of action requires characterizing host cell receptors. For example, a difference of a few amino acid residues between mouse and human precursors of a heparin-binding epidermal growth factor-like growth factor (HB-EGF) defines resistance in mice to diphtheria toxin (Naglich et al. 1992). The use of new screening methods to define host factors involved in cell intoxication will certainly accelerate the identification of host factor receptors. Notably, the use of gene-trap retrovirus-driven insertional mutagenesis to produce a collection of null-allele cell mutants for screening (Carette et al. 2009) will help identify toxin receptors and also help define host factors involved in facilitating toxin enzymatic domains' translocation and activation into the host cell cytosol, as well as toxin cofactors. Such an elegant strategy identified TMEM181 as the host cell receptor of HdCDT, together with sphingomyelin synthase-1, an enzymatic activity likely required to maintain proper organization of lipid rafts (Carette et al. 2009). Similarly, the lipolysis-stimulated lipoprotein receptor has also been recently identified as the receptor of the binary actin ADP-ribosylating toxin of hypervirulent strains of *C. difficile* (Papatheodorou et al. 2011).

An approach of major interest, although probably highly counterintuitive, is to define conditions in which bacterial toxins can trigger host antimicrobial responses and thus exert local or cell-type-specific "avirulence" effects. This likely will help decipher how the host senses and reacts properly to pathogenic bacteria rather than to commensals (Sansonetti 2011). One such interesting example is the infection of macrophages by LT of *B. anthracis* (Ali et al. 2011). The lethal factor enzymatic component of this toxin penetrates cells and cleaves the amino-terminal part of MAP kinase kinases, interrupting these signaling pathways (Duesbery et al. 1998). This induces a leakage of ATP through connexin channels as a result of impaired p38 MAP kinase kinase and AKT signaling pathways (Ali et al. 2011). Next, the sensing of extracellular ATP via P_2X_7 purinergic receptors leads to activation of the inflammasome system and the production of interleukin-1β, a critical factor in mounting antibacterial responses (Ali et al. 2011). Sensing of ATP leakage and potassium efflux induced by several pore-forming toxins are likely critical elements allowing host cells to properly decipher between commensal and pathogenic bacteria-producing toxins (Ali et al. 2011; Hamon and Cossart 2011). Also in line with this, a recent study conducted in *Drosophila* revealed that direct activation of Rac activity by the CNF1 toxin of uropathogenic *E. coli* triggers an antimicrobial response via Rip kinase (IMD in flies) (Boyer et al. 2011). Importantly, recombinant insects expressing CNF1 become resistant to infection by *Pseudomonas* entomopathogenic bacteria. This indicates that some bacterial toxins may favor infection of the host at specific stages, while triggering antimicrobial responses in other contexts.

CONCLUDING REMARKS

Protein toxins provide the bacterium with an advantage in host–pathogen interactions. Each protein toxin appears unique and provides the producing bacterial pathogen with a selective advantage in these host–pathogen interactions. Continued studies on these protein toxins will extend our understanding of bacterial pathogenesis and may identify novel applications of these toxins as therapies to treat human disease.

ACKNOWLEDGMENTS

E.L. is supported by institutional funding from INSERM and grants from Foundation Infectiopôle Sud, the Agence Nationale de la Recherche (ANR 11BSV3 004 01), and the Association pour la Recherche sur le Cancer (ARC SFI20 111203659). J.T.B. is supported by a grant from the National Institutes of Health (AI30 162) and is a member of the Great Lakes Regions Center of Excellence (AIU54–057153). We apologize to colleagues whose important contributions could not be cited because of space limitations.

REFERENCES

Ahuja N, Kumar P, Bhatnagar R. 2004. The adenylate cyclase toxins. *Crit Rev Microbiol* **30:** 187–196.

Aktories K. 2011. Bacterial protein toxins that modify host regulatory GTPases. *Nat Rev Microbiol* **9:** 487–498.

Aktories K, Barmann M, Ohishi I, Tsuyama S, Jakobs KH, Habermann E. 1986. Botulinum C2 toxin ADP-ribosylates actin. *Nature* **322:** 390–392.

Aktories K, Lang AE, Schwan C, Mannherz HG. 2011. Actin as target for modification by bacterial protein toxins. *FEBS J* **278:** 4526–4543.

Ali SR, Timmer AM, Bilgrami S, Park EJ, Eckmann L, Nizet V, Karin M. 2011. Anthrax toxin induces macrophage death by p38 MAPK inhibition but leads to inflammasome activation via ATP leakage. *Immunity* **35:** 34–44.

Alouf JE, Popoff MR, ed. 2006. *The comprehensive sourcebook of bacterial protein toxins*, 3rd ed. Elsevier, Amsterdam, The Netherlands.

Bischofberger M, Gonzalez MR, van der Goot FG. 2009. Membrane injury by pore-forming proteins. *Curr Opin Cell Biol* **21:** 589–595.

Bohnert S, Schiavo G. 2005. Tetanus toxin is transported in a novel neuronal compartment characterized by a specialized pH regulation. *J Biol Chem* **280:** 42336–42344.

Bokoch GM. 2005. Regulation of innate immunity by Rho GTPases. *Trends Cell Biol* **15:** 163–171.

Boquet P, Lemichez E. 2003. Bacterial virulence factors targeting Rho GTPases: Parasitism or symbiosis? *Trends Cell Biol* **13:** 238–246.

Boyer L, Doye A, Rolando M, Flatau G, Munro P, Gounon P, Clement R, Pulcini C, Popoff MR, Mettouchi A, et al. 2006. Induction of transient macroapertures in endothelial cells through RhoA inhibition by *Staphylococcus aureus* factors. *J Cell Biol* **173:** 809–819.

Boyer L, Magoc L, Dejardin S, Cappillino M, Paquette N, Hinault C, Charriere GM, Ip WK, Fracchia S, Hennessy E, et al. 2011. Pathogen-derived effectors trigger protective immunity via activation of the Rac2 enzyme and the IMD or Rip kinase signaling pathway. *Immunity* **35:** 536–549.

Brunger AT, Jin R, Breidenbach MA. 2008. Highly specific interactions between botulinum neurotoxins and synaptic vesicle proteins. *Cell Mol Life Sci* **65:** 2296–2306.

Byres E, Paton AW, Paton JC, Lofling JC, Smith DF, Wilce MC, Talbot UM, Chong DC, Yu H, Huang S, et al. 2008. Incorporation of a non-human glycan mediates human susceptibility to a bacterial toxin. *Nature* **456:** 648–652.

Carette JE, Guimaraes CP, Varadarajan M, Park AS, Wuethrich I, Godarova A, Kotecki M, Cochran BH, Spooner E, Ploegh HL, et al. 2009. Haploid genetic screens in human cells identify host factors used by pathogens. *Science* **326:** 1231–1235.

Chen C, Fu Z, Kim JJ, Barbieri JT, Baldwin MR. 2009. Gangliosides as high affinity receptors for tetanus neurotoxin. *J Biol Chem* **284:** 26569–26577.

Collier RJ, Young JA. 2003. Anthrax toxin. *Annu Rev Cell Dev Biol* **19:** 45–70.

Collins D, Hogan AM, Winter DC. 2011. Microbial and viral pathogens in colorectal cancer. *Lancet Oncol* **12:** 504–512.

Cossart P. 2011. Illuminating the landscape of host-pathogen interactions with the bacterium *Listeria monocytogenes. Proc Natl Acad Sci* **108:** 19484–19491.

Cruz-Migoni A, Hautbergue GM, Artymiuk PJ, Baker PJ, Bokori-Brown M, Chang CT, Dickman MJ, Essex-Lopresti A, Harding SV, Mahadi NM, et al. 2011. A *Burkholderia pseudomallei* toxin inhibits helicase activity of translation factor eIF4A. *Science* **334:** 821–824.

Cui J, Yao Q, Li S, Ding X, Lu Q, Mao H, Liu L, Zheng N, Chen S, Shao F. 2010. Glutamine deamidation and dysfunction of ubiquitin/NEDD8 induced by a bacterial effector family. *Science* **329:** 1215–1218.

Dikic I, Wakatsuki S, Walters KJ. 2009. Ubiquitin-binding domains—From structures to functions. *Nat Rev Mol Cell Biol* **10:** 659–671.

Dong M, Yeh F, Tepp WH, Dean C, Johnson EA, Janz R, Chapman ER. 2006. SV2 is the protein receptor for botulinum neurotoxin A. *Science* **312:** 592–596.

Doye A, Mettouchi A, Bossis G, Clement R, Buisson-Touati C, Flatau G, Gagnoux L, Piechaczyk M, Boquet P, Lemichez E. 2002. CNF1 exploits the ubiquitin-proteasome machinery to restrict Rho GTPase activation for bacterial host cell invasion. *Cell* **111:** 553–564.

Duesbery NS, Webb CP, Leppla SH, Gordon VM, Klimpel KR, Copeland TD, Ahn NG, Oskarsson MK, Fukasawa K, Paull KD, et al. 1998. Proteolytic inactivation of MAP-kinase-kinase by anthrax lethal factor. *Science* **280:** 734–737.

Elwell CA, Dreyfus LA. 2000. DNase I homologous residues in CdtB are critical for cytolethal distending toxin-mediated cell cycle arrest. *Mol Microbiol* **37:** 952–963.

Fabbri A, Travaglione S, Falzano L, Fiorentini C. 2008. Bacterial protein toxins: Current and potential clinical use. *Curr Med Chem* **15:** 1116–1125.

Flatau G, Lemichez E, Gauthier M, Chardin P, Paris S, Fiorentini C, Boquet P. 1997. Toxin-induced activation of the G protein p21 Rho by deamidation of glutamine. *Nature* **387:** 729–733.

Galan JE. 2009. Common themes in the design and function of bacterial effectors. *Cell Host Microbe* **5:** 571–579.

Guerra L, Cortes-Bratti X, Guidi R, Frisan T. 2011. The biology of the cytolethal distending toxins. *Toxins (Basel)* **3:** 172–190.

Guichard A, McGillivray SM, Cruz-Moreno B, van Sorge NM, Nizet V, Bier E. 2010. Anthrax toxins cooperatively inhibit endocytic recycling by the Rab11/Sec15 exocyst. *Nature* **467:** 854–858.

Hamon MA, Cossart P. 2011. K^+ efflux is required for histone H3 dephosphorylation by *Listeria monocytogenes* listeriolysin O and other pore-forming toxins. *Infect Immun* **79:** 2839–2846.

Hill KK, Smith TJ, Helma CH, Ticknor LO, Foley BT, Svensson RT, Brown JL, Johnson EA, Smith LA, Okinaka RT, et al. 2007. Genetic diversity among botulinum neurotoxin-producing clostridial strains. *J Bacteriol* **189:** 818–832.

Hoffmann C, Pop M, Leemhuis J, Schirmer J, Aktories K, Schmidt G. 2004. The *Yersinia pseudotuberculosis* cytotoxic necrotizing factor (CNFY) selectively activates RhoA. *J Biol Chem* **279:** 16026–16032.

Jaffe AB, Hall A. 2005. Rho GTPases: Biochemistry and biology. *Annu Rev Cell Dev Biol* **21:** 247–269.

Johannes L, Romer W. 2010. Shiga toxins—From cell biology to biomedical applications. *Nat Rev Microbiol* **8:** 105–116.

Just I, Selzer J, Wilm M, von Eichel-Streiber C, Mann M, Aktories K. 1995. Glucosylation of Rho proteins by *Clostridium difficile* toxin B. *Nature* **375:** 500–503.

Kim C, Wilcox-Adelman S, Sano Y, Tang WJ, Collier RJ, Park JM. 2008. Antiinflammatory cAMP signaling and cell migration genes co-opted by the anthrax bacillus. *Proc Natl Acad Sci* **105:** 6150–6155.

Landsberg MJ, Jones SA, Rothnagel R, Busby JN, Marshall SD, Simpson RM, Lott JS, Hankamer B, Hurst MR. 2011. 3D structure of the *Yersinia entomophaga* toxin complex and implications for insecticidal activity. *Proc Natl Acad Sci* **108:** 20544–20549.

Lang AE, Schmidt G, Schlosser A, Hey TD, Larrinua IM, Sheets JJ, Mannherz HG, Aktories K. 2010. *Photorhabdus luminescens* toxins ADP-ribosylate actin and RhoA to force actin clustering. *Science* **327:** 1139–1142.

Lara-Tejero M, Galan JE. 2000. A bacterial toxin that controls cell cycle progression as a deoxyribonuclease I-like protein. *Science* **290:** 354–357.

Lemichez E, Lecuit M, Nassif X, Bourdoulous S. 2010. Breaking the wall: Targeting of the endothelium by pathogenic bacteria. *Nat Rev Microbiol* **8:** 93–104.

Lemonnier M, Landraud L, Lemichez E. 2007. Rho GTPase-activating bacterial toxins: From bacterial virulence regulation to eukaryotic cell biology. *FEMS Microbiol Rev* **31:** 515–534.

Leppla SH. 1982. Anthrax toxin edema factor: A bacterial adenylate cyclase that increases cyclic AMP concentrations of eukaryotic cells. *Proc Natl Acad Sci* **79:** 3162–3166.

Maddugoda MP, Stefani C, Gonzalez-Rodriguez D, Saarikangas J, Torrino S, Janel S, Munro P, Doye A, Prodon F, Aurrand-Lions M, et al. 2011. cAMP signaling by anthrax edema toxin induces transendothelial cell tunnels, which are resealed by MIM via Arp2/3-driven actin polymerization. *Cell Host Microbe* **10:** 464–474.

Munro P, Flatau G, Anjuere F, Hofman V, Czerkinsky C, Lemichez E. 2005. The Rho GTPase activators CNF1 and DNT bacterial toxins have mucosal adjuvant properties. *Vaccine* **23:** 2551–2556.

Munro P, Flatau G, Lemichez E. 2007. Bacteria and the ubiquitin pathway. *Curr Opin Microbiol* **10:** 39–46.

Murphy JR. 2011. Mechanism of diphtheria toxin catalytic domain delivery to the eukaryotic cell cytosol and the cellular factors that directly participate in the process. *Toxins (Basel)* **3:** 294–308.

Naglich JG, Metherall JE, Russell DW, Eidels L. 1992. Expression cloning of a diphtheria toxin receptor: Identity with a heparin-binding EGF-like growth factor precursor. *Cell* **69:** 1051–1061.

Nesic D, Hsu Y, Stebbins CE. 2004. Assembly and function of a bacterial genotoxin. *Nature* **429:** 429–433.

Oberoi TK, Dogan T, Hocking JC, Scholz RP, Mooz J, Anderson CL, Karreman C, Meyer Zu Heringdorf D, Schmidt G, Ruonala M, et al. 2011. IAPs regulate the plasticity of cell migration by directly targeting Rac1 for degradation. *EMBO J* **31:** 14–28.

Papatheodorou P, Carette JE, Bell GW, Schwan C, Guttenberg G, Brummelkamp TR, Aktories K. 2011. Lipolysis-stimulated lipoprotein receptor (LSR) is the host receptor for the binary toxin *Clostridium difficile* transferase (CDT). *Proc Natl Acad Sci* **108:** 16422–16427.

Ribet D, Hamon M, Gouin E, Nahori MA, Impens F, Neyret-Kahn H, Gevaert K, Vandekerckhove J, Dejean A, Cossart P. 2010. *Listeria monocytogenes* impairs SUMOylation for efficient infection. *Nature* **464:** 1192–1195.

Rolando M, Stefani C, Flatau G, Auberger P, Mettouchi A, Mhlanga M, Rapp U, Galmiche A, Lemichez E. 2010. Transcriptome dysregulation by anthrax lethal toxin plays a key role in induction of human endothelial cell cytotoxicity. *Cell Microbiol* **12:** 891–905.

Sandvig K, Garred O, Prydz K, Kozlov JV, Hansen SH, van Deurs B. 1992. Retrograde transport of endocytosed Shiga toxin to the endoplasmic reticulum. *Nature* **358:** 510–512.

Sansonetti PJ. 2011. To be or not to be a pathogen: That is the mucosally relevant question. *Mucosal Immunol* **4:** 8–14.

Schmidt G, Sehr P, Wilm M, Selzer J, Mann M, Aktories K. 1997. Gln 63 of Rho is deamidated by *Escherichia coli* cytotoxic necrotizing factor-1. *Nature* **387:** 725–729.

Shenker BJ, Dlakic M, Walker LP, Besack D, Jaffe E, LaBelle E, Boesze-Battaglia K. 2007. A novel mode of action for a microbial-derived immunotoxin: The cytolethal distending toxin subunit B exhibits phosphatidylinositol 3,4,5-triphosphate phosphatase activity. *J Immunol* **178:** 5099–5108.

Taieb F, Nougayrede JP, Oswald E. 2011. Cycle inhibiting factors (Cifs): Cyclomodulins that usurp the ubiquitin-dependent degradation pathway of host cells. *Toxins (Basel)* **3:** 356–368.

Torrino S, Visvikis O, Doye A, Boyer L, Stefani C, Munro P, Bertoglio J, Gacon G, Mettouchi A, Lemichez E. 2011. The E3 ubiquitin-ligase HACE1 catalyzes the ubiquitylation of active Rac1. *Dev Cell* **21:** 959–965.

Visvikis O, Maddugoda MP, Lemichez E. 2010. Direct modifications of Rho proteins: Deconstructing GTPase regulation. *Biol Cell* **102:** 377–389.

Weissman AM. 2001. Themes and variations on ubiquitylation. *Nat Rev Mol Cell Biol* **2:** 169–178.

Worby CA, Mattoo S, Kruger RP, Corbeil LB, Koller A, Mendez JC, Zekarias B, Lazar C, Dixon JE. 2009. The Fic domain: Regulation of cell signaling by adenylylation. *Mol Cell* **34:** 93–103.

Yarbrough ML, Li Y, Kinch LN, Grishin NV, Ball HL, Orth K. 2009. AMPylation of Rho GTPases by *Vibrio* VopS disrupts effector binding and downstream signaling. *Science* **323:** 269–272.

Zhang Y, Zhu X, Torelli AT, Lee M, Dzikovski B, Koralewski RM, Wang E, Freed J, Krebs C, Ealick SE, et al. 2010. Diphthamide biosynthesis requires an organic radical generated by an iron-sulphur enzyme. *Nature* **465:** 891–896.

Epigenetics and Bacterial Infections

Hélène Bierne[1,2,3], Mélanie Hamon[1,2,3], and Pascale Cossart[1,2,3]

[1]Institut Pasteur, Unité des Interactions Bactéries-Cellules, Paris F-75015, France

[2]Inserm, U604, Paris F-75015, France

[3]INRA, USC2020, Paris F-75015, France

Correspondence: helene.bierne@pasteur.fr

Epigenetic mechanisms regulate expression of the genome to generate various cell types during development or orchestrate cellular responses to external stimuli. Recent studies highlight that bacteria can affect the chromatin structure and transcriptional program of host cells by influencing diverse epigenetic factors (i.e., histone modifications, DNA methylation, chromatin-associated complexes, noncoding RNAs, and RNA splicing factors). In this article, we first review the molecular bases of the epigenetic language and then describe the current state of research regarding how bacteria can alter epigenetic marks and machineries. Bacterial-induced epigenetic deregulations may affect host cell function either to promote host defense or to allow pathogen persistence. Thus, pathogenic bacteria can be considered as potential epimutagens able to reshape the epigenome. Their effects might generate specific, long-lasting imprints on host cells, leading to a memory of infection that influences immunity and might be at the origin of unexplained diseases.

Upon a microbial attack, host cells undergo massive changes in their transcriptional program, mobilizing genes involved in key processes (e.g., immunity, cell death/survival, and adhesion/motility) to trigger an appropriate response (Jenner and Young 2005). It is thus not surprising that successful pathogens have developed specific mechanisms to deregulate the expression levels and/or kinetics of these defense genes. Host transcription factors are first obvious targets to reprogram the genome and bacteria use diverse tricks to alter their function. For instance, bacterial factors can hijack cellular signaling pathways that activate or sequester transcription factors (e.g., NF-κB, IRF/STATs, or AP-1) in the cytosol of targeted cells, or manipulate their half-lives via posttranslational modifications (Bhavsar et al. 2007; Ribet and Cossart 2010; Perrett et al. 2011). Some bacteria, such as the phytopathogen *Xanthomonas*, even produce transcriptional activators that function as eukaryotic transcription factors (Kay et al. 2007).

However, selective activation or silencing of specific genes not only depends on transcription factors, but also on their cross talk with epigenetic modulators, which regulate DNA accessibility by controlling the chromatin structure. Epigenetic modifications of chromatin during development and in response to distinct environmental factors contribute to adult phenotypic variability and susceptibility to a

Cite this article as *Cold Spring Harb Perspect Med* doi: 10.1101/cshperspect.a010272

number of diseases, including cancers and metabolic and autoimmune disorders (van Vliet et al. 2007; Liu et al. 2008; Wilson 2008; Portela and Esteller 2010). This article uncovers the most recent data highlighting that epigenetic changes can also contribute to and/or result from bacterial infectious diseases. After an overview of the complex mechanisms governing chromatin dynamics, we will describe how they may contribute to host response to infection and/or are hijacked by bacteria to impose a transcriptional program beneficial for infection. Only animal pathogenic bacteria will be addressed here; for chromatin targeting by plant pathogenic bacteria, we refer to recent reviews (Ma et al. 2011b; Rivas 2011; Bierne and Cossart 2012).

CHROMATIN REGULATION OF GENE EXPRESSION AND THE EPIGENETIC MEMORY

In eukaryotic cells, DNA is packaged with histones into chromatin, allowing its confinement into the tight space of the nucleus. The state of compaction of chromatin not only organizes the genome but also plays a major role in nuclear processes requiring access to DNA, i.e., transcription, replication, recombination, and DNA repair. The structure of chromatin has several levels of organization (Woodcock and Dimitrov 2001). On the lowest level, the structure is based on repeating units, the nucleosomes, which consist of octamers of histone proteins (two copies of each of the core histones H2A, H2B, H3, and H4) around which ~147 base pairs of DNA is wound. Nucleosomes line up along the DNA in nucleosomal arrays and associate with linker histone H1 (or its isoforms), nonhistone proteins, and RNAs, forming higher-order chromatin structures, such as the loosely packaged and transcriptionally active euchromatin, or the highly condensed and transcriptionally silent heterochromatin. Chromatin-remodeling and modifying mechanisms function together to dynamically control the state of chromatin compaction and open or close regions of the genome for appropriate outcomes.

ATP-Dependent Remodeling of Nucleosomes

Sequence-dependent physical properties of DNA, thermal motion, and binding of transcription factors to DNA influence nucleosome mobility. Yet, the movement of histone octamers relative to DNA is mainly catalyzed by ATP-dependent remodeling enzymes that use energy from ATP hydrolysis to move, destabilize, evict, or reassemble nucleosomes (Hota et al. 2011). These ATPases are encoded by ~27 genes in humans and are usually associated with several other proteins within multiprotein complexes, grouped into four families: SWI/SNF, ISWI, CHD, INO80/SWR (all abbreviations are listed in Table 1) (Gangaraju and Bartholomew 2007; Hargreaves and Crabtree 2011). These complexes contribute either to activation or repression of transcription or both, depending on their interaction with histones and other chromatin-binding proteins.

Posttranslational Modifications of Histones and DNA Methylation: The Epigenetic Code

Nucleosome stability is also deeply influenced by changes occurring in the nucleosome itself. The histone octamer can be modified by exchange of histone variants and by a plethora of posttranslational modifications (PTMs), such as acetylation, methylation, phosphorylation, ubiquitylation, sumoylation, ADP ribosylation, and others, mainly in the histone tails that protrude from the nucleosome (Kouzarides 2007; Suganuma and Workman 2011), but also, as shown more recently, in globular domains (Tropberger and Schneider 2011). DNA can also be modified by methylation of the 5 position of the pyrimidine ring of cytosine (5mC), mainly in the context of CpG dinucleotides in somatic mammalian cells (Klose and Bird 2006; Chen and Riggs 2011).

These covalent modifications of chromatin, the "chromatin marks," are added or removed by a wide range of enzymes termed as "writers," such as histone acetyltransferases (HATs), methyltransferases (HMTs), and kinases, and

Table 1. Abbreviations and acronyms

Name	Full name	Function
5mC/5mU	5-Methylcytosine/5-methyluracyl	Modified DNA base
5hmC/ 5hmU	5-Hydroxymethylcytosine/5′-hydroxymethyluracyl	Modified DNA base
5fC	5-Formylcytosine	Modified DNA base
5caC	5-Carboxylcytosine	Modified DNA base
AID/ APOBEC	Activation-induced cytidine deaminase/apolipoprotein B mRNA-editing enzyme catalytic polypeptide	Nucleic acid mutator
BER/NER	Base excision repair /nucleotide excision repair	DNA repair
BAHD1	Bromo adjacent homology domain-containing 1	Chromatin-repressor complex subunit
CHD3/4	Chromodomain-helicase-DNA-binding protein 3/4 (Mi-2α/Mi-2β)	Nucleosome remodeler
CIITA	Class II, major histocompatibility complex, transactivator	Transcription factor
DNMT	DNA methyltransferase	DNA writer
G9a (EHMT2)	Euchromatic histone-lysine *N*-methyltransferase 2 (EHMT2)	Histone writer
H2AK119ub	Histone H2A ubiquitylated at lysine 119	Modified histone
H3K9me 2/3	Histone H3 di/trimethylated at lysine 9	Modified histone
H3K27me3	Histone H3 trimethylated at lysine 27	Modified histone
H3K9ac	Histone H3 acetylated at lysine 9	Modified histone
H3K14ac	Histone H3 acetylated at lysine 14	Modified histone
H3K23ac	Histone H3 acetylated at lysine 23	Modified histone
H3T3	Histone H3 phosphorylated at threonine 3	Modified histone
H3K9ac	Histone H3 acetylated at lysine 9	Modified histone
H3S10p	Histone H3 phosphorylated at serine 10	Modified histone
H4K9me	Histone H4 methylated at lysine 9	Modified histone
HMT	Histone methylatransferase	Modified histone
HDM	Histone demethylase	Histone writer
HDAC	Histone deacetylase	Histone writer
HMT	Histone methyltransferase	Histone writer
HP1	Heterochromatin protein 1	Chromatin-silencing factor reader
IFN-γ	Interferon γ	Cytokine
INO80	INOsitol requiring protein 80	Nucleosome remodeler
IKKα	IκB kinase α	Kinase
ISG	Interferon-stimulated gene	Gene
ISWI	Imitation switch (ISWI)	Nucleosome remodeler
JNK	c-Jun amino-terminal kinases	Kinase
LPS	Lipopolysaccharide	Major component of the outer membrane of gram-negative bacteria
LSD1	Lysine (K)-specific demethylase 1A (KDM1A)	Histone writer
MAMPs	Microbe-associated molecular patterns	Bacterial molecules
MAPK	Mitogen-activated protein (MAP) kinases	Kinase
MBD1	Methyl-CpG-binding protein 1	Chromatin-silencing factor; 5mC reader
MSK1/2	Mitogen- and stress-activated kinase 1/2	Kinase-histone writer
MTA1/2/3	Metastasis-associated gene 1/2/3	Chromatin-remodeling complex subunit
NF-κB	Nuclear factor κ-light-chain enhancer of activated B cells	Transcription factor

Continued

Table 1. *Continued*

Name	Full name	Function
NuRD	Nucleosome remodeling and histone deacetylase	Chromatin-remodeling complex
p300/CBP	E1A binding protein p300/CREB-binding protein	Histone writer, reader
PTB		Splicing factor
PRC1/PRC2	Polycomb repressive complex 1/2	Chromatin-repressor complex
SETDB1	SET domain, bifurcated 1	Histone writer
SIN3A	SIN3 transcription regulator homolog A; histone deacetylase complex subunit Sin3a	Chromatin-repressor complex subunit
SWI/SNF	Switch/Sucrose Nonfermentable	Nucleosome remodeler
TET	Ten-eleven translocation	Enzyme converting 5-meC to 5hmC
TNF-α	Tumor necrosis factor-α	Cytokine
TLR	Toll-like receptor	Receptor for MAMPs
TNF-R	Tumor necrosis factor-receptor	Cytokine receptor
TSIX	TSIX transcript, XIST antisense RNA	Noncoding RNA
VRK1	Vaccinia-related kinase 1	Kinase
XIST	X-inactive specific transcript	Noncoding RNA

CREB, cyclic AMP response element binding protein; PTB, phosphotyrosine binding; MAMPs, microbe-associated molecular patterns.

"erasers," such as histone deacetylases (HDACs), demethylases (HDMs), and phosphatases (Fig. 1) (Zhou et al. 2011). In the case of DNA methylation, writers are DNA methyltransferases (DNMTs), either involved in establishing methylation (i.e., the "de novo" methyltransferases DNMT3a and DNMT3b) or copying methylation patterns to the newly synthesized DNA strand during replication (i.e., the "maintenance" methyltransferase DNMT1). 5mC is a relatively stable mark and its erasure is more complex than that of histone PTMs. Intense recent research has nevertheless started to identify enzymes capable of modifying preexisting methylation patterns. DNA methylation is reversible either by a passive way, when 5mC is not copied during DNA replication, or by active mechanisms involving intermediate chemical modifications of 5mC, followed by passive demethylation or DNA repair, as detailed in Figure 1 (Bhutani et al. 2011; Chen and Riggs 2011; Wu and Zhang 2011).

DNA methylation, especially at promoters and enhancers, is mainly coupled with transcriptional silencing, whereas histone modifications are involved in both positive and negative regulation of transcription. For instance, H3 methylated at lysine 4 (H3K4me), phosphorylated at serine 10 (H3S10p), and acetylated at lysine 14 (H3K14ac) are linked to transcriptional activation, whereas deacetylated histones and H3 methylated at lysine 9 (H3K9me) (Fig. 1) and lysine 27 (H3K27me) are associated with repression. Some genes can have both repressive and active histone marks, and constitutively bind RNA polymerase II at their proximal promoters, thus being prepared for efficient future transcription while remaining silent (Mikkelsen et al. 2007). Such so-called "poised" genes can operate in cells in which a rapid change in expression levels is required, enabling cell plasticity (Cuddapah et al. 2010).

Chromatin modifications affect gene expression in two nonexclusive ways. First, they can directly alter the nucleosome structure. In particular, acetylation and phosphorylation of histones are proposed to attenuate the electrostatic interactions between basic histones and negatively charged DNA, thereby increasing nucleosome mobility and access of transcription factors to DNA (Fenley et al. 2010). Second, histone PTMs and 5mC serve as "signaling platforms" that recruit chromatin-modifying and/or regulatory proteins, termed "readers," which themselves interact or stabilize other chromatin components (Fig. 1). Readers are docked to histone PTMs or methylated DNA via chromatin-binding modules, such as methyl-CpG

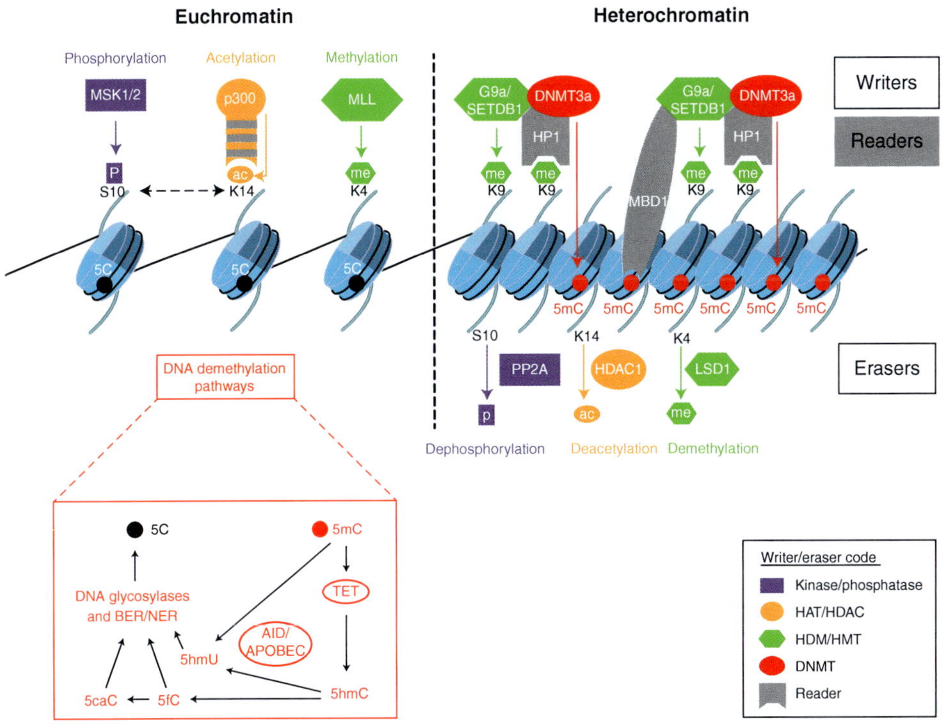

Figure 1. Chromatin modifications and regulators. Chromatin modifications open or close the chromatin structure, thereby activating or repressing gene expression. Serine phosphorylation on lysine 10 of histone H3 (S10p), acetylation on lysine 14 of H3 (K14ac), and methylation on lysine 4 of H4 (K4me) are examples of activating marks. Conversely, dephosphorylation, deacetylation, and demethylation of the same residues are associated with repression. Methylation of lysine 9 of H3 (K9me) and 5-cytosine methylation of DNA (5mC) are also repressive marks. These modifications are catalyzed or reversed by different enzymes known as "writers" or "erasers," respectively. An example of writer and eraser for each modification is shown in the color of the modification. Erasure of DNA methylation involves intermediate chemical modifications of 5mC, followed by passive demethylation or DNA repair. 5mC can be hydroxylated by TET dioxygenases (Tahiliani et al. 2009) to form 5hmC, and further oxidized to produce 5fC and 5caC (Wu and Zhang 2011). 5hmC is poorly recognized by DNMT1 and thus can lead to passive demethylation. In addition, 5mC and 5hmC can both be deaminated by AID/APOBEC deaminases to form modified uracyls, 5mU or 5hmU. Excision of modified bases (5mU, 5hmU, 5fC, and 5caC) by DNA glycosylases followed by repair via BER (base excision repair)/NER (nucleotide excision repair) is proposed to regenerate the unmethylated cytosine (Bhutani et al. 2011; Wu and Zhang 2011). Modified residues are recognized and interpreted by different protein modules, known as "readers." For instance, the bromodomain of HAT p300 binds H3K14ac, the chromodomain of HP1 binds H3K9me, and the methyl-CpG-binding domain (MBD) of MBD1 binds 5meC. Examples of epigenetic cross talks are shown: H3S10p is often associated with H3K14ac and H3K9ac (not shown). H3K9me is often associated with 5mC, on cross-interactions between the H3K9 methyltransferase (G9a/SETDB1), DNMT, HP1, and MBD1. Overall, the chromatin compaction, loss of histone activation marks, and removal of transcription factors accompany gene silencing. HAT, histone acetyltransferase; HDAC, histone deacetylase; HMT, histone methyltransferase; HDM, histone demethyltransferase; DNMT, DNA methyltransferases. Abbreviations are listed in Table 1.

binding domains of MBD proteins that interact with methylated DNA (Sasai and Defossez 2009), or bromodomains, chromodomain, 14-3-3, Tudor, plant homeodomains (PHDs), etc., which interact with specific histone PTMs (Taverna et al. 2007). For instance, the chromodomain of heterochromatin protein 1 (HP1) binds H3K9me2/3, which initiates heterochromatin formation and gene silencing (Bannister et al. 2001; Lachner et al. 2001).

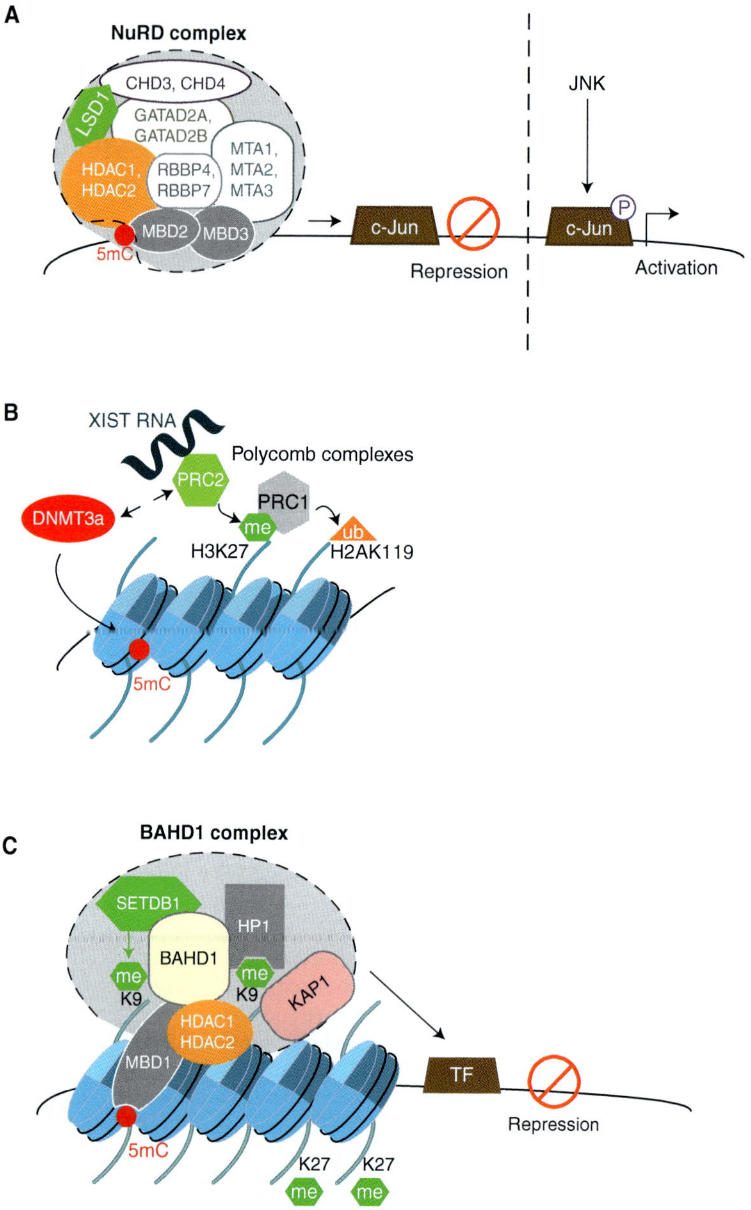

Figure 2. Three examples of chromatin-modifying complexes and their mode of action. (*A*) The nucleosome remodeling and histone deacetylase NuRD (or Mi-2) complex contains different subunits involved in nucleosome remodeling (CHD3/CHD4), histone deacetylation (HDAC1/HDAC2), histone demethylation (LSD1), and binding to transcription factors and histones. Additionally, MBD2 recruits NuRD to methylated DNA, whereas MBD3 (whose MBD region is mutated and does not bind methylated DNA) interacts with transcription factors. The function and targeting of the complex to specific loci depends on the combinatorial assembly of the different subunit isoforms. Most often, recruitment of the NuRD complex by a tissue-specific transcription factor to gene promoters mediates transcriptional repression. In the case of the transcription factor c-Jun of the AP1 family, recent studies suggest that its recruitment to the NuRD complex involves MBD3 interaction with unphosphorylated c-Jun. On activation of the JNK pathway, c-Jun phosphorylation mediates MBD3/c-Jun dissociation, resulting in derepression of transcription of target genes (e.g., cell-cycle and cell-differentiation genes) (Aguilera et al. 2011). Abbreviations are listed in Table 1. (*Legend continues on facing page.*)

Conversely, the bromodomain of the HAT p300/CBP binds acetylated-lysine histone residues, and is important for the enzyme activity and transcriptional activation (Fig. 1) (Chen et al. 2010).

A Ballet of Chromatin-Modifying and -Remodeling Complexes: The Epigenetic Language

Writers, readers, and erasers are often modular proteins displaying several properties. This is well exemplified by p300/CBP, which is a writer (via its HAT module), a reader (via its bromodomain), and an adaptor (via other modules, such as transcription factor and steroid receptor interaction domains), thus functioning as a transcriptional integrator. In addition, chromatin-modifying enzymes usually assemble into multisubunit complexes, together with nonenzymatic subunits acting as readers and/or as linkers for sequence-specific DNA-binding factors. For instance, the NuRD complex (Fig. 2A) contains subunits involved in nucleosome remodeling (CHD3 and CHD4), histone deacetylation (HDAC1 and HDAC2), histone demethylation (LSD1), and in binding to other subunits and histones (RBBP4, RBBP7, GATAD2A, and GATAD2B), to methylated DNA (MBD2) and to transcription factors (MTA1, MTA2, MTA3, and MBD3) (Hayakawa and Nakayama 2011; Lai and Wade 2011). Combinatorial assembly of these subunits determines the function of NuRD in genomic targeting and mediating cell type-specific transcriptional regulations, such as transcriptional repression of tumor-suppressor genes.

These macromolecular complexes may function alone or with others to orchestrate, in time and space, a specific chromatin state. The vast array and different combinations of histone PTMs coordinate the sequential recruitment of complexes, a stepwise regulation that reinforces or reverses existing histone PTMs (Latham and Dent 2007; Lee et al. 2010; Suganuma and Workman 2011). For instance, H3S10p is required for acetylation at lysine 14 on the same histone by the HAT Gcn5. Another example is the sequential recruitment of the polycomb repressive complexes. PRC2 "writes" H3K27me3, while PRC1 "reads" it and induces monoubiquitylation of histone H2A, ultimately leading to chromatin compaction at target genes (Fig. 2B) (Margueron and Reinberg 2011). Such cross talk also takes place between histone PTMs and DNA methylation. Interaction of H3K9 HMT (e.g., SETDB1, G9a) or H3K27 HMT (PRC2) with DNMT3A/B is proposed to promote DNA methylation at specific regions (Figs. 1 and 2B) (Li et al. 2006; Vire et al. 2006). Once formed, 5mC sites recruit the methyl-CpG binding proteins, such as MBD1, which can interact with both H3K9 HMTs and HP1 (Fujita et al. 2003; Ichimura et al. 2005). Such sequence of events may explain the spreading of repressive marks at targeted loci.

Figure 2. (*Continued*) (*B*) Mammalian PcG proteins form two multiprotein complexes, polycomb repressive complexes 1 and 2 (PRC1 and PRC2). PRC2 contains a histone-methyltransferase subunit (EZH2, not shown) that generates H3K27me3. PRC1 binds H3K27me3 and catalyzes ubiquitinylation of H2AK119. An early event in chromosome X inactivation in mammalian females is the recruitment of PRC2 by the XIST long coding RNA, inducing H3K27me3 and H2AK119ub along the inactive X chromosome. PRC2 has the ability to recruit DNMTs to some of its target genes, thereby stimulating de novo DNA methylation. (*C*) The BAHD1-associated chromatin-silencing complex has been discovered by the study of the bacterial protein LntA from *L. monocytogenes*. Two-hybrid screen, coimmunoprecipitation, and colocalization experiments (Bierne et al. 2009), as well as tandem-affinity purification (Lebreton et al. 2011), identified components associated to BAHD1, i.e., HP1, MBD1, KAP1, histone methyltransferase (SETDB1), and deacetylases (HDAC1/2). Our data suggest that BAHD1 acts as a silencer by tethering chromatin regulators and modifying enzymes to sequence-specific transcription factors (TF), enabling local chromatin compaction. BAHD1-associated heterochromatin sites are enriched in H3K27me3, but the relationship with PRC2 is unknown. Like for NuRD, the function and targeting of the BAHD1 complex to specific genes likely depend on the combinatorial assembly of the different subunits, in response to signals to which cells are submitted. On *Listeria* infection, the BAHD1 complex assembles at promoters of a set of interferon-stimulated genes (ISGs), as shown in Figure 4.

The recruitment, regulation, and interactions of the complexes implementing chromatin marks trigger dynamic combinatorial patterns that are proposed to form a language (Lee et al. 2010). The "grammar" of this language also involves modification of nonhistone substrates by chromatin-modifying enzymes (Xu et al. 2007; Rathert et al. 2008), as well as a wide range of PTMs induced by cell signaling pathways that alter interaction or stability of subunits of chromatin complexes and transcription factors. This translates signal transduction information to the chromatin structure in response to different environmental signals (Mohammad and Baylin 2010). Phosphorylation and sumoylation are important modifications in the function of a variety of repressor complexes (Garcia-Dominguez and Reyes 2009; Baek 2011) and, for instance, play a central role in the targeting of HP1α to chromatin (Hiragami-Hamada et al. 2011; Maison et al. 2011). Phosphorylation of the transcription factor c-Jun by JNK kinase impairs the binding of c-Jun to the MBD3 subunit of the NuRD complex, thereby relieving repression of target genes (Fig. 2A) (Aguilera et al. 2011). JNK, MSK1/2, IKKα, and several other kinases of signal transduction pathways can also directly phosphorylate histones in the nucleus (Yamamoto et al. 2003; Baek 2011, Tiwari et al. 2011). Thus, signaling molecules activated in cells in response to external stimuli have key effects on the chromatin syntax.

Other Epigenetic Regulators: Noncoding RNAs and RNA Splicing Factors

Additional complexity was recently added to the epigenetic landscape by the discovery of a crucial role of RNAs and RNA-binding proteins in dynamic changes of the chromatin structure and spatial organization of the genome inside the nucleus (for reviews, see Bernstein and Allis 2005; Faghihi and Wahlestedt 2009; Chen and Carmichael 2010; Grewal 2010; Turner and Morris 2010; Kaikkonen et al. 2011).

Eukaryotes transcribe most of their genomic DNA, albeit only 1%–2% of all cellular transcripts encode proteins. The vast majority of RNAs are thus noncoding RNAs (ncRNA) of diverse size (from small 17–30 bp to long >200 bp ncRNAs), which can be generated from intergenic regions, exons, introns, or enhancers, in sense or antisense orientation (Mattick and Makunin 2005; Tisseur et al. 2011). Most reported data link ncRNAs to gene silencing. Well-known examples are two long noncoding transcripts, XIST and TSIX, which regulate inactivation of one of the two X chromosomes in female mammals (Nora and Heard 2010; Morey and Avner 2011). XIST transcripts coat the inactive X chromosome in *cis* and recruit chromatin-remodeling complexes, such as PRC2 and PRC1, that trigger repressive marks and heterochomatin formation (Fig. 2B) (Pontier and Gribnau 2011; Wutz 2011). The action of XIST is antagonized by TSIX, a noncoding antisense RNA to the *Xist* gene that represses *Xist* transcription through chromatin modifications. The mechanism at play might involve XIST/TSIX RNA duplexes processed by the RNAi machinery and subsequent recruitment of chromatin-remodeling complexes by small interfering RNAs (siRNAs; see below) (Ogawa et al. 2008). Numerous cases of sense–antisense transcript pairs have now been discovered in mammalian genomes and increasing evidence suggests that antisense RNAs might play a general role in epigenetic changes by recruiting DNA methyltransferase and/or histone-modifying enzymes (Faghihi and Wahlestedt 2009; Kaikkonen et al. 2011). Additionally, long ncRNAs were recently shown to act as scaffolds, localizing genes to specific subnuclear regions for either repressive or activating effects (Yang et al. 2011).

Mammalian cells also produce small ncRNAs that were originally characterized for their role in posttranscriptional gene silencing in the cytoplasm, by targeting specific mRNAs for degradation or translational inhibition. However, recent data indicate that small ncRNAs also play an important role in chromatin dynamics in the nucleus. These RNAs are generated by cleavage of double-stranded RNAs (dsRNAs) by dsRNA-specific RNase III-type enzymes: microRNAs (miRNAs) derive from unique hairpin-shaped precursors in long transcripts, whereas endogenous small interfering RNAs (endo-siRNAs)

 Cite this article as *Cold Spring Harb Perspect Med* doi: 10.1101/cshperspect.a010272

are formed by cleavage of long dsRNA generated at repeated sequences (e.g., centromere, telomere, and transposons) or by antisense transcription (Turner and Morris 2010; Rother and Meister 2011). Current models suggest that the regulatory functions of nuclear small ncRNAs and associated factors in gene expression are to serve as guides for chromatin-modifying complexes and target them at specific sites, in *cis* or *trans*, possibly through base pairing or formation of an RNA/DNA triple helix (Schmitz et al. 2010).

Last, but not least, mounting evidence suggests that chromatin modifications also contribute to pre-mRNA alternative splicing, which in eukaryotes allows for one gene to encode functionally distinct proteins (Allemand et al. 2008; Hnilicova and Stanek 2011). The spliceosome, a complex of small nuclear RNA (snRNA) and protein subunits (snRNPs, small nuclear ribonucleoprotein particles) that removes introns, seems to be influenced by chromatin modifications. For instance, HDAC inhibition changes alternative splicing of ~700 genes in human cells (Hnilicova et al. 2011). In addition, splicing regulators, such as PTB or hnRNPs, have been shown to interact with chromatin regulators, such as MRG15 (Luco et al. 2010) or HP1, respectively (Piacentini and Pimpinelli 2010). Interestingly H3K9me2/3 marks are enriched in introns, in contrast to exons, and a reader of such marks, HP1γ, was recently shown to modulate alternative splicing (Saint-Andre et al. 2011). Furthermore, siRNA-mediated heterochromatin formation seems to contribute to splicing events (Allo et al. 2009). Splicing decisions are thus likely assisted by chromatin factors and ncRNAs.

The Epigenetic Memory

Chromatin modifications can promote mitotically heritable changes in gene expression without altering the DNA sequence, especially during cell differentiation. This property to persist through cell division, despite the fact that the initiating signal is no longer present, and thus allowing a cell to "remember" its transcriptional profile and identity, defines epigenetic regulation (Russo et al. 1996). Chromatin modifications are thus commonly referred to as "epigenetic" modifications. However, many marks induced by cell signaling, DNA repair, or cell-cycle transitions, are short-lived and do not give rise to long-term memory storage. To account for both transient and stable modifications of the epigenetic language, we thus prefer another definition of epigenetic events: "the structural adaptation of chromosomal regions so as to register, signal, or perpetuate altered activity states," recently proposed by Adrian Bird (Bird 2007). When inherited, chromatin modifications generate an epigenetic memory.

BACTERIAL CYTOSOLIC SIGNALING TO CHROMATIN

Mechanisms controlling gene expression at the chromatin level show high levels of complexity. Various bacterial products can affect them in many ways (Table 2), through activation of signaling cascades or directly in the nucleus, as we will review now.

Histone Modifications

So far, most of the reported chromatin modifications induced by bacteria are histone acetylation/deacetylation and phosphorylation/dephosphorylation events generated through activation of host cell signaling cascades by bacterial components (e.g., microbe-associated molecular patterns, metabolites, and virulence factors) (Fig. 3). The effects are complex, because they differ according to the bacterial agonist, cell type, and kinetics parameters. Among the host signaling pathways that a number of bacteria activate, MAPKs (e.g., ERK and p38), NF-κB, and PI3K pathways are known to activate the kinases that phosphorylate H3S10 in the nucleus (i.e., MSK1/2, IKKα, and AKT, respectively) (Yamamoto et al. 2003; Baek 2011). Any bacterial stimulus activating these pathways has therefore the potential to induce H3S10p and associated acetylated histones. Examples of these are listed below.

In endothelial HUVEC cells, the causative agent of listeriosis, *Listeria monocytogenes*, rap-

Table 2. Bacterial species, factors, and effects

Bacterial species	Bacterial factor	Effect
Anaplasma phagocytophilum	?	Histone deacetylation and silencing of host defense genes
	AnkA	Binding to chromatin at AT-rich sequences; silencing of *CYBB*
Bacillus anthracis	LT	Inhibition of MAPK and of H3S10p and down-regulation of *IL-8* and *KC* genes
Campylobacter rectus	?	Hypermethylation in the promoter region P0 of the *IGF2* gene in the murine-infected placenta
Chlamydia trachomatis	NUE	Methylation of mammalian histones
Escherichia coli (uropathogenic)	?	DNA methylation and down-regulation of *CDKN2A*
Ehrlichia chaffeensis		Binding to chromatin at Alu-Sx elements
Helicobacter pylori		Induction of DNA methylation in gastric mucosa
		Induction of H3 modifications
		Induction of miRNA expression
		Induction of RNA splicing factors
Legionella pneumophila	Flagellin	Histone acetylation in infected lung epithelial cells
Listeria	?	Acetylation of H4 and phosphorylation/acetylation of H3, in particular at the *IL-8* promoter and decreased binding of HDAC1 at the *IL8* promoter in *Listeria*-infected endothelial cells
	LLO	Histone modifications and modulation of host defense genes by a signaling pathway involving K^+ efflux
	?	Down-regulation of ISG by BAHD1-associated complex in epithelial cells
	LntA	Up-regulation of ISG by LntA-mediated inhibition of BAHD1-associated complex
	?	Change in expression of a subset of host miRNAs
Moraxella catarrhalis		Reduction in the global expression and activity of HDAC1/2 in airway epithelial cells
Mycobacterium tuberculosis		Control of chromatin complex at ISG downstream from IFN-γ
Porphyromonas gingivalis		Reactivation of latent viruses via chromatin modification induced by butyric acid
Salmonella		Change in expression of a subset of host miRNAs
Shigella flexneri	OspF	Down-regulation of MAP kinase in the nucleus by eliminylation; inhibition of phosphorylation of H3S10; down-regulation of immune response genes inhibition of NF-κB chromatin access leading to the inactivation of *IL-8* and other genes essential for innate immune responses
	OspB	Binding to Rb
	IpaH9.3	Ubiquitin ligase, which targets a splicing factor for degradation and impairs splicing
Bacterial product	LPS	Inducer of innate immunity via activation of TLR4-mediated responses and the production of proinflammatory cytokines; induces immunosuppression by chromatin modifications on repeated challenge
	Butyrate	Inhibition of HDAC activity

LT, lethal toxin; LLO, listeriolysin O; LPS, lipopolysaccharide.

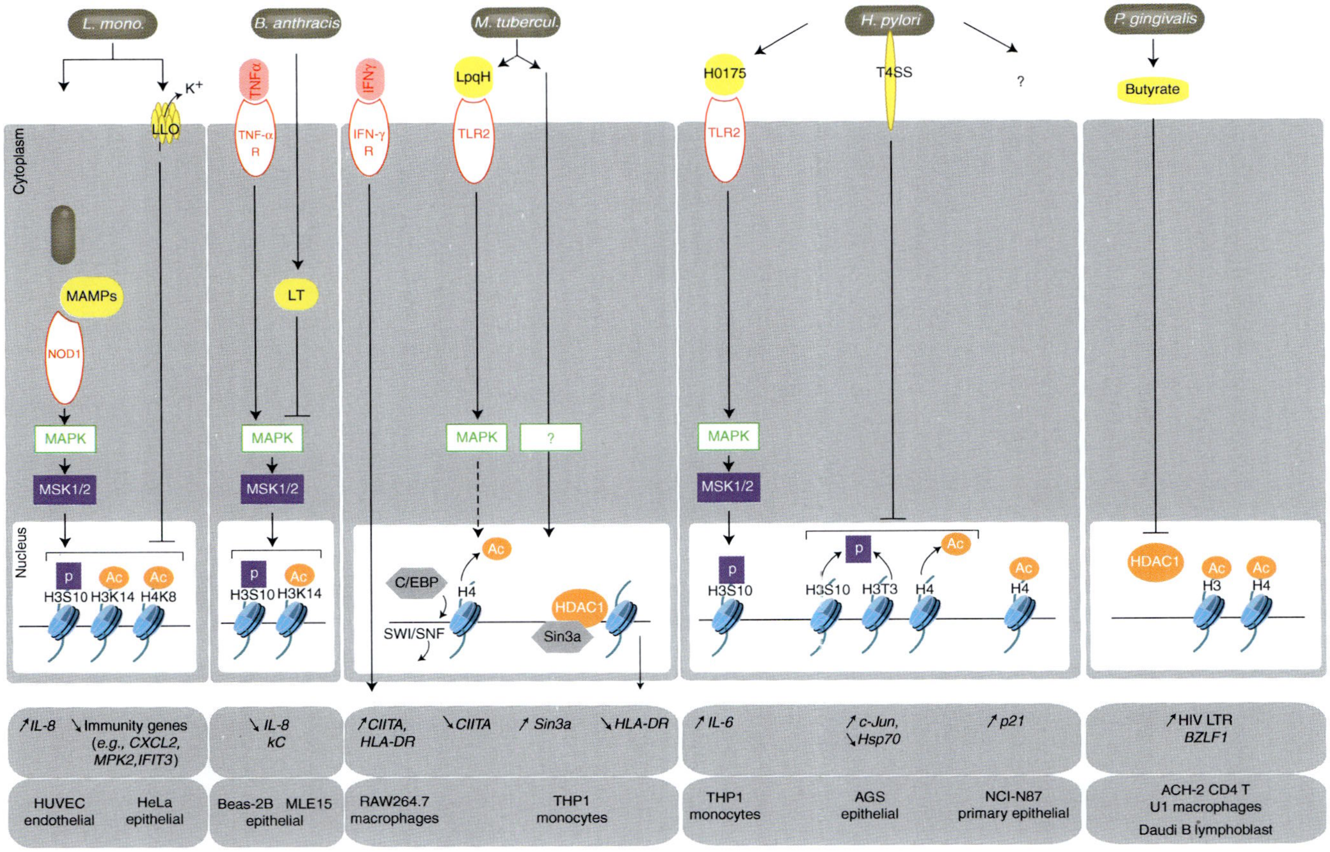

Figure 3. Bacterial signaling to histones and downstream effects. Schematic representation of *Listeria monocytogenes-*, *Bacillus anthracis-*, *Mycobacterium tuberculosis-*, *Helicobacter pylori-*, or *Porphyromonas gingivalis*-induced signaling pathways leading to histone modifications, as detailed in the text. Bacterial products inducing host cellular signaling are in yellow. Membrane (TNF-α-R, IFN-γ, TLR2, TLR4) or cytosolic (NOD1) receptors are indicated by a red oval. Effects on target genes are indicated by arrows (up for activation and down for repression). Cell types in which studies have been performed are listed below. Abbreviations are listed in Table 1.

idly increases phosphorylation and acetylation of histones (H3S10p, H3K14ac, and H4K8ac), and specifically at the promoter of the proinflammatory gene *IL-8* (Schmeck et al. 2005; Opitz et al. 2006). This effect involves NOD1-mediated sensing of cytosolic bacteria and activation of MAP kinases (MAPK), followed by enhanced binding of HAT p300/CBP and decreased binding of HDAC1 to the *IL-8* promoter (Fig. 3). Flagellin of *Legionella pneumophila*, the causative agent of Legionnaire disease and bacterial lipopolysaccharide (LPS), has similar effects on *IL-8* in lung epithelial cells and dendritic cells, respectively (Saccani et al. 2002; Schmeck et al. 2008). Not only pathogenic but also commensal bacteria, such as *Moraxella catarrhalis*, a saprophytic bacterium of the respiratory tract, or *Bacteroides vulgatus*, a commensal of the intestinal flora, can induce phosphorylation/acetylation of H3 via induction of an inflammatory signaling cascade (Haller et al. 2003; Slevogt et al. 2006).

However, with such an important role on inflammation, MAPK signaling is often counteracted by bacterial-induced mechanisms. In the case of *B. vulgatus*, this is performed by bacterial induction of the TGF-β1 anti-inflammatory pathway, which in turn induces H3 deacetylation via HDAC recruitment at proinflammatory gene promoters (Haller et al. 2003). This prevents *B. vulgatus* from eliciting an inflammatory response in the gut and contributes to its tolerance by the host. Bacterial toxins also dampen the host innate immune responses by inhibiting H3 phosphorylation/acetylation. Once endocytosed into the host cell, lethal toxin (LT) from *Bacillus anthracis*, the agent of anthrax, cleaves and inactivates MAPKKs leading to disruption in MAPK signaling (Bardwell et al. 2004). In lung epithelial cells activated by TNF-α, LT-mediated MAPK inhibition promotes a decrease in the levels of H3S10p and H3K14ac at the promoter of *IL-8* and *KC* genes (Fig. 3) (Raymond et al. 2009). Pore-forming toxins that "intoxicate" cells by disruption of membrane permeability use another signaling pathway to change histone marks. In epithelial HeLa cells, *L. monocytogenes* listeriolysin O (LLO) promotes deacetylation and dephosphorylation of histones, in particular at a subset of immune genes, such as *CXCL2*, *MKP2*, or *IFIT3*, which become repressed (Fig. 3) (Hamon et al. 2007). The mechanism at play involves a K^+ efflux through pore formation at the membrane (Hamon and Cossart 2011). Other pore-forming toxins, such as PFO of *Clostridium perfringens*, PLY of *Streptococcus pneumonia*, and Aerolysin from *Aeromonas hydrophila* have similar effects, suggesting that all of these toxins activate an intracellular K^+ sensor leading to a pathway that modifies histones and subsequent gene expression (Hamon and Cossart 2011).

Of note, MAP kinases have several downstream targets other than MSK1/2, some of which are involved in histone deacetylation and transcriptional repression at later time points (Yang et al. 2001). This might explain why bacterial-induced MAPK signaling does not always lead to transcriptional activation of immune genes, as exemplified by *Mycobacterium tuberculosis* (Mtb), the causative agent of tuberculosis. Control of Mtb infection requires IFN-γ, but some IFN-γ-induced genes are repressed in macrophages infected with Mtb or exposed to Mtb components, such as the lipoprotein LpqH (Wang et al. 2005; Pennini et al. 2006). These genes include *CIITA*, coding for the master regulator of MHC class II genes, as well as some of its targets (e.g., *HLA-DR*). Activation of the TLR2/MAPK-dependent pathway on Mtb infection stimulates recruitment of the transcriptional repressor C/EBP and histone deacetylation at the promoter of *CIITA*, antagonizing the nucleosome-remodeling activity of the SWI/SNF complex and down-regulating *CIITA* expression (Fig. 3) (Pathak et al. 2006; Pennini et al. 2007). Additionally, mycobacterial infection up-regulates expression of *Sin3A* encoding a corepressor that acts in concert with HDACs to repress a number of genes, including MHC class II genes (Wang et al. 2005). Thus, to counteract IFN-γ-induced pathways, Mtb not only silences *CIITA*, but also CIITA-regulated genes, such as *HLA-DR*, on increased recruitment of Sin3A-HDACs to their promoters (Fig. 3).

The carcinogenic bacterium *Helicobacter pylori* is also capable of modifying histones by diverse means, including the modulation of the

 Cite this article as *Cold Spring Harb Perspect Med* doi: 10.1101/cshperspect.a010272

MAPK pathway. The peptidyl prolyl *cis-*, *trans-*isomerase HP0175 secreted by *H. pylori* binds the innate immunity receptor TLR4, which in turn activates MAPKs and the downstream MSK1-dependent phosphorylation of H3, in particular, at the *IL-6* promoter in THP-1 monocytes (Fig. 3) (Pathak et al. 2006). In contrast, in gastric epithelial cells *H. pylori* has an opposite effect by inducing dephosphorylation of H3S10 and H3T3, as well as deacetylation of H3K23 (Fehri et al. 2009; Ding et al. 2010a). *H. pylori*-mediated chromatin modifications are in this case dependent on a functional type IV secretion system, suggesting involvement of an injected bacterial product. *H. pylori*-induced dephosphorylation of H3S10 is transient and impacts both the cell cycle (Fehri et al. 2009) and transcription of oncogene *c-Jun* (positively) and heat shock gene *hsp70* (negatively) (Ding et al. 2010a). *H. pylori*-mediated premitotic arrest involves dephosphorylation of H3S10 upon deregulation of the mitotic histone kinase VRK1, followed by rephosphorylation of H3S10 by an IKKα-dependent pathway. Another study showed that exposure of *H. pylori* to gastric epithelial cells promotes release of HDAC1 from the promoter of the cell-cycle regulator gene $p21^{WAF}$, hyperacetylation of H4, and increased expression of $p21^{WAF}$. Chromatin alterations might contribute to the effects of *H. pylori* on cell-cycle progression, cellular proliferation, and cell death (Xia et al. 2008).

Changes in the levels of chromatin-modifying components are another possible outcome of bacterial infections. As discussed above, Mtb infection up-regulates expression of *Sin3A* (Wang et al. 2005), whereas the tick-transmitted pathogen *Anaplasma phagocytophilum* up-regulates the expression of HDACs in infected cells (Garcia-Garcia et al. 2009a), an effect that contributes to silencing of host defense genes by deacetylation of histones. In contrast, the levels of HDAC transcripts are decreased in gingival epithelial cells treated with oral pathogen *Porphyromonas gingivalis* and nonpathogenic *Fusobacterium nucleatum* (Yin and Chung 2011).

Bacteria can also produce metabolites acting as inhibitors of chromatin-modifying enzymes. One such product is butyric acid, a short-chain fatty acid acting as a potent inhibitor of HDACs (Riggs et al. 1977). Interestingly, the adverse effect of *Porphyromanas gingivalis* in reactivating latent viruses, such as human immunodeficiency virus (HIV) and Epstein–Barr syndrome (EBV), seems to result from production of butyrate by this bacterium (Imai et al. 2009; Imai et al. 2011). It is proposed that viral genes maintained silent by HDAC-containing complexes are reactivated following inhibition of HDACs by butyric acid (Fig. 3). Thus, *P. gingivalis* infection might be a risk factor for viral diseases, such as AIDS or herpes. Butyric acid also exerts beneficial anti-inflammatory effects on the host, at least in animal models, possibly via epigenetic up-regulation of anti-inflammatory genes. Such observations open up the interesting possibility to use butyrate-producing probiotic bacteria as immunosuppressors (Licciardi et al. 2010).

DNA Methylation

The importance of DNA methylation events associated with bacterial infections is also becoming increasingly appreciated. The best documented example is *H. pylori* infection that induces aberrant DNA methylation in the human gastric mucosa, strikingly at promoters of genes found methylated in gastric cancer cells (Maekita et al. 2006; Ding et al. 2010b; Ushijima and Hattori 2012). *H. pylori*-associated hypermethylation occurs, for instance, at the E-cadherin gene *CDH1* (Chan et al. 2003), tumor-suppressor genes (e.g., *USF1/2* and *WWOX* [Bussiere et al. 2010; Yan et al. 2011]), DNA repair genes (e.g., *MLH1* [Yao et al. 2006]), as well as to CpG islands of miRNA genes (Ando et al. 2009). The ability of *H. pylori* to induce DNA methylation in gastric mucosa was confirmed in the gerbil animal model and, interestingly, this effect was diminished on treatment with the immunosuppressor cyclosporin A (Niwa et al. 2010). Indeed, in contrast to ethanol or NaCl stimuli that induce neutrophil infiltration in the stomach, *H. pylori*-mediated inflammation triggered lymphocyte and macrophage infiltration, which appears to have a key role in induction of methylation (Hur et al. 2011). Thus, although the mechanisms by which *H. pylori* induces DNA hypermethylation are still unclear, the infection-associated inflammatory response is a tempting explanation (Ding et al. 2010b; Ushijima and Hattori 2012). Among signals resulting from chronic inflammation, elevated levels of IL1β and nitric oxide (NO) are proposed to contribute to influence the recruitment of DNMTs at specific loci.

In fact, evidence for an epigenetic control of the inflammatory response is increasingly recognized (Medzhitov and Horng 2009; Bayarsaihan 2011). In the intestine, some PRC2 target genes are subject to aberrant DNA methylation following chronic inflammation (Hahn et al. 2008). In a noninflammatory context, the LPS-sensing receptor encoding gene *TLR4* is down-regulated in intestinal epithelial cells and there is evidence for a role of commensal bacteria in *TLR4* methylation (Takahashi et al. 2011). This is proposed to maintain intestinal homeostasis by preventing an excessive inflammatory reaction toward the gut microbiota. In the oral cavity, similar mechanisms may dampen uncontrolled inflammatory response triggered by bacterial-induced chronic infection. In particular, periodontally inflamed gingival biopsies showed a significant increase in promoter methylation of the gene encoding the proinflammatory enzyme COX-2, when compared with noninflamed biopsy samples (Zhang et al. 2010). This would allow a chronic inflammatory stimulus to be tolerated, preventing unrestricted tissue destruction. Whether this is a bacteria-triggered phenomenon is unknown, but it is of note that resident bacteria, such as *P. gingivalis*, can induce hypermethylation of specific genes in gingival epithelial cells (Yin and Chung 2011).

Bacteria-induced DNA methylation can also affect genes involved in cell proliferation, as illustrated by uropathogenic *Escherichia coli* (UPEC). In human uroepithelial cells, infection with UPEC results in the up-regulation of DNA methyltransferase (DNMT) activity and *DNMT1* expression and induces CpG methylation and down-regulation of *CDKN2A*, a G_1-cell-cycle inhibitor regulator (Tolg et al. 2011). This may increase uroepithelial cell proliferation and pathogen persistence, by counteracting infection-stimulated host cell apoptosis. Other organs can be targeted by bacteria-mediated epigenetic changes, including placenta. Indeed, maternal infection with *Campylobacter rectus* induces hypermethylation of the imprinted *IGF2* gene promoter in murine placental tissue (Bobetsis et al. 2007). This result suggests that bacterial infections during pregnancy might epigenetically affect genes that play an important role in fetal development.

ncRNA and RNA Splicing

It is now well established that miRNAs are important regulators of immune responses (O'Connell et al. 2011). Not surprisingly, several of them are induced in response to pathogenic bacteria, such as *H. pylori* (Fassi Fehri et al. 2010; Matsushima et al. 2011), *Salmonella typhimurium* (Schulte et al. 2011), *L. monocytogenes*, and *Mycobacterium bovis* BCG (Ma et al. 2011a; Izar et al. 2012), and even by microbiota (Dalmasso et al. 2011). However, in all of the reported studies, miRNAs function in posttranscriptional repression of mRNAs in the cytosol. Whether a bacterial stimulus induces expression of miRNA or endogenous siRNA acts at the chromatin level in the nucleus is unknown.

Likewise, whereas numerous immunologically relevant genes undergo alternative splicing (Lynch 2004), it remains to be explored whether bacteria could affect alternative splicing by changes in chromatin to deregulate the function of the immune system. A recent study analyzing changes in the phosphoproteome of gastric cells on *H. pylori* infection described an enrichment of RNA processing and splicing factors in infected cells (Holland et al. 2011). Besides this study and the reported targeting of a splicing factor by the *Shigella flexneri* protein IpaH9.8 (see below) (Okuda et al. 2005), information of the link between splicing and bacterial infection is lacking.

As we have seen, bacteria-mediated signaling pathways affect epigenetic processes in many ways. However, it is not always easy to distinguish what is the contribution of the host response to infection and what is a specific

 Cite this article as *Cold Spring Harb Perspect Med* doi: 10.1101/cshperspect.a010272

bacteria-targeted mechanism's on the chromatin changes. The existence of bacterial factors that act directly in the nucleus, which we recently referred to as the "nucleomodulins" (Bierne and Cossart 2012), highlights that bacteria have evolved mechanisms to actively manipulate the chromatin-regulatory machinery in the nucleus and reprogram host gene expression to their advantage (Fig. 4).

BACTERIAL PROTEINS MIMICKING OR CONTROLLING HOST CHROMATIN-REGULATORY FACTORS

Histone-Modifying Bacterial Enzymes

Several components of the eukaryotic protein methylation system might have their origins in bacteria. In particular, SET domain-containing HMT (e.g., G9a, SETDB1) may have first emerged in prokaryotes from the SAF superfamily of carbohydrate-binding domains (Aravind et al. 2011). Conversely, a subset of the bacterial SET domain-containing protein might have evolved the chromatin-related role of their eukaryotic counterparts. Evidence for this hypothesis emerged from studies of chlamydial SET proteins. A SET domain protein (cpnSET) from *Chlamydophila pneumoniae*, which causes acute respiratory diseases in humans, was first shown to methylate the chlamydial histone-like proteins HC1/HC2 (Murata et al. 2007). Its homolog nuclear effector (NUE) from *Chlamydia trachomatis* was subsequently identified as a T3SS effector that enters the nucleus of infected cells, where it associates with chromatin (Fig. 4) (Pennini et al. 2010). Importantly, NUE can methylate mammalian histones H2B, H3, and H4 in vitro, confirming its histone methyltransferase activity. Additionally, nuclear effector (NUE) can methylate itself, a property shared by eukaryotic G9a. However, experimental evidence for a direct relationship between NUE activity in the nucleus and host gene expression is still lacking. Nonetheless, with regard to the number of SET domain proteins present in many bacterial species that interact with eukaryotes, it is tempting to speculate that several bacteria might use this strategy.

Ankyrin Repeat (Ank)-Containing Bacterial Proteins

Several effectors of human intracellular bacterial pathogens, such as *Anaplasma*, *Ehrlichia*, *Ricketssia*, *Orientia*, *Coxiella*, and *Legionella* species contain proteins with Ank eukaryotic motifs. These motifs mediate protein–protein interactions in a multitude of host processes, including transcriptional regulation (Mosavi et al. 2004). Ank-containing proteins, AnkA and p200, from the related pathogens *Anaplasma phagocytophilum* and *Ehrlichia chaffeensis*, respectively, bind chromatin in the nucleus of infected cells (Fig. 4) (Park et al. 2004; Zhu et al. 2009). AnkA preferentially targets AT-rich chromatin sites and decreases expression of a set of genes. In particular, AnkA represses *CYBB* that encodes Cytochrome b-245, a component of the phagocyte oxidase that influences *A. phagocytophilum* survival (Garcia-Garcia et al. 2009b). This is concomitant to deacetylation of H3 at the *CYBB* locus. *E. chaffeensis* p200 binds to chromatin at Alu-Sx elements located in promoters and introns of various human genes (Zhu et al. 2009). Several p200 target genes are strongly up-regulated during infection, suggesting that p200 may affect gene transcription globally through mechanisms associated with Alu element gene regulation.

Bacterial Proteins Targeting or Modifying Chromatin-Remodeling Complexes

Our recent studies identified a bacterial protein that targets a subunit of a chromatin-repressive complex, highlighting that bacteria can directly hijack the heterochromatin machinery. The nuclear-targeted factor LntA from *L. monocytogenes* interacts with BAHD1, a previously unknown protein that we characterized as a novel heterochromatinization factor in vertebrates. BAHD1 acts in complex with other chromatin factors (e.g., HP1, MBD1, SETDB1, HDACs, and KAP1) to silence gene expression (Fig. 2C) (Bierne et al. 2009; Lebreton et al. 2011). The set of genes repressed by the BAHD1-associated complex depends on the signal to which cells are submitted and on the cell type. In

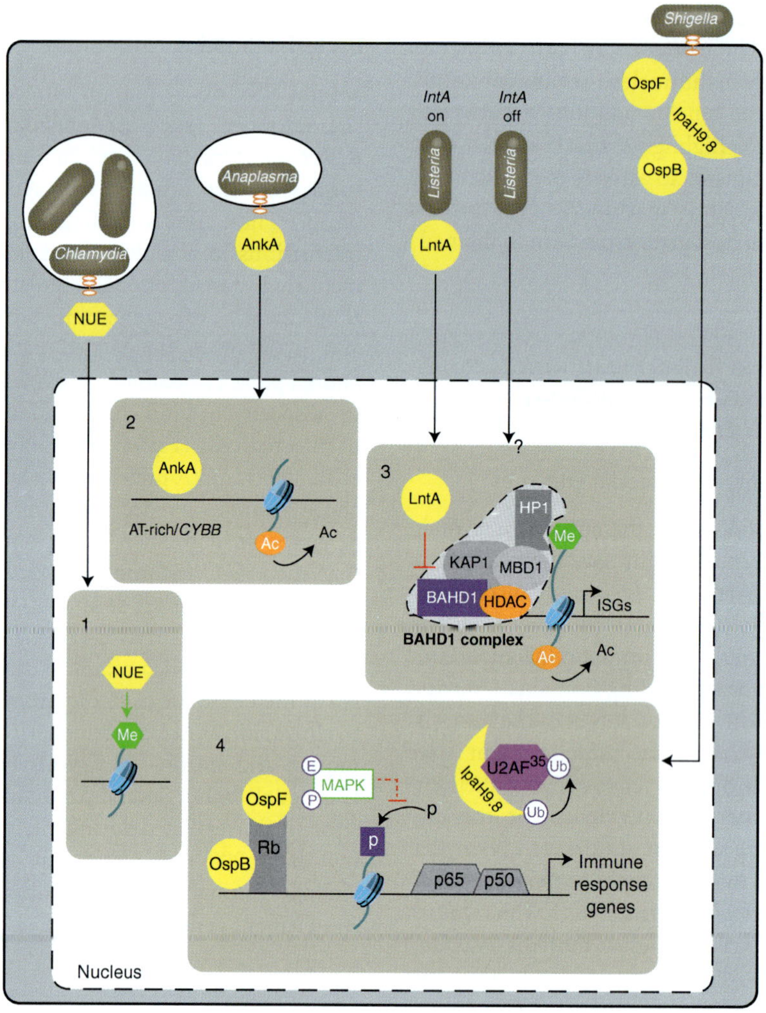

Figure 4. Bacterial nucleomodulins targeting chromatin. Schematic representation of *Chlamydia, Anaplasma, Listeria*, and *Shigella* secreted factors involved in the control of gene expression in the nucleus of host cells, as detailed in the text. The bacterial nucleomodulins are in yellow. 1. *Chlamydia* histone-methyltransferase NUE methylate's mammalian histones. However, target genes are unknown. 2. Binding to AT-rich sequences and silencing of *CYBB* expression by *Anaplasma* Ank effector AnkA. 3. Inhibition of the BAHD1-associated heterochromatic complex and induction of interferon-stimulated genes by *Listeria* LntA. On *Listeria* infection, an unknown signaling pathway drives the BAHD1-associated chromatin complex (see Fig. 2C) to repress interferon-stimulated genes. When *Listeria* produces and secretes LntA, this factor enters the nucleus and interacts with BAHD1, destabilizes the silencing complex, restores histone acetylation (Ac), and enhances the expression of ISGs. 4. Control of a set of NF-κB (p65-p50) regulated genes by *Shigella* posttranslational modifiers OspF, which eliminylates MAP kinases preventing phosphorylation of histone H3, and IpaH9.8, which ubiquitinylates and promotes degradation of the splicing factor $U2AF^{35}$. OspF and another effector, OspB, bind the retinoblastoma protein (Rb), which potentially recruits several chromatin-remodeling enzymes (not shown). Ac, Acetylation; Me, methylation; P, phosphorylation; E, eliminylation; UB, ubiquitinylation.

 Cite this article as *Cold Spring Harb Perspect Med* doi: 10.1101/cshperspect.a010272

particular, BAHD1 represses interferon-stimulated genes (ISG) in epithelial cells infected with *L. monocytogenes* (Lebreton et al. 2011). Upon an unknown signal, *L. monocytogenes* switches on expression of the *lntA* gene. The secreted LntA factor enters the nucleus and alleviates BAHD1 binding to ISG promoters, thereby up-regulating their expression. By preventing recruitment of BAHD1 and HDAC1/2 at ISG promoters, LntA is thought to trigger chromatin unwinding (Fig. 4). Consistent with this model, acetylation of histone H3 increases at ISG promoters in cells infected with LntA-producing bacteria. We propose that the LntA–BAHD1 interplay spatiotemporally modulates the interferon response to control bacterial colonization of the host (Lebreton et al. 2012). However, questions remain regarding how BAHD1 is specifically recruited to a subset of infection-dependent targets and what signal determines the switch between a *lntA* nonexpressing and expressing state of *Listeria* during the infection process.

Another strategy used by bacteria to target chromatin is to modify locally and specifically the activity of key chromatin/transcriptional regulatory factors. This is used by several T3SS effectors from gram-negative pathogens, such as OspF of *S. flexneri.* OspF is a phosphothreonine lyase that irreversibly modifies host MAPKs by a reaction called eliminylation (Li et al. 2007; Brennan and Barford 2009). This enzymatic reaction converts a phosphothreonine residue into a dehydrobutyrine residue that can no longer be phosphorylated, and hence locks the substrate in an inactive form. Inhibition of MAPK signaling by OspF abrogates phosphorylation of histone H3 at a set of NF-κB-regulated promoters and blocks the activation of a pool of proinflammatory genes (Arbibe et al. 2007). In addition, OspF and another nuclear-targeted effector, OspB, interact with the human retinoblastoma protein Rb, which is known to bind several chromatin-remodeling factors (Zurawski et al. 2009). *Shigella* likely uses OspF–OspB synergy to down-regulate host innate immunity via alteration of the chromatin structure at specific genes (Fig. 4).

S. flexneri secretes another effector, IpaH9.8, which also contributes to *Shigella* pathogenesis by impairing host inflammatory responses. IpaH9.8 is an E3 ubiquitin ligase that binds several host proteins, likely by its LRR domain, and targets them for proteasome-dependent degradation (Rohde et al. 2007). IpaH9.8 affects the NF-κB-dependent pathway in the cytosol (Ashida et al. 2010), as well as the activity of an mRNA splicing factor, U2AF^{35}, in the nucleus (Fig. 4), thus interfering with U2AF^{35}-dependent splicing (Toyotome 2001; Okuda et al. 2005; Seyedarabi et al. 2011). This *Shigella* IpaH9.8 effector is a member of a novel family of bacterial E3 ubiquitin ligases found in several animal and plant pathogens (Rohde et al. 2007; Singer et al. 2008; Zhu et al. 2008). Orthologs of IpaH9.8, such as SspH1 of *Salmonella enterica* and YopM of *Yersinia pestis*, have also been shown to migrate to the host-cell nucleus (Haraga and Miller 2003; Benabdillah et al. 2004) but their effect on chromatin regulation has not yet been examined.

THE EPIGENETIC MEMORY OF INFECTION: THE BACTERIAL IMPRINTS

As chromatin modifications may be transmitted to daughter cells during cell division, leading to heritable changes in gene function, it is possible that a bacterial infection could generate heritable marks after pathogen eradication. Deciphering whether histone modification and/or DNA methylation fingerprints imposed by bacterial components are maintained over time is thus a key issue. The following two examples support this idea.

The Epigenetic Component of Severe Systemic Inflammation and LPS Tolerance

Aberrant inflammatory reactions in response to sustained exposure to microbes and microbial products, such as LPS, lead to tissue damage, multiorgan dysfunction, septic shock, and death. To compensate these adverse effects, the immune system has developed postseptic immunosuppression (PSI) mechanisms that enable hematopoietic cells to become temporary hyporesponsive. This compensatory anti-inflammatory response counteracts the harmful effects of sepsis, but leaves individuals more susceptible to opportunistic infections for extended periods of time (weeks to years). Although PSI is a complex multifactorial process, the contribution of epigenetic regulations is increasingly recognized, as reviewed recently (McCall et al. 2010; Carson et al. 2011). One of the facets of PSI is LPS tolerance, in which LPS-elicited TLR4 responses are reprogrammed toward silencing of proinflammatory cytokine genes and expression of anti-inflammatory or antimicrobial mediators. LPS activation of TLR4 first elicits transcription of poised proinflammatory genes, which are rapidly derepressed and then returned to basal state within hours. As mentioned previously, opening the chromatin at target genes during this acute phase involves histone phosphorylation and acetylation. However, sustained exposure to LPS or subsequent LPS challenge activates a pathway leading to permanent gene repression, as characterized for *TNF-α* and *IL1-β* (El Gazzar et al. 2008; Chen et al. 2009). This relies on a change in the composition of NF-κB transcription factor at the proximal promoters of *TNF-α* and *IL1-β*, with a shift from activating p65-p50 to repressive RelB-p50, owing to the up-regulation of *RelB* expression by p65-p50. RelB interacts with H3K9 HMT G9a, leading to H3K9me2 and subsequent recruitment of HP1. The repressive complex formed by G9a and HP1 recruits DNMT3A/B, which induces de novo CpG methylation (McCall et al. 2010). This leads to assembly of silent, facultative heterochromatin in a similar cross talk as depicted in Figure 1.

LPS tolerance can last for weeks in humans, but whether this memory is passed through cell division is not yet proven. Additionally, even if imprinted hematopoietic cells divide, why new cells from progenitors in the bone marrow do not restore an efficient immune system is an open question. A tempting hypothesis would be that epigenetic imprinting also occurs at the level of stem cells. This hypothesis needs to be investigated by analyzing the epigenome of stem cells isolated from animal models of sepsis. The reversal of heterochromatin to euchromatin at genes targeted for LPS-mediated repression is also a key issue to understand how "imprinted" immune cells return to homeostasis.

Bacterial Infection-Associated Cancers

As mentioned above, *H. pylori* is an important acquired risk factor for gastric cancer. Besides *H. pylori*-mediated effects on cell proliferation and DNA integrity, *H. pylori*-induced aberrant DNA methylation emerges as an important mechanism in stomach carcinogenesis (Ding et al. 2010b; Ushijima and Hattori 2012). Remarkably, it has been shown that eradication of *H. pylori* infection in human patients leads to a decrease but not to full disappearance in methylation of promoter CpG islands of genes closely correlated with the risk of gastric cancer development (Nakajima et al. 2010). This is a strong indication that a bacterial infection may leave epigenetic imprints in a tissue enabling permanent changes in gene expression. Given the fact that cancer must arise from a cell that has the potential to divide, this bacterial reprogramming might be induced in long-living cells, such as stem cells or progenitors, thereby being propagated to daughter cells. It may also target mature epithelial cells for dedifferentiation by silencing components of the stem cell signaling network, leading to increased proliferation and survival (Katoh 2007).

The "*Helicobacter* paradigm" is transposable to bacteria targeting other tissues. It is speculated that *E. coli* infection may be linked with bladder carcinoma risk (Tolg et al. 2011) and intestinal bacteria might predispose to colon cancer (Sun 2010). More generally, deregulation of tumor-suppressor and/or stem cell-associated pathways (e.g., Wnt, JAK-STAT, JNK, and Notch) on genetic alteration and epigenetic reprogramming induced by bacteria is a possible cause of cancer development in epithelial niches.

CONCLUDING REMARKS

There is growing evidence for a role of bacterial infection in modulating the epigenetic information of host cells by diverse mechanisms. This is particularly well illustrated with the pathogen *L. monocytogenes*: on the one hand, this pathogen induces signaling events in host cells leading to activation (via activation of MAP kinases) or repression (via activation of

 Cite this article as *Cold Spring Harb Perspect Med* doi: 10.1101/cshperspect.a010272

the BAHD1 silencing complex) of immunity genes; on the other hand, this bacterium produces at least two proteins, LLO toxin and LntA nucleomodulin, to counteract these pathways. How these events are spatiotemporally coordinated, in particular, how expression and secretion of bacterial factors is controlled by *Listeria* during the infectious process, and how chromatin writers and erasers become specifically localized at specific regions of the genome, at a given time and in a specific cell, are important questions to be answered. In this regard, deciphering how chromatin modifications are spread throughout the genome (the "pathoepigenome") is likely to provide important clues. Technological advances now make human genome-wide mapping of DNA methylation and histone modification profiles feasible. Not only *Listeria*, but all bacteria presented here, as well as others eliciting chromatin modification in mammalian cells, will undoubtedly benefit from these new technologies, highlighting genes epigenetically deregulated on bacterial infection. Epigenetics has been discussed here for its role in regulating transcription; however, it has additional roles in DNA repair, DNA replication, and cell division. Loss of epigenetic control of these other DNA-based processes may also become relevant in bacterial infectious diseases. Interestingly, studying bacterial modulation of epigenetics processes can help in understanding the fundamental principles of their occurrence and regulation, as we have shown with the discovery of the BAHD1-chromatin-silencing complex (Fig. 2C).

Understanding the role of chromatin modifications and their regulators in the physiopathology of infectious diseases is another challenging perspective. Investigations will have to be performed at the tissue and cell levels, with the objective of analyzing cells with long lifespans, to determine the "chromatin signature" of infection, not only during infection but also following recovery. Results of these investigations will probably depend on the cell type and host genetic background. With the better understanding of the connections between bacterial infectious diseases and the epigenome, opportunities will arise for therapeutic applications, especially as epigenetic processes can be reverted. Rapid elimination of microbe-induced pathoepigenetic changes may prevent chronic or latent infections, some cancers, or autoimmune diseases. This opens avenues for future research in the field of bacterial pathogenesis and chromatin-based regulation of defense genes.

ACKNOWLEDGMENTS

Studies in our laboratory are funded by Institut Pasteur, INRA, INSERM, French Ligue Nationale Contre le Cancer (LNCC RS10/75-76 Bierne), European Research Council (ERC Advanced Grant 233348), French National Research Agency (ANR, grant EPILIS), and Fondation Louis Jeantet, Fondation Le Roch les Mousquetaires.

REFERENCES

Aguilera C, Nakagawa K, Sancho R, Chakraborty A, Hendrich B, Behrens A. 2011. c-Jun N-terminal phosphorylation antagonises recruitment of the Mbd3/NuRD repressor complex. *Nature* **469:** 231–235.

Allemand E, Batsche E, Muchardt C. 2008. Splicing, transcription, and chromatin: A menage a trois. *Curr Opin Genet Dev* **18:** 145–151.

Allo M, Buggiano V, Fededa JP, Petrillo E, Schor I, de la Mata M, Agirre E, Plass M, Eyras E, Elela SA, et al. 2009. Control of alternative splicing through siRNA-mediated transcriptional gene silencing. *Nat Struct Mol Biol* **16:** 717–724.

Ando T, Yoshida T, Enomoto S, Asada K, Tatematsu M, Ichinose M, Sugiyama T, Ushijima T. 2009. DNA methylation of microRNA genes in gastric mucosae of gastric cancer patients: Its possible involvement in the formation of epigenetic field defect. *Int J Cancer* **124:** 2367–2374.

Aravind L, Abhiman S, Iyer LM. 2011. Natural history of the eukaryotic chromatin protein methylation system. *Prog Mol Biol Transl Sci* **101:** 105–176.

Arbibe L, Kim DW, Batsche E, Pedron T, Mateescu B, Muchardt C, Parsot C, Sansonetti PJ. 2007. An injected bacterial effector targets chromatin access for transcription factor NF-κB to alter transcription of host genes involved in immune responses. *Nat Immunol* **8:** 47–56.

Ashida H, Kim M, Schmidt-Supprian M, Ma A, Ogawa M, Sasakawa C. 2010. A bacterial E3 ubiquitin ligase IpaH9.8 targets NEMO/IKKγ to dampen the host NF-κB-mediated inflammatory response. *Nat Cell Biol* **12:** 61–69.

Baek SH. 2011. When signaling kinases meet histones and histone modifiers in the nucleus. *Mol Cell* **42:** 274–284.

Bannister AJ, Zegerman P, Partridge JF, Miska EA, Thomas JO, Allshire RC, Kouzarides T. 2001. Selective recognition of methylated lysine 9 on histone H3 by the HP1 chromo domain. *Nature* **410:** 120–124.

Bardwell AJ, Abdollahi M, Bardwell L. 2004. Anthrax lethal factor-cleavage products of MAPK (mitogen-activated protein kinase) kinases exhibit reduced binding to their cognate MAPKs. *Biochem J* **378:** 569–577.

Bayarsaihan D. 2011. Epigenetic mechanisms in inflammation. *J Dent Res* **90:** 9–17.

Benabdillah R, Jaime Mota Ls, Lützelschwab S, Demoinet E, Cornelis GR. 2004. Identification of a nuclear targeting signal in YopM from Yersinia spp. *Microbial Pathogenesis* **36:** 247–261.

Bernstein E, Allis CD. 2005. RNA meets chromatin. *Genes Dev* **19:** 1635–1655.

Bhavsar AP, Guttman JA, Finlay BB. 2007. Manipulation of host-cell pathways by bacterial pathogens. *Nature* **449:** 827–834.

Bhutani N, Burns DM, Blau HM. 2011. DNA demethylation dynamics. *Cell* **146:** 866–872.

Bierne H, Cossart P. 2012. When bacteria target the nucleus: The emerging family of nucleomodulins. *Cell Microbiol* **14:** 622–633.

Bierne H, Tham TN, Batsche E, Dumay A, Leguillou M, Kerneis-Golsteyn S, Regnault B, Seeler JS, Muchardt C, Feunteun J, et al. 2009. Human BAHD1 promotes heterochromatic gene silencing. *Proc Natl Acad Sci* **106:** 13826–13831.

Bird A. 2007. Perceptions of epigenetics. *Nature* **447:** 396–398.

Bobetsis YA, Barros SP, Lin DM, Weidman JR, Dolinoy DC, Jirtle RL, Boggess KA, Beck JD, Offenbacher S. 2007. Bacterial infection promotes DNA hypermethylation. *J Dent Res* **86:** 169–174.

Brennan DF, Barford D. 2009. Eliminylation: A post-translational modification catalyzed by phosphothreonine lyases. *Trends Biochem Sci* **34:** 108–114.

Bussiere FI, Michel V, Memet S, Ave P, Vivas JR, Huerre M, Touati E. 2010. *H. pylori*-induced promoter hypermethylation downregulates USF1 and USF2 transcription factor gene expression. *Cell Microbiol* **12:** 1124–1133.

Carson WF, Cavassani KA, Dou Y, Kunkel SL. 2011. Epigenetic regulation of immune cell functions during post-septic immunosuppression. *Epigenetics* **6:** 273–283.

Chan AO, Lam SK, Wong BC, Wong WM, Yuen MF, Yeung YH, Hui WM, Rashid A, Kwong YL. 2003. Promoter methylation of E-cadherin gene in gastric mucosa associated with *Helicobacter pylori* infection and in gastric cancer. *Gut* **52:** 502–506.

Chen LL, Carmichael GG. 2010. Long noncoding RNAs in mammalian cells: What, where, and why? *Wiley Interdiscip Rev RNA* **1:** 2–21.

Chen ZX, Riggs AD. 2011. DNA methylation and demethylation in mammals. *J Biol Chem* **286:** 18347–18353.

Chen X, Yoza BK, El Gazzar M, Hu JY, Cousart SL, McCall CE. 2009. RelB sustains IκBα expression during endotoxin tolerance. *Clin Vaccine Immunol* **16:** 104–110.

Chen J, Ghazawi FM, Li Q. 2010. Interplay of bromodomain and histone acetylation in the regulation of p300-dependent genes. *Epigenetics* **5:** 509–515.

Cuddapah S, Barski A, Zhao K. 2010. Epigenomics of T cell activation, differentiation, and memory. *Curr Opin Immunol* **22:** 341–347.

Dalmasso G, Nguyen HT, Yan Y, Laroui H, Charania MA, Ayyadurai S, Sitaraman SV, Merlin D. 2011. Microbiota modulate host gene expression via microRNAs. *PLoS ONE* **6:** e19293.

Ding SZ, Fischer W, Kaparakis-Liaskos M, Liechti G, Merrell DS, Grant PA, Ferrero RL, Crowe SE, Haas R, Hatakeyama M, et al. 2010a. *Helicobacter pylori*-induced histone modification, associated gene expression in gastric epithelial cells, and its implication in pathogenesis. *PLoS ONE* **5:** e9875.

Ding SZ, Goldberg JB, Hatakeyama M. 2010b. *Helicobacter pylori* infection, oncogenic pathways and epigenetic mechanisms in gastric carcinogenesis. *Future Oncol* **6:** 851–862.

El Gazzar M, Yoza BK, Chen X, Hu J, Hawkins GA, McCall CE. 2008. G9a and HP1 couple histone and DNA methylation to TNFα transcription silencing during endotoxin tolerance. *J Biol Chem* **283:** 32198–32208.

Faghihi MA, Wahlestedt C. 2009. Regulatory roles of natural antisense transcripts. *Nat Rev Mol Cell Biol* **10:** 637–643.

Fassi Fehri L, Koch M, Belogolova E, Khalil H, Bolz C, Kalali B, Mollenkopf HJ, Beigier-Bompadre M, Karlas A, Schneider T, et al. 2010. *Helicobacter pylori* induces miR-155 in T cells in a cAMP-Foxp3-dependent manner. *PLoS ONE* **5:** e9500.

Fehri LF, Rechner C, Janssen S, Mak TN, Holland C, Bartfeld S, Bruggemann H, Meyer TF. 2009. *Helicobacter pylori*-induced modification of the histone H3 phosphorylation status in gastric epithelial cells reflects its impact on cell cycle regulation. *Epigenetics* **4:** 577–586.

Fenley AT, Adams DA, Onufriev AV. 2010. Charge state of the globular histone core controls stability of the nucleosome. *Biophys J* **99:** 1577–1585.

Fujita N, Watanabe S, Ichimura T, Tsuruzoe S, Shinkai Y, Tachibana M, Chiba T, Nakao M. 2003. Methyl-CpG binding domain 1 (MBD1) interacts with the Suv39h1-HP1 heterochromatic complex for DNA methylation-based transcriptional repression. *J Biol Chem* **278:** 24132–24138.

Gangaraju VK, Bartholomew B. 2007. Mechanisms of ATP dependent chromatin remodeling. *Mutat Res* **618:** 3–17.

Garcia-Dominguez M, Reyes JC. 2009. SUMO association with repressor complexes, emerging routes for transcriptional control. *Biochim Biophys Acta* **1789:** 451–459.

Garcia-Garcia JC, Barat NC, Trembley SJ, Dumler JS. 2009a. Epigenetic silencing of host cell defense genes enhances intracellular survival of the rickettsial pathogen *Anaplasma phagocytophilum*. *PLoS Pathog* **5:** e1000488.

Garcia-Garcia JC, Rennoll-Bankert KE, Pelly S, Milstone AM, Dumler JS. 2009b. Silencing of host cell CYBB gene expression by the nuclear effector AnkA of the intracellular pathogen *Anaplasma phagocytophilum*. *Infect Immun* **77:** 2385–2391.

Grewal SI. 2010. RNAi-dependent formation of heterochromatin and its diverse functions. *Curr Opin Genet Dev* **20:** 134–141.

Hahn MA, Hahn T, Lee DH, Esworthy RS, Kim BW, Riggs AD, Chu FF, Pfeifer GP. 2008. Methylation of polycomb target genes in intestinal cancer is mediated by inflammation. *Cancer Res* **68:** 10280–10289.

Haller D, Holt L, Kim SC, Schwabe RF, Sartor RB, Jobin C. 2003. Transforming growth factor-β1 inhibits non-pathogenic Gram negative bacteria-induced NF-κB recruitment to the interleukin-6 gene promoter in intestinal epithelial cells through modulation of histone acetylation. *J Biol Chem* **278:** 23851–23860.

Hamon MA, Cossart P. 2011. K^+ efflux is required for histone H3 dephosphorylation by *Listeria monocytogenes* listeriolysin O and other pore-forming toxins. *Infect Immun* **79:** 2839–2846.

Hamon MA, Batsche E, Regnault B, Tham TN, Seveau S, Muchardt C, Cossart P. 2007. Histone modifications induced by a family of bacterial toxins. *Proc Natl Acad Sci* **104:** 13467–13472.

Haraga A, Miller SI. 2003. A *Salmonella enterica* serovar typhimurium translocated leucine-rich repeat effector protein inhibits NF-κB-dependent gene expression. *Infect Immun* **71:** 4052–4058.

Hargreaves DC, Crabtree GR. 2011. ATP-dependent chromatin remodeling: Genetics, genomics and mechanisms. *Cell Res* **21:** 396–420.

Hayakawa T, Nakayama J. 2011. Physiological roles of class I HDAC complex and histone demethylase. *J Biomed Biotechnol* **2011:** 129383.

Hiragami-Hamada K, Shinmyozu K, Hamada D, Tatsu Y, Uegaki K, Fujiwara S, Nakayama J. 2011. N-terminal phosphorylation of HP1α promotes its chromatin binding. *Mol Cell Biol* **31:** 1186–1200.

Hnilicova J, Stanek D. 2011. Where splicing joins chromatin. *Nucleus* **2:** 182–188.

Hnilicova J, Hozeifi S, Duskova E, Icha J, Tomankova T, Stanek D. 2011. Histone deacetylase activity modulates alternative splicing. *PLoS ONE* **6:** e16727.

Holland C, Schmid M, Zimny-Arndt U, Rohloff J, Stein R, Jungblut PR, Meyer TF. 2011. Quantitative phosphoproteomics reveals link between *Helicobacter pylori* infection and RNA splicing modulation in host cells. *Proteomics* **11:** 2798–2811.

Hota SK, Dechassa ML, Prasad P, Bartholomew B. 2011. Mapping protein-DNA and protein-protein interactions of ATP-dependent chromatin remodelers. *Methods Mol Biol* **809:** 381–409.

Hur K, Niwa T, Toyoda T, Tsukamoto T, Tatematsu M, Yang HK, Ushijima T. 2011. Insufficient role of cell proliferation in aberrant DNA methylation induction and involvement of specific types of inflammation. *Carcinogenesis* **32:** 35–41.

Ichimura T, Watanabe S, Sakamoto Y, Aoto T, Fujita N, Nakao M. 2005. Transcriptional repression and heterochromatin formation by MBD1 and MCAF/AM family proteins. *J Biol Chem* **280:** 13928–13935.

Imai K, Ochiai K, Okamoto T. 2009. Reactivation of latent HIV-1 infection by the periodontopathic bacterium *Porphyromonas gingivalis* involves histone modification. *J Immunol* **182:** 3688–3695.

Imai K, Inoue H, Tamura M, Cueno ME, Takeichi O, Kusama K, Saito I, Ochiai K. 2011. The periodontal pathogen *Porphyromonas gingivalis* induces the Epstein-Barr virus lytic switch transactivator ZEBRA by histone modification. *Biochimie* **94:** 839–846.

Izar B, Mannala GK, Mraheil MA, Chakraborty T, Hain T. 2012. microRNA response to *Listeria monocytogenes* infection in epithelial cells. *Int J Mol Sci* **13:** 1173–1185.

Jenner RG, Young RA. 2005. Insights into host responses against pathogens from transcriptional profiling. *Nat Rev Microbiol* **3:** 281–294.

Kaikkonen MU, Lam MT, Glass CK. 2011. Non-coding RNAs as regulators of gene expression and epigenetics. *Cardiovasc Res* **90:** 430–440.

Katoh M. 2007. Dysregulation of stem cell signaling network due to germline mutation, SNP, *Helicobacter pylori* infection, epigenetic change and genetic alteration in gastric cancer. *Cancer Biol Ther* **6:** 832–839.

Kay S, Hahn S, Marois E, Hause G, Bonas U. 2007. A bacterial effector acts as a plant transcription factor and induces a cell size regulator. *Science* **318:** 648–651.

Klose RJ, Bird AP. 2006. Genomic DNA methylation: The mark and its mediators. *Trends Biochem Sci* **31:** 89–97.

Kouzarides T. 2007. Chromatin modifications and their function. *Cell* **128:** 693–705.

Lachner M, O'Carroll D, Rea S, Mechtler K, Jenuwein T. 2001. Methylation of histone H3 lysine 9 creates a binding site for HP1 proteins. *Nature* **410:** 116–120.

Lai AY, Wade PA. 2011. Cancer biology and NuRD: A multifaceted chromatin remodelling complex. *Nat Rev Cancer* **11:** 588–596.

Latham JA, Dent SY. 2007. Cross-regulation of histone modifications. *Nat Struct Mol Biol* **14:** 1017–1024.

Lebreton A, Lakisic G, Job V, Fritsch L, Tham TN, Camejo A, Mattei P-J, Regnault B, Nahori M-A, Cabanes D, et al. 2011. A bacterial protein targets the BAHD1 chromatin complex to stimulate type III interferon response. *Science* **3:** 1319–1321.

Lebreton A, Cossart P, Bierne H. 2012. Bacteria tune interferon responses by playing with chromatin. *Virulence* **3:** 87–91.

Lee JS, Smith E, Shilatifard A. 2010. The language of histone crosstalk. *Cell* **142:** 682–685.

Li H, Rauch T, Chen ZX, Szabo PE, Riggs AD, Pfeifer GP. 2006. The histone methyltransferase SETDB1 and the DNA methyltransferase DNMT3A interact directly and localize to promoters silenced in cancer cells. *J Biol Chem* **281:** 19489–19500.

Li H, Xu H, Zhou Y, Zhang J, Long C, Li S, Chen S, Zhou J-M, Shao F. 2007. The phosphothreonine lyase activity of a bacterial type III effector family. *Science* **315:** 1000–1003.

Licciardi PV, Wong SS, Tang ML, Karagiannis TC. 2010. Epigenome targeting by probiotic metabolites. *Gut Pathog* **2:** 24.

Liu L, Li Y, Tollefsbol TO. 2008. Gene-environment interactions and epigenetic basis of human diseases. *Curr Issues Mol Biol* **10:** 25–36.

Luco RF, Pan Q, Tominaga K, Blencowe BJ, Pereira-Smith OM, Misteli T. 2010. Regulation of alternative splicing by histone modifications. *Science* **327:** 996–1000.

Lynch KW. 2004. Consequences of regulated pre-mRNA splicing in the immune system. *Nat Rev Immunol* **4:** 931–940.

Ma F, Xu S, Liu X, Zhang Q, Xu X, Liu M, Hua M, Li N, Yao H, Cao X. 2011a. The microRNA miR-29 controls innate and adaptive immune responses to intracellular bacterial infection by targeting interferon-γ. *Nat Immunol* **12:** 861–869.

Ma KW, Flores C, Ma W. 2011b. Chromatin configuration as a battlefield in plant-bacteria interactions. *Plant Physiol* **57:** 535–543.

Maekita T, Nakazawa K, Mihara M, Nakajima T, Yanaoka K, Iguchi M, Arii K, Kaneda A, Tsukamoto T, Tatematsu M, et al. 2006. High levels of aberrant DNA methylation in *Helicobacter pylori*-infected gastric mucosae and its possible association with gastric cancer risk. *Clin Cancer Res* **12:** 989–995.

Maison C, Bailly D, Roche D, Montes de Oca R, Probst AV, Vassias I, Dingli F, Lombard B, Loew D, Quivy JP, et al. 2011. SUMOylation promotes de novo targeting of HP1α to pericentric heterochromatin. *Nat Genet* **43:** 220–227.

Margueron R, Reinberg D. 2011. The Polycomb complex PRC2 and its mark in life. *Nature* **469:** 343–349.

Matsushima K, Isomoto H, Inoue N, Nakayama T, Hayashi T, Nakayama M, Nakao K, Hirayama T, Kohno S. 2011. MicroRNA signatures in *Helicobacter pylori*-infected gastric mucosa. *Int J Cancer J* **128:** 361–370.

Mattick JS, Makunin IV. 2005. Small regulatory RNAs in mammals. *Hum Mol Genet* **14** (Spec No 1)**:** R121–R132.

McCall CE, Yoza B, Liu T, El Gazzar M. 2010. Gene-specific epigenetic regulation in serious infections with systemic inflammation. *J Innate Immun* **2:** 395–405.

Medzhitov R, Horng T. 2009. Transcriptional control of the inflammatory response. *Nat Rev Immunol* **9:** 692–703.

Mikkelsen TS, Ku M, Jaffe DB, Issac B, Lieberman E, Giannoukos G, Alvarez P, Brockman W, Kim TK, Koche RP, et al. 2007. Genome-wide maps of chromatin state in pluripotent and lineage-committed cells. *Nature* **448:** 553–560.

Mohammad HP, Baylin SB. 2010. Linking cell signaling and the epigenetic machinery. *Nat Biotechnol* **28:** 1033–1038.

Morey C, Avner P. 2011. The demoiselle of X-inactivation: 50 years old and as trendy and mesmerising as ever. *PLoS Genet* **7:** e1002212.

Mosavi LK, Cammett TJ, Desrosiers DC, Peng ZY. 2004. The ankyrin repeat as molecular architecture for protein recognition. *Protein Sci* **13:** 1435–1448.

Murata M, Azuma Y, Miura K, Rahman MA, Matsutani M, Aoyama M, Suzuki H, Sugi K, Shirai M. 2007. Chlamydial SET domain protein functions as a histone methyltransferase. *Microbiology* **153:** 585–592.

Nakajima T, Enomoto S, Yamashita S, Ando T, Nakanishi Y, Nakazawa K, Oda I, Gotoda T, Ushijima T. 2010. Persistence of a component of DNA methylation in gastric mucosae after *Helicobacter pylori* eradication. *J Gastroenterol* **45:** 37–44.

Niwa T, Tsukamoto T, Toyoda T, Mori A, Tanaka H, Maekita T, Ichinose M, Tatematsu M, Ushijima T. 2010. Inflammatory processes triggered by *Helicobacter pylori* infection cause aberrant DNA methylation in gastric epithelial cells. *Cancer Res* **70:** 1430–1440.

Nora EP, Heard E. 2010. Chromatin structure and nuclear organization dynamics during X-chromosome inactivation. *Cold Spring Harb Symp Quant Biol* **75:** 333–344.

O'Connell RM, Rao DS, Baltimore D. 2011. MicroRNA regulation of inflammatory responses. *Ann Rev Immunol* **30:** 295–312.

Ogawa Y, Sun BK, Lee JT. 2008. Intersection of the RNA interference and X-inactivation pathways. *Science* **320:** 1336–1341.

Okuda J, Toyotome T, Kataoka N, Ohno M, Abe H, Shimura Y, Seyedarabi A, Pickersgill R, Sasakawa C. 2005. Shigella effector IpaH9.8 binds to a splicing factor U2AF35 to modulate host immune responses. *Biochem Biophys Res Commun* **333:** 531–539.

Opitz B, Puschel A, Beermann W, Hocke AC, Forster S, Schmeck B, van Laak V, Chakraborty T, Suttorp N, Hippenstiel S. 2006. *Listeria monocytogenes* activated p38 MAPK and induced IL-8 secretion in a nucleotide-binding oligomerization domain 1-dependent manner in endothelial cells. *J Immunol* **176:** 484–490.

Park J, Kim KJ, Choi K-S, Grab DJ, Dumler JS. 2004. *Anaplasma phagocytophilum* AnkA binds to granulocyte DNA and nuclear proteins. *Cell Microbiol* **6:** 743–751.

Pathak SK, Basu S, Bhattacharyya A, Pathak S, Banerjee A, Basu J, Kundu M. 2006. TLR4-dependent NF-κB activation and mitogen- and stress-activated protein kinase 1-triggered phosphorylation events are central to *Helicobacter pylori* peptidyl prolyl *cis-*, *trans*-isomerase (HP0175)-mediated induction of IL-6 release from macrophages. *J Immunol* **177:** 7950–7958.

Pennini ME, Pai RK, Schultz DC, Boom WH, Harding CV. 2006. *Mycobacterium tuberculosis* 19-kDa lipoprotein inhibits IFN-γ-induced chromatin remodeling of MHC2TA by TLR2 and MAPK signaling. *J Immunol* **176:** 4323–4330.

Pennini ME, Liu Y, Yang J, Croniger CM, Boom WH, Harding CV. 2007. CCAAT/enhancer-binding protein β and δ binding to CIITA promoters is associated with the inhibition of CIITA expression in response to *Mycobacterium tuberculosis* 19-kDa lipoprotein. *J Immunol* **179:** 6910–6918.

Pennini MSP, Dautry-Varsat A, Subtil A. 2010. Histone methylation by NUE, a novel nuclear effector of the intracellular pathogen *Chlamydia trachomatis. PLoS Pathog* **6:** e1000995.

Perrett CA, Lin DY, Zhou D. 2011. Interactions of bacterial proteins with host eukaryotic ubiquitin pathways. *Front Microbiol* **2:** 143.

Piacentini L, Pimpinelli S. 2010. Positive regulation of euchromatic gene expression by HP1. *Fly (Austin)* **4:** 299–301.

Pontier DB, Gribnau J. 2011. Xist regulation and function explored. *Hum Genet* **130:** 223–236.

Portela A, Esteller M. 2010. Epigenetic modifications and human disease. *Nat Biotechnol* **28:** 1057–1068.

Rathert P, Dhayalan A, Murakami M, Zhang X, Tamas R, Jurkowska R, Komatsu Y, Shinkai Y, Cheng X, Jeltsch A. 2008. Protein lysine methyltransferase G9a acts on nonhistone targets. *Nat Chem Biol* **4:** 344–346.

Raymond B, Batsche E, Boutillon F, Wu YZ, Leduc D, Balloy V, Raoust E, Muchardt C, Goossens PL, Touqui

L. 2009. Anthrax lethal toxin impairs IL-8 expression in epithelial cells through inhibition of histone H3 modification. *PLoS Pathog* **5:** e1000359.

Ribet D, Cossart P. 2010. SUMOylation and bacterial pathogens. *Virulence* **6:** 532–534.

Riggs MG, Whittaker RG, Neumann JR, Ingram VM. 1977. n-Butyrate causes histone modification in HeLa and Friend erythroleukaemia cells. *Nature* **268:** 462–464.

Rivas S. 2011. Nuclear dynamics during plant innate immunity. *Plant Physiol* **158:** 87–94.

Rohde JR, Breitkreutz A, Chenal A, Sansonetti PJ, Parsot C. 2007. Type III secretion effectors of the IpaH family are E3 ubiquitin ligases. *Cell Host Microbe* **1:** 77–83.

Rother S, Meister G. 2011. Small RNAs derived from longer non-coding RNAs. *Biochimie* **93:** 1905–1915.

Russo VEA, Martienssen RA, Riggs AD. 1996. *Epigenetic mechanisms of gene regulation.* Cold Spring Harbor Laboratory Press, Cold Spring Harbor, NY.

Saccani S, Pantano S, Natoli G. 2002. p38-Dependent marking of inflammatory genes for increased NF-κB recruitment. *Nat Immunol* **3:** 69–75.

Saint-Andre V, Batsche E, Rachez C, Muchardt C. 2011. Histone H3 lysine 9 trimethylation and HP1γ favor inclusion of alternative exons. *Nat Struct Mol Biol* **18:** 337–344.

Sasai N, Defossez PA. 2009. Many paths to one goal? The proteins that recognize methylated DNA in eukaryotes. *Int J Dev Biol* **53:** 323–334.

Schmeck B, Beermann W, van Laak V, Zahlten J, Opitz B, Witzenrath M, Hocke AC, Chakraborty T, Kracht M, Rosseau S, et al. 2005. Intracellular bacteria differentially regulated endothelial cytokine release by MAPK-dependent histone modification. *J Immunol* **175:** 2843–2850.

Schmeck B, Lorenz J, N'Guessan PD, Opitz B, van Laak V, Zahlten J, Slevogt H, Witzenrath M, Flieger A, Suttorp N, et al. 2008. Histone acetylation and flagellin are essential for *Legionella pneumophila*-induced cytokine expression. *J Immunol* **181:** 940–947.

Schmitz KM, Mayer C, Postepska A, Grummt I. 2010. Interaction of noncoding RNA with the rDNA promoter mediates recruitment of DNMT3b and silencing of rRNA genes. *Genes Dev* **24:** 2264–2269.

Schulte LN, Eulalio A, Mollenkopf HJ, Reinhardt R, Vogel J. 2011. Analysis of the host microRNA response to Salmonella uncovers the control of major cytokines by the let-7 family. *EMBO J* **30:** 1977–1989.

Seyedarabi A, Sullivan JA, Sasakawa C, Pickersgill RW. 2011. A disulfide driven domain swap switches off the activity of Shigella IpaH9.8 E3 ligase. *FEBS Lett* **584:** 4163–4168.

Singer AU, Rohde JR, Lam R, Skarina T, Kagan O, Dileo R, Chirgadze NY, Cuff ME, Joachimiak A, Tyers M, et al. 2008. Structure of the Shigella T3SS effector IpaH defines a new class of E3 ubiquitin ligases. *Nat Struct Mol Biol* **15:** 1293–1301.

Slevogt H, Schmeck B, Jonatat C, Zahlten J, Beermann W, van Laak V, Opitz B, Dietel S, N'Guessan PD, Hippenstiel S, et al. 2006. *Moraxella catarrhalis* induces inflammatory response of bronchial epithelial cells via MAPK and NF-κB activation and histone deacetylase activity reduction. *Am J Physiol Lung Cell Mol Physiol* **290:** L818–L826.

Suganuma T, Workman JL. 2011. Signals and combinatorial functions of histone modifications. *Annu Rev Biochem* **80:** 473–499.

Sun J. 2010. Enteric bacteria and cancer stem cells. *Cancers (Basel)* **3:** 285–297.

Tahiliani M, Koh KP, Shen Y, Pastor WA, Bandukwala H, Brudno Y, Agarwal S, Iyer LM, Liu DR, Aravind L, et al. 2009. Conversion of 5-methylcytosine to 5-hydroxymethylcytosine in mammalian DNA by MLL partner TET1. *Science* **324:** 930–935.

Takahashi K, Sugi Y, Nakano K, Tsuda M, Kurihara K, Hosono A, Kaminogawa S. 2011. Epigenetic control of the host gene by commensal bacteria in large intestinal epithelial cells. *J Biol Chem* **286:** 35755–35762.

Taverna SD, Li H, Ruthenburg AJ, Allis CD, Patel DJ. 2007. How chromatin-binding modules interpret histone modifications: Lessons from professional pocket pickers. *Nat Struct Mol Biol* **14:** 1025–1040.

Tisseur M, Kwapisz M, Morillon A. 2011. Pervasive transcription—Lessons from yeast. *Biochimie* **93:** 1889–1896.

Tiwari VK, Stadler MB, Wirbelauer C, Paro R, Schubeler D, Beisel C. 2011. A chromatin-modifying function of JNK during stem cell differentiation. *Nat Genet* **44:** 94–100.

Tolg C, Sabha N, Cortese R, Panchal T, Ahsan A, Soliman A, Aitken KJ, Petronis AB, Auml Gli DJ. 2011. Uropathogenic *E. coli* infection provokes epigenetic downregulation of CDKN2A (p16INK4A) in uroepithelial cells. *Lab Invest* **91:** 825–836.

Toyotome T. 2001. Shigella protein IpaH9.8 is secreted from bacteria within mammalian cells and transported to the nucleus. *J Biol Chem* **276:** 32071–32079.

Tropberger P, Schneider R. 2011. Going global: Novel histone modifications in the globular domain of H3. *Epigenetics* **5:** 112–117.

Turner AM, Morris KV. 2010. Controlling transcription with noncoding RNAs in mammalian cells. *Biotechniques* **48:** x–xvi.

Ushijima T, Hattori N. 2012. Molecular pathways: Involvement of *Helicobacter pylori*-triggered inflammation in the formation of an epigenetic field defect, and its usefulness as cancer risk and exposure markers. *Clin Cancer Res* **18:** 923–929.

van Vliet J, Oates NA, Whitelaw E. 2007. Epigenetic mechanisms in the context of complex diseases. *Cell Mol Life Sci* **64:** 1531–1538.

Vire E, Brenner C, Deplus R, Blanchon L, Fraga M, Didelot C, Morey L, Van Eynde A, Bernard D, Vanderwinden JM, et al. 2006. The Polycomb group protein EZH2 directly controls DNA methylation. *Nature* **439:** 871–874.

Wang Y, Curry HM, Zwilling BS, Lafuse WP. 2005. Mycobacteria inhibition of IFN-γ induced HLA-DR gene expression by up-regulating histone deacetylation at the promoter region in human THP-1 monocytic cells. *J Immunol* **174:** 5687–5694.

Wilson AG. 2008. Epigenetic regulation of gene expression in the inflammatory response and relevance to common diseases. *J Periodontol* **79:** 1514–1519.

Woodcock CL, Dimitrov S. 2001. Higher-order structure of chromatin and chromosomes. *Curr Opin Genet Dev* **11:** 130–135.

Wu H, Zhang Y. 2011. Mechanisms and functions of Tet protein-mediated 5-methylcytosine oxidation. *Genes Dev* **25:** 2436–2452.

Wutz A. 2011. Gene silencing in X-chromosome inactivation: Advances in understanding facultative heterochromatin formation. *Nat Rev Genet* **12:** 542–553.

Xia G, Schneider-Stock R, Diestel A, Habold C, Krueger S, Roessner A, Naumann M, Lendeckel U. 2008. *Helicobacter pylori* regulates p21(WAF1) by histone H4 acetylation. *Biochem Biophys Res Commun* **369:** 526–531.

Xu WS, Parmigiani RB, Marks PA. 2007. Histone deacetylase inhibitors: Molecular mechanisms of action. *Oncogene* **26:** 5541–5552.

Yamamoto Y, Verma UN, Prajapati S, Kwak YT, Gaynor RB. 2003. Histone H3 phosphorylation by IKK-α is critical for cytokine-induced gene expression. *Nature* **423:** 655–659.

Yan J, Zhang M, Zhang J, Chen X, Zhang X. 2011. *Helicobacter pylori* infection promotes methylation of WWOX gene in human gastric cancer. *Biochem Biophys Res Commun* **408:** 99–102.

Yang SH, Vickers E, Brehm A, Kouzarides T, Sharrocks AD. 2001. Temporal recruitment of the mSin3A-histone deacetylase corepressor complex to the ETS domain transcription factor Elk-1. *Mol Cell Biol* **21:** 2802–2814.

Yang L, Lin C, Liu W, Zhang J, Ohgi KA, Grinstein JD, Dorrestein PC, Rosenfeld MG. 2011. ncRNA- and Pc2 methylation-dependent gene relocation between nuclear structures mediates gene activation programs. *Cell* **147:** 773–788.

Yao Y, Tao H, Park DI, Sepulveda JL, Sepulveda AR. 2006. Demonstration and characterization of mutations induced by *Helicobacter pylori* organisms in gastric epithelial cells. *Helicobacter* **11:** 272–286.

Yin L, Chung WO. 2011. Epigenetic regulation of human β-defensin 2 and CC chemokine ligand 20 expression in gingival epithelial cells in response to oral bacteria. *Mucosal Immunol* **4:** 409–419.

Zhang S, Barros SP, Niculescu MD, Moretti AJ, Preisser JS, Offenbacher S. 2010. Alteration of PTGS2 promoter methylation in chronic periodontitis. *J Dent Res* **89:** 133–137.

Zhou Y, Kim J, Yuan X, Braun T. 2011. Epigenetic modifications of stem cells: A paradigm for the control of cardiac progenitor cells. *Circ Res* **109:** 1067–1081.

Zhu Y, Li H, Hu L, Wang J, Zhou Y, Pang Z, Liu L, Shao F. 2008. Structure of a Shigella effector reveals a new class of ubiquitin ligases. *Nat Struct Mol Biol* **15:** 1302–1308.

Zhu B, Nethery KA, Kuriakose JA, Wakeel A, Zhang X, McBride JW. 2009. Nuclear translocated *Ehrlichia chaffeensis* ankyrin protein interacts with a specific adenine-rich motif of host promoter and intronic *Alu* elements. *Infect Immun* **77:** 4243–4255.

Zurawski DV, Mumy KL, Faherty CS, Mccormick BA, Maurelli AT. 2009. *Shigella flexneri* type III secretion system effectors OspB and OspF target the nucleus to downregulate the host inflammatory response via interactions with retinoblastoma protein. *Mol Microbiol* **71:** 350–368.

Concepts and Mechanisms: Crossing Host Barriers

Kelly S. Doran[1,2], Anirban Banerjee[1], Olivier Disson[3,4], and Marc Lecuit[3,4,5]

[1]Department of Biology and Center for Microbial Sciences, San Diego State University, San Diego, California 92182

[2]Department of Pediatrics, University of California, San Diego, La Jolla, California 92123

[3]Institut Pasteur, Biology of Infection Unit, French National Reference Center, and WHO Collaborating Center for *Listeria*, Paris 75015, France

[4]INSERM, U1117 Paris, France

[5]Université Paris Descartes, Sorbonne Paris Cité, Institut Imagine, Centre d'Infectiologie Necker-Pasteur, Hôpital Universitaire Necker-Enfants Malades, Paris 75015, France

Correspondence: kdoran@mail.sdsu.edu; marc.lecuit@pasteur.fr

The human body is bordered by the skin and mucosa, which are the cellular barriers that define the frontier between the internal milieu and the external nonsterile environment. Additional cellular barriers, such as the placental and the blood–brain barriers, define protected niches within the host. In addition to their physiological roles, these host barriers provide both physical and immune defense against microbial infection. Yet, many pathogens have evolved elaborated mechanisms to target this line of defense, resulting in a microbial invasion of cells constitutive of host barriers, disruption of barrier integrity, and systemic dissemination and invasion of deeper tissues. Here we review representative examples of microbial interactions with human barriers, including the intestinal, placental, and blood–brain barriers, and discuss how these microbes adhere to, invade, breach, or compromise these barriers.

INVASIVE PATHOGENS AND HOST BARRIER

A critical event in systemic infection is the crossing by its causative agent of one or multiple host barriers (Fig. 1). It can schematically occur according to two nonexclusive scenarios: (i) the barrier function of a tissue can be compromised either by physical means (wound, catheter, surgery, arthropod bite) or in the context of a host condition that disrupts the integrity of tissues (inflammation, dysfunction of a host gene product implicated in barrier function), and this barrier damage allows microbial invasion; (ii) barrier(s) can also be crossed actively, via the direct action of microbial gene products that mediates microbial adhesion to and translocation across cells constitutive of host barriers.

ENTEROPATHOGENS AND THE INTESTINAL BARRIER

The skin and mucosa form the barriers that separate the host from its external nonsterile

Cite this article as *Cold Spring Harb Perspect Med* doi: 10.1101/cshperspect.a010090

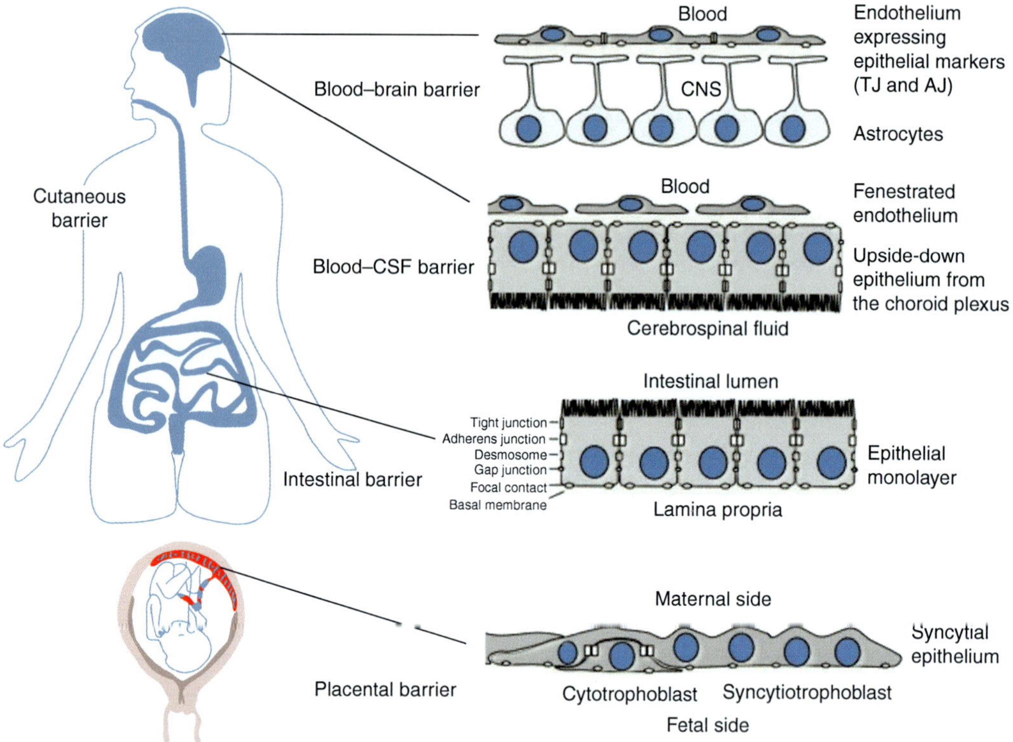

Figure 1. Example of host barriers. From *top* to *bottom*: The blood–brain barrier is made by brain endothelial cells that express tight junctions. The blood–CSF barrier is constituted by a monolayer of epithelial cells from the choroid plexus, which separates the blood located in a fenestrated endothelium from the cerebrospinal fluid. The intestinal barrier is a mucosal barrier made by a monolayer of epithelial cells. The placental barrier is constituted by both syncytial trophoblastic cells and mononuclear cytotrophoblasts.

environment. The intestinal barrier is made of a monolayer of polarized epithelial cells called enterocytes. Tight junctions (TJs) and adherens junctions (AJs) hold together enterocytes. TJs are the most apical structures and are made of transmembrane proteins (occludins, claudins, junction adhesion molecules [JAMs], and the coxsackievirus and adenovirus receptor [CAR]), connected to intracellular proteins (such as ZO-1, -2, -3, MAGUK, and PAR family proteins). TJs prevent paracellular diffusion of molecules or luminal microorganisms (Anderson et al. 2004). AJs are required for the integrity of TJs and are located directly below them (Harris and Tepass 2010). They are made of E-cadherin, which is connected to the actin cytoskeleton. AJs and TJs are located at the apical and luminal side of enterocytes. Their basolateral side seats on a basement membrane made of extracellular matrix proteins, which separate them from the lamina propria, that contains connective tissue, stromal cells, blood and lymphatic capillaries, and cellular effectors of the immune system, such as lymphocytes, dendritic cells, or resident macrophages, some of which are in direct contact with enterocytes and the intestinal lumen via extending dendrites (Rescigno and Di Sabatino 2009).

The intestinal epithelium apical surface delineates the interface between the host and the intestinal lumen. The intestinal barrier is in continuous contact and crosstalk with up to 10^{14} microorganisms, which constitute the intestinal microbiota. It represents a major potential portal entry for microbes into the host. Microbial pathogens have evolved specific capacities to adhere to, invade, or disrupt the intestinal barrier (Sousa et al. 2005; Guttman and Finlay

2009; Yoshida et al. 2011). These pathogens have to counteract host defenses and competition from the resident microflora in the gut. Microbes can cross epithelial barriers by disrupting intercellular junctions or by interacting with epithelial cell receptors, leading to the traversal in a transcellular or paracellular route. Pathogens may also take advantage of the luminal sampling property of host cells to get access to the lamina propria.

Disruption of Tight Junctions by Enteropathogens

Intestinal pathogens may adhere to the apical surface of enterocytes, remain extracellular, and modify the barrier properties of the epithelium. This event can lead to local invasion of the lamina propria. Enteropathogenic and enterohemorrhagic *E. coli* (EPEC and EHEC) are diarrheagenic bacteria (Wong et al. 2011) that affect TJs (occluding and ZO-1) through type-III secretion system (TTSS) effectors EspF, Map, and EspG (Dean and Kenny 2004; Matsuzawa et al. 2005), leading to a decreased transepithelial resistance and aberrant TJs. *Helicobacter pylori* expresses several adhesins that mediate bacterial attachment to gastric epithelial glycan receptors, and possesses a type-IV secretion system that injects the multifunctional bacterial effector CagA into the host cell cytoplasm (Rieder et al. 2005). CagA alters the composition and the function of TJs (Amieva et al. 2003) and leads to a disruption of the epithelial polarity, which could have a role in *H. pylori*-associated carcinogenesis (Saadat et al. 2007).

Toxin A and toxin B from *Clostridium difficile* are both implicated in the pathogenesis of pseudomembranous colitis (Kuehne et al. 2010). They dissociate TJ's proteins occludin, ZO-1 and ZO-2, thus increasing intestinal epithelium permeability (Voth and Ballard 2005). The bifunctional enterotoxin CPE of *Clostridium perfringens*, which can cause diarrhea, forms pores in the host cell plasma membrane, and binds claudin-3 and claudin-4 with high affinity (Sonoda et al. 1999; Winkler et al. 2009; Robertson et al. 2010), inducing their degradation, a decrease in transepithelial resistance and disruption of TJs. The role of claudins in the barrier function of TJs was discovered via these studies (Sonoda et al. 1999).

Rotavirus, member of the Reovirus family causing gastroenteritis, also colonizes the gut and disrupts tight junctions. VP8 subunit of VP4, an outer rotavirus protein, moves ZO-1, claudin-3, and occludin from TJs, then disrupting the barrier integrity (Nava et al. 2004; Beau et al. 2007). Similar effects are produced by the viral enterotoxin NSP4, which blocks the formation of TJs by preventing lateral recruitment of ZO-1 (Tafazoli et al. 2001). Reoviruses use the TJ protein JAM-1 as a cellular receptor (Barton et al. 2001). The viral surface protein $\sigma 1$ directly interacts with the amino-terminal domain of JAM-1 and disrupts the homodimers of this junction protein (Forrest et al. 2003; Campbell et al. 2005; Guglielmi et al. 2007).

Active Microbial Crossing of the Intestinal Barrier

Translocation across Enterocytes

Listeria monocytogenes is a facultative intracellular bacterium that crosses the intestinal, blood–brain and placental barriers, inducing gastroenteritis, meningitis/encephalitis, and materno-fetal infection. InlA and InlB, two surface proteins of *L. monocytogenes*, which interact with their respective host receptors E-cadherin and Met, induce internalization of *L. monocytogenes* into cultured epithelial cells. Importantly, in contrast to human E-cadherin, mouse E-cadherin is unable to interact with InlA and to promote *L. monocytogenes* entry into cells (Lecuit et al. 1999, 2001). This species specificity relies on the sixteenth amino acid of the mature E-cadherin, a proline in humans, and a glutamic acid in mice. E-cadherin from guinea pig is also recognized by InlA and harbors a proline at position 16. In this species, contrary to the mouse, *L. monocytogenes* is able to cross the intestinal barrier, disseminate systemically, and induce a dose- and InlA-dependent lethality (Lecuit et al. 2001). To prove the role of the InlA–E-cadherin interaction in the ability of *L. monocytogenes* to cross the intestinal barrier, a transgenic mouse model expressing human E-cadherin in postmitotic nonproliferative small intestinal enterocytes was designed (Lecuit et al. 2001). In contrast to nontransgenic mice, and similarly to guinea pigs and humans, transgenic mice expressing human E-cadherin are highly permissive to orally acquired listeriosis, demonstrating a critical role for InlA in the ability of *L. monocytogenes* to cross the intestinal barrier. Epidemiological study of human cases of listeriosis evidence supports this result, because clinical strains express a functional InlA significantly more often (96%) than food isolates do (65%) (Jacquet et al. 2004), which is in favor of a role of InlA in crossing the intestinal barrier in humans. In contrast to the InlA–E-cadherin interaction, the InlB–Met interaction is not necessary for crossing the intestinal barrier in transgenic mice, as well as in gerbil, which is permissive for InlA and InlB (Khelef et al. 2006; Disson et al. 2008). The way by which *L. monocytogenes* gets access to the adherens junction protein E-cadherin and the mechanism by which bacteria cross the intestinal barrier were not known until recently. It had been proposed that *L. monocytogenes* access to E-cadherin may be facilitated when extruding epithelial cells detach from the tips of intestinal microvilli, thereby exposing E-cadherin to the luminal side of the intestine (Pentecost et al. 2006). This was shown in vitro and correlated with a colocalization of bacteria with extruding cells in rabbit ileal loop in vivo. It has now been shown that E-cadherin is luminally accessible not only around extruding enterocytes at the tip and lateral sides of villi, but also particularly around mucus-expelling goblet cells (GCs) and in villus epithelial folds. *L. monocytogenes* preferentially adheres to accessible E-cadherin on GCs in a ligated loop of transgenic mice expressing human E-cadherin, is internalized, rapidly transcytosed across the intestinal epithelium, and released in the lamina propria by exocytosis from where it disseminates systemically (Nikitas et al. 2011). *L. monocytogenes* is thus able to exploit intrinsic tissue heterogeneity to access its receptor. These results highlight transcytosis as an unexpected pathway that is hijacked by *L. monocytogenes* to breach the intestinal barrier and cause systemic infection. The precise mechanism by which *L. monocytogenes* transcytoses across enterocytes is currently being investigated.

Some viral species also use basolateral cell adhesion molecules as receptors to infect epithelia. The coxsackievirus and adenovirus receptor (CAR) is a component of the tight junction, which is located under the classical TJ proteins (occludin and claudin) and is thus inaccessible to viruses that contact host cells via their apical surface (Freimuth et al. 2008). As indicated by its name, it is both targeted by adenovirus and coxsackieviruses, leading to a decrease in epithelial impermeability (Bergelson et al. 1997). Coxsackieviruses B are members of the Enterovirus family and the causative agents of meningitis and myocarditis. Coxsackievirus-B infects polarized cell layers through binding to the decay-accelerating factor (DAF) proteins on the apical cell surface (Coyne and Bergelson 2006). Virus attachment to DAF leads to the activation of the src kinase Abl, inducing a Rac-dependent reorganization of the actin cytoskeleton that enables the viruses to relocalize to tight junctions. Once in the junction, viral particles interact with CAR, and are internalized in caveolar vesicles, a mechanism dependent on the activation of Fyn kinase by DAF (Coyne and Bergelson 2006). Internalization of coxsackievirus-B also requires the TJ protein occludin (Coyne et al. 2007). The subsequent release from epithelial cells has been suggested to occur through a calcium signal and calpain-dependent necrosis of the infected cells (Bozym et al. 2011).

Host Cell Sensing of Intestinal Lumen Content as an Achilles' Heel Exploited by Microbes

M cells, which are specialized epithelial cells located at the Peyer's patch level and in villi and $CD11c^{+}CX3CR1^{high}$ cells, continuously sample the intestinal lumen and transport antigens to the mucosal lymphoid tissues (Rescigno 2011). This process is required to either induce tolerance or to initiate immune responses. However, many microbes exploit M cells as an Achilles' heel to enter the lamina propria, including enteroviruses, *Yersinia enterocolitica*, *Shigella flexneri*, and *Salmonella enterica* (Corr et al. 2008).

 Cite this article as *Cold Spring Harb Perspect Med* doi: 10.1101/cshperspect.a010090

S. enterica causes severe diseases in humans, ranging from gastroenteritis to systemic infections (typhoid fever). Systemic infections because of *Salmonella* serovars are species-specific: *S. enterica* serovar Typhi (*S.* Typhi) induce typhoid fever, a systemic infection in human, whereas *S. enterica* serovar Typhimurium (*S.* Typhimurium), which induces a self-limited gastroenteritis in immunocompetent humans, but causes a typhoid-like systemic infection in mice. For this reason, mice have been widely used to study typhoid fever (Monack et al. 2004), although they are generally more resistant than humans to intestinal colonization by *S. enterica*. To circumvent this limitation, a streptomycin pretreated mouse model of enterocolitis has been developed that allows probing the host and bacterial factors contributing to intestinal immunopathology (Grassl and Finlay 2008). After oral ingestion, *S.* Typhimurium invades intestinal tissues and induces mucosal inflammation. It may enter the blood circulation and infect downstream organs such as liver and spleen (Watson and Holden 2010). *S.* Typhimurium exploits both dendritic cells and M cells to access the lamina propria. It was shown that, in the terminal ileum of mice, CD11c^{+} (mainly CX3CR1^{high}) lamina propria monocytic phagocytes extend protrusions across the epithelium, which facilitates pathogen transport from the gut lumen into the lamina propria (Rescigno et al. 2001; Niess et al. 2005; Chieppa et al. 2006). DCs may then allow the clearance of the pathogen, but in an immunocompromised host may lead to the subsequent colonization of deeper organs. However, sampling by DCs alone does not account for the translocation of *S.* Typhimurium to the lamina propria (Hapfelmeier et al. 2008). *S.* Typhimurium is also able to translocate through M cells to access Peyer's patches lamina propria (Jones et al. 1994). The glycoprotein 2 (GP2), specifically expressed on the apical plasma membrane of M cells, recognizes FimH, a component of type-I pili expressed by *Enterobacteriaceae*, including *Escherichia coli* and *S.* Typhimurium, which allows the internalization of these bacteria (Hase et al. 2009). Another protein, encoded by *lpf* from the fimbrial operon, interacts with mouse M cells (Baumler et al. 1996). Once attached, *Salmonella* pathogenicity island 1 (SPI-1), which codes for bacterial proteins translocated into target cells by the SPI-1 type-III secretion system (TTSS-1), promotes bacterial invasion of mouse M cells (Clark et al. 1998). *S.* Typhimurium may also cross the intestinal barrier through the epithelial cells in an SPI-1 dependent manner. Recently, the mechanism by which *S.* Typhimurium crosses the epithelial barrier at the cecal level and is released in the lamina propria has been studied in detail (Muller et al. 2012). It has been shown by use of in vivo confocal microscopy in a murine *Salmonella*–diarrhea model that the breaching of the epithelial barrier relies on both TTSS-1-mediated invasion and TTSS-2-dependent transcytosis to the basolateral side. Once in the lamina propria, the bacterium is transiently extracellular before being phagocytized, among others by CD11c^{+}CX3CR1^{high} monocytic phagocytes, which enables it to colonize the lamina propria. *S.* Typhimurium may then exploit the immune surveillance after crossing the epithelial barrier to infect its host.

Paracellular Pathway

Pathogens such as parasites (*Entamoeba histolytica* and *Toxoplasma gondii*) (Yoshida et al. 2011) or bacteria (*Streptococcus agalactiae*) have been proposed to cross the intestinal barrier in a paracellular route. This serves to target intercellular junctions and to actively propel between two cell membranes. *Toxoplasma gondii* (*Tg*) is a human parasite with the capacity to actively cross not only the intestinal barrier but also the blood–brain and placental barriers (Barragan and Sibley 2003). *Tg* parasites have been shown to accumulate around intercellular junctions. They may use the paracellular route to cross epithelial barriers without altering the barrier permeability. Crossing may occur through the interaction of the *Tg* surface adhesin MIC2 with the intercellular adhesion molecule 1 (ICAM-1), which is present in the apical surfaces of epithelial barriers (Barragan et al. 2005) and requires the motility property of the parasite (Barragan and Sibley 2002). *Tg* has to be internalized by host cells to replicate, and its entry and the formation of a parasitophorous vacuole depend on the association of its apical tip with the host cell through the parasite transmembrane protein AMA1, and a subsequent rearrangement of the host and *Tg* cytoskeletons. This allows the parasite to replicate at the intestinal level, and could also be the first step for the invasion of downstream organs (Tyler and Boothroyd 2011).

Streptococcus agalactiae (group B *Streptococcus* [GBS]) is a normal constituent of the intestinal microflora and the major cause of human neonatal meningitis. Among the GBS clones, ST-17 is strongly associated with meningitis in infants. It has been shown that a ST-17-specific surface-anchored protein called hypervirulent GBS adhesin (HvgA) is required for intestinal colonization and translocation across the intestinal barrier and the blood–brain barrier (BBB), leading to meningitis (Tazi et al. 2010). The mechanism by which the ST-17 clone crosses the intestinal barrier is still unknown. However, previous data suggest that the paracellular route could be a way for GBS to cross epithelial barriers. Using differentiated epithelial cells grown on Transwell inserts as a model, it has been shown by transmission electron microscopy that GBS interacts with intercellular junctions and has the capacity to cross the monolayer by a paracellular mechanism via an active and transient opening of cell junctions (Soriani et al. 2006). This process depends on the pilus backbone of GBS, but the host receptors are still not known (Pezzicoli et al. 2008). However, the mechanism by which the ST-17 clone crosses the intestinal barrier is currently unknown.

MICROBES AND THE PLACENTAL BARRIER

The human maternofetal barrier contains two anatomically distinct components, the chorioallantoic placenta and the chorioamnion. The barrier is formed at the placental level by specialized epithelial cells, the villous syncytiotrophoblasts, which are differentiated from underlying mononuclear cytotrophoblasts. Multinucleated syncytiotrophoblasts are in direct contact with maternal blood circulating in the intervillous space. A basement membrane separates these trophoblastic cells from a connective tissue that contains fetal capillaries. The amniotic epithelium forms the maternofetal interface in the chorioamnion. The apical surface of this epithelium is exposed to amniotic fluid, whereas its basal surface sits on a basement membrane that overlies the amniotic mesoderm (Cross et al. 1994; Huppertz 2008; Maltepe et al. 2010).

A number of pathogens invade the placenta but few are known to actively cross the placental barrier and disseminate to the fetus. Infection of placenta can lead to abortion, premature delivery, and fetal and neonatal infection. *Coxiella burnettii*, the agent of Q fever, is an intracellular bacterium which may infect trophoblast (Baumgartner and Bachmann 1992; Ben Amara et al. 2010) and, in rare cases, the fetus (Carcopino et al. 2009) leading to abortion and premature delivery. *Plasmodium falciparum*, the etiological agent of the malaria, has long been known to affect pregnancy (Brabin et al. 2004). Parasitized erythrocytes bind to chondroitin sulfate A expressed at the syncytiotrophoblast apical surface, where they accumulate, impeding the physiological functions of the placenta (Scherf et al. 2001). Adverse outcomes are low birth weight babies, fetal loss, as well as increased perinatal and maternal mortality. Several Chlamydia spp, such as *Chlamydia trachomatis*, *Chlamydophila abortus*, *Chlamydophila psittaci*, and *Chlamydophila pneumoniae*, are involved in adverse pregnancy outcomes in humans and/or animals (Baud et al. 2008). These obligate intracellular bacteria have been shown to infect trophoblasts in a mouse model of infection, probably via infected maternal neutrophils (Buendia et al. 1998).

The exact mechanism by which pathogens may cross the placental barrier and infect the fetus remain often elusive. The mechanisms by which some of the microorganisms responsible for maternofetal infections in human may cross this barrier are described below.

Tg is responsible for congenital infection in up to 30% of infected pregnant women, when primary infection is acquired during pregnancy. *Tg* has been shown to infect directly primary human trophoblasts (Abbasi et al. 2003) and extravillous trophoblasts of human ex vivo inoculated placental explants (Robbins et al. 2012), as well as trophoblasts in a rodent animal model (*Calomys callosus*) (Ferro et al. 2002). This invasion may be the critical event leading to the infection of the fetus. Another nonexclusive hypothesis is that *Tg* crosses the placental barrier in a paracellular way through the interaction between the parasite adhesin MIC2 and ICAM-1 (Barragan et al. 2005), ICAM-1 being up-regulated during the infection (Juliano et al. 2006).

Brucella abortus (Ba) is an intracellular bacterium that infects animals and humans, inducing a fetal abortion. *Ba* is thought to be transmitted in cows via the oro-pharyngeal route after exposure to infected placenta released on abortion. It colonizes placenta in its natural hosts (Carvalho Neta et al. 2010) as well as giant trophoblasts in a mouse animal model (Tobias et al. 1993; Kim et al. 2005b). It could be transmitted to the fetus, indicating a crossing of the placental barrier. A surface protein of *Ba* (SP41) is associated with adherence to and invasion of nonphagocytic HeLa cells (Castaneda-Roldan et al. 2006). However, the receptor is not known and trophoblasts may express other receptors recognized by *Ba*. The exact mechanism by which *Ba* may cross the placental barrier is not known.

Coxsackieviruses B (CVB) may be transmitted to the fetus during delivery (Moore 1982) but the rate of transplacental transmission although present (Konstantinidou et al. 2007), is not known. In a study conducted by Satosar et al. (2004) it was shown that CVB was present in 35% of infected placentae obtained from cases with fetal or neonatal death or severe neonatal morbidity, indicating a potential strong role of coxsackieviruses in placental infection. It is not known, however, if the mechanism by which CVB crosses the fetal barrier has similarities with crossing of other host barriers (intestinal barrier and BBB).

Rubella virus, which belongs to the *Togaviridae* family, is associated with congenital rubella syndrome (CRS), a major cause of developmental abnormalities, including blindness and deafness (Banatvala and Brown 2004; Duszak 2009). Despite efficient vaccination programs, this disease still affects some areas in developing countries. Infection by Rubella virus occurs by inhalation, the placenta being subsequently infected, leading to focal necrosis of the epithelium in the chorionic villi and in the endothelial cells of its capillaries (Tondury and Smith 1966). Despite the existence of animal models of vertical transmission (Cotlier et al. 1968; Kono et al. 1969), the pathophysiology of the infection remains mostly elusive. It has been proposed that infected cells in the placenta detach into the lumen of vessels, resulting in infection and damage of fetal organs. As classically observed for vertically transmitted microbes, symptoms associated with CRS are more severe when the infection occurs during the first trimester of gestation but the infection can occur throughout the pregnancy (Grillner et al. 1983; Miller 1991).

The recent reemergence of another member of the *Togaviridae* family that belongs to the *Alphavirus* genus, Chikungunya virus, led to the demonstration that it is associated with severe neonatal infections, including encephalopathy. As mother-to-child transmission is rare in antepartum but frequent in the context of intrapartum maternal viremia, and the placenta is not a site of viral replication, this favors a model of vertical transmission via placental breaches that forms around the term during parturition, as described for a number of other viruses associated with elevated maternal viremia during delivery, such as HIV, HBV, and arboviruses such as Dengue virus or West Nile virus (Couderc et al. 2008; Gerardin et al. 2008).

L. monocytogenes actively crosses the placental barrier and causes severe maternofetal infections in humans (Mylonakis et al. 2002; Lecuit et al. 2007). The ability of *L. monocytogenes* to infect cytotrophoblastic and syncytiotrophoblastic cells has been shown from immunohistopathological studies of placentas from women with listeriosis, in conjunction with the use of human primary trophoblasts and third trimester human placental explants (Lecuit et al. 2004). Importantly, and in contrast to what is observed at the intestinal barrier level, both InlA and InlB interactions with their receptors on syncytiotrophoblasts are necessary for efficient invasion (Lecuit et al. 2004). These findings suggest that *L. monocytogenes* crossing of the human fetoplacental unit follows a transplacental route, after attachment to and invasion of the syncytiotrophoblasts in an InlA- and InlB-dependent manner, respectively. It has been recently proposed in an ex vivo model of human first trimester placental explant that *L. monocytogenes* enter the placenta in an InlA-dependent manner mostly through the extravillous trophoblasts, which anchor the placenta in the decidua, and not through the syncytiotrophoblasts (Robbins et al. 2010). This apparent difference may be because of intrinsic anatomical differences in first and third trimester explants, the former presenting with more anchoring villi and extravillous trophobalsts than the latter. Of note, infection of fetus and abortion because of *L. monocytogenes* occur far more frequently during later stages of pregnancy (~80% of maternofetal listeriosis occur after 22 weeks of amenorrhea; data from the French National Reference Center for *Listeria*). The requirement of InlA in *L. monocytogenes* crossing of the placental barriers has been independently shown epidemiologically, because *L. monocytogenes* isolates expressing a functional InlA are very significantly associated with their fetoplacental origin (Jacquet et al. 2004; Disson et al. 2008). The mechanism of infection has also been studied in animal models in vivo. As previously described, a limitation to the study of *L. monocytogenes* in vivo is that InlA and InlB interactions with their respective receptors are species-specific: InlA interacts with guinea pig E-cadherin, as in human, but not mouse and rat, E-cadherin (Lecuit et al. 1999), whereas InlB interacts with mouse Met, as in human but not guinea pig Met (Khelef et al. 2006). In apparent contradiction with the above epidemiological, ex vivo and in vitro results, InlA plays no role in *L. monocytogenes* fetoplacental infection in the guinea pig, despite its permissiveness to the InlA pathway (Bakardjiev et al. 2004). Similarly, InlB plays no role in *L. monocytogenes* fetoplacental infection in the mouse, despite its permissiveness to the InlB pathway (Le Monnier et al. 2007). To investigate the role of both InlA and InlB in vivo, studies were conducted in two complementary animal models permissive for these two pathways: the gerbil, a natural host for *L. monocytogenes*, and a knockin mouse line that expresses a humanized version of E-cadherin in place of endogenous mouse E-cadherin. Using these two models, it has been established that *L. monocytogenes* specifically targets the placenta in vivo, in an InlA and InlB interdependent manner, and interacts with its receptors E-cadherin and Met, thereby deciphering the molecular mechanism underlying the ability of *L. monocytogenes* to target and cross the placental barrier (Disson et al. 2008).

CENTRAL NERVOUS SYSTEM PATHOGENS AND THE BLOOD–BRAIN BARRIER

Infectious meningitis is a severe and potentially lethal infection of the meninges, the protective membranes covering the central nervous system (CNS), even in otherwise healthy children or adults. Clinical and neuropathologic studies have clearly documented the complications of bacterial meningitis, which are brain edema, increased intracranial pressure, seizure, arterial and venous cerebral vascular insults, and other motor, sensitive, sensorial, and cognitive sequelae (Scheld et al. 2002). Host immune responses are frequently inadequately tuned to control infection within the CNS, either insufficient and unable to control microbial burden, or excessive and associated with excessive inflammation and irreversible tissue damages (Tunkel and Scheld 1995; Pfister and Scheld 1997). In this context, disruption of the BBB is a hallmark event in the pathophysiology of meningitis and subsequent CNS disease, which is critical not only for microbial access to the CNS, but also in the control of the infection.

The BBB maintains the homeostasis of the CNS microenvironment by restricting access of macromolecules, cells, and pathogens. The BBB *sensu stricto* is anatomically represented by the cerebral microvascular endothelium. Epithelial cells of the choroid plexus and meningeal vessels are two additional potential sites of crossing the blood–cerebrospinal fluid barrier. Brain microvascular endothelial cells (BMEC) are held together by tight junctions, display a paucity of pinocytosis, and thus form a barrier (Betz 1985, 1992). Yet, despite its highly restrictive nature, certain pathogens are able to penetrate the BBB and gain entry into the CNS. The exact mechanisms by which these microorganisms pass through the BBB result from a complex interplay between host and microbial factors. Two principal mechanisms, operating individually or in tandem, are an increased permeability of the barrier and the direct invasion of the microorganism. The reasons for increased permeability of the BBB during infections are not fully understood. A dose- and time-dependent increase in permeability was observed in various animal models of bacterial meningitis (Temesvari et al. 1993; Kim et al. 1997; Mayhan 1998). These data suggest a role for bacterial factors, like cell wall components and toxins, in promoting BBB permeability either directly or indirectly through induction of host factors. Several reports have linked increased BBB permeability to a parallel increase in levels of circulating cytokines in the blood and CSF during infection (Lo et al. 1994). Tumor necrosis factor α (TNF-α), IL-1, and IL-6 are major early response cytokines that trigger a cascade of inflammatory mediators, including other cytokines, chemokines, arachidonic acid metabolites, and reactive nitrogen and oxygen intermediates (Dinarello 2000).

Blood–Brain Barrier Composition

The BBB, composed primarily of a specialized layer of BMEC, separates the brain and its surrounding tissues from the circulating blood, tightly regulating the flow of nutrients and molecules and thereby maintaining the proper biochemical conditions for normal brain function (Betz 1985, 1992). In addition, other periendothelial structures, such as pericytes, astrocytes, and a basement membrane, contribute to the BBB function (Abbott et al. 2010). These cells function in a coordinated fashion to regulate the passage of nutrients and other small lipophlic molecules while restricting the access of other macromolecules (Pfister et al. 1992; Rubin and Staddon 1999). The brain endothelial cells are mainly characterized by the presence of tight intercellular junctions that promote high trans-endothelial electrical resistance and therefore impede paracellular flux of macromolecules (Ballabh et al. 2004). Similar to epithelial cells, endothelial cells express adherens junctions (AJ) and tight junctions (TJ) proteins at sites of intercellular contacts (Schulze and Firth 1993; Wolburg and Lippoldt 2002). In BMEC, TJ are composed of four integral membrane proteins: occludin, claudins, junctional adhesion molecules, and cell-selective adhesion molecules, which are linked through cytoplasmic proteins (zonala-occludin-1 (ZO-1), -2, -3, cingulin) to the actin cytoskeleton. AJ are composed of VE-cadherins, which are connected to actin cytoskeleton through catenins (Furuse 2010; Lampugnani et al. 1995; Hirase et al. 1997; Martin-Padura et al. 1998; Rudini and Dejana 2008). Claudin 1, -3, -5, and -12 have been identified at the BBB with claudin-5 being especially critical for BBB formation (Nitta et al. 2003), whereas occludin specifically regulates calcium flux between the blood and brain (Liebner et al. 2008). In addition to maintaining endothelial integrity for normal homeostasis these junctional complexes also play important roles in the maintenance of cell polarity, signaling, and modulation of transcription (Bazzoni and Dejana 2004; Dejana 2004).

Pathogen Invasion of the BBB

Significant progress has been made in understanding the molecular interactions between the BBB and microbial pathogens with the availability of in vitro tissue culture models of human BMEC (hBMEC) (Stins et al. 1994; Weksler et al. 2005) and in vivo animal models (Smith et al. 1973; Doran et al. 2002; Wickham et al. 2007; Uchiyama et al. 2009; Sheen et al. 2010). Microbial interactions with the BBB may involve crossing of BMEC in a vacuole (transcytosis), through the intercellular junctional spaces (paracytosis), or while inside a host cell (e.g., infected phagocyte) using it as a vehicle to cross the barrier (Trojan horse) (Kim 2008, 2010) (Fig. 2).

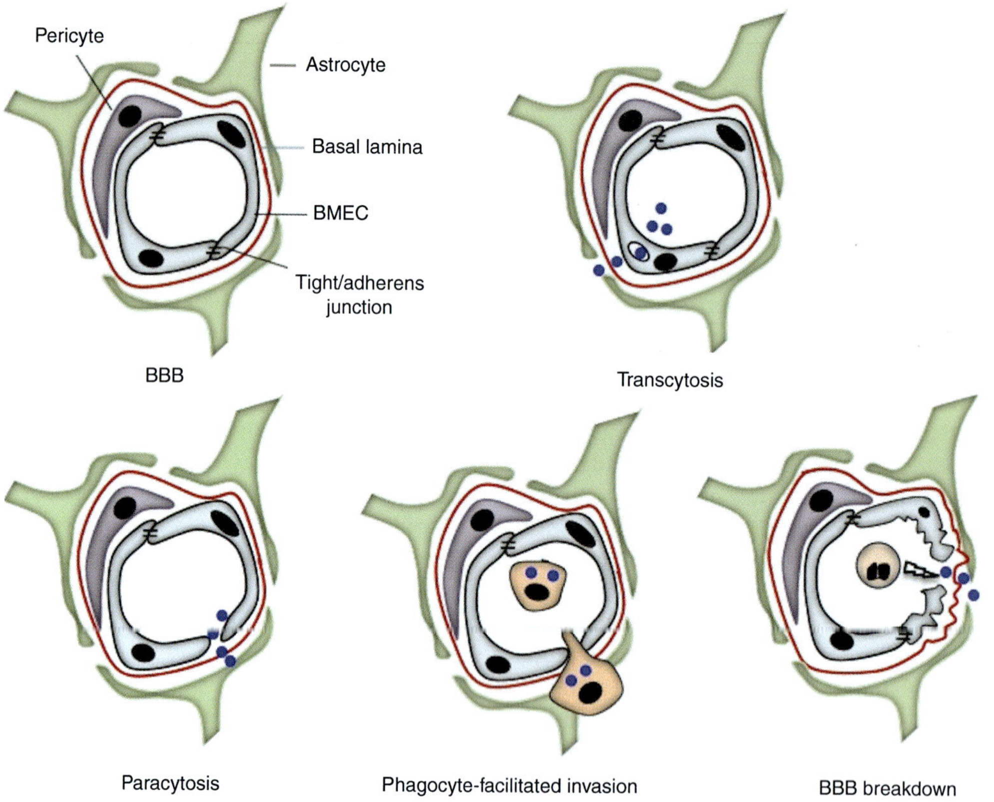

Figure 2. (*Top left*) The blood–brain barrier (BBB) is composed of brain microvascular endothelial cells (BMEC), expressing tight junctions and surrounded by pericytes and astrocytes. From *top* to *bottom* and from *left* to *right*: Pathogens crossing of the BBB may occur in a vacuole (trancytosis), through intercellular junctional spaces (paracytosis), through an infected blood circulating cell (Trojan's Horse mechanism), or following the barrier disruption (BBB breakdown).

Microbial Adhesion to BMEC and Transcytosis across the BBB

Many meningeal pathogens are able to enter or "invade" brain endothelium apically and exit the cell on the basolateral side, thereby crossing the BBB cellularly. The engagement of host cell receptors results in signaling processes that drive massive host actin cytoskeletal rearrangement to promote bacterial uptake into the cell. Electron microscopic (EM) studies have shown the presences of meningeal pathogens in membrane bound vacuoles within BMEC, suggesting the involvement of endocytic pathways as well as avoidance of lysosomal fusion for BBB traversal (Nizet et al. 1997; van Sorge et al. 2008, 2011). Many studies have focused on the identification of bacterial and host components involved in microbial interactions with the BBB (Kim 2008). Critical bacterial components have been discovered by screening random mutant libraries (Badger et al. 2000; Doran et al. 2005), analyzing bacterial transcription profiles during infection (Dietrich et al. 2003; Teng et al. 2005), and using bioinformatic approaches for whole genome comparisons (Uchiyama et al. 2009; van Sorge et al. 2009; Tazi et al. 2010). Bacterial pili, or fimbriae, have emerged as a common class of adhesins used by many meningeal pathogens, such as *E. coli* K1(Teng et al. 2005), GBS (Maisey et al. 2007; van Sorge et al. 2009), and *N. meningitides* (Kirchner and Meyer 2005) to initiate attachment to brain endothelium. Other well-characterized or recently described bacterial components such as CbpA and NanA, a surface anchored sialidase, in *S. pneumonia* (Ring et al. 1998; Uchiyama et al. 2009), lipoteichoic acid and HvgA in GBS (Doran et al. 2005; Tazi et al. 2010), type1 fimbriae, OmpA and ibeA/B in *E. coli* K1 (Prasadarao et al. 1999a,b; Teng et al. 2005), promote brain endothelial cell attachment and subsequent cellular penetration. Separately, bacterial toxins play an important role in microbial interactions with the BBB. In case of *E. coli* K1, cytotoxic necrotizing factor-1 (CNF-1) activates RhoA resulting in increased invasion of BMEC in vitro and increased penetration of the BBB in vivo (Khan et al. 2002, 2003). In contrast, β hemolysin (β-h/c) of GBS (Doran et al. 2003; Lembo et al. 2010), pneumolysin of *S. pneumonia* (Zysk et al. 2001) and HiB lipopolysaccharide (Patrick et al. 1992) all cause apoptosis in brain endothelial cells, which lead to increased BBB permeability and bacterial penetration in vivo.

A broad range of host cell surface receptors have been reported to mediate interaction with pathogens that lead to subsequent cellular penetration in the CNS. The laminin receptor (LR) and platelet activating factor receptor (PAFr) have been identified as common portals of CNS entry for the leading meningeal pathogens such as *E. coli* K1 (Chung et al. 2003; Kim et al. 2005a), *Neisseria meningitidis* (Virji et al. 1993), *Streptococcus pneumoniae* (Ring et al. 1998; Orihuela et al. 2009), and *Haemophilus influenzae* (Swords et al. 2001). LR was identified to function as a receptor for CNF-1 of *E. coli* K1 (Chung et al. 2003; Kim et al. 2005a), PilQ and PorA of *N. meningitidis,* CbpA of *S. pneumoniae,* and OmpP2 of *H. influenzae* (Orihuela et al. 2009). The PAFr binds to phosphorylcholine on the bacterial surface, which results in β-arrestin-mediated invasion of brain endothelial cells (Radin et al. 2005). Inflammatory activation of host cells shifts the targeting of *S. pneumoniae* to this G-protein-coupled receptor (Cundell et al. 1995). This results in promoting bacterial transcytosis across the cell, whereas non-PAFr-mediated entry shunts bacteria for exit and reentry on the apical surface in a novel recycling pathway (Ring et al. 1998). Integrins are also emerging as common receptors for many meningeal pathogens. Bacteria may engage components of an extracellular matrix to bridge to integrins and initiate integrin signaling pathways. For example, human collagen bridges GBS pili adhesin, PilA, and $\alpha_2\beta_1$ integrin on BMEC, resulting in bacterial attachment, immune activation, and ultimately penetration of the CNS (Banerjee et al. 2011). Similarly, fibronectin bridges *N. meningitidis* to $\alpha_5\beta_1$ integrin on BMEC surface enabling bacterial internalization (Unkmeir et al. 2002). Glycan structures, specially glucosaminoglycans (GAG), that interact with integrin signaling machinery, have been shown to interact with bacterial proteins such as the GBS α C protein (Chang et al. 2011). Given that many of these host cell factors preferentially localize to the basolateral surface of polarized endothelium, disruption of junctional protein complexes on infection may result in a nonpolarized distribution of proteins on the BBB plasma membrane and their luminal accessibility (see below).

Paracytosis

Several meningeal pathogens are capable of either cleaving or remodeling these intercellular junctional proteins and therefore affect BBB integrity. Disruption of these cellular contacts leads to increased BBB permeability and also facilitates bacterial translocation into the brain. Bacterial toxins such as pneumolysin and β-hemolysin, of *S. pneumoniae* and GBS, respectively, are implicated in directly affecting the BBB integrity by causing endothelial cell death (Zysk et al. 2001; Lembo et al. 2010). For GBS, higher toxin production has been associated with increased capacity to cause meningitis (Doran et al. 2003). In addition, toxin production results in increased production of proinflammatory cytokines/chemokines, especially TNF-α, which negatively impact the BBB function (Sharief et al. 1992; Kim et al. 1997; Barichello et al. 2011).

Pathogens may also bind to endothelial cells at or near the TJs (Grab et al. 2004, 2005) and induce the secretion of host proteases that can cleave TJ or AJ proteins. For example, *N. meningitidis* induces specific cleavage of the TJ component occludin through the release of host matrix metalloproteinase 8, resulting in endothelial cell detachment and increased paracellular permeability (Schubert-Unkmeir et al. 2010). A similar strategy is used by *Borrelia*, which was shown to degrade TJ proteins and allow passage of monocytes (Reijerkerk et al. 2006). Some pathogens may adopt a more sophisticated strategy of remodeling the junctional complexes. *E. coli* K1 OmpA initiates a signaling process that leads to the dissociation of β-catenins from cadherins which further results in disruption of barrier integrity (Sukumaran and Prasadarao 2003). The type-IV pili of *N. meningitides*, on the other hand, stimulates the β2-adrenoceptor/β-arrestin signaling pathway in endothelial cells, leading to recruitment of the Par3/Par6/PKCζ polarity complex and delocalization of junctional proteins which result in anatomical gaps that are used by the bacteria to penetrate into the CNS (Coureuil et al. 2009, 2010). Human immunodeficiency virus (HIV) envelope glycoprotein gp120 has also been shown to increase BBB permeability by interacting with endothelial *trans*-membrane coreceptor CCR5 and CXCR4 (Kanmogne et al. 2007).

Phagocyte-Facilitated Invasion

The use of phagocytic cells for BBB invasion (Trojan horse mechanism) appears to be mainly used by intracellular pathogens that are able to survive and eventually multiply in professional phagocytes. According to this scenario, a pathogen first infects a bone-marrow-derived cell, such as a macrophage, monocyte, or dendritic cell, which may then penetrate the BBB, acting as a shuttle for the microbe. For example, CD11b expressing cells are infected and used by *T. gondii* to breach the BBB (Lachenmaier et al. 2011). Similarly, monocytes or macrophages may be used by *C. neoformans*, *L. monocytogenes*, *M. tuberculosis*, and *Brucella* and *Salmonella* spp for entry into the CNS (Greiffenberg et al. 1998; Drevets et al. 2004; Join-Lambert et al. 2005; Charlier et al. 2009). West Nile virus infection of BMEC modulates TJ proteins in endothelial cells and facilitates migration of virus-infected cells through that space into CNS (Verma et al. 2009). HIV infection of monocytes increases surface expression of LFA-1, an ICAM1 ligand that increases the likelihood of adhesion to CNS endothelial cells (Stent and Crowe 1997). After penetration, these infected monocytes and macrophages can spread the pathogen to microglia, perivascular macrophages, and astrocytes, creating a reservoir for the organism in the CNS. These parenchymal pathogen reservoirs primarily result in microglial activation, leukocyte infiltration into the CNS, and neuronal damage.

Inflammation and BBB Permeability

Finally, host inflammatory factors may also contribute to loss of BBB integrity, resulting in pathogen penetration into the CNS. It was recently shown that polymorphonuclear neutrophils contribute to BBB impermeability loss during the GBS meningitis (Banerjee et al. 2011). Although neutrophils are critically important to fight against bacterial sepsis, depletion of peripheral neutrophils in a murine hematogenous model of GBS meningitis prolonged survival, decreased bacterial CNS load, and decreased BBB leakage in GBS-infected mice (Banerjee et al. 2011). Correspondingly, prevention of leukocyte infiltration into the CNS using anti-CD18 antibody has been shown to improve pneumococcal and HiB meningitis outcomes (Tuomanen et al. 1989; Saez-Llorens et al. 1991).

Different bacterial components initiate inflammatory activation of brain endothelial cells, resulting in secretion of cytokines and chemokines such as TNF-α, IL-1, IL-8, or TGF-β, the levels of which are typically elevated in meningitis patients (Waage et al. 1989; Fida et al. 2006; Nagesh Babu et al. 2008). These cytokines may have an autocrine effect on BMEC, leading to increased expression of host receptors that further promote bacterial penetration. For example, cells stimulated with TNF-α, IL-8, or TGF-β resulted in an increased bacterial uptake in vitro (Cundell et al. 1995; Zhang et al. 2002; Banerjee et al. 2010). OmpA of *E. coli* K1 binds to gp96 glycoprotein on BMEC surface and this interaction in turn increases the expression of

 Cite this article as *Cold Spring Harb Perspect Med* doi: 10.1101/cshperspect.a010090

type-1 fimbriae that bind to CD48 (Prasadarao et al. 2003; Khan et al. 2007). Pathogen interaction with BMEC also results in activation of several signal transduction pathways that either leads to actin cytoskeleton rearrangement and subsequent bacterial uptake or proinflammatory chemokine/cytokine secretion and localized inflammation. In the majority of instances actin cytoskeleton rearrangement signaling events involve activation of focal adhesion kinase (FAK) (Prasadarao et al. 1999b; Banerjee et al. 2011), phosphatidylinositol 3-kinase (PI3K), ρ GTPases (Khan et al. 2003) and cytosolic phospholipase A2 (Das et al. 2001) ($cPLA_2$), and src kinases (Slanina et al. 2010). Additionally, activation of mitogen-activated protein (MAP) kinase signaling pathways promotes proinflammatory cytokine secretion (Sokolova et al. 2004; Banerjee et al. 2010, 2011).

Transcriptional analysis of the BMEC on GBS infection revealed a specific gene-activation program acting to orchestrate neutrophil recruitment (IL-8, CXCL1, CXCL2), activation (IL-6, IL-8), survival (GM-CSF), and extravasation (ICAM-1). This response was dependent on the GBS β-h/c toxin, the transcriptional regulator of virulence, CovR and PilA expression (Doran et al. 2003; Lembo et al. 2010; Banerjee et al. 2011). Additional in vitro studies with *S. pneumoniae*, HiB, *S.* typhimurium, *E. coli* K1, *Streptococcus suis*, *L. monocytogenes*, and *N. meningitidis* reveal a similar core BMEC transcriptional response that results in the induction of neutrophil signaling pathways (Wilson and Drevets 1998; Vadeboncoeur et al. 2003; Galanakis et al. 2006; Schubert-Unkmeir et al. 2007; Banerjee et al. 2010; van Sorge et al. 2011). Continued exposure and invasion of the pathogen may result in overactivation of BBB endothelium and lead to increased inflammation that may ultimately compromise BBB integrity.

IN SUMMARY

Microbial interactions with host barriers are key events in invasive infectious diseases. The deciphering of the molecular mechanisms underlying microbial attachment to, invasion of, translocation across, and disruption of host barriers is not only important in a biomedical perspective but also on a basic perspective. Future studies will hopefully help improve the severe prognosis of invasive infection, and also contribute to a better understanding of the physiology of host barriers.

ACKNOWLEDGMENTS

The authors thank their respective laboratory members for their help and support and apologize to researchers whose work has not been discussed in detail or reviewed here because of space constraints. Work on the BBB and bacterial meningitis in K.S.D.'s laboratory is supported by funding from the National Institutes of Health/National Institute of Neurological Disorders and Stroke (grant No. RO1NS051247). Work in M.L's laboratory is supported by Institut Pasteur, INSERM, ERC, FRM, Ville de Paris, Labex IBEID, Listress, and ICRES EU Programs and Fondation BNP-Paribas. The authors declare no competing financial interests.

REFERENCES

Abbasi M, Kowalewska-Grochowska K, Bahar MA, Kilani RT, Winkler-Lowen B, Guilbert LJ. 2003. Infection of placental trophoblasts by *Toxoplasma gondii*. *J Infect Dis* **188:** 608–616.

Abbott NJ, Patabendige AA, Dolman DE, Yusof SR, Begley DJ. 2010. Structure and function of the blood–brain barrier. *Neurobiol Dis* **37:** 13–25.

Amieva MR, Vogelmann R, Covacci A, Tompkins LS, Nelson WJ, Falkow S. 2003. Disruption of the epithelial apical-junctional complex by *Helicobacter pylori* CagA. *Science* **300:** 1430–1434.

Anderson JM, Van Itallie CM, Fanning AS. 2004. Setting up a selective barrier at the apical junction complex. *Curr Opin Cell Biol* **16:** 140–145.

Badger JL, Wass CA, Kim KS. 2000. Identification of *Escherichia coli* K1 genes contributing to human brain microvascular endothelial cell invasion by differential fluorescence induction. *Mol Microbiol* **36:** 174–182.

Bakardjiev AI, Stacy BA, Fisher SJ, Portnoy DA. 2004. Listeriosis in the pregnant guinea pig: A model of vertical transmission. *Infect Immun* **72:** 489–497.

Ballabh P, Braun A, Nedergaard M. 2004. The blood–brain barrier: an overview: Structure, regulation, and clinical implications. *Neurobiol Dis* **16:** 1–13.

Banatvala JE, Brown DW. 2004. Rubella. *Lancet* **363:** 1127–1137.

Banerjee A, Van Sorge NM, Sheen TR, Uchiyama S, Mitchell TJ, Doran KS. 2010. Activation of brain endothelium by pneumococcal neuraminidase NanA promotes bacterial internalization. *Cell Microbiol* **12:** 1576–1588.

Banerjee A, Kim BJ, Carmona EM, Cutting AS, Gurney MA, Carlos C, Feuer R, Prasadarao NV, Doran KS. 2011. Bacterial Pili exploit integrin machinery to promote immune activation and efficient blood–brain barrier penetration. *Nat Commun* **2:** 462.

Barichello T, Pereira JS, Savi GD, Generoso JS, Cipriano AL, Silvestre C, Petronilho F, Dal-Pizzol F, Vilela MC, Teixeira AL. 2011. A kinetic study of the cytokine/chemokines levels and disruption of blood–brain barrier in infant rats after pneumococcal meningitis. *J Neuroimmunol* **233:** 12–17.

Barragan A, Sibley LD. 2002. Transepithelial migration of *Toxoplasma gondii* is linked to parasite motility and virulence. *J Exp Med* **195:** 1625–1633.

Barragan A, Sibley LD. 2003. Migration of *Toxoplasma gondii* across biological barriers. *Trends Microbiol* **11:** 426–430.

Barragan A, Brossier F, Sibley LD. 2005. Transepithelial migration of *Toxoplasma gondii* involves an interaction of intercellular adhesion molecule 1 (ICAM-1) with the parasite adhesin MIC2. *Cell Microbiol* **7:** 561–568.

Barton ES, Forrest JC, Connolly JL, Chappell JD, Liu Y, Schnell FJ, Nusrat A, Parkos CA, Dermody TS. 2001. Junction adhesion molecule is a receptor for reovirus. *Cell* **104:** 441–451.

Baud D, Regan L, Greub G. 2008. Emerging role of Chlamydia and Chlamydia-like organisms in adverse pregnancy outcomes. *Curr Opin Infect Dis* **21:** 70–76.

Baumgartner W, Bachmann S. 1992. Histological and immunocytochemical characterization of *Coxiella burnetii*-associated lesions in the murine uterus and placenta. *Infect Immun* **60:** 5232–5241.

Baumler AJ, Tsolis RM, Heffron F. 1996. The lpf fimbrial operon mediates adhesion of *Salmonella typhimurium* to murine Peyer's patches. *Proc Natl Acad Sci* **93:** 279–283.

Bazzoni G, Dejana E. 2004. Endothelial cell-to-cell junctions: Molecular organization and role in vascular homeostasis. *Physiol Rev* **84:** 869–901.

Beau I, Cotte-Laffitte J, Amsellem R, Servin AL. 2007. A protein kinase A-dependent mechanism by which rotavirus affects the distribution and mRNA level of the functional tight junction-associated protein, occludin, in human differentiated intestinal Caco-2 cells. *J Virol* **81:** 8579–8586.

Ben Amara A, Ghigo E, Le Priol Y, Lepolard C, Salcedo SP, Lemichez E, Bretelle F, Capo C, Mege JL. 2010. *Coxiella burnetii*, the agent of Q fever, replicates within trophoblasts and induces a unique transcriptional response. *PLoS ONE* **5:** e15315.

Bergelson JM, Cunningham JA, Droguett G, Kurt-Jones EA, Krithivas A, Hong JS, Horwitz MS, Crowell RL, Finberg RW. 1997. Isolation of a common receptor for Coxsackie B viruses and adenoviruses 2 and 5. *Science* **275:** 1320–1323.

Betz AL. 1985. Epithelial properties of brain capillary endothelium. *Fed Proc* **44:** 2614–2615.

Betz AL. 1992. An overview of the multiple functions of the blood–brain barrier. *NIDA Res Monogr* **120:** 54–72.

Bozym RA, Patel K, White C, Cheung KH, Bergelson JM, Morosky SA, Coyne CB. 2011. Calcium signals and calpain-dependent necrosis are essential for release of coxsackievirus B from polarized intestinal epithelial cells. *Mol Biol Cell* **22:** 3010–3021.

Brabin BJ, Romagosa C, Abdelgalil S, Menendez C, Verhoeff FH, McGready R, Fletcher KA, Owens S, D'Alessandro U, Nosten F, et al. 2004. The sick placenta—the role of malaria. *Placenta* **25:** 359–378.

Buendia AJ, Sanchez J, Martinez MC, Camara P, Navarro JA, Rodolakis A, Salinas J. 1998. Kinetics of infection and effects on placental cell populations in a murine model of *Chlamydia psittaci*-induced abortion. *Infect Immun* **66:** 2128–2134.

Campbell JA, Schelling P, Wetzel JD, Johnson EM, Forrest JC, Wilson GA, Aurrand-Lions M, Imhof BA, Stehle T, Dermody TS. 2005. Junctional adhesion molecule a serves as a receptor for prototype and field-isolate strains of mammalian reovirus. *J Virol* **79:** 7967–7978.

Carcopino X, Raoult D, Bretelle F, Boubli L, Stein A. 2009. Q Fever during pregnancy: A cause of poor fetal and maternal outcome. *Ann NY Acad Sci* **1166:** 79–89.

Carvalho Neta AV, Mol JP, Xavier MN, Paixao TA, Lage AP, Santos RL. 2010. Pathogenesis of bovine brucellosis. *Vet J* **184:** 146–155.

Castaneda-Roldan EI, Ouahrani-Bettache S, Saldana Z, Avelino F, Rendon MA, Dornand J, Giron JA. 2006. Characterization of SP41, a surface protein of *Brucella* associated with adherence and invasion of host epithelial cells. *Cell Microbiol* **8:** 1877–1887.

Chang YC, Wang Z, Flax LA, Xu D, Esko JD, Nizet V, Baron MJ. 2011. Glycosaminoglycan binding facilitates entry of a bacterial pathogen into central nervous systems. *PLoS Pathog* **7:** e1002082.

Charlier C, Nielsen K, Daou S, Brigitte M, Chretien F, Dromer F. 2009. Evidence of a role for monocytes in dissemination and brain invasion by Cryptococcus neoformans. *Infect Immun* **77:** 120–127.

Chieppa M, Rescigno M, Huang AY, Germain RN. 2006. Dynamic imaging of dendritic cell extension into the small bowel lumen in response to epithelial cell TLR engagement. *J Exp Med* **203:** 2841–2852.

Chung JW, Hong SJ, Kim KJ, Goti D, Stins MF, Shin S, Dawson VL, Dawson TM, Kim KS. 2003. 37-kDa laminin receptor precursor modulates cytotoxic necrotizing factor 1-mediated RhoA activation and bacterial uptake. *J Biol Chem* **278:** 16857–16862.

Clark MA, Hirst BH, Jepson MA. 1998. Inoculum composition and *Salmonella* pathogenicity island 1 regulate M-cell invasion and epithelial destruction by *Salmonella typhimurium*. *Infect Immun* **66:** 724–731.

Corr SC, Gahan CC, Hill C. 2008. M-cells: Origin, morphology and role in mucosal immunity and microbial pathogenesis. *FEMS Immunol Med Microbiol* **52:** 2–12.

Cotlier E, Fox J, Bohigian G, Beaty C, Du Pree A. 1968. Pathogenic effects of rubella virus on embryos and newborn rats. *Nature* **217:** 38–40.

Couderc T, Chretien F, Schilte C, Disson O, Brigitte M, Guivel-Benhassine F, Touret Y, Barau G, Cayet N, Schuffenecker I, et al. 2008. A mouse model for Chikungunya: Young age and inefficient type-I interferon signaling are risk factors for severe disease. *PLoS Pathog* **4:** e29.

Coureuil M, Mikaty G, Miller F, Lecuyer H, Bernard C, Bourdoulous S, Dumenil G, Mege RM, Weksler BB, Romero IA, et al. 2009. Meningococcal type IV pili recruit the polarity complex to cross the brain endothelium. *Science* **325:** 83–87.

Coureuil M, Lecuyer H, Scott MG, Boularan C, Enslen H, Soyer M, Mikaty G, Bourdoulous S, Nassif X, Marullo S. 2010. Meningococcus hijacks a β2-adrenoceptor/β-Arrestin pathway to cross brain microvasculature endothelium. *Cell* **143:** 1149–1160.

Coyne CB, Bergelson JM. 2006. Virus-induced Abl and Fyn kinase signals permit coxsackievirus entry through epithelial tight junctions. *Cell* **124:** 119–131.

Coyne CB, Shen L, Turner JR, Bergelson JM. 2007. Coxsackievirus entry across epithelial tight junctions requires occludin and the small GTPases Rab34 and Rab5. *Cell Host Microbe* **2:** 181–192.

Cross JC, Werb Z, Fisher SJ. 1994. Implantation and the placenta: Key pieces of the development puzzle. *Science* **266:** 1508–1518.

Cundell DR, Gerard NP, Gerard C, Idanpaan-Heikkila I, Tuomanen EI. 1995. *Streptococcus pneumoniae* anchor to activated human cells by the receptor for platelet-activating factor. *Nature* **377:** 435–438.

Das A, Asatryan L, Reddy MA, Wass CA, Stins MF, Joshi S, Bonventre JV, Kim KS. 2001. Differential role of cytosolic phospholipase A2 in the invasion of brain microvascular endothelial cells by *Escherichia coli* and *Listeria monocytogenes*. *J Infect Dis* **184:** 732–737.

Dean P, Kenny B. 2004. Intestinal barrier dysfunction by enteropathogenic *Escherichia coli* is mediated by two effector molecules and a bacterial surface protein. *Mol Microbiol* **54:** 665–675.

Dejana E. 2004. Endothelial cell–cell junctions: Happy together. *Nat Rev Mol Cell Biol* **5:** 261–270.

Dietrich G, Kurz S, Hubner C, Aepinus C, Theiss S, Guckenberger M, Panzner U, Weber J, Frosch M. 2003. Transcriptome analysis of *Neisseria meningitidis* during infection. *J Bacteriol* **185:** 155–164.

Dinarello CA. 2000. Proinflammatory cytokines. *Chest* **118:** 503–508.

Disson O, Grayo S, Huillet E, Nikitas G, Langa-Vives F, Dussurget O, Ragon M, Le Monnier A, Babinet C, Cossart P, et al. 2008. Conjugated action of two species-specific invasion proteins for fetoplacental listeriosis. *Nature* **455:** 1114–1118.

Doran KS, Chang JC, Benoit VM, Eckmann L, Nizet V. 2002. Group B streptococcal β-hemolysin/cytolysin promotes invasion of human lung epithelial cells and the release of interleukin-8. *J Infect Dis* **185:** 196–203.

Doran KS, Liu GY, Nizet V. 2003. Group B streptococcal β-hemolysin/cytolysin activates neutrophil signaling pathways in brain endothelium and contributes to development of meningitis. *J Clin Invest* **112:** 736–744.

Doran KS, Engelson EJ, Khosravi A, Maisey HC, Fedtke I, Equils O, Michelsen KS, Arditi M, Peschel A, Nizet V. 2005. Blood–brain barrier invasion by group B *Streptococcus* depends upon proper cell-surface anchoring of lipoteichoic acid. *J Clin Invest* **115:** 2499–2507.

Drevets DA, Dillon MJ, Schawang JS, Van Rooijen N, Ehrchen J, Sunderkotter C, Leenen PJ. 2004. The Ly-6Chigh monocyte subpopulation transports *Listeria monocytogenes* into the brain during systemic infection of mice. *J Immunol* **172:** 4418–4424.

Duszak RS. 2009. Congenital rubella syndrome—Major review. *Optometry* **80:** 36–43.

Ferro EA, Silva DA, Bevilacqua E, Mineo JR. 2002. Effect of *Toxoplasma gondii* infection kinetics on trophoblast cell population in Calomys callosus, a model of congenital toxoplasmosis. *Infect Immun* **70:** 7089–7094.

Fida NM, Al-Mughales J, Farouq M. 2006. Interleukin-1α, interleukin-6 and tumor necrosis factor-α levels in children with sepsis and meningitis. *Pediatr Int* **48:** 118–124.

Forrest JC, Campbell JA, Schelling P, Stehle T, Dermody TS. 2003. Structure-function analysis of reovirus binding to junctional adhesion molecule 1. Implications for the mechanism of reovirus attachment. *J Biol Chem* **278:** 48434–48444.

Freimuth P, Philipson L, Carson SD. 2008. The coxsackievirus and adenovirus receptor. *Curr Top Microbiol Immunol* **323:** 67–87.

Furuse M. 2010. Molecular basis of the core structure of tight junctions. *Cold Spring Harb Perspect Biol* **2:** a002907.

Galanakis E, Di Cello F, Paul-Satyaseela M, Kim KS. 2006. *Escherichia coli* K1 induces IL-8 expression in human brain microvascular endothelial cells. *Eur Cytokine Netw* **17:** 260–265.

Gerardin P, Barau G, Michault A, Bintner M, Randrianaivo H, Choker G, Lenglet Y, Touret Y, Bouveret A, Grivard P, et al. 2008. Multidisciplinary prospective study of mother-to-child chikungunya virus infections on the island of La Reunion. *PLoS Med* **5:** e60.

Grab DJ, Nikolskaia O, Kim YV, Lonsdale-Eccles JD, Ito S, Hara T, Fukuma T, Nyarko E, Kim KJ, Stins MF, et al. 2004. African trypanosome interactions with an in vitro model of the human blood–brain barrier. *J Parasitol* **90:** 970–979.

Grab DJ, Perides G, Dumler JS, Kim KJ, Park J, Kim YV, Nikolskaia O, Choi KS, Stins MF, Kim KS. 2005. Borrelia burgdorferi, host-derived proteases, and the blood–brain barrier. *Infect Immun* **73:** 1014–1022.

Grassl GA, Finlay BB. 2008. Pathogenesis of enteric *Salmonella* infections. *Curr Opin Gastroenterol* **24:** 22–26.

Greiffenberg L, Goebel W, Kim KS, Weiglein I, Bubert A, Engelbrecht F, Stins M, Kuhn M. 1998. Interaction of *Listeria monocytogenes* with human brain microvascular endothelial cells: InlB-dependent invasion, long-term intracellular growth, and spread from macrophages to endothelial cells. *Infect Immun* **66:** 5260–5267.

Grillner L, Forsgren M, Barr B, Bottiger M, Danielsson L, De Verdier C. 1983. Outcome of rubella during pregnancy with special reference to the 17th–24th weeks of gestation. *Scand J Infect Dis* **15:** 321–325.

Guglielmi KM, Kirchner E, Holm GH, Stehle T, Dermody TS. 2007. Reovirus binding determinants in junctional adhesion molecule-A. *J Biol Chem* **282:** 17930–17940.

Guttman JA, Finlay BB. 2009. Tight junctions as targets of infectious agents. *Biochim Biophys Acta* **1788:** 832–841.

Hapfelmeier S, Muller AJ, Stecher B, Kaiser P, Barthel M, Endt K, Eberhard M, Robbiani R, Jacobi CA, Heikenwalder M, et al. 2008. Microbe sampling by mucosal dendritic cells is a discrete, MyD88-independent step in Δ*invG S.* Typhimurium colitis. *J Exp Med* **205:** 437–450.

Harris TJ, Tepass U. 2010. Adherens junctions: From molecules to morphogenesis. *Nat Rev Mol Cell Biol* **11:** 502–514.

Hase K, Kawano K, Nochi T, Pontes GS, Fukuda S, Ebisawa M, Kadokura K, Tobe T, Fujimura Y, Kawano S, et al. 2009. Uptake through glycoprotein 2 of FimH$^+$ bacteria by M cells initiates mucosal immune response. *Nature* **462:** 226–230.

Hirase T, Staddon JM, Saitou M, Ando-Akatsuka Y, Itoh M, Furuse M, Fujimoto K, Tsukita S, Rubin LL. 1997. Occludin as a possible determinant of tight junction permeability in endothelial cells. *J Cell Sci* **110:** 1603–1613.

Huppertz B. 2008. The anatomy of the normal placenta. *J Clin Pathol* **61:** 1296–1302.

Jacquet C, Doumith M, Gordon JI, Martin PM, Cossart P, Lecuit M. 2004. A molecular marker for evaluating the pathogenic potential of foodborne *Listeria monocytogenes. J Infect Dis* **189:** 2094–2100.

Join-Lambert OF, Ezine S, Le Monnier A, Jaubert F, Okabe M, Berche P, Kayal S. 2005. *Listeria monocytogenes*-infected bone marrow myeloid cells promote bacterial invasion of the central nervous system. *Cell Microbiol* **7:** 167–180.

Jones BD, Ghori N, Falkow S. 1994. *Salmonella typhimurium* initiates murine infection by penetrating and destroying the specialized epithelial M cells of the Peyer's patches. *J Exp Med* **180:** 15–23.

Juliano PB, Blotta MH, Altemani AM. 2006. ICAM-1 is overexpressed by villous trophoblasts in placentitis. *Placenta* **27:** 750–757.

Kanmogne GD, Schall K, Leibhart J, Knipe B, Gendelman HE, Persidsky Y. 2007. HIV-1 gp120 compromises blood–brain barrier integrity and enhances monocyte migration across blood–brain barrier: Implication for viral neuropathogenesis. *J Cereb Blood Flow Metab* **27:** 123–134.

Khan NA, Wang Y, Kim KJ, Chung JW, Wass CA, Kim KS. 2002. Cytotoxic necrotizing factor-1 contributes to *Escherichia coli* K1 invasion of the central nervous system. *J Biol Chem* **277:** 15607–15612.

Khan NA, Shin S, Chung JW, Kim KJ, Elliott S, Wang Y, Kim KS. 2003. Outer membrane protein A and cytotoxic necrotizing factor-1 use diverse signaling mechanisms for *Escherichia coli* K1 invasion of human brain microvascular endothelial cells. *Microb Pathog* **35:** 35–42.

Khan NA, Kim Y, Shin S, Kim KS. 2007. FimH-mediated *Escherichia coli* K1 invasion of human brain microvascular endothelial cells. *Cell Microbiol* **9:** 169–178.

Khelef N, Lecuit M, Bierne H, Cossart P. 2006. Species specificity of the *Listeria monocytogenes* InlB protein. *Cell Microbiol* **8:** 457–470.

Kim KS. 2008. Mechanisms of microbial traversal of the blood–brain barrier. *Nat Rev Microbiol.*

Kim KS. 2010. Acute bacterial meningitis in infants and children. *Lancet Infect Dis* **10:** 32–42.

Kim KS, Wass CA, Cross AS. 1997. Blood–brain barrier permeability during the development of experimental bacterial meningitis in the rat. *Exp Neurol* **145:** 253–257.

Kim KJ, Chung JW, Kim KS. 2005a. 67-kDa laminin receptor promotes internalization of cytotoxic necrotizing factor 1-expressing *Escherichia coli* K1 into human brain microvascular endothelial cells. *J Biol Chem* **280:** 1360–1368.

Kim S, Lee DS, Watanabe K, Furuoka H, Suzuki H, Watarai M. 2005b. Interferon-γ promotes abortion due to *Brucella* infection in pregnant mice. *BMC Microbiol* **5:** 22.

Kirchner M, Meyer TF. 2005. The PilC adhesin of the *Neisseria* type IV pilus-binding specificities and new insights into the nature of the host cell receptor. *Mol Microbiol* **56:** 945–957.

Kono R, Hayakawa Y, Hibi M, Ishii K. 1969. Experimental vertical transmission of rubella virus in rabbits. *Lancet* **1:** 343–347.

Konstantinidou A, Anninos H, Spanakis N, Kotsiakis X, Syridou G, Tsakris A, Patsouris E. 2007. Transplacental infection of coxsackievirus B3 pathological findings in the fetus. *J Med Virol* **79:** 754–757.

Kuehne SA, Cartman ST, Heap JT, Kelly ML, Cockayne A, Minton NP. 2010. The role of toxin A and toxin B in *Clostridium difficile* infection. *Nature* **467:** 711–713.

Lachenmaier SM, Deli MA, Meissner M, Liesenfeld O. 2011. Intracellular transport of *Toxoplasma gondii* through the blood–brain barrier. *J Neuroimmunol* **232:** 119–130.

Lampugnani MG, Corada M, Caveda L, Breviario F, Ayalon O, Geiger B, Dejana E. 1995. The molecular organization of endothelial cell to cell junctions: Differential association of plakoglobin, β-catenin, and α-catenin with vascular endothelial cadherin (VE-cadherin). *J Cell Biol* **129:** 203–217.

Lecuit M, Dramsi S, Gottardi C, Fedor-Chaiken M, Gumbiner B, Cossart P. 1999. A single amino acid in E-cadherin responsible for host specificity towards the human pathogen *Listeria monocytogenes. EMBO J* **18:** 3956–3963.

Lecuit M, Vandormael-Pournin S, Lefort J, Huerre M, Gounon P, Dupuy C, Babinet C, Cossart P. 2001. A transgenic model for listeriosis: Role of internalin in crossing the intestinal barrier. *Science* **292:** 1722–1725.

Lecuit M, Nelson DM, Smith SD, Khun H, Huerre M, Vacher-Lavenu MC, Gordon JI, Cossart P. 2004. Targeting and crossing of the human maternofetal barrier by *Listeria monocytogenes*: Role of internalin interaction with trophoblast E-cadherin. *Proc Natl Acad Sci* **101:** 6152–6157.

Lecuit M, Sonnenburg JL, Cossart P, Gordon JI. 2007. Functional genomic studies of the intestinal response to a foodborne enteropathogen in a humanized gnotobiotic mouse model. *J Biol Chem* **282:** 15065–15072.

Lembo A, Gurney MA, Burnside K, Banerjee A, bfde los Reyes M, Connelly JE, Lin WJ, Jewell KA, Vo A, Renken CW, et al. 2010. Regulation of CovR expression in Group B *Streptococcus* impacts blood–brain barrier penetration. *Mol Microbiol* **77:** 431–443.

Le Monnier A, Autret N, Join-Lambert OF, Jaubert F, Charbit A, Berche P, Kayal S. 2007. ActA is required for crossing of the fetoplacental barrier by *Listeria monocytogenes. Infect Immun* **75:** 950–957.

Liebner S, Corada M, Bangsow T, Babbage J, Taddei A, Czupalla CJ, Reis M, Felici A, Wolburg H, Fruttiger M, et al. 2008. Wnt β-catenin signaling controls development of the blood–brain barrier. *J Cell Biol* **183:** 409–417.

Lo WD, Wolny A, Boesel C. 1994. Blood–brain barrier permeability in staphylococcal cerebritis and early brain abscess. *J Neurosurg* **80:** 897–905.

Maisey HC, Hensler M, Nizet V, Doran KS. 2007. Group B streptococcal pilus proteins contribute to adherence to and invasion of brain microvascular endothelial cells. *J Bacteriol* **189:** 1464–1467.

Maltepe E, Bakardjiev AI, Fisher SJ. 2010. The placenta: Transcriptional, epigenetic, and physiological integration during development. *J Clin Invest* **120:** 1016–1025.

Martin-Padura I, Lostaglio S, Schneemann M, Williams L, Romano M, Fruscella P, Panzeri C, Stoppacciaro A, Ruco L, Villa A, et al. 1998. Junctional adhesion molecule, a novel member of the immunoglobulin superfamily that distributes at intercellular junctions and modulates monocyte transmigration. *J Cell Biol* **142:** 117–127.

Matsuzawa T, Kuwae A, Abe A. 2005. Enteropathogenic *Escherichia coli* type III effectors EspG and EspG2 alter epithelial paracellular permeability. *Infect Immun* **73:** 6283–6289.

Mayhan WG. 1998. Effect of lipopolysaccharide on the permeability and reactivity of the cerebral microcirculation: Role of inducible nitric oxide synthase. *Brain Res* **792:** 353–357.

Miller E. 1991. Rubella in the United Kingdom. *Epidemiol Infect* **107:** 31–42.

Monack DM, Mueller A, Falkow S. 2004. Persistent bacterial infections: The interface of the pathogen and the host immune system. *Nat Rev Microbiol* **2:** 747–765.

Moore M. 1982. Centers for Disease Control. Enteroviral disease in the United States, 1970–1979. *J Infect Dis* **146:** 103–108.

Muller AJ, Kaiser P, Dittmar KE, Weber TC, Haueter S, Endt K, Songhet P, Zellweger C, Kremer M, Fehling HJ, et al. 2012. Salmonella gut invasion involves TTSS-2-dependent epithelial traversal, basolateral exit, and uptake by epithelium-sampling lamina propria phagocytes. *Cell Host Microbe* **11:** 19–32.

Mylonakis E, Paliou M, Hohmann EL, Calderwood SB, Wing EJ. 2002. Listeriosis during pregnancy: A case series and review of 222 cases. *Medicine (Baltimore)* **81:** 260–269.

Nagesh Babu G, Kumar A, Kalita J, Misra UK. 2008. Proinflammatory cytokine levels in the serum and cerebrospinal fluid of tuberculous meningitis patients. *Neurosci Lett* **436:** 48–51.

Nava P, Lopez S, Arias CF, Islas S, Gonzalez-Mariscal L. 2004. The rotavirus surface protein VP8 modulates the gate and fence function of tight junctions in epithelial cells. *J Cell Sci* **117:** 5509–5519.

Niess JH, Brand S, Gu X, Landsman L, Jung S, McCormick BA, Vyas JM, Boes M, Ploegh HL, Fox JG, et al. 2005. CX3CR1-mediated dendritic cell access to the intestinal lumen and bacterial clearance. *Science* **307:** 254–258.

Nikitas G, Deschamps C, Disson O, Niault T, Cossart P, Lecuit M. 2011. Transcytosis of *Listeria monocytogenes* across the intestinal barrier upon specific targeting of goblet cell accessible E-cadherin. *J Exp Med* **208:** 2263–2277.

Nitta T, Hata M, Gotoh S, Seo Y, Sasaki H, Hashimoto N, Furuse M, Tsukita S. 2003. Size-selective loosening of the blood–brain barrier in claudin-5-deficient mice. *J Cell Biol* **161:** 653–660.

Nizet V, Kim KS, Stins M, Jonas M, Chi EY, Nguyen D, Rubens CE. 1997. Invasion of brain microvascular endothelial cells by group B streptococci. *Infect Immun* **65:** 5074–5081.

Orihuela CJ, Mahdavi J, Thornton J, Mann B, Wooldridge KG, Abouseada N, Oldfield NJ, Self T, Ala'Aldeen DA, Tuomanen EI. 2009. Laminin receptor initiates bacterial contact with the blood brain barrier in experimental meningitis models. *J Clin Invest* **119:** 1638–1646.

Patrick D, Betts J, Frey EA, Prameya R, Dorovini-Zis K, Finlay BB. 1992. *Haemophilus influenzae* lipopolysaccharide disrupts confluent monolayers of bovine brain endothelial cells via a serum-dependent cytotoxic pathway. *J Infect Dis* **165:** 865–872.

Pentecost M, Otto G, Theriot JA, Amieva MR. 2006. *Listeria monocytogenes* invades the epithelial junctions at sites of cell extrusion. *PLoS Pathog* **2:** e3.

Pezzicoli A, Santi I, Lauer P, Rosini R, Rinaudo D, Grandi G, Telford JL, Soriani M. 2008. Pilus backbone contributes to group B *Streptococcus* paracellular translocation through epithelial cells. *J Infect Dis* **198:** 890–898.

Pfister HW, Scheld WM. 1997. Brain injury in bacterial meningitis: Therapeutic implications. *Curr Opin Neurol* **10:** 254–259.

Pfister HW, Borasio GD, Dirnagl U, Bauer M, Einhaupl KM. 1992. Cerebrovascular complications of bacterial meningitis in adults. *Neurology* **42:** 1497–1504.

Prasadarao NV, Wass CA, Huang SH, Kim KS. 1999a. Identification and characterization of a novel Ibe10 binding protein that contributes to *Escherichia coli* invasion of brain microvascular endothelial cells. *Infect Immun* **67:** 1131–1138.

Prasadarao NV, Wass CA, Stins MF, Shimada H, Kim KS. 1999b. Outer membrane protein A-promoted actin condensation of brain microvascular endothelial cells is required for *Escherichia coli* invasion. *Infect Immun* **67:** 5775–5783.

Prasadarao NV, Srivastava PK, Rudrabhatla RS, Kim KS, Huang SH, Sukumaran SK. 2003. Cloning and expression of the *Escherichia coli* K1 outer membrane protein A receptor, a gp96 homologue. *Infect Immun* **71:** 1680–1688.

Radin JN, Orihuela CJ, Murti G, Guglielmo C, Murray PJ, Tuomanen EI. 2005. β-Arrestin 1 participates in platelet-activating factor receptor-mediated endocytosis of *Streptococcus pneumoniae*. *Infect Immun* **73:** 7827–7835.

Reijerkerk A, Kooij G, van der Pol SM, Khazen S, Dijkstra CD, de Vries HE. 2006. Diapedesis of monocytes is associated with MMP-mediated occludin disappearance in brain endothelial cells. *FASEB J* **20:** 2550–2552.

Rescigno M. 2011. Dendritic cells in bacteria handling in the gut. *J Leukoc Biol* **90:** 669–672.

Rescigno M, Di Sabatino A. 2009. Dendritic cells in intestinal homeostasis and disease. *J Clin Invest* **119:** 2441–2450.

Rescigno M, Urbano M, Valzasina B, Francolini M, Rotta G, Bonasio R, Granucci F, Kraehenbuhl JP, Ricciardi-Castagnoli P. 2001. Dendritic cells express tight junction proteins and penetrate gut epithelial monolayers to sample bacteria. *Nat Immunol* **2:** 361–367.

Rieder G, Fischer W, Haas R. 2005. Interaction of *Helicobacter pylori* with host cells: Function of secreted and translocated molecules. *Curr Opin Microbiol* **8:** 67–73.

Ring A, Weiser JN, Tuomanen EI. 1998. Pneumococcal trafficking across the blood–brain barrier. Molecular analysis of a novel bidirectional pathway. *J Clin Invest* **102:** 347–360.

Robbins JR, Skrzypczynska KM, Zeldovich VB, Kapidzic M, Bakardjiev AI. 2010. Placental syncytiotrophoblast constitutes a major barrier to vertical transmission of *Listeria monocytogenes*. *PLoS Pathog* **6:** e1000732.

Robbins JR, Zeldovich VB, Poukchanski A, Boothroyd JC, Bakardjiev AI. 2012. Tissue barriers of the human placenta to infection with *Toxoplasma gondii*. *Infect Immun* **80:** 418–428.

Robertson SL, Smedley JG III, McClane BA. 2010. Identification of a claudin-4 residue important for mediating the host cell binding and action of *Clostridium perfringens* enterotoxin. *Infect Immun* **78:** 505–517.

Rubin LL, Staddon JM. 1999. The cell biology of the blood–brain barrier. *Annu Rev Neurosci* **22:** 11–28.

Rudini N, Dejana E. 2008. Adherens junctions. *Curr Biol* **18:** R1080–1082.

Saadat I, Higashi H, Obuse C, Umeda M, Murata-Kamiya N, Saito Y, Lu H, Ohnishi N, Azuma T, Suzuki A, et al. 2007. *Helicobacter pylori* CagA targets PAR1/MARK kinase to disrupt epithelial cell polarity. *Nature* **447:** 330–333.

Saez-Llorens X, Jafari HS, Severien C, Parras F, Olsen KD, Hansen EJ, Singer II, McCracken GH Jr. 1991. Enhanced attenuation of meningeal inflammation and brain edema by concomitant administration of anti-CD18 monoclonal antibodies and dexamethasone in experimental *Haemophilus meningitis*. *J Clin Invest* **88:** 2003–2011.

Satosar A, Ramirez NC, Bartholomew D, Davis J, Nuovo GJ. 2004. Histologic correlates of viral and bacterial infection of the placenta associated with severe morbidity and mortality in the newborn. *Hum Pathol* **35:** 536–545.

Scheld WM, Koedel U, Nathan B, Pfister HW. 2002. Pathophysiology of bacterial meningitis: Mechanism(s) of neuronal injury. *J Infect Dis* **186:** S225–233.

Scherf A, Pouvelle B, Buffet PA, Gysin J. 2001. Molecular mechanisms of *Plasmodium falciparum* placental adhesion. *Cell Microbiol* **3:** 125–131.

Schubert-Unkmeir A, Sokolova O, Panzner U, Eigenthaler M, Frosch M. 2007. Gene expression pattern in human brain endothelial cells in response to *Neisseria meningitidis*. *Infect Immun* **75:** 899–914.

Schubert-Unkmeir A, Konrad C, Slanina H, Czapek F, Hebling S, Frosch M. 2010. *Neisseria meningitidis* induces brain microvascular endothelial cell detachment from the matrix and cleavage of occludin: A role for MMP-8. *PLoS Pathog* **6:** e1000874.

Schulze C, Firth JA. 1993. Immunohistochemical localization of adherens junction components in blood–brain barrier microvessels of the rat. *J Cell Sci* **104:** 773–782.

Sharief MK, Ciardi M, Thompson EJ. 1992. Blood–brain barrier damage in patients with bacterial meningitis: Association with tumor necrosis factor-α but not interleukin-1 β. *J Infect Dis* **166:** 350–358.

Sheen TR, Ebrahimi CM, Hiemstra IH, Barlow SB, Peschel A, Doran KS. 2010. Penetration of the blood–brain barrier by *Staphylococcus* aureus: Contribution of membrane-anchored lipoteichoic acid. *J Mol Med* **88:** 633–639.

Slanina H, Konig A, Hebling S, Hauck CR, Frosch M, Schubert-Unkmeir A. 2010. Entry of *Neisseria meningitidis* into mammalian cells requires the Src family protein tyrosine kinases. *Infect Immun* **78:** 1905–1914.

Smith AL, Smith DH, Averill DR Jr, Marino J, Moxon ER. 1973. Production of *Haemophilus influenzae* b meningitis in infant rats by intraperitoneal inoculation. *Infect Immun* **8:** 278–290.

Sokolova O, Heppel N, Jagerhuber R, Kim KS, Frosch M, Eigenthaler M, Schubert-Unkmeir A, Turner DP, Marietou AG, Johnston L, et al. 2004. Interaction of *Neisseria meningitidis* with human brain microvascular endothelial cells: Role of MAP- and tyrosine kinases in invasion and inflammatory cytokine release. *Cell Microbiol* **6:** 1153–1166.

Sonoda N, Furuse M, Sasaki H, Yonemura S, Katahira J, Horiguchi Y, Tsukita S. 1999. Clostridium perfringens enterotoxin fragment removes specific claudins from tight junction strands: Evidence for direct involvement of claudins in tight junction barrier. *J Cell Biol* **147:** 195–204.

Soriani M, Santi I, Taddei A, Rappuoli R, Grandi G, Telford JL. 2006. Group B *Streptococcus* crosses human epithelial cells by a paracellular route. *J Infect Dis* **193:** 241–250.

Sousa S, Lecuit M, Cossart P. 2005. Microbial strategies to target, cross or disrupt epithelia. *Curr Opin Cell Biol* **17:** 489–498.

Stent G, Crowe SM. 1997. Effects of HIV-1 on the surface expression of LFA-1 on cultured monocytes. *J Acquir Immune Defic Syndr Hum Retrovirol* **15:** 95–103.

Stins MF, Prasadarao NV, Ibric L, Wass CA, Luckett P, Kim KS. 1994. Binding characteristics of S fimbriated *Escherichia coli* to isolated brain microvascular endothelial cells. *Am J Pathol* **145:** 1228–1236.

Sukumaran SK, Prasadarao NV. 2003. *Escherichia coli* K1 invasion increases human brain microvascular endothelial cell monolayer permeability by disassembling vascular-endothelial cadherins at tight junctions. *J Infect Dis* **188:** 1295–1309.

Swords WE, Ketterer MR, Shao J, Campbell CA, Weiser JN, Apicella MA. 2001. Binding of the non-typeable *Haemophilus influenzae* lipooligosaccharide to the PAF receptor initiates host cell signalling. *Cell Microbiol* **3:** 525–536.

Tafazoli F, Zeng CQ, Estes MK, Magnusson KE, Svensson L. 2001. NSP4 enterotoxin of rotavirus induces paracellular leakage in polarized epithelial cells. *J Virol* **75:** 1540–1546.

Tazi A, Disson O, Bellais S, Bouaboud A, Dmytruk N, Dramsi S, Mistou MY, Khun H, Mechler C, Tardieux I, et al. 2010. The surface protein HvgA mediates group B *Streptococcus* hypervirulence and meningeal tropism in neonates. *J Exp Med* **207:** 2313–2322.

Temesvari P, Abraham CS, Speer CP, Kovacs J, Megyeri P. 1993. *Escherichia coli* 0111 B4 lipopolysaccharide given intracisternally induces blood–brain barrier opening during experimental neonatal meningitis in piglets. *Pediatr Res* **34:** 182–186.

Teng CH, Cai M, Shin S, Xie Y, Kim KJ, Khan NA, Di Cello F, Kim KS. 2005. *Escherichia coli* K1 RS218 interacts with human brain microvascular endothelial cells via type 1 fimbria bacteria in the fimbriated state. *Infect Immun* **73:** 2923–2931.

Tobias L, Cordes DO, Schurig GG. 1993. Placental pathology of the pregnant mouse inoculated with *Brucella abortus* strain 2308. *Vet Pathol* **30:** 119–129.

Tondury G, Smith DW. 1966. Fetal rubella pathology. *J Pediatr* **68:** 867–879.

Tunkel AR, Scheld WM. 1995. Acute bacterial meningitis. *Lancet* **346:** 1675–1680.

Tuomanen EI, Saukkonen K, Sande S, Cioffe C, Wright SD. 1989. Reduction of inflammation, tissue damage, and mortality in bacterial meningitis in rabbits treated with monoclonal antibodies against adhesion-promoting receptors of leukocytes. *J Exp Med* **170:** 959–969.

Tyler JS, Boothroyd JC. 2011. The carboxyl terminus of Toxoplasma RON2 provides the crucial link between AMA1 and the host-associated invasion complex. *PLoS Pathog* **7:** e1001282.

Uchiyama S, Carlin AF, Khosravi A, Weiman S, Banerjee A, Quach D, Hightower G, Mitchell TJ, Doran KS, Nizet V. 2009. The surface-anchored NanA protein promotes pneumococcal brain endothelial cell invasion. *J Exp Med* **206:** 1845–1852.

Unkmeir A, Latsch K, Dietrich G, Wintermeyer E, Schinke B, Schwender S, Kim KS, Eigenthaler M, Frosch M. 2002. Fibronectin mediates Opc-dependent internalization of Neisseria meningitidis in human brain microvascular endothelial cells. *Mol Microbiol* **46:** 933–946.

Vadeboncoeur N, Segura M, Al-Numani D, Vanier G, Gottschalk M. 2003. Pro-inflammatory cytokine and chemokine release by human brain microvascular endothelial cells stimulated by *Streptococcus suis* serotype 2. *FEMS Immunol Med Microbiol* **35:** 49–58.

van Sorge NM, Ebrahimi CM, McGillivray SM, Quach D, Sabet M, Guiney DG, Doran KS. 2008. Anthrax toxins inhibit neutrophil signaling pathways in brain endothelium and contribute to the pathogenesis of meningitis. *PLoS ONE* **3:** e2964.

van Sorge NM, Quach D, Gurney MA, Sullam PM, Nizet V, Doran KS. 2009. The group B streptococcal serine-rich repeat 1 glycoprotein mediates penetration of the blood–brain barrier. *J Infect Dis* **199:** 1479–1487.

van Sorge NM, Zialcita PA, Browne SH, Quach D, Guiney DG, Doran KS. 2011. Penetration and activation of brain endothelium by *Salmonella enterica* serovar Typhimurium. *J Infect Dis* **203:** 401–405.

Verma S, Lo Y, Chapagain M, Lum S, Kumar M, Gurjav U, Luo H, Nakatsuka A, Nerurkar VR. 2009. West Nile virus infection modulates human brain microvascular endothelial cells tight junction proteins and cell adhesion molecules: Transmigration across the in vitro blood–brain barrier. *Virology* **385:** 425–433.

Virji M, Saunders JR, Sims G, Makepeace K, Maskell D, Ferguson DJ. 1993. Pilus-facilitated adherence of *Neisseria meningitidis* to human epithelial and endothelial cells: Modulation of adherence phenotype occurs concurrently with changes in primary amino acid sequence and the glycosylation status of pilin. *Mol Microbiol* **10:** 1013–1028.

Voth DE, Ballard JD. 2005. *Clostridium difficile toxins*: Mechanism of action and role in disease. *Clin Microbiol Rev* **18:** 247–263.

Waage A, Halstensen A, Shalaby R, Brandtzaeg P, Kierulf P, Espevik T. 1989. Local production of tumor necrosis factor α, interleukin 1, and interleukin 6 in meningococcal meningitis. Relation to the inflammatory response. *J Exp Med* **170:** 1859–1867.

Watson KG, Holden DW. 2010. Dynamics of growth and dissemination of *Salmonella* in vivo. *Cell Microbiol* **12:** 1389–1397.

Weksler BB, Subileau EA, Perriere N, Charneau P, Holloway K, Leveque M, Tricoire-Leignel H, Nicotra A, Bourdoulous S, Turowski P, et al. 2005. Blood–brain barrier-specific properties of a human adult brain endothelial cell line. *FASEB J* **19:** 1872–1874.

Wickham ME, Brown NF, Provias J, Finlay BB, Coombes BK. 2007. Oral infection of mice with *Salmonella enterica* serovar Typhimurium causes meningitis and infection of the brain. *BMC Infect Dis* **7:** 65.

Wilson SL, Drevets DA. 1998. *Listeria monocytogenes* infection and activation of human brain microvascular endothelial cells. *J Infect Dis* **178:** 1658–1666.

Winkler L, Gehring C, Wenzel A, Muller SL, Piehl C, Krause G, Blasig IE, Piontek J. 2009. Molecular determinants of the interaction between *Clostridium perfringens* enterotoxin fragments and claudin-3. *J Biol Chem* **284:** 18863–18872.

Wolburg H, Lippoldt A. 2002. Tight junctions of the blood–brain barrier: Development, composition and regulation. *Vascul Pharmacol* **38:** 323–337.

Wong AR, Pearson JS, Bright MD, Munera D, Robinson KS, Lee SF, Frankel G, Hartland EL. 2011. Enteropathogenic and enterohaemorrhagic *Escherichia coli*: Even more subversive elements. *Mol Microbiol* **80:** 1420–1438.

Yoshida N, Tyler KM, Llewellyn MS. 2011. Invasion mechanisms among emerging food-borne protozoan parasites. *Trends Parasitol* **27:** 459–466.

Zhang WG, Khan AN, Kim KJ, Stins M, Kim KS. 2002. Transforming growth factor-β increases *Escherichia coli* K1 adherence, invasion, and transcytosis in human brain microvascular endothelial cells. *Cell Tissue Res* **309:** 281–286.

Zysk G, Schneider-Wald BK, Hwang JH, Bejo L, Kim KS, Mitchell TJ, Hakenbeck R, Heinz HP. 2001. Pneumolysin is the main inducer of cytotoxicity to brain microvascular endothelial cells caused by *Streptococcus pneumoniae*. *Infect Immun* **69:** 845–852.

Host Specificity of Bacterial Pathogens

Andreas Bäumler[1] and Ferric C. Fang[2,3]

[1]Department of Medical Microbiology and Immunology, University of California, Davis School of Medicine, Davis, California 95616

[2]Department of Laboratory Medicine, University of Washington School of Medicine, Seattle, Washington 98195-7735

[3]Department Microbiology, University of Washington School of Medicine, Seattle, Washington 98195-7735

Correspondence: fcfang@u.washington.edu

Most pathogens are able to infect multiple hosts but some are highly adapted to a single-host species. A detailed understanding of the basis of host specificity can provide important insights into molecular pathogenesis, the evolution of pathogenic microbes, and the potential for pathogens to cross the species barrier to infect new hosts. Comparative genomics and the development of humanized mouse models have provided important new tools with which to explore the basis of generalism and specialism. This review will examine host specificity of bacterial pathogens with a focus on generalist and specialist serovars of *Salmonella enterica*.

AN EVOLUTIONARY PERSPECTIVE ON HOST SPECIFICITY

Definitions

In host–pathogen interactions, pathogens called *generalists* are capable of infecting a wide range of host species, whereas others referred to as *specialists* establish an intimate relationship with only a single-host species. Most pathogens are capable of infecting multiple hosts. Genetic diversity and ecological opportunities for cross-species transmission favor generalism (Woolhouse et al. 2001). The evolutionary dynamic leading to specialism has been long debated. The potential to expand into new niches is implicated as an important driving force in specialism, and some studies suggest that coexistence with other specialists can increase resource exploitation within an ecosystem. Some genetic changes that enhance adaptation to a particular niche may reduce fitness for another niche. This so-called *antagonistic pleiotropy* can lead to evolutionary trade-offs, which are not necessarily a requirement for specialization but may reinforce tendencies toward specialization. Although often discussed as a dichotomy, host specialization is often a continuum, and many pathogens are able to infect a limited number of hosts (Kirchner and Roy 2002). For example, humans and other primates are susceptible to *Shigella*, whereas rabbits are resistant but develop inflammation following the injection of *Shigella* into ileal loops, and mice are resistant to intestinal shigellosis (Phalipon and Sansonetti 2007).

Cite this article as *Cold Spring Harb Perspect Med* doi: 10.1101/cshperspect.a010041

The Host Species Barrier

As pathogens establish defined host ranges, the barrier that must be overcome to make a transition to a new host species is an important determinant of the ecology of infectious diseases. Zoonotic pathogens are able to overcome the species barrier to be transmitted from other vertebrate animals to humans. Crossing the species barrier has important implications for pathogenesis, as pathogens may show increased virulence in hosts to which they are not adapted.

Gene-for-Gene or Multifactorial?

Studies in plants have led to the gene-for-gene model of host pathogen interactions, in which plants that possess a specific resistance gene show resistance to pathogens possessing a corresponding avirulence (*avr*) gene (Flor 1971). However, the relevance of this model for most animal host–pathogen interactions is limited. Attempts to convert host-specific *Salmonella* Typhi to a murine pathogen by transduction of DNA from the generalist serovar *S.* Typhimurium, for example, have led to the conclusion that host specificity in *Salmonella* is multifactorial (Zahrt 1998). This increases the challenge of understanding the mechanisms underlying host specialization. This review will examine host specificity in bacterial pathogens, with a particular emphasis on the enteric pathogen *Salmonella.*

EXAMPLES OF SPECIALIST AND GENERALIST BACTERIAL PATHOGENS

Salmonella enterica Serovars

One approach to exploring the genetic basis of host specificity is to compare microbes with a narrow host range (specialists) with close relatives that display broad host specificity (generalists). The genus *Salmonella* consists of a group of pathogens that are particularly well suited for such a comparison. *Salmonella* serovars are Gram-negative pathogens belonging to the family Enterobacteriaceae in the phylum Proteobacteria. Most of the >2000 *Salmonella* serovars are associated with gastroenteritis in humans and have animal reservoirs in a broad range of reptilian, avian, and/or mammalian species (Fig. 1) (Kelterborn 1967). Gastroenteritis is a diarrheal disease that remains localized to the intestine and mesenteric lymph nodes in immunocompetent individuals (Santos et al. 2001). A typical representative generalist serovar associated with gastroenteritis is *Salmonella enterica* serovar Typhimurium (*S.* Typhimurium). However, the genus *Salmonella* also contains a small number of specialists with a narrow host range that are no longer associated with gastroenteritis in humans. Instead, these specialists are associated with disseminated septicemic infections in humans (e.g., typhoid fever) or other animal species (e.g., fowl typhoid). Humans are the only known reservoir for typhoidal *Salmonella* serovars, which developed in four phylogenetically unrelated clonal lineages within the genus *Salmonella*, presumably from ancestral organisms that were originally associated with gastroenteritis (Selander et al. 1990). *S. enterica* serovar Typhi (*S.* Typhi), the causative agent of typhoid fever, forms one of these clonal lineages, which has been estimated to be between 10,000 and 71,000 years old (Roumagnac et al. 2006). *S. enterica* serovars Paratyphi C (*S.* Paratyphi C) and Paratyphi B (*S.* Paratyphi B) each form single lineages, whereas a fourth lineage is formed by the *S. enterica* serovars Paratyphi A (*S.* Paratyphi A) and Sendai (*S.* Sendai). *S.* Paratyphi A, *S.* Paratyphi B, *S.* Paratyphi C, and *S.* Sendai cause paratyphoid fever, which is milder in its course but otherwise clinically indistinguishable from typhoid fever.

In addition to typhoidal serovars, the genus *Salmonella* contains a few specialists associated with septicemic illnesses in other animal species. For example, *S. enterica* serovar Gallinarum (*S.* Gallinarum) causes pullorum disease (biotype Pullorum) and fowl typhoid (biotype Gallinarum) in poultry, but is avirulent for other animal species (reviewed in Shivaprasad 2000). *S. enterica* serovars Dublin (*S.* Dublin) and Choleraesuis (*S.* Choleraesuis) cause bacteremia in their bovine and porcine hosts, respectively (Hughes et al. 1971; Sojka et al. 1977). However, these pathogens are not fully host restricted, because consumption of unpasteurized

Cite this article as *Cold Spring Harb Perspect Med* doi: 10.1101/cshperspect.a010041

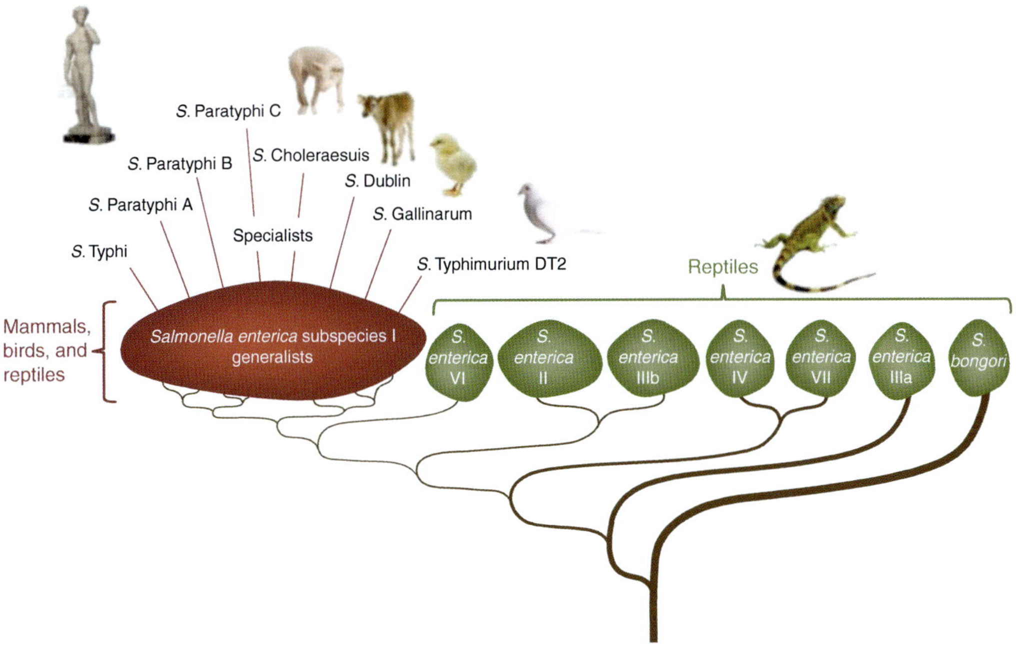

Figure 1. Host range of members of the genus *Salmonella*. The genus *Salmonella* consists of two species, *S. enterica* and *S. bongori*. *S. enterica* is further subdivided into seven subspecies, designated I, II, IIIa, IIIb, IV, VI, and VII. Serovars of *S. bongori* and *S. enterica* subspecies II, IIIa, IIIb, IV, VI, and VII are largely reptile associated and can be occasionally transmitted from this reservoir to humans. The majority of serovars belonging to *S. enterica* subspecies I are generalists that have reservoirs in mammalian, avian, and reptilian species. Specialists with a more restricted host range have independently evolved from this group several times.

milk, undercooked beef, or undercooked pork products can result in invasive bloodstream infections in humans (Saphra and Winter 1957; Fang and Fierer 1991; Threlfall et al. 1992). Closely related specialist/generalist pairs can be found even within a particular *Salmonella* serovar. For example, whereas the majority of S. Typhimurium isolates are prototypical generalists, a clonal group designated S. Typhimurium phage type DT2 represents a specialist associated with septicemic infections in pigeons (reviewed in Rabsch et al. 2002).

Yersinia Species

A second interesting group of organisms for comparative analysis within the family Enterobacteriaceae is the genus *Yersinia*, which contains three species pathogenic for humans: *Yersinia enterocolitica*, *Y. pseudotuberculosis*, and *Y. pestis*. The enteropathogenic species *Y. enterocolitica* and *Y. pseudotuberculosis* are associated with diarrhea in humans. Both of these pathogens are generalists associated with reservoirs in pigs and wild animals (Kapperud 1991; Fredriksson-Ahomaa 2009) and can also be found in the environment. *Y. pestis*, the causative agent of plague, branched an estimated 1500 to 20,000 years ago from the closely related *Y. pseudotuberculosis* serotype O:1b (Achtman et al. 1999). All contemporary *Y. pestis* isolates arose by clonal diversification from strains associated with the European black death of the 14th century (Bos et al. 2011). *Y. pestis* circulates in its rodent reservoir by flea-borne transmission, a dramatic change from the fecal–oral transmission of its *Y. pseudotuberculosis*-like ancestor. The very recent transition from an enteropathogenic lifestyle to that of a vector-borne septicemic infection makes the comparison of the *Y. pseudotuberculosis* and *Y. pestis* genomes particularly interesting. Although *Y. enterocolitica*,

Y. pseudotuberculosis, and *Y. pestis* are able to infect a wide range of mammalian hosts, another *Yersinia* species called *Y. ruckeri* is a specialist pathogen of salmonid fish (Sulakvelidze 2000).

IMPLICATIONS FOR VIRULENCE

The Host as Driver of Pathogen Variation

As different hosts represent different environmental niches, the fitness of a pathogen may vary from host to host. Different strains of a pathogen that are highly adapted to specific hosts may evolve in what is referred to as *polymorphism*. Alternatively, a single pathogen may develop into a generalist with the ability to use different strategies to survive and replicate in different hosts, referred to as *polyphenism*. In this regard, variation in host phenotypes is an important driver of the evolution of virulence. It has been proposed that pathogens that evolve within a host are more likely to be polymorphic, whereas those evolving outside a host are more likely to be polyphenic (Pfennig 2001). Another potential determinant of pathogen specialization is the strategy used by the host to resist infection (Miller et al. 2005). Theoretical models suggest that attempts by a host to control pathogen replication are more likely to select for polymorphic genotypes than strategies aimed at tolerating pathogen-induced damage.

Effects of Specialization on Virulence

During the coevolution of pathogens and hosts, adaptation is thought to result in the gradual attenuation of virulence, as the survival of the pathogen is dependent on the survival of the host (Dineen 1963; Mims 1975). However, exceptions to this tendency are observed for pathogens in which debility or death of the host are required for transmission. Anderson and May (1982) emphasized that the critical end point of evolution is not necessarily commensalism, but rather maximal transmission of a pathogen, which will tend to be proportional to the pathogen's reproductive rate. Garnick (1992) further refined this concept by noting that the immune response of the host to a pathogen can limit reproduction, such that optimal transmissibility is dependent on both host and pathogen factors, and the consequences for the host might differ considerably depending on whether the infection is short-lived or chronic. Therefore, the relationship between host specialization and virulence is complex. Human-adapted *S.* Typhi is generally regarded as more virulent than the generalist serovars, although both specialist and generalist *Salmonella* serovars can cause a wide range of clinical manifestations ranging from subclinical infection to life-threatening sepsis.

MOLECULAR MECHANISMS OF HOST SPECIFICITY

Microbial Adherence

In some instances, the ability of a pathogen to infect a host has been correlated with its ability to adhere to cells from that species. For example, the enteropathogenic RDEC-1 strain of *Escherichia coli* is able to cause diarrhea in rabbits but not in rats or guinea pigs and is able to adhere to intestinal brush borders of rabbits but not of the other two species (Cheney et al. 1980). The Opa protein of *Neiserria gonorrhoeae* is able to bind to the human CEACAM1 glycoprotein but not its canine, bovine, or murine orthologs (Voges et al. 2010), correlating with the specificity of this pathogen for humans. A type-3 secretory system-1 (T3SS-1)-associated lipoprotein called InvH promotes the ability of *S. enterica* to adhere to and invade cultured epithelial cells (Altmeyer et al. 1993). The effects of InvH on *Salmonella* interactions with host-intestinal mucosa appear to be host specific, as an *invH* mutation reduces the severity of *Salmonella* enteritis in calves but does not affect intestinal colonization or virulence in chicks (Watson et al. 1995, 1998; Porter and Curtiss 1997). The host-specific pathogen *S.* Typhi also requires InvH for efficient interaction with epithelial cells (Altmeyer et al. 1993) but enters human cells via an interaction between its type-IVB pili and the CFTR (cystic fibrosis transmembrane conductance regulator) protein (Pier et al. 1998; Tsui et al. 2003). Resistance to typhoid has therefore been proposed as a selection pressure that may have maintained the cystic fibrosis mutation at high frequencies in the population (van de Vosse et al. 2005).

Host-specific interactions during entry limit the host range of *Listeria monocytogenes* (family Listeriaceae, class Bacilli, phylum Firmicutes), an opportunistic food-borne pathogen of humans. The invasion protein internalin (InlA) enables *L. monocytogenes* to bind E-cadherin on epithelial cells to induce its internalization and crossing of the intestinal barrier. Internalin interacts with E-cadherin from humans but not E-cadherin from mice, owing to a single amino acid difference between the human and murine proteins (Lecuit et al. 1999). Although mice are naturally resistant to oral challenge with *L. monocytogenes*, transgenic mice expressing human E-cadherin in the intestine become susceptible to infection by this route (Lecuit et al. 2001). The interaction between InlA and human E-cadherin also plays an important role in the ability of *L. monocytogenes* to cross the placental barrier and cause severe in utero infections (Lecuit et al. 2004). A second invasion protein of *L. monocytogenes*, InlB, mediates binding to the hepatocyte growth factor receptor (Met) of murine and human cells, but not to Met from rabbits or guinea pigs (Khelef et al. 2006), which represents another example of species specificity. Both InlA and InlB have been implicated in species-specific interactions involved in crossing the placental barrier in vivo (Lecuit et al. 2004; Disson et al. 2008).

Replication in Host Cells

Adherence is only one of the possible mechanisms underlying host specificity. Viruses may show host tropism at the level of adherence and entry or at the level of downstream intracellular events (McFadden 2005). For pathogens able to colonize both cold-blooded and warm-blooded vertebrates, the complex organization of the adaptive immune system in birds and mammals creates unique challenges. In particular, pathogens must confront a broad repertoire of immunoglobulins and highly ordered lymphoid tissues containing mixed populations of lymphocytes, macrophages, and dendritic cells. The ability to survive in macrophages from different types of host has been proposed to be an important factor in *Salmonella* host specificity (Bäumler et al. 1998). For instance, human-adapted *S.* Typhi has been reported to display enhanced replication in human macrophages in comparison to the generalist *S.* Typhimurium, whereas the reverse is true in murine macrophages (Vladoianu et al. 1990; Pascopella et al. 1995; Ishibashi and Arai 1996; Schwan 2000; Xu 2009). The absence of the T3SS effector protein GtgE, a protease that targets the host-cell protein Rab32, has recently been proposed as an explanation for the poor growth of *S.* Typhi in murine cells (Spanò and Galán 2012).

Toxins

The ability to cause disease in a particular host species can also be affected by a species-specific interaction of toxins with their host-cell targets. *Actinobacillus suis* and *Actinobacillus equuli* (family Pasteurellaceae, phylum Proteobacteria) are host-adapted pathogens affecting suckling pigs and newborn foals (sleepy foal disease), respectively. The ability to cause disease in their respective host species is at least, in part, mediated by the elaboration of host-specific pore-forming toxins. Apxl/II, a pore-forming toxin produced by *A. suis*, shows specific cytotoxicity for porcine leukocytes, whereas Aqx, a pore-forming toxin released by *A. equuli*, specifically lyses equine leukocytes (reviewed in Frey 2011).

Iron Acquisition and Complement Resistance

Host-specific interactions relating to metal acquisition have been described for the human-adapted pathogens *Neisseria gonorrhoeae* and *Neisseria meningitidis* (family Neisseriaceae, phylum Proteobacteria). Like many pathogenic microbes, *N. gonorrhoeae* and *N. meningitidis* require iron during extracellular growth in the host. One strategy for iron acquisition is to strip iron from the human iron-binding protein transferrin using transferrin-binding protein B (TbpB) (reviewed in Schryvers and Stojiljkovic

1999). This process is host-specific, because TbpB of *N. meningitidis* binds human transferrin but not murine transferrin (Renauld-Mongenie et al. 1997). *N. meningitidis* can persist at least 2 d in intraperitoneally challenged transgenic mice expressing human transferrin, whereas bacteria are rapidly cleared from human transferrin transgenic mice challenged with a *N. meningitidis tbpB* mutant or from wild-type mice challenged with wild-type *N. meningitidis* (Zarantonelli et al. 2007). However, mice expressing human transferrin eventually clear even wild-type *N. meningitidis*, suggesting that additional species-specific interactions may be required for infection. During extracellular growth in the human body, *N. meningitidis* is exposed to complement. A mechanism that confers complement resistance is the recruitment of a molecule that down-regulates complement activation, complement factor H (fH), to the bacterial surface by factor H-binding protein (fHbp) (Schneider et al. 2006; Ngampasutadol et al. 2008). *N. meningitidis* fHbp binds fH from humans, but not fH from primates or mice (Granoff et al. 2009). Interestingly, differences between human and primate fH sequences cluster to the precise site of interaction with fHbp (Schneider et al. 2009), which likely represents another interaction contributing to the host specificity of *N. meningitidis* (Everett et al. 2011).

IMPLICATIONS OF HOST SPECIFICITY FOR TRANSMISSION

Host restriction limits the number of susceptible hosts available for transmission. Although this is rarely a limiting factor for transmission of human-adapted microbes in urban settings, it must have posed a formidable challenge for the persistence of pathogens within the small groups of human hunters and gatherers that existed before the agricultural revolution of the Neolithic age. One way for a microbe to meet this challenge is the development of a chronic persistent state. The transition from a broad-host-range pathogen (generalist) that is cleared from the body within a few weeks to a host-restricted pathogen (specialist) that chronically persists within a host for decades is illustrated by comparison of the lifestyles of broad-host-range and host-restricted *Salmonella* serovars.

Generalist *Salmonella* Serovars

Upon ingestion, the broad-host-range pathogen *S.* Typhimurium uses its virulence factors, two type-III secretion systems (T3SS-1 and T3SS-2), to invade epithelial cells (Galán and Curtiss III 1989) and survive in macrophages (Ochman et al. 1996), which elicits acute intestinal inflammation (Tsolis et al. 1999), producing symptoms of gastroenteritis within 24 hours of ingestion (Glynn and Palmer 1992). During the ensuing host response, neutrophils migrate into the intestinal lumen and produce reactive oxygen species in an attempt to kill bacteria. A by-product of oxyradical production is the generation of a new respiratory electron acceptor, tetrathionate, which enables *S.* Typhimurium to out-compete obligate anaerobic microbes in the gut lumen (Winter et al. 2010a) by utilizing carbon sources that cannot be fermented (Thiennimitr et al. 2011). The resulting bloom of *S.* Typhimurium in the intestinal lumen enhances its transmission to the next host by the fecal–oral route (Lawley et al. 2008). With the onset of adaptive immunity, the pathogen is cleared from intestinal tissues, and bacterial shedding in the feces subsides within a few weeks.

Host-Specific *Salmonella* Serovars

Similar to *S.* Typhimurium, the host-restricted *S.* Typhi uses its two type-III secretion systems (T3SS-1 and T3SS-2), to invade epithelial cells (Elsinghorst et al. 1989) and survive in macrophages (Forest et al. 2010). However, *S.* Typhi does not elicit overt host responses during the initial invasion of the intestinal mucosa, as indicated by a 2-wk incubation period of typhoid fever (Olsen et al. 2003). *S.* Typhi avoids the generation of responses during the initial phase of infection because it evades innate immunity by expressing a capsular polysaccharide (Raffatellu et al. 2005; Wilson et al. 2008; Haneda et al. 2009) and by rapidly repressing expression of flagella and T3SS-1 during the transition

 Cite this article as *Cold Spring Harb Perspect Med* doi: 10.1101/cshperspect.a010041

through the epithelial lining (Winter et al. 2010b). These mechanisms enable *S.* Typhi to slip past the host's defenses to cause an invasive bloodstream infection. Systemic dissemination results in colonization of internal organs and enables the pathogen to establish chronic carriage in the gallbladder in a fraction of infected individuals (e.g., Typhoid Mary). Chronic carriers are important reservoirs for transmission within the human population, intermittently releasing bacteria through the bile duct into the intestine (Putnam 1927; Leavitt 1992).

Other host-restricted *Salmonella* serovars also cause systemic infections that provide access to new routes of transmission. For example, the most important route of transmission for poultry-adapted *S.* Gallinarum is infection of the ovaries, which results in vertical spread via the egg to the chick or poult (Shivaprasad 2000). In cattle, long-term *S.* Dublin carrier animals contribute to transmission of the pathogen within infected herds by periodically shedding bacteria into milk (Nielsen et al. 2004). These examples help to illustrate that within the genus *Salmonella*, the transition from a pathogen with a broad host range to a pathogen with a restricted host range commonly involves a change in the route of transmission.

Yersinia Species and the Importance of Vector-Borne Transmission

The transition from free-living *Y. pseudotuberculosis* to host-dependent *Y. pestis* also involved a dramatic change in the route of transmission. Although *Y. pseudotuberculosis* is an enteric pathogen that is transmitted via the fecal–oral route, the vector-borne transmission of *Y. pestis* exchanges a transition through the mammalian intestine for a transition through the flea gut. However, the selective forces that selected for this change in the route of transmission during the emergence of *Y. pestis* from an ancestral *Y. pseudotuberculosis*-like organism remain unknown.

Dependence on an arthropod vector for transmission may restrict host availability and therefore impose constraints on the opportunity for a pathogen to encounter novel hosts. For example, Lyme disease is an important zoonotic infection in the Northeastern United States, where a tick vector transmits the *Borellia burgdorferi* spirochete to *Peromyscus leucopus*, a highly competent and ubiquitous rodent reservoir host (LoGiudice et al. 2003). In contrast, Lyme disease is nonendemic in Northern Colorado where the primary rodent reservoir is *Neotoma mexicana*, a solitary species with little human contact (Maupin et al. 1994).

GENOMIC SIGNATURES OF HOST SPECIFICITY

Genomic Decay

Host specificity is associated with a number of genomic signatures, including genomic decay, genomic rearrangements, and gene acquisition by lateral gene transfer. The genomes of specialists commonly show signs of genomic decay, as indicated by gene deletion or gene inactivation through point mutation (pseudogene formation). For example, 80 genes have been deleted from the genome of *Y. pestis* since this pathogen diverged from *Y. pseudotuberculosis* and became host dependent (Chain et al. 2004). Furthermore, the genome of *Y. pestis* contains ~208 pseudogenes, compared with only 62 in *Y. pseudotuberculosis* (Parkhill et al. 2001; Chain et al. 2004). The specialist *Y. ruckeri* has the smallest genome of any *Yersinia* species, although this does not appear to be a result of reductive evolution, as its genome contains relatively few pseudogenes. Both *Y. pestis* and *Y. ruckeri* lack genes for urease, vitamin B12 metabolism, and the methionine salvage pathway, perhaps because these pathogens do not colonize the mammalian intestine (Chen et al. 2010). The genome of the generalist *S.* Typhimurium (strain LT2) contains 25 pseudogenes (McClelland et al. 2001), far fewer than the genomes of host-restricted pathogens such as *S.* Typhi (204 pseudogenes), *S.* Paratyphi A (173 pseudogenes), *S.* Paratyphi C (152 pseudogenes), *S.* Gallinarum (309 pseudogenes), *S.* Dublin (177 pseudogenes), or *S.* Choleraesuis (151 pseudogenes) (Parkhill et al. 2001; McClelland et al. 2004; Chiu et al. 2005; Thomson et al. 2008; Liu et al. 2009; Betancor et al. 2012). The pigeon-associated *S.* Typhimurium DT2 strain 94-213 has 84 pseudogenes, including 21 loci that are intact in other closely related *S.* Typhimurium strains, and shows distinctive patterns of gene expression (Kingsley et al. 2013). Similarly, the presence of 77 pseudogenes in the multidrug-resistant *S.* Typhimurium ST313 genotype has led to the suggestion that this lineage is becoming host-adapted to humans (Kingsley et al. 2009; Feasey et al. 2012).

One reason for genomic decay in specialists is related to their divergence from a generalist ancestor along with a change in their route of transmission. When genes necessary for transmission of the ancestral organism are no longer under selection, their functions become dispensable. As a result, such genes accumulate mutations or are deleted. For example, the *shdA* and *misL* genes encode nonfimbrial adhesins that enhance intestinal colonization of *S.* Typhimurium (Kingsley et al. 2000, 2002; Dorsey et al. 2005). Pseudogene formation in host-restricted *Salmonella* serovars has inactivated both *shdA* (*S.* Typhi, *S.* Paratyphi A, *S.* Paratyphi C, *S.* Dublin, and *S.* Gallinarum) and *misL* (*S.* Typhi) (Parkhill et al. 2001; McClelland et al. 2004; Thomson et al. 2008; Liu et al. 2009; Betancor et al. 2012). Another group of intestinal colonization factors of *S.* Typhimurium is encoded by fimbrial operons of the chaperone usher assembly class (Weening et al. 2005). Genomic decay of genes encoding these intestinal colonization factors is evident in host-restricted *Salmonella* serovars, as pseudogenes are present in 7 of 11 fimbrial operons in *S.* Typhi strain CT18 (Townsend et al. 2001), 8 of 12 fimbrial operons in *S.* Gallinarium (Thomson et al. 2008), and 3 of 12 fimbrial operons in *S.* Paratyphi A (McClelland et al. 2004). It has been postulated that mutations that alter adhesion-binding specificity or affinity can also contribute to adaptation to different host environments (Kisiela et al. 2012). The generalist *S.* Typhimurium uses tetrathionate respiration encoded by the *ttrRS*/*ttrBCA* gene cluster to out-compete microbes in the gut during its fecal–oral transmission cycle (Winter et al. 2010a). Pseudogenes in this pathway are found in *S.* Typhi strain CT18 (*ttrS*) and *S.* Gallinarum (*ttrB*, *ttrC*) (Parkhill et al. 2001; Thomson et al. 2008). Tetrathionate respiration supports growth of *S.* Typhimurium on ethanolamine as a carbon source in the gut lumen during gastroenteritis (Thiennimitr et al. 2011). This pathway is degraded in specialists *S.* Paratyphi C (*eutA*, *eutC*, *eutK*, and *eutN*) and *S.* Choleraesuis (*eutN*) (Liu et al. 2009). Growth on ethanolamine under anaerobic conditions necessitates biosynthesis of vitamin B12 by the *cob*/*cbi* gene cluster (Roof and Roth 1988), which contains pseudogenes in *S.* Typhi (*cbiM*, *cbiK*, *cbiJ*, and *cbiC*), *S.* Paratyphi A (*cbiA*), and *S.* Gallinarum (*cobD*, *cbiD*, and *cbiC*) (Parkhill et al. 2001; McClelland et al. 2004; Thomson et al. 2008).

Many of the pseudogenes found in the genomes of host-restricted *Salmonella* serovars encode functions that aid intestinal colonization and competition with the intrinsic microbiota during gastroenteritis. These genes are intact in broad-host-range *Salmonella* serovars because they confer properties that enhance transmission of these pathogens by the fecal–oral route during gastroenteritis (Lawley et al. 2008), which imposes selective constraints that prevent loss of the encoded functions by mutation. However, these genes are no longer under selection in host-restricted pathogens that cause septicemic infections and transmit from reservoirs in the gallbladder (e.g., *S.* Typhi), the udder (e.g., *S.* Dublin), or the ovaries (e.g., *S.* Gallinarum). Thus, one driving force responsible for pseudogene formation in host-restricted *Salmonella* serovars is that functions required for transmission from an intestinal reservoir are no longer under selection. The resulting genomic decay of these pathways in host-restricted *Salmonella* serovars represent an experiment of nature, giving rise to a nonsaturation mutagenesis of genes necessary for fecal–oral transmission of generalists associated with gastroenteritis.

Genomic Rearrangements

Genomic rearrangements are another genomic signature associated with specialists but absent from generalists. The chromosomes of *E. coli* and *S.* Typhimurium display a high degree of conservation with regard to their gene order (Riley and Anilionis 1978), suggesting that this feature may be under selection. Large rearrangements that alter chromosomal gene order are commonly associated with a lower growth rate, because inversions may alter gene dosage, replication-transcription conflicts, or chromosome symmetry (Hill and Gray 1988; Rebollo et al. 1988; Campo et al. 2004). The generalist *S.* Typhimurium needs to multiply at a maximum growth rate in the intestine because successful transmission requires bacterial numbers in the feces to reach a sufficient threshold (Lawley et al. 2008). Thus, the maximum growth rate needed for transmission might place the gene order on the chromosomes of broad-host-range *Salmonella* serovars under selection.

In contrast, host-restricted *Salmonella* serovars no longer depend on multiplication at a maximal rate to successfully compete with other microbes in the intestine because their transmission success is aided by chronic carriage, a state of relative dormancy that is likely associated with slow growth. Therefore, selective constraints that prevent chromosomal rearrangements may no longer apply to host-restricted *Salmonella* serovars. This might explain why major genomic rearrangements owing to homologous recombination between the *rrn* operons are a characteristic feature of the chromosomes of *S.* Typhi isolates (Liu and Sanderson 1995, 1996; Kothapalli et al. 2005; Matthews et al. 2010). Similar genome rearrangements are also found in other host-restricted lineages within the genus *Salmonella*, including *S.* Paratyphi C, *S.* Gallinarum, and *S.* Typhimurium phage type DT2 (Liu and Sanderson 1998; Helm et al. 2004; Wu et al. 2005; Liu et al. 2007). Thus, rearrangements are a genomic signature of host-restricted members of the genus *Salmonella* that rely on chronic persistence in the host for transmission.

Lateral Gene Transfer

A third genomic signature associated with specialists is the acquisition of genetic material by lateral gene transfer, but this genomic signature is also commonly found in genomes of generalists. In some cases, laterally acquired genomic regions help explain how host-restricted pathogens adapted to their new lifestyle as they diverged from their generalist ancestors. For example, the generalist *S.* Typhimurium triggers innate inflammatory responses that prevent bacterial dissemination from the intestine (reviewed in Santos et al. 2009). In contrast, the specialist *S.* Typhi evades innate immunity through a rapid change in virulence gene expression when the pathogen enters the intestinal mucosa (Winter et al. 2010b). These changes in virulence gene expression are mediated by the TviA protein, which activates expression of a capsular polysaccharide, the Vi antigen, and represses expression of flagella and T3SS-1 (Winter et al. 2009). The *tviA* gene and genes involved in the biosynthesis of the capsular polysaccharide (*tviBCDE vexABCDE*) are encoded by a horizontally acquired DNA region designated *Salmonella* pathogenicity island (SPI)-7 (Virlogeux et al. 1995; Parkhill et al. 2001). SPI-7 is present in two other host-restricted serovars associated with disseminated septicemic infections, *S.* Dublin and *S.* Paratyphi C, suggesting that convergence by horizontal transfer of this DNA region occurred in their evolutionary history (Hashimoto and Khan 1997; Morris et al. 2003; Pickard et al. 2003). *S.* Typhi also produces a cytolethal distending toxin known as the "typhoid toxin," which is encoded by a pathogenicity islet (Spanò et al. 2008; Hodak and Galán 2013), although recent genomic studies have shown the presence of the islet in some nontyphoidal serovars as well (den Bakker et al. 2011).

Similarly, acquisition of the 10-kb pPst plasmid and the 100-kb pFra plasmid by lateral gene transfer enabled *Y. pestis* to adapt to the vector-borne route of transmission after its divergence from a *Y. pseudotuberculosis*-like ancestor. For instance, the *ymt* gene located on the pFra plasmid encodes phospholipase D, which enhances the ability of *Y. pestis* to colonize the flea midgut (Hinnebusch et al. 2002). The pFra plasmid also carries the *caf1M1A1* operon, which encodes a protein capsule that is required for infection of mice by the flea-borne route (Sebbane et al.

2009). The *pla* gene carried on the pPst plasmid encodes a plasminogen activator that is specifically required for inducing the bubonic form of plague in the mouse after a flea bite (Sebbane et al. 2006). Thus, acquisition of two *Y. pestis*-specific plasmids by lateral gene transfer facilitated the rapid evolutionary transition of *Y. pestis* to flea-borne transmission.

MODELS TO STUDY HOST SPECIFICITY

Cell Culture Models

Comparing infectivity in cultured cells from permissive and nonpermissive hosts has been a useful approach to understanding the molecular mechanisms of host specificity shown by viral pathogens, but has been generally less useful for bacteria. One exception has been the avian-specific *S. enterica* serovar Gallinarum biotypes Gallinarum and Pullorum, which have lost their ability to show mannose-sensitive hemagglutination and also show reduced adherence to and invasion of human HEp-2 epithelial cells owing to changes in their type-1 fimbriae. Expression of type-1 fimbriae from generalist S. Typhimurium restores hemagglutination and dramatically enhances HEp-2 invasion by S. Gallinarum and S. Pullorum. As mentioned previously, comparative survival in human and murine macrophages correlates with host specificity in S. Typhi and S. Typhimurium (Vladoianu et al. 1990; Pascopella et al. 1995; Ishibashi and Arai 1996; Schwan 2000; Xu 2009). However, comparative survival in chicken macrophages does not correlate with the avian virulence of S. Gallinarum (Chadfield et al. 2003). Cell culture has been useful for understanding the basis of host specificity in hemotropic *Bartonella* spp. The ability to adhere to and invade erythrocytes from humans, cats, mice, or rats recapitulates the host specificity of *B. quintana*, *B. henseleae*, *B. tribocorum*, and *B. birtlesii* observed in vivo (Vayssier-Taussat et al. 2010). The ability of *Bartonella* to infect specific hosts appears to be encoded within the Trw type-IV secretory system of these pathogens. In contrast, the host specificity of rabbit enteropathogenic *E. coli* strains is not mirrored by in vitro studies of ileal brush border adherence or erythrocyte agglutination (Robins-Browne et al. 1994). In summary, cell culture can be useful for exploring the mechanistic basis of host specificity but in many cases fails to recapitulate the complexity of in vivo host-bacterial interactions.

Comparative Zoology

Where permissive and nonpermissive animal hosts have been available for study, the comparative zoology of host–pathogen interactions has been informative. It has been reported, for instance, that type-1 fimbriae are not the sole factor responsible for the avian host specificity of S. Gallinarum, because S. Typhimurium is better able to proliferate in deep tissues and cause lethal infection in mice after intravenous inoculation, which bypasses the mucosal stage of infection, whereas the opposite is true in chickens (Barrow et al. 1994). The investigators making these observations concluded that host specificity for *Salmonella* is primarily expressed at the level of the reticuloendothelial system. Both bovine-adapted *Salmonella* serovar Dublin and swine-adapted S. Choleraesuis are virulent when administered orally to cattle, whereas avian S. Gallinarum and ovine S. Abortusovis are not. The initial interactions of these serovars with the intestinal mucosa do not appear to be appreciably different, but only S. Dublin is able to persist within the cattle intestine and translocate to mesenteric lymph nodes (Paulin et al. 2002).

Mice fail to develop diarrhea following oral *Salmonella* administration, representing an important difference between murine and human hosts. However, the administration of streptomycin in drinking water disrupts the endogenous intestinal flora and renders mice susceptible to *Salmonella* colitis (Bohnhoff and Miller 1962; Barthel et al. 2003). Specialist and generalist *Salmonella* strains differ in their ability to cause enteritis in the streptomycin-treated mouse model, although substantial strain-to-strain differences within serovars are observed (Suar et al. 2006). The human-adapted serovars S. Typhi and S. Paratyphi do not elicit inflammation in this model (Suar et al. 2006; Raffatellu et al. 2007; Haneda et al. 2009), correlating with their reduced propensity to cause diarrhea in humans. Introduction of the *viaB* capsule locus into *S.* Typhimurium suppresses the inflammatory pathology elicited by this generalist serovar, indicating that *S.* Typhi uses a specific mechanism to reduce intestinal inflammation.

Humanized Mice

Human-adapted pathogens present a special challenge for experimentalists. Until recently, researchers have had to extrapolate from observations in related animal-adapted pathogens, an approach with obvious limitations. For example, *S.* Typhimurium infection in mice is often touted as a model to understand *S.* Typhi infection in humans, but "murine typhoid" and human typhoid are different in a number of fundamental respects. As discussed earlier, typhoidal and nontyphoidal *Salmonella* serovars differ with regard to virulence factors, pseudogene content, and interactions with the innate immune system. Studies in mice have shown that IFN-γ/IL-12 signaling and the Nox2 phagocyte oxidase are critically important for host resistance to *Salmonella* (Hess et al. 1996; Mastroeni et al. 1996, 2000). Yet, although humans deficient in these host defenses show enhanced susceptibility to nontyphoidal *Salmonella* infections, they do not appear to be appreciably more susceptible to *S.* Typhi (de Jong et al. 1998; Winkelstein et al. 2000; Ottenhoff et al. 2002). Similarly, mice lacking $CD4^+$ T cells are highly susceptible to *S.* Typhimurium, as are humans with low $CD4^+$ T-cell counts owing to HIV infection (Fischl et al. 1986; Hess et al. 1996; Gordon et al. 2002), but HIV infection does not confer an increased susceptibility to typhoid fever (Gordon 2008). To render mice susceptible to an *S.* Typhi challenge, hog gastric mucin must be coadministered to incapacitate host phagocytes, or massive bacterial inocula must be given (Collins and Carter 1978; Hone et al. 1991). Such models provide limited insight into the pathogenesis of actual human typhoid.

It is generally believed that one of the reasons for the host restriction of *S.* Typhi is that mice possess some host defenses that are absent from humans. Consistent with this idea, a recent study suggests that inactivation of the murine Toll-like receptor 11 (TLR11), which is absent in humans, renders mice susceptible to *S.* Typhi infection (Mathur et al. 2012). TLR11-deficient mice thus represent a potential model to study aspects of typhoid fever pathogenesis (Fang and Bäumler 2012). Conversely, it is thought that *L. monocytogenes* does not infect mice by the oral route because receptors for epithelial invasion are not expressed in this species. The finding that transgenic expression in mice of the InlA receptor human E-cadherin renders these animals susceptible to invasion of enterocytes and crossing of the intestinal barrier supports this idea (Lecuit et al. 2001). Similarly, *Streptococcus pyogenes* secretes an enzyme that specifically activates human plasminogen and increases virulence in transgenic mice expressing human plasminogen (Sun et al. 2004). *N. meningitidis* expresses an iron acquisition system (TbpA and TbpB) that specifically strips iron from human transferrin and enhances growth in transgenic mice expressing human transferrin (Zarantonelli et al. 2007). Collectively, these studies show that the targeted deletion or addition of genes can increase the susceptibility of mice to selected human-specific pathogens.

Recent technological advances have created "humanized" chimeric mice that reconstitute specific anatomic and functional elements of human biology in laboratory mice (Mosier et al. 1988; Traggiai et al. 2004). Although studies of infection in humanized mice have only recently begun, this approach promises to provide unprecedented insights into mechanisms of host interaction with human pathogens including hepatitis viruses, HIV, CMV, Epstein-Barr virus, dengue virus, *N. meningitidis*, and *Plasmodium falciparum* (e.g., Bente 2005; Jiménez-Díaz 2009; Smith et al. 2010; Berges and Rowan 2011; Sato et al. 2011; Zeisel et al. 2011; Melican et al. 2013).

The generation of humanized mice requires genetic modifications that render mice receptive to the engraftment of human tissues, hematopoietic stem cells, or peripheral blood mononuclear cells. These may include a *scid* (severe combined immunodeficiency) mutation, which results in sensitivity to radiation and impaired B- and T-cell maturation, a *NOD* (nonobese diabetic) mutation, which confers reduced natural killer (NK) cell activity and additional innate immune defects, and targeted mutation of the *Il2rg* (interleukin-2 receptor γ-chain) locus, which results in severe impairment of T, B, and NK cell development (Shultz et al. 2007; Bernard et al. 2008; Brehm et al. 2010). A combination of mutations is required to support high levels of foreign cell engraftment, and transgenic human cytokine expression may enhance the subsequent survival, development, and differentiation of human cells in the murine host (Manz 2007). Following γ-irradiation of immunodeficient mice at birth, human cells or tissues may be introduced at various sites, such as the blood, liver, heart, peritoneal cavity, bone marrow, or kidney

Despite their severe deficiencies in both innate and adaptive immunity, NOD-*scid* IL2rγ^{null} mice are completely resistant to *S.* Typhi administered by intraperitoneal inoculation. However, engraftment with human hematopoietic stem cells, to create hu-SRC-SCID mice, renders the animals susceptible to progressive lethal infection (Fig. 2) (Libby et al. 2010). This suggests that *S.* Typhi requires human hematopoietic cells to proliferate and cause progressive infection in vivo. The pathology observed in hu-SRC-SCID mice infected with *S.* Typhi resembles human typhoid in a number of respects, including hepatic Kupffer cell swelling and splenic granulomatous inflammation with epithelioid macrophages and multinucleated giant cells (Mallory 1898; Ayhan et al. 1973; Bharadwaj et al. 2009). Elevated serum levels of the cytokines IL-6, IFN-γ, and TNF-α are also observed, as in patients with typhoid fever (Butler et al. 1993; Keuter et al. 1994). Engraftment of human hematopoietic stem cells into a different mouse background, Rag2 IL2rγ^{null} mice, has also been shown to render mice susceptible to *S.* Typhi infection. However, although the humanized Rag2 IL2rγ^{null} mice show high organism burdens indicative of bacterial replication, their infections do not appear to be as lethal as those observed in hu-SRC-SCID mice (Song et al. 2010; Firoz Mian et al. 2011). It is not clear

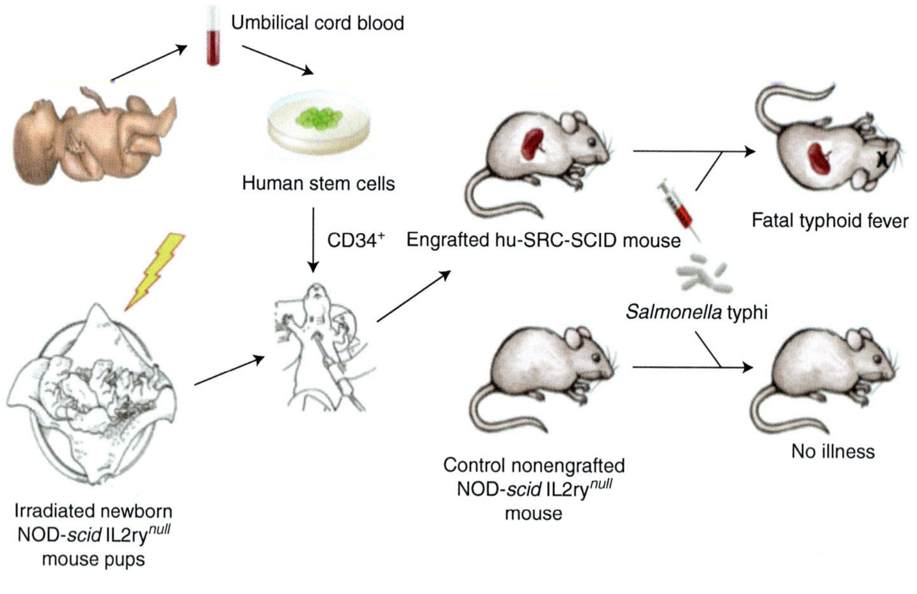

Figure 2. Humanized mouse model of typhoid fever. Newborn immunodeficient NOD-*scid* IL2rγ^{null} mice are irradiated before transplantation of human CD34^{+} hematopoietic stem cells from T-cell-depleted umbilical cord blood. Engrafted mice that have been repopulated with human SCID repopulating cells (hu-SRC) are now susceptible to lethal systemic infection with the human-adapted pathogen *Salmonella* Typhi.

whether this is owing to different levels or types of human cell engraftment, or perhaps to contributions of the NOD mutation to typhoid susceptibility. Preliminary observations indicate that at least some *S.* Typhi virulence determinants are required for virulence in humanized mice (Libby et al. 2010; Song et al. 2010), and the production of human antibodies to *S.* Typhi was observed in some infected Rag2 IL2rγ^{null} mice, suggesting that humanized mice may become useful for the preclinical evaluation of typhoid vaccine candidates (Song et al. 2010). Collectively, these initial observations indicate that humanized mice can recapitulate certain aspects of human typhoid and will be useful to obtain new insights into typhoid pathogenesis, host specificity, and immunity.

Despite their considerable promise, humanized mice at the present time have a number of significant limitations for the study of infectious diseases, including cost, various defects in reconstituted immunity (including mucosal immunity in the intestinal tract), mouse-to-mouse variability, and graft-versus-host disease. Further refinements to optimize engraftment can be anticipated as humanized mouse models continue to be developed and refined.

CONCLUSIONS

Importance of Understanding Pathogen Host Specificity

Host specialism represents an avenue of evolutionary development that takes a pathogen down an irreversible path of adaptation to a specific host. Specialism may be viewed as a gambit by the pathogen, in which exploitation of a specific niche is achieved at the potential expense of a trade-off in versatility (Bäumler et al. 1998). The extinction of the host can now result in the elimination of the pathogen as well. Yet the substantial number of host-specific pathogens attests to the fact that the benefits of this strategy can outweigh the risks. For the researcher, host specificity provides a valuable opportunity to study the complexities of coevolution between two organisms whose fates have become inextricably intertwined.

Crossing the Species Barrier

Another important reason to understand the basis of host specificity is that host barriers are relative, not absolute. Zoonotic pathogens, which are able to overcome the species barrier to infect humans, comprise a majority of the emerging and reemerging microbial challenges that afflict mankind (Taylor et al. 2001; Woolhouse et al. 2001). By exploring the mechanistic basis of specialism, one can gain essential insights into the pathogenesis of human-specific infections and anticipate future threats from pathogens waiting to make the species jump. As the playwright Arthur Miller wrote, "You specialize in something until one day you find it is specializing in you" (Miller 1968).

ACKNOWLEDGMENTS

The authors acknowledge research support from the National Institutes of Health (AI39557, AI40124, AI44170, AI44486, AI76246, AI77629, AI88122, AI91966, and AI96528).

REFERENCES

Achtman M, Zurth K, Morelli G, Torrea G, Guiyoule A, Carniel E. 1999. *Yersinia pestis*, the cause of plague, is a recently emerged clone of *Yersinia pseudotuberculosis*. *Proc Natl Acad Sci* **96:** 14043–14048.

Altmeyer RM, McNern JK, Bossio JC, Rosenshine I, Finlay BB, Galan JE. 1993. Cloning and molecular characterization of a gene involved in *Salmonella* adherence and invasion of cultured epithelial cells. *Mol Microbiol* **7:** 89–98.

Anderson RM, May RM. 1982. Coevolution of hosts and parasites. *Parasitology* **85:** 411–426.

Ayhan A, Gokoz A, Karacadag S, Telatar H. 1973. The liver in typhoid fever. *Am J Gastroenterol* **59:** 141–146.

Barrow PA, Huggins MB, Lovell MA. 1994. Host specificity of *Salmonella* infection in chickens and mice is expressed in vivo primarily at the level of the reticuloendothelial system. *Infect Immun* **62:** 4602–4610.

Barthel M, Hapfelmeier S, Quintanilla-Martinez L, Kremer M, Rohde M, Hogardt M, Pfeffer K, Russmann H, Hardt WD. 2003. Pretreatment of mice with streptomycin provides a *Salmonella enterica* serovar Typhimurium colitis model that allows analysis of both pathogen and host. *Infect Immun* **71:** 2839–2858.

Bäumler AJ, Tsolis RM, Ficht TA, Adams LG. 1998. Evolution of host adaptation in *Salmonella enterica*. *Infect Immun* **66:** 4579–4587.

Bente DA, Melkus MW, Garcia JV, Rico-Hesse R. 2005. Dengue fever in humanized NOD/SCID mice. *J Virol* **79:** 13797–13799.

Berges BK, Rowan MR. 2011. The utility of the new generation of humanized mice to study HIV-1 infection: Transmission, prevention, pathogenesis, and treatment. *Retrovirology* **8:** 65.

Bernard D, Peakman M, Hayday AC. 2008. Establishing humanized mice using stem cells: Maximizing the potential. *Clin Exp Immunol* **152:** 406–414.

Betancor L, Yim L, Martínez A, Fookes M, Sasias S, Schelotto F, Thomson N, Maskell D, Chabalgoity JA. 2012. Genomic comparison of the closely related *Salmonella enterica* serovars Enteritidis and Dublin. *Open Microbiol J* **6:** 5–13.

Bharadwaj S, Anim JT, Ebrahim F, Aldahham A. 2009. Granulomatous inflammatory response in a case of typhoid fever. *Med Princ Pract* **18:** 239–241.

Bohnhoff M, Miller CP. 1962. Enhanced susceptibility to *Salmonella* infection in streptomycin-treated mice. *J Infect Dis* **111:** 117–127.

Bos KI, Schuenemann VJ, Golding GB, Burbano HA, Waglechner N, Coombes BK, McPhee JB, DeWitte SN, Meyer M, Schmedes S, et al. 2011. A draft genome of *Yersinia pestis* from victims of the Black Death. *Nature* **478:** 506–510.

Brehm MA, Shultz LD, Greiner DL. 2010. Humanized mouse models to study human diseases. *Curr Opin Endocrinol Diabetes Obes* **17:** 120–125.

Butler T, Ho M, Acharya G, Tiwari M, Gallati H. 1993. Interleukin-6, γ interferon, and tumor necrosis factor receptors in typhoid fever related to outcome of antimicrobial therapy. *Antimicrob Agents Chemother* **37:** 2418–2421.

Campo N, Dias MJ, Daveran-Mingot ML, Ritzenthaler P, Le Bourgeois P. 2004. Chromosomal constraints in Gram-positive bacteria revealed by artificial inversions. *Mol Microbiol* **51:** 511–522.

Chadfield MS, Brown DJ, Aabo S, Christensen JP, Olsen JE. 2003. Comparison of intestinal invasion and macrophage response of *Salmonella* Gallinarum and other host-adapted *Salmonella* enterica serovars in the avian host. *Vet Microbiol* **92:** 49–64.

Chain PS, Carniel E, Larimer FW, Lamerdin J, Stoutland PO, Regala WM, Georgescu AM, Vergez LM, Land ML, Motin VL, et al. 2004. Insights into the evolution of *Yersinia pestis* through whole-genome comparison with *Yersinia pseudotuberculosis*. *Proc Natl Acad Sci* **101:** 13826–13831.

Chen PE, Cook C, Stewart AC, Nagarajan N, Sommer DD, Pop M, Thomason B, Kiley Thomason MP, Lentz S, Nolan N, et al. 2010. Genomic characterization of the *Yersinia* genus. *Genome Biol* **11:** R1.

Cheney CP, Schad PA, Formal SB, Boedeker EC. 1980. Species specificity of in vitro *Escherichia coli* adherence to host intestinal cell membranes and its correlation with in vivo colonization and infectivity. *Infect Immun* **28:** 1019–1027.

Chiu CH, Tang P, Chu C, Hu S, Bao Q, Yu J, Chou YY, Wang HS, Lee YS. 2005. The genome sequence of *Salmonella enterica* serovar Choleraesuis, a highly invasive and resistant zoonotic pathogen. *Nucleic Acids Res* **33:** 1690–1698.

Collins FM, Carter PB. 1978. Growth of salmonellae in orally infected germfree mice. *Infect Immun* **21:** 41–47.

de Jong R, Altare F, Haagen IA, Elferink DG, Boer T, van Breda Vriesman PJ, Kabel PJ, Draaisma JM, van Dissel JT, Kroon FP, et al. 1998. Severe mycobacterial and *Salmonella* infections in interleukin-12 receptor-deficient patients. *Science* **280:** 1435–1438.

den Bakker HC, Moreno Switt AI, Govoni G, Cummings CA, Ranieri ML, Degoricija L, Hoelzer K, Rodriguez-Rivera LD, Brown S, Bolchakova E, et al. 2011. Genome sequencing reveals diversification of virulence factor content and possible host adaptation in distinct subpopulations of *Salmonella enterica*. *BMC Genomics* **12:** 425.

Dineen JK. 1963. Immunological aspects of parasitism. *Nature* **197:** 268–269.

Disson O, Grayo S, Huillet E, Nikitas G, Langa-Vives F, Dussurget O, Ragon M, Le Monnier A, Babinet C, Cossart P, et al. 2008. Conjugated action of two species-specific invasion proteins for fetoplacental listeriosis. *Nature* **455:** 1114–1118.

Dorsey CW, Laarakker MC, Humphries AD, Weening EH, Bäumler AJ. 2005. *Salmonella enterica* serotype Typhimurium MisL is an intestinal colonization factor that binds fibronectin. *Mol Microbiol* **57:** 196–211.

Elsinghorst EA, Baron LS, Kopecko DJ. 1989. Penetration of human intestinal epithelial cells by *Salmonella*: Molecular cloning and expression of *Salmonella typhi* invasion determinants in *Escherichia coli*. *Proc Natl Acad Sci* **86:** 5173–5177.

Everett J, Lea SM, Caesar JJE, Johnson S, Tang CM. 2011. Investigating host specificity of *Neisseria meningitidis*. *Mol Immunol* **48:** 1673.

Fang FC, Bäumler AJ. 2012. A tollgate for typhoid. *Cell* **151:** 473–475.

Fang FC, Fierer J. 1991. Human infection with *Salmonella dublin*. *Medicine* **70:** 198–207.

Feasey NA, Dougan G, Kingsley RA, Heyderman RS, Gordon MA. 2012. Invasive non-typhoidal *Salmonella* disease: An emerging and neglected tropical disease in Africa. *Lancet* **379:** 2489–2499.

Firoz Mian M, Pek EA, Chenoweth MJ, Ashkar AA. 2011. Humanized mice are susceptible to *Salmonella typhi* infection. *Cell Mol Immunol* **8:** 83–87.

Fischl MA, Dickinson GM, Sinave C, Pitchenik AE, Cleary TJ. 1986. *Salmonella* bacteremia as manifestation of acquired immunodeficiency syndrome. *Arch Intern Med* **146:** 113–115.

Flor HH. 1971. Current status of the gene-for-gene concept. *Ann Rev Phytopathol* **9:** 275–296.

Forest CG, Ferraro E, Sabbagh SC, Daigle F. 2010. Intracellular survival of *Salmonella enterica* serovar Typhi in human macrophages is independent of *Salmonella* pathogenicity island (SPI)-2. *Microbiology* **156:** 3689–3698.

Fredriksson-Ahomaa M. 2009. Epidemiology of human *Yersinia pseudotuberculosis* infection. *Archiv Fur Lebensmittelhygiene* **60:** 82–87.

Frey J. 2011. The role of RTX toxins in host specificity of animal pathogenic Pasteurellaceae. *Vet Microbiol* **153:** 51–58.

Galán JE, Curtiss R III. 1989. Cloning and molecular characterization of genes whose products allow *Salmonella typhimurium* to penetrate tissue culture cells. *Proc Natl Acad Sci* **86:** 6383–6387.

Garnick E. 1992. Parasite virulence and parasite-host coevolution: A reappraisal. *J Parasitol* **78:** 381–386.

Glynn JR, Palmer SR. 1992. Incubation period, severity of disease, and infecting dose: Evidence from a *Salmonella* outbreak. *Am J Epidemiol* **136:** 1369–1377.

Gordon MA. 2008. *Salmonella* infections in immunocompromised adults. *J Infect* **56:** 413–422.

Gordon MA, Banda HT, Gondwe M, Gordon SB, Boeree MJ, Walsh AL, Corkill JE, Hart CA, Gilks CF, Molyneux ME. 2002. Non-typhoidal *Salmonella* bacteraemia among HIV-infected Malawian adults: High mortality and frequent recrudescence. *AIDS* **16:** 1633–1641.

Granoff DM, Welsch JA, Ram S. 2009. Binding of complement factor H (fH) to *Neisseria meningitidis* is specific for human fH and inhibits complement activation by rat and rabbit sera. *Infect Immun* **77:** 764–769.

Haneda T, Winter SE, Butler BP, Wilson RP, Tukel C, Winter MG, Godinez I, Tsolis RM, Bäumler AJ. 2009. The capsule-encoding *viaB* locus reduces intestinal inflammation by a *Salmonella* pathogenicity island 1-independent mechanism. *Infect Immun* **77:** 2932–2942.

Hashimoto Y, Khan AQ. 1997. Comparison of ViaB regions of Vi-positive organisms. *FEMS Microbiol Lett* **157:** 55–57.

Helm RA, Porwollik S, Stanley AE, Maloy S, McClelland M, Rabsch W, Eisenstark A. 2004. Pigeon-associated strains of *Salmonella enterica* serovar Typhimurium phage type DT2 have genomic rearrangements at rRNA operons. *Infect Immun* **72:** 7338–7341.

Hess J, Ladel C, Miko D, Kaufmann SH. 1996. *Salmonella typhimurium aroA*-infection in gene-targeted immunodeficient mice: Major role of $CD4^+$ TCR-$\alpha\beta$ cells and IFN-γ in bacterial clearance independent of intracellular location. *J Immunol* **156:** 3321–3326.

Hill CW, Gray JA. 1988. Effects of chromosomal inversion on cell fitness in *Escherichia coli* K-12. *Genetics* **119:** 771–778.

Hinnebusch BJ, Rudolph AE, Cherepanov P, Dixon JE, Schwan TG, Forsberg A. 2002. Role of *Yersinia* murine toxin in survival of *Yersinia pestis* in the midgut of the flea vector. *Science* **296:** 733–735.

Hodak H, Galán JE. 2013. A *Salmonella* Typhi homologue of bacteriophage muramidases controls typhoid toxin secretion. *EMBO Rep* **14:** 95–102.

Hone DM, Harris AM, Chatfield S, Dougan G, Levine MM. 1991. Construction of genetically defined double *aro* mutants of *Salmonella typhi*. *Vaccine* **9:** 810–816.

Hughes LE, Gibson EA, Roberts HE, Davies ET, Davies G, Sojka WJ. 1971. Bovine salmonellosis in England and Wales. *Br Vet J* **127:** 225–238.

Ishibashi Y, Arai T. 1996. A possible mechanism for host-specific pathogenesis of *Salmonella* serovars. *Microb Pathog* **21:** 435–446.

Jiménez-Díaz MB, Mulet T, Viera S, Gómez V, Garuti H, Ibáñez J, Alvarez-Doval A, Shultz LD, Martínez A, Gargallo-Viola D, Angulo-Barturen I. 2009. Improved murine model of malaria using *Plasmodium falciparum* competent strains and non-myelodepleted NOD-scid $IL2R\gamma^{null}$ mice engrafted with human erythrocytes. *Antimicrob Agents Chemother* **53:** 4533–4536.

Kapperud G. 1991. *Yersinia enterocolitica* in food hygiene. *Int J Food Microbiol* **12:** 53–65.

Kelterborn E. 1967. Salmonella-species. First isolations, names and occurrence. Springer-Verlag, New York.

Keuter M, Dharmana E, Gasem MH, van der Ven-Jongekrijg J, Djokomoeljanto R, Dolmans WM, Demacker P, Sauerwein R, Gallati H, van der Meer JW. 1994. Patterns of proinflammatory cytokines and inhibitors during typhoid fever. *J Infect Dis* **169:** 1306–1311.

Khelef N, Lecuit M, Bierne H, Cossart P. 2006. Species specificity of the *Listeria monocytogenes* InlB protein. *Cell Microbiol* **8:** 457–470.

Kingsley RA, van Amsterdam K, Kramer N, Bäumler AJ. 2000. The *shdA* gene is restricted to serotypes of *Salmonella enterica* subspecies I and contributes to efficient and prolonged fecal shedding. *Infect Immun* **68:** 2720–2727.

Kingsley RA, Santos RL, Keestra AM, Adams LG, Bäumler AJ. 2002. *Salmonella enterica* serotype Typhimurium ShdA is an outer membrane fibronectin-binding protein that is expressed in the intestine. *Mol Microbiol* **43:** 895–905.

Kingsley RA, Msefula CL, Thomson NR, Kariuki S, Holt KE, Gordon MA, Harris D, Clarke L, Whitehead S, Sangal V, et al. 2009. Epidemic multiple drug resistant *Salmonella* Typhimurium causing invasive disease in sub-Saharan Africa have a distinct genotype. *Genome Res* **19:** 2279–2287.

Kingsley RA, Kay S, Connor T, Barquist L, Salt L, Holt KE, Sivaraman K, Wileman T, Goulding D, Clare S, et al. 2013. Genome and transcriptome adaptation accompanying emergence of the definitive type 2 host-restricted *Salmonella enterica* serovar Typhimurium pathovar. *mBio* **4:** e00565.

Kirchner JW, Roy BA. 2002. Evolutionary implications of host-pathogen specificity: Fitness consequences of pathogen virulence traits. *Evol Ecol Res* **4:** 27–48.

Kisiela DI, Chattopadhyay S, Libby SJ, Karlinsey JE, Fang FC, Tchesnokova V, Kramer JJ, Beskhlebnaya V, Samadpour M, Grzymajlo K, et al. 2012. Evolution of *Salmonella enterica* virulence via point mutations in the fimbrial adhesin. *PLoS Pathog* **8:** e1002733.

Kothapalli S, Nair S, Alokam S, Pang T, Khakhria R, Woodward D, Johnson W, Stocker BA, Sanderson KE, Liu SL. 2005. Diversity of genome structure in *Salmonella enterica* serovar Typhi populations. *J Bacteriol* **187:** 2638–2650.

Lawley TD, Bouley DM, Hoy YE, Gerke C, Relman DA, Monack DM. 2008. Host transmission of *Salmonella enterica* serovar Typhimurium is controlled by virulence factors and indigenous intestinal microbiota. *Infect Immun* **76:** 403–416.

Leavitt JW. 1992. "Typhoid Mary" strikes back. Bacteriological theory and practice in early twentieth-century public health. *Isis* **83:** 608–629.

Lecuit M, Dramsi S, Gottardi C, Fedor-Chaiken M, Gumbiner B, Cossart P. 1999. A single amino acid in E-cadherin responsible for host specificity towards the human pathogen *Listeria monocytogenes*. *EMBO J* **18:** 3956–3963.

Lecuit M, Vandormael-Pournin S, Lefort J, Huerre M, Gounon P, Dupuy C, Babinet C, Cossart P. 2001. A transgenic model for listeriosis: Role of internalin in crossing the intestinal barrier. *Science* **292:** 1722–1725.

Lecuit M, Nelson DM, Smith SD, Khun H, Huerre M, Vacher-Lavenu MC, Gordon JI, Cossart P. 2004. Targeting and crossing of the human maternofetal barrier by *Listeria monocytogenes*: Role of internalin interaction with trophoblast E-cadherin. *Proc Natl Acad Sci* **101:** 6152–6157.

Libby SJ, Brehm MA, Greiner DL, Shultz LD, McClelland M, Smith KD, Cookson BT, Karlinsey JE, Kinkel TL, Porwollik S, et al. 2010. Humanized nonobese diabetic-*scid* $IL2r\gamma^{null}$ mice are susceptible to lethal *Salmonella* Typhi infection. *Proc Natl Acad Sci* **107:** 15589–15594.

Liu SL, Sanderson KE. 1995. Rearrangements in the genome of the bacterium *Salmonella typhi*. *Proc Natl Acad Sci* **92:** 1018–1022.

Liu SL, Sanderson KE. 1996. Highly plastic chromosomal organization in *Salmonella typhi*. *Proc Natl Acad Sci* **93:** 10303–10308.

Liu SL, Sanderson KE. 1998. Homologous recombination between *rrn* operons rearranges the chromosome in host-specialized species of *Salmonella*. *FEMS Microbiol Lett* **164:** 275–281.

Liu WQ, Liu GR, Li JQ, Xu GM, Qi D, He XY, Deng J, Zhang FM, Johnston RN, Liu SL. 2007. Diverse genome structures of *Salmonella paratyphi* C. *BMC Genomics* **8:** 290.

Liu WQ, Feng Y, Wang Y, Zou QH, Chen F, Guo JT, Peng YH, Jin Y, Li YG, Hu SN, et al. 2009. *Salmonella paratyphi* C: Genetic divergence from *Salmonella choleraesuis* and pathogenic convergence with *Salmonella typhi*. *PLoS ONE* **4:** e4510.

LoGiudice K, Ostfield RS, Schmidt KA, Keesing F. 2003. The ecology of infectious disease: Effects of host diversity and community composition on Lyme disease risk. *Proc Natl Acad Sci* **100:** 567–571.

Mallory FB. 1898. A histological study of typhoid fever. *J Exp Med* **3:** 611–638.

Manz MG. 2007. Human-hemato-lymphoid-system mice: Opportunities and challenges. *Immunity* **26:** 537–541.

Mastroeni P, Harrison JA, Chabalgoity JA, Hormaeche CE. 1996. Effect of interleukin 12 neutralization on host resistance and γ interferon production in mouse typhoid. *Infect Immun* **64:** 189–196.

Mastroeni P, Vazquez-Torres A, Fang FC, Xu Y, Khan S, Hormaeche CE, Dougan G. 2000. Antimicrobial actions of the NADPH phagocyte oxidase and inducible nitric oxide synthase in experimental salmonellosis. II. Effects on microbial proliferation and host survival in vivo. *J Exp Med* **192:** 237–248.

Mathur R, Oh H, Zhang D, Park SG, Seo J, Koblansky A, Hayden MS, Ghosh S. 2012. A mouse model of *Salmonella* Typhi infection. *Cell* **151:** 590–602.

Matthews TD, Edwards R, Maloy S. 2010. Chromosomal rearrangements formed by *rrn* recombination do not improve replichore balance in host-specific *Salmonella enterica* serovars. *PLoS ONE* **5:** e13503.

Maupin GO, Gage KL, Piesman J, Montenieri J, Sviat SL, VanderZanden L, Happ CM, Dolan M, Johnson BJ. 1994. Discovery of an enzootic cycle of *Borrelia burgdorferi* in *Neotoma mexicana* and *Ixodes spinipalpis* from northern Colorado, an area where Lyme disease is nonendemic. *J Infect Dis* **170:** 636–643.

McClelland M, Sanderson KE, Spieth J, Clifton SW, Latreille P, Courtney L, Porwollik S, Ali J, Dante M, Du F, et al. 2001. Complete genome sequence of *Salmonella enterica* serovar Typhimurium LT2. *Nature* **413:** 852–856.

McClelland M, Sanderson KE, Clifton SW, Latreille P, Porwollik S, Sabo A, Meyer R, Bieri T, Ozersky P, McLellan M, et al. 2004. Comparison of genome degradation in Paratyphi A and Typhi, human-restricted serovars of *Salmonella enterica* that cause typhoid. *Nat Genet* **36:** 1268–1274.

McFadden G. 2005. Poxvirus tropism. *Nat Rev Microbiol* **3:** 201–213.

Melican K, Michea Veloso P, Martin T, Bruneval P, Duménil G. 2013. Adhesion of *Neisseria meningitidis* to dermal vessels leads to local vascular damage and purpura in a humanized mouse model. *PLoS Pathog* **9:** e1003139.

Miller A. 1968. *The price: A play.* Secker and Warburg, London.

Miller MR, White A, Boots M. 2005. The evolution of host resistance: Tolerance and control as distinct strategies. *J Theor Biol* **236:** 198–207.

Mims CA. 1975. The meaning of persistent infections in nature. *Bull World Health Org* **52:** 745.

Morris C, Tam CK, Wallis TS, Jones PW, Hackett J. 2003. *Salmonella enterica* serovar Dublin strains, which are Vi antigen-positive use type IVB pili for bacterial self-association and human intestinal cell entry. *Microb Pathog* **35:** 279–284.

Mosier DE, Gulizia RJ, Baird SM, Wilson DB. 1988. Transfer of a functional human immune system to mice with severe combined immunodeficiency. *Nature* **335:** 256–259.

Ngampasutadol J, Ram S, Gulati S, Agarwal S, Li C, Visintin A, Monks B, Madico G, Rice PA. 2008. Human factor H interacts selectively with *Neisseria gonorrhoeae* and results in species-specific complement evasion. *J Immunol* **180:** 3426–3435.

Nielsen LR, Schukken YH, Grohn YT, Ersboll AK. 2004. *Salmonella* Dublin infection in dairy cattle: Risk factors for becoming a carrier. *Prev Vet Med* **65:** 47–62.

Ochman H, Soncini FC, Solomon F, Groisman EA. 1996. Identification of a pathogenicity island for *Salmonella* survival in host cells. *Proc Natl Acad Sci* **93:** 7800–7804.

Olsen SJ, Bleasdale SC, Magnano AR, Landrigan C, Holland BH, Tauxe RV, Mintz ED, Luby S. 2003. Outbreaks of typhoid fever in the United States, 1960–99. *Epidemiol Infect* **130:** 13–21.

Ottenhoff TH, Verreck FA, Lichtenauer-Kaligis EG, Hoeve MA, Sanal O, van Dissel JT. 2002. Genetics, cytokines and human infectious disease: Lessons from weakly pathogenic mycobacteria and salmonellae. *Nat Genet* **32:** 97–105.

Parkhill J, Dougan G, James KD, Thomson NR, Pickard D, Wain J, Churcher C, Mungall KL, Bentley SD, Holden MT, et al. 2001. Complete genome sequence of a multiple drug resistant *Salmonella enterica* serovar Typhi CT18. *Nature* **413:** 848–852.

Pascopella L, Raupach B, Ghori N, Monack D, Falkow S, Small PL. 1995. Host restriction phenotypes of *Salmonella typhi* and *Salmonella gallinarum*. *Infect Immun* **63:** 4329–4335.

Paulin SM, Watson PR, Benmore AR, Stevens MP, Jones PW, Villarreal-Ramos B, Wallis TS. 2002. Analysis of *Salmonella enterica* serotype-host specificity in calves: Avirulence of *S. enterica* serotype gallinarum correlates with bacterial dissemination from mesenteric lymph nodes and persistence in vivo. *Infect Immun* **70:** 6788–6797.

Pfennig KS. 2001. Evolution of pathogen virulence: The role of variation in host phenotype. *Proc Biol Sci* **268:** 755–760.

Phalipon A, Sansonetti PJ. 2007. *Shigella*'s ways of manipulating the host intestinal innate and adaptive immune system: A tool box for survival? *Immunol Cell Biol* **85:** 119–129.

Pickard D, Wain J, Baker S, Line A, Chohan S, Fookes M, Barron A, Gaora PO, Chabalgoity JA, Thanky N, et al. 2003. Composition, acquisition, and distribution of the Vi exopolysaccharide-encoding *Salmonella enterica* pathogenicity island SPI-7. *J Bacteriol* **185:** 5055–5065.

Pier GB, Grout M, Zaidi T, Meluleni G, Mueschenborn SS, Banting G, Ratcliff R, Evans MJ, Colledge WH. 1998. *Salmonella typhi* uses CFTR to enter intestinal epithelial cells. *Nature* **393:** 79–82.

Porter SB, Curtiss RR. 1997. Effect of inv mutations on *Salmonella* virulence and colonization in 1-day-old White Leghorn chicks. *Avian Dis* **41:** 45–57.

Putnam P. 1927. The trend of typhoid fever mortality in the United States. *Am J Hyg* **7:** 762–781.

Rabsch W, Andrews HL, Kingsley RA, Prager R, Tschape H, Adams LG, Bäumler AJ. 2002. *Salmonella enterica* serotype Typhimurium and its host-adapted variants. *Infect Immun* **70:** 2249–2255.

Raffatellu M, Chessa D, Wilson RP, Dusold R, Rubino S, Bäumler AJ. 2005. The Vi capsular antigen of *Salmonella enterica* serotype Typhi reduces Toll-like receptor-dependent interleukin-8 expression in the intestinal mucosa. *Infect Immun* **73:** 3367–3374.

Raffatellu M, Santos RL, Chessa D, Wilson RP, Winter SE, Rossetti CA, Lawhon SD, Chu H, Lau T, Bevins CL, et al. 2007. The capsule encoding the *viaB* locus reduces interleukin-17 expression and mucosal innate responses in the bovine intestinal mucosa during infection with *Salmonella enterica* serotype Typhi. *Infect Immun* **75:** 4342–4350.

Rebollo JE, Francois V, Louarn JM. 1988. Detection and possible role of two large nondivisible zones on the *Escherichia coli* chromosome. *Proc Natl Acad Sci* **85:** 9391–9395.

Renauld-Mongenie G, Poncet D, von Olleschik-Elbheim L, Cournez T, Mignon M, Schmidt MA, Quentin-Millet MJ. 1997. Identification of human transferrin-binding sites within meningococcal transferrin-binding protein B. *J Bacteriol* **179:** 6400–6407.

Riley M, Anilionis A. 1978. Evolution of the bacterial genome. *Ann Rev Microbiol* **32:** 519–560.

Robins-Browne RM, Tokhi AM, Adams LM, Bennett-Wood V. 1994. Host specificity of enteropathogenic *Escherichia coli* from rabbits: Lack of correlation between adherence in vitro and pathogenicity for laboratory animals. *Infect Immun* **62:** 3329–3336.

Roof DM, Roth JR. 1988. Ethanolamine utilization in *Salmonella typhimurium*. *J Bacteriol* **170:** 3855–3863.

Roumagnac P, Weill FX, Dolecek C, Baker S, Brisse S, Chinh NT, Le TA, Acosta CJ, Farrar J, Dougan G, et al. 2006. Evolutionary history of *Salmonella typhi*. *Science* **314:** 1301–1304.

Santos RL, Zhang S, Tsolis RM, Kingsley RA, Adams LG, Bäumler AJ. 2001. Animal models of *Salmonella* infections: Enteritis vs. typhoid fever. *Microb Infect* **3:** 1335–1344.

Santos RL, Raffatellu M, Bevins CL, Adams LG, Tukel C, Tsolis RM, Baumler AJ. 2009. Life in the inflamed intestine, *Salmonella* style. *Trends Microbiol* **17:** 498–506.

Saphra I, Winter JW. 1957. Clinical manifestations of salmonellosis in man—An evaluation of 7779 human infections identified at the New York *Salmonella* Center. *N Engl J Med* **256:** 1128–1134.

Sato K, Misawa N, Nie C, Satou Y, Iwakiri D, Matsuoka M, Takahashi R, Kuzushima K, Ito M, Takada K, et al. 2011. A novel animal model of Epstein–Barr virus-associated hemophagocytic lymphohistiocytosis in humanized mice. *Blood* **117:** 5663–5673.

Schneider MC, Exley RM, Chan H, Feavers I, Kang YH, Sim RB, Tang CM. 2006. Functional significance of factor H binding to *Neisseria meningitidis*. *J Immunol* **176:** 7566–7575.

Schneider MC, Prosser BE, Caesar JJE, Kugelberg E, Li S, Zhang Q, Quoraishi S, Lovett JE, Deane JE, Sim RB, et al. 2009. *Neisseria meningitidis* recruits factor H using protein mimicry of host carbohydrates. *Nature* **458:** 890–899.

Schryvers AB, Stojiljkovic I. 1999. Iron acquisition systems in the pathogenic *Neisseria*. *Mol Microbiol* **32:** 1117–1123.

Schwan WR, Huang XZ, Hu L, Kopecko DJ. 2000. Differential bacterial survival, replication, and apoptosis-inducing ability of *Salmonella* serovars within human and murine macrophages. *Infect Immun* **68:** 1005–1013.

Sebbane F, Jarrett CO, Gardner D, Long D, Hinnebusch BJ. 2006. Role of the *Yersinia pestis* plasminogen activator in the incidence of distinct septicemic and bubonic forms of flea-borne plague. *Proc Natl Acad Sci* **103:** 5526–5530.

Sebbane F, Jarrett C, Gardner D, Long D, Hinnebusch BJ. 2009. The *Yersinia pestis caf1M1A1* fimbrial capsule operon promotes transmission by flea bite in a mouse model of bubonic plague. *Infect Immun* **77:** 1222–1229.

Selander RK, Beltran P, Smith NH, Helmuth R, Rubin FA, Kopecko DJ, Ferris K, Tall BD, Cravioto A, Musser JM. 1990. Evolutionary genetic relationships of clones of *Salmonella* serovars that cause human typhoid and other enteric fevers. *Infect Immun* **58:** 2262–2275.

Shivaprasad HL. 2000. Fowl typhoid and pullorum disease. *Rev Sci Tech* **19:** 405–424.

Shultz LD, Ishikawa F, Greiner DL. 2007. Humanized mice in translational biomedical research. *Nat Rev Immunol* **7:** 118–130.

Smith MS, Goldman DC, Bailey AS, Pfaffle DL, Kreklywich CN, Spencer DB, Othieno FA, Streblow DN, Garcia JV, Fleming WH, et al. 2010. Granulocyte-colony stimulating factor reactivates human cytomegalovirus in a latently infected humanized mouse model. *Cell Host Microbe* **8:** 284–291.

Sojka WJ, Wray C, Shreeve J, Benson AJ. 1977. Incidence of *Salmonella* infection in animals in England and Wales 1968–1974. *J Hyg* **78:** 43–56.

Song J, Willinger T, Rongvaux A, Eynon EE, Stevens S, Manz MG, Flavell RA, Galan JE. 2010. A mouse model for the human pathogen *Salmonella* Typhi. *Cell Host Microbe* **8:** 369–376.

Spanò S, Galán JE. 2012. A Rab32-dependent pathway contributes to *Salmonella* Typhi host restriction. *Science* **338:** 960–963.

Spanò S, Ugalde JE, Galán JE. 2008. Delivery of a *Salmonella* Typhi exotoxin from a host intracellular compartment. *Cell Host Microbe* **3:** 30–38.

Suar M, Jantsch J, Hapfelmeier S, Kremer M, Stallmach T, Barrow PA, Hardt WD. 2006. Virulence of broad- and narrow-host-range *Salmonella enterica* serovars in the streptomycin-pretreated mouse model. *Infect Immun* **74:** 632–644.

Sulakvelidze A. 2000. Yersiniae other than *Y. enterocolitica, Y. pseudotuberculosis,* and *Y. pestis*: The ignored species. *Microbes Infect* **2:** 297–513.

Sun H, Ringdahl U, Homeister JW, Fay WP, Engelberg NC, Yang AY, Rozek LS, Wang X, Sjöbring U, Ginsburg D. 2004. Plasminogen is a critical host pathogenicity factor for group A streptococcal infection. *Science* **305:** 1283–1286.

Taylor LH, Latham SM, Woolhouse ME. 2001. Risk factors for human disease emergence. *Philos Trans R Soc Lond B Biol Sci* **356:** 983–989.

Thiennimitr P, Winter SE, Winter MG, Xavier MN, Tolstikov V, Huseby DL, Sterzenbach T, Tsolis RM, Roth JR, Bäumler AJ. 2011. Intestinal inflammation allows *Salmonella* to use ethanolamine to compete with the microbiota. *Proc Natl Acad Sci* **108:** 17480–17485.

Thomson NR, Clayton DJ, Windhorst D, Vernikos G, Davidson S, Churcher C, Quail MA, Stevens M, Jones MA, Watson M, et al. 2008. Comparative genome analysis of *Salmonella* Enteritidis PT4 and *Salmonella* Gallinarum 287/91 provides insights into evolutionary and host adaptation pathways. *Genome Res* **18:** 1624–1637.

Threlfall EJ, Hall ML, Rowe B. 1992. *Salmonella* bacteraemia in England and Wales, 1981–1990. *J Clin Pathol* **45:** 34–36.

Townsend SM, Kramer NE, Edwards R, Baker S, Hamlin N, Simmonds M, Stevens K, Maloy S, Parkhill J, Dougan G, et al. 2001. *Salmonella enterica* serotype Typhi possesses a unique repertoire of fimbrial gene sequences. *Infect Immun* **69:** 2894–2901.

Traggiai E, Chicha L, Mazzucchelli L, Bronz L, Piffaretti JC, Lanzavecchia A, Manz MG. 2004. Development of a human adaptive immune system in cord blood cell-transplanted mice. *Science* **304:** 104–107.

Tsolis RM, Adams LG, Ficht TA, Bäumler AJ. 1999. Contribution of *Salmonella typhimurium* virulence factors to diarrheal disease in calves. *Infect Immun* **67:** 4879–4885.

Tsui IS, Yip CM, Hackett J, Morris C. 2003. The type IVB pili of *Salmonella enterica* serovar Typhi bind to the cystic fibrosis transmembrane conductance regulator. *Infect Immun* **71:** 6049–6050.

van de Vosse E, Ali S, de Visser AW, Surjadi C, Widjaja S, Vollaard AM, van Dissel JT. 2005. Susceptibility to typhoid fever is associated with a polymorphism in the cystic fibrosis transmembrane conductance regulator (CFTR). *Hum Genet* **118:** 138–140.

Vayssier-Taussat M, Le Rhun D, Deng HK, Biville F, Cescau S, Danchin A, Marignac G, Lenaour E, Boulouis HJ, Mavris M, et al. 2010. The Trw type IV secretion system of *Bartonella* mediates host-specific adhesion to erythrocytes. *PLoS Pathog* **6:** e1000946.

Virlogeux I, Waxin H, Ecobichon C, Popoff MY. 1995. Role of the *viaB* locus in synthesis, transport and expression of *Salmonella typhi* Vi antigen. *Microbiology* **141:** 3039–3047.

Vladoianu IR, Chang HR, Pechere JC. 1990. Expression of host resistance to *Salmonella typhi* and *Salmonella typhimurium*: Bacterial survival within macrophages of murine and human origin. *Microb Pathog* **8:** 83–90.

Voges M, Bachmann V, Kammerer R, Gophna U, Hauck CR. 2010. CEACAM1 recognition by bacterial pathogens is species specific. *BMC Microbiol* **10:** 117.

Watson PR, Paulin SM, Bland AP, Jones PW, Wallis TS. 1995. Characterization of intestinal invasion by *Salmonella typhimurium* and *Salmonella dublin* and effect of a mutation in the invH gene. *Infect Immun* **63:** 2743–2754.

Watson PR, Galyov EE, Paulin SM, Jones PW, Wallis TS. 1998. Mutation of *invH*, but not *stn*, reduces *Salmonella*-induced enteritis in cattle. *Infect Immun* **66:** 1432–1438.

Weening EH, Barker JD, Laarakker MC, Humphries AD, Tsolis RM, Bäumler AJ. 2005. The *Salmonella enterica* serotype Typhimurium *lpf, bcf, stb, stc, std*, and *sth* fimbrial operons are required for intestinal persistence in mice. *Infect Immun* **73:** 3358–3366.

Wilson RP, Raffatellu M, Chessa D, Winter SE, Tukel C, Bäumler AJ. 2008. The Vi-capsule prevents Toll-like receptor 4 recognition of *Salmonella*. *Cell Microbiol* **10:** 876–890.

Winkelstein JA, Marino MC, Johnston RBJ, Boyle J, Curnutte J, Gallin JI, Malech HL, Holland SM, Ochs H, Quie P, et al. 2000. Chronic granulomatous disease. Report on a national registry of 368 patients. *Medicine (Baltimore)* **79:** 155–169.

Winter SE, Winter MG, Thiennimitr P, Gerriets VA, Nuccio SP, Russmann H, Bäumler AJ. 2009. The TviA auxiliary protein renders the *Salmonella enterica* serotype Typhi RcsB regulon responsive to changes in osmolarity. *Mol Microbiol* **74:** 175–193.

Winter SE, Thiennimitr P, Winter MG, Butler BP, Huseby DL, Crawford RW, Russell JM, Bevins CL, Adams LG, Tsolis RM, et al. 2010a. Gut inflammation provides a respiratory electron acceptor for *Salmonella*. *Nature* **467:** 426–429.

Winter SE, Winter MG, Godinez I, Yang H-J, Russmann H, Andrews-Polymenis HL, Bäumler AJ. 2010b. A rapid change in virulence gene expression during the transition from the intestinal lumen into tissue promotes systemic dissemination of *Salmonella*. *PLoS Pathog* **6:** e1001060.

Woolhouse ME, Taylor LH, Haydon DT. 2001. Population biology of multihost pathogens. *Science* **292:** 1109–1112.

Wu KY, Liu GR, Liu WQ, Wang AQ, Zhan S, Sanderson KE, Johnston RN, Liu SL. 2005. The genome of *Salmonella enterica* serovar Gallinarum: Distinct insertions/deletions and rare rearrangements. *J Bacteriol* **187:** 4720–4727.

Xu T, Maloy S, McGuire KL. 2009. Macrophages influence *Salmonella* host-specificity in vivo. *Microb Pathog* **47:** 212–222.

Zahrt TC. 1998. "A genetic analysis of the *Salmonella typhi* host range". PhD thesis, University of Illinois, Urbana-Champaign.

Zarantonelli ML, Szatanik M, Giorgini D, Hong E, Huerre M, Guillou F, Alonso JM, Taha MK. 2007. Transgenic mice expressing human transferrin as a model for meningococcal infection. *Infect Immun* **75:** 5609–5614.

Zeisel MB, Da Costa D, Baumert TF. 2011. Opening the door for hepatitis C virus infection in genetically humanized mice. *Hepatology* **54:** 1873–1875.

A Genome-Wide Perspective of Human Diversity and Its Implications in Infectious Disease

Jérémy Manry[1,2,3] and Lluis Quintana-Murci[1,2]

[1]Institut Pasteur, Unit of Human Evolutionary Genetics, Department of Genomes and Genetics, F-75015 Paris, France

[2]Centre National de la Recherche Scientifique, CNRS URA3012, F-75015 Paris, France

[3]Université Pierre et Marie Curie (UPMC), Cellule Pasteur, F-75015 Paris, France

Correspondence: lluis.quintana-murci@pasteur.fr

Progress in genomic technologies, such as DNA arrays and next-generation sequencing, is allowing systematic characterization of the degree of human genetic variation at the scale of individual genomes. Public efforts, such as the International HapMap Project and the 1000 Genomes Project, have provided a realistic picture of the levels of genetic diversity in individuals and populations. These genomic techniques are also making it possible to evaluate the contribution of host genetic diversity to differences in susceptibility to both rare and common infectious diseases. Recent studies have revealed the power of whole-exome sequencing for dissecting the immunological mechanisms underlying the pathogenesis of severe, rare infectious diseases. Likewise, genome-wide association studies on common viral, bacterial, and parasitic infections have shed light on the host genetic basis of susceptibility to infectious diseases and, in some cases, of disease progression and drug responses.

Dissection of the genetic basis of the phenotypic diversity observed in the human host is essential if we are to understand our relationship to disease, a major goal in biology and medicine. The release of the reference human genome sequence in the early 2000s (Lander et al. 2001; Venter et al. 2001) provided the first foundations for studies of the genetics of the human host, but provided little insight into the extent of naturally occurring genetic variation between the genomes of different individuals and populations that could account for their phenotypic variation. The advent of new genomic technologies, such as DNA microarrays and next-generation sequencing, has allowed comparative studies of variation at the level of the entire genome, between populations of different geographic origins and ethnic backgrounds, thereby increasing our understanding of the level of naturally occurring variation of the human genome and its relationship to disease. These genome-wide studies have yielded important insight into population histories and patterns of migration and admixture, into the genetic mechanisms underlying the adaptation of our species to changing environments, such as climate, nutritional resources, and pathogen pressures, and into the contribution of human genetics to disease susceptibility, severity, and response to treatment.

Cite this article as *Cold Spring Harb Perspect Med* doi: 10.1101/cshperspect.a012450

THE DIVERSITY OF THE HUMAN GENOME

Genome-Wide Data Sets for Human Genetic Variation

The Human Genome Project (Lander et al. 2001), the SNP Consortium (Sachidanandam et al. 2001), the International HapMap Project (The International HapMap Consortium 2005; Frazer et al. 2007; Altshuler et al. 2010), and, more recently, the 1000 Genomes Project (The 1000 Genomes Project Consortium 2010) have collectively identified ~15 million common DNA variants, mostly single-nucleotide polymorphisms (SNPs), constituting a key tool for investigations of the extent of genetic variation in human populations. The first large-scale studies of human genetic variation at the genome-wide level were based on genotyping technologies, such as analyses of variation at hundreds of thousands of SNPs surveyed only in the genomic regions probed by the array. The International HapMap Project has been the most popular and successful project of this type. It was launched in 2002, with the aim of providing genotype and haplotype data for common SNPs in 270 individuals from four geographically diverse populations (Yoruba from Nigeria, individuals of Western European descent from Utah, Han Chinese, and Japanese). Phase I of the project involved the genotyping of ~1.3 million SNPs (The International HapMap Consortium 2005), a figure extended to >3.1 million SNPs during phase II (Frazer et al. 2007). The goal of phase III has been to increase the number of samples and populations investigated, rather than the number of SNPs, by genotyping 1.6 million SNPs in 1184 individuals from 11 global populations, and sequencing 10 100-kb regions in 692 of these individuals (Altshuler et al. 2010). The International HapMap Project has thus provided a high-resolution catalog of the allele frequencies of hundreds of thousands of SNPs distributed throughout the genome, across human populations. These data have provided important insight into the fine-scale structure of linkage disequilibrium (LD; i.e., the pattern of correlation between SNPs located close together on the chromosome) in the genome (Conrad et al. 2006b; Frazer et al. 2007), the distribution and causes of recombination hotspots (McVean et al. 2004; Myers et al. 2005), the prevalence of structural variation such as copy number polymorphisms (CNPs) (Conrad et al. 2006a; Redon et al. 2006), and the identity of genes that have been targeted by different forms of natural selection in the human genome (Voight et al. 2006; Frazer et al. 2007; Sabeti et al. 2007; Barreiro et al. 2008).

These genotyping resources have had a profound impact on the design and interpretation of genome-wide association studies (GWAs), by making it possible for the vast majority of common variants—defined as having a frequency >5%—to be tested for association with various traits or diseases. However, the down side of these genotyping approaches is that they largely ignore low-frequency (from 0.5% to 5%) and rare (<0.5% frequency) variants, which make up a large proportion of the human genome and may also contribute to disease susceptibility (Manolio et al. 2009). The 1000 Genomes Project was launched in 2008, to improve our understanding of the genetic contribution to human phenotypes through a consideration of the entire allelic spectrum of sequence-based variation (The 1000 Genomes Project Consortium 2010). Its aim was to provide a large number of complete human genome sequences and, therefore, accurate information on all forms of DNA polymorphism in major population groups from West Africa, Europe, East and South Asia, and the Americas. More specifically, the goal was to characterize >95% of the variants with an allele frequency of 1% or higher (the accepted, conventional definition of polymorphism), or a frequency of 0.5% or lower within genes (the probability to find deleterious variants in this frequency range is the highest). The results of the pilot phase of the project have recently been released, providing information about the location, allele frequency, and local haplotype structure of ~15 million SNPs, 1 million short insertions and deletions, and 20,000 structural variants, most of which were newly identified in this project (Fig. 1). Overall ~8.4 million SNPs and ~840,000 insertions/deletions were described for the first time (The 1000 Genomes Project Consortium 2010). Collectively, these

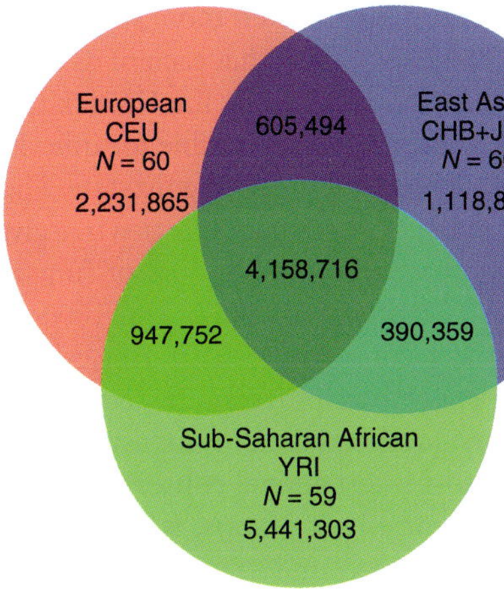

Figure 1. Venn diagram showing the numbers of SNPs identified in the low-coverage pilot project on the 1000 Genomes Project (The 1000 Genomes Project Consortium 2010), in each population, including the already known (present in dbSNP release 129) and new SNPs. *N* is the number of individuals sequenced. CEU, European ancestry in Utah; CHB + JPT, Han Chinese individuals in Beijing and Japanese individuals in Tokyo; YRI, Yoruba from Ibadan, Nigeria.

sequencing data, together with the genotyping data obtained from the International HapMap Project (Altshuler et al. 2010) and the HGDP-CEPH panel (Li et al. 2008), have provided a realistic picture of the existing levels of genetic variation in our species, including the identification of rare variants, both at the interindividual and interpopulation levels.

Human Genome Diversity and Population History

The level of genetic variation of a population reflects its demographic history. Characterization of the patterns of genetic diversity in large numbers of human populations from diverse ethnic backgrounds should therefore provide important information about human origins and migration patterns. In this context, genome-wide data have confirmed that African populations have the highest levels of genetic diversity and that non-African diversity is essentially a subgroup of the diversity observed in Africans (Frazer et al. 2007; Li et al. 2008; Altshuler et al. 2010; The 1000 Genomes Project Consortium 2010). This provides conclusive support for a recent African origin of modern humans ~200,000 years ago, as suggested by other pieces of evidence from studies of uniparentally inherited mitochondrial DNA and Y-chromosome DNA and sets of neutrally evolving regions of the genome (Ingman et al. 2000; Thomson et al. 2000; Cavalli-Sforza and Feldman 2003; Voight et al. 2005; Fagundes et al. 2007; Campbell and Tishkoff 2008; Laval et al. 2010). The 1000 Genomes Project has shown, for example, that populations with African ancestry have the highest proportion of novel variants, reflecting the greater diversity of African populations (The 1000 Genomes Project Consortium 2010). However, as Asian and European populations have been expanding more rapidly in the recent past than the African population, it has been suggested that the sequencing of larger numbers of individuals from each population might yield a larger number of polymorphic sites in Eurasians than in Africans (Gravel et al. 2011). All the existing genome-wide data sets for human genetic variation have increased our understanding of the demographic history of human populations and provided insight into the degree of population genetic structure and levels of admixture in many populations worldwide. This information is essential for studies of association with disease.

Human Genome Diversity and Natural Selection

The observed patterns of population diversity or population structure can be explained by random genetic drift in most cases, reflecting the demographic history of the population under study. However, natural selection also influences our genomic diversity (Sabeti et al. 2006; Nielsen et al. 2007). Indeed, many genome-wide scans for selection have shown that the observed genetic patterns are incompatible with neutrality and have identified a number of genomic regions that have been subject to positive selection (Voight et al. 2006; Frazer et al. 2007; Sabeti et al. 2007; Barreiro et al. 2008; Akey 2009; Pickrell et al. 2009; Hancock et al. 2011). The detection of genes subject to positive selection facilitates the identification of genes harboring functional variants that have conferred a selective advantage on the host. Studies of this type have identified genes and gene variants potentially accounting for differences in human phenotypes, which may be benign (e.g., skin color, height, hair development) or disease related. Collectively, these massive genome-wide data sets have shed light on the intensity and types of natural selection acting on the human genome, providing information complementary to that generated by clinical and epidemiological genetic approaches, through the identification of genes playing an important role in disease susceptibility and progression and, more generally, in host survival (Sabeti et al. 2006; Nielsen et al. 2007; Quintana-Murci et al. 2007; Gilad et al. 2009; Barreiro and Quintana-Murci 2010).

The Burden of Deleterious Mutations in the Human Population

Both demographic factors and natural selection shape the patterns of genetic diversity in populations, and the cumulative effects of these two forces have an impact on the burden of potentially deleterious mutations present in the human population. For example, the 1000 Genomes Project has estimated that each of us carries 190–210 in-frame insertions/deletions, 80–100 premature stop codons, 40–50 splice-site-disrupting variants, and 220–250 frameshift deletions, most of which lead to premature stop codons (Table 1) (The 1000 Genomes Project Consortium 2010). Even more surprisingly, this project has estimated that each of us carries 50–100 variants classified by the Human Gene Mutation Database as responsible for inherited disorders, albeit in the heterozygous state. Thus, 250–300 of the ~20,500 genes carried by a given individual harbor putative loss-of-function mutations (Clamp et al. 2007; The 1000 Genomes Project Consortium 2010). However, despite the enormous progress made in recent years, many questions remain unanswered. For example, most international consortia have focused on major ethnic groups, whereas ethnic minorities are largely underrepresented in these studies. The inclusion of adequate numbers of individuals from these minorities is of crucial importance for the identification of population-specific or region-specific functional variants associated with disease or drug response (Tishkoff et al. 2009). In addition, most of the signals of natural selection detected to date remain unexplained. The functional impact of the selected variants is usually postulated on the basis of predictive algorithms, such as PolyPhen (Adzhubei et al. 2010), but the actual role of most of the variants of interest remains unknown (Lohmueller et al. 2008).

THE INFLUENCE OF HOST GENETIC DIVERSITY ON HUMAN DISEASE

Studies of the degree of genetic variation at the level of the genome are making it possible to test blindly for genetic associations with disease

Table 1. Different classes of genic variants, estimated by the 1000 Genomes Project low-coverage data set (The 1000 Genomes Project Consortium 2010)

			Population range[a]		
Genic variants	Total	Total range[a]	African (YRI)	European (CEU)	East-Asian (CHB + JPT)
Synonymous SNPs	55,217	10,572–12,126	12,334–12,591	10,533–10,652	10,357–10,471
Nonsynonymous SNPs	61,284	9966–10,819	11,048–11,422	9690–9776	9932–10,115
Stop-introducing SNPs	951	88–101	93–100	74–78	102–108
Splice-site-disrupting SNPs	500	41–49	41–47	34–38	53–58
Small frameshift indels	890	227–242	247–255	217–223	226–230
Genes disrupted by large deletions	143	28–36	32–35	34–37	33–37

SNP, single-nucleotide polymorphism.

[a] These ranges correspond to the interquartile range of the number of variants of specified type per individual.

susceptibility, pathogenesis, or response to treatment, without having to make prior assumptions as in candidate-gene studies. Such studies have been done for both rare ("Mendelian") and common (mostly "complex") diseases. The dividing line between Mendelian and complex diseases may be considered vague and simplistic, but Mendelian diseases are usually rare and mostly result from highly penetrant, deleterious alleles that segregate in families (Zwick et al. 2000; Pritchard and Cox 2002). However, most phenotypic differences in morphology, physiology, and disease are quantitative rather than discrete and follow a continuous, almost normal distribution (Pritchard and Cox 2002). These so-called "complex traits" are generally caused by a number of genetic and environmental factors and they therefore do not follow simple Mendelian rules of inheritance (Badano and Katsanis 2002; Botstein and Risch 2003). Genes involved in human disease may vary considerably in the number and population frequency of the corresponding disease-predisposing alleles (Fig. 2) (Reich and Lander 2001). For severe Mendelian diseases, the frequency of susceptibility mutations is usually very low ($<<1\%$) and is inversely proportional to the severity of the disease (Pritchard and Cox 2002). Such simple rules cannot be applied to complex susceptibility to disease. In this context, two principal nonmutually exclusive hypotheses have been put forward. According to the *common disease–rare variant* hypothesis, a significant proportion of the inherited susceptibility to disease results from the additive effects of a number of low-frequency, independent variants of different

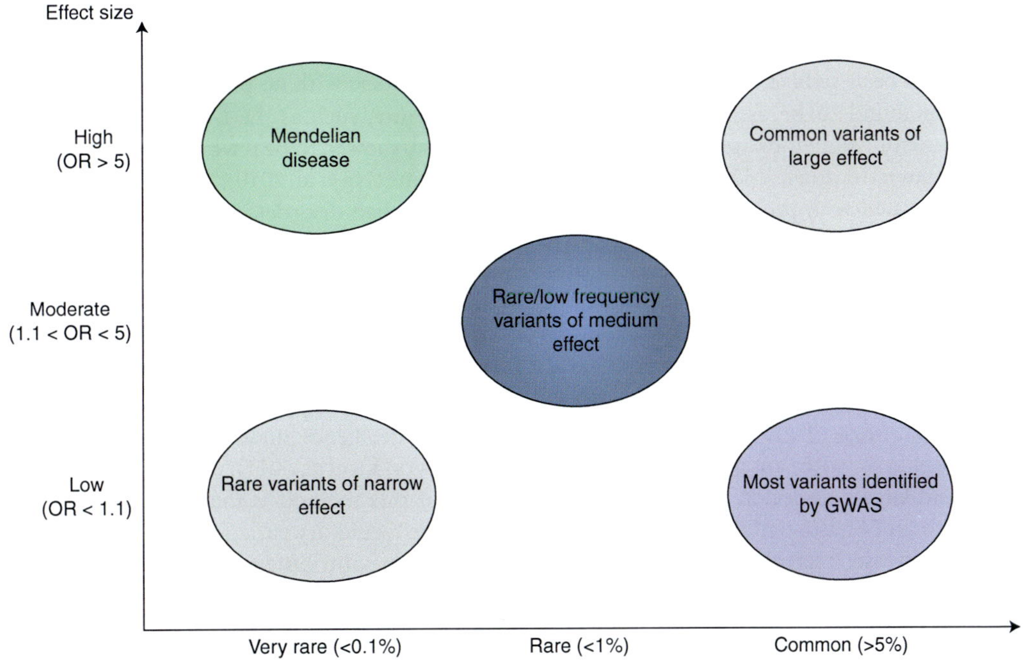

Figure 2. Scheme representing the frequency of the variants and their effects in the framework of genome-wide association studies based on genotyping by next-generation sequencing. Most of the variants identified by GWAs are common variants with a moderate effect but, with advances in next-generation sequencing, it is now possible to identify rarer variants, as reported in recent exome studies detecting very rare variants responsible for Mendelian diseases. Rare variants with weak effects are likely to be difficult to identify genetically, as a number of such variants will probably be required to account for a disease or trait and this will require high levels of coverage of the entire genome. Common variants with a large effect are unusual as they would be expected to be eliminated by natural selection and should therefore not be present at such high frequencies. (From McCarthy et al. 2008; adapted, with permission.)

genes, with each variant making a moderate contribution to disease risk (Bodmer and Bonilla 2008). According to the *common disease–common variant* hypothesis, disease susceptibility is often owing to variants with relatively high population frequencies (Reich and Lander 2001).

In recent years, genome-wide studies of disease susceptibility have involved the genotyping of hundreds of thousands of SNPs, using array technologies such as those of Illumina or Affymetrix, and have focused principally on common, complex diseases or traits. By comparing the genome-wide frequencies of SNPs between affected and unaffected individuals, it is possible to identify genomic regions, and specific SNPs in particular, that are associated with the disease or trait studied, through the use of various statistical thresholds (i.e., the *P* values used to define a SNP as "robustly" associated with a given disease may differ in the extent to which they are conservative). By January 2012, >1000 GWAs had been published and >5000 SNPs had been found to be associated, to various extents, with numerous phenotypes (www.genome.gov). In total, 237 phenotypes have been associated with precise genomic regions, at P values $<5 \times 10^{-8}$. The diseases, syndromes, or traits overrepresented in these studies are complex traits or late-onset disorders, such as asthma (21 studies), Alzheimer's disease (27), breast cancer (20), height (19), prostate cancer (19), schizophrenia (25), and type 2 diabetes (29). Most of these GWAs investigated polymorphisms with an allele frequency of >5% and largely neglected rare variants. This has led to an overassociation of common variants with disease. It has also been suggested that some GWA signals may reflect the effects of multiple, rare variants that have been credited to common variants (Dickson et al. 2010).

From a genome-wide perspective, studies of Mendelian disorders, which are expected to result from low-frequency variants, have progressed much more slowly than those focusing on complex traits or diseases. High-throughput sequencing is now making it possible to take into account these rare, sometimes population-specific variants. In this context, the whole-genome sequencing of affected individuals would be the most effective method, because it can capture causal variants located both within genes and elsewhere in the genome (e.g., regulatory variants). However, the technology required remains costly. In contrast, whole-exome sequencing, in which only exons, accounting for ~1% of the genome, are resequenced, has become technically feasible and affordable. It is now the most widely used genome-wide method in research into Mendelian disorders. In 2009, Ng and colleagues published the first exome-based study, providing proof of concept that causal genetic variants could be identified by whole-exome sequencing (Ng et al. 2009). A comparison of the exomes of individuals with Freeman-Sheldon syndrome (FSS), a rare autosomal-dominant disorder, with those of eight HapMap individuals led to the identification of *MYH3*, which had previously been implicated as the causal gene of FSS. Subsequent studies have identified variants or genes involved in numerous Mendelian disorders with no previously identified genetic cause, such as the *DHODH* gene, which Ng et al. showed to be the causal agent of Miller syndrome (Ng et al. 2010). More than 40 studies of Mendelian disorders based on whole-exome sequencing or the sequencing of whole genomes for a small number of individuals have been published. Ku and coworkers recently reviewed many of the exome studies done to date, highlighting the power and efficiency of whole-exome sequencing as an approach to identification of the genes underlying rare Mendelian disorders (Ku et al. 2011). One of the shortcomings of this method is that it cannot pick up possible regulatory variation outside genes that might also contribute to disease.

GENOME-WIDE STUDIES OF MENDELIAN DEFECTS IN IMMUNITY AND INFECTION

The Mendelian inheritance of numerous severe, rare infectious diseases or immunological defects has been dissected mostly by candidate-gene approaches. As this topic has been extensively reviewed elsewhere (Casanova and Abel 2005; Hill 2006; Fellay et al. 2009), we will focus here exclusively on studies based on genome-

 Cite this article as *Cold Spring Harb Perspect Med* doi: 10.1101/cshperspect.a012450

wide approaches. The first identification of a Mendelian defect related to an infectious disease by a genome-wide approach was reported in 2010 (Byun et al. 2010). By whole-exome sequencing in a single child, the investigators identified a mutation responsible for fatal Kaposi sarcoma (KS), which although not infectious in itself, is directly related to human herpesvirus-8 (HHV-8) infection. After quality control and the exclusion of variants corresponding to simple polymorphisms, a splice-site mutation in the *STIM* gene, which is known to be involved in a primary T-cell immunodeficiency (Picard et al. 2009), was the only remaining potential disease-causing variant. *STIM1* mRNA splicing, protein production, and Ca^{2+} influx were abolished in cell lines from the patient, but rescued by the expression of wild-type *STIM1*. This work showed that childhood KS could result from a single-gene defect and provided proof of concept that whole-exome sequencing in single patients could be used to decipher the genetic basis of rare inborn errors of immunity (Byun et al. 2010).

Since this pioneering study, other studies on Mendelian susceptibility to infectious diseases have been published. For example, a combination of genome-wide linkage analysis and whole-exome sequencing led to the identification of a homozygous missense mutation in the *FADD* gene of patients with autoimmune lymphoproliferative syndrome (ALPS) (Bolze et al. 2010). In this case, ALPS was observed in a context of severe bacterial and viral diseases, recurrent liver disease and encephalopathy, together with cardiac malformations. Supporting their genetic observations with functional analyses, the investigators determined not only the genetic basis of this complex disorder, but also the mechanisms underlying its pathogenesis. They showed that the observed bacterial infections resulted partly from functional hyposplenism, and that the observed viral infections resulted from impaired interferon (IFN) immunity. A similar approach has been used to show that germline mutations of *STAT1* are responsible for chronic mucocutaneous disease (CMCD) (Liu et al. 2011). Loss-of-function *STAT1* mutations had previously been implicated in predisposition to intracellular bacterial and viral diseases, impairing the STAT1-dependent response to IFNs (Zhang et al. 2008). However, the more recent exome-based study showed that some patients have a gain-of-function *STAT1* mutation that leads to an impairment of IL-17 immunity, causing autosomal-dominant CMCD (Liu et al. 2011).

A recent study based on whole-exome sequencing in a patient suffering from a Crohn disease-like illness not only identified the causal mutation but also highlighted the potential use of this technology in a clinical setting (Worthey et al. 2011). The exome study identified a missense mutation in the X-linked inhibitor of apoptosis gene (XIAP). The XIAP protein plays a key role in the proinflammatory response and in bacterial sensing via the NOD signaling pathway. The results of functional analyses were consistent with aberrant XIAP function, with impaired IL-8 production in response to the stimulation of the patient's cells with muramyl dipeptide, and enhanced apoptosis after the CD3 stimulation of peripheral blood mononuclear cells (PBMC). Based on these findings, allogeneic hematopoietic progenitor cell transplantation was performed, in accordance with recommendations for the treatment of XIAP deficiency. An assessment more than 42 days after transplantation showed that the child had recovered, and no recurrence of gastrointestinal disease was observed (Worthey et al. 2011).

Collectively, these studies show that whole-exome sequencing in a small number of individuals, or even an isolated individual, is an efficient and cost-effective strategy for investigating the mechanisms underlying the pathogenesis of severe infectious diseases or immunity-related disorders, and for improving disease diagnosis, potentially leading to better treatment and personalized medicine.

GENOME-WIDE STUDIES OF COMPLEX SUSCEPTIBILITY TO INFECTIOUS DISEASES

GWAs of susceptibility to infectious diseases have progressed much more slowly than similar studies of other diseases, such as cancer, psychiatric disorders, or metabolic diseases, with only seven complex infectious diseases having been studied in this way to date (for an extensive review, see Vannberg et al. 2011). However various GWAs for viral, bacterial, and parasitic infectious diseases have, in some cases, shed light on the contribution of host genetic diversity to the differences in response to both treatment and vaccine administration between individuals.

Viral Infections

The first reported GWAs of an infectious disease focused on HIV-1 infection and found that various SNPs in the human leukocyte antigen (HLA) region were associated with viral load (Fellay et al. 2007). They investigated HIV-1 viral load set point during the asymptomatic period of infection or the viral load "resting state" after the peak of viremia following acute infection. The strongest correlation was obtained for HLA-B and HLA-C, which were independently associated with a lower viral load set point (Fellay et al. 2007). A subsequent study by the same investigators confirmed these association signals, with some variants in this genomic region showing the strongest associations with both viral set point and progression (Fellay et al. 2009). The most likely causal variants appear to be the HLA-B*5701 allele and the HLA-C promoter SNP-35C, which has been shown to regulate the production of HLA-C mRNA (Fellay et al. 2007, 2009). The investigators estimated that human genetic variation and demographic variables could account for up to 22% of the variability in HIV-1 viral load, at least in the European population (Fellay et al. 2009).

Since this initial study, other GWAs of viral infections have been conducted. The association between hepatitis C virus (HCV) infection and the *IL28B* locus, which encodes interferon-λ3, merits particular attention. Several SNPs within or close to *IL28B* have been found to be associated with spontaneous HCV clearance or a better response to pegylated IFNα-ribavirin treatment for chronic HCV infection in various human populations (Ge et al. 2009; Suppiah et al. 2009; Tanaka et al. 2009; Thomas et al. 2009; Rauch et al. 2010). The frequency of rs12979860, a SNP located 3 kb upstream of *IL28B* and associated with both spontaneous viral clearance and better response to drug treatment, varies significantly between populations, with the protective C allele being most prevalent in Asians, present at an intermediate frequency in Europeans and least prevalent in individuals of African descent (Fig. 3) (Ge et al. 2009; Thomas et al. 2009). Furthermore, it has recently been shown that this high frequency of the protective allele in Eurasian populations is the result of positive selection (Manry et al. 2011). These studies also pave the way for improvements in the treatment of HCV, to minimize adverse effects. Indeed, ribavirin-induced hemolytic anemia is a major side effect of pegylated IFNα-ribavirin treatment, necessitating dose modification in up to 15% of patients. It has recently been shown that *ITPA* gene variants protect against hemolytic anemia in patients with hepatitis C treated with ribavirin (Fellay et al. 2010), again highlighting the importance of host genotype in response to treatment.

Another study identified 11 SNPs in a region including *HLA-DPA1* and *HLA-DPB1* as associated with hepatitis B virus (HBV) infection in a Japanese cohort (Kamatani et al. 2009). These associations were replicated in additional cohorts, and specific haplotypes associated with risk and protection were identified (Kamatani et al. 2009; Nishida et al. 2012). These findings suggest that antigen presentation on HLA-DP molecules may be critical for virus elimination, the pathogenesis of chronic hepatitis B. In the context of public health, the standard course of vaccination against HBV fails to elicit protective antibody levels in almost 10% of adult recipients. A GWAS of antibody titers in a cohort of hepatitis B vaccine recipients from Indonesia has recently been conducted, to identify genetic variants explaining such differences in vaccine responses (Png et al. 2011). The investigators identified a number of hits within the HLA region, including HLA-DR, HLA-DP, and a gene-rich HLA class III interval. The concordance between the alleles associated with protective effects against chronic HBV infection and those associated with a response to the vaccine (i.e., HLA-

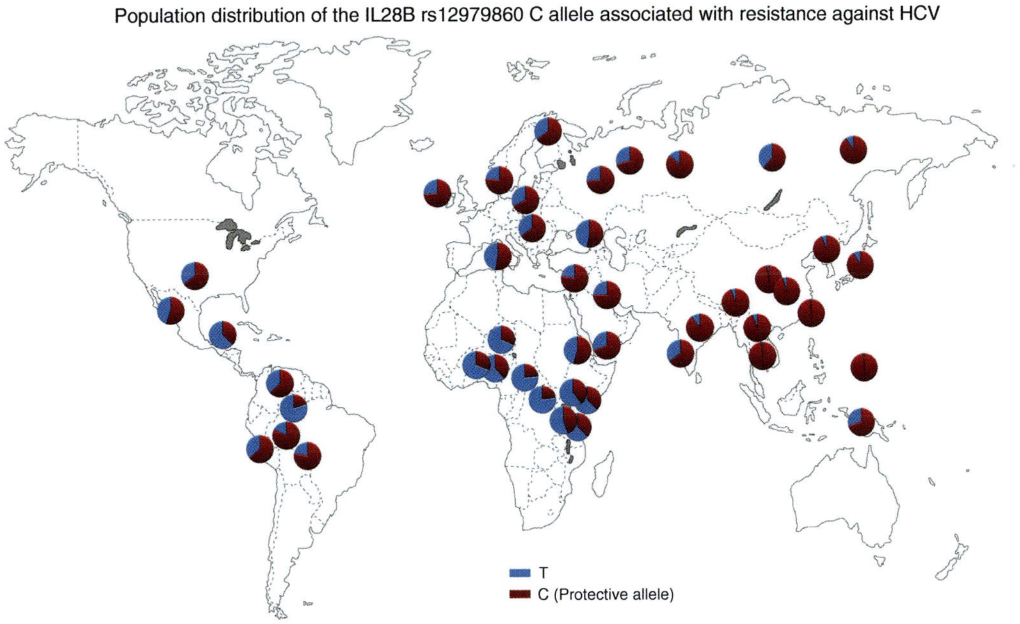

Figure 3. Worldwide distribution of the *IL28B* SNP rs12979860 allele variation. The pie charts show the frequency of the T (blue) and C (red) alleles in each population sampled in the ALFRED database. The "protective" C variant has been associated with both a better response to HCV drug treatment and enhanced spontaneous HCV viral clearance (Ge et al. 2009; Thomas et al. 2009).

DP) suggests that the genetic bases of the two phenotypes may have major aspects in common.

More recently, another GWAs has identified several SNPs associated with variation in cytokine responses in smallpox vaccine recipients, suggesting the complexity of the host response to vaccinia virus involving multiple genes and gene families (Kennedy et al. 2012a,b; Ovsyannikova et al. 2012). The most significant result they obtained was an intronic SNP in the *WDR92* gene, which is associated with variation in IFN-α secretion. This, together with the reported associations of HLA-DP variations with HBV infection (Kamatani et al. 2009; Nishida et al. 2012) and of *IL28B* variants with both spontaneous viral clearance and response to treatment in HCV infection (Ge et al. 2009; Suppiah et al. 2009; Tanaka et al. 2009; Thomas et al. 2009; Rauch et al. 2010), suggests that the identification of host genetic variants associated with different susceptibilities to infection may generally provide useful information for improving both the prevention and treatment of infectious diseases.

Bacterial Infections

Three major bacterial infections have been studied from a genome-wide perspective: tuberculosis, leprosy, and meningococcal disease. The initial tuberculosis study combined two GWAs from Ghana and the Gambia and found weak to moderate associations with several genes, including those of the HLA-DQ region (Thye et al. 2010). The most significant hit concerned a non-HLA region located in a gene-poor region on chromosome 18q11.2. After imputation analysis of the data from the 1000 Genomes Project on the same Ghanian cohort, a possible resistance locus on chromosome 11p13 has been more recently identified (Thye et al. 2012), but the findings remain overall conflicting (Mahasirimongkol et al. 2012; Png et al. 2012). The apparent lack of success of these studies may conceal an interesting and more promising result. Gene-poor regions, such as the one on chromosome 18q11.2 associated with tuberculosis (Thye et al. 2010), may harbor regulatory elements (The ENCODE Project Consortium

2012) that may play a major role in disease susceptibility in addition to that of classical missense mutations in genes.

A first genome-wide association for leprosy detected significant associations with SNPs in the *CCDC122*, *C13orf31* (LACC1), *NOD2*, *TNFSF15*, *HLA-DR*, *RIPK2* and, possibly, *LRRK2* genes in a cohort from China (Zhang et al. 2009). Subsequent studies achieved replication of these findings for *CCDC122* and *C13orf31* (LACC1) (Wong et al. 2010b) or obtained signals for other loci, such as *HLA-DRB1*, *HLA-DQA1*, and *TLR1* (Wong et al. 2010a) in the Indian population. New susceptibility genes for leprosy have recently been identified and include regions encompassing the *IL23R* and *RAB32* genes (Zhang et al. 2011). The *RAB32* gene has been implicated in the response to *Mycobacterium tuberculosis* infection (Seto et al. 2011), suggesting that there may be genetic variants involved in susceptibility to both of these mycobacterial infections (Zhang et al. 2011). The conflicting results obtained in leprosy studies may reflect genuine biological differences in leprosy susceptibility between populations, or more technical issues relating to study design, such as the small sample sizes and low power of some studies.

Finally, a GWAS of meningococcal disease in various European cohorts found and replicated an association with a cluster of complement factor SNPs, including a missense mutation in the complement factor H gene (*CFH*) and an intronic mutation in the CFH-related protein 3 gene (*CFHR3*) (Davila et al. 2010). These results suggest that genetic variation in complement genes may play an important role in determining whether a subject presents invasive disease or asymptomatic *Neisseria meningitides* colonization, opening up new avenues of research relating to possible treatment of the disease by pharmacological modulations of CFH protein levels in vivo.

Parasitic Infections

Malaria, particularly its severe forms, is the only parasitic disease to have been studied from a genome-wide perspective. A large GWAS of severe malaria in a cohort of children from the Gambia identified the strongest signal of association as being located close to the *HBB* gene on chromosome 11p15 (Jallow et al. 2009), to which the well-known HbS polymorphism (rs334) maps. HbS homozygotes suffer from sickle-cell disease, but heterozygotes have a 10-fold-reduced risk of severe malaria, which explains the high frequencies (up to 30%) of this allele in Africa (Allison 1954).

By combining genome-wide studies with the targeted resequencing of the *HBB* genomic region in a subset of individuals, the investigators identified the causal HbS polymorphism as the SNP displaying the strongest association. This study provided proof of principle that the causal polymorphism within a GWA signal can be identified by regional sequencing of the corresponding genomic region. In addition to the *HBB* hit, the investigators identified 19 additional but weaker association signals, including regions encompassing or located close to genes such as *SPATA3*, *LOC257407*, *PSMD1*, and *GPR55*. Some of these variants may be authentic loci with a modest genetic effect on malaria susceptibility, but further validation is required.

INSIGHT FROM THE EVOLUTIONARY APPROACH TO INFECTIOUS DISEASE RESEARCH

The evolutionary genetics of immunity is a field of research investigating whether and how natural selection has shaped the variability of host defense or immunity-related genes in present-day human populations. This approach has proved an indispensable complement to clinical and epidemiological genetics for understanding the biological relevance of host genes in host survival (Casanova and Abel 2007; Nielsen et al. 2007; Quintana-Murci et al. 2007; Sabeti et al. 2007; Barreiro and Quintana-Murci 2010). Indeed, the burden of infectious diseases has been massive throughout history, with life expectancy not exceeding 20 to 25 years of age until the advent of Pasteur's microbial theory of disease and the resulting control of infections by hygiene, vaccines, and antibiotics (Casanova and Abel 2005). It is therefore hardly

 Cite this article as *Cold Spring Harb Perspect Med* doi: 10.1101/cshperspect.a012450

surprising that immunity-related genes appear to be among those with biological functions most strongly targeted by natural selection in the human genome, both with respect to the chimpanzee and within different human populations (Barreiro and Quintana-Murci 2010). For example, several genome-wide scans for positive selection have detected more than 5000 loci presenting at least one genomic signature of positive selection (see Akey 2009 and Barreiro and Quintana-Murci 2010 for reviews). Within the genomic regions targeted by positive selection, 360 genes with immune-related functions have been identified, including 186 genes identified as targets of positive selection in at least two studies (Barreiro and Quintana-Murci 2010). This group of "selected genes" may display functional variation that is differentially distributed between human populations and is therefore highly likely to be involved in the differences in susceptibility to infectious diseases or, more generally, to immunity-related disorders, such as autoimmune, inflammatory, or allergic disorders, observed today in human populations. The most obvious source of selection pressure on these genes is the presence of the pathogen (e.g., pathogen-driven selection). Proof of concept is provided by HLA class I, blood group antigen, and interleukin genes, for which genetic variation has been shown to be associated with pathogen species richness and diversity (Prugnolle et al. 2005; Fumagalli et al. 2009a,b). Balancing selection, which is considered to be rare in humans, is known to have acted on several immune-related genes; the best known example is major histocompatibility complex (MHC) genes, which display an extraordinary amount of variation in the population because heterozygous individuals are able to recognize a wider variety of microbes (Prugnolle et al. 2005; Cagliani et al. 2008; Andres et al. 2009; Fumagalli et al. 2009a,b).

Past selection, leading to an increase in resistance to the presence of the pathogen may lead, in some cases, to higher susceptibility to autoimmune or inflammatory disorders, as predicted by the hygiene hypothesis (Sironi and Clerici 2010). For example, several interleukin/interleukin receptor genes involved in the pathogenesis of inflammatory bowel disease (IBD) and celiac disease (CeD) have been subjected to pathogen-driven selective pressure. Specifically, the IBD/CeD risk alleles of *IL18-RAP*, *IL18R1*, *IL23*, *IL18R1*, and the intergenic region between *IL2* and *IL21* are more frequent in populations exposed to high microbial/viral loads, suggesting that these variants play a protective role in the host response to these pathogens (Fumagalli et al. 2009b; Sironi and Clerici 2010). Furthermore, several alleles or haplotypes associated with greater susceptibility to autoimmune disorders, including type I diabetes (T1D), CeD, and multiple sclerosis (MS), have been shown to have been subject to recent positive selection (Barreiro and Quintana-Murci 2010). This suggests that some unknown selective pressure, possibly pathogen driven, has resulted in an increase in the frequency of risk alleles/haplotypes in the general population or even in specific populations. For example, European populations display high frequencies of variants conferring susceptibility to celiac disease (Abadie et al. 2011).

CONCLUDING REMARKS

The relatively slow pace at which genome-wide research into infectious diseases has progressed reflects, at least partially, the complex nature of infectious diseases, which have diverse genetic bases, extending from single genetic defects to the presence of complex susceptibility alleles with various degrees of penetrance and much more complex situations in which many common or rare variants are involved, each having a modest effect. In the next few years, whole-genome sequencing technology is likely to become more affordable, and this will probably lead to a huge increase in the number of studies based on these techniques. This technology will be used not only in the context of rare infectious diseases or immunological defects probably resulting from single-gene lesions, but also in the context of complex susceptibility to infectious diseases, which may be caused by numerous low-frequency variants. The genetic data obtained from these whole-genome sequence-based association studies will then be combined with other independent sources of data, in large cohorts of individuals presenting clinically well-defined infectious phenotypes. Multidisciplinary approaches, as those used by the ENCODE Project Consortium (The ENCODE Project Consortium 2012), integrating whole-genome sequencing data, expression, and proteomic profiles obtained with arrays, RNA-Seq and mass spectrometry, epigenetic markers such as genome-wide methylation profiles, and evolutionary and population genetics data should facilitate the identification of functionally important genes and variants involved in host immunity to infection and uncover the immunological mechanisms favoring our past and present survival in the natural setting. This, together with improvements in statistical methods and modeling, will facilitate the dissection and more precise quantification of the contribution of host genetics to the considerable heterogeneity between individuals and populations in terms of susceptibility to infection, disease progression, and drug responses.

ACKNOWLEDGMENTS

This work is supported by the Institut Pasteur, the Agence Nationale de la Recherche (ANR-08-MIEN-009-01), the Fondation pour la Recherche Médicale (FRM), the Centre National de la Recherche Scientifique, and an EPFL Debiopharm Life Sciences Award to L.Q.-M. J.M. is supported by an FRM fellowship.

REFERENCES

Abadie V, Sollid LM, Barreiro LB, Jabri B. 2011. Integration of genetic and immunological insights into a model of celiac disease pathogenesis. *Annu Rev Immunol* **29:** 493–525.

Adzhubei IA, Schmidt S, Peshkin L, Ramensky VE, Gerasimova A, Bork P, Kondrashov AS, Sunyaev SR. 2010. A method and server for predicting damaging missense mutations. *Nat Methods* **7:** 248–249.

Akey JM. 2009. Constructing genomic maps of positive selection in humans: Where do we go from here? *Genome Res* **19:** 711–722.

Allison AC. 1954. Protection afforded by sickle-cell trait against subtertian malareal infection. *Br Med J* **1:** 290–294.

Altshuler DM, Gibbs RA, Peltonen L, Dermitzakis E, Schaffner SF, Yu F, Bonnen PE, de Bakker PI, Deloukas P, Gabriel SB, et al. 2010. Integrating common and rare genetic variation in diverse human populations. *Nature* **467:** 52–58.

Andres AM, Hubisz MJ, Indap A, Torgerson DG, Degenhardt JD, Boyko AR, Gutenkunst RN, White TJ, Bustamante CD, Clark AG, et al. 2009. Targets of balancing selection in the human genome. *Mol Biol Evol* **26:** 2755–2764.

Badano JL, Katsanis N. 2002. Beyond Mendel: An evolving view of human genetic disease transmission. *Nat Rev Genet* **3:** 779–789.

Barreiro LB, Quintana-Murci L. 2010. From evolutionary genetics to human immunology: How selection shapes host defence genes. *Nat Rev Genet* **11:** 17–30.

Barreiro LB, Laval G, Quach H, Patin E, Quintana-Murci L. 2008. Natural selection has driven population differentiation in modern humans. *Nat Genet* **40:** 340–345.

Bodmer W, Bonilla C. 2008. Common and rare variants in multifactorial susceptibility to common diseases. *Nat Genet* **40:** 695–701.

Bolze A, Byun M, McDonald D, Morgan NV, Abhyankar A, Premkumar L, Puel A, Bacon CM, Rieux-Laucat F, Pang K, et al. 2010. Whole-exome-sequencing-based discovery of human FADD deficiency. *Am J Hum Genet* **87:** 873–881.

Botstein D, Risch N. 2003. Discovering genotypes underlying human phenotypes: Past successes for mendelian disease, future approaches for complex disease. *Nat Genet* **33:** 228–237.

Byun M, Abhyankar A, Lelarge V, Plancoulaine S, Palanduz A, Telhan L, Boisson B, Picard C, Dewell S, Zhao C, et al. 2010. Whole-exome sequencing-based discovery of STIM1 deficiency in a child with fatal classic Kaposi sarcoma. *J Exp Med* **207:** 2307–2312.

Cagliani R, Fumagalli M, Riva S, Pozzoli U, Comi GP, Menozzi G, Bresolin N, Sironi M. 2008. The signature of long-standing balancing selection at the human defensin β-1 promoter. *Genome Biol* **9:** R143.

Campbell MC, Tishkoff SA. 2008. African genetic diversity: Implications for human demographic history, modern human origins, and complex disease mapping. *Annu Rev Genomics Hum Genet* **9:** 403–433.

Casanova JL, Abel L. 2005. Inborn errors of immunity to infection: The rule rather than the exception. *J Exp Med* **202:** 197–201.

Casanova JL, Abel L. 2007. Human genetics of infectious diseases: A unified theory. *EMBO J* **26:** 915–922.

Cavalli-Sforza LL, Feldman MW. 2003. The application of molecular genetic approaches to the study of human evolution. *Nat Genet* **33:** 266–275.

Clamp M, Fry B, Kamal M, Xie X, Cuff J, Lin MF, Kellis M, Lindblad-Toh K, Lander ES. 2007. Distinguishing protein-coding and noncoding genes in the human genome. *Proc Natl Acad Sci* **104:** 19428–19433.

Conrad DF, Andrews TD, Carter NP, Hurles ME, Pritchard JK. 2006a. A high-resolution survey of deletion polymorphism in the human genome. *Nat Genet* **38:** 75–81.

Conrad DF, Jakobsson M, Coop G, Wen X, Wall JD, Rosenberg NA, Pritchard JK. 2006b. A worldwide survey of haplotype variation and linkage disequilibrium in the human genome. *Nat Genet* **38:** 1251–1260.

Davila S, Wright VJ, Khor CC, Sim KS, Binder A, Breunis WB, Inwald D, Nadel S, Betts H, de Groot R, et al. 2010. Genome-wide association study identifies variants in the CFH region associated with host susceptibility to meningococcal disease. *Nat Genet* **42:** 772–776.

Dickson SP, Wang K, Krantz I, Hakonarson H, Goldstein DB. 2010. Rare variants create synthetic genome-wide associations. *PLoS Biol* **8:** e1000294.

Fagundes NJ, Ray N, Beaumont M, Neuenschwander S, Salzano FM, Bonatto SL, Excoffier L. 2007. Statistical evaluation of alternative models of human evolution. *Proc Natl Acad Sci* **104:** 17614–17619.

Fellay J, Shianna KV, Ge D, Colombo S, Ledergerber B, Weale M, Zhang K, Gumbs C, Castagna A, Cossarizza A, et al. 2007. A whole-genome association study of major determinants for host control of HIV-1. *Science* **317:** 944–947.

Fellay J, Ge D, Shianna KV, Colombo S, Ledergerber B, Cirulli ET, Urban TJ, Zhang K, Gumbs CE, Smith JP, et al. 2009. Common genetic variation and the control of HIV-1 in humans. *PLoS Genet* **5:** e1000791.

Fellay J, Thompson AJ, Ge D, Gumbs CE, Urban TJ, Shianna KV, Little LD, Qiu P, Bertelsen AH, Watson M, et al. 2010. ITPA gene variants protect against anaemia in patients treated for chronic hepatitis C. *Nature* **464:** 405–408.

Frazer KA, Ballinger DG, Cox DR, Hinds DA, Stuve LL, Gibbs RA, Belmont JW, Boudreau A, Hardenbol P, Leal SM, et al. 2007. A second generation human haplotype map of over 3.1 million SNPs. *Nature* **449:** 851–861.

Fumagalli M, Cagliani R, Pozzoli U, Riva S, Comi GP, Menozzi G, Bresolin N, Sironi M. 2009a. Widespread balancing selection and pathogen-driven selection at blood group antigen genes. *Genome Res* **19:** 199–212.

Fumagalli M, Pozzoli U, Cagliani R, Comi GP, Riva S, Clerici M, Bresolin N, Sironi M. 2009b. Parasites represent a major selective force for interleukin genes and shape the genetic predisposition to autoimmune conditions. *J Exp Med* **206:** 1395–1408.

Ge D, Fellay J, Thompson AJ, Simon JS, Shianna KV, Urban TJ, Heinzen EL, Qiu P, Bertelsen AH, Muir AJ, et al. 2009. Genetic variation in IL28B predicts hepatitis C treatment-induced viral clearance. *Nature* **461:** 399–401.

Gilad Y, Pritchard JK, Thornton K. 2009. Characterizing natural variation using next-generation sequencing technologies. *Trends Genet* **25:** 463–471.

Gravel S, Henn BM, Gutenkunst RN, Indap AR, Marth GT, Clark AG, Yu F, Gibbs RA, Bustamante CD. 2011. Demographic history and rare allele sharing among human populations. *Proc Natl Acad Sci* **108:** 11983–11988.

Hancock AM, Witonsky DB, Alkorta-Aranburu G, Beall CM, Gebremedhin A, Sukernik R, Utermann G, Pritchard JK, Coop G, Di Rienzo A. 2011. Adaptations to climate-mediated selective pressures in humans. *PLoS Genet* **7:** e1001375.

Hill AV. 2006. Aspects of genetic susceptibility to human infectious diseases. *Annu Rev Genet* **40:** 469–486.

Ingman M, Kaessmann H, Paabo S, Gyllensten U. 2000. Mitochondrial genome variation and the origin of modern humans. *Nature* **408:** 708–713.

Jallow M, Teo YY, Small KS, Rockett KA, Deloukas P, Clark TG, Kivinen K, Bojang KA, Conway DJ, Pinder M, et al. 2009. Genome-wide and fine-resolution association analysis of malaria in West Africa. *Nat Genet* **41:** 657–665.

Kamatani Y, Wattanapokayakit S, Ochi H, Kawaguchi T, Takahashi A, Hosono N, Kubo M, Tsunoda T, Kamatani N, Kumada H, et al. 2009. A genome-wide association study identifies variants in the HLA-DP locus associated with chronic hepatitis B in Asians. *Nat Genet* **41:** 591–595.

Kennedy RB, Ovsyannikova IG, Pankratz VS, Haralambieva IH, Vierkant RA, Jacobson RM, Poland GA. 2012a. Genome-wide genetic associations with IFNγ response to smallpox vaccine. *Human Genet* **131:** 1433–1451.

Kennedy RB, Ovsyannikova IG, Shane Pankratz V, Haralambieva IH, Vierkant RA, Poland GA. 2012b. Genome-wide analysis of polymorphisms associated with cytokine responses in smallpox vaccine recipients. *Human Genet* **131:** 1403–1421.

Ku CS, Naidoo N, Pawitan Y. 2011. Revisiting Mendelian disorders through exome sequencing. *Hum Genet* **129:** 351–370.

Lander ES, Linton LM, Birren B, Nusbaum C, Zody MC, Baldwin J, Devon K, Dewar K, Doyle M, FitzHugh W, et al. 2001. Initial sequencing and analysis of the human genome. *Nature* **409:** 860–921.

Laval G, Patin E, Barreiro LB, Quintana-Murci L. 2010. Formulating a historical and demographic model of recent human evolution based on resequencing data from noncoding regions. *PLoS ONE* **5:** e10284.

Li JZ, Absher DM, Tang H, Southwick AM, Casto AM, Ramachandran S, Cann HM, Barsh GS, Feldman M, Cavalli-Sforza LL, et al. 2008. Worldwide human relationships inferred from genome-wide patterns of variation. *Science* **319:** 1100–1104.

Liu L, Okada S, Kong XF, Kreins AY, Cypowyj S, Abhyankar A, Toubiana J, Itan Y, Audry M, Nitschke P, et al. 2011. Gain-of-function human STAT1 mutations impair IL-17 immunity and underlie chronic mucocutaneous candidiasis. *J Exp Med* **208:** 1635–1648.

Lohmueller KE, Indap AR, Schmidt S, Boyko AR, Hernandez RD, Hubisz MJ, Sninsky JJ, White TJ, Sunyaev SR, Nielsen R, et al. 2008. Proportionally more deleterious genetic variation in European than in African populations. *Nature* **451:** 994–997.

Mahasirimongkol S, Yanai H, Mushiroda T, Promphittayarat W, Wattanapokayakit S, Phromjai J, Yuliwulandari R, Wichukchinda N, Yowang A, Yamada N, et al. 2012. Genome-wide association studies of tuberculosis in Asians identify distinct at-risk locus for young tuberculosis. *J Hum Genet* **57:** 363–367.

Manolio TA, Collins FS, Cox NJ, Goldstein DB, Hindorff LA, Hunter DJ, McCarthy MI, Ramos EM, Cardon LR, Chakravarti A, et al. 2009. Finding the missing heritability of complex diseases. *Nature* **461:** 747–753.

Manry J, Laval G, Patin E, Fornarino S, Itan Y, Fumagalli M, Sironi M, Tichit M, Bouchier C, Casanova JL, et al. 2011.

Evolutionary genetic dissection of human interferons. *J Exp Med* **208:** 2747–2759.

McCarthy MI, Abecasis GR, Cardon LR, Goldstein DB, Little J, Ioannidis JP, Hirschhorn JN. 2008. Genome-wide association studies for complex traits: Consensus, uncertainty and challenges. *Nat Rev Genet* **9:** 356–369.

McVean GA, Myers SR, Hunt S, Deloukas P, Bentley DR, Donnelly P. 2004. The fine-scale structure of recombination rate variation in the human genome. *Science* **304:** 581–584.

Myers S, Bottolo L, Freeman C, McVean G, Donnelly P. 2005. A fine-scale map of recombination rates and hotspots across the human genome. *Science* **310:** 321–324.

Ng SB, Turner EH, Robertson PD, Flygare SD, Bigham AW, Lee C, Shaffer T, Wong M, Bhattacharjee A, Eichler EE, et al. 2009. Targeted capture and massively parallel sequencing of 12 human exomes. *Nature* **461:** 272–276.

Ng SB, Buckingham KJ, Lee C, Bigham AW, Tabor HK, Dent KM, Huff CD, Shannon PT, Jabs EW, Nickerson DA, et al. 2010. Exome sequencing identifies the cause of a mendelian disorder. *Nat Genet* **42:** 30–35.

Nielsen R, Hellmann I, Hubisz M, Bustamante C, Clark AG. 2007. Recent and ongoing selection in the human genome. *Nat Rev Genet* **8:** 857–868.

Nishida N, Sawai H, Matsuura K, Sugiyama M, Ahn SH, Park JY, Hige S, Kang JH, Suzuki K, Kurosaki M, et al. 2012. Genome-wide association study confirming association of HLA-DP with protection against chronic hepatitis B and viral clearance in Japanese and Korean. *PLoS ONE* **7:** e39175.

Ovsyannikova IG, Kennedy RB, O'Byrne M, Jacobson RM, Pankratz VS, Poland GA. 2012. Genome-wide association study of antibody response to smallpox vaccine. *Vaccine* **30:** 4182–4189.

Picard C, McCarl CA, Papolos A, Khalil S, Luthy K, Hivroz C, LeDeist F, Rieux-Laucat F, Rechavi G, Rao A, et al. 2009. STIM1 mutation associated with a syndrome of immunodeficiency and autoimmunity. *N Engl J Med* **360:** 1971–1980.

Pickrell JK, Coop G, Novembre J, Kudaravalli S, Li JZ, Absher D, Srinivasan BS, Barsh GS, Myers RM, Feldman MW, et al. 2009. Signals of recent positive selection in a worldwide sample of human populations. *Genome Res* **19:** 826–837.

Png E, Thalamuthu A, Ong RT, Snippe H, Boland GJ, Seielstad M. 2011. A genome-wide association study of hepatitis B vaccine response in an Indonesian population reveals multiple independent risk variants in the HLA region. *Hum Mol Genet* **20:** 3893–3898.

Png E, Alisjahbana B, Sahiratmadja E, Marzuki S, Nelwan R, Balabanova Y, Nikolayevskyy V, Drobniewski F, Nejentsev S, Adnan I, et al. 2012. A genome wide association study of pulmonary tuberculosis susceptibility in Indonesians. *BMC Med Genet* **13:** 5.

Pritchard JK, Cox NJ. 2002. The allelic architecture of human disease genes: Common disease-common variant ... or not? *Hum Mol Genet* **11:** 2417–2423.

Prugnolle F, Manica A, Charpentier M, Guegan JF, Guernier V, Balloux F. 2005. Pathogen-driven selection and worldwide HLA class I diversity. *Curr Biol* **15:** 1022–1027.

Quintana-Murci L, Alcais A, Abel L, Casanova JL. 2007. Immunology in natura: Clinical, epidemiological and evolutionary genetics of infectious diseases. *Nat Immunol* **8:** 1165–1171.

Rauch A, Kutalik Z, Descombes P, Cai T, Di Iulio J, Mueller T, Bochud M, Battegay M, Bernasconi E, Borovicka J, et al. 2010. Genetic variation in IL28B is associated with chronic hepatitis C and treatment failure: A genome-wide association study. *Gastroenterology* **138:** 1338–1345.

Redon R, Ishikawa S, Fitch KR, Feuk L, Perry GH, Andrews TD, Fiegler H, Shapero MH, Carson AR, Chen W, et al. 2006. Global variation in copy number in the human genome. *Nature* **444:** 444–454.

Reich DE, Lander ES. 2001. On the allelic spectrum of human disease. *Trends Genet* **17:** 502–510.

Sabeti PC, Schaffner SF, Fry B, Lohmueller J, Varilly P, Shamovsky O, Palma A, Mikkelsen TS, Altshuler D, Lander ES. 2006. Positive natural selection in the human lineage. *Science* **312:** 1614–1620.

Sabeti PC, Varilly P, Fry B, Lohmueller J, Hostetter E, Cotsapas C, Xie X, Byrne EH, McCarroll SA, Gaudet R, et al. 2007. Genome-wide detection and characterization of positive selection in human populations. *Nature* **449:** 913–918.

Sachidanandam R, Weissman D, Schmidt SC, Kakol JM, Stein LD, Marth G, Sherry S, Mullikin JC, Mortimore BJ, Willey DL, et al. 2001. A map of human genome sequence variation containing 1.42 million single nucleotide polymorphisms. *Nature* **409:** 928–933.

Seto S, Tsujimura K, Koide Y. 2011. Rab GTPases regulating phagosome maturation are differentially recruited to mycobacterial phagosomes. *Traffic* **12:** 407–420.

Sironi M, Clerici M. 2010. The hygiene hypothesis: An evolutionary perspective. *Microbes Infect* **12:** 421–427.

Suppiah V, Moldovan M, Ahlenstiel G, Berg T, Weltman M, Abate ML, Bassendine M, Spengler U, Dore GJ, Powell E, et al. 2009. IL28B is associated with response to chronic hepatitis C interferon-α and ribavirin therapy. *Nat Genet* **41:** 1100–1104.

Tanaka Y, Nishida N, Sugiyama M, Kurosaki M, Matsuura K, Sakamoto N, Nakagawa M, Korenaga M, Hino K, Hige S, et al. 2009. Genome-wide association of IL28B with response to pegylated interferon-α and ribavirin therapy for chronic hepatitis C. *Nat Genet* **41:** 1105–1109.

The 1000 Genomes Project Consortium. 2010. A map of human genome variation from population-scale sequencing. *Nature* **467:** 1061–1073.

The ENCODE Project Consortium. 2012. An integrated encyclopedia of DNA elements in the human genome. *Nature* **489:** 57–74.

The International HapMap Consortium. 2005. A haplotype map of the human genome. *Nature* **437:** 1299–1320.

Thomas DL, Thio CL, Martin MP, Qi Y, Ge D, O'Huigin C, Kidd J, Kidd K, Khakoo SI, Alexander G, et al. 2009. Genetic variation in IL28B and spontaneous clearance of hepatitis C virus. *Nature* **461:** 798–801.

Thomson R, Pritchard JK, Shen P, Oefner PJ, Feldman MW. 2000. Recent common ancestry of human Y chromosomes: Evidence from DNA sequence data. *Proc Natl Acad Sci* **97:** 7360–7365.

Thye T, Vannberg FO, Wong SH, Owusu-Dabo E, Osei I, Gyapong J, Sirugo G, Sisay-Joof F, Enimil A, Chinbuah MA, et al. 2010. Genome-wide association analyses identifies a susceptibility locus for tuberculosis on chromosome 18q11.2. *Nat Genet* **42:** 739–741.

Thye T, Owusu-Dabo E, Vannberg FO, van Crevel R, Curtis J, Sahiratmadja E, Balabanova Y, Ehmen C, Muntau B, Ruge G, et al. 2012. Common variants at 11p13 are associated with susceptibility to tuberculosis. *Nat Genet* **44:** 257–259.

Tishkoff SA, Reed FA, Friedlaender FR, Ehret C, Ranciaro A, Froment A, Hirbo JB, Awomoyi AA, Bodo JM, Doumbo O, et al. 2009. The genetic structure and history of Africans and African Americans. *Science* **324:** 1035–1044.

Vannberg FO, Chapman SJ, Hill AV. 2011. Human genetic susceptibility to intracellular pathogens. *Immunol Rev* **240:** 105–116.

Venter JC, Adams MD, Myers EW, Li PW, Mural RJ, Sutton GG, Smith HO, Yandell M, Evans CA, Holt RA, et al. 2001. The sequence of the human genome. *Science* **291:** 1304–1351.

Voight BF, Adams AM, Frisse LA, Qian Y, Hudson RR, Di Rienzo A. 2005. Interrogating multiple aspects of variation in a full resequencing data set to infer human population size changes. *Proc Natl Acad Sci* **102:** 18508–18513.

Voight BF, Kudaravalli S, Wen X, Pritchard JK. 2006. A map of recent positive selection in the human genome. *PLoS Biol* **4:** e72.

Wong SH, Gochhait S, Malhotra D, Pettersson FH, Teo YY, Khor CC, Rautanen A, Chapman SJ, Mills TC, Srivastava A, et al. 2010a. Leprosy and the adaptation of human toll-like receptor 1. *PLoS Pathog* **6:** e1000979.

Wong SH, Hill AV, Vannberg FO. 2010b. Genomewide association study of leprosy. *N Engl J Med* **362:** 1447–1448.

Worthey EA, Mayer AN, Syverson GD, Helbling D, Bonacci BB, Decker B, Serpe JM, Dasu T, Tschannen MR, Veith RL, et al. 2011. Making a definitive diagnosis: Successful clinical application of whole exome sequencing in a child with intractable inflammatory bowel disease. *Genet Med* **13:** 255–262.

Zhang SY, Boisson-Dupuis S, Chapgier A, Yang K, Bustamante J, Puel A, Picard C, Abel L, Jouanguy E, Casanova JL. 2008. Inborn errors of interferon (IFN)-mediated immunity in humans: Insights into the respective roles of IFN-α/β, IFN-γ, and IFN-λ in host defense. *Immunol Rev* **226:** 29–40.

Zhang FR, Huang W, Chen SM, Sun LD, Liu H, Li Y, Cui Y, Yan XX, Yang HT, Yang RD, et al. 2009. Genomewide association study of leprosy. *N Engl J Med* **361:** 2609–2618.

Zhang F, Liu H, Chen S, Low H, Sun L, Cui Y, Chu T, Li Y, Fu X, Yu Y, et al. 2011. Identification of two new loci at IL23R and RAB32 that influence susceptibility to leprosy. *Nat Genet* **43:** 1247–1251.

Zwick ME, Cutler DJ, Chakravarti A. 2000. Patterns of genetic variation in Mendelian and complex traits. *Annu Rev Genomics Hum Genet* **1:** 387–407.

Vaccines, Reverse Vaccinology, and Bacterial Pathogenesis

Isabel Delany, Rino Rappuoli, and Kate L. Seib

Novartis Vaccines and Diagnostics, 53100 Siena, Italy

Correspondence: Rino.rappuoli@novartis.com

Advances in genomics and innovative strategies such as reverse vaccinology have changed the concepts and approaches to vaccine candidate selection and design. Genome mining and blind selection of novel antigens provide a novel route to investigate the mechanisms that underpin pathogenesis. The resulting lists of novel candidates are revealing new aspects of pathogenesis of target organisms, which in turn drives the rational design of optimal vaccine antigens. Here we use the discovery, characterization, and exploitation of fHbp, a vaccine candidate and key virulence factor of meningococcus, as an illustrative case in point. Applying genomic approaches to study both the pathogen and host will ultimately increase our fundamental understanding of pathogen biology, mechanisms responsible for the development of protective immunity, and guide next-generation vaccine design.

Along with improved sanitation and the discovery and use of antibiotics, vaccination is the intervention that has had the greatest impact on human health and the standard of living in recent history. Vaccines also represent the most cost-effective way of improving health and saving lives (Levine and Lagos 2004). In the past, the development of new vaccines and therapeutics was aided, and largely driven, by increased understanding of the pathogenesis of infectious agents. The antigens used in vaccines do not necessarily have to be virulence factors; however, understanding and interrupting the cycle of infection by directing the immune responses toward key virulence determinants has historically been a successful rationale in eliciting protective immunity. However, the development of vaccines for many pathogens remains elusive, and there is a growing requirement for the fast development of effective vaccines for emerging diseases (Morens et al. 2008). During the last three decades, the vaccine field has been transformed by new technologies such as recombinant DNA and chemical conjugation. More recently, new methods of antigen discovery and design as well as investigation of vaccine responses have been applied, including reverse vaccinology, structural biology, and systems biology (Rinaudo et al. 2009). Genome-based technologies have enabled functionally blind selection of vaccine candidates, and have not only led to the discovery of novel protective antigens but have also revealed new virulence factors of several pathogens. Consequently, the pathogenesis-to-vaccine paradigm has been reversed in several situations, and vaccine development frequently leads to a better understanding of pathogenesis, which has in turn led to novel

Cite this article as *Cold Spring Harb Perspect Med* doi: 10.1101/cshperspect.a012476

approaches in studying not only the organism itself but also the strategies for the design of more successful vaccines (Fig. 1). The continuing advancement of genome-based technologies will hopefully lead to new vaccines for unmet diseases, as well as increasingly efficient development pathways for these vaccines (Rappuoli et al. 2011).

HOW OUR UNDERSTANDING OF BACTERIAL PATHOGENESIS HAS LED TO NEW VACCINES

Since the first observations of Jenner at the turn of the nineteenth century, that milkmaids exposed to cowpox appeared to be protected from the more severe human smallpox disease, the understanding of pathogenesis has been fundamental to the development of vaccines. Once it became evident in the early 1800s that diseases were caused by microbes, Louis Pasteur started the rational development of vaccines and established the basic rules of vaccinology. Most of the vaccines licensed to date have been developed based on Pasteur's principles of "isolate, inactivate and inject" the causative agent of disease (Plotkin 2009). One could say that they have been developed empirically, with little or no understanding of the complex immunological mechanisms by which they induce protective immunity. However, the strategy was based on a unifying rationale, that by reducing the virulence of the disease-causing organism or by inhibiting its ability to replicate, the consequent

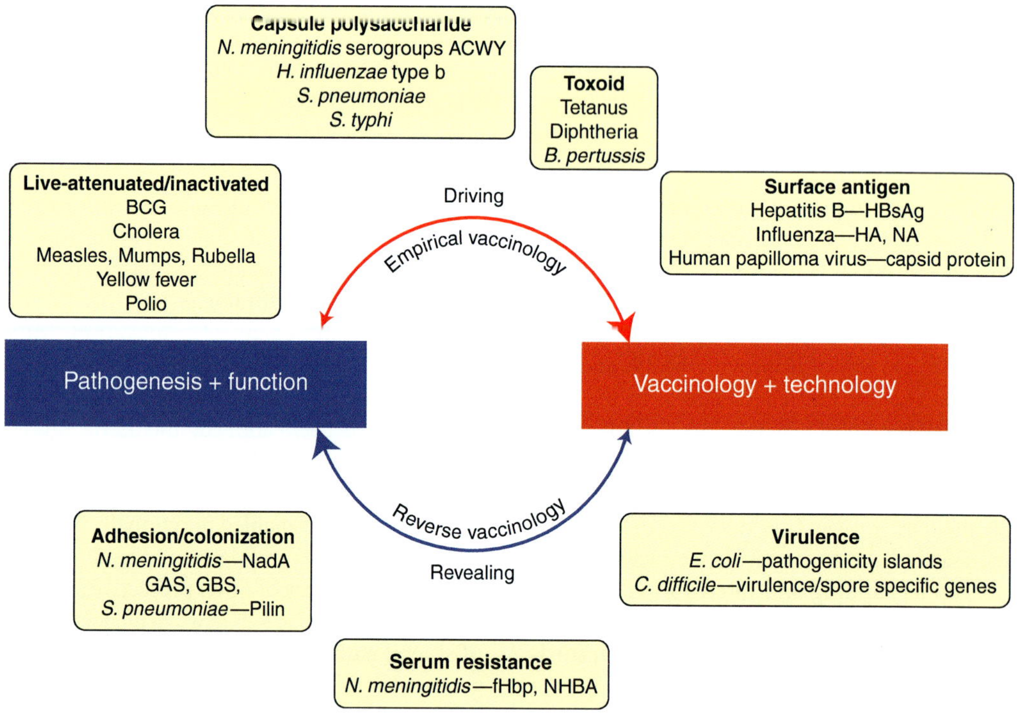

Figure 1. Interactions between microbial pathogenesis and vaccine development. Historically, the understanding of microbial mechanisms of pathogenesis has driven vaccine development, with empirical vaccinology approaches being based on inactivating, attenuating, and/or targeting and interrupting the function of virulence factors, such as capsule polysaccharides, toxins, and other surface proteins. However, with new technologies and vaccine approaches, the focus has shifted to functionally blind antigen discovery based on high-throughput screening of the pathogen's genome and proteome. These approaches have identified many promising vaccine candidates, which have subsequently been found to be involved in the pathogenic process, such as colonization of the host or serum resistance.

attenuation or inactivation robs infectious organisms of their pathogenicity while preserving their immunogenicity. The mechanism of action of vaccines relies entirely on activation of the human body's own protective immunological mechanisms, and live-attenuated or inactivated vaccines basically mimic the kind of protective immunity induced in people who survive a normal infection.

Pasteur applied the principles of "isolate, inactivate and inject" in the late 1800s to both viruses (dried rabies virus isolated from rabbits) and bacteria (heat-inactivated anthrax bacilli). These strategies have been successful for those diseases that result in natural immunity to reinfection, in particular for many viral diseases that have invariant antigens (e.g., smallpox, yellow fever, polio, rabies, mumps, measles, rubella, varicella, herpes zoster, hepatitis A, and Japanese encephalitis) (Rappuoli 2007). Furthermore, inactivated or live-attenuated vaccines against some bacterial infectious diseases have also been developed with varying degrees of success; for instance, Bacille Calmette–Guérin (BCG; live-attenuated *Mycobacterium bovis*) is possibly the most widely used vaccine worldwide and currently the only vaccine available against tuberculosis (Parida and Kaufmann 2010). Whole-cell vaccines against enteric diseases such as typhoid and cholera, and a whole-cell pertussis vaccine are currently licensed and in use, although there is a general need and trend to replace these with newer improved vaccines.

A further development to Pasteur's strategy was undertaken when essential components of the organism, usually virulence determinants, were isolated and inactivated, for use as "subunit" vaccines. This approach was first used for diphtheria and tetanus in which the pathology of the disease is largely owing to the toxins produced by the bacterial pathogen, and initial toxoid vaccines were chemically inactivated toxins isolated from the bacteria (Glenny and Hopkins 1923; Ramon 1924). Similarly, capsule polysaccharides of Gram-negative bacteria are common virulence factors and subunit vaccines based on purified polysaccharides were initially developed for *Streptococcus pneumoniae* in the 1940s (MacLeod et al. 1945), *Neisseria meningitidis* in the 1960s (Gotschlich et al. 1969), *Haemophilus influenzae* in the 1970s (Peltola et al. 1977), and *Salmonella typhi* Vi in the 1980s (Acharya et al. 1987). With advances in new technologies, such as recombinant DNA and chemical conjugation of proteins to polysaccharides, subunit vaccines have advanced notably and are able to provide improved memory responses and increased protection in infants. These vaccines are safe and efficacious, and have also been developed in multivalent forms to provide protection against numerous pneumococcal serotypes and meningococcal serogroups. For the three major causes of bacteremic disease—*H. influenzae* type b (Hib), pneumococci, and meningococci—the use of polysaccharide conjugate vaccines has led to successful control of invasive diseases caused by many serogroups in various countries worldwide (Kelly et al. 2004; Gill et al. 2010; Jefferies et al. 2011).

Genetic inactivation by targeted mutagenesis as well as modern recombinant techniques for the expression and purification of protein-based subunit vaccines have led to the development of safer live-attenuated vaccines and subunit vaccines through rational design. For example, recombinant DNA technology enabled the generation of the detoxified *Bordetella pertussis* toxin (Pizza et al. 1989), and the recombinant hepatitis B surface antigen (HBsAg) (Buynak et al. 1976; Ellis 1990). Ultimately, recombinant DNA technology has formed the basis of a wide range of new platforms for vaccine design and delivery.

Subunit vaccine antigens are usually better tolerated than inactivated or live-attenuated pathogens, and this has led to a push for the development of new and improved vaccines using recombinant techniques. Acellular pertussis combination vaccines containing between one and five pertussis virulence factors are substantially less reactogenic than the original whole-cell vaccines and have increasingly replaced whole-cell-containing vaccines in developed countries (Kitchin 2011). However, a limitation of subunit vaccines is that they are generally less immunogenic and often require the addition of an adjuvant to elicit protective immune responses (Mbow et al. 2010). Unfortunately, few adjuvants are currently licensed and widely used (e.g., aluminum salts, Alum, that have been used for more than 80 years). However, newer adjuvants (e.g., MF59 [O'Hagan et al. 2011] and AS04 [Garcon et al. 2011]) have been developed, each of which have specific properties designed to enhance the magnitude and quality of the immune response to target specific diseases.

Although the virtual disappearance of diseases like diphtheria, tetanus, pertussis, and invasive *H. influenzae* B are landmarks to the success of vaccines, there still remain many bacterial diseases for which vaccination strategies have so far been unsuccessful. Often this is owing to the hypervariability of their antigens, an incomplete understanding of pathogenic mechanisms, and/or a need for cell-mediated immune responses (Rappuoli 2007). For instance, for those diseases that do not induce immunity after natural infection (e.g., malaria, or respiratory syncytial virus [RSV]) or those that cause persistent or latent infection (e.g., HIV and hepatitis C virus), a vaccine-induced protective immune response must go one step beyond the mechanisms that nature has evolved and vaccine development must go beyond the traditional empirical methodologies.

REVERSING THE PARADIGM: REVERSE VACCINOLOGY APPROACHES LEAD TO INCREASED UNDERSTANDING OF BACTERIAL PATHOGENESIS

With the arrival of whole-genome sequencing, genome-based antigen selection has played a major role in antigen discovery and vaccine design. One approach that has been used to mine pathogenic bacterial genomes has been coined "reverse vaccinology," and allows the investigation of the complete potential antigenic repertoire of an organism from its genome sequence. Reverse vaccinology involves the cloning and expression of all the proteins in an organism's genome sequence that are predicted in silico to be surface exposed or secreted. Then each protein is screened, through high-throughput immunization, for their ability to elicit antibodies in mice that can kill or neutralize the target organism. The first pathogen addressed using reverse vaccinology was *N. meningitidis* serogroup B (MenB) (Pizza et al. 2000), for which no broadly protective vaccine exists owing to the similarity of its capsular polysaccharide to a self-antigen and the hypervariability of its major outer membrane protein antigens. This approach led to the identification of 29 novel antigens that can elicit bactericidal antibodies against the pathogen in vitro. After successful preclinical studies, the vaccine deriving from the genome approach entered phase I testing in 2002 (Giuliani et al. 2006). In November 2012, the European Medicinal Agency recommended the granting of a marketing authorization for Bexsero, the first vaccine to provide broad coverage against meningococcal serogroup B. This was followed by the approval of the European Commission in January 2013. The three main antigens identified by reverse vaccinology that are formulated as components of the MenB vaccine, *Neisserial* heparin-binding antigen (NHBA), factor H-binding protein (fHbp), and *N. meningitidis* adhesin A (NadA), have all subsequently been implicated as playing important roles in meningococcal virulence (Capecchi et al. 2005; Madico et al. 2006; Schneider et al. 2006; Serruto et al. 2010) (see below for more details on fHbp).

Since the first application of reverse vaccinology, based on the single genome of a MenB strain (Pizza et al. 2000), various advancements have been made to this genomic approach. A multigenome or pan-genome reverse vaccinology approach was applied to Group B streptococcus (GBS) to identify antigens from the extended gene repertoire of the species rather than from a single organism (Maione et al. 2005), and the subtractive reverse vaccinology approach has been used to identify antigens present in pathogenic but not commensal strains of *Escherichia coli* (Rasko et al. 2008; Moriel et al. 2010). Reverse vaccinology has been applied to a wide range of bacterial pathogens and has provided a long list of promising antigens from functionally blind interrogation of their genomes, and the subsequent studies on antigen function are leading to increased understanding of the biology of the pathogens (Fig. 1; Table 1).

Table 1. Microbial virulence factors identified during vaccine development

Organism	Disease	Vaccine approach/genomic-based technology	Antigen/virulence factor	Role	References
Neisseria meningitidis	Meningococcal meningitis and sepsis	Reverse Vaccinology: genome analysis of serogroup B strain Structural Vaccinology: chimeric fHbp engineered based on variant 1 backbone-containing epitopes from variants 2 and 3	Factor H-binding protein (fHbp)	Binding of complement factor H (fH), serum resistance	Pizza et al. 2000, 2008; Giuliani et al. 2006; Madico et al. 2006; Scarselli et al. 2011
		Reverse Vaccinology: genome analysis of serogroup B strain	Neisserial adhesin A (NadA)	Adhesion, colonization	Comanducci et al. 2002; Capecchi et al. 2005
		Reverse Vaccinology: genome analysis of serogroup B strain	Neisserial heparin-binding antigen (NHBA)	Binding of heparin, serum resistance	Serruto et al. 2010
Group B *Streptococcus* (*Streptococcus agalactiae*)	Sepsis, pneumonia, and neonatal meningitis	Pan-Genome Reverse Vaccinology: genome analysis of 8 GBS strains Comparative Genomics and Genome Hybridization: *Streptococcus pneumoniae*, GAS genomes, 19 GBS strains	Pili (BP-1, BP-2a, AP-1, AP-2a, GBS67)	Adhesion, colonization, biofilm formation, translocation through epithelial cells	Lauer et al. 2005; Maione et al. 2005; Tettelin et al. 2005; Dramsi et al. 2006; Rosini et al. 2006; Pezzicoli et al. 2008; Margarit et al. 2009; Rinaudo et al. 2010; Nuccitelli et al. 2011
Group A *Streptococcus* (*Streptococcus pyogenes*)	Pharyngitis, necrotizing fasciitis, pneumonia, sepsis, arthritis, myositis	Pan-Genome Reverse Vaccinology/Comparative Genomics: analysis of five GAS strains for pilus encoding (LPXTG motif) genes	Pili (Lancefield antigens)	Adhesion, colonization, biofilm formation	Mora et al. 2005; Abbot et al. 2007; Manetti et al. 2007
Streptococcus pneumoniae	Pneumonia, meningitis, otitis media, sepsis	Reverse Vaccinology/Comparative Genomics: analysis of T4 and INV104 genomes for putative virulence factors and pilus genes	Pili (RrgB, RrgA, RrgC encoded by *rlrA* pathogenicity Islet; PitB, pilus islet 2)	Adhesion, colonization, biofilm formation	Hava and Camilli 2002; Barocchi et al. 2006; LeMieux et al. 2006; Bagnoli et al. 2008; Hilleringmann et al. 2009; Donati et al. 2010

Continued

Cite this article as *Cold Spring Harb Perspect Med* doi: 10.1101/cshperspect.a012476

Table 1. *Continued*

Organism	Disease	Vaccine approach/genomic-based technology	Antigen/virulence factor	Role	References
Escherichia coli	Intestinal (diarrhea) and extraintestinal (urinary tract infection, sepsis, meningitis, hemolytic-uremic syndrome) disease	Comparative/Subtractive Reverse Vaccinology: comparison of genomes of commensal and pathogenic strains Proteomics: analysis of OMV proteins	Pathogenicity islands, pathogen-specific OMPs (e.g., ECOK1_0290)	Adhesion, colonization	Berlanda Scorza et al. 2008; Moriel et al. 2010; Nesta et al. 2012
Staphylococcus aureus	Nosocomial, skin, wound infections, endocarditis, pneumonia, bacteremia, osteomyelitis, toxic shock syndrome	Reverse Vaccinology: genome analysis of eight strains for the presence of sortase substrate genes (based on loss of pathogenesis of sortase mutant) Comparative Genomics: genome analysis of 58 strains for variation of putative surface proteins	Conserved virulence-associated surface proteins (e.g., IsdA, IsdB, SdrD, SdrE)	Pathogenesis in murine renal abscess model	Mazmanian et al. 2000; Stranger-Jones et al. 2006; McCarthy and Lindsay 2010
Clostridium difficile	Nosocomial diarrhea	Comparative Genomics: genome analysis of historic nonepidemic and newly emerged hypervirulent epidemic strains, and evolutionary dynamics of 30 isolates Proteomics: investigation of core spore gene, spore–host interactions	Genetic regions specific for epidemic strain (e.g., flagella biosynthesis and glycosylation proteins) 29 *C. difficile* spore specific proteins	Motility, autoagglutination, sporulation	Stabler et al. 2006, 2009; Lawley et al. 2009; He et al. 2010
Chlamydia trachomatis	Most prevalent sexually transmitted infection worldwide (causes ectopic pregnancy, infertility), trachoma	Reverse Vaccinology: genome analysis Comparative Genomics: genome analysis of five chlamydial species Proteomics: investigation of elementary bodies	Outer membrane complex and elementary body proteins	Host–pathogen interactions	Montigiani et al. 2002; Thorpe et al. 2007; Heinz et al. 2009; Liu et al. 2010

OMV, outer membrane vesicle; OMPs, outer membrane proteins.

Cite this article as *Cold Spring Harb Perspect Med* doi: 10.1101/cshperspect.a012476

Although a vaccine antigen does not have to be a virulence factor, focusing on proteins with a crucial function in bacterial pathogenesis or survival is a useful selection criterion, because antibodies elicited to these antigens may block their function and result in neutralization. For example, the use of protective antigens involved in adhesion is particularly attractive, as there may be important implications for providing herd immunity through reduction of the pathogen's colonization and circulation. Indeed, functional studies of many protective antigens have subsequently revealed them to be novel adhesins involved in colonization of the pathogen (Table 1) (Capecchi et al. 2005; Rosini et al. 2006). The pili of gram-negative bacteria are well-described virulence factors, however, little was known about pili in gram-positive bacteria before their discovery as a result of reverse vaccinology on GBS (Lauer et al. 2005; Maione et al. 2005), and subsequent studies on *Streptococcus pyogenes,* and *S. pneumoniae* (reviewed in Telford et al. 2006). The presence of pili that contain protective antigens in all three principal streptococcal pathogens indicates that these structures play an important role in virulence. Genome-based comparison of pathogenic and nonpathogenic strains has also been applied to enable vaccine targets to be identified on the basis of proteins that are specifically present in pathogenic strains and therefore implicated in virulence. Recently, subtractive reverse vaccinology has been applied to *E. coli* and five out of the nine protective antigens identified are located on putative pathogenicity islands specific for pathogenic *E. coli* strains (Moriel et al. 2010).

A CASE STUDY OF HOW VACCINE DEVELOPMENT AND UNDERSTANDING OF BACTERIAL PATHOGENESIS ARE LINKED: DISCOVERY, CHARACTERIZATION, AND EXPLOITATION OF fHbp

Reverse vaccinology, based on functionally blind selection of candidate antigens from genome screening, led to the discovery and characterization of factor H-binding protein (fHbp), which is a novel protective antigen and virulence determinant of meningococcal infection. The continuing study of fHbp has led to significant advances in the understanding of meningococcal pathogenesis, contributed to the development of improved animal models for *N. meningitidis*, and provided greater clarity for host susceptibility studies, which has in turn led to novel approaches in studying the organism itself as well as the design of more successful vaccines (Fig. 2).

Identification of fHbp as a Vaccine Antigen

fHbp was initially identified as the genome-derived *Neisseria* antigen 1870 (GNA1870), and was described as a *Neisseria*-specific putative surface lipoprotein of unknown function (Pizza et al. 2000). An unrelated vaccine discovery project also identified this antigen (referred to as lipoprotein [LP] 2086) using a proteomic approach to analyze soluble outer membrane proteins (Fletcher et al. 2004). fHbp induces high levels of bactericidal antibodies in mice (Fletcher et al. 2004; Giuliani et al. 2006) and humans (Jacobsson et al. 2009; Plested et al. 2009; Findlow et al. 2010; Giuliani et al. 2010; Halperin et al. 2010; Toneatto et al. 2011a,b). Protection against meningococcal disease correlates with the level of bactericidal antibodies in the serum bactericidal activity (SBA) assay (Borrow et al. 2006; Frasch et al. 2009), indicating that fHbp is an important vaccine antigen. Antibodies to fHbp also confer passive protection in the infant rat model of meningococcal bacteremia (Masignani et al. 2003). Development of fHbp as a vaccine antigen has greatly progressed and it is currently a component of the Novartis multivalent 4CMenB vaccine (Giuliani et al. 2006) that was submitted for licensure to the European Medicines Agency in 2010. Another vaccine contains two alleles of the recombinant fHbp protein (rLP2086) and is in late stages of development by Pfizer (Fletcher et al. 2004). In another strategy for the production of outer membrane vesicle (OMV)-based vaccines, meningococcal strains have been engineered to overexpress variants of fHbp, which has shown to provide broad protection

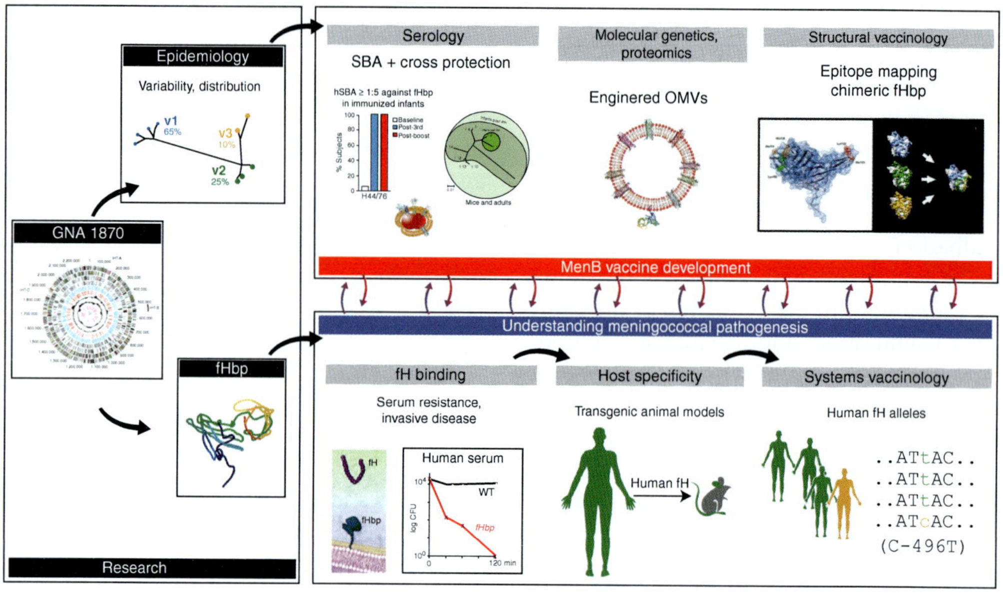

Figure 2. The meningococcal fHbp antigen as an example of how vaccine development has led to an increased understanding of pathogenesis. Genome-derived *Neisseria* antigen 1870 (GNA1870) was identified as a novel vaccine antigen from the genome sequence of *N. meningitidis* by reverse vaccinology. Since then it has been studied extensively for vaccine development (*top* panel) in terms of its distribution and sequence variation, as well as its ability to induce cross-protective bactericidal antibodies. This information has also been used to engineer meningococcal strains overexpressing fHbp for outer membrane vesicle (OMV) vaccines. In addition, structural studies have been used for epitope mapping and the generation of a chimeric fHbp antigen that is able to induce broad cross protection. Parallel to this work, the understanding of meningococcal pathogenesis has greatly advanced by studying this antigen (*bottom* panel). GNA1870 was renamed as factor H-binding protein (fHbp) owing to the discovery of its functional role in binding human factor H (fH), which increases serum resistance. This role has also led to increased understanding of meningococcal host specificity and the development of transgenic mice models expressing human fH. Furthermore, human fH alleles have been identified that increase host susceptibility to meningococcal disease.

against several MenB strains (Keiser et al. 2011; Koeberling et al. 2011a,b).

fHbp Function and Role in Pathogenesis

Before its identification as a vaccine candidate the function of fHbp was unknown. Two independent laboratories subsequently found that this antigen binds human complement factor H (fH) (Madico et al. 2006; Schneider et al. 2006), which is a key regulatory protein that acts as an inhibitor of the alternative complement pathway (reviewed in Meri et al. 2008). fHbp is an important meningococcal survival factor in human serum and blood (Madico et al. 2006; Welsch et al. 2008; Seib et al. 2009), because recruitment of fH to the bacterial surface via fHbp binding enables the pathogen to evade alternative complement-mediated killing by the host innate immune system. This has implications for fHbp not only as a virulence factor in vivo but also as a target antigen for vaccination as interestingly, anti-fHbp antibodies can also increase the susceptibility of the bacterium to killing by the alternative complement pathway by blocking fH binding to the bacteria (Madico et al. 2006). Additional functions for fHbp have also been proposed, including resistance to the antimicrobial LL-37 (Seib et al. 2009) and binding of the siderophore enterobactin (Veggi et al. 2012). Investigation of *fHbp* transcription and regulation revealed that fHbp expression is induced under oxygen limitation and implicates an important role in oxygen-poor microenvironments, such as the submucosa or intracellular locations (Bartolini et al. 2006; Oriente et al. 2010).

Meningococcal Epidemiology and Typing

fHbp is expressed by almost all *N. meningitidis* strains studied to date, although levels of expression vary between isolates (Masignani et al. 2003; Fletcher et al. 2004; McNeil et al. 2009; Murphy et al. 2009). Furthermore, a significant degree of fHbp sequence variability exists in MenB strains circulating globally, and sequencing of *fHbp* in ~2000 MenB strains led to the identification of ~300 subvariant polypeptides (Bambini et al. 2009; Murphy et al. 2009). Different sequence variants can be classified into three groups (variants 1, 2, and 3) (Masignani et al. 2003; Brehony et al. 2009) or two subfamilies (A and B: corresponding to variants 2/3 and 1, respectively) (Fletcher et al. 2004; McNeil et al. 2009; Murphy et al. 2009; Jiang et al. 2010) that do not induce cross-protective immunity. However, all variants studied to date are able to bind fH, mediate serum resistance, and induce bactericidal antibodies (Dunphy et al. 2011; Seib et al. 2011). Members of the variant 1, 2, and 3 families are present in 65%, 25%, and 10% of the MenB global population, respectively (Beernink et al. 2007; Bambini et al. 2009; Murphy et al. 2009).

Analysis of the 4CMenB vaccine antigens fHbp, NadA, and NHBA variant distribution in meningococcal population (Bambini et al. 2009) indicated that vaccine coverage could not be predicted on the basis of multilocus sequence typing (MLST), the gold standard for meningococcal classification (Brehony et al. 2007). As a result of this, a high-throughput meningococcal antigen typing system (MATS) was developed based on a unique vaccine antigen-specific enzyme-linked immunosorbent assay (ELISA), which measures both immunologic cross reactivity and quantity of fHbp, NHBA, and NadA, and correlates with bactericidal activity (Donnelly et al. 2010). MATS analysis of large panels of strains has predicted that adults and infants immunized with the 4CMenB vaccine would be protected from disease caused by 86% and 77% of the diverse MenB strains circulating globally, respectively. This typing approach could also be applied for vaccines against other variable pathogens, including nontypable *H. influenzae*, *Streptococcus* groups A and B, and *S. pneumoniae*.

fHbp and Structural Vaccinology

Analysis of the three-dimensional structure of fHbp, by nuclear magnetic resonance (NMR) of the carboxy-terminal portion (Cantini et al. 2009) and the crystal structure of the entire fHbp protein in complex with a fragment of fH (Schneider et al. 2009) has led to identification of residues important for binding of bactericidal antibodies and fH, respectively. Using this data a structural vaccinology approach has enabled the rational design of a chimeric fHbp antigen, which combines the antigenic repertoire of the three major fHbp variant groups into a single molecule, using the variant 1 scaffold carrying patches of amino acids from the surfaces of variants 2 and 3. The chimeric fHbp protein G1 is capable of inducing antibodies that are bactericidal against a diverse panel of strains of meningococcus B, suggesting that it could be used to produce a broadly protective vaccine (Scarselli et al. 2011).

Understanding fHbp-fH Binding Leads to Better Animal Models and Improved Vaccine Candidates

Meningococcus is an obligate human pathogen that has no known reservoir outside the human host, and one of the biggest limitations in studying the pathogenesis of the disease is the lack of good animal models of infection. The finding that fHbp specifically binds human fH, but not rat or mouse fH, provided a key element to understanding meningococcal species specificity (Granoff et al. 2009). Indeed, the administration of human fH to infant rats challenged with MenB greatly increased meningococcal survival (>10-fold) (Granoff et al. 2009). Recently, a human fH transgenic mouse model has been developed by microinjection of mouse embryos with human fH cDNA, leading to concentrations of human fH in the sera of the transgenic mice comparable to that seen in human sera (Beernink et al. 2011). This model was used to test the immunogenicity of an fHbp mutant with a single amino acid substitution, R41S, that is no longer able to bind human fH, to investigate whether epitopes may be obscured when human fH is bound to the wild-type fHbp vaccine given to humans. The fHbp R41S mutant induced higher serum bactericidal antibody responses than the wild-type fHbp and these antibodies had increased ability to block fH binding to wild-type fHbp (Beernink et al. 2011).

Host SNPs and fH Levels Are Involved in Susceptibility to Meningococcal Diseases

The increased incidence and recurrence of infection and disease in people with deficiencies (Nielsen et al. 1989) or overproduction of fH (Haralambous et al. 2006) indicated the importance of host fH in controlling meningococcal disease. A single-nucleotide polymorphism (SNP; C-496T C/C) in the promoter region of the gene encoding fH (*CFH*) is associated with slightly increased fH concentrations, reduced bactericidal activity against meningococci, and an increased risk of disease (Haralambous et al. 2006). Genome-wide association studies have also revealed a set of strongly correlated SNPs in *CFH* and the gene encoding CFH-related protein 3 (*CFHR3*) that are associated with decreased susceptibility to meningococcal disease for minor allele carriers. Interestingly, the protective alleles of the identified SNPs are rare in Africa, suggesting that these variants may play a role in the increased disease rate seen in the meningitis belt (Davila et al. 2010). However, it has been highlighted that association studies typically do not provide sufficient coverage of African populations, although new tools such as the "The 1000 Genomes Project" may help gain a better understanding of the link between host fH polymorphisms and meningococcal disease in high-risk African populations (de Bakker and Telenti 2010).

In conclusion, the discovery of fHbp through the reverse vaccinology approach and its characterization and study as a vaccine antigen has considerably advanced the understanding of meningococcal pathogenesis, and furthermore, this enhanced knowledge has led to advancements in the use of fHbp as a vaccine candidate (Fig. 2).

THE WAY AHEAD FOR VACCINOLOGY

The use of reverse vaccinology has enabled identification of numerous promising vaccine candidates against meningococcus, GBS, group A streptococcus, pneumococcus, pathogenic *E. coli*, and also for antibiotic-resistant bacteria such as *Staphylococcus aureus* (Table 1). The subsequent challenge is the refinement of these long lists of new antigens, enabling detection of the best candidate antigens and formulations to take forward for clinical development. This selection of the best candidates will largely depend on the use of novel approaches, such as structural vaccinology to engineer optimal antigens, and systems biology to gain a better understanding of signatures of immunogenicity and correlates of protection.

STRUCTURAL VACCINOLOGY

Although the mining of bacterial genome sequences has identified excellent targets that could be used in a vaccine, the sequence variability of antigens is often a major challenge to their use and development as broadly protective vaccine components. Traditionally, vaccines have been developed using natural antigens, however, rational optimization of antigens could improve their properties as immunogens by combining, exposing, and/or improving the immunogenicity of epitopes. After all, native surface proteins of pathogens have adapted to evade, not induce, immunity (Dormitzer et al. 2008).

Structural biological studies enable the atomic resolution of an antigen, and structure-based design allows the engineering of multiple immunodominant epitopes in one molecule to induce broad immune responses against different protein variants. Analogous to three musketeers, who together make a formidable team (all for one), one can combine all immunogenic epitopes of non-cross-reacting variants of an antigen into a rationally designed individual molecular champion (one for all), as achieved for the meningococcal fHbp (Scarselli et al. 2011) and a group B streptococcal pilus antigen (Nuccitelli et al. 2011). Structure-based antigen design has also been used in HIV vaccine development to render conserved epitopes immunogenic (Tobin et al. 2008; Burton 2010). Furthermore, stabilizing the structure of difficult antigens can improve their value as vaccine components, by optimizing their efficient production and storage stability.

SYSTEMS VACCINOLOGY

Systems biology is an interdisciplinary approach that systematically investigates the structure and complex interactions between all parts of a biological system with the ultimate goal of predicting the behavior of the system (Kitano 2002). A systems-based approach is increasingly being applied to vaccinology with two main goals in mind, first to understand the mechanisms by which vaccines stimulate protective immunity, and secondly to identify markers or molecular signatures that can be used to predict the immunogenicity or efficacy of vaccines (Pulendran et al. 2010; Nakaya et al. 2011).

Systems vaccinology, which relies on next-generation sequencing and postgenomic technologies for the high-throughput acquisition and analysis of data, can be used to define and monitor the global architecture of the human immune response and the changes that occur following vaccination. For example, analysis of vaccine-induced changes in blood, including patterns of global gene expression or alterations in immune cells, with respect to several immunological parameters, has successfully been used to identify molecular signatures induced early after vaccination with the live-attenuated yellow fever vaccine (YF-17D) that could be used to predict immunogenicity of the vaccine (Gaucher et al. 2008; Querec et al. 2009). Such signatures of immunogenicity, when identified, would also allow early identification of unsafe vaccine candidates and guide selection of the most effective vaccine composition, in terms of optimal candidate combinations, formulations, delivery systems, and immunization schemes, thus bridging the gap between antigen discovery and clinical research. Furthermore, such signatures would enable monitoring of suboptimal immune responses in certain individuals (e.g., elderly, infants, or immunocompromised populations) in whom vaccination may not confer protective immunity and may remain at risk of infection (Nakaya et al. 2011).

Other relevant data that can be gained from this field of study is the characterization of human genetic variation, at the individual or population level, and how this impacts on vaccine success. The application of state-of-the-art next-generation sequencing technology and tools that allow high-throughput detection of gene variations can provide genome-wide information, which when integrated with immune-response signature profiles will further fill the gap in our understanding of the relationship between the phenotype–genotype of an immune response. A large and growing family of polymorphisms in genes of the immune system have been identified that lead to significant variations in immune responses and that are critical to vaccine efficacy (Poland et al. 2007, 2009; Haralambieva and Poland 2010). For example, many studies have revealed associations with human leukocyte antigen (HLA) haplotypes and immune outcomes after vaccination with the MMR (measles, mumps, and rubella) vaccine (Ovsyannikova et al. 2007) hepatitis B vaccine (Kruskall et al. 1992) and the rubella vaccine (Dhiman et al. 2010). This raises the notion of "personalized vaccinology" in which a prevaccination genome-wide screen would allow individualized vaccination approaches that can be tailored and guided by the genetic uniqueness of each patient, maximizing immunogenicity and minimizing the risk of either vaccine failure or adverse events (Poland et al. 2008; Dhiman et al. 2009).

Systems vaccinology can be used to define the signatures of protection that are required to elicit a protective immune response, which will increase our knowledge of the mechanisms of action of currently successful vaccines,

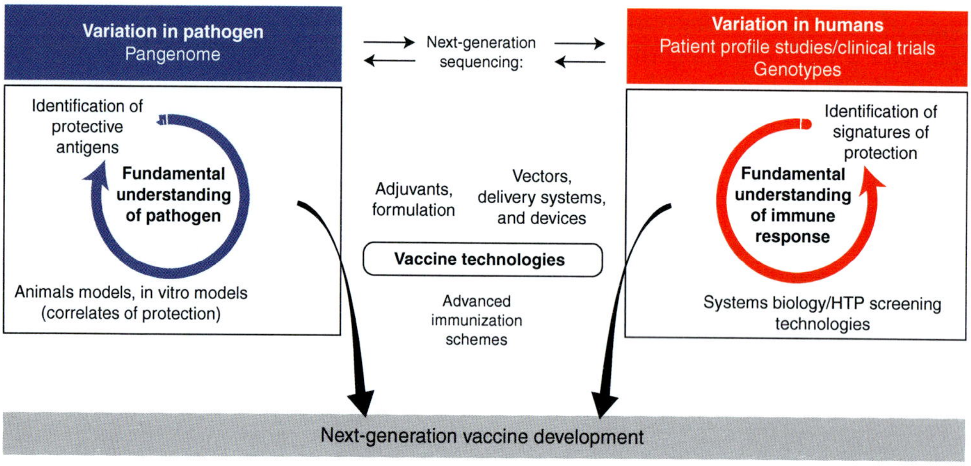

Figure 3. Key areas of next-generation vaccine development. High-throughput (HTP) DNA sequencing and screening technologies are leading to a better understanding of the genetic variation of both pathogens and the human host, which in turn is improving the fundamental understanding of the pathogen and the immune response, respectively. This information can be exploited to enable identification of protective antigens from the pathogen as well as the signatures in the host immune response that lead to protection. These two parallel fields of vaccinology, combined with improving vaccine formulation and delivery technologies, are key components of next-generation vaccine development.

as well as enable the improved identification, rational design, and testing of novel vaccine antigens (Pulendran et al. 2010; Bernstein et al. 2011).

CONCLUSIONS

With the growing problem of antimicrobial resistance, and newly emerging or reemerging pathogens (for review see Morens et al. 2008), we are increasingly looking to the development and use of vaccines to control infectious diseases. Furthermore, new and improved vaccines are needed to replace several vaccines that are suboptimal in terms of efficacy or safety. Advancing technologies continue to transform the field of vaccinology, and we are now able to use genomic-based approaches to aid selection of vaccine candidates and structure-based design to optimize the chosen immunogens. In parallel, the increasing use of systems biology will provide essential insights into the immune response elicited by vaccines, and should help identify signatures of immunogenicity and correlates of protection. This improved understanding of both the host and the pathogen is aiding the development of new vaccine technologies, including the use of small molecule adjuvants to target specific immune responses, as well as new delivery systems and immunization schemes to optimize vaccine efficacy, which are essential components for the next generation of vaccine development (Fig. 3). The knowledge gained through systems biology will also increase fundamental understanding of microbial pathogenesis, enabling continued advancements in vaccine development.

ACKNOWLEDGMENTS

We thank Giorgio Corsi for artwork.

REFERENCES

Abbot EL, Smith WD, Siou GP, Chiriboga C, Smith RJ, Wilson JA, Hirst BH, Kehoe MA. 2007. Pili mediate specific adhesion of *Streptococcus pyogenes* to human tonsil and skin. *Cell Microbiol* **9:** 1822–1833.

Acharya IL, Lowe CU, Thapa R, Gurubacharya VL, Shrestha MB, Cadoz M, Schulz D, Armand J, Bryla DA, Trollfors B, et al. 1987. Prevention of typhoid fever in Nepal with the Vi capsular polysaccharide of *Salmonella* typhi. A preliminary report. *N Engl J Med* **317:** 1101–1104.

Bagnoli F, Moschioni M, Donati C, Dimitrovska V, Ferlenghi I, Facciotti C, Muzzi A, Giusti F, Emolo C, Sinisi A, et al. 2008. A second pilus type in *Streptococcus pneumoniae* is prevalent in emerging serotypes and mediates adhesion to host cells. *J Bacteriol* **190:** 5480–5492.

Bambini S, Muzzi A, Olcen P, Rappuoli R, Pizza M, Comanducci M. 2009. Distribution and genetic variability of three vaccine components in a panel of strains representative of the diversity of serogroup B meningococcus. *Vaccine* **27:** 2794–2803.

Barocchi MA, Ries J, Zogaj X, Hemsley C, Albiger B, Kanth A, Dahlberg S, Fernebro J, Moschioni M, Masignani V, et al. 2006. A pneumococcal pilus influences virulence and host inflammatory responses. *Proc Natl Acad Sci* **103:** 2857–2862.

Bartolini E, Frigimelica E, Giovinazzi S, Galli G, Shaik Y, Genco C, Welsch JA, Granoff DM, Grandi G, Grifantini R. 2006. Role of FNR and FNR-regulated, sugar fermentation genes in *Neisseria meningitidis* infection. *Mol Microbiol* **60:** 963–972.

Beernink PT, Welsch JA, Harrison LH, Leipus A, Kaplan SL, Granoff DM. 2007. Prevalence of factor H-binding protein variants and NadA among meningococcal group B isolates from the United States: Implications for the development of a multicomponent group B vaccine. *J Infect Dis* **195:** 1472–1479.

Beernink PT, Shaughnessy J, Braga EM, Liu Q, Rice PA, Ram S, Granoff DM. 2011. A meningococcal factor H binding protein mutant that eliminates factor H binding enhances protective antibody responses to vaccination. *J Immunol* **186:** 3606–3614.

Berlanda Scorza F, Doro F, Rodriguez-Ortega MJ, Stella M, Liberatori S, Taddei AR, Serino L, Gomes Moriel D, Nesta B, Fontana MR, et al. 2008. Proteomics characterization of outer membrane vesicles from the extraintestinal pathogenic *Escherichia coli* ΔtolR IHE3034 mutant. *Mol Cell Proteomics* **7:** 473–485.

Bernstein A, Pulendran B, Rappuoli R. 2011. Systems vaccinomics: The road ahead for vaccinology. *OMICS* **15:** 529–531.

Borrow R, Carlone GM, Rosenstein N, Blake M, Feavers I, Martin D, Zollinger W, Robbins J, Aaberge I, Granoff DM, et al. 2006. *Neisseria meningitidis* group B correlates of protection and assay standardization—International meeting report Emory University, Atlanta, Georgia, United States, 16–17 March 2005. *Vaccine* **24:** 5093–5107.

Brehony C, Jolley KA, Maiden MC. 2007. Multilocus sequence typing for global surveillance of meningococcal disease. *FEMS Microbiol Rev* **31:** 15–26.

Brehony C, Wilson DJ, Maiden MC. 2009. Variation of the factor H-binding protein of *Neisseria meningitidis*. *Microbiology* **155:** 4155–4169.

Burton DR. 2010. Scaffolding to build a rational vaccine design strategy. *Proc Natl Acad Sci* **107:** 17859–17860.

Buynak EB, Roehm RR, Tytell AA, Bertland AU II, Lampson GP, Hilleman MR. 1976. Vaccine against human hepatitis B. *JAMA* **235:** 2832–2834.

Cantini F, Veggi D, Dragonetti S, Savino S, Scarselli M, Romagnoli G, Pizza M, Banci L, Rappuoli R. 2009. Solution structure of the factor H binding protein, a survival factor and protective antigen of *Neisseria meningitidis*. *J Biol Chem* **284:** 9022–9026.

Capecchi B, Adu-Bobie J, Di Marcello F, Ciucchi L, Masignani V, Taddei A, Rappuoli R, Pizza M, Arico B. 2005. *Neisseria meningitidis* NadA is a new invasin which promotes bacterial adhesion to and penetration into human epithelial cells. *Mol Microbiol* **55:** 687–698.

Comanducci M, Bambini S, Brunelli B, Adu-Bobie J, Arico B, Capecchi B, Giuliani MM, Masignani V, Santini L, Savino S, et al. 2002. NadA, a novel vaccine candidate of *Neisseria meningitidis*. *J Exp Med* **195:** 1445–1454.

Davila S, Wright VJ, Khor CC, Sim KS, Binder A, Breunis WB, Inwald D, Nadel S, Betts H, Carrol ED, et al. 2010. Genome-wide association study identifies variants in the CFH region associated with host susceptibility to meningococcal disease. *Nat Genet* **42:** 772–776.

de Bakker PI, Telenti A. 2010. Infectious diseases not immune to genome-wide association. *Nat Genet* **42:** 731–732.

Dhiman N, Smith DI, Poland GA. 2009. Next-generation sequencing: A transformative tool for vaccinology. *Expert Rev Vaccines* **8:** 963–967.

Dhiman N, Haralambieva IH, Kennedy RB, Vierkant RA, O'Byrne MM, Ovsyannikova IG, Jacobson RM, Poland GA. 2010. SNP/haplotype associations in cytokine and cytokine receptor genes and immunity to rubella vaccine. *Immunogenetics* **62:** 197–210.

Donati C, Hiller NL, Tettelin H, Muzzi A, Croucher NJ, Angiuoli SV, Oggioni M, Dunning Hotopp JC, Hu FZ, Riley DR, et al. 2010. Structure and dynamics of the pangenome of *Streptococcus pneumoniae* and closely related species. *Genome Biol* **11:** R107.

Donnelly J, Medini D, Boccadifuoco G, Biolchi A, Ward J, Frasch C, Moxon ER, Stella M, Comanducci M, Bambini S, et al. 2010. Qualitative and quantitative assessment of meningococcal antigens to evaluate the potential strain coverage of protein-based vaccines. *Proc Natl Acad Sci* **107:** 19490–19495.

Dormitzer PR, Ulmer JB, Rappuoli R. 2008. Structure-based antigen design: A strategy for next generation vaccines. *Trends Biotechnol* **26:** 659–667.

Dramsi S, Caliot E, Bonne I, Guadagnini S, Prevost MC, Kojadinovic M, Lalioui L, Poyart C, Trieu-Cuot P. 2006. Assembly and role of pili in group B streptococci. *Mol Microbiol* **60:** 1401–1413.

Dunphy KY, Beernink PT, Brogioni B, Granoff DM. 2011. Effect of factor H-binding protein sequence variation on factor H binding and survival of *Neisseria meningitidis* in human blood. *Infect Immun* **79:** 353–359.

Ellis RW. 1990. New and improved vaccines against hepatitis B. In *New generation vaccines* (ed. Woodrow G, Levine M), pp. 439–447. Marcel Dekker, New York.

Findlow J, Borrow R, Snape MD, Dawson T, Holland A, John TM, Evans A, Telford KL, Ypma E, Toneatto D, et al. 2010. Multicenter, open-label, randomized phase II controlled trial of an investigational recombinant Meningococcal serogroup B vaccine with and without outer membrane vesicles, administered in infancy. *Clin Infect Dis* **51:** 1127–1137.

Fletcher LD, Bernfield L, Barniak V, Farley JE, Howell A, Knauf M, Ooi P, Smith RP, Weise P, Wetherell M, et al.

2004. Vaccine potential of the *Neisseria meningitidis* 2086 lipoprotein. *Infect Immun* **72:** 2088–2100.

Frasch CE, Borrow R, Donnelly J. 2009. Bactericidal antibody is the immunologic surrogate of protection against meningococcal disease. *Vaccine* **27** (Suppl 2): B112–B116.

Garcon N, Wettendorff M, Van Mechelen M. 2011. Role of AS04 in human papillomavirus vaccine: Mode of action and clinical profile. *Expert Opin Biol Ther* **11:** 667–677.

Gaucher D, Therrien R, Kettaf N, Angermann BR, Boucher G, Filali-Mouhim A, Moser JM, Mehta RS, Drake DR III, Castro E, et al. 2008. Yellow fever vaccine induces integrated multilineage and polyfunctional immune responses. *J Exp Med* **205:** 3119–3131.

Gill CJ, Baxter R, Anemona A, Ciavarro G, Dull P. 2010. Persistence of immune responses after a single dose of Novartis meningococcal serogroup A, C, W-135 and Y CRM-197 conjugate vaccine (Menveo) or Menactra among healthy adolescents. *Hum Vaccin* **6:** 881–887.

Giuliani MM, Adu-Bobie J, Comanducci M, Arico B, Savino S, Santini L, Brunelli B, Bambini S, Biolchi A, Capecchi B, et al. 2006. A universal vaccine for serogroup B meningococcus. *Proc Natl Acad Sci* **103:** 10834–10839.

Giuliani MM, Biolchi A, Serruto D, Ferlicca F, Vienken K, Oster P, Rappuoli R, Pizza M, Donnelly J. 2010. Measuring antigen-specific bactericidal responses to a multicomponent vaccine against serogroup B meningococcus. *Vaccine* **28:** 5023–5030.

Glenny AT, Hopkins BE. 1923. Diphtheria toxoid as an immunizing agent. *Br J Exp Pathol* **4:** 283–288.

Gotschlich EC, Goldschneider I, Artenstein MS. 1969. Human immunity to the meningococcus. IV. Immunogenicity of group A and group C meningococcal polysaccharides in human volunteers. *J Exp Med* **129:** 1367–1384.

Granoff DM, Welsch JA, Ram S. 2009. Binding of complement factor H (fH) to *Neisseria meningitidis* is specific for human fH and inhibits complement activation by rat and rabbit sera. *Infect Immun* **77:** 764–769.

Halperin SA, Diaz-Mitoma F, Dull P, Anemona A, Ceddia F. 2010. Safety and immunogenicity of an investigational quadrivalent meningococcal conjugate vaccine after one or two doses given to infants and toddlers. *Eur J Clin Microbiol Infect Dis* **29:** 259–267.

Haralambieva IH, Poland GA. 2010. Vaccinomics, predictive vaccinology and the future of vaccine development. *Future Microbiol* **5:** 1757–1760.

Haralambous E, Dolly SO, Hibberd ML, Litt DJ, Udalova IA, O'Dwyer C, Langford PR, Simon Kroll J, Levin M. 2006. Factor H, a regulator of complement activity, is a major determinant of meningococcal disease susceptibility in UK Caucasian patients. *Scand J Infect Dis* **38:** 764–771.

Hava DL, Camilli A. 2002. Large-scale identification of serotype 4 *Streptococcus pneumoniae* virulence factors. *Mol Microbiol* **45:** 1389–1406.

He M, Sebaihia M, Lawley TD, Stabler RA, Dawson LF, Martin MJ, Holt KE, Seth-Smith HM, Quail MA, Rance R, et al. 2010. Evolutionary dynamics of *Clostridium difficile* over short and long time scales. *Proc Natl Acad Sci* **107:** 7527–7532.

Heinz E, Tischler P, Rattei T, Myers G, Wagner M, Horn M. 2009. Comprehensive in silico prediction and analysis of chlamydial outer membrane proteins reflects evolution and life style of the Chlamydiae. *BMC Genomics* **10:** 634.

Hilleringmann M, Ringler P, Muller SA, De Angelis G, Rappuoli R, Ferlenghi I, Engel A. 2009. Molecular architecture of *Streptococcus pneumoniae* TIGR4 pili. *EMBO J* **28:** 3921–3930.

Jacobsson S, Molling P, Olcen P. 2009. Seroprevalence of antibodies against fHbp and NadA, two potential vaccine antigens for *Neisseria meningitidis*. *Vaccine* **27:** 5755–5759.

Jefferies JM, Macdonald E, Faust SN, Clarke SC. 2011. 13-valent pneumococcal conjugate vaccine (PCV13). *Hum Vaccin* **7:** 1012–1018.

Jiang HQ, Hoiseth SK, Harris SL, McNeil LK, Zhu D, Tan C, Scott AA, Alexander K, Mason K, Miller L, et al. 2010. Broad vaccine coverage predicted for a bivalent recombinant factor H binding protein based vaccine to prevent serogroup B meningococcal disease. *Vaccine* **28:** 6086–6093.

Keiser PB, Biggs-Cicatelli S, Moran EE, Schmiel DH, Pinto VB, Burden RE, Miller LB, Moon JE, Bowden RA, Cummings JF, et al. 2011. A phase 1 study of a meningococcal native outer membrane vesicle vaccine made from a group B strain with deleted lpxL1 and synX, over-expressed factor H binding protein, two PorAs and stabilized OpcA expression. *Vaccine* **29:** 1413–1420.

Kelly DF, Moxon ER, Pollard AJ. 2004. *Haemophilus influenzae* type b conjugate vaccines. *Immunology* **113:** 163–174.

Kitano H. 2002. Systems biology: A brief overview. *Science* **295:** 1662–1664.

Kitchin NR. 2011. Review of diphtheria, tetanus and pertussis vaccines in clinical development. *Expert Rev Vaccines* **10:** 605–615.

Koeberling O, Delany I, Granoff DM. 2011a. A critical threshold of meningococcal factor H binding protein expression is required for increased breadth of protective antibodies elicited by native outer membrane vesicle vaccines. *Clin Vaccine Immunol* **18:** 736–742.

Koeberling O, Seubert A, Santos G, Colaprico A, Ugozzoli M, Donnelly J, Granoff DM. 2011b. Immunogenicity of a meningococcal native outer membrane vesicle vaccine with attenuated endotoxin and over-expressed factor H binding protein in infant rhesus monkeys. *Vaccine* **29:** 4728–4734.

Kruskall MS, Alper CA, Awdeh Z, Yunis EJ, Marcus-Bagley D. 1992. The immune response to hepatitis B vaccine in humans: Inheritance patterns in families. *J Exp Med* **175:** 495–502.

Lauer P, Rinaudo CD, Soriani M, Margarit I, Maione D, Rosini R, Taddei AR, Mora M, Rappuoli R, Grandi G, et al. 2005. Genome analysis reveals pili in Group B Streptococcus. *Science* **309:** 105.

Lawley TD, Croucher NJ, Yu L, Clare S, Sebaihia M, Goulding D, Pickard DJ, Parkhill J, Choudhary J, Dougan G. 2009. Proteomic and genomic characterization of highly infectious *Clostridium difficile* 630 spores. *J Bacteriol* **191:** 5377–5386.

LeMieux J, Hava DL, Basset A, Camilli A. 2006. RrgA and RrgB are components of a multisubunit pilus encoded by the *Streptococcus pneumoniae* rlrA pathogenicity islet. *Infect Immun* **74:** 2453–2456.

Levine MM, Lagos R. 2004. Vaccines and vaccination in historical perspective. In *New generation vaccines* (ed. Levine MM, Kaper JB, Rappouli R, et al.), pp. 1–10. Marcel Dekker, New York.

Liu X, Afrane M, Clemmer DE, Zhong G, Nelson DE. 2010. Identification of *Chlamydia trachomatis* outer membrane complex proteins by differential proteomics. *J Bacteriol* **192:** 2852–2860.

MacLeod CM, Hodges GR, Heidelberger M, Bernhard WG. 1945. Prevention of pneumococcal pneumonia by immunization with specific capsular polysaccharides. *J Exp Med* **82:** 445–465.

Madico G, Welsch JA, Lewis LA, McNaughton A, Perlman DH, Costello CE, Ngampasutadol J, Vogel U, Granoff DM, Ram S. 2006. The meningococcal vaccine candidate GNA1870 binds the complement regulatory protein factor H and enhances serum resistance. *J Immunol* **177:** 501–510.

Maione D, Margarit I, Rinaudo CD, Masignani V, Mora M, Scarselli M, Tettelin H, Brettoni C, Iacobini ET, Rosini R, et al. 2005. Identification of a universal Group B streptococcus vaccine by multiple genome screen. *Science* **309:** 148–150.

Manetti AG, Zingaretti C, Falugi F, Capo S, Bombaci M, Bagnoli F, Gambellini G, Bensi G, Mora M, Edwards AM, et al. 2007. Streptococcus pyogenes pili promote pharyngeal cell adhesion and biofilm formation. *Mol Microbiol* **64:** 968–983.

Margarit I, Rinaudo CD, Galeotti CL, Maione D, Ghezzo C, Buttazzoni E, Rosini R, Runci Y, Mora M, Buccato S, et al. 2009. Preventing bacterial infections with pilus-based vaccines: The group B streptococcus paradigm. *J Infect Dis* **199:** 108–115.

Masignani V, Comanducci M, Giuliani MM, Bambini S, Adu-Bobie J, Arico B, Brunelli B, Pieri A, Santini L, Savino S, et al. 2003. Vaccination against *Neisseria meningitidis* using three variants of the lipoprotein GNA1870. *J Exp Med* **197:** 789–799.

Mazmanian SK, Liu G, Jensen ER, Lenoy E, Schneewind O. 2000. Staphylococcus aureus sortase mutants defective in the display of surface proteins and in the pathogenesis of animal infections. *Proc Natl Acad Sci* **97:** 5510–5515.

Mbow ML, De Gregorio E, Valiante NM, Rappuoli R. 2010. New adjuvants for human vaccines. *Curr Opin Immunol* **22:** 411–416.

McCarthy AJ, Lindsay JA. 2010. Genetic variation in *Staphylococcus aureus* surface and immune evasion genes is lineage associated: Implications for vaccine design and host-pathogen interactions. *BMC Microbiol* **10:** 173.

McNeil LK, Murphy E, Zhao XJ, Guttmann S, Harris SL, Scott AA, Tan C, Mack M, DaSilva I, Alexander K, et al. 2009. Detection of LP2086 on the cell surface of *Neisseria meningitidis* and its accessibility in the presence of serogroup B capsular polysaccharide. *Vaccine* **27:** 3417–3421.

Meri S, Jordens M, Jarva H. 2008. Microbial complement inhibitors as vaccines. *Vaccine* **26:** I113–I117.

Montigiani S, Falugi F, Scarselli M, Finco O, Petracca R, Galli G, Mariani M, Manetti R, Agnusdei M, Cevenini R, et al. 2002. Genomic approach for analysis of surface proteins in *Chlamydia pneumoniae*. *Infect Immun* **70:** 368–379.

Mora M, Bensi G, Capo S, Falugi F, Zingaretti C, Manetti AG, Maggi T, Taddei AR, Grandi G, Telford JL. 2005. Group A streptococcus produce pilus-like structures containing protective antigens and Lancefield T antigens. *Proc Natl Acad Sci* **102:** 15641–15646.

Morens DM, Folkers GK, Fauci AS. 2008. Emerging infections: A perpetual challenge. *Lancet Infect Dis* **8:** 710–719.

Moriel DG, Bertoldi I, Spagnuolo A, Marchi S, Rosini R, Nesta B, Pastorello I, Corea VA, Torricelli G, Cartocci E, et al. 2010. Identification of protective and broadly conserved vaccine antigens from the genome of extraintestinal pathogenic *Escherichia coli*. *Proc Natl Acad Sci* **107:** 9072–9077.

Murphy E, Andrew L, Lee KL, Dilts DA, Nunez L, Fink PS, Ambrose K, Borrow R, Findlow J, Taha MK, et al. 2009. Sequence diversity of the factor H binding protein vaccine candidate in epidemiologically relevant strains of serogroup B *Neisseria meningitidis*. *J Infect Dis* **200:** 379–389.

Nakaya HI, Li S, Pulendran B. 2011. Systems vaccinology: Learning to compute the behavior of vaccine induced immunity. *Wiley Interdiscip Rev Syst Biol Med* **4:** 193–205.

Nesta B, Spraggon G, Alteri C, Moriel DG, Rosini R, Veggi D, Smith S, Bertoldi I, Pastorello I, Ferlenghi I, et al. 2012. FdeC, a novel broadly conserved Escherichia coli adhesin eliciting protection against urinary tract infections. *MBio* **3:** e00010–12.

Nielsen HE, Christensen KC, Koch C, Thomsen BS, Heegaard NH, Tranum-Jensen J. 1989. Hereditary, complete deficiency of complement factor H associated with recurrent meningococcal disease. *Scand J Immunol* **30:** 711–718.

Nuccitelli A, Cozzi R, Gourlay LJ, Donnarumma D, Necchi F, Norais N, Telford JL, Rappuoli R, Bolognesi M, Maione D, et al. 2011. Structure-based approach to rationally design a chimeric protein for an effective vaccine against Group B Streptococcus infections. *Proc Natl Acad Sci* **108:** 10278–10283.

O'Hagan DT, Rappuoli R, De Gregorio E, Tsai T, Del Giudice G. 2011. MF59 adjuvant: The best insurance against influenza strain diversity. *Expert Rev Vaccines* **10:** 447–462.

Oriente F, Scarlato V, Delany I. 2010. Expression of factor H binding protein of meningococcus responds to oxygen limitation through a dedicated FNR-regulated promoter. *J Bacteriol* **192:** 691–701.

Ovsyannikova IG, Jacobson RM, Vierkant RA, Pankratz VS, Poland GA. 2007. HLA supertypes and immune responses to measles-mumps-rubella viral vaccine: Findings and implications for vaccine design. *Vaccine* **25:** 3090–3100.

Parida SK, Kaufmann SH. 2010. Novel tuberculosis vaccines on the horizon. *Curr Opin Immunol* **22:** 374–384.

Peltola H, Kayhty H, Sivonen A, Makela H. 1977. *Haemophilus influenzae* type b capsular polysaccharide vaccine in children: A double-blind field study of 100,000 vaccinees 3 months to 5 years of age in Finland. *Pediatrics* **60:** 730–737.

Pezzicoli A, Santi I, Lauer P, Rosini R, Rinaudo D, Grandi G, Telford JL, Soriani M. 2008. Pilus backbone contributes to group B *Streptococcus* paracellular translocation through epithelial cells. *J Infect Dis* **198:** 890–898.

Pizza M, Covacci A, Bartoloni A, Perugini M, Nencioni L, De Magistris MT, Villa L, Nucci D, Manetti R, Bugnoli M, et al. 1989. Mutants of pertussis toxin suitable for vaccine development. *Science* **246:** 497–500.

Pizza M, Scarlato V, Masignani V, Giuliani MM, Arico B, Comanducci M, Jennings GT, Baldi L, Bartolini E, Capecchi B, et al. 2000. Identification of vaccine candidates against serogroup B meningococcus by whole-genome sequencing. *Science* **287:** 1816–1820.

Pizza M, Donnelly J, Rappuoli R. 2008. Factor H-binding protein, a unique meningococcal vaccine antigen. *Vaccine* **26** (Suppl 8): 146–148.

Plested JS, Welsch JA, Granoff DM. 2009. Ex vivo model of meningococcal bacteremia using human blood for measuring vaccine-induced serum passive protective activity. *Clin Vaccine Immunol* **16:** 785–791.

Plotkin SA. 2009. Vaccines: The fourth century. *Clin Vaccine Immunol* **16:** 1709–1719.

Poland GA, Ovsyannikova IG, Jacobson RM, Smith DI. 2007. Heterogeneity in vaccine immune response: The role of immunogenetics and the emerging field of vaccinomics. *Clin Pharmacol Ther* **82:** 653–664.

Poland GA, Ovsyannikova IG, Jacobson RM. 2008. Personalized vaccines: The emerging field of vaccinomics. *Expert Opin Biol Ther* **8:** 1659–1667.

Poland GA, Ovsyannikova IG, Jacobson RM. 2009. Application of pharmacogenomics to vaccines. *Pharmacogenomics* **10:** 837–852.

Pulendran B, Li S, Nakaya HI. 2010. Systems vaccinology. *Immunity* **33:** 516–529.

Querec TD, Akondy RS, Lee EK, Cao W, Nakaya HI, Teuwen D, Pirani A, Gernert K, Deng J, Marzolf B, et al. 2009. Systems biology approach predicts immunogenicity of the yellow fever vaccine in humans. *Nat Immunol* **10:** 116–125.

Ramon G. 1924. Sur la toxine et sur l'anatoxine diphtériques. Pouvoir floculant et propriétés immunisantes. *Ann Inst Pasteur* **38:** 1–106.

Rappuoli R. 2007. Bridging the knowledge gaps in vaccine design. *Nat Biotechnol* **25:** 1361–1366.

Rappuoli R, Black S, Lambert PH. 2011. Vaccine discovery and translation of new vaccine technology. *Lancet* **378:** 360–368.

Rasko DA, Rosovitz MJ, Myers GS, Mongodin EF, Fricke WF, Gajer P, Crabtree J, Sebaihia M, Thomson NR, Chaudhuri R, et al. 2008. The pangenome structure of *Escherichia coli*: Comparative genomic analysis of *E. coli* commensal and pathogenic isolates. *J Bacteriol* **190:** 6881–6893.

Rinaudo CD, Telford JL, Rappuoli R, Seib KL. 2009. Vaccinology in the genome era. *J Clin Invest* **119:** 2515–2525.

Rinaudo CD, Rosini R, Galeotti CL, Berti F, Necchi F, Reguzzi V, Ghezzo C, Telford JL, Grandi G, Maione D. 2010. Specific involvement of pilus type 2a in biofilm formation in group B Streptococcus. *PLoS ONE* **5:** e9216.

Rosini R, Rinaudo CD, Soriani M, Lauer P, Mora M, Maione D, Taddei A, Santi I, Ghezzo C, Brettoni C, et al. 2006. Identification of novel genomic islands coding for antigenic pilus-like structures in *Streptococcus agalactiae*. *Mol Microbiol* **61:** 126–141.

Scarselli M, Arico B, Brunelli B, Savino S, Di Marcello F, Palumbo E, Veggi D, Ciucchi L, Cartocci E, Bottomley MJ, et al. 2011. Rational design of a meningococcal antigen inducing broad protective immunity. *Sci Transl Med* **3:** 91ra62.

Schneider MC, Exley RM, Chan H, Feavers I, Kang YH, Sim RB, Tang CM. 2006. Functional significance of factor H binding to *Neisseria meningitidis*. *J Immunol* **176:** 7566–7575.

Schneider MC, Prosser BE, Caesar JJ, Kugelberg E, Li S, Zhang Q, Quoraishi S, Lovett JE, Deane JE, Sim RB, et al. 2009. *Neisseria meningitidis* recruits factor H using protein mimicry of host carbohydrates. *Nature* **458:** 890–893.

Seib KL, Serruto D, Oriente F, Delany I, Adu-Bobie J, Veggi D, Arico B, Rappuoli R, Pizza M. 2009. Factor H-binding protein is important for meningococcal survival in human whole blood and serum and in the presence of the antimicrobial peptide LL-37. *Infect Immun* **77:** 292–299.

Seib KL, Brunelli B, Brogioni B, Palumbo E, Bambini S, Muzzi A, Dimarcello F, Marchi S, van der Ende A, Ariço B, et al. 2011. Characterisation of diverse sub-variants of the meningococcal factor H binding protein (fHbp) for their ability to bind fH, mediate serum resistance and induce bactericidal antibodies. *Infect Immun* **79:** 970–981.

Serruto D, Spadafina T, Ciucchi L, Lewis LA, Ram S, Tontini M, Santini L, Biolchi A, Seib KL, Giuliani MM, et al. 2010. *Neisseria meningitidis* GNA2132, a heparin-binding protein that induces protective immunity in humans. *Proc Natl Acad Sci* **107:** 3770–3775.

Stabler RA, Gerding DN, Songer JG, Drudy D, Brazier JS, Trinh HT, Witney AA, Hinds J, Wren BW. 2006. Comparative phylogenomics of *Clostridium difficile* reveals clade specificity and microevolution of hypervirulent strains. *J Bacteriol* **188:** 7297–7305.

Stabler RA, He M, Dawson L, Martin M, Valiente E, Corton C, Lawley TD, Sebaihia M, Quail MA, Rose G, et al. 2009. Comparative genome and phenotypic analysis of *Clostridium difficile* 027 strains provides insight into the evolution of a hypervirulent bacterium. *Genome Biol* **10:** R102.

Stranger-Jones YK, Bae T, Schneewind O. 2006. Vaccine assembly from surface proteins of *Staphylococcus aureus*. *Proc Natl Acad Sci* **103:** 16942–16947.

Telford JL, Barocchi MA, Margarit I, Rappuoli R, Grandi G. 2006. Pili in gram-positive pathogens. *Nat Rev Microbiol* **4:** 509–519.

Tettelin H, Masignani V, Cieslewicz MJ, Donati C, Medini D, Ward NL, Angiuoli SV, Crabtree J, Jones AL, Durkin AS, et al. 2005. Genome analysis of multiple pathogenic isolates of *Streptococcus agalactiae*: Implications for the microbial "pan-genome." *Proc Natl Acad Sci* **102:** 13950–13955.

Thorpe C, Edwards L, Snelgrove R, Finco O, Rae A, Grandi G, Guilio R, Hussell T. 2007. Discovery of a vaccine antigen that protects mice from *Chlamydia pneumoniae* infection. *Vaccine* **25:** 2252–2260.

Tobin GJ, Trujillo JD, Bushnell RV, Lin G, Chaudhuri AR, Long J, Barrera J, Pena L, Grubman MJ, Nara PL. 2008.

Deceptive imprinting and immune refocusing in vaccine design. *Vaccine* **26:** 6189–6199.

Toneatto D, Ismaili S, Ypma E, Vienken K, Oster P, Dull P. 2011a. The first use of an investigational multicomponent meningococcal serogroup B vaccine (4CMenB) in humans. *Hum Vaccin* **7:** 646–653.

Toneatto D, Oster P, Deboer AC, Emerson A, Santos GF, Ypma E, Detora L, Pizza M, Kimura A, Dull P. 2011b. Early clinical experience with a candidate meningococcal B recombinant vaccine (rMenB) in healthy adults. *Hum Vaccin* **7:** 781–791.

Veggi D, Gentile MA, Cantini F, Lo Surdo P, Nardi-Dei V, Seib KL, Pizza M, Rappuoli R, Banci L, Savino S, et al. 2012. The factor H binding protein of *Neisseria meningitidis* interacts with xenosiderophores in vitro. *Biochemistry* **51:** 9384–9393.

Welsch JA, Ram S, Koeberling O, Granoff DM. 2008. Complement-dependent synergistic bactericidal activity of antibodies against factor H-binding protein, a sparsely distributed meningococcal vaccine antigen. *J Infect Dis* **197:** 1053–1061.

Probiotics: Properties, Examples, and Specific Applications

Judith Behnsen, Elisa Deriu, Martina Sassone-Corsi, and Manuela Raffatellu

Department of Microbiology and Molecular Genetics, Institute for Immunology, University of California, Irvine, California 92697

Correspondence: manuelar@uci.edu

Probiotics are beneficial components of the microbiota that have been used for centuries because of the health benefits they confer to the host. Only recently, however, has the contribution of probiotics to modulation of immunological, respiratory, and gastrointestinal functions started to be fully appreciated and scientifically evaluated. Probiotics such as *Escherichia coli* Nissle 1917 and lactic acid bacteria are currently used to, or have been evaluated for use to, prevent or treat a range of intestinal maladies including inflammatory bowel disease, constipation, and colon cancer. Engineering these natural probiotics to produce immunomodulatory molecules may help to further increase the benefit to the host. In this article, we will discuss some of the mechanisms of action of probiotics as well as advances in the rational design of probiotics.

Probiotics are defined as living bacteria that, when administered in adequate amounts, confer a health benefit on the host (FAO/WHO 2001). The original observation of the beneficial properties conferred by some bacteria is attributed to the Nobel Prize winner Eli Metchnikoff, who is regarded as the grandfather of modern probiotics. In the early 20th century, Metchnikoff discovered that "healthy bacteria," especially lactic acid bacteria (LAB), can have a positive influence on digestion and the immune system (Anukam and Reid 2008). Most microorganisms recognized to date as probiotics are Gram-positive, with *Lactobacillus* and *Bifidobacterium* being the main species used as treatments of intestinal dysfunctions (Marco et al. 2006). However, some Gram-negatives are also used as probiotics. The best example of this group is *Escherichia coli* Nissle 1917 (EcN) (Nissle 1959), also known as "Mutaflor," which has been used in Germany for many years in the treatment of chronic constipation (Mollenbrink and Bruckschen 1994) and colitis (Schutz 1989).

In the last century, many studies have reported probiotic bacteria to play important roles in the modulation of immunological, respiratory, and gastrointestinal functions (Floch et al. 2011). Furthermore, probiotics have been shown to play a protective role by directly competing with intestinal pathogens through the release of antibacterial substances such as bacteriocins (Cotter et al. 2005) or metabolites such as acetic acid and lactic acid (Servin 2004). Although most studies on probiotics have been empirical, new advancements may originate from research on the interactions between com-

Cite this article as *Cold Spring Harb Perspect Med* doi: 10.1101/cshperspect.a010074

mensal microorganisms (termed the microbiota), pathogens, and the host. Understanding the mechanisms of gut colonization in both normal and inflammatory conditions is essential to designing probiotics for a specific use. In this article, we will discuss some of the recent developments on the mechanisms of action of probiotics and their utilization for delivering molecules to specific sites within the host.

THE HUMAN DIGESTIVE TRACT AND THE MICROBIOTA

The human digestive tract comprises a variety of ecological niches populated by several bacterial species that have established a symbiotic relationship with the host. This bacterial population, also called the intestinal microbiota, plays an important role in the development of the gut immune system, digestion of food, production of short-chain fatty acids and essential vitamins, and resistance to colonization from pathogenic microorganisms (Hooper and Gordon 2001). The human gut microbiome consists of many different species of bacteria, some of which are nonculturable and therefore not well known or characterized. Indeed, it has only been through the advent of deep sequencing, genomics, and metagenomics in the last decade that the complexity of the microbiota has been fully appreciated.

The distribution of the intestinal microbiota varies along three main locations in the digestive tract: (1) the stomach, populated by $<10^2$ colony-forming units (cfu)/mL, including lactobacilli and streptococci; (2) the ileum and distal ileum, populated by 10^2-10^3 cfu/mL of bacteria, including *E. coli*, *Klebsiella*, *Enterococcus*, and *Bacteroides*; and (3) the large intestine, which constitutes the largest microbial population of the body, with $10^{10}-10^{12}$ cfu/mL (DiBaise et al. 2008). Remarkably, each individual organism presents a specific "bacterial fingerprint," which is influenced by a variety of factors including the maternal environment, host genotype, diet, and antibiotic treatment (Spor et al. 2011). But even though the composition of the microbiota differs from person to person, it clusters in three distinct groups, so-called enterotypes. These human enterotypes are enriched in *Bacteroides*, *Prevotella*, or *Ruminococcus* and use different routes to generate energy from fermentable substrates available in the colon: *Bacteroides* uses carbohydrates; *Prevotella*, mucins; and *Ruminococcus*, mucins and sugars (Arumugam et al. 2011). Enterotypes have also been associated with long-term diets (Wu et al. 2011). Despite this heterogeneity, Firmicutes and Bacteroidetes are the most common intestinal phyla across all vertebrates, representing >90% of the microbiota, followed by Actinobacteria and Proteobacteria (Mahowald et al. 2009). Members of the microbiota from phylum Bacteroidetes are represented by a variety of species, including *Bacteroides fragilis*, which was recently shown to possess immunomodulatory capabilities via its polysaccharide capsule (Cobb et al. 2004; Coyne et al. 2005; Mazmanian et al. 2005; Liu et al. 2008). In contrast, phylum Firmicutes is mainly represented by species belonging to class Clostridia, which are known for their abilities to metabolize fiber and produce butyrate, a short-chain fatty acid with immunomodulatory activity (Gophna et al. 2006; Vinolo et al. 2011). Bacteria belonging to class Bacilli, including *Lactobacillus acidophilus* and *Enterococcus faecalis*, constitute the rest of the Firmicutes phylum.

The advent of germ-free mice gave rise to a better understanding of the impact the intestinal microbiota has on the host. Studies have shown that these mice exhibit a thin intestinal epithelium (Jervis and Biggers 1964), loss of short-chain fatty acid production (Høverstad and Midtvedt 1986), and alterations to the immune system (Maslowski et al. 2009; Tlaskalova-Hogenova et al. 2011). Remarkably, administration of the microbiota restores full functionality of the gut (Aureli et al. 2011). Moreover, some components of the microbiota have been shown to induce the development of T-cell subsets (Ivanov et al. 2009; Feng et al. 2010; Atarashi et al. 2011; Shaw et al. 2012) and the release of defensins (Bevins and Salzman 2011). Another notable aspect is that in normal individuals, the host establishes tolerance to commensal bacteria while maintaining its capacity to mount an immune response

 Cite this article as *Cold Spring Harb Perspect Med* doi: 10.1101/cshperspect.a010074

against pathogens. This balance between tolerance and immunity, when disrupted, can lead to the development of a number of intestinal pathologies including inflammatory bowel disease (IBD) and intestinal cancer (Karin et al. 2006; Artis 2008). Components of the immune response that normally orchestrate the mucosal barrier to the microbiota can, if altered, also contribute to the development of IBD. For example, an inflammatory response mounted against the microbiota can lead to a change in the host environment that in turn affects the composition and quantity of the microbiota, which further exacerbates intestinal inflammation (Fava and Danese 2011). Intestinal inflammation results in dramatic alterations to the microbiota, with loss of the most abundant species (Bacteroidetes and Clostridiales) and enhanced growth of Enterobacteriaceae (Fava and Danese 2011; Mukhopadhya et al. 2012). Similarly, in many cases of self-limiting gastrointestinal infections, antibiotic treatment is not recommended because of its limited efficacy and notable side effects, including alterations to the microbiota that can result in antibiotic-associated colitis (Hogenauer et al. 1998).

In light of these observations, it becomes apparent that restoring balance to or augmenting the microbiota can potentially provide beneficial resolution of a variety of diseases. Moreover, as we discuss later, probiotics can be engineered to better mitigate the conditions arising from a particular intestinal pathology by cytokine and enzyme delivery or through direct competition with pathogenic microorganisms.

THE BEST-CHARACTERIZED PROBIOTICS

Many probiotics are culturable components of the microbiota that have been used for their beneficial functions since long before the term "probiotic" was coined. The most commonly used probiotic strains include the lactic acid bacteria (LAB), Gram-positive microbes that have been used for centuries in food production processes (yogurt, cheese, pickles). Members of the LAB such as *Lactococcus* and *Streptococcus* are also important components of the endogenous microbiota in the human ileum and jejunum and, at more moderate densities, in the colon (Hayashi et al. 2005). The approval for use in food combined with the absence of lipopolysaccharides (LPS) and lack of secreted proteases make many LAB ideal for use as probiotics. Similarly, as these organisms are Gram-positive, recombinant proteins can be secreted without getting trapped in the periplasm, making LAB species attractive vehicles for food-grade production of proteins and enzymes (Le Loir et al. 2005; Mierau and Kleerebezem 2005; Peterbauer et al. 2011).

The most widely used LAB species for cytokine delivery is *Lactococcus lactis*. This bacterium, unlike other *Lactococcus* species, is only transiently present in the human gut and is hence classified as a noncommensal (Nouaille et al. 2003). This transient colonization provides an advantage for the utilization of this probiotic strain as a delivery vehicle for protein vaccines and even, more recently, DNA vaccines. In these cases, a sustained colonization with a strain would be a disadvantage, as it may lead to overstimulation of the targeted pathway. The possibility to deliver proteins directly to the mucosa opens up new methods of therapeutic treatment in which traditional routes of medication fail as a result of, for example, low local availability of the therapeutic substance.

Regarding Gram-negative probiotics, the most commonly used strain is *Escherichia coli* Nissle 1917 (EcN) The army surgeon Alfred Nissle originally isolated this strain in 1917 from the feces of a soldier during the First World War who, in contrast to his comrades, did not develop infectious diarrhea during an outbreak of the highly contagious organism *Shigella*. Nissle's observation suggested that EcN might provide colonization resistance to mucosal pathogens. Consistent with this hypothesis, the probiotic effect and biosafety of EcN has since been extensively shown in numerous trials and underlined by its long medical history in Central Europe as a microbial remedy (Kruis et al. 1997; Westendorf et al. 2005; Henker et al. 2007). At present, EcN is contained in a drug called Mutaflor, which is used for the treatment of both infectious diarrheal diseases and IBD (Schutz 1989). Furthermore, EcN has been administered to neonates to prevent the colonization of their digestive tract by multidrug-resistant pathogens (Lodinova-Zadnikova and Sonnenborn 1997; Boudeau et al. 2003; Kruis et al. 2004; Grabig et al. 2006; Henker et al. 2007).

LAB AND THEIR ROLE IN DELIVERING CYTOKINES AND OTHER MOLECULES TO RECONSTITUTE BARRIER DEFECTS

One of the recent advances in probiotics research involves the use of LAB to deliver cytokines directly to target sites within a host. LAB-delivered cytokines can be applied to treat diseases that weaken the mucosal barrier, such as IBD subtypes Crohn's disease and ulcerative colitis. IBD is a major health concern in the Western world and manifests as chronic inflammation of the intestine that results in diarrhea, abdominal pain, and weight loss (Steidler et al. 2000). Although the etiology of IBD is unknown, it is clear that both genetic alterations in host pattern recognition receptors and proinflammatory genes as well as the microbiota play a role in causing the sustained intestinal inflammation seen in IBD patients (Nagalingam and Lynch 2012). Therefore, treatments aimed at reducing intestinal inflammation in these patients are highly desirable.

One way to reduce the chronic inflammation of IBD patients is through the administration of anti-inflammatory cytokines such as interleukin 10 (IL-10). This cytokine plays a central role in down-regulating inflammatory cascades and in the establishment of tolerance in the mucosa (Huibregtse et al. 2011). However, clinical trials of subcutaneous administration of IL-10 were disappointing because of the low efficacy and side effects (van Deventer et al. 1997; Colombel et al. 2001; Tilg et al. 2002). Owing to the acid sensitivity of IL-10, direct oral administration was not deemed a viable option. Instead, in an attempt to use the oral route of administration but protect IL-10 from degradation, Schotte et al. (2000) engineered an IL-10-producing *L. lactis* strain (*LL-mIL10*). They subsequently showed in a landmark study in 2000 that murine colitis significantly improved following treatment with this IL-10-producing strain (Fig. 1) (Steidler et al. 2000). Furthermore, the onset of colitis was inhibited in IL-10-deficient mice and the amount of IL-10 needed for the observed reduction in colitis was several orders of magnitudes lower than what was needed to reduce it by systemic administration (Steidler et al. 2000). Thus, delivery to the mucosa via LAB was shown to be a key element in the effectiveness of an IL-10-based treatment.

A major issue to be resolved for clinical probiotic applications is the biological containment of genetically modified organisms. To achieve this in *L. lactis*, the thymidylate synthase gene (*thyA*) was replaced by the human IL-10 gene. Mutants in *thyA* require the presence of thymidine or thymine in the media to replicate, and therefore their growth is restricted to the human body and the accumulation of the strain in the environment is prevented (Steidler et al. 2003). The IL-10-producing *L. lactis* strain has since been administered to 10 Crohn's disease patients during a phase 1 clinical trial. Out of 10 patients, eight had a clinical benefit and five went into complete clinical remission (Braat et al. 2006). These exciting findings have paved the way for subsequent clinical trials (Steidler et al. 2009).

Given the anti-inflammatory properties of IL-10, other strategies have been used to induce secretion of this cytokine in the gut mucosa. One such strategy involved a protein from a pathogen. Pathogenic yersiniae produce an anti-inflammatory protein called LcrV, which mediates evasion of the host's immune response by enhancing IL-10 production (Fig. 1) (Foligne et al. 2007; Depaolo et al. 2008). Remarkably, an *L. lactis* strain producing LcrV was able to reduce inflammation in a trinitrobenzene sulfonic acid mouse model of colitis as efficiently as an *L. lactis* strain secreting IL-10 (Foligne et al. 2007).

Opposite to IL-10 on the cytokine spectrum is tumor necrosis factor α (TNF-α), a cytokine that mediates many of the clinical symptoms of IBD. Although systemic administration of anti-TNF-α antibodies has become an established therapy for Crohn's disease and ulcerative colitis, it has drawbacks, similar to those seen with the systemic administration of IL-10. The

 Cite this article as *Cold Spring Harb Perspect Med* doi: 10.1101/cshperspect.a010074

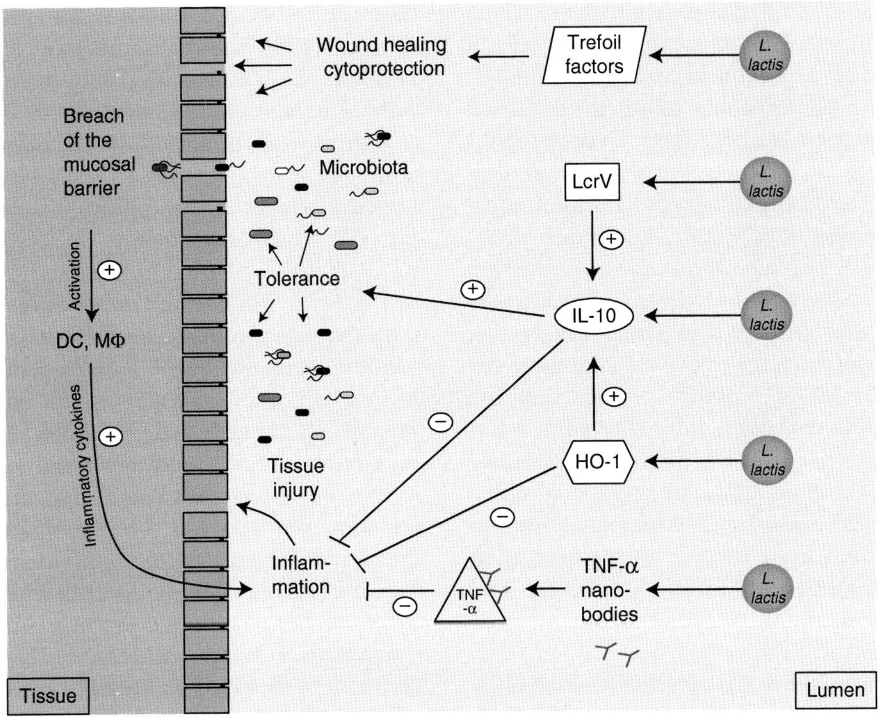

Figure 1. *Lactococcus lactis.* The probiotic *L. lactis* was engineered to produce different cytokines: interleukin 10 (IL-10), heme oxygenase 1 (HO-1), the *Yersinia* protein LcrV, TNF-α nanobodies, or trefoil factors. After oral administration, *L. lactis* secretes these proteins in direct proximity to mucosal surfaces. These proteins can modulate the immune system in different ways to dampen the immune response and establish tolerance to the microbiota in the case of mucosal injury during IBD or cancer treatment.

therapy is costly and has poor patient compliance and several adverse side effects (Vandenbroucke et al. 2010). It was believed that oral administration and local production of the antibody at the site of inflammation in the gut could alleviate these problems and partially restore normal mucosal function. Thus, *L. lactis* was engineered to produce anti-TNF-α nanobodies (Fig. 1), small and stable single-domain antibody fragments that were derived from heavy-chain camelid antibodies. As expected, daily administration of *L. lactis* producing these nanobodies reduced colonic inflammation in both dextran sodium sulfate (DSS)-treated and IL-10-deficient mice. Remarkably, the effect of the anti-TNF-α nanobodies produced by *L. lactis* was restricted to the gut and did not interfere with systemic function of TNF-α (Vandenbroucke et al. 2010). Therefore, LAB-mediated delivery of IL-10 and anti-TNF-α antibodies appears to be more beneficial than systemic delivery to control mucosal inflammation.

Another host factor that improves the integrity of the gut mucosal barrier is the enzyme heme oxygenase 1 (HO-1), which provides protection against oxidative stress and has anti-inflammatory and other immunomodulatory functions (Fig. 1) (Vijayan et al. 2010). Notably, an *L. lactis* strain secreting this enzyme proved to be effective in reducing morbidity and mortality in a model of LPS-induced mucosal injury (Pang et al. 2008) and in a model of hemorrhagic shock (Pang et al. 2009) in rats. Possible mechanisms for the protective effects in the gut mucosa of the *L. lactis* strain secreting HO-1 are the observed reductions in TNF-α and myeloperoxidase levels accompanied by increased IL-10 production.

Other molecules that have a broad protective effect on the mucosa are peptides of the trefoil factor family (TFF): Human TFF-1 and -2 are produced by mucus-producing cells in the stomach and duodenum, whereas TFF-3 is highly expressed in goblet cells in the small and large intestine (Kjellev 2009). TFFs are cytoprotective and promote epithelial wound healing and reconstitution of the gastrointestinal tract (Vandenbroucke et al. 2004) and are therefore excellent candidates for mucosa restoration (Fig. 1). However, one major drawback to the use of trefoil factors as therapeutic agents is that they do not reach the colon when administered orally. Because TFFs bind to mucus and are absorbed in the cecum, intrarectal administration has proven effective (Tran et al. 1999; Vandenbroucke et al. 2004). Intragastric administration of TFF-secreting *L. lactis* to mice has also been proven to be effective in the prevention and healing of acute DSS-induced colitis and chronic colitis in IL-10-deficient mice (Vandenbroucke et al. 2004).

Because TFF-1 and -3 are also secreted by human salivary glands and thus present in saliva, an *L. lactis* strain secreting TFF-1 was recently formulated into a mouthwash for treatment of oral mucositis, a very common and painful complication of radio- or chemotherapy in cancer patients (Caluwaerts et al. 2010). Cytotoxic anticancer drugs that affect fast-growing cancer cells also affect mucosal cells with their rapid mitotic rate, leading to atrophy, swelling, erythema, and ulceration (Raber-Durlacher et al. 2010). The TFF-1-secreting *L. lactis* strain was highly efficacious in alleviating oral mucositis in a hamster model (Caluwaerts et al. 2010) and in patients in a phase 1b clinical trial, leading to a clinical phase 2/3 trial to begin in 2013 (http://www.actogenix.com).

Taken together, these studies underline the potential of using probiotics to deliver molecules directly at the target site in order to restore the mucosal barrier function without interfering with systemic immunity.

THE PROBIOTIC *E. COLI* NISSLE 1917

Although engineering Gram-positive probiotic strains like LAB to deliver molecules of interest has been a relatively uncomplicated affair because of the comparatively simple nature of their cell walls, Gram-negative probiotics like EcN have also been engineered to secrete molecules (Rao et al. 2005; Choi et al. 2012). In an attempt to block infection with the human immunodeficiency virus (HIV), EcN was engineered to secrete a hybrid peptide comprising the HIV protein gp41 (which catalyzes receptor-mediated membrane fusion) and the EcN hemolysin A (Hly), which allows direct export from the EcN cytoplasm into the extracellular medium. Remarkably, the secreted peptide inhibited HIV fusion and entry into the host cells. Moreover, the engineered EcN colonized mice for weeks to months, indicating that secretion of microbicides by this commensal may be a way to prevent HIV entry (Rao et al. 2005). In this scenario, the persistent colonization of EcN constitutes an advantage over LAB, which only transiently colonize the gut mucosa. In another recent study, EcN was engineered to deliver epidermal growth factor as a means to enhance wound healing (Choi et al. 2012). Therefore, although LAB have been the probiotics of choice for targeted delivery of molecules, EcN may provide distinct advantages when persistent colonization is desirable. For this reason, understanding the mechanisms by which EcN colonizes and persists in the gut is critical to enhancing its efficacy and potentially developing other Gram-negative bacteria for use as probiotics.

Since its serendipitous discovery, EcN has been widely used to shorten the duration of diarrhea in children and to alleviate intestinal inflammation in patients with IBD, and in particular ulcerative colitis. Although the molecular mechanisms by which EcN exerts its beneficial effects are largely unknown (Schultz and Lindstrom 2008), several studies have tried to understand what makes EcN a probiotic. Because EcN administration alleviates gastrointestinal tract inflammatory disorders and is highly protective against pathogenic bacteria and fungi including *Listeria monocytogenes*, *Candida albicans*, and *Salmonella enterica* serovar Typhimurium (Hockertz 1991, 1997; Mandel et al. 1995), one might expect that this probiotic strain would down-regulate inflammation. However, there is strong evidence that this bacterium instead enhances the host cell-mediated response, leading to a modulation of the balance between both pro- and anti-inflammatory local cytokines (Fig. 2) (Cross et al. 2004; Ukena et al. 2005).

One of the mechanisms by which EcN limits intestinal inflammation while maintaining intestinal immunological homeostasis is through the induction of naïve and memory peripheral blood $CD4^+$-T-cell clonal expansion without the activation of mucosal and lamina propria T cells (Fig. 2) (Sturm et al. 2005). EcN also increases the activation, cell cycle progression, and cytokine secretion of γδ T cells, which play an important role in the immune response to bacterial antigens (Fig. 2). Remarkably, γδ-T-cell activation is followed by apoptosis of these cells, probably as a way to limit intestinal inflammation (Guzy et al. 2008). EcN has also been shown to mitigate experimental colitis in mice while also reducing expression of the proinflammatory cytokines and interferon γ (IFN-γ), which are activated through Toll-like receptor 4 (TLR4) and TLR5 (Fig. 2) (Grabig et al. 2006). EcN administration can also induce systemic humoral immunity in infants as well as induce specific IgA and IgM antibodies in the mucosa (Fig. 2) (Cukrowska et al. 2002; Ouwehand et al. 2002).

The immunomodulatory effects of EcN have also been observed in colonic epithelial cells, where both cell debris and cell extract fractions of EcN induced the secretion of the proinflammatory cytokine IL-8 (Lammers et al. 2002) while decreasing the expression of cyclooxygenase-2 (COX-2) and the secretion of prostaglandin E2 (PGE2), two molecules that have been implicated in colorectal carcinogenesis (Fig. 2) (Otte et al. 2009). Furthermore, EcN has been shown to be a potent activator of the antimicrobial peptide human β-defensin-2 through flagellin stimulation of NF-κB- and AP-1-mediated signaling, thereby enhancing the colonic epithelial chemical defense system (Fig. 2) (Fellermann and Stange 2001; Ganz 2003; Wehkamp et al. 2004; Splichal et al. 2005; Schlee et al. 2007). Another mechanism by which EcN enhances the mucosal barrier is through up-regulation of the tight junction-associated protein zonula occludens 2 (ZO-2) in intestinal epithelial cells (Fig. 2) (Schulze and Downward 2001; Ukena et al. 2007). Remarkably, EcN counteracted the reduced expression of ZO-2 resulting from enteropathogenic *E. coli* infection (Zyrek et al. 2007). Furthermore, oral administration of EcN to DSS-treated mice reduced loss of body weight and colon shortening and also conferred protection from the DSS colitis-associated increase in mucosal permeability to lumenal substances (Ukena et al. 2007).

Taken together, these studies point to an immunomodulatory role for EcN, with a balanced

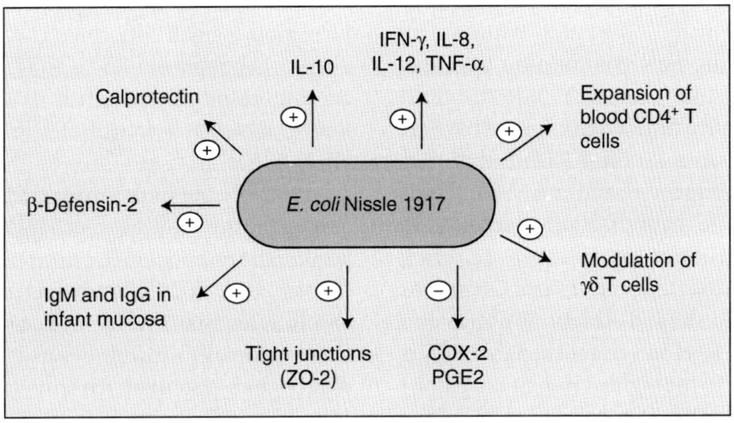

Figure 2. Immune modulation by *E. coli* Nissle 1917. The probiotic *E. coli* Nissle 1917 modulates the host immune system in multiple ways (summary of data from in vitro and in vivo experiments).

activation of the immune response counteracted by inactivation mechanisms. However, these studies do not explain the molecular mechanisms by which EcN modulates the immune response in vivo and ameliorates intestinal inflammation.

Important insight into the mechanisms of probiotic activity exhibited by EcN may be found by comparing its genome to those of other *E. coli* strains. Although all *E. coli* strains colonize the gut, differences in the genome content explain why some *E. coli* strains do not cause disease while others are pathogenic (Hacker and Kaper 2000; Hentschel and Hacker 2001). Many differentiating genetic determinants are acquired by horizontal gene transfer and oftentimes cluster on the chromosome in genomic islands that carry factors that enhance the fitness of a strain in a given environment (Kaper and Hacker 1999; Hacker and Carniel 2001). Therefore, horizontal gene transfer plays an important role in the adaptation of bacteria to specialized niches and may explain why EcN is a probiotic (Koonin et al. 2001). Indeed, EcN represents an excellent example of bacterial genome evolution within the pathogenic *E. coli* serotype O6 lineage. Although EcN serotype O6:K5:H1 is typical of *E. coli* strains associated with urinary tract infections, EcN is completely nonpathogenic (Gunzer et al. 2002). Although EcN lacks prominent virulence genes, it exhibits several fitness factors that contribute to its colonization efficiency and survival within the host (Fig. 3) (Reid et al. 2001; Sanders 2003). However, the contribution of many of the putative fitness factors that promote EcN colonization and survival in the intestine is not well understood.

Several putative fitness determinants of EcN are localized on four genomic islands that have been partially sequenced and analyzed. Comparative genomic hybridization studies with the available genomes of *E. coli* K-12 strain MG1655 and uropathogenic *E. coli* O6 strains CFT073 and 536 showed structural similarities on the genomic level and established that EcN is strongly related to the highly virulent uropathogenic strain CFT073 (Grozdanov et al. 2004; Vejborg et al. 2010). Some fitness factors encoded by both CFT073 and EcN are curli and both type 1 and F1C fimbriae (Fig. 3). In particular, F1C fimbrie of EcN are required for biofilm formation, adherence to epithelial cells, intestinal colonization, and persistence in the gut of infant mice (Blum et al. 1995; Stentebjerg-Olesen et al. 1999; Lasaro et al. 2009). An important difference between EcN and CFT073 is in the structure of LPS (Fig. 3). A point mutation introducing a stop codon in the gene for the O6 antigen polymerase makes the O6 polysaccharide side chain very short, consisting of only a single "repeating unit" of the oligosaccharide building block typical of the O6 antigen. This peculiar characteristic is believed to contribute to EcN serum complement sensitivity and to be responsible for the semirough phenotypic aspect of the colonies grown on solid nutrient media (Fig. 3). This change could also play a role in the special immunomodulating properties exhibited by EcN, which is free of immunotoxic side effects in patients (Grozdanov et al. 2002). Another characteristic of EcN is the presence of an extracellular capsule of the K5 serotype, which is found in only 1% of *E. coli* isolates and is important for adhesion and colonization (Fig. 3) (Herias et al. 1997; Burns and Hull 1998). Despite this, and in contrast to other extraintestinal pathogenic capsule-forming *E. coli*, K5 does not contribute to serum resistance, and EcN is rapidly killed in the classic serum resistance test (Hughes et al. 1982). Moreover, the EcN K5 capsule was shown to stimulate TLR5 and to increase the induction of chemokines in both intestinal epithelial cells and ex vivo mouse small intestine (Hafez et al. 2009, 2010). Although the K5 capsule seems to play an important role in vitro, to date it is not known whether it contributes to the probiotic effect of EcN in vivo.

Probably the most striking feature of EcN's genome is the multiple mechanisms present to acquire the essential metal nutrient iron (Fig. 3) (Crosa et al. 2004). Most bacteria display an absolute requirement for iron, and its acquisition is generally difficult because of its low solubility and potential toxicity (Andrews and Schmidt 2007). Bacteria can survive in iron-limiting conditions using specialized iron transport mechanisms, which are usually induced by low

 Cite this article as *Cold Spring Harb Perspect Med* doi: 10.1101/cshperspect.a010074

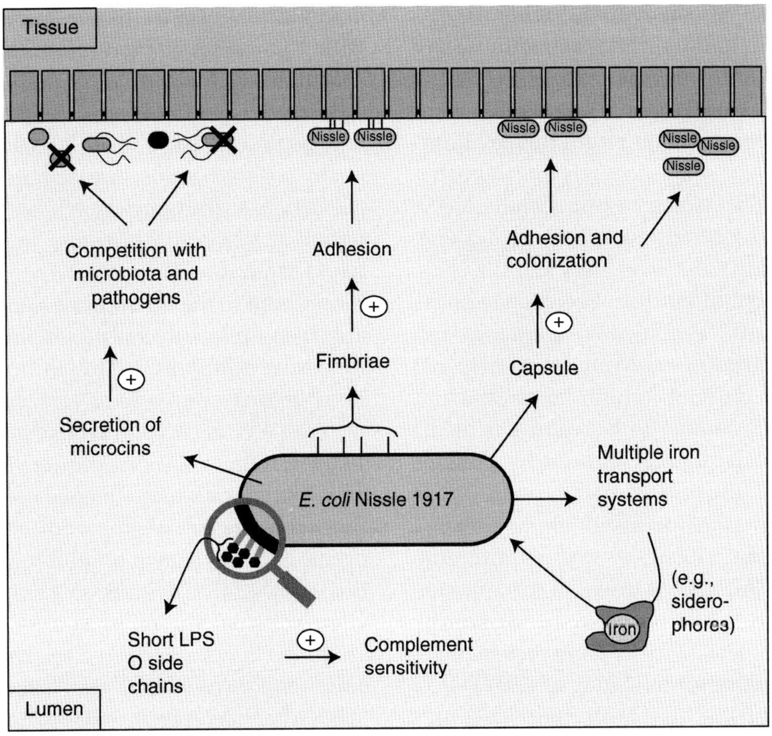

Figure 3. Fitness factors of *E. coli* Nissle 1917. The probiotic *E. coli* Nissle 1917 possesses multiple fitness factors that enable it to colonize the gut and compete with the resident microbiota and pathogenic bacteria.

iron availability and provide specificity for alternative sources of this metal (Skaar 2010). One of the mechanisms used by Enterobacteriaceae to acquire iron is to secrete small chelators termed siderophores. EcN produces several siderophores, including enterobactin, salmochelin, aerobactin, and yersiniabactin (Bäumler et al. 1998; Hantke et al. 2003; Valdebenito et al. 2006). Moreover, EcN has the hemin- and citrate-dependent iron acquisition systems, as well as the Iha siderophore receptor, which was initially identified as a putative nonfimbrial adherence-conferring molecule in the uropathogenic strain CFT073 (Tarr et al. 2000; Torres et al. 2001; Welch et al. 2002; Léveillé et al. 2006; Hancock et al. 2010). Besides the ferric iron uptake transporters, EcN also produces EfeU, an elemental ferrous iron uptake system of the oxidase-dependent iron transporters (OFeT) family, which is a homolog of the yeast iron permease Ftr1p. Notably, the *efeU* gene has been shown to be inactivated in *E. coli* K-12 by a frameshift mutation (Grosse et al. 2006). When the mutation is repaired, EfeU is functional and alleviates iron starvation in a strain defective in all other iron transporters (Cao et al. 2007).

Overall, it appears that EcN has maintained the same redundancy in iron uptake systems as its closest relative, the uropathogenic strain CFT073. Although iron uptake promotes CFT 073's colonization of the bladder and the kidney (Garcia et al. 2011), it is probable that the many specialized iron uptake systems in EcN contribute to its colonization of the intestine and promote competition for a niche with the resident microbiota (Crosa et al. 2004), particularly if the intestine is inflamed.

One feature of the inflammatory response is that it releases antimicrobial proteins that sequester metal ions to further limit their availability to pathogens, a process known as nutritional immunity (Kehl-Fie and Skaar 2010). One mechanism is the release of the antimicrobial proteins lipocalin-2 and calprotectin by both epithelial cells and the neutrophils recruited to the site of infection. Lipocalin-2 binds to a subset of catecholate siderophores including enterochelin, the siderophore secreted by all Enterobacteriaceae to acquire iron, thereby limiting the growth of strains that rely on enterochelin as the sole scavenger of iron (Smith 2007). Calprotectin sequesters zinc and manganese, thereby limiting their availability and thus limiting the growth of sensitive bacteria (Kehl-Fie and Skaar 2010). Remarkably, *S. enterica* serovar Typhimurium (and likely other intestinal pathogens) is resistant to both lipocalin-2 and calprotectin. Resistance to lipocalin-2 is mediated by the siderophore salmochelin (Müller et al. 2009), a glycosylated enterochelin that is too big to fit in the lipocalin-2 binding pocket (Fischbach et al. 2006). A high-affinity zinc transporter, ZnuABC, is essential for zinc uptake when this element is limited and contributes to the low sensitivity to calprotectin exhibited by *S.* Typhimurium (Liu et al. 2012). It is thus not surprising that both lipocalin-2 and calprotectin provide this pathogen with an advantage in growing in the inflamed gut and competing with the microbiota (Raffatellu et al. 2009; Liu et al. 2012).

Although the mechanisms by which EcN ameliorates diarrhea are not completely understood, fitness factors enhancing its survival in the inflamed gut similar to those used by *S.* Typhimurium will likely play an important role. Notably, EcN was shown to induce a significant increase of calprotectin in the small intestine of gnotobiotic piglets. Contrary to this, calprotectin did not increase in the gut after infection with the nonpathogenic *E. coli* strain O86 or with the enteropathogenic *E. coli* strain O55 (Splichal et al. 2005), suggesting that calprotectin may play a role in the probiotic activity of EcN. Building on this, further studies are necessary to assess whether EcN colonization and its probiotic function are enhanced in the inflamed gut when antimicrobials like lipocalin-2 and calprotectin are highly expressed. In this scenario, high-affinity metal transporters may boost EcN colonization of the inflamed gut and may provide a means to compete for metals with other organisms, including pathogens.

Other factors that may help EcN to compete with bacteria in the gastrointestinal tract are the microcins MccH47 and MccM (Fig. 3). Microcins are low-molecular-weight antimicrobial peptides that, similar to bacteriocins of Gram-positive strains, display potent bactericidal activity against phylogenetically related bacteria that lack complementary immunity proteins (Baquero and Moreno 2006). MccH47 and MccM bind to the siderophore salmochelin and are taken up by catecholate siderophore receptors, thus exhibiting a "Trojan horse" mechanism of entry into strains (Patzer et al. 2003; Duquesne et al. 2007; Vassiliadis et al. 2010). In light of these observations, the MccH47 and MccM microcins may enhance EcN's competition with the microbiota and bacterial pathogens. EcN was shown to inhibit human intestinal epithelial cell invasion of adherent and invasive *E. coli*, *S.* Typhimurium, *Yersinia enterocolitica*, *Shigella flexneri*, *Legionella pneumophila*, and *L. monocytogenes* (Boudeau et al. 2003; Altenhoefer et al. 2004). The mechanism of this inhibition is not known, but it is possible that at least in some conditions EcN secretes anti-invasive components that counteract pathogens, which likely include the MccH47 and MccM microcins (Baquero and Moreno 2006). Studies in animal models are needed to determine the contribution of metal transporters, microcins, and other factors to the probiotic effect of EcN.

CONCLUDING REMARKS

Our understanding of the intestinal mucosal barrier function and its alteration in IBD patients has dramatically improved in recent years. In light of this, production of cytokines, enzymes, and other molecules by gut commensals in order to strengthen the mucosa has become a rapidly expanding field of research. Engineered probiotics like LAB are an excellent tool to deliver molecules directly to the mucosa without the side effects and shortcomings of systemic delivery. Notably, they are safe to use in humans and have been closely associated with humans for centuries. Moreover, those in use do not permanently colonize the gut and have been

 Cite this article as *Cold Spring Harb Perspect Med* doi: 10.1101/cshperspect.a010074

engineered to be environmentally safe, which ensures a tightly controlled administration of the desired molecules. Because of this, there has been an expansion in clinical applications of LAB to deliver molecules. Some products, for example, the TFF-1-producing *L. lactis* mouthwash, are now in phase 2/3 clinical trials, and more will hopefully follow soon.

With the advent of deep sequencing, we have also gained new knowledge on the complexity of the microbiota that colonizes our intestine and its alterations in a variety of pathologies including IBD and infections. We now appreciate that inflammation shapes the microbial communities of the gut and that only the fittest survive in an inflamed environment. Intestinal pathogens like *S.* Typhimurium have acquired several mechanisms to thrive in the inflamed gut and compete with the microbiota. By understanding how pathogens survive in the inflamed gut, we can isolate or engineer probiotic strains that share similar traits to colonize the inflamed gut and compete with pathogens for a niche; EcN, which shares many fitness factors of intestinal pathogens, is a natural example of such probiotics. Moreover, the persistent residence of Gram-negative bacteria like EcN in the inflamed gut mucosa may make these organisms attractive candidates for molecule delivery when long-term colonization of relatively harsh host environments is desirable.

ACKNOWLEDGMENTS

We acknowledge Sean-Paul Nuccio for help with editing this article. Work in M.R.'s lab is supported by National Institutes of Health Public Health Service grant AI083619 and by the Pacific Southwest Regional Center of Excellence. J.B. is supported by a postdoctoral fellowship of the American Heart Association (11POST7090006).

REFERENCES

Altenhoefer A, Oswald S, Sonnenborn U, Enders C, Schulze J, Hacker J, Oelschlaeger TA. 2004. The probiotic *Escherichia coli* strain Nissle 1917 interferes with invasion of human intestinal epithelial cells by different enteroinvasive bacterial pathogens. *FEMS Immunol Med Microbiol* **40:** 223–229.

Andrews NC, Schmidt PJ. 2007. Iron homeostasis. *Annu Rev Physiol* **69:** 69–85.

Anukam KC, Reid G. 2008. Probiotics: 100 years (1907–2007) after Elie Metchnikoff's observations. In *Communicating current research and educational topics and trends in applied microbiology*, 2007 ed. (ed. Mendez-vilas A), pp. 466–474. Formatex.org, Spain.

Artis D. 2008. Epithelial-cell recognition of commensal bacteria and maintenance of immune homeostasis in the gut. *Nat Rev Immunol* **8:** 411–420.

Arumugam M, Raes J, Pelletier E, Le Paslier D, Yamada T, Mende DR, Fernandes GR, Tap J, Bruls T, Batto JM, et al. 2011. Enterotypes of the human gut microbiome. *Nature* **473:** 174–180.

Atarashi K, Tanoue T, Shima T, Imaoka A, Kuwahara T, Momose Y, Cheng G, Yamasaki S, Saito T, Ohba Y, et al. 2011. Induction of colonic regulatory T cells by indigenous *Clostridium* species. *Science* **331:** 337–341.

Aureli P, Capurso L, Castellazzi AM, Clerici M, Giovannini M, Morelli L, Poli A, Pregliasco F, Salvini F, Zuccotti GV. 2011. Probiotics and health: An evidence-based review. *Pharmacol Res* **63:** 366–376.

Baquero F, Moreno F. 2006. The microcins. *FEMS Microbiol Lett* **23:** 117–124.

Bäumler AJ, Norris TL, Lasco T, Voight W, Reissbrodt R, Rabsch W, Heffron F. 1998. IroN, a novel outer membrane siderophore receptor characteristic of *Salmonella enterica*. *J Bacteriol* **180:** 1446–1453.

Bevins CL, Salzman NH. 2011. Paneth cells, antimicrobial peptides and maintenance of intestinal homeostasis. *Nat Rev Microbiol* **9:** 356–368.

Blum G, Marre R, Hacker J. 1995. Properties of *Escherichia coli* strains of serotype O6. *Infection* **23:** 234–236.

Boudeau J, Glasser AL, Julien S, Colombel JF, Darfeuille-Michaud A. 2003. Inhibitory effect of probiotic *Escherichia coli* strain Nissle 1917 on adhesion to and invasion of intestinal epithelial cells by adherent-invasive *E. coli* strains isolated from patients with Crohn's disease. *Aliment Pharmacol Ther* **18:** 45–56.

Braat H, Rottiers P, Hommes DW, Huyghebaert N, Remaut E, Remon JP, van Deventer SJ, Neirynck S, Peppelenbosch MP, Steidler L. 2006. A phase I trial with transgenic bacteria expressing interleukin-10 in Crohn's disease. *Clin Gastroenterol Hepatol* **4:** 754–759.

Burns SM, Hull SI. 1998. Comparison of loss of serum resistance by defined lipopolysaccharide mutants and an acapsular mutant of uropathogenic *Escherichia coli* O75:K5. *Infect Immun* **66:** 4244–4253.

Caluwaerts S, Vandenbroucke K, Steidler L, Neirynck S, Vanhoenacker P, Corveleyn S, Watkins B, Sonis S, Coulie B, Rottiers P. 2010. AG013, a mouth rinse formulation of *Lactococcus lactis* secreting human Trefoil Factor 1, provides a safe and efficacious therapeutic tool for treating oral mucositis. *Oral Oncol* **46:** 564–570.

Cao J, Woodhall MR, Alvarez J, Cartron ML, Andrews SC. 2007. EfeUOB (YcdNOB) is a tripartite, acid-induced and CpxAR-regulated, low-pH Fe^{2+} transporter that is cryptic in *Escherichia coli* K-12 but functional in *E. coli* O157:H7. *Mol Microbiol* **65:** 857–875.

Choi HJ, Ahn JH, Park SH, Do KH, Kim J, Moon Y. 2012. Enhanced wound healing by recombinant *Escherichia coli* Nissle 1917 via human epidermal growth factor receptor in human intestinal epithelial cells: Therapeutic implication using recombinant probiotics. *Infect Immun* **80:** 1079–1087.

Cobb BA, Wang Q, Tzianabos AO, Kasper DL. 2004. Polysaccharide processing and presentation by the MHCII pathway. *Cell* **117:** 677–687.

Colombel JF, Rutgeerts P, Malchow H, Jacyna M, Nielsen OH, Rask-Madsen J, Van Deventer S, Ferguson A, Desreumaux P, Forbes A, et al. 2001. Interleukin 10 (Tenovil) in the prevention of postoperative recurrence of Crohn's disease. *Gut* **49:** 42–46.

Cotter PD, Hill C, Ross RP. 2005. Bacteriocins: Developing innate immunity for food. *Nat Rev Microbiol* **3:** 777–788.

Coyne MJ, Reinap B, Lee MM, Comstock LE. 2005. Human symbionts use a host-like pathway for surface fucosylation. *Science* **307:** 1778–1781.

Crosa JH, Mey AR, Payne SM. 2004. *Iron transport in bacteria*. ASM Press, Washington, DC.

Cross ML, Ganner A, Teilab D, Fray LM. 2004. Patterns of cytokine induction by gram-positive and gram-negative probiotic bacteria. *FEMS Immunol Med Microbiol* **42:** 173–180.

Cukrowska B, Lodinova-Zadnikova R, Enders C, Sonnenborn U, Schulze J, Tlaskalova-Hogenova H. 2002. Specific proliferative and antibody responses of premature infants to intestinal colonization with nonpathogenic probiotic *E. coli* strain Nissle 1917. *Scand J Immunol* **55:** 204–209.

Depaolo RW, Tang F, Kim I, Han M, Levin N, Ciletti N, Lin A, Anderson D, Schneewind O, Jabri B. 2008. Toll-like receptor 6 drives differentiation of tolerogenic dendritic cells and contributes to LcrV-mediated plague pathogenesis. *Cell Host Microbe* **4:** 350–361.

DiBaise JK, Zhang H, Crowell MD, Krajmalnik-Brown R, Decker GA, Rittmann BE. 2008. Gut microbiota and its possible relationship with obesity. *Mayo Clin Proc* **83:** 460–469.

Duquesne S, Destoumieux-Garzon D, Peduzzi J, Rebuffat S. 2007. Microcins, gene-encoded antibacterial peptides from enterobacteria. *Nat Prod Rep* **24:** 708–734.

FAO/WHO. 2001. Health and nutritional properties of probiotics in food including powder milk with live lactic acid bacteria. Cordoba, Argentina, October 1–4.

Fava F, Danese S. 2011. Intestinal microbiota in inflammatory bowel disease: Friend of foe? *World J Gastroenterol* **17:** 557–566.

Fellermann K, Stange EF. 2001. Defensins—Innate immunity at the epithelial frontier. *Eur J Gastroenterol Hepatol* **13:** 771–776.

Feng T, Wang L, Schoeb TR, Elson CO, Cong Y. 2010. Microbiota innate stimulation is a prerequisite for T cell spontaneous proliferation and induction of experimental colitis. *J Exp Med* **207:** 1321–1332.

Fischbach MA, Lin H, Zhou L, Yu Y, Abergel RJ, Liu DR, Raymond KN, Wanner BL, Strong RK, Walsh CT, et al. 2006. The pathogen-associated *iroA* gene cluster mediates bacterial evasion of lipocalin 2. *Proc Natl Acad Sci* **103:** 16502–16507.

Floch MH, Walker WA, Madsen K, Sanders ME, Macfarlane GT, Flint HJ, Dieleman LA, Ringel Y, Guandalini S, Kelly CP, et al. 2011. Recommendations for probiotic use—2011 update. *J Clin Gastroenterol* **45** (Suppl): S168–S171.

Foligne B, Dessein R, Marceau M, Poiret S, Chamaillard M, Pot B, Simonet M, Daniel C. 2007. Prevention and treatment of colitis with *Lactococcus lactis* secreting the immunomodulatory *Yersinia* LcrV protein. *Gastroenterology* **133:** 862–874.

Ganz T. 2003. Defensins: Antimicrobial peptides of innate immunity. *Nat Rev Immunol* **3:** 710–720.

Garcia EC, Brumbaugh AR, Mobley HL. 2011. Redundancy and specificity of *Escherichia coli* iron acquisition systems during urinary tract infection. *Infect Immun* **79:** 1225–1235.

Gophna U, Sommerfeld K, Gophna S, Doolittle WF, Veldhuyzen van Zanten SJ. 2006. Differences between tissue-associated intestinal microfloras of patients with Crohn's disease and ulcerative colitis. *J Clin Microbiol* **44:** 4136–4141.

Grabig A, Paclik D, Guzy C, Dankof A, Baumgart DC, Erckenbrecht J, Raupach B, Sonnenborn U, Eckert J, Schumann RR, et al. 2006. *Escherichia coli* strain Nissle 1917 ameliorates experimental colitis via Toll-like receptor 2- and Toll-like receptor 4-dependent pathways. *Infect Immun* **74:** 4075–4082.

Grosse C, Scherer J, Koch D, Otto M, Taudte N, Grass G. 2006. A new ferrous iron-uptake transporter, EfeU (YcdN), from *Escherichia coli*. *Mol Microbiol* **62:** 120–131.

Grozdanov L, Zahringer U, Blum-Oehler G, Brade L, Henne A, Knirel YA, Schombel U, Schulze J, Sonnenborn U, Gottschalk G, et al. 2002. A single nucleotide exchange in the *wzy* gene is responsible for the semirough O6 lipopolysaccharide phenotype and serum sensitivity of *Escherichia coli* strain Nissle 1917. *J Bacteriol* **184:** 5912–5925.

Grozdanov L, Raasch C, Schulze J, Sonnenborn U, Gottschalk G, Hacker J, Dobrindt U. 2004. Analysis of the genome structure of the nonpathogenic probiotic *Escherichia coli* strain Nissle 1917. *J Bacteriol* **186:** 5432–5441.

Gunzer F, Hennig-Pauka I, Waldmann KH, Sandhoff R, Grone HJ, Kreipe HH, Matussek A, Mengel M. 2002. Gnotobiotic piglets develop thrombotic microangiopathy after oral infection with enterohemorrhagic *Escherichia coli*. *Am J Clin Pathol* **118:** 364–375.

Guzy C, Paclik D, Schirbel A, Sonnenborn U, Wiedenmann B, Sturm A. 2008. The probiotic *Escherichia coli* strain Nissle 1917 induces γδ T cell apoptosis via caspase- and FasL-dependent pathways. *Int Immunol* **20:** 829–840.

Hacker J, Carniel E. 2001. Ecological fitness, genomic islands and bacterial pathogenicity. A Darwinian view of the evolution of microbes. *EMBO Rep* **2:** 376–381.

Hacker J, Kaper JB. 2000. Pathogenicity islands and the evolution of microbes. *Annu Rev Microbiol* **54:** 641–679.

Hafez M, Hayes K, Goldrick M, Warhurst G, Grencis R, Roberts IS. 2009. The K5 capsule of *Escherichia coli* strain Nissle 1917 is important in mediating interactions with intestinal epithelial cells and chemokine induction. *Infect Immun* **77:** 2995–3003.

Hafez M, Hayes K, Goldrick M, Grencis RK, Roberts IS. 2010. The K5 capsule of *Escherichia coli* strain Nissle 1917 is important in stimulating expression of Toll-like receptor 5, CD14, MyD88, and TRIF together with the induction of interleukin-8 expression via the mitogen-activated protein kinase pathway in epithelial cells. *Infect Immun* **78:** 2153–2162.

Hancock V, Vejborg RM, Klemm P. 2010. Functional genomics of probiotic *Escherichia coli* Nissle 1917 and 83972, and UPEC strain CFT073: Comparison of transcriptomes, growth and biofilm formation. *Mol Genet Genomics* **284:** 437–454.

Hantke K, Nicholson G, Rabsch W, Winkelmann G. 2003. Salmochelins, siderophores of *Salmonella enterica* and uropathogenic *Escherichia coli* strains, are recognized by the outer membrane receptor IroN. *Proc Natl Acad Sci* **100:** 3677–3682.

Hayashi H, Takahashi R, Nishi T, Sakamoto M, Benno Y. 2005. Molecular analysis of jejunal, ileal, caecal and recto-sigmoidal human colonic microbiota using 16S rRNA gene libraries and terminal restriction fragment length polymorphism. *J Med Microbiol* **54** (Pt 11): 1093–1101.

Henker J, Laass M, Blokhin BM, Bolbot YK, Maydannik VG, Elze M, Wolff C, Schulze J. 2007. The probiotic *Escherichia coli* strain Nissle 1917 (EcN) stops acute diarrhoea in infants and toddlers. *Eur J Pediatr* **166:** 311–318.

Hentschel U, Hacker J. 2001. Pathogenicity islands: The tip of the iceberg. *Microbes Infect* **3:** 545–548.

Herias MV, Midtvedt T, Hanson LA, Wold AE. 1997. *Escherichia coli* K5 capsule expression enhances colonization of the large intestine in the gnotobiotic rat. *Infect Immun* **65:** 531–536.

Hockertz S. 1991. Immunomodulating effect of killed, apathogenic *Escherichia coli*, strain Nissle 1917, on the macrophage system. *Arzneimittelforschung* **41:** 1108–1112.

Hockertz S. 1997. Augmentation of host defence against bacterial and fungal infections of mice pretreated with the non-pathogenic *Escherichia coli* strain Nissle 1917. *Arzneimittelforschung* **47:** 793–796.

Hogenauer C, Hammer HF, Krejs GJ, Reisinger EC. 1998. Mechanisms and management of antibiotic-associated diarrhea. *Clin Infect Dis* **27:** 702–710.

Hooper LV, Gordon JI. 2001. Commensal host-bacterial relationships in the gut. *Science* **292:** 1115–1118.

Høverstad T, Midtvedt T. 1986. Short-chain fatty acids in germfree mice and rats. *J Nutr* **116:** 1772–1776.

Hughes C, Phillips R, Roberts AP. 1982. Serum resistance among *Escherichia coli* strains causing urinary tract infection in relation to O type and the carriage of hemolysin, colicin, and antibiotic resistance determinants. *Infect Immun* **35:** 270–275.

Huibregtse IL, Zaat SA, Kapsenberg ML, Sartori da Silva MA, Peppelenbosch MP, van Deventer SJ, Braat H. 2011. Genetically modified *Lactococcus lactis* for delivery of human interleukin-10 to dendritic cells. *Gastroenterol Res Pract* **2012:** 639291.

Ivanov II, Atarashi K, Manel N, Brodie EL, Shima T, Karaoz U, Wei D, Goldfarb KC, Santee CA, Lynch SV, et al. 2009. Induction of intestinal Th17 cells by segmented filamentous bacteria. *Cell* **139:** 485–498.

Jervis HR, Biggers DC. 1964. Mucosal enzymes in the cecum of conventional and germfree mice. *Anat Rec* **148:** 591–597.

Kaper JB, Hacker J. 1999. *Pathogenicity islands and other mobile virulence elements.* ASM Press, Washington, DC.

Karin M, Lawrence T, Nizet V. 2006. Innate immunity gone awry: Linking microbial infections to chronic inflammation and cancer. *Cell* **124:** 823–835.

Kehl-Fie TE, Skaar EP. 2010. Nutritional immunity beyond iron: A role for manganese and zinc. *Curr Opin Chem Biol* **14:** 218–224.

Kjellev S. 2009. The trefoil factor family—Small peptides with multiple functionalities. *Cell Mol Life Sci* **66:** 1350–1369.

Koonin EV, Makarova KS, Aravind L. 2001. Horizontal gene transfer in prokaryotes: Quantification and classification. *Annu Rev Microbiol* **55:** 709–742.

Kruis W, Schutz E, Fric P, Fixa B, Judmaier G, Stolte M. 1997. Double-blind comparison of an oral *Escherichia coli* preparation and mesalazine in maintaining remission of ulcerative colitis. *Aliment Pharmacol Ther* **11:** 853–858.

Kruis W, Fric P, Pokrotnieks J, Lukas M, Fixa B, Kascak M, Kamm MA, Weismueller J, Beglinger C, Stolte M, et al. 2004. Maintaining remission of ulcerative colitis with the probiotic *Escherichia coli* Nissle 1917 is as effective as with standard mesalazine. *Gut* **53:** 1617–1623.

Lammers KM, Helwig U, Swennen E, Rizzello F, Venturi A, Caramelli E, Kamm MA, Brigidi P, Gionchetti P, Campieri M. 2002. Effect of probiotic strains on interleukin 8 production by HT29/19A cells. *Am J Gastroenterol* **97:** 1182–1186.

Lasaro MA, Salinger N, Zhang J, Wang Y, Zhong Z, Goulian M, Zhu J. 2009. F1C fimbriae play an important role in biofilm formation and intestinal colonization by the *Escherichia coli* commensal strain Nissle 1917. *Appl Environ Microbiol* **75:** 246–251.

Le Loir Y, Azevedo V, Oliveira SC, Freitas DA, Miyoshi A, Bermudez-Humaran LG, Nouaille S, Ribeiro LA, Leclercq S, Gabriel JE, et al. 2005. Protein secretion in *Lactococcus lactis*: An efficient way to increase the overall heterologous protein production. *Microb Cell Fact* **4:** 2.

Léveillé S, Caza M, Johnson JR, Clabots C, Sabri M, Dozois CM. 2006. Iha from an *Escherichia coli* urinary tract infection outbreak clonal group A strain is expressed in vivo in the mouse urinary tract and functions as a catecholate siderophore receptor. *Infect Immun* **74:** 3427–3436.

Liu CH, Lee SM, Vanlare JM, Kasper DL, Mazmanian SK. 2008. Regulation of surface architecture by symbiotic bacteria mediates host colonization. *Proc Natl Acad Sci* **105:** 3951–3956.

Liu JZ, Jellbauer S, Poe AJ, Ton V, Pesciaroli M, Kehl-Fie TE, Restrepo NA, Hosking MP, Edwards RA, Battistoni A, et al. 2012. Zinc sequestration by the neutrophil protein calprotectin enhances *Salmonella* growth in the inflamed gut. *Cell Host Microbe* **11:** 227–239.

Lodinova-Zadnikova R, Sonnenborn U. 1997. Effect of preventive administration of a nonpathogenic *Escherichia coli* strain on the colonization of the intestine with microbial pathogens in newborn infants. *Biol Neonate* **71:** 224–232.

Mahowald MA, Rey FE, Seedorf H, Turnbaugh PJ, Fulton RS, Wollam A, Shah N, Wang C, Magrini V, Wilson RK, et al. 2009. Characterizing a model human gut microbiota composed of members of its two dominant bacterial phyla. *Proc Natl Acad Sci* **106:** 5859–5864.

Mandel L, Trebichavsky I, Splichal I, Schulze J. 1995. Stimulation of intestinal immune cells by *E. coli* in gnotobiotic piglets. *Adv Exp Med Biol* **371A:** 463–464.

Marco ML, Pavan S, Kleerebezem M. 2006. Towards understanding molecular modes of probiotic action. *Curr Opin Biotechnol* **17:** 204–210.

Maslowski KM, Vieira AT, Ng A, Kranich J, Sierro F, Yu D, Schilter HC, Rolph MS, Mackay F, Artis D, et al. 2009. Regulation of inflammatory responses by gut microbiota and chemoattractant receptor GPR43. *Nature* **461:** 1282–1286.

Mazmanian SK, Liu CH, Tzianabos AO, Kasper DL. 2005. An immunomodulatory molecule of symbiotic bacteria directs maturation of the host immune system. *Cell* **122:** 107–118.

Mierau I, Kleerebezem M. 2005. 10 years of the nisin-controlled gene expression system (NICE) in *Lactococcus lactis*. *Appl Microbiol Biotechnol* **68:** 705–717.

Mollenbrink M, Bruckschen E. 1994. Treatment of chronic constipation with physiologic *Escherichia coli* bacteria. Results of a clinical study of the effectiveness and tolerance of microbiological therapy with the *E. coli* Nissle 1917 strain (Mutaflor). *Med Klin (Munich)* **89:** 587–593.

Mukhopadhya I, Hansen R, El-Omar EM, Hold GL. 2012. IBD—What role do Proteobacteria play? *Nat Rev Gastroenterol Hepatol* **9:** 219–230.

Müller SI, Valdebenito M, Hantke K. 2009. Salmochelin, the long-overlooked catecholate siderophore of *Salmonella*. *Biometals* **22:** 691–695.

Nagalingam NA, Lynch SV. 2012. Role of the microbiota in inflammatory bowel diseases. *Inflamm Bowel Dis* **18:** 968–984.

Nissle A. 1959. Explanations of the significance of colonic dysbacteria & the mechanism of action of *E. coli* therapy (mutaflor). *Medizinische* **4:** 1017–1022.

Nouaille S, Ribeiro LA, Miyoshi A, Pontes D, Le Loir Y, Oliveira SC, Langella P, Azevedo V. 2003. Heterologous protein production and delivery systems for *Lactococcus lactis*. *Genet Mol Res* **2:** 102–111.

Otte JM, Mahjurian-Namari R, Brand S, Werner I, Schmidt WE, Schmitz F. 2009. Probiotics regulate the expression of COX-2 in intestinal epithelial cells. *Nutr Cancer* **61:** 103–113.

Ouwehand AC, Salminen S, Isolauri E. 2002. Probiotics: An overview of beneficial effects. *Antonie Van Leeuwenhoek* **82:** 279–289.

Pang Q, Ji Y, Li Y, Bermudez-Humaran LG, Hu G, Zeng Y. 2008. Intragastric administration with recombinant *Lactococcus lactis* producing heme oxygenase-1 prevents lipopolysaccharide-induced endotoxemia in rats. *FEMS Microbiol Lett* **283:** 62–68.

Pang QF, Ji Y, Bermudez-Humaran LG, Zhou QM, Hu G, Zeng Y. 2009. Protective effects of a heme oxygenase-1-secreting *Lactococcus lactis* on mucosal injury induced by hemorrhagic shock in rats. *J Surg Res* **153:** 39–45.

Patzer SI, Baquero MR, Bravo D, Moreno F, Hantke K. 2003. The colicin G, H and X determinants encode microcins M and H47, which might utilize the catecholate siderophore receptors FepA, Cir, Fiu and IroN. *Microbiology* **149:** 2557–2570.

Peterbauer C, Maischberger T, Haltrich D. 2011. Food-grade gene expression in lactic acid bacteria. *Biotechnol J* **6:** 1147–1161.

Raber-Durlacher JE, Elad S, Barasch A. 2010. Oral mucositis. *Oral Oncol* **46:** 452–456.

Raffatellu M, George MD, Akiyama Y, Hornsby MJ, Nuccio SP, Paixao TA, Butler BP, Chu H, Santos RL, Berger T, et al. 2009. Lipocalin-2 resistance confers an advantage to *Salmonella enterica* serotype Typhimurium for growth and survival in the inflamed intestine. *Cell Host Microbe* **5:** 476–486.

Rao S, Hu S, McHugh L, Lueders K, Henry K, Zhao Q, Fekete RA, Kar S, Adhya S, Hamer DH. 2005. Toward a live microbial microbicide for HIV: Commensal bacteria secreting an HIV fusion inhibitor peptide. *Proc Natl Acad Sci* **102:** 11993–11998.

Reid G, Howard J, Gan BS. 2001. Can bacterial interference prevent infection? *Trends Microbiol* **9:** 424–428.

Sanders ME. 2003. Probiotics: Considerations for human health. *Nutr Rev* **61:** 91–99.

Schlee M, Wehkamp J, Altenhoefer A, Oelschlaeger TA, Stange EF, Fellermann K. 2007. Induction of human β-defensin 2 by the probiotic *Escherichia coli* Nissle 1917 is mediated through flagellin. *Infect Immun* **75:** 2399–2407.

Schotte L, Steidler L, Vandekerckhove J, Remaut E. 2000. Secretion of biologically active murine interleukin-10 by *Lactococcus lactis*. *Enzyme Microb Technol* **27:** 761–765.

Schultz M, Lindstrom AL. 2008. Rationale for probiotic treatment strategies in inflammatory bowel disease. *Expert Rev Gastroenterol Hepatol* **2:** 337–355.

Schulze A, Downward J. 2001. Navigating gene expression using microarrays—A technology review. *Nat Cell Biol* **3:** E190–E195.

Schutz E. 1989. The treatment of intestinal diseases with Mutaflor. A multicenter retrospective study. *Fortschr Med* **107:** 599–602.

Servin AL. 2004. Antagonistic activities of lactobacilli and bifidobacteria against microbial pathogens. *FEMS Microbiol Rev* **28:** 405–440.

Shaw MH, Kamada N, Kim YG, Nunez G. 2012. Microbiota-induced IL-1β, but not IL-6, is critical for the development of steady-state T_H17 cells in the intestine. *J Exp Med* **209:** 251–258.

Skaar EP. 2010. The battle for iron between bacterial pathogens and their vertebrate hosts. *PLoS Pathog* **6:** e1000949.

Smith KD. 2007. Iron metabolism at the host pathogen interface: Lipocalin 2 and the pathogen-associated *iroA* gene cluster. *Int J Biochem Cell Biol* **39:** 1776–1780.

Splichal I, Fagerhol MK, Trebichavsky I, Splichalova A, Schulze J. 2005. The effect of intestinal colonization of germ-free pigs with *Escherichia coli* on calprotectin levels in plasma, intestinal and bronchoalveolar lavages. *Immunobiology* **209:** 681–687.

Spor A, Koren O, Ley R. 2011. Unravelling the effects of the environment and host genotype on the gut microbiome. *Nat Rev Microbiol* **9:** 279–290.

Steidler L, Hans W, Schotte L, Neirynck S, Obermeier F, Falk W, Fiers W, Remaut E. 2000. Treatment of murine colitis by *Lactococcus lactis* secreting interleukin-10. *Science* **289:** 1352–1355.

Steidler L, Neirynck S, Huyghebaert N, Snoeck V, Vermeire A, Goddeeris B, Cox E, Remon JP, Remaut E. 2003. Biological containment of genetically modified *Lactococcus lactis* for intestinal delivery of human interleukin 10. *Nat Biotechnol* **21:** 785–789.

Steidler L, Rottiers P, Coulie B. 2009. Actobiotics as a novel method for cytokine delivery. *Ann NY Acad Sci* **1182:** 135–145.

Stentebjerg-Olesen B, Chakraborty T, Klemm P. 1999. Type 1 fimbriation and phase switching in a natural *Escherichia coli fimB* null strain, Nissle 1917. *J Bacteriol* **181:** 7470–7478.

Sturm A, Rilling K, Baumgart DC, Gargas K, Abou-Ghazale T, Raupach B, Eckert J, Schumann RR, Enders C, Sonnenborn U, et al. 2005. *Escherichia coli* Nissle 1917 distinctively modulates T-cell cycling and expansion via Toll-like receptor 2 signaling. *Infect Immun* **73:** 1452–1465.

Tarr PI, Bilge SS, Vary JC Jr, Jelacic S, Habeeb RL, Ward TR, Baylor MR, Besser TE. 2000. Iha: A novel *Escherichia coli* O157:H7 adherence-conferring molecule encoded on a recently acquired chromosomal island of conserved structure. *Infect Immun* **68:** 1400–1407.

Tilg H, van Montfrans C, van den Ende A, Kaser A, van Deventer SJ, Schreiber S, Gregor M, Ludwiczek O, Rutgeerts P, Gasche C, et al. 2002. Treatment of Crohn's disease with recombinant human interleukin 10 induces the proinflammatory cytokine interferon γ. *Gut* **50:** 191–195.

Tlaskalova-Hogenova H, Stepankova R, Kozakova H, Hudcovic T, Vannucci L, Tuckova L, Rossmann P, Hrncir T, Kverka M, Zakostelska Z, et al. 2011. The role of gut microbiota (commensal bacteria) and the mucosal barrier in the pathogenesis of inflammatory and autoimmune diseases and cancer: Contribution of germ-free and gnotobiotic animal models of human diseases. *Cell Mol Immunol* **8:** 110–120.

Torres AG, Redford P, Welch RA, Payne SM. 2001. TonB-dependent systems of uropathogenic *Escherichia coli*: Aerobactin and heme transport and TonB are required for virulence in the mouse. *Infect Immun* **69:** 6179–6185.

Tran CP, Cook GA, Yeomans ND, Thim L, Giraud AS. 1999. Trefoil peptide TFF2 (spasmolytic polypeptide) potently accelerates healing and reduces inflammation in a rat model of colitis. *Gut* **44:** 636–642.

Ukena SN, Westendorf AM, Hansen W, Rohde M, Geffers R, Coldewey S, Suerbaum S, Buer J, Gunzer F. 2005. The host response to the probiotic *Escherichia coli* strain Nissle 1917: Specific up-regulation of the proinflammatory chemokine MCP-1. *BMC Med Genet* **6:** 43.

Ukena SN, Singh A, Dringenberg U, Engelhardt R, Seidler U, Hansen W, Bleich A, Bruder D, Franzke A, Rogler G, et al. 2007. Probiotic *Escherichia coli* Nissle 1917 inhibits leaky gut by enhancing mucosal integrity. *PLoS ONE* **2:** e1308.

Valdebenito M, Crumbliss AL, Winkelmann G, Hantke K. 2006. Environmental factors influence the production of enterobactin, salmochelin, aerobactin, and yersiniabactin in *Escherichia coli* strain Nissle 1917. *Int J Med Microbiol* **296:** 513–520.

Vandenbroucke K, Hans W, Van Huysse J, Neirynck S, Demetter P, Remaut E, Rottiers P, Steidler L. 2004. Active delivery of trefoil factors by genetically modified *Lactococcus lactis* prevents and heals acute colitis in mice. *Gastroenterology* **127:** 502–513.

Vandenbroucke K, de Haard H, Beirnaert E, Dreier T, Lauwereys M, Huyck L, Van Huysse J, Demetter P, Steidler L, Remaut E, et al. 2010. Orally administered *L. lactis* secreting an anti-TNF Nanobody demonstrate efficacy in chronic colitis. *Mucosal Immunol* **3:** 49–56.

van Deventer SJ, Elson CO, Fedorak RN. 1997. Multiple doses of intravenous interleukin 10 in steroid-refractory Crohn's disease. Crohn's Disease Study Group. *Gastroenterology* **113:** 383–389.

Vassiliadis G, Destoumieux-Garzon D, Lombard C, Rebuffat S, Peduzzi J. 2010. Isolation and characterization of two members of the siderophore-microcin family, microcins M and H47. *Antimicrob Agents Chemother* **54:** 288–297.

Vejborg RM, Friis C, Hancock V, Schembri MA, Klemm P. 2010. A virulent parent with probiotic progeny: Comparative genomics of *Escherichia coli* strains CFT073, Nissle 1917 and ABU 83972. *Mol Genet Genomics* **283:** 469–484.

Vijayan V, Mueller S, Baumgart-Vogt E, Immenschuh S. 2010. Heme oxygenase-1 as a therapeutic target in inflammatory disorders of the gastrointestinal tract. *World J Gastroenterol* **16:** 3112–3119.

Vinolo MA, Rodrigues HG, Nachbar RT, Curi R. 2011. Regulation of inflammation by short chain fatty acids. *Nutrients* **3:** 858–876.

Wehkamp J, Harder J, Wehkamp K, Wehkamp-von Meissner B, Schlee M, Enders C, Sonnenborn U, Nuding S, Bengmark S, Fellermann K, et al. 2004. NF-κB- and AP-1-mediated induction of human beta defensin-2 in intestinal epithelial cells by *Escherichia coli* Nissle 1917: A novel effect of a probiotic bacterium. *Infect Immun* **72:** 5750–5758.

Welch RA, Burland V, Plunkett G III, Redford P, Roesch P, Rasko D, Buckles EL, Liou SR, Boutin A, Hackett J, et al. 2002. Extensive mosaic structure revealed by the complete genome sequence of uropathogenic *Escherichia coli*. *Proc Natl Acad Sci* **99:** 17020–17024.

Westendorf AM, Gunzer F, Deppenmeier S, Tapadar D, Hunger JK, Schmidt MA, Buer J, Bruder D. 2005. Intestinal immunity of *Escherichia coli* NISSLE 1917: A safe carrier for therapeutic molecules. *FEMS Immunol Med Microbiol* **43:** 373–384.

Wu GD, Chen J, Hoffmann C, Bittinger K, Chen YY, Keilbaugh SA, Bewtra M, Knights D, Walters WA, Knight R, et al. 2011. Linking long-term dietary patterns with gut microbial enterotypes. *Science* **334:** 105–108.

Zyrek AA, Cichon C, Helms S, Enders C, Sonnenborn U, Schmidt MA. 2007. Molecular mechanisms underlying the probiotic effects of *Escherichia coli* Nissle 1917 involve ZO-2 and PKCζ redistribution resulting in tight junction and epithelial barrier repair. *Cell Microbiol* **9:** 804–816.

Therapeutic and Prophylactic Applications of Bacteriophage Components in Modern Medicine

Sankar Adhya[1], Carl R. Merril[2], and Biswajit Biswas[3]

[1]National Cancer Institute, National Institutes of Health, Bethesda, Maryland 20892

[2]National Institute of Mental Health, National Institutes of Health, Bethesda, Maryland 20892

[3]Department of Genomics, Biological Defense Research Directorate, Naval Medical Research Center, Fort Detrick, Maryland 21702

Correspondence: sadhya@helix.nih.gov

As the interactions of phage with mammalian innate and adaptive immune systems are better delineated and with our ability to recognize and eliminate toxins and other potentially harmful phage gene products, the potential of phage therapies is now being realized. Early efforts to use phage therapeutically were hampered by inadequate phage purification and limited knowledge of phage–bacterial and phage–human relations. However, although use of phage as an antibacterial therapy in countries that require controlled clinical studies has been hampered by the high costs of patient trials, their use as vaccines and the use of phage components such as lysolytic enzymes or lysozymes has progressed to the point of commercial applications. Recent studies concerning the intimate associations between mammalian hosts and bacterial and phage microbiomes should hasten this progress.

The human superorganism is made up of eukaryotic and prokaryotic cells, in which the prokaryotic cells far outnumber the eukaryotic cells. Given our current limited understanding of ourselves as super organisms it is remarkable that we have achieved successes, albeit limited, in developing therapies for infectious diseases. This was especially true in the 19th and 20th centuries when specific strains of bacteria were first being discovered and in some cases associated with animal or human disease states.

The therapeutic advances realized at that time were based on limited observations and heroic empirical efforts, such as the discovery of the antisyphilitic compound Salvarsan or 606 (in reference to the hundreds of compounds screened) by S. Hata in Paul Ehrlich's laboratory in the early 1900s (Sneade 2005).

Given the state of knowledge at the beginning of the 20th century concerning microbiology and the immune defense mechanisms, the discovery of bacteriophage (phage) viruses that kill specific strains of bacteria was welcomed. However, soon after their discovery a number of issues arose, including a controversy as to whether the bacteriophages were self-replicating particles or a lytic enzymatic activity activated

Cite this article as *Cold Spring Harb Perspect Med* doi: 10.1101/cshperspect.a012518

in the bacteria. One of the pioneers associated with their discovery, Félix d'Herelle, consulted with Albert Einstein who reassured d'Herelle that his experiments designed to prove that they were self-replicating particles were valid (Summers 1999). Perhaps more serious were the problems that arose owing to a lack of appreciation of the narrow host range of most phage strains, so that a phage isolated from an infected patient did not appear to be effective in treating another patient with a seemingly similar infection. In addition, there was a lack of knowledge of bacterial toxins and methods to purify phage preparations so that they were not contaminated with such toxins. These problems coupled with inadequately designed animal and clinical experiments were some of the factors that led, in the United States, to the development of a government agency to oversee therapeutic developments and claims (Merril et al. 2003, 2006).

In addition to these early problems, the discovery of the broad host range antibiotics—beginning with penicillin—more than 85 years ago, resulted in the virtual abandonment of efforts to continue in the development of the phage as a therapeutic antibacterial agent in the United States. However, the increasing incidence of antibiotic-resistant bacterial strains, particularly the bacterial strains commonly found in hospital settings, has prompted a reexamination of the potentials of phage antibacterial therapies. In addition, there is now an increased awareness of clinical problems associated with the extended bacterial host range of most commonly used antibiotics. Although an extended bacterial host range was one of the criteria used in searching for clinically relevant antibiotics, as this property partially alleviated the need to accurately identify infecting bacterial species before therapy was initiated, it is now recognized that such extended host range antibiotics often result in collateral damage to the natural human microbiome (Jernberg et al. 2010). Even before the medical community recognized the extent of this problem, it was clear that individuals treated successfully for one species of infectious agent would often succumb to an infection by another secondary foreign bacterial strain.

As most phage strains have a narrow bacterial host range, they are expected to cause minimal perturbations of the normal human microbiome when they are used to treat bacterial infections. Also, as their mechanisms of antibacterial activity of phage are not related to the mode of action of the antibiotic therapies, it should be possible to find effective phage therapies for the antibiotic-resistant disease-associated bacterial strains.

In addition our ability to develop molecular tools that permit manipulations of phage genomes and our knowledge of the microbiology and physiology of phage along with some knowledge of their interactions with the human immune system, they are beginning to be adapted to serve as vaccines. The phage strains are also being used to produce antibacterial proteins that are encoded in their genomes. By manipulating the promoters for these genes, it is possible to produce antibacterial proteins such as the phage-based lysozymes in quantity. The clinical applications of phage, including the development of phage-based vaccines and phage products are now beginning to enter commercial development.

PHAGE ANTIBACTERIAL THERAPY

One of the discoverers of phage, Félix d'Herelle, clearly recognized the clinical potential of phage as an antibacterial agent (Summers 1999). The ability of phage to kill bacteria and in addition to replicate exponentially suggests some of the potential advantages of phage antibacterial therapy. Although d'Herelle's reported successes in his first clinical applications of antibacterial phage therapies, others had difficulties when they tried to scale up the therapies to larger and/or different patient populations (Merril et al. 2003, 2006). Despite these problems, numerous phage strains were isolated and used clinically in Europe, the Middle East, and Asia (Alisky et al. 1998; Sulakvelidze et al. 2001). Unfortunately, most of the phage data in these studies are presented in a qualitative manner with inadequate clinical details and controls.

Cite this article as *Cold Spring Harb Perspect Med* doi: 10.1101/cshperspect.a012518

Potential Problems for Use of Phage as Antibacterial Therapy

Oral and topical antibacterial applications of phage should present few problems other than an occasional inflammation owing to an immune response or a reaction to bacterial byproducts that have not been sufficiently reduced in concentration during the phage preparation. However, system applications require greater care in reducing bacterial contaminations including endo and exotoxins. In addition, phage pharmacokinetic data is rudimentary at best, as many phage strains have not been administered systemically to animals and almost none have been studied carefully in humans. Animal studies have clearly shown that most phage are taken up by the liver with some trapping by the spleen (Merril et al. 2003). In some cases, it has been possible to select phage strains that can remain in the circulation, and remain available for interaction with possible systemic bacterial infectious agents by genetic selection methods (Merril et al. 1996; Vitiello et al. 2005). In a similar manner it should be possible to select phage strains that have a reduced immunogenicity. Reduced immunogenicity, to minimize adverse reactions to phage therapy, may be also accomplished by developing phage strains displaying epitopes that match surface proteins on normal human blood cells. In this regard it might be useful to develop a "platform" phage that is free of toxin and antibiotic resistance genes, but with the ability to stay in the circulation longer, and with reduced immunogenicity. Such a platform phage could be engineered so that it could be easily modified for enhancement of its host range for specific bacterial strains. For example, this could be accomplished with a phage in which the tail proteins could be specified in a manner similar to that found in coliphage K1-5. This phage is a "dual" specificity phage that encodes two different tail proteins allowing it to attack and replicate on both K1 and K5 strains of *Escherichia coli* (Scholl et al. 2001). Alternatively one could use a site-specific recombination system that permits the phage to switch between alternative tail fiber proteins, or by using a reverse transcriptase controlled tropism switch to generate alternate tail fibers (Sandmeier 1994; Liu et al. 2002). Phage strains considered for therapeutic purposes should be screened for toxins, antibiotic resistance, and genes that increase bacterial pathogenicity (Merril et al. 2003). With the growth of the genomic data banks and the development of bioinformatics, tools such as screens will be become more effective (Pauwels et al. 2009; Verheust et al. 2010).

Animal Models for Treatment of Infectious Diseases

At the beginning of international hostilities in the 1940s, the U.S. Department of Defense sponsored carefully designed phage therapy experiments for animal models of infectious diseases. Despite the encouraging results obtained in these experiments this approach was virtually abandoned in the United States following the discovery and development of broad-spectrum antibiotics (Merril et al. 2003, 2006). However, with the advent of antibiotic-resistant pathogenic bacterial strains, research on the potential of phage therapy has been rejuvenated. In addition, with the expanded knowledge of phage molecular biology and interactions with mammalian immune systems it is possible to genetically engineer phage that might be more efficacious than the wild types found in nature. This potential was shown in the development of long circulating phage strains that could stay in the circulatory system longer than the laboratory strains from which they were derived. There have been a number of successful demonstrations of the effectiveness of phage in animal models of bacterial infectious diseases (e.g., see Dubos et al. 1943; Smith and Huggins 1982; Merril et al. 1996; Biswas et al. 2002; Bull et al. 2002; Gu et al. 2012) including a recent success in the treatment of rats using an animal model system for sepsis and meningitis caused by S242, a fatal neonatal meningitis *E. coli* strain (Pouillot et al. 2012).

Clinical Studies

One of the first problems encountered by the infectious disease physician is to determine what agent would offer an efficacious treatment for the patient's disease. Although the narrow host range of phage strains would appear to offer an impediment for the clinical adaptation of phage therapy, it should be possible to address this problem by developing libraries of clinically relevant phage strains carrying marker genes. The use of such strains could provide both an identification of the infecting bacterial strains along with information as to which strains of phage might offer an effective therapy for the patient's infection (Merril et al. 2003, 2006). One of the major problems confronting phage therapy is the need and cost of adequate clinical trials. This is of course a problem for any new antibacterial therapy. The early clinical applications often lacked the new standard double-blind statistically valid experimental designs (Carlton 1999). The years of experience devoted to developing an understanding of the molecular biology, physiology, and methods for the purification of phage should be an asset in helping to design clinical applications and evaluations. Clinical trials needed to meet the full requirements of the U.S. Food and Drug Administration (FDA) for therapeutic phage have not yet been fulfilled for any of the phage strains that have been shown to be effective in animal models of infectious diseases. Such trials are expensive; however, given the need for such agents, for the treatment of antibiotic-resistant bacterial infectious organisms, approval should be gained for some in the not-too-distant future.

USE OF PHAGE COMPONENTS

As Prophylactic Agents

Despite reasonable success of phage therapy in diagnosis and treatment of bacterial infections by bacteriophages as discussed above, parallel research has been progressing in the use of bacteriophage components mostly as prophylaxis. Lysozymes, originally discovered by Alexander Fleming (Fleming 1922), are ubiquitous in nature, being associated with phage in mammals. Lysozymes have been used for decades as a prophylactic agent to kill Gram-positive bacteria because of their ability to degrade cell walls from outside by hydrolysis of one of the four major bonds in the peptidoglycan, which results in hypotonic bursting of the inner membrane and leaking of intracellular components and thereby causing cell death. Lysozymes from different sources including phage have been used as an adduct in the brewing industry and in yogurt and other milk products. As discussed below, phage-encoded lysozymes are of two kinds: (1) endolysin, which is made by phage during phage lytic growth after infection of bacterial host, and which lyses the cell wall from inside to help release phage progeny particles; and (2) phage tail-associated murein lytic enzymes (TAME), which can hydrolyze cell wall bonds from outside after phage adsorption to the host.

Brewing industries face problems with their fermentation systems, as airborne bacteria frequently contaminate them. Bacterial contamination can make a beer turbid, aromatic, odious, or ropy. The most common contaminations are lactic-acid-producing Gram-positive bacteria (*Lactobacillus* and *Pediococcus*) and acetic-acid-producing Gram-negative bacteria (*Acetobacter*). The spoilage comes from the fact that acetic acid and lactic acids, among other by-products, are made by bacteria from sugar. Some measured amount of specific strains of lactic-acid-producing bacteria are allowed during fermentation to keep the pH low, thus facilitating beer production as well as lowering acidity by "malolactic acid" production. In wine making, unwanted contaminations are routinely taken care of by sanitation practices. In fact, direct lysozyme addition acts as a preservative for storage of foods like yogurt, tofu, cheese, and sake (Larson 2005).

More recently, the addition of phage endolysins has been shown to protect against a broad spectrum of Gram-positive bacteria. If microbial contamination is envisioned, lysozyme is used to kill the lactic acid bacteria. In fact, lysozyme can be added to the final product. Depending on factors like temperature, pH, etc., lysozyme starts working immediately after addition and kills bacteria at once. Lysozyme in beer remains 50% active for at least a 6-month period. Lysozyme present in beer has minimal effects on the physical and sensory properties of the beer (Feeney and Nagy 1952; Daeschel et al. 1991; Landschoot 2005; Waite and Daeschel 2007).

As Curative Agents

More recently, however, lysozymes are being developed for treatment of mammalian infections. Endolysin, used by phage to release progeny particles from bacterial hosts, acts by breaking up the cell from inside. This occurs after another phage product, called holin, creates pores in the inner membrane through which lysozymes can pass through the membrane and reach the cell wall (Wang et al. 2003). However, peptidoglycan-degrading enzymes have the ability to digest bacterial cell walls. The phage-encoded endolysins precisely do that except in a species- or genus-specific manner (Nelson et al. 2001; Fischetti 2010). Phage-encoded endolysin usually can lyse from outside only the corresponding host. Recently, cell wall lysis-mediated killing activity of phage-encoded pure lysozymes has been successfully used to its maximal level by, among others, Fischetti and colleagues, and Loessner and colleagues. Like the current surge in bacteriophage therapy, development of endolysin-mediated therapy of infection was motivated by the emergence of antibiotic-resistant bacteria in clinics. Endolysins create holes in the cell from outside by peptidoglycan digestion and expansion of the inner cytoplasmic membrane and subsequent hypotonic lysis.

Most human infections begin at the mucosal membranes followed by their colonization, which are usually a reservoir of many pathogens. Very few anti-infective agents are known that prevent mucosal colonization. Nonetheless, if one can reduce the bacterial load of the mucous membranes in the community, in hospitals, and in nursing homes, the incidence of the diseases would very likely reduce. Precisely, endolysins are expected to be very effective in such cases. Specific endolysins have now been identified to very effectively kill Gram-positive bacteria (Nelson et al. 2001; Loeffler et al. 2003). Interestingly, the endolysins bind very tightly to their cell wall substrates and thus and do not have a turnover number requiring multiple molecules to bind and hydrolyze several cell wall bonds to make effective cell wall lysis (Loessner et al. 2002; Jervis et al. 2005). An oral colonization animal model has been developed with *Streptococcus pyogenes*, a nasal model with pneumococci, and a vaginal model with group B streptococci. These studies showed, for example, that nanogram quantities of endolysin kill *S. pyogenes* 106-fold in 2–4 h after lysozyme treatments. All current success in dealing with endolysin-mediated cell killing are only with Gram-positive bacteria simply because the enzyme when added from outside can access the cell wall peptidoglycan. Gram-negative bacteria, however, have an outer membrane that sterically interferes with lysozyme action thus making lysozyme-mediated cell killing generally ineffective. However, recently, a structurally engineered phage lysozyme containing the FyuA-binding domain of pesticin fused to the amino terminus of T4 lysozyme has given encouraging results against Gram-negative pathogens (Lukacik et al. 2012).

Immunity and Resistance

Both in bacteriophage therapy and in direct lysozyme (of phage or any other source) therapy, it has always been feared that such foreign objects or proteins when used systematically would develop neutralizing antibodies and thus hinder their antibacterial action in the future. However, it has been found that highly immune serum slows down but does not block bacteriolytic activities of pneumococcal-specific endolysins (Loeffler et al. 2001). Another concern of endolysin treatment was its short circulation time after administration (Loeffler et al. 2001). But this issue did not affect the treatment because of the very rapid action time of the lysozyme molecules. Both in antibiotic and phage therapy, another major concern is the phenomenon of the increase of resistant pathogenic strains. Interestingly, every attempt to generate resistant mutants against endolysins in the laboratory setup has so far failed (Loeffler et al. 2001). Although this gives more credence to the potential of lysozyme therapy, we note that most antibiotic-resistance elements that have infiltrated into antibiotic therapy are because of bacteria-acquiring resistance elements from natural environments by lateral gene transfer.

Use of Lysozyme for Treating Bacterial Contamination in Plant Cell Cultures

Even egg white lysozyme has been successfully used to reduce *Bacillus circulans* and *Sphingomonas paucimobilis* infection of in vitro shoot cultures quince and hybrid rootstock (Marino et al. 2003). Although lysozyme did not have a negative effect on shoot growth, under optimal conditions it was effective in eliminating *B. circulans* in quince shoot and hybrid rootstock cultures, and has a bacteriostatic effect on *S. paucimobilis*. These results suggest that lysozyme may be able to replace antibiotic treatments of in vitro shoot cultures although it may not be effective against every plant bacterial infection.

Use of Bacteriophage Tail-Associated Lytic Enzymes as Antibacterial Therapeutic Agents

It has been known for a while that peptidoglycan-degrading murein hydrolases cleave bacterial cell walls very efficiently in a species-specific manner. Thus, they have the potential of being therapeutic agents against specific pathogens. Recently, this idea has been successfully exploited. These hydrolases are different from the phage-encoded endolysins, which are made and used by phage to come out of host cells by cell wall degradation from inside as discussed above. Unlike the endolysins the murein hydrolases are present at the tip of tails of phage virions. It is believed that the tail-attached hydrolases help DNA injection after phage adsorption. If a bacterial cell is infected by phage particles at very high multiplicities, then the cell lyses before phage DNA starts replication. It has been called "lysis from without" (Delbruck 1940). These tail hydrolases are more or less ubiquitous among tailed phages. Motivated by the emergence of drug-resistant human pathogen *Staphylococcus aureus*, Paul et al. (2011) identified the muralytic activity of the tail of the broad host range staphylococcal-specific phage, called phage K. These investigators showed the efficacy of the K-encoded purified hydrolase as a bacteriocidal agent in cell culture. The efficacy of this protein was further enhanced by making a hybrid protein comprising the catalytic domain of the hydrolase and the staphylococcal cell wall binding domain of a bacteriocin called lysostaphin (Baba and Schneewind 1998) to generate a protein called P128. The bacteriolytic activity of P128 was more than two orders of magnitude higher than the catalytic domain of the hydrolase alone because of increased substrate specificity. The hybrid has already been tested in experimental nasal colonization of methicillin-resistant *S. aureus* (MRSA) in experimental rats; the protein has been shown extremely effective in decolonizing the animals. Thus it is an excellent prospective therapeutic against infection.

PHAGE-MEDIATED VACCINE

Prevention of infectious diseases by direct inoculation of a minute quantity of pathogens was shown in China during the 15th century, where healthy children were inoculated with a minute quantity of exudates collected from the pustule of smallpox-infected persons. It was observed that the process of this crude inoculation technique protected the children from future epidemics of smallpox (Tizard 1984, Needham 1999). Despite the fact that the science behind the treatment of infectious diseases through vaccination was first shown by Edward Jenner more than three centuries ago (Riedel 2005) and that use of vaccine in treatment of diseases is widespread now, there are still problems and limitations in preparing vaccines by traditional means. This set the stage to explore other approaches for producing better vaccines (Burdin et al. 2004; O'Hagan and Rappuoli 2004). The advancement of recombinant DNA technology led us to produce vaccine candidates that are recombinant protein molecules displaying subunits of pathogens. However, such vaccine candidates lack many immunogenic features of the original pathogens (Petrovsky and Aguilar 2004). Although many of these recombinant vaccines contain succession of immunodominant epitopes, they fail to stimulate the production of neutralizing antibodies. To overcome this situation, currently many of these recombinant vaccines are mixed with adjuvants to enhance immunostimulatory effects. Unfortunately, the adjuvants that have been tried are either questionable or the process is time consuming and thus not an ideal method of developing vaccines for rapidly evolving diseases like swine flu (Barrett and Stanberry 2008).

Previously immunostimulatory effects of synthetic-immunodominant peptides (epitopes) of antigen and recombinant viruses or nucleotide acid-based (DNA) vaccines were evaluated through various clinical trials. Results of these trials indicated that these vaccines were lacking an effective delivery system for proper activation of the immune system (Petrovsky and Aguilar 2004). These activation deficiencies were reflected in the inability of the vaccines to elicit both humoral and cell-mediated immunity against targeted pathogens. Thus, development of a proper delivery system for any type of subunit or DNA vaccine is of paramount importance (Burdin et al. 2004).

Vaccinologists are now evaluating bacteriophages (phages) as vaccine delivery agents, and the idea is gaining popularity owing to the inherited character of phages to stimulate both arms of the immune system (humoral and cell-mediated immunity).

The ability of phage to deliver gene(s) to mammalian cells was discovered in studies in which lambda (λ) phage carrying a galactose transferase gene was shown capable of inducing galactose transferase enzyme activity in human fibroblast cells from a child lacking the capacity to make the Gal-transferase enzyme (Merril et al. 1971). This ability of phage to deliver genes to human cells was later independently confirmed with a phage carrying the β-galactosidase gene in human cells deficient in that enzyme activity (Horst et al. 1975).

A concern that was expressed concerning these experiments is that the DNA used to correct the genetic defects in the eukaryotic cells is bacterial or prokaryotic. However, it has now been clearly shown that prokaryotic genes, such as the genes in a segment of the Ti plasmid of the bacteria *Agrobacterium tumefaciens* can function in eukaryotes. These genes have been shown to be capable of transforming eukaryotic cells in plants and they are the cause of the commonly observed plant crown gall tumors. In addition, the bacterial Ti plasmid genes have also been shown to be capable of affecting yeast, filamentous fungi, cultivated mushrooms, and human cells in culture (Lacroix et al. 2006).

Currently, two separate approaches of antigen delivery systems are attempted with phage vaccines. These systems are designated as "phage display vaccines" and "phage DNA vaccines" (Clark and March 2004a). Broadly, phage display vaccines are mainly referred for producing immunogenic phage particles that display foreign antigen on phage surface. This generally is achieved by expressing antigens as fusion products of one of the major surface proteins of phage virion; phage DNA vaccines, on the other hand, are produced by incorporation of foreign antigen genes in phage genome under the control of strong eukaryotic promoters. In the latter case, phage acts as a passive carrier to transfer the foreign DNA in mammalian cells where the antigen gene is expressed. These two methods are often combined to produce phages that carry a foreign antigen gene along with a display protein or peptide on the surface. The display proteins or peptides are generally selected for their specific binding affinity of antigen-presenting cells. Thus, such phage constructs pose the additional ability of delivering the antigen gene directly to the immunoreactive cells like dendritic cells, Kupffer cells, etc. (Zanghi et al. 2007).

Phage Display Vaccine

A filamentous phage M13 display system was first used for expression of antibody fragments and their affinity maturation (Benhar 2001). Although M13 is successful in several cases in displaying melanoma-specific tumor antigen to produce cancer vaccines effective in reducing tumor growth in animal models (Fang et al. 2005), the M13 display system has limitations (Gupta et al. 2003). The system is not efficient in displaying larger proteins when using the high copy coat protein gp VIII. The other coat protein, gpIII, which can effectively display foreign proteins or peptides, has a low copy number in the M13 virion (Smith 1985).

Soc and Hoc, two of the virion proteins of phage T4, are more effective major display platforms for foreign antigens. Unlike M13, T4 is able to display peptides of various sizes and sequences without much problem. Hoc and Soc are also highly immunogenic and therefore act as good adjuvants. Thus, T4 phage is considered as a better display platform than filamentous phages (Li et al. 2006a). Currently, several vaccines for infectious diseases are prepared by using the T4 phage display system, which has shown promising results in animal models (Jiang et al. 1997).

T7 is another phage capable of displaying proteins and peptides including antigens. Fusion products are generated by cloning the coding sequences for the peptides and proteins at the carboxyl end of the capsid protein, 10B (Lindner et al. 2011). Lewis lung cancer vaccine prepared by T7 phage display of vascular endothelial growth factor (VEGF) has been successfully used to break immunologic tolerance. When injected in mice, T7-VEGF display produced strong immunogenic response against cancer cells and inhibited cancer growth (Li et al. 2006b). The comparative aspects of different phages for antigen display are shown in Table 1.

Recently, the common cloning vector phage lambda (λ) is also gaining popularity as a phage display platform. The complex cloning method of λ phage display is considerably simplified by using a Cre-*lox*P site-specific recombination process to develop a λ phage display system that facilitates rapid high efficiency cloning of foreign DNA in λ genome (Gupta et al. 2003). The method allows displaying foreign genes fused to λ capsid protein gene D; nearly about 450 copies of foreign antigen can display as D fusion on λ surface (Gupta et al. 2003). The ability of λ to display multiple copies of the same antigen on the surface of a single phage particle is more effective in eliciting a strong immune response than low copy display (Clark and March 2004a). Production of successful λ display antigen to produce neutralizing antibodies against infectious porcine circovirus infection has been shown by Hayes et al. (2010), using gpD fusion antigen on phage λ. Activation of immune system by phage display vaccine is described in Figure 1.

Phage DNA Vaccine

Successful delivery and expression of foreign genes in eukaryotic cells by phage-mediated gene transfer is also in common use now. The use of strong eukaryotic promoters has extended the capacity of phage to deliver foreign antigen genes to mammalian cells. Generally for production of vaccine, the antigen genes under the control of a eukaryotic promoter are cloned inside of a nonessential region of phage genome. When injected in a mammalian system,

Table 1. List of phages and their virion proteins, which are used for display

Vector phage	Phage virion proteins for fusion display	Copy number of display	Size of display shown
M13	gpVIII	≈2700	Small peptide (6–8 amino acids)
	gpIII	≈5	Large protein with reduced viability
T4	Soc	≈960	Large protein (up to 837 amino acids)
	Hoc	≈160	Protein (up to 183 amino acids)
T7	10A	≈415	Protein (40–50 amino acids)
	10B	≈1	Protein (up to 1200 amino acids)
Lambda (λ)	gpD	≈420	Large protein (1024 amino acids)

Data were taken from Smith (1985), Markland et al. (1991), Mikawa et al. (1996), Ren et al. (1996, 1997), Ren and Black (1998), Iwasaki et al. (2000), Savinov and Austin (2001), Sche et al. (2001), Gupta et al. (2003), Pacheco et al. (2006), and Li et al. (2007).

The above phages are used for presentation of displayed antigens to mammalian cells; the λ system was also developed as a DNA delivery platform to antigen-presenting cells (APC).

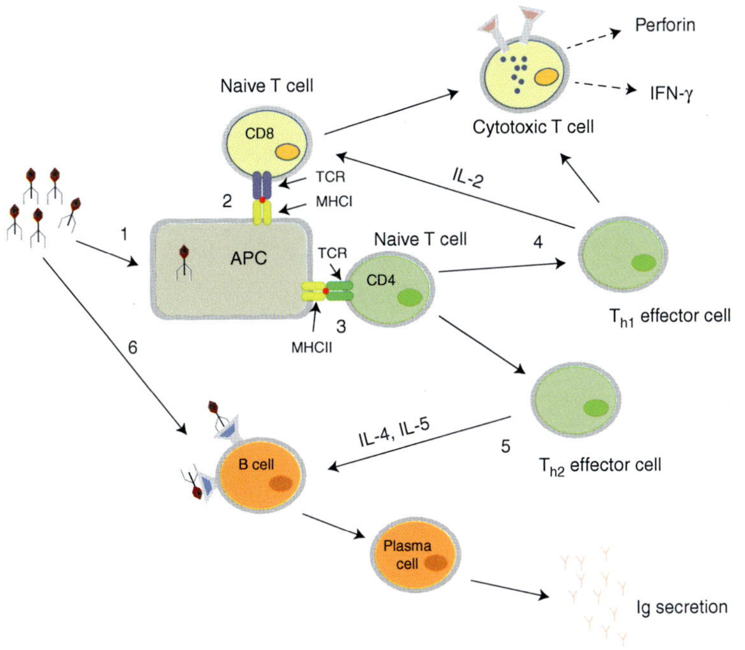

Figure 1. Immune activation of mammalian system with phage display antigens. (1) Particulate nature of phage activates the antigen-presenting cells (APC), which process the antigens for immune presentation. (2) Presentation of processed antigens by major histocompatibility complexes (MHC) class I molecules to CD8 T cells, which leads to T-cell activation. (3) Antigens are presented by MHC class II molecules to CD4 T cell, which in turn activates T_{h1} and T_{h2} effector cells. (4) T_{h1} cells generate cytotoxic T-cell (CTL) responses and help produce interferon γ (IFN-γ). (5) T_{h2} cells activate B cells to make antibodies. (6) Direct activation of B cells by phage vaccines also leads to massive antibody response. T-cell receptors are denoted by TCR.

this phage vaccine acting as a DNA vaccine can induce potent immune response by expressing foreign antigen inside of APCs (Fig. 2) (Clark and March 2006). Several λ-based DNA vaccines for infectious diseases have been prepared that have shown promising results in animal models (Clark and March 2004b; March et al. 2004).

Advantages of Phage-Mediated Vaccine Delivery

Recent global surveillance indicates that infectious microorganisms are emerging at an alarming rate (Barrett and Stanberry 2008). There is also a significant concern about the potential of generating highly virulent microorganisms using genetic engineering and synthetic biology techniques. The outbreak caused by these types of natural or man-made variants can spread rapidly and has to be stopped before diseases become epidemic. The quarantine and rapid vaccination against targeted pathogens is the only way to prevent epidemics during massive disease outbreaks. One of the best advantages of phage vaccine is that it can be produced inexpensively and deployed rapidly during disease outbreaks.

Utilizing advanced knowledge in bioinformatics, gene sequencing, and phage genetics, it is possible to create and manufacture vaccines in record time. Genomic and proteomic data of any old or newly emerging pathogenic bacteria or virus can be scanned using existing algorithms and database management systems to identify novel antigenic motifs that can be cloned into phage genome to produce phage vaccine. The resulting phage vaccine can propagate rapidly using bacteriological media and does not need cell culture or a chicken-egg-

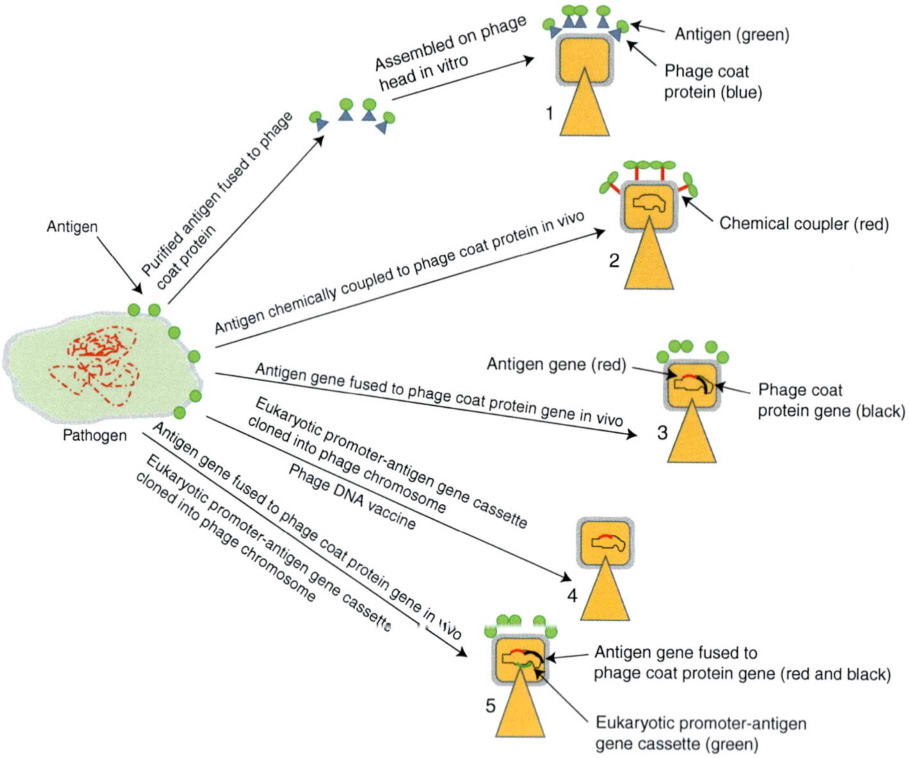

Figure 2. Schematic representations of several phage-mediated vaccine delivery systems. (1) Production of virus-like particles (VLPs) by adding purified phage capsid proteins fused to antigens during assembly of phage particles in vitro. (2) Antigens are attached by chemical conjugation on preassembled phage head. (3) Antigens are displayed on phage surface by fusion of antigen genes with phage capsid protein genes. (4) Phage as a DNA delivery vehicle where antigen genes are cloned in phage genome under the control of eukaryotic promoters. (5) Phage for DNA vaccine, where phage carries antigen genes under the control of eukaryotic promoters. The phage also displays foreign proteins on its surface as fusion of phage capsid proteins. This protein targets the phage to antigen-presenting cells (APC).

based system for manufacturing sufficient quantities of vaccine to protect against a new pathogen; such preparations can be made in weeks rather than months. The phage particles are nonpathogenic to humans and the resultant vaccines do not require adjuvants. By using phage vaccine, all relevant information can be delivered to the immune system to allow for an immune response that includes antibody production and the development of immunity to a pathogen (Fig. 1). The particulate nature of phage vaccine attracts APCs that engulf, process, and present phage-mediated antigens through MHCs (class I and class II pathways) (Manoutcharian et al. 1999; Gaubin et al. 2003) to evoke both cell-mediated and humoral immunities (Fig. 1). It was expected that phage as an extracellular antigen will only activate antibody production through MHC class II bias pathway, but it has been shown that phage as a particulate antigen can access the MHC I pathway through cross priming, which activates a cellular immune response (Gaubin et al. 2003). In addition, the immunostimulatory unmethylated CpG motifs of the phage genome are recognized by the innate immune cells through Toll-like receptors (TLRs) (Krieg et al. 1995; Mason et al. 2005), which further enhance immunity. Immunostimulatory effect of phage vaccination through oral route has been evaluated by Clark et al. They have succeeded in making a phage DNA vaccine against *Yersina pestis* by cloning

V-antigen in phage λ genome. Upon oral administration, it produced a significant immune response in mouse (Clark and March 2004b). Others have also explored the oral delivery route of phage vaccination, and showed antigen-specific response (Delmastro et al. 1997; Zuercher et al. 2000). These results indicate the possibility of oral delivery of phage vaccine in developing countries.

In summary, phage particles pose several intrinsic characters that make them lucrative for developing vaccine-delivery platforms. The natural stability of phage particles provides easy processing and cheaper manufacturing capability for large-scale production of vaccines. The particulate nature of phage makes them easy to purify by simple centrifugation steps, which are far easier and cheaper than the methods applied for purifying soluble recombinant protein vaccines. Additionally, phage itself acts as a strong adjuvant and thus enhances excellent immune response against any antigens present along with phage (Frenkel et al. 2000; Manoutcharian et al. 2004). In fact, the response against phage-displayed antigens can break immunological tolerance against self-proteins and thus is an ideal vehicle for delivering cancer antigens (Fang et al. 2005).

PERSPECTIVES

The ever-increasing list of antibiotic-resistant bacterial strains and newly emerging viruses is forcing a reexamination of alternate modes of therapies for infectious diseases. Of these, the phage and their molecular components and products offer a number of new therapeutic alternatives. Although the development of phage-based applications will require new directions, there are advantages over the prior art of treating infectious disease. A useful aspect is the ability to develop therapies for bacterial infection that minimally perturb the human microbiome. As recent research of the microbiome has shown, even small alterations can have effects that range from alterations in the perception of hunger to increased susceptibility to lethal infections by foreign bacterial strains. These advances in our ability to work with phage extend from the treatment of human infections to increasing the efficiency of industrial processes, particularly those that rely on fermentation. The development of phage-based delivery of antigens and genes will result in new vaccines for protection from numerous diseases and new therapeutic approaches to both infectious and noninfectious diseases, such as some forms of cancer. The newly developed genomic sequencing methods along with other high-throughput technologies including cell sorters and multi-well metabolic monitoring robots should result in the rapid development of the needed information for phage-based therapies.

REFERENCES

Alisky J, Iczkowski K, Rapoport A, Troitsky N. 1998. Bacteriophages show promise as antimicrobial agents. *J Infect* **36:** 5–15.

Baba T, Schneewind O. 1998. Instruments of microbial warfare: Bacteriocin synthesis, toxicity and immunity. *Trends Microbiol* **6:** 66–71.

Barrett ADT, Stanberry LR. 2008. *Vaccines for biodefence and emerging and neglected diseases*. Academic, Waltham, MA.

Benhar I. 2001. Biotechnological applications of phage and cell display. *Biotechnol Adv* **19:** 1–33.

Biswas B, Adhya S, Washart P, Paul B, Trostel AN, Powell B, Carlton R, Merril CR. 2002. Bacteriophage therapy rescues mice bacteremic from a clinical isolate of vancomycin-resistant *Enterococcus faecium*. *Infect Immun* **70:** 204–210.

Bull JJ, Levin BR, DeRouin T, Walker N, Bloch CA. 2002. Dynamics of success and failure in phage and antibiotic therapy in experimental infections. *BMC Microbiol* **2:** 35.

Burdin N, Guy B, Moingeon P. 2004. Immunological foundations to the quest for new vaccine adjuvants. *BioDrugs* **18:** 79–93.

Carlton RM. 1999. Phage therapy: Past history and future prospects. *Arch Immunol Ther Exp* **47:** 267–274.

Clark JR, March JB. 2004a. Bacterial viruses as human vaccines? *Expert Rev Vaccines* **3:** 463–476.

Clark JR, March JB. 2004b. Bacteriophage-mediated nucleic acid immunisation. *FEMS Immunol Med Microbiol* **40:** 21–26.

Clark JR, March JB. 2006. Bacteriophages and biotechnology: Vaccines, gene therapy and antibacterials. *Trends Biotechnol* **24:** 212–218.

Daeschel MA, Jung DS, Watson BT. 1991. Controlling wine malolactic fermentation with nisin and nisin-resistant strains of *Leuconostoc oenos*. *Appl Environ Microbiol* **57:** 601–603.

Delbruck M. 1940. The growth of bacteriophage and lysis of the host. *J Gen Physiol* **23:** 643–660.

Delmastro P, Meola A, Monaci P, Cortese R, Galfre G. 1997. Immunogenicity of filamentous phage displaying peptide mimotopes after oral administration. *Vaccine* **15:** 1276–1285.

Dubos RJ, Straus JH, Pierce C. 1943. The multiplication of bacteriophage in vivo and its protective effect against an experimental infection with *Shigella dysenteriae*. *J Exp Med* **78:** 161–168.

Fang J, Wang G, Yang Q, Song J, Wang Y, Wang L. 2005. The potential of phage display virions expressing malignant tumor specific antigen MAGE-A1 epitope in murine model. *Vaccine* **23:** 4860–4866.

Feeney RE, Nagy DA. 1952. The antibacterial activity of the egg white protein conalbumin. *J Bacteriol* **64:** 629–643.

Fischetti VA. 2010. Bacteriophage endolysins: A novel anti-infective to control Gram-positive pathogens. *Int J Med Microbiol* **300:** 357–362.

Fleming A. 1922. On a remarkable bacteriolytic element found in tissues and secretions. *P Roy Soc London B* **93:** 306–317.

Frenkel D, Katz O, Solomon B. 2000. Immunization against Alzheimer's β-amyloid plaques via EFRH phage administration. *Proc Natl Acad Sci* **97:** 11455–11459.

Gaubin M, Fanutti C, Mishal Z, Durrbach A, De Berardinis P, Sartorius R, Del Pozzo G, Guardiola J, Perham RN, Piatier-Tonneau D. 2003. Processing of filamentous bacteriophage virions in antigen-presenting cells targets both HLA class I and class II peptide loading compartments. *DNA Cell Biol* **22:** 11–18.

Gu J, Liu X, Li Y, Han W, Lei L, Yang Y, Zhao H, Gao Y, Song J, Lu R, et al. 2012. A method for generation phage cocktail with great therapeutic potential. *PloS ONE* **7:** e31698.

Gupta A, Onda M, Pastan I, Adhya S, Chaudhary VK. 2003. High-density functional display of proteins on bacteriophage λ. *J Mol Biol* **334:** 241–254.

Hayes S, Gamage LN, Hayes C. 2010. Dual expression system for assembling phage λ display particle (LDP) vaccine to porcine Circovirus 2 (PCV2). *Vaccine* **28:** 6789–6799.

Horst J, Kluge JF, Beyreuther K, Gerok W. 1975. Gene transfer to human cells: Transducing phage λ plac gene expression in GM1-gangliosidosis fibroblasts. *Proc Natl Acad Sci* **72:** 3531–3535.

Iwasaki K, Trus BL, Wingfield PT, Cheng N, Campusano G, Rao VB, Steven AC. 2000. Molecular architecture of bacteriophage T4 capsid: Vertex structure and bimodal binding of the stabilizing accessory protein, Soc. *Virology* **271:** 321–333.

Jernberg C, Lofmark S, Edlund C, Jansson JK. 2010. Long-term impacts of antibiotic exposure on the human intestinal microbiota. *Microbiology* **156:** 3216–3223.

Jervis EJ, Guarna MM, Doheny JG, Haynes CA, Kilburn DG. 2005. Dynamic localization and persistent stimulation of factor-dependent cells by a stem cell factor/cellulose binding domain fusion protein. *Biotechnol Bioeng* **91:** 314–324.

Jiang J, AbuShilbayeh L, Rao VB. 1997. Display of a PorA peptide from *Neisseria meningitidis* on the bacteriophage T4 capsid surface. *Infect Immun* **65:** 4770–4777.

Krieg AM, Yi AK, Matson S, Waldschmidt TJ, Bishop GA, Teasdale R, Koretzky GA, Klinman DM. 1995. Cpg motifs in bacterial-DNA trigger direct B-cell activation. *Nature* **374:** 546–549.

Lacroix B, Tzfira T, Vainstein A, Citovsky V. 2006. A case of promiscuity: Agrobacterium's endless hunt for new partners. *Trends Genet* **22:** 29–37.

Landschoot AV. 2005. Antibacterial properties of hen egg white lysozyme against beer spoilage bacteria and effect of lysozyme on yeast fermentation. *Cerevisia* **32:** 219–225.

Larson AE. 2005. *Lysozyme in antimicrobial in food*. Taylor and Francis, London.

Li Q, Shivachandra SB, Leppla SH, Rao VB. 2006a. Bacteriophage T4 capsid: A unique platform for efficient surface assembly of macromolecular complexes. *J Mol Biol* **363:** 577–588.

Li XH, Tang L, Liu D, Sun HM, Zhou CC, Tan LS, Wang LP, Zhang PD, Zhang SQ. 2006b. Antitumor effect of recombinant T7 phage vaccine expressing xenogenic vascular endothelial growth factor on Lewis lung cancer in mice. *Ai Zheng* **25:** 1221–1226.

Li Q, Shivachandra SB, Zhang Z, Rao VB. 2007. Assembly of the small outer capsid protein, Soc, on bacteriophage T4: A novel system for high density display of multiple large anthrax toxins and foreign proteins on phage capsid. *J Mol Biol* **370:** 1006–1019.

Lindner T, Kolmar H, Haberkorn U, Mier W. 2011. DNA libraries for the construction of phage libraries: Statistical and structural requirements and synthetic methods. *Molecules* **16:** 1625–1641.

Liu M, Deora R, Doulatov SR, Gingery M, Eiserling FA, Preston A, Maskell DJ, Simons RW, Cotter PA, Parkhill J, et al. 2002. Reverse transcriptase-mediated tropism switching in *Bordetella bacteriophage*. *Science* **295:** 2091–2094.

Loeffler JM, Nelson D, Fischetti VA. 2001. Rapid killing of *Streptococcus pneumoniae* with a bacteriophage cell wall hydrolase. *Science* **294:** 2170–2172.

Loeffler JM, Djurkovic S, Fischetti VA. 2003. Phage lytic enzyme Cpl-1 as a novel antimicrobial for pneumococcal bacteremia. *Infect Immun* **71:** 6199–6204.

Loessner MJ, Kramer K, Ebel F, Scherer S. 2002. C-terminal domains of *Listeria monocytogenes* bacteriophage murein hydrolases determine specific recognition and high-affinity binding to bacterial cell wall carbohydrates. *Mol Microbiol* **44:** 335–349.

Lukacik P, Barnard TJ, Keller PW, Chaturvedi KS, Seddiki N, Fairman JW, Noinaj N, Kirby TL, Henderson JP, Steven AC, et al. 2012. *Proc Natl Acad Sci* **109:** 9857–9862.

Manoutcharian K, Terrazas LI, Gevorkian G, Acero G, Petrossian P, Rodriguez M, Govezensky T. 1999. Phage-displayed T-cell epitope grafted into immunoglobulin heavy-chain complementarity-determining regions: An effective vaccine design tested in murine cysticercosis. *Infect Immun* **67:** 4764–4770.

Manoutcharian K, Diaz-Orea A, Gevorkian G, Fragoso G, Acero G, Gonzalez E, De Aluja A, Villalobos N, Gomez-Conde E, Sciutto E. 2004. Recombinant bacteriophage-based multiepitope vaccine against *Taenia solium* pig cysticercosis. *Vet Immunol Immunopathol* **99:** 11–24.

March JB, Clark JR, Jepson CD. 2004. Genetic immunisation against hepatitis B using whole bacteriophage λ particles. *Vaccine* **22:** 1666–1671.

Marino G, Ferrarini V, Giardini S, Biavati B. 2003. Use of lysozyme for treatment of bacterial contamination in in vitro shoot cultures of fruit plants. *In Vitro Cell Dev Biol* **39:** 327–331.

Markland W, Roberts BL, Saxena MJ, Guterman SK, Ladner RC. 1991. Design, construction and function of a multicopy display vector using fusions to the major coat protein of bacteriophage M13. *Gene* **109:** 13–19.

Mason KA, Ariga H, Neal R, Valdecanas D, Hunter N, Krieg AM, Whisnant JK, Milas L. 2005. Targeting toll-like receptor 9 with CpG oligodeoxynucleotides enhances tumor response to fractionated radiotherapy. *Clin Cancer Res* **11:** 361–369.

Merril CR, Geier MR, Petricciani JC. 1971. Bacterial virus gene expression in human cells. *Nature* **233:** 398–400.

Merril CR, Biswas B, Carlton R, Jensen NC, Creed GJ, Zullo S, Adhya S. 1996. Long-circulating bacteriophage as antibacterial agents. *Proc Natl Acad Sci* **93:** 3188–3192.

Merril CR, Scholl D, Adhya SL. 2003. The prospect for bacteriophage therapy in Western medicine. *Nat Rev Drug Discov* **2:** 489–497.

Merril CR, Scholl D, Adhya S. 2006. *The bacteriophage*. Oxford University Press, London.

Mikawa YG, Maruyama IN, Brenner S. 1996. Surface display of proteins on bacteriophage λ heads. *J Mol Biol* **262:** 21–30.

Needham J. 1999. *Science and civilisation in China*. Cambridge University Press, Cambridge.

Nelson D, Loomis L, Fischetti VA. 2001. Prevention and elimination of upper respiratory colonization of mice by group A streptococci by using a bacteriophage lytic enzyme. *Proc Natl Acad Sci* **98:** 4107–4112.

O'Hagan DT, Rappuoli R. 2004. Novel approaches to vaccine delivery. *Pharm Res* **21:** 1519–1530.

Pacheco S, Gomez I, Sato R, Bravo A, Soberon M. 2006. Functional display of *Bacillus thuringiensis* Cry1Ac toxin on T7 phage. *J Invertebr Pathol* **92:** 45–49.

Paul VD, Sundarrajan S, Rajagopalan SS, Hariharan S, Kempashanaiah N, Padmanabhan S, Sriram B, Ramachandran J. 2011. Lysis-deficient phages as novel therapeutic agents for controlling bacterial infection. *BMC Microbiol* **11:** 195.

Pauwels K, Gijsbers R, Toelen J, Schambach A, Willard-Gallo K, Verheust C, Debyser Z, Herman P. 2009. State-of-the-art lentiviral vectors for research use: Risk assessment and biosafety recommendations. *Curr Gene Ther* **9:** 459–474.

Petrovsky N, Aguilar JC. 2004. Vaccine adjuvants: Current state and future trends. *Immunol Cell Biol* **82:** 488–496.

Pouillot F, Chomton M, Blois H, Courroux C, Noelig J, Bidet P, Bingen E, Bonacorsi S. 2012. Efficacy of bacteriophage therapy in experimental sepsis and meningitis caused by a clone O25b:H4-ST131 *Escherichia coli* strain producing CTX-M-15. *Antimicrob Agents Chemother* **56:** 3568–3575.

Ren Z, Black LW. 1998. Phage T4 SOC and HOC display of biologically active, full-length proteins on the viral capsid. *Gene* **215:** 439–444.

Ren ZJ, Lewis GK, Wingfield PT, Locke EG, Steven AC, Black LW. 1996. Phage display of intact domains at high copy number: A system based on SOC, the small outer capsid protein of bacteriophage T4. *Protein Sci* **5:** 1833–1843.

Ren ZJ, Baumann RG, Black LW. 1997. Cloning of linear DNAs in vivo by overexpressed T4 DNA ligase: Construction of a T4 phage hoc gene display vector. *Gene* **195:** 303–311.

Riedel S. 2005. Edward Jenner and the history of smallpox and vaccination. *Proc (Bayl Univ Med Cent)* **18:** 21–25.

Sandmeier H. 1994. Acquisition and rearrangement of sequence motifs in the evolution of bacteriophage tail fibres. *Mol Microbiol* **12:** 343–350.

Savinov SN, Austin DJ. 2001. The cloning of human genes using cDNA phage display and small-molecule chemical probes. *Comb Chem High T Scr* **4:** 593–597.

Sche PP, McKenzie KM, White JD, Austin DJ. 2001. Corrigendum to: "Display cloning: Functional identification of natural product receptors using cDNA-phage display" [Chemistry & Biology 6 (1999) 707–716]. *Chem Biol* **8:** 399–400.

Scholl D, Rogers S, Adhya S, Merril CR. 2001. Bacteriophage K1-5 encodes two different tail fiber proteins, allowing it to infect and replicate on both K1 and K5 strains of *Escherichia coli*. *J Virol* **75:** 2509–2515.

Smith GP. 1985. Filamentous fusion phage: Novel expression vectors that display cloned antigens on the virion surface. *Science* **228:** 1315–1317.

Smith HW, Huggins MB. 1982. Successful treatment of experimental *Escherichia coli* infections in mice using phage: Its general superiority over antibiotics. *J Gen Microbiol* **128:** 307–318.

Sneade W. 2005. *Drug discovery: A history*. Wiley-Interscience, Hoboken, NJ.

Sulakvelidze A, Alavidze Z, Morris JG Jr. 2001. Bacteriophage therapy. *Antimicrob Agents Chemother* **45:** 649–659.

Summers WC. 1999. *Félix ďHerelle and the origins of molecular biology*. Yale University Press, New Haven and London.

Tizard IR. 1984. *Immunology: An introduction*. Cengage Learning, Stamford, CT.

Verheust C, Pauwels K, Mahillon JD, Helinski R, Herman P. 2010. Contained use of bacteriophages: Risk assessment and biosafety recommendations. *Appl Biosafety* **15:** 32–44.

Vitiello CL, Merril CR, Adhya S. 2005. An amino acid substitution in a capsid protein enhances phage survival in mouse circulatory system more than a 1000-fold. *Virus Res* **114:** 101–103.

Waite JG, Daeschel MA. 2007. Contribution of wine components to inactivation of food-borne pathogens. *J Food Sci* **72:** M286–M291.

Wang IN, Deaton J, Young R. 2003. Sizing the holin lesion with an endolysin-β-galactosidase fusion. *J Bacteriol* **185:** 779–787.

Zanghi CN, Sapinoro R, Bradel-Tretheway B, Dewhurst S. 2007. A tractable method for simultaneous modifications to the head and tail of bacteriophage λ and its application to enhancing phage-mediated gene delivery. *Nucleic Acids Res* **35:** e59.

Zuercher AW, Miescher SM, Vogel M, Rudolf MP, Stadler MB, Stadler BM. 2000. Oral anti-IgE immunization with epitope-displaying phage. *Eur J Immunol* **30:** 128–135.

Index

I

R

S

T

U

V